Introductory Algebra

FIFTH EDITION

Introductory Algebra

FIFTH EDITION

MARGARET L. LIAL
American River College

E. JOHN HORNSBY, JR.
University of New Orleans

CHARLES D. MILLER

HarperCollinsCollegePublishers

Sponsoring Editor: Karin E. Wagner
Developmental Editor: Sandi Goldstein
Project Editor: Ann-Marie Buesing
Design Administrator: Jess Schaal
Text Design: Lesiak/Crampton Design: Lucy Lesiak
Cover Design: Lesiak/Crampton Design: Lucy Lesiak
Cover Photo: Joyce P. Lopez
Photo Researcher: Nina Page
Production Administrator: Randee Wire
Compositor: Interactive Composition Corporation
Printer and Binder: R. R. Donnelley & Sons Company
Cover Printer: The Lehigh Press, Inc.

Introductory Algebra, Fifth Edition

94 95 96 97 9 8 7 6 5 4 3 2 1

Content Overview

Contents

Preface

This fifth edition of *Introductory Algebra* is designed for college students who have not been exposed to algebra or who require further review before taking additional courses in mathematics, science, business, or computer science. The primary objective of the course is for students to gain familiarity with mathematical symbols and operations in order to formulate and solve first- and second-degree equations.

This text retains the successful features of previous editions: learning objectives for each section, careful exposition, fully developed examples, margin problems, cautions and notes, and boxes that set off important definitions, formulas, rules, and procedures. In this new edition, we have made several changes in content. Many of these changes follow the guidelines set forth in the *Curriculum and Evaluation Standards for School Mathematics*, published by the National Council of Teachers of Mathematics.

CHANGES IN CONTENT

- For continuous review, we have added problems involving fractions and decimals throughout the text.
- Every exercise set has been extensively revised, with more than 80 percent of the exercises being new. Nine accuracy checkers were hired to ensure that the highest level of accuracy has been maintained.
- Cumulative reviews now appear after each chapter, beginning with Chapter 2, and cover material learned up to that point.
- In this text, we view calculators as a means of allowing students to spend more time on the conceptual nature of mathematics and less time on computation with paper and pencil. We have included an introduction to scientific calculators, and the use of the scientific calculator is discussed wherever it is appropriate throughout the book. Some sections include exercises that require a calculator. We also discuss the role of estimation in deciding whether an answer is reasonable.
- The review material on fractions, decimals, and percents has been grouped into Chapter R, a new review chapter, that can be covered in class or left for individual review.
- New problems on percent and geometry applications of algebra have been added in Chapter 2.
- In Chapter 8, we have included some work with cube roots and fourth roots.

FEATURES

Definitions, Formulas, Rules, and Procedures *These items are outlined in color boxes to stress the importance of this material.*

Quest for Numeracy *In these pages, we address various concerns of the NCTM Standards: technology, graph reading, estimation skills, statistical techniques, group discussions, and so on.*

6.3 THE SLOPE OF A LINE

OBJECTIVES

1. Find the slope of a line given two points.
2. Find the slope from the equation of a line.
3. Use slopes to determine whether two lines are parallel, perpendicular, or neither.

FOR EXTRA HELP

Tape 9 | SSM pp. 229–233 | MAC: A IBM: A

We can graph a straight line if at least two different points on the line are known. We can also graph a line by using just one point on the line, along with the "steepness" of the line.

1 One way to measure the steepness of a line is to compare the vertical change in the line (the rise) to the horizontal change (the run) while moving along the line from one fixed point to another. This measure of steepness is called the *slope* of the line.

Figure 13 shows a line with the points (x_1, y_1) and (x_2, y_2). (Read x_1 as "x-sub-one" and x_2 as "x-sub-two.") Moving along the line from the point (x_1, y_1) to the point (x_2, y_2) causes y to change by $y_2 - y_1$ units. This is the vertical change. Similarly, x changes by $x_2 - x_1$ units, the horizontal change. The ratio of the change in y to the change in x gives the slope of the line. We usually denote slope with the letter m. The slope of a line is defined as follows.

SLOPE FORMULA

The **slope** of the line through the points (x_1, y_1) and (x_2, y_2) is

$$m = \frac{\text{change in } y}{\text{change in } x} = \frac{y_2 - y_1}{x_2 - x_1} \quad \text{if } x_2 \neq x_1.$$

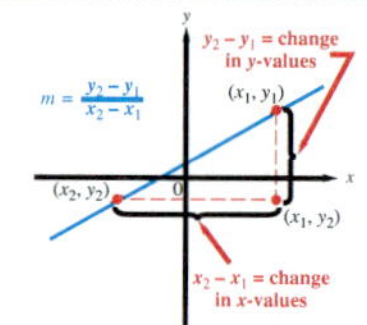

FIGURE 13

WORK PROBLEM 1 AT THE SIDE.

1. Evaluate $\frac{y_2 - y_1}{x_2 - x_1}$ for the following values.

(a) $y_2 = 4, y_1 = -1$ $x_2 = 3, x_1 = 4$

(b) $x_1 = 3, x_2 = -5,$ $y_1 = 7, y_2 = -9$

(c) $x_1 = 2, x_2 = 7,$ $y_1 = 4, y_2 = 9$

The slope of a line tells how fast y changes for each unit of change in x; that is, the slope gives the rate of change in y for each unit of change in x.

The idea of slope is useful in many everyday situations. For example, a highway with a 10% or $\frac{1}{10}$ grade (or slope) rises 1 meter for every 10 meters horizontally. Architects specify the pitch of a roof by indicating the slope; a $\frac{5}{12}$ roof means that the roof rises 5 feet for every 12 feet in the horizontal direction. The slope of a stairwell also indicates the ratio of the vertical rise to the horizontal run. See Figure 14.

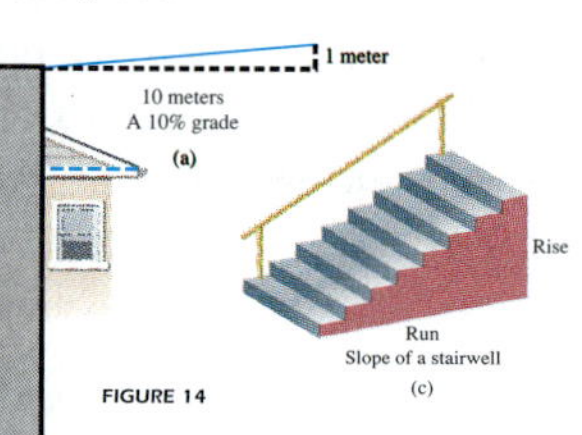

FIGURE 14

ANSWERS
1. (a) −5 (b) 2 (c) 1

397

QUEST FOR NUMERACY

Inquiring Minds Want to Know . . . What's So Curious and Interesting About Numbers?

The theory of numbers is the branch of mathematics that deals with the study of the properties of the natural numbers. One of the most remarkable books on number theory is *The Penguin Dictionary of Curious and Interesting Numbers* (1986) by David Wells. This book contains fascinating numbers and their properties, some of which are given here.

- 113 and 199 and 337 are the only three-digit numbers that are prime and all rearrangements of the digits are prime.
- Find the sum of the cubes of the digits of 136: $1^3 + 3^3 + 6^3 = 244$. Repeat the process with the digits of 244: $2^3 + 4^3 + 4^3 = 136$. We're back to where we started.
- 635,318,657 is the smallest number that can be expressed as the sum of two fourth powers in two ways: $635{,}318{,}657 = 59^4 + 158^4 = 133^4 + 134^4$.
- The number 24,678,050 has an interesting property: $24{,}678{,}050 = 2^8 + 4^8 + 6^8 + 7^8 + 8^8 + 0^8 + 5^8 + 0^8$.
- The number 54,748 has a similar interesting property: $54{,}748 = 5^5 + 4^5 + 7^5 + 4^5 + 8^5$.
- The number 3435 has this property: $3435 = 3^3 + 4^4 + 3^3 + 5^5$.

If your curiosity is piqued by such facts, this book is for you!

FOR GROUP DISCUSSION Discover some curious and interesting properties of numbers by going through the following processes.

1. Choose a three-digit number that is a multiple of 3. Add the cubes of the digits. Repeat the process until the same number is obtained over and over. Compare your results with another student.
2. Take any three-digit number whose digits are not all the same. Arrange the digits in decreasing order, and then arrange them in increasing order. Now subtract. Repeat the process, using a 0 if necessary in the event that the difference consists of only two digits. Continue until the same number repeats over and over. What is the number?
3. Repeat Item 2 for a four-digit number. What result do you get?
4. When told that the number 1729 was rather dull, the Indian mathematician Srinivasa Ramanujan commented that, on the contrary, it was interesting because it is the smallest number that can be expressed as the sum of two cubes in exactly two ways. What are these two ways? Can you express 85 as the sum of two squares in two ways? (For more on Ramanujan, read *The Man Who Knew Infinity*, by Robert Kanigel.)

152

Student Resources *This new feature cross-references relevant student supplements to the respective text section. They are located at the head of each section.*

(c) $-\sqrt{\frac{16}{49}}$

Because $\frac{16}{49} = \frac{4}{7} \cdot \frac{4}{7}$, $-\sqrt{\frac{16}{49}} = -\frac{4}{7}$. ■

WORK PROBLEM 3 AT THE SIDE.

3. Find each square root.

(a) $\sqrt{16}$

(b) $-\sqrt{169}$

(c) $-\sqrt{225}$

(d) $\sqrt{729}$

(e) $\sqrt{\frac{36}{25}}$

2 A number that is not a perfect square has a square root that is not a rational number. For example, $\sqrt{5}$ is not a rational number because it cannot be written as the ratio of two integers. However, $\sqrt{5}$ is a real number and corresponds to a point on the number line. As mentioned in Chapter 1, a real number that is not rational is called an *irrational number*. The number $\sqrt{5}$ is irrational. Many square roots of integers are irrational.

Not every number has a *real number* square root. For example, there is no real number that can be squared to get -36. (The square of a real number can never be negative.) Because of this, $\sqrt{-36}$ is not a real number. (A calculator will show an error message in a case like this.)

If a is a negative real number, $\sqrt{a}$ is not a real number.

■ **EXAMPLE 4** *Identifying Types of Square Roots*

Tell whether each square root is rational, irrational, or not a real number.

(a) $\sqrt{17}$

Because 17 is not a perfect square, $\sqrt{17}$ is irrational.

(b) $\sqrt{64}$

The number 64 is a perfect square, 8^2, so $\sqrt{64} = 8$, a rational number.

(c) $\sqrt{-25}$

There is no real number whose square is -25. Therefore, $\sqrt{-25}$ is not a real number. ■

WORK PROBLEM 4 AT THE SIDE.

4. Tell whether each square root is *rational, irrational,* or *not a real number.*

(a) $\sqrt{9}$

(b) $\sqrt{7}$

(c) $\sqrt{\frac{4}{9}}$

(d) $\sqrt{72}$

(e) $\sqrt{-43}$

Note Not all irrational numbers are square roots of integers. For example, π (approximately 3.14159) is an irrational number that is not a square root of any integer.

3 Even if a number is irrational, a decimal that approximates the number can be found by using a calculator.

For example, if we use a calculator to find $\sqrt{10}$, the display will show 3.16227766, which is only a rational approximation of $\sqrt{10}$.

ANSWERS
3. (a) 4 (b) −13 (c) −15 (d) 27 (e) $\frac{6}{5}$
4. (a) rational (b) irrational (c) rational (d) irrational (e) not a real number

Calculators *Passages discussing use of the calculator as well as special calculator exercises are included wherever appropriate throughout the text.*

...TS AND RADICALS

...ball, Texas, at the same ...rth at 25 miles per hour, ... west at 60 miles per hour. ...ey after 3 hours?

Tomball

56. A boat is being pulled toward a dock with a rope attached at water level. When the boat is 24 feet from the dock, 30 feet of rope is extended. What is the height of the dock above the water?

57. What is the value of x in the figure?

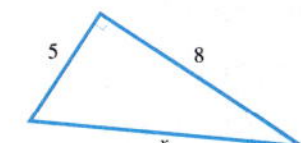

58. What is the value of y in the figure?

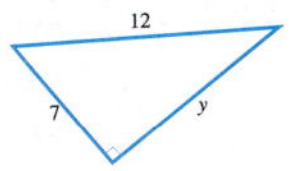

59. Use specific values for a and b different from those given in the "Caution" following Example 6 to show that $\sqrt{a^2 + b^2} \neq a + b$.

60. Why would the values $a = 0$ and $b = 1$ *not* be satisfactory in Exercise 59?

Find each of the following roots that are real numbers.

61. $\sqrt[3]{1000}$ **62.** $\sqrt[3]{8}$ **63.** $\sqrt[3]{125}$ **64.** $\sqrt[3]{216}$

65. $\sqrt[3]{-27}$ **66.** $\sqrt[3]{-64}$ **67.** $\sqrt[4]{625}$ **68.** $\sqrt[4]{10{,}000}$

69. $\sqrt[4]{-1}$ **70.** $\sqrt[4]{-625}$ **71.** $-\sqrt[5]{243}$ **72.** $-\sqrt[5]{100{,}000}$

PREVIEW EXERCISES

Write each number in prime factored form. See Section 4.1.

73. 72 **74.** 100 **75.** 40 **76.** 242

Margin Exercises *Practice exercises appear in the margins throughout the text. They parallel each example and keep students involved with the presentation by allowing them to practice new concepts immediately. Answers conveniently appear at the bottom of the page.*

Cautionary Remarks *Common student errors and difficulties are highlighted graphically and identified with the heading "Caution." Important comments are similarly highlighted with the heading "Note."*

Examples *More than 650 examples include detailed, step-by-step solutions and descriptive side comments in color. Each example includes a brief descriptive title to help students understand the purpose of the example and to aid students in studying for the examinations.*

348 CHAPTER 5 RATIONAL EXPRESSIONS

2. Solve each problem.

(a) In 1989, Emerson Fitipaldi won the Indianapolis 500 (mile) race in 2.984 hours. What was his average speed?

(b) The winner of the 1988 Indianapolis 500 (mile) race was Rick Mears, who drove his Penske-Chevy V8 at an average speed of 144.8 miles per hour. What was Mears' driving time?

(c) A small plane flew from Warsaw to Rome averaging 164 miles per hour. The trip took 2 hours. What is the distance between Warsaw and Rome?

EXAMPLE 2 *Finding Distance, Rate, or Time*

(a) The speed of sound is 1088 feet per second at sea level at 32°F. In 5 seconds under these conditions, sound travels

$$\underset{\text{Rate}}{1088} \times \underset{\text{Time}}{5} = \underset{\text{Distance}}{5440} \text{ feet.}$$

Here, we found distance given rate and time, using $d = rt$.

(b) Over a short distance, an elephant can travel at a rate of 25 miles per hour. In order to travel $\frac{1}{4}$ mile, it would take an elephant

$$\frac{\frac{1}{4}}{25} = \frac{1}{4} \times \frac{1}{25} = \frac{1}{100} \text{ hour.}$$

Distance → (numerator); Rate → (denominator); ← Time

Here, we find time given rate and distance, using $t = \frac{d}{r}$. To convert $\frac{1}{100}$ hour to minutes, multiply $\frac{1}{100}$ by 60 to get $\frac{60}{100}$ or $\frac{3}{5}$ minute. To convert $\frac{3}{5}$ minute to seconds, multiply $\frac{3}{5}$ by 60 to get 36 seconds.

(c) In the 1988 Olympic Games, the USSR won the 400-meter relay with a time of 38.19 seconds. The rate of the team was

$$\frac{400}{38.19} = 10.47 \text{ (rounded) meters per second.}$$

Distance → 400; Time → 38.19; ← Rate

This answer was obtained using a calculator. Here, we found rate given distance and time, using $r = \frac{d}{t}$. ■

WORK PROBLEM 2 AT THE SIDE.

Many applied problems use the formulas just discussed. The next example shows how to solve a typical application of the formula $d = rt$. A strategy for solving such problems involves two major steps:

SOLVING MOTION PROBLEMS

Step 1 Set up a sketch showing what is happening in the problem.
Step 2 Make a chart using the information given in the problem, along with the unknown quantities.

The chart will help you organize the information, and the sketch will help you set up the equation.

EXAMPLE 3 *Solving a Motion Problem*

Two cars leave Baton Rouge, Louisiana, at the same time and travel east on Interstate 12. One travels at a constant speed of 55 miles per hour and the other travels at a constant speed of 63 miles per hour. In how many hours will the distance between them be 24 miles?

Since we are looking for time, let t = the number of hours until the distance between them is 24 miles. The sketch in Figure 1 shows what is happening in the problem. Now, construct a chart like the one that follows. Fill in the information given in the problem, and use t for the time traveled by each car. Multiply rate by time to get the expressions for distances traveled.

150 CHAPTER 2 SOLVING EQUATIONS AND INEQUALITIES

5. Solve each equation.

(a) $\frac{z}{2} = \frac{z+1}{3}$

(b) $\frac{p+3}{3} = \frac{p-5}{4}$

6. On a map, 12 inches represents 500 miles. How many miles would be represented by 30 inches?

ANSWERS
5. (a) 2 (b) −27
6. 1250 miles

Caution The cross product method cannot be used directly if there is more than one term on either side.

EXAMPLE 5 *Solving an Equation Using Cross Products*

Solve the equation

$$\frac{m-2}{5} = \frac{m+1}{3}.$$

Find the cross products.

$3(m-2) = 5(m+1)$	Be sure to use parentheses.
$3m - 6 = 5m + 5$	Distributive property
$3m = 5m + 11$	Add 6.
$-2m = 11$	Subtract 5m.
$m = -\frac{11}{2}$	Divide by −2. ■

WORK PROBLEM 5 AT THE SIDE.

Note Remember, when you set cross products equal, you are really multiplying each ratio in the proportion by a common denominator.

4 Proportions are useful in many practical applications.

EXAMPLE 6 *Applying Proportions*

A local drugstore is offering 3 packs of toothpicks for $.87. How much would be charged for 10 packs?

Let x represent the cost of 10 packs of toothpicks.

Set up a proportion. One ratio in the proportion can involve the number of packs, and the other can involve the costs. Make sure that the corresponding numbers appear in the numerator and denominator.

$$\frac{\text{Cost of 3}}{\text{Cost of 10}} = \frac{3}{10}$$

$\frac{.87}{x} = \frac{3}{10}$	
$3x = .87(10)$	Cross products
$3x = 8.7$	
$x = 2.90$	Divide by 3.

The 10 packs should cost $2.90. As we saw earlier, the proportion could also be written as

$$\frac{3}{.87} = \frac{10}{x},$$

which would give the same cross products. ■

WORK PROBLEM 6 AT THE SIDE.

EXERCISES

As a key feature of the text, approximately 5800 exercises are provided—including approximately 1100 review exercises, 400 conceptual and writing exercises, and 900 margin problems. Every exercise set has been completely revised, with more than 80 percent of the exercises being new. Care has been taken to pair exercises (evens with odds) and to grade the exercises with regard to increasing difficulty. Exercises now include a number of special types.

Conceptual and writing exercises are designed to require a deeper understanding of concepts. Nearly 150 of these exercises require the student to respond by writing a few sentences. Answers are not given for the writing exercises because they are open-ended, and instructors may use them in different ways.

Challenging exercises require the student to go beyond the examples in the text.

Applications now include more realistic and interesting examples, in many cases using actual data from current events, sports, etc.

Calculator exercises are included in selected sections where use of the scientific calculator is appropriate. The first calculator exercise in a group is marked with ▦. Subsequent exercises are identified with a colored exercise number.

Cumulative reviews end each chapter after Chapter 1 with a set of exercises that test students on the topics covered from the beginning of the text up to that point.

Preview Exercises, formerly called *Skill Sharpeners*, are intended to review the basic skills needed to do the work in the next section. These exercises also help to show how earlier material connects with and is needed for later topics.

SUPPLEMENTS

Our extensive supplemental package includes an annotated instructor's edition, testing materials, solutions, software, and videotapes.

For the Instructor

Annotated Instructor's Edition
This edition provides instructors with immediate access to the answers to every exercise in the text, with the exception of writing exercises. Each answer is printed in color next to the corresponding text exercise. Symbols are used to identify the conceptual (●) and writing (✎) exercises to assist in making homework assignments. Calculator (▦) and challenging (▲) exercises are also marked for the instructor for this purpose.

Instructor's Resource Manual
The Instructor's Resource Manual includes suggestions for using the textbook in a mathematics laboratory; short-answer and multiple-choice versions of a placement test; six forms of chapter tests for each chapter, including four open-response and two multiple-choice forms; short-answer and multiple-choice forms of a final examination; and an extensive set of additional exercises, providing 10 to 20 exercises for each textbook objective, which instructors can use as an additional source of questions for tests, quizzes, or student review of difficult topics. In addition, a new section containing teaching tips has been added for the instructor's convenience. This manual also includes a list of all conceptual, writing, challenging, and calculator exercises.

Instructor's Solution Manual
Available at no charge to instructors, this book includes solutions to the even-numbered section exercises. The two solution manuals plus the solutions given at the back of the textbook provide detailed, worked-out solutions to each exercise and margin problem in the book. This manual also includes a list of all conceptual, writing, challenging, and calculator exercises.

Instructor's Answer Manual
This manual includes answers to all exercises and a list of conceptual, writing, challenging, and calculator exercises.

HarperCollins Test Generator/Editor for Mathematics with QuizMaster
Available in IBM and Macintosh versions, the test generator is fully networkable. The test generator enables instructors to select questions by objective, section, or chapter, or to use a ready-made test for each chapter. The editor enables instructors to edit any preexisting data or to easily create their own questions. The software is algorithm driven, allowing the instructor to generate constants while maintaining problem type, providing a nearly unlimited number of available test or quiz items in multiple-choice and/or open-response formats for one or more test forms. The system features printed graphics and accurate mathematics symbols. **QuizMaster** enables instructors to create tests and quizzes using the Test Generator/Editor and save them to disk so students can take the test or quiz on a stand-alone computer or network. **QuizMaster** then grades the test or quiz and allows the instructor to create reports on individual students or entire classes. CLAST and TASP versions of this package are also available for IBM and Mac machines.

For the Student

Student's Solution Manual
This book contains solutions to every other odd-numbered section exercise (those not included at the back of the textbook) as well as solutions to all margin problems, chapter review exercises, chapter tests, and cumulative review exercises. (ISBN 0-673-99062-1)

Interactive Mathematics Tutorial Software with Management System
This innovative package is available in DOS, Windows, and Macintosh versions and is fully networkable. As with the Test Generator/Editor, this software is algorithm driven, which automatically regenerates constants, so students will not see the numbers repeat in a problem type if they revisit any particular section. The tutorial is objective-based, self-paced, and provides unlimited opportunities to review lessons and to practice problem solving. If students give a wrong answer, they can ask to see the problem worked out and get a textbook page reference. The program is menu-driven for ease of use, and on-screen help can be obtained at any time with a single keystroke. Students' scores are automatically recorded and can be printed for a permanent record. The optional **Management System** lets instructors record student scores on disk and print diagnostic reports for individual students or classes. CLAST and TASP versions of this tutorial are also available for both IBM and Mac machines. This software may also be purchased by students for use outside the classroom or lab.

Videotapes
A new videotape series has been developed to accompany *Introductory Algebra*, Fifth Edition. In a separate lesson for each section in the book, the series covers all objectives, topics, and problem-solving techniques discussed within the text.

Overcoming Math Anxiety
This book, written by Randy Davidson and Ellen Levitov, includes step-by-step guides to problem solving, note taking, and applied problems. Discover the reasons behind math anxiety and ways to overcome those obstacles. Learn relaxation techniques, build better math skills, and improve study habits. This is the answer for anyone who is nervous with numbers! (ISBN 0-06-501651-3)

ACKNOWLEDGMENTS

We appreciate the many contributions of users of the fourth edition of the book. We also wish to thank our reviewers for their insightful comments and suggestions:

Carla K. Ainsworth, *Salt Lake Community College*
Lisa G. Angelo, *Bucks County Community College*
T. Patrick Burke, *Naugatuck Valley Community Technical College*
Dennis Carrie, *Golden West College*
Sandra Letha Cox, *Rend Lake College*
Mimi Elwell, *Lake Michigan College*
Mavis Escondel, *Nicholls State University*
Patsy J. Fagan, *Drake University*
Kathy Fenimore, *Frederick Community College*
Dorothy M. Fitzgerald, *Golden West College*
Beth Fraser, *Middlesex Community College*
John J. Gaudio, *Waubonsee Community College*
Gail C. Gonyo, *Adirondack Community College*
Newton Grant, *Delgado Community College*
D. Michael Hamm, *Brookhaven College*
James E. Harrington, *Adirondack Community College*
Bruce H. Hoelter, *Raritan Valley Community College*
Winfield A. Ihlow, *Miami Dade Community College*
Phyllis H. Jore, *Valencia Community College*
Margie Keener, *Oklahoma State University*
Harvey A. Leboff, *Massachusetts Bay Community College*
Timothy E. McKenna, *University of Michigan at Dearborn*
Susan R. Mills, *Riverside Community College*
Elmo Moore, *Humboldt State University*
Nancy Myers, *University of Southern Indiana*
Vicki Norwich, *Brevard Community College*
Frank W. Post, *South Seattle Community College*
David Pruis, *Grand Rapids Community College*
Jerry Shipman, *Alabama A & M University*
Janet M. Sibol, *Hillsborough Community College–Dale Mabry Campus*
Lori Smellegar, *Manatee Community College*

As always, Paul Elderveld, *College of DuPage*, has done an outstanding job of coordinating all the print supplements for us.

We also wish to thank those who did an excellent job checking all the answers for us: T. Patrick Burke, *Naugatuck Valley Community Technical College*; Mimi Elwell, *Lake Michigan College*; Kevin Hopkins, *Southwest Baptist University*; Ken Johnston, *Hinds Community College*; Margie Keener, *Oklahoma State University*; Thomas Nichols, *North Central Technical College*; Jeannette O'Rourke, *Middlesex County College*; Kathleen L. Pellissier; and Kathy Timblin, *Waubonsee Community College*.

We would like to thank Tracey Hoy, *College of Lake County*, and Gerald Krusinski, *College of DuPage*, for writing the solutions at the back of the text, and Theresa McGinnis for editing those solutions.

Our appreciation also goes to Tommy Thompson, *Cedar Valley College*, for his suggestions for the feature "To the Student: Success in Algebra." Special thanks go to the dedicated staff at HarperCollins who have worked so long and hard to make this book a success: Karin Wagner, Sandi Goldstein, Ann-Marie Buesing, Anne Kelly, Linda Youngman, Kevin Connors, and Ed Moura.

Margaret L. Lial
E. John Hornsby, Jr.

An Introduction to Scientific Calculators

The emphasis placed on paper-and-pencil computation, which has long been a part of the school mathematics curriculum, is not as important as it once was. The search for easier ways to calculate and compute has culminated in the development of hand-held calculators and computers. Professional organizations devoted to the teaching and learning of mathematics recommend that this technology be incorporated throughout the mathematics curriculum. In light of this recommendation, this text assumes that all students have access to calculators, and the authors recommend that students become computer literate to the extent their academic resources and individual finances allow.

In this text we view calculators as a means of allowing students to spend more time on the conceptual nature of mathematics and less time on the drudgery of computation with paper and pencil. Calculators come in a large array of different types, sizes, and prices, making it difficult to decide which machine best fits your needs. For the general population, a calculator that performs the operations of arithmetic and a few other functions is sufficient. These are known as **four-function calculators.** Students who take higher mathematics courses (engineers, for example) usually need the added power of **scientific calculators. Programmable calculators,** which allow short programs to be written and performed, and **graphing calculators,** which actually plot graphs on small screens, are also available. Keep in mind that calculators differ from one manufacturer to the other. For this reason, remember the following.

Always refer to your owner's manual if you need assistance in performing an operation with your calculator.

Because of their relatively inexpensive cost and features that far exceed those of four-function calculators, scientific calculators probably provide students the best value for their money. For this reason, we will give a short synopsis of the major functions of scientific calculators.

Most scientific calculators use *algebraic logic.* (Models sold by Texas Instruments, Sharp, Casio, and Radio Shack, for example, use algebraic logic.) A notable exception is Hewlett Packard, a company whose calculators use *Reverse Polish Notation* (RPN). In this introduction, we explain how to use calculators with algebraic logic.

Arithmetic Operations

To perform an operation of arithmetic, simply enter the first number, touch the operation key (+ , − , × , or ÷), enter the second number, and then touch the = key. For example, to add 4 and 3, use the following keystrokes.

(The final answer is displayed in color.)

Change Sign Key

The key marked ± allows you to change the sign of a display. This is particularly useful when you wish to enter a negative number. For example, to enter −3, use the following keystrokes.

Parentheses Keys

These keys, (and), are designed to allow you to group numbers in arithmetic operations as you desire. For example, if you wish to evaluate $4 \times (6 + 3)$, use the following keystrokes.

Memory Key

Scientific calculators can hold a number in memory for later use. The label of the memory key varies among models; two versions are M and STO. M+ and M− allow you to add to or subtract from the value currently in memory. The memory recall key, labeled MR, RM, or RCL, allows you to retrieve the value stored in memory.

Suppose that you wish to store the number 5 in memory. Enter 5, then touch the key for memory. You can then perform other calculations. When you need to retrieve the 5, touch the key for memory recall.

If a calculator has a constant memory feature, the value in memory will be retained even after the power is turned off. Some advanced calculators have more than one memory. It is best to read the owner's manual for your model to see exactly how memory is activated.

Clearing/Clear Entry Keys

These keys allow you to clear the display or clear the last entry entered into the display. They are usually marked C and CE. In some models, touching the C key once will clear the last entry, while touching it twice will clear the entire operation in progress.

Second Function Key

This key is used in conjunction with another key to activate a function that is printed *above* an operation key (and not on the key itself). It is usually marked 2nd. For example, suppose you wish to find the square of a number, and the squaring function (explained in more detail later) is printed above another key. You would need to touch 2nd before the desired squaring function can be activated.

Some newer models of scientific calculators (the TI-35X by Texas Instruments, for example) even provide a third function key, marked 3rd, which is used in a manner similar to the one described for the second function key.

Square Root and Cube Root Keys

Touching the square root key, $\sqrt{x}$, will give the square root (or an approximation of the square root) of the number in the display. For example, to find the square root of 36, use the following keystrokes.

The square root of 2 is an example of an irrational number (See Section 8.1). The calculator will give an approximation of its value, since the decimal for $\sqrt{2}$ never terminates and never repeats. The number of digits shown will vary among models. To find an approximation of $\sqrt{2}$, use the following keystrokes.

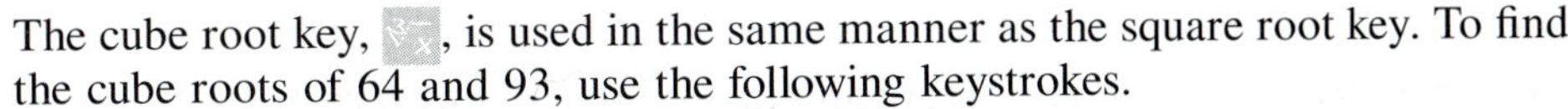

The cube root key, $\sqrt[3]{x}$, is used in the same manner as the square root key. To find the cube roots of 64 and 93, use the following keystrokes.

Note: Calculators differ in the number of digits provided in the display. For example, one calculator gives the approximation of the cube root of 93 as 4.530654896, showing more digits than shown above.

Squaring and Cubing Keys

The squaring key, x^2, allows you to square the entry in the display. For example, to square 35.7, use the following keystrokes.

The squaring key and the square root key are often found on the same key, with one of them being a second function (that is, activated by the second function key, described above).

The cubing key, x^3, allows you to cube the entry in the display. To cube 3.5, follow these keystrokes.

Reciprocal Key

The key marked $1/x$ (or x^{-1}) is the reciprocal key. (When two numbers have a product of 1, they are called *reciprocals.*) Suppose you wish to find the reciprocal of 5. Use the following keystrokes.

Inverse Key

Some calculators have an inverse key, marked INV. Inverse operations are operations that "undo" each other. For example, the operations of squaring and taking the square root are inverse operations. The use of the INV key varies among different models of calculators, so read your owner's manual carefully.

Exponential Key

This key, marked x^y or y^x, allows you to raise a number to a power. For example, if you wish to raise 4 to the fifth power (that is, find 4^5), use the following keystrokes.

Root Key

Some calculators have this key specifically marked $\sqrt[y]{x}$ or $\sqrt[x]{y}$; with others, the operation of taking roots is accomplished by using the inverse key in conjunction with the exponential key. Suppose, for example, your calculator is of the latter type, and you wish to find the fifth root of 1024. Use the following keystrokes.

Notice how this "undoes" the operation explained in the exponential key discussion earlier.

Pi Key

The number π is important in mathematics. It occurs, for example, in the area and circumference formulas for a circle. By touching the π key, you can get the display of the first few digits of π. (Because π is irrational, the display shows only an approximation.) One popular model gives the following display when the π key is activated: 3.1415927. As mentioned before, calculators will vary in the number of digits they give for π in their displays.

OTHER CONSIDERATIONS

When decimal approximations are shown on scientific calculators, they are either *truncated* or *rounded.* To see which of these a particular model is programmed to do, evaluate 1/18 as an example. If the display shows .0555555 (last digit 5), it truncates the display. If it shows .0555556 (last digit 6), it rounds off the display.

When very large or very small numbers are obtained as answers, scientific calculators often express these numbers in scientific notation. For example, if you multiply 6,265,804 by 8,980,591, the display might look like this:

5.6270623 13

The "13" at the far right means that the number on the left is multiplied by 10^{13}. This means that the decimal point must be moved 13 places to the right if the answer is to be expressed in its usual form. Even then, the value obtained will only be an approximation: 56,270,623,000,000.

Two features of advanced scientific calculators are programmability and graphing capability. A programmable calculator has the capability of running small programs, much like a mini-computer. A graphing calculator can be used to plot graphs of functions on a small screen. One of the issues in mathematics education today deals with how graphing calculators should be incorporated into the curriculum. Their availability in the 1990s parallels the availability of scientific calculators in the 1980s, and they will no doubt be a major component in mathematics education as we move into the twenty-first century.

To The Student: Success in Algebra

The main reason students have difficulty with mathematics is that they don't know how to study it. Studying mathematics *is* different from studying subjects like English or history. The key to success is regular practice.

This should not be surprising. After all, can you learn to play the piano or to ski well without a lot of regular practice? The same thing is true for learning mathematics. Working problems nearly every day is the key to becoming successful. Here is a list of things that will help you succeed in studying algebra.

1. *Attend class regularly*. Pay attention to what your teacher says and does in class, and take careful notes. In particular, note the problems the teacher works on the board and copy the complete solutions. Keep these notes separate from your homework to avoid confusion when you read them over later.
2. Don't hesitate to ask questions in class. It is not a sign of weakness, but of strength. There are always other students with the same question who are too shy to ask.
3. *Read your text carefully*. Many students read only enough to get by, usually only the examples. Reading the complete section will help you solve the homework problems. Most exercises are keyed to specific examples or objectives that will explain the procedures for working them.
4. Before you start on your homework assignment, rework the problems the teacher worked in class. This will reinforce what you have learned. Many students say, "I understand it perfectly when you do it, but I get stuck when I try to work the problem myself."
5. Do your homework assignment only *after* reading the text and reviewing your notes from class. Check your work against the answers in the back of the book. If you get a problem wrong and are unable to understand why, mark that problem and ask your instructor about it. Then practice working additional problems of the same type to reinforce what you have learned.
6. Work as neatly as you can. Write your symbols clearly, and make sure the problems are clearly separated from each other. Working neatly will help you to think clearly and also make it easier to review the homework before a test.
7. After you complete a homework assignment, look over the text again. Try to identify the main ideas that are in the lesson. Often they are clearly highlighted or boxed in the text.
8. Use the chapter test at the end of each chapter as a practice test. Work through the problems under test conditions, without referring to the text or the answers until you are finished. You may want to time yourself to see how long it takes you. When you finish, check your answers against those in the back of the book, and study the problems you missed. Answers are keyed to the appropriate sections of the text.
9. Keep any quizzes and tests that are returned to you, and use them when you study for future tests and the final exam. These quizzes and tests indicate what your instructor considers most important. Be sure to correct any problems on these tests that you missed, so you will have the corrected work to study.
10. Don't worry if you do not understand a new topic right away. As you read more about it and work through the problems, you will gain understanding. Each time you review a topic you will understand it a little better. No one understands each topic completely right from the start.

R Prealgebra Review

R.1 FRACTIONS

OBJECTIVES

1. Identify prime numbers.
2. Write numbers in prime factored form.
3. Write fractions in lowest terms.
4. Multiply and divide fractions.
5. Add and subtract fractions.

FOR EXTRA HELP

Tape 1

SSM pp. 1–5

MAC: B IBM: B

Studying algebra requires good arithmetic skills. Most people do not get much practice using fractions, so we review the rules for fractions in this section.

The numbers used most often in everyday life are the **whole numbers,**

$$0, 1, 2, 3, 4, 5, \ldots$$

and the **fractions,** such as

$$\frac{1}{3}, \frac{5}{4}, \quad \text{and} \quad \frac{11}{12}.$$

The parts of a fraction are named as follows.

$$\frac{4}{7} \begin{array}{l} \leftarrow \textbf{numerator} \\ \leftarrow \textbf{denominator} \end{array}$$

If the numerator of a fraction is smaller than the denominator, we call it a **proper fraction.** If the reverse is true, the fraction is an **improper fraction.** An improper fraction is often written as a **mixed number.** For example, $\frac{12}{5}$ may be written as $2\frac{2}{5}$. In algebra, we prefer to use the improper form.

1 In work with fractions, we will need to write the numerators and denominators as products. When 12 is written as $2 \cdot 6$, for example, 2 and 6 are called **factors** of 12. Other factors of 12 are 1, 3, 4, and 12. A whole number is **prime** if it has exactly two different factors (itself and 1). The first dozen primes are listed here.

$$2, 3, 5, 7, 11, 13, 17, 19, 23, 29, 31, 37$$

A whole number that is not prime is called a **composite** number. The number 1 is neither prime nor composite.

1. Tell whether each number is prime or composite.

(a) 12

(b) 13

(c) 27

(d) 59

(e) 1806

ANSWERS
1. (b) and (d) are prime; the others are composite.

EXAMPLE 1 *Distinguishing between Prime and Composite Numbers*
Decide which of the following numbers is prime.

(a) 33

33 has factors of 3 and 11 as well as 1 and 33, so it is composite.

(b) 43

Divide 43 by consecutive prime numbers until one divides evenly, or the quotient is smaller than the number you are dividing by. A calculator will speed up the work. If the quotient is smaller than the divisor, no larger prime number will be a factor. Since none of the primes up to and including 7 are factors, and the quotient is $6\frac{1}{7}$, which is smaller than 7, 43 is prime.

(c) 9832

9832 can be divided by 2, so it is composite. ■

WORK PROBLEM 1 AT THE SIDE.

2 As mentioned earlier, to factor a number means to write it as the product of two or more numbers. Factoring is just the reverse of multiplying two numbers to get the product.

Multiplication	Factoring
$6 \cdot 3 = 18$	$18 = 6 \cdot 3$
↑ ↑ ↑	↑ ↑ ↑
factors product	product factors

In algebra raised dots are used instead of the × symbol to indicate multiplication. Each composite number can be written as the product of prime numbers. A number written as the product of prime numbers is in **prime factored form.**

EXAMPLE 2 *Writing Numbers in Prime Factored Form*
Write each number in prime factored form.

(a) 35

Factor 35 as the product of the prime factors 5 and 7, or as

$$35 = 5 \cdot 7.$$

(b) 24

A handy way to keep track of the factors is to use a tree, as shown at the side. The prime factors are circled.

Divide by the smallest prime, 2, to get $24 = 2 \cdot 12$.

Now divide 12 by 2 to find factors of 12. $24 = 2 \cdot 2 \cdot 6$

Since 6 can be factored as $2 \cdot 3$, $24 = 2 \cdot 2 \cdot 2 \cdot 3$

where all factors are prime. ■

24
(2) · 12
(2) · 6
(2) · (3)

WORK PROBLEM 2 AT THE SIDE.

3 We use prime factors to simplify fractions to **lowest terms.** A fraction is in lowest terms when the numerator and denominator have no factors in common (other than 1). We can write a fraction in this form by using the following fact.

PROPERTY OF 1

A nonzero number divided by itself is equal to 1.

For example,

$$\frac{3}{3}=1, \quad \frac{8}{8}=1, \quad \text{and} \quad \frac{17}{17}=1.$$

WRITING A FRACTION IN LOWEST TERMS

Step 1 Write the numerator and denominator in prime factored form.

Step 2 Replace each pair of factors common to the numerator and denominator with 1.

Step 3 Multiply the remaining factors in the numerator and in the denominator.

EXAMPLE 3 *Writing Fractions in Lowest Terms*

Write each fraction in lowest terms.

(a) $\frac{10}{15}=\frac{2\cdot 5}{3\cdot 5}=\frac{2}{3}\cdot\frac{5}{5}=\frac{2}{3}\cdot 1=\frac{2}{3}$

Since 5 is a common factor of 10 and 15, we use the property of 1 to replace $\frac{5}{5}$ with 1.

(b) $\frac{15}{45}=\frac{3\cdot 5}{3\cdot 3\cdot 5}=\frac{1\cdot 3\cdot 5}{3\cdot 3\cdot 5}=\frac{1}{3}\cdot\frac{3}{3}\cdot\frac{5}{5}=\frac{1}{3}\cdot 1\cdot 1=\frac{1}{3}$

Multiplying by 1 in the numerator does not change the value of the numerator and makes it possible to rewrite the expression as the product of three fractions in the next step.

(c) $\frac{150}{200}$

It is not always necessary to factor into *prime* factors in Step 1. Here, if you see that 50 is a common factor of the numerator and the denominator, factor as follows:

$$\frac{150}{200}=\frac{3\cdot 50}{4\cdot 50}=\frac{3}{4}\cdot 1=\frac{3}{4}. \quad ■$$

2. Write each number in prime factored form.

(a) 70

(b) 72

(c) 693

(d) 97

ANSWERS

2. (a) $2\cdot 5\cdot 7$ (b) $2\cdot 2\cdot 2\cdot 3\cdot 3$ (c) $3\cdot 3\cdot 7\cdot 11$ (d) 97 is prime.

3. Write each fraction in lowest terms.

(a) $\frac{8}{14}$

(b) $\frac{35}{42}$

(c) $\frac{120}{72}$

4. Find each product, and write it in lowest terms.

(a) $\frac{5}{8} \cdot \frac{2}{10}$

(b) $\frac{1}{10} \cdot \frac{12}{5}$

(c) $\frac{7}{9} \cdot \frac{12}{14}$

ANSWERS

3. (a) $\frac{4}{7}$ (b) $\frac{5}{6}$ (c) $\frac{5}{3}$

4. (a) $\frac{1}{8}$ (b) $\frac{6}{25}$ (c) $\frac{2}{3}$

Note When you are writing a fraction in lowest terms, look for the largest common factor in the numerator and denominator. If none is obvious, factor the numerator and the denominator into prime factors. *Any* common factor can be used, and the fraction reduced in stages. For example,

$$\frac{150}{200} = \frac{15 \cdot 10}{20 \cdot 10} = \frac{3 \cdot 5 \cdot 10}{4 \cdot 5 \cdot 10} = \frac{3}{4}.$$

WORK PROBLEM 3 AT THE SIDE.

4 We are now ready to review multiplication of fractions.

MULTIPLYING FRACTIONS

To multiply two fractions, we multiply the numerators to get the numerator of the product, and multiply the denominators to get the denominator of the product.

In practice, we often simplify before performing the multiplication as shown in the next example.

EXAMPLE 4 *Multiplying Fractions*

Find the product of $\frac{3}{8}$ and $\frac{4}{9}$, and write it in lowest terms.

$$\frac{3}{8} \cdot \frac{4}{9} = \frac{3 \cdot 4}{8 \cdot 9} \quad \text{Multiply numerators. Multiply denominators.}$$

$$= \frac{3 \cdot 4}{2 \cdot 4 \cdot 3 \cdot 3} \quad \text{Factor.}$$

$$= \frac{1}{2 \cdot 3} = \frac{1}{6} \quad \text{Write in lowest terms.}$$

WORK PROBLEM 4 AT THE SIDE.

Two fractions are **reciprocals** of each other if their product is 1. For example, $\frac{3}{4}$ and $\frac{4}{3}$ are reciprocals because

$$\frac{3}{4} \cdot \frac{4}{3} = 1.$$

The numbers $\frac{7}{11}$ and $\frac{11}{7}$ are reciprocals also. We use reciprocals to divide fractions.

DIVIDING FRACTIONS

To divide two fractions, we multiply the first fraction and the reciprocal of the second.

The reason this method works will be explained later. The answer to a division problem is called the **quotient.** For example, the quotient of 20 and 10 is 2, since $20 \div 10 = 2$. (Note that the quotient of 10 and 20 is $10 \div 20 = \frac{1}{2}$.)

EXAMPLE 5 *Dividing Fractions*

Find the following quotients, and write them in lowest terms.

(a) $\frac{3}{4} \div \frac{8}{5} = \frac{3}{4} \cdot \frac{5}{8} = \frac{3 \cdot 5}{4 \cdot 8} = \frac{15}{32}$

↑ Multiply by the reciprocal of second fraction.

(b) $\frac{3}{4} \div \frac{5}{8} = \frac{3}{4} \cdot \frac{8}{5} = \frac{3 \cdot 8}{4 \cdot 5} = \frac{6}{5}$

(c) $\frac{5}{8} \div 10 = \frac{5}{8} \div \frac{10}{1} = \frac{5}{8} \cdot \frac{1}{10} = \frac{1}{16}$ ■

↑ Write 10 as $\frac{10}{1}$.

WORK PROBLEM 5 AT THE SIDE.

5. Find each quotient, and write it in lowest terms.

(a) $\frac{3}{10} \div \frac{2}{7}$

(b) $\frac{3}{4} \div \frac{7}{16}$

(c) $\frac{4}{3} \div 6$

5 The result of adding two numbers is called the *sum* of the numbers. For example, since $2 + 3 = 5$, the sum of 2 and 3 is 5.

ADDING FRACTIONS

To find the **sum** of two fractions with the same denominator, we add their numerators, keeping the same denominator.

EXAMPLE 6 *Adding Fractions with the Same Denominator*

Add. Write sums in lowest terms.

(a) $\frac{3}{7} + \frac{2}{7} = \frac{3+2}{7} = \frac{5}{7}$ Denominator does not change.

(b) $\frac{2}{10} + \frac{3}{10} = \frac{2+3}{10} = \frac{5}{10} = \frac{1}{2}$ Write in lowest terms. ■

WORK PROBLEM 6 AT THE SIDE.

6. Add. Write sums in lowest terms.

(a) $\frac{3}{5} + \frac{4}{5}$

(b) $\frac{5}{14} + \frac{3}{14}$

If the fractions to be added have different denominators, we use the property of 1 to write them with a common denominator. To do this, we first find the **least common denominator** (or **LCD**), the smallest number that has both denominators as factors.

EXAMPLE 7 *Adding Fractions with Different Denominators*

Add each of the following.

(a) $\frac{4}{15} + \frac{5}{9}$

To find the LCD, we first factor both denominators.

$$15 = 5 \cdot 3 \quad \text{and} \quad 9 = 3 \cdot 3$$

The LCD must have each factor of each denominator. In this example, the LCD needs one factor of 5 and two factors of 3, because the second denominator has two factors of 3. The LCD should be the smallest multiple of both denominators. Here,

$$\text{LCD} = 5 \cdot 3 \cdot 3 = 45.$$

ANSWERS

5. (a) $\frac{21}{20}$ (b) $\frac{12}{7}$ (c) $\frac{2}{9}$

6. (a) $\frac{7}{5}$ (b) $\frac{4}{7}$

We can use the property of 1 to write each fraction with the LCD as denominator.

$$\frac{4}{15} = \frac{4}{15} \cdot \frac{3}{3} = \frac{12}{45} \quad \text{and} \quad \frac{5}{9} = \frac{5}{9} \cdot \frac{5}{5} = \frac{25}{45}$$

Now we add the two equivalent fractions to get the required sum.

$$\frac{4}{15} + \frac{5}{9} = \frac{12}{45} + \frac{25}{45} = \frac{37}{45}$$

(b) $3\frac{1}{2} + 2\frac{3}{4}$

Change both mixed numbers to improper fractions, as follows.

$$3\frac{1}{2} = 3 + \frac{1}{2} = \frac{3}{1} + \frac{1}{2} = \frac{6}{2} + \frac{1}{2} = \frac{6+1}{2} = \frac{7}{2}$$

$$2\frac{3}{4} = 2 + \frac{3}{4} = \frac{8}{4} + \frac{3}{4} = \frac{8+3}{4} = \frac{11}{4}$$

Now add. By inspection, the least common denominator is 4.

$$3\frac{1}{2} + 2\frac{3}{4} = \frac{7}{2} + \frac{11}{4} = \frac{14}{4} + \frac{11}{4} = \frac{25}{4} \quad \text{or} \quad 6\frac{1}{4}$$ ■

WORK PROBLEM 7 AT THE SIDE.

The *difference* between two numbers is found by subtracting the numbers. For example, $9 - 5 = 4$, so the difference between 9 and 5 is 4. We find the difference between two fractions as follows.

SUBTRACTING FRACTIONS

To find the **difference** between two fractions with the same denominator, we subtract their numerators, keeping the same denominator.

■ EXAMPLE 8 *Subtracting Fractions*

Subtract. Write differences in lowest terms.

(a) $\frac{15}{8} - \frac{3}{8} = \frac{15-3}{8} = \frac{12}{8} = \frac{3}{2}$ Lowest terms

(b) $\frac{15}{16} - \frac{4}{9}$

Since $16 = 2 \cdot 2 \cdot 2 \cdot 2$ and $9 = 3 \cdot 5$, with no common factors, the LCD is $16 \cdot 9 = 144$.

$$\frac{15}{16} - \frac{4}{9} = \frac{15 \cdot 9}{16 \cdot 9} - \frac{4 \cdot 16}{9 \cdot 16}$$ Get a common denominator.

$$= \frac{135}{144} - \frac{64}{144}$$ Use a calculator.

$$= \frac{71}{144}$$ Subtract.

(c) $2\frac{1}{2} - 1\frac{3}{4}$

7. Add.

(a) $\frac{7}{30} + \frac{2}{45}$

(b) $\frac{17}{10} + \frac{8}{27}$

(c) $2\frac{1}{8} + 1\frac{2}{3}$

ANSWERS

7. (a) $\frac{5}{18}$ (b) $\frac{539}{270}$ (c) $\frac{91}{24}$ or $3\frac{19}{24}$

First, change the mixed numbers $2\frac{1}{2}$ and $1\frac{3}{4}$ into improper fractions.

$$2\frac{1}{2} = 2 + \frac{1}{2} = \frac{4}{2} + \frac{1}{2} = \frac{5}{2}$$

$$1\frac{3}{4} = 1 + \frac{3}{4} = \frac{4}{4} + \frac{3}{4} = \frac{7}{4}$$

$$2\frac{1}{2} - 1\frac{3}{4} = \frac{5}{2} - \frac{7}{4} \quad \text{Write as improper fractions.}$$

$$= \frac{10}{4} - \frac{7}{4} \quad \text{Get a common denominator.}$$

$$= \frac{3}{4} \quad \text{Subtract.}$$

WORK PROBLEM 8 AT THE SIDE.

We often see mixed numbers used in applications of mathematics.

EXAMPLE 9 *Solving an Applied Problem with Fractions*

The diagram below appears in the book *Woodworker's 39 Sure-Fire Projects*. It is the front view of a corner bookcase/desk. Add the fractions shown in the diagram to find the approximate height of the bookcase/desk.

We must find the following sum (in inches).

$$\frac{3}{4} + 4\frac{1}{2} + 9\frac{1}{2} + \frac{3}{4} + 9\frac{1}{2} + \frac{3}{4} + 4\frac{1}{2}$$

Change the mixed numbers to improper fractions.

$$\frac{3}{4} + \frac{9}{2} + \frac{19}{2} + \frac{3}{4} + \frac{19}{2} + \frac{3}{4} + \frac{9}{2}$$

The LCD is 4. Change all fractions to fourths.

$$\frac{3}{4} + \frac{18}{4} + \frac{38}{4} + \frac{3}{4} + \frac{38}{4} + \frac{3}{4} + \frac{18}{4}$$

Now we can add and simplify the answer.

$$\frac{3}{4} + \frac{18}{4} + \frac{38}{4} + \frac{3}{4} + \frac{38}{4} + \frac{3}{4} + \frac{18}{4} = \frac{121}{4} = 30\frac{1}{4}$$

The approximate height is $30\frac{1}{4}$ inches.

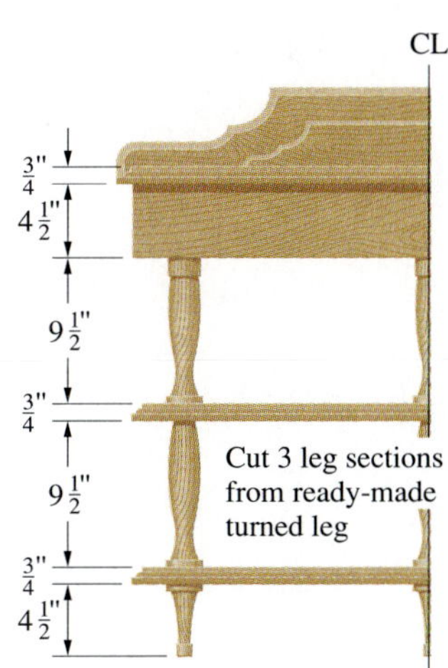

FRONT VIEW

WORK PROBLEM 9 AT THE SIDE.

8. Subtract.

(a) $\frac{9}{11} - \frac{3}{11}$

(b) $\frac{13}{15} - \frac{5}{6}$

(c) $2\frac{3}{8} - 1\frac{1}{2}$

9. If an upholsterer needs $2\frac{1}{4}$ yards of fabric to re-cover a chair, how many chairs can be re-covered with $23\frac{2}{3}$ yards of fabric?

ANSWERS

8. (a) $\frac{6}{11}$ (b) $\frac{1}{30}$ (c) $\frac{7}{8}$

9. 10 chairs (Some fabric will be left over.)

QUEST FOR NUMERACY

What is the Meaning of Numeracy?

Letter is to *number* as *literacy* is to *numeracy*. In recent years much has been written about how important it is that the general population be "numerate." The essay "Quantity" by James T. Fey in *On the Shoulders of Giants: New Approaches to Numeracy* contains the following description of an approach to numeracy.

> Given the fundamental role of quantitative reasoning in application of mathematics as well as the innate human attraction to numbers, it is not surprising that number concepts and skills form the core of school mathematics. In the earliest grades all children start on a mathematical path designed to develop computational procedures of arithmetic together with corresponding conceptual understanding that is required to solve quantitative problems and make informed decisions. Children learn many ways to describe quantitative data and relationships using numerical, graphic, and symbolic representations; to plan arithmetic and algebraic operations and to execute those plans using effective procedures; and to interpret quantitative information, to draw inferences, and to test the conclusions for reasonableness.

FOR GROUP DISCUSSION With calculator in hand, each student in the class is to attempt to fill in the boxes with the digits 3, 4, 5, 6, 7, or 8, with each digit used at most once. Then take a poll to see who was able to come up with the closest number to the "goal number." You are allowed one minute per round. Good luck!

Round I	□ × □□□□ = 30,000
Round II	□ × □□□□ = 40,000
Round III	□ × □□□□ = 50,000
Round IV	□□ × □□□ = 30,000
Round V	□□ × □□□ = 60,000

R.1 EXERCISES

NAME DATE HOUR

1. In the fraction $\frac{3}{8}$, _____ is the numerator and _____ is the denominator.

2. How may $\frac{15}{7}$ be written as a mixed number?

3. How may $3\frac{4}{5}$ be written as an improper fraction in lowest terms?

4. What is the reciprocal of $\frac{9}{8}$?

5. The answer in a multiplication problem is called the __________, and the answer in a division problem is called the __________.

6. The answer in an addition problem is called the __________, and the answer in a subtraction problem is called the __________.

Identify each number as prime, composite, or neither. See Example 1.

7. 17 **8.** 23 **9.** 54 **10.** 88

11. 3458 **12.** 2895 **13.** 1 **14.** 10

* *Write each number in prime factored form. See Example 2.*

15. 30 **16.** 40 **17.** 500 **18.** 700

19. 124 **20.** 120 **21.** 29 **22.** 31

* Color exercise numbers are used to indicate exercises designed for calculator use.

Write each fraction in lowest terms. See Example 3.

23. $\frac{8}{16}$ **24.** $\frac{4}{12}$ **25.** $\frac{15}{18}$ **26.** $\frac{16}{20}$

27. $\frac{15}{45}$ **28.** $\frac{16}{64}$ **29.** $\frac{144}{120}$ **30.** $\frac{132}{77}$

31. One of the following is the correct way to write $\frac{16}{24}$ in lowest terms. Which one is it?

(a) $\frac{16}{24} = \frac{8 + 8}{8 + 16} = \frac{8}{16} = \frac{1}{2}$

(b) $\frac{16}{24} = \frac{4 \cdot 4}{4 \cdot 6} = \frac{4}{6}$

(c) $\frac{16}{24} = \frac{8 \cdot 2}{8 \cdot 3} = \frac{2}{3}$

(d) $\frac{16}{24} = \frac{14 + 2}{21 + 3} = \frac{2}{3}$

32. For the fractions p/q and r/s, which one of the following can serve as a common denominator?

(a) $q \cdot s$

(b) $q + s$

(c) $p \cdot r$

(d) $p + r$

Find each product or quotient, and write it in lowest terms. See Examples 4 and 5.

33. $\frac{4}{5} \cdot \frac{6}{7}$ **34.** $\frac{5}{9} \cdot \frac{10}{7}$ **35.** $\frac{1}{10} \cdot \frac{12}{5}$ **36.** $\frac{6}{11} \cdot \frac{2}{3}$

37. $\frac{15}{4} \cdot \frac{8}{25}$ **38.** $\frac{4}{7} \cdot \frac{21}{8}$ **39.** $2\frac{2}{3} \cdot 5\frac{4}{5}$ **40.** $3\frac{3}{5} \cdot 7\frac{1}{6}$

41. $\frac{5}{4} \div \frac{3}{8}$ **42.** $\frac{7}{6} \div \frac{9}{10}$ **43.** $\frac{32}{5} \div \frac{8}{15}$ **44.** $\frac{24}{7} \div \frac{6}{21}$

45. $\frac{3}{4} \div 12$

46. $\frac{2}{5} \div 30$

47. $2\frac{5}{8} \div 1\frac{15}{32}$

48. $2\frac{3}{10} \div 7\frac{4}{5}$

49. In your own words, explain how to divide two fractions.

50. In your own words, explain how to add two fractions that have different denominators.

Find each sum or difference, and write it in lowest terms. See Examples 6–8.

51. $\frac{7}{12} + \frac{1}{12}$

52. $\frac{3}{16} + \frac{5}{16}$

53. $\frac{5}{9} + \frac{1}{3}$

54. $\frac{4}{15} + \frac{1}{5}$

55. $3\frac{1}{8} + \frac{1}{4}$

56. $5\frac{3}{4} + \frac{2}{3}$

57. $\frac{7}{12} - \frac{1}{9}$

58. $\frac{11}{16} - \frac{1}{12}$

59. $6\frac{1}{4} - 5\frac{1}{3}$

60. $8\frac{4}{5} - 7\frac{4}{9}$

61. $\frac{5}{3} + \frac{1}{6} - \frac{1}{2}$

62. $\frac{7}{15} + \frac{1}{6} - \frac{1}{10}$

63. A cent is equal to $\frac{1}{100}$ of one dollar. Two dimes added to three dimes gives an amount equal to that of one half-dollar. Give an arithmetic problem using fractions that describes this equality, using 100 as a denominator throughout.

64. Three nickels added to twelve nickels gives an amount equal to that of three quarters. Give an arithmetic problem using fractions that describes this equality, using 100 as a denominator throughout.

Solve each applied problem. See Example 9.

65. A motel owner has decided to expand his business by buying a piece of property next to the motel. The property has an irregular shape, with five sides as shown in the figure. Find the total distance around the piece of property. This is called the *perimeter* of the figure.

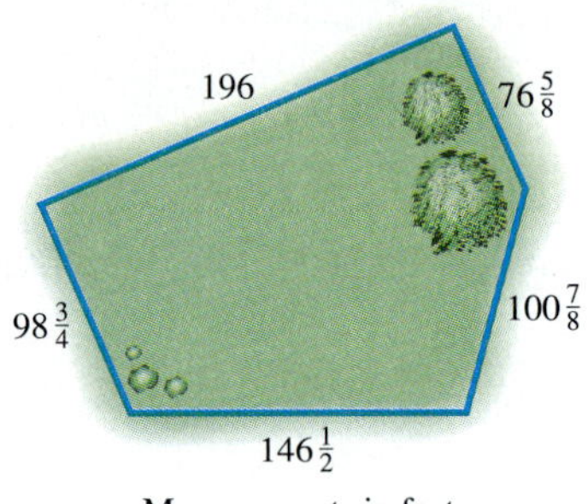

Measurements in feet

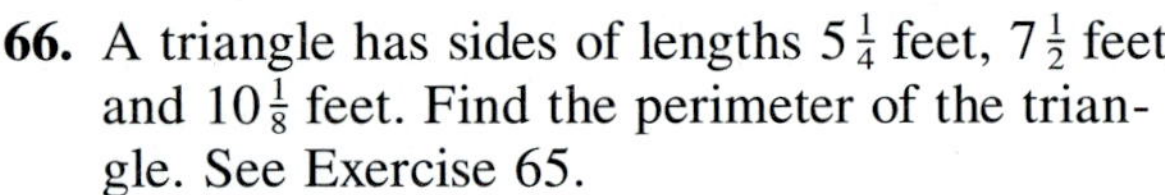

66. A triangle has sides of lengths $5\frac{1}{4}$ feet, $7\frac{1}{2}$ feet, and $10\frac{1}{8}$ feet. Find the perimeter of the triangle. See Exercise 65.

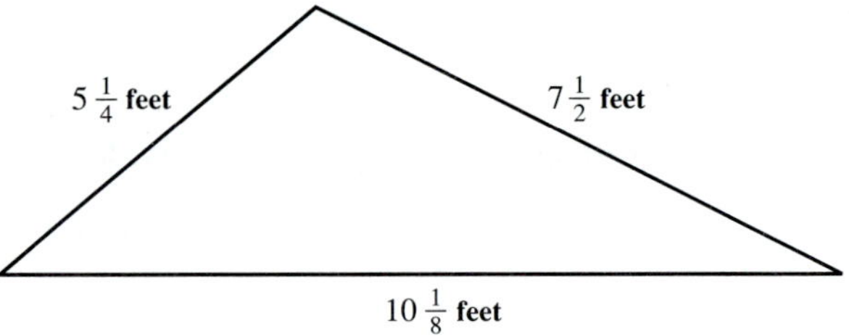

67. A hardware store sells a 40-piece socket wrench set. The measure of the largest socket is $\frac{3}{4}$ inch, while the measure of the smallest socket is $\frac{3}{16}$ inch. What is the difference between these measures?

68. Two sockets in a socket wrench set have measures of $\frac{9}{16}$ inch and $\frac{3}{8}$ inch. What is the difference between these two measures?

69. Tex's favorite recipe for barbecue sauce calls for $2\frac{1}{3}$ cups of tomato sauce. The recipe makes enough barbecue sauce to serve 7 people. How much tomato sauce is needed for 1 serving?

70. A cake recipe calls for $1\frac{3}{4}$ cups of sugar. A caterer has $15\frac{1}{2}$ cups of sugar on hand. How many cakes can the caterer make?

R.2 DECIMALS AND PERCENTS

Fractions are one way to represent parts of a whole. Another way is with a **decimal fraction** or **decimal,** a number written with a decimal point, such as 9.4. Each digit in a decimal number has a place value, as shown below.

4,896,328.9721

Each successive place value is ten times larger than the place value to its right and is one-tenth as large as the place value to its left.

Prices are often written as decimals. The price \$14.75 means 14 dollars and 75 cents, or 14 dollars and $\frac{75}{100}$ of a dollar.

1 Place value is used to write a decimal number as a fraction. For example, since the last digit of .67 is in the *hundredths* place,

$$.67 = \frac{67}{100}.$$

Similarly,

$$.9 = \frac{9}{10}, \qquad .25 = \frac{25}{100}, \qquad .342 = \frac{342}{1000}.$$

These examples suggest the following rule.

CONVERTING A DECIMAL TO A FRACTION

Read the name using the correct place value. Write it in fraction form just as you read it. The denominator will be a power of 10.

For example, we read .16 as "sixteen hundredths" and write it in fraction form as $\frac{16}{100}$.

EXAMPLE 1 *Writing a Decimal as a Fraction*

Write each decimal as a fraction. Do not write in lowest terms.

(a) .95

There are two places to the right of the decimal, so there will be two zeros in the denominator.

$$.95 = \frac{95}{100}$$

2 places 2 zeros

(b) $.056 = \frac{56}{1000}$

3 places 3 zeros

(c) $4.2095 = \frac{42{,}095}{10{,}000}$ ■

4 places 4 zeros

WORK PROBLEM 1 AT THE SIDE.

OBJECTIVES

1. Write decimals as fractions.
2. Add and subtract decimals.
3. Multiply and divide decimals.
4. Write fractions as decimals.
5. Convert percents to decimals and decimals to percents.

FOR EXTRA HELP

Tape 1 | SSM pp. 5–9 | MAC: B IBM: B

1. Write each decimal as a fraction. Do not write in lowest terms.

(a) .8

(b) .431

(c) 20.58

ANSWERS

1. (a) $\frac{8}{10}$ (b) $\frac{431}{1000}$ (c) $\frac{2058}{100}$

2 In the next example, we explain addition and subtraction of decimals.

EXAMPLE 2 *Adding and Subtracting Decimals*
Add or subtract as indicated.

(a) $6.92 + 14.8 + 3.217$

Place the numbers in a column, with decimal points lined up so that tenths are in one column, hundredths in another column, and so on.

$$\begin{array}{r} 6.92 \\ 14.8 \\ +\ 3.217 \\ \hline 24.937 \end{array}$$

Decimal points lined up

A good way to avoid errors is to attach zeros to make all the numbers the same length. For example,

$$\begin{array}{r} 6.92 \\ 14.8 \\ +\ 3.217 \\ \hline \end{array} \quad \text{becomes} \quad \begin{array}{r} 6.920 \\ 14.800 \\ +\ 3.217 \\ \hline 24.937. \end{array}$$

Attach zeros.

(b) $47.6 - 32.509$

Write the numbers in a column, attaching zeros to 47.6.

$$\begin{array}{r} 47.6 \\ -\ 32.509 \\ \hline \end{array} \quad \text{becomes} \quad \begin{array}{r} 47.600 \\ -\ 32.509 \\ \hline 15.091 \end{array}$$

(c) $3 - .253$

A whole number is assumed to have the decimal at the right of the number. Write 3 as 3.000; then subtract.

$$\begin{array}{r} 3.000 \\ -\ \ .253 \\ \hline 2.747 \end{array}$$ ■

WORK PROBLEM 2 AT THE SIDE.

A calculator may be used to add and subtract decimal numbers. If you use a calculator, it is a good idea to estimate the answer to be sure you did not enter numbers incorrectly or press a wrong key. To get an easy estimate, round the numbers so that only the first digit is not zero. Use the following rule for rounding.

RULE FOR ROUNDING

If the first digit to become 0 or be dropped is 5 or more, round the last digit to be kept to the next highest number.

If the digit to become 0 or be dropped is 4 or less, do not round up.

To estimate the answer to Example 2(a), round

6.92 to 7, 14.8 to 10, and 3.217 to 3.

↑ 5 or more ↑ 4 or less ↑ 4 or less

Since $7 + 10 + 3 = 20$, the answer of 24.937 is reasonable.

2. Add or subtract as indicated.

(a) $$\begin{array}{r} 68.9 \\ 42.72 \\ +\ 8.973 \\ \hline \end{array}$$

(b) $$\begin{array}{r} 32.5 \\ -\ 21.72 \\ \hline \end{array}$$

(c) $42.83 + 71.629 + 3.074$

(d) $351.8 - 2.706$

ANSWERS
2. (a) 120.593 (b) 10.78
(c) 117.533 (d) 349.094

3 We multiply decimals by slightly modifying multiplication of whole numbers.

MULTIPLYING DECIMALS

Ignore the decimal points and multiply as if the numbers were whole numbers. Then add together the number of **decimal places** (digits after the decimal point) in each number being multiplied. Locate the decimal point in the answer that many digits from the right.

EXAMPLE 3 *Multiplying Decimals*

Multiply.

(a) 29.3 × 4.52

Multiply as if the numbers were whole numbers.

```
     29.3     1 decimal place in top number
   × 4.52     2 decimal places in second number
     586      | 1 + 2 = 3
    1465      |
   1172       ↓
  132.436     3 decimal places in answer
```

(b) 7.003 × 55.8

```
     7.003    3 decimal places
  ×   55.8    1 decimal place
    56024     | 3 + 1 = 4
   35015      |
  35015       ↓
  390.7674    4 decimal places
```

(c) 31.42 × 65

```
    31.42     2 decimal places
  ×    65     0 decimal places
    15710     | 2 + 0 = 2
   18852      ↓
  2042.30     2 decimal places ■
```

WORK PROBLEM 3 AT THE SIDE.

3. Multiply.

(a) 2.13 × .05

(b) 69.32 × 1.4

(c) 397.12 × .152

(d) 42,980 × .012

To divide decimals, we convert the divisor to a whole number.

DIVIDING DECIMALS

Change the **divisor** (the number you are dividing *by*) into a whole number by moving the decimal point as many places as necessary to the right. Move the decimal point in the **dividend** (the number you are dividing *into*) to the right by the same number of places. Finally, bring the decimal point straight up and divide as with whole numbers.

```
                 5 ← quotient
divisor → 25)125
              ↑
          dividend
```

ANSWERS
3. (a) .1065 (b) 97.048 (c) 60.36224 (d) 515.76

4. Divide.

(a) $32.3\overline{)481.27}$

(b) $.37\overline{)5.476}$

(c) $375.125 \div 3.001$

ANSWERS
4. (a) 14.9 (b) 14.8 (c) 125

■ **EXAMPLE 4** *Dividing Decimals*

Divide.

(a) $279.45 \div 24.3$

Step 1 Write the problem as follows.

$$24.3\overline{)279.45}$$

Step 2 To change 24.3 into a whole number, move the decimal point one place to the right. Move the decimal point in 279.45 the same number of places to the right to get 2794.5.

$$24.3.\overline{)279.4.5}$$ Move one decimal place to the right.

To see why this works, write the division in fraction form and multiply the numerator and denominator by $\frac{10}{10}$ or 1.

$$\frac{279.45}{24.3} \cdot \frac{10}{10} = \frac{2794.5}{243}$$

The result is the same as when we moved the decimal point one place to the right in the divisor and the dividend.

Step 3 Bring the decimal point straight up and divide as with whole numbers.

```
        11.5
243)2794.5     Move decimal point straight up.
    243
     364
     243
     1215
     1215
        0
```

(b) $73.82\overline{)1852.882}$

Move the decimal point two places to the right in 73.82, to get 7382. Do the same thing with 1852.882, to get 185288.2.

$$73.82.\overline{)1852.88.2}$$

Bring the decimal point straight up and divide as with whole numbers.

```
          25.1
7382)185288.2
     14764
      37648
      36910
        7382
        7382
           0
```
■

◀◀ WORK PROBLEM 4 AT THE SIDE.

Note In working with decimals, it is helpful to check the work by estimating the answer. For example, in Example 3(a) the answer should be about $30 \times 5 = 150$. The answer, 132.436, is the correct size. Estimate the answers to Example 4 by rounding both numbers to whole numbers with only the first digit nonzero. (*Hint:* In Example 4(a) use $300 \div 20$.)

4 A fraction is written in decimal form as follows.

Write a fraction as a decimal by dividing the denominator into the numerator.

EXAMPLE 5 *Writing a Fraction as a Decimal*

Write each fraction as a decimal.

(a) $\frac{3}{8}$

$$\begin{array}{r} .375 \\ 8\overline{)3.000} \\ \underline{24} \\ 60 \\ \underline{56} \\ 40 \\ \underline{40} \\ 0 \end{array} \qquad \frac{3}{8} = .375$$

(b) $\frac{2}{3}$

$$\begin{array}{r} .6666 \\ 3\overline{)2.000\ldots} \\ \underline{18} \\ 20 \\ \underline{18} \\ 20 \\ \underline{18} \\ 20 \end{array}$$

The remainder in this division is never 0. Because a 2 is always left after the subtraction, this quotient is a **repeating decimal**. A convenient notation for a repeating decimal is a bar over the digit (or digits) that repeats. For instance, we can write .66666 . . . as $.\overline{6}$. In applications we often round repeating decimals to as many places as needed. For example, rounding to the nearest thousandth,

$$\frac{2}{3} = .667. \quad ■$$

Caution When rounding, be careful to distinguish between *thousandths* and *thousands* or between *hundredths* and *hundreds*, and so on.

WORK PROBLEM 5 AT THE SIDE.

5 An important application of decimals is in work with percents. The word **percent** means "per one hundred." Percent is written with the sign %. One percent means "one per one hundred" or "one one-hundredth."

$$1\% = .01 \quad \text{or} \quad 1\% = \frac{1}{100}$$

5. Convert to decimals. For repeating decimals, write the answer two ways: using the bar notation, when applicable, and rounding to the nearest thousandth.

(a) $\frac{2}{9}$

(b) $\frac{17}{20}$

(c) $\frac{5}{8}$

(d) $\frac{1}{7}$

ANSWERS

5. (a) $.\overline{2}$, .222 (b) .85 (c) .625 (d) $.\overline{142857}$, .143

6. Convert as indicated.

(a) 23% to a decimal

(b) 310% to a decimal

(c) .71 to a percent

(d) 1.32 to a percent

ANSWERS
6. (a) .23 **(b)** 3.10 **(c)** 71% **(d)** 132%

■ EXAMPLE 6 *Converting Percents and Decimals*

(a) Write 73% as a decimal.
Since 1% = .01,

$$73\% = 73 \cdot \mathbf{1\%} = 73 \cdot \mathbf{.01} = .73.$$

Also, 73% can be written as a decimal using the fraction form $1\% = \frac{1}{100}$.

$$73\% = 73 \cdot \mathbf{1\%} = 73 \cdot \left(\frac{\mathbf{1}}{\mathbf{100}}\right) = \frac{73}{100} = .73$$

(b) Write 125% as a decimal.

$$125\% = 125 \cdot 1\% = 125 \cdot .01 = 1.25$$

(c) Write .32 as a percent.
Since .32 means 32 hundredths, write .32 as 32 × .01. Finally, replace .01 with 1%.

$$.32 = 32 \cdot \mathbf{.01} = 32 \cdot \mathbf{1\%} = 32\%$$

(d) Write 2.63 as a percent.

$$2.63 = 263 \cdot .01 = 263 \cdot 1\% = 263\%$$ ■

Note To change a decimal to a percent or a percent to a decimal, we move the decimal point two places.
Move to the left to convert a percent to a decimal.
Move to the right to convert a decimal to a percent.
Keeping in mind the meaning of percent (per hundred) will help you remember which way to move the decimal point.

■ EXAMPLE 7 *Converting Percents and Decimals by Moving the Decimal Point*

Convert each decimal to a percent and each percent to a decimal.

(a) 45% = .45

(b) 250% = 2.50

(c) .57 = 57%

(d) 1.5 = 150% ■

◀◀ WORK PROBLEM 6 AT THE SIDE.

R.2 EXERCISES

NAME DATE HOUR

1. In the decimal 367.9412, name the digit that is in each place value.
(a) tens **(b)** tenths **(c)** thousandths **(d)** ones or units **(e)** hundredths

2. Write a numeral that has 5 in the thousands place, 0 in the tenths place, and 4 in the ten thousandths place.

3. For the sum 35.89 + 24.1, which one of the following is the best estimate?
(a) 40 **(b)** 50 **(c)** 60 **(d)** 70

4. For the difference 149.83 − 99.4, which one of the following is the best estimate?
(a) 40 **(b)** 50 **(c)** 60 **(d)** 70

5. For the product 84.9 × 98.3, which one of the following is the best estimate?
(a) 7200 **(b)** 8500 **(c)** 85,000 **(d)** 72,000

6. For the quotient 9845.3 ÷ 97.2, which one of the following is the best estimate?
(a) 10 **(b)** 1000 **(c)** 100 **(d)** 10,000

Write each decimal as a fraction. Do not write in lowest terms. See Example 1.

7. .4 **8.** .6 **9.** .64 **10.** .82

11. .138 **12.** .104 **13.** 3.805 **14.** 5.166

Add or subtract as indicated. Make sure that your answer is reasonable by using estimation skills. See Example 2.

15. 25.32 + 10.92 + 85.74 + 29.826 **16.** 90.527 + 32.43 + 589.83 + 399.327

17. 28.73 − 3.12 **18.** 46.88 − 13.45 **19.** 43.5 − 28.17 **20.** 345.1 − 56.31

21. $\begin{array}{r} 32.56 \\ 47.356 \\ +\ \ 1.8 \\ \hline \end{array}$

22. $\begin{array}{r} 75.22 \\ 123.96 \\ +\ \ 3.897 \\ \hline \end{array}$

23. $\begin{array}{r} 7.56 \\ -\ 2.789 \\ \hline \end{array}$

24. $\begin{array}{r} 9.24 \\ -\ 8.582 \\ \hline \end{array}$

Multiply or divide as indicated. Make sure that your answer is reasonable by using estimation skills. See Examples 3 and 4.

25. 44.3 × 5.6 **26.** 12.8 × 9.1 **27.** 34.045 × .56 **28.** 57.116 × .97

29. 57.2 ÷ 8 **30.** 73.36 ÷ 14 **31.** 24.837 ÷ 9.74 **32.** 44.4788 ÷ 5.27

33. Explain in your own words the procedure for adding or subtracting decimals.

34. Explain in your own words the procedures for (a) multiplying decimals and (b) dividing decimals.

35. For the decimal number 46.249, round to the place value indicated.
(a) hundredths **(b)** tenths **(c)** ones or units **(d)** tens

36. Round each of the following decimals to the nearest thousandth.
(a) $.\overline{8}$ **(b)** $.\overline{5}$ **(c)** .9762 **(d)** .8642

Write each fraction as a decimal. For repeating decimals, write the answer two ways: using the bar notation and rounding to the nearest thousandth. See Example 5.

37. $\frac{1}{8}$ **38.** $\frac{7}{8}$ **39.** $\frac{1}{4}$ **40.** $\frac{3}{4}$

41. $\frac{3}{16}$ **42.** $\frac{7}{16}$ **43.** $\frac{3}{7}$ **44.** $\frac{6}{7}$

45. $\frac{5}{9}$ **46.** $\frac{8}{9}$ **47.** $\frac{1}{6}$ **48.** $\frac{5}{6}$

49. In your own words, explain how you would convert a decimal to a percent.

50. In your own words, explain how you would convert a percent to a decimal.

Convert the following percents to decimals. See Examples 6, 7(a) and 7(b).

51. 54% **52.** 39% **53.** 117% **54.** 189%

55. 2.4% **56.** 3.1% **57.** .8% **58.** .9%

Convert the following decimals to percents. See Examples 6, 7(c) and 7(d).

59. .75 **60.** .83 **61.** .004 **62.** .00

63. 1.28 **64.** 2.35 **65.** .3 **66.** .6

One method of converting a fraction to a percent is to first convert the fraction to a decimal, as shown in Example 5, and then convert the decimal to a percent, as shown in Examples 6 and 7. Convert each of the following fractions to a percent in this way.

67. $\frac{3}{4}$ **68.** $\frac{1}{4}$ **69.** $\frac{3}{2}$ **70.** $\frac{5}{4}$

71. $\frac{5}{6}$ **72.** $\frac{11}{16}$ **73.** $\frac{5}{16}$ **74.** $\frac{7}{15}$

1 The Real Number System

1.1 EXPONENTS, ORDER OF OPERATIONS, AND INEQUALITY

OBJECTIVES

1. Use exponents.
2. Use the order of operations rules.
3. Use more than one grouping symbol.
4. Know the meanings of $\neq$, $<$, $>$, $\leq$, and $\geq$.
5. Translate word statements to symbols.
6. Reverse the direction of inequality statements.

FOR EXTRA HELP

Tape 1

SSM pp. 10–12

MAC: A IBM: A

1 Earlier, in Chapter R, when we found prime factorizations, we often found products that had the same factor appearing many times. As mentioned there, we use a raised dot to indicate multiplication. For example, when 81 is written in prime factored form as

$$81 = 3 \cdot 3 \cdot 3 \cdot 3,$$

the factor 3 appears four times. Repeated factors can be written in a short form by using an *exponent*. For example, the factorization of 81 is written with an exponent as

$$\underbrace{3 \cdot 3 \cdot 3 \cdot 3}_{\text{4 factors of 3}} = 3^4$$

(4 is labeled **exponent**; 3 is labeled **base**.)

The number 4 is the **exponent** and 3 is the **base** in the **exponential expression** 3^4. Exponents are also called **powers.** We read 3^4 as "3 to the fourth power" or simply "3 to the fourth."

EXAMPLE 1 *Finding the Value of an Exponential Expression*

Find the values of the following.

(a) 5^2

$$\underbrace{5 \cdot 5}= 25$$

5 is used as a factor 2 times.

Read 5^2 as "5 squared."

(b) 6^3

$$\underbrace{6 \cdot 6 \cdot 6} = 216$$

6 is used as a factor 3 times.

Read 6^3 as "6 cubed."

(c) 2^5

$$2 \cdot 2 \cdot 2 \cdot 2 \cdot 2 = 32 \quad \text{2 is used as a factor 5 times.}$$

Read 2^5 as "2 to the fifth power."

1. Find the value of each exponential expression.

(a) 6^2

(b) 3^5

(c) $\left(\frac{3}{4}\right)^2$

(d) $\left(\frac{1}{2}\right)^4$

(e) $(.4)^3$

2. Find the value of each expression.

(a) $3 \cdot 8 + 7$

(b) $9 + 12 \cdot 6$

ANSWERS

1. (a) 36 (b) 243 (c) $\frac{9}{16}$ (d) $\frac{1}{16}$ (e) .064

2. (a) 31 (b) 81

(d) 7^4

$$7 \cdot 7 \cdot 7 \cdot 7 = 2401 \qquad \text{7 is used as a factor 4 times.}$$

Read 7^4 as "7 to the fourth power."

(e) $\left(\frac{2}{3}\right)^3$

$$\frac{2}{3} \cdot \frac{2}{3} \cdot \frac{2}{3} = \frac{8}{27} \qquad \tfrac{2}{3} \text{ is used as a factor 3 times.}$$ ■

WORK PROBLEM 1 AT THE SIDE.

2 Many problems involve more than one operation. For example, in finding the value of

$$5 + 2 \cdot 3,$$

which should be done first—multiplication or addition? The following **order of operations** has been agreed on as the most reasonable. (This is the order used by most calculators and computers.)

ORDER OF OPERATIONS

If possible, simplify within parentheses and above and below fraction bars.

Step 1 Apply all exponents.
Step 2 Do any multiplications or divisions in the order in which they occur, working from left to right.
Step 3 Do any additions or subtractions in the order in which they occur, working from left to right.

■ **EXAMPLE 2** *Using the Order of Operations*

Find the value of $5 + 2 \cdot 3$.

Using the order of operations given above, first multiply 2 and 3, and then add 5.

$$\begin{aligned} 5 + 2 \cdot 3 &= 5 + 6 && \text{Multiply.} \\ &= 11 && \text{Add.} \end{aligned}$$ ■

WORK PROBLEM 2 AT THE SIDE.

A dot has been used to show multiplication; another way to show multiplication is with parentheses. For example, 3(7) means $3 \cdot 7$ or 21. Also $3(4 + 5)$ means 3 times the sum of 4 and 5. By the order of operations, the sum must be found first, then the product. The next example shows the use of parentheses for multiplication and parentheses and fraction bars for grouping.

■ **EXAMPLE 3** *Using the Order of Operations*

Find the value of each of the following.

(a) $9(6 + 11)$

Work first inside the parentheses.

$$\begin{aligned} 9(6 + 11) &= 9(17) && \text{Add inside parentheses.} \\ &= 153 && \text{Multiply.} \end{aligned}$$

(b) $2(5 + 6) + 7 \cdot 3 = 2(11) + 7 \cdot 3$ Add inside parentheses.
$= 22 + 21$ Multiply.
$= 43$ Add.

(c) $\dfrac{4(5 + 3) + 3}{2(3) - 1}$

Simplify the numerator and denominator separately.

$$\frac{4(5 + 3) + 3}{2(3) - 1} = \frac{4(8) + 3}{2(3) - 1} \quad \text{Add inside parentheses.}$$
$$= \frac{32 + 3}{6 - 1} \quad \text{Multiply.}$$
$$= \frac{35}{5} \quad \text{Add and subtract.}$$
$$= 7 \quad \text{Divide.}$$

(d) $9 + 2^3 - 5$

Following the order of operations, we calculate 2^3 first.

$9 + 2^3 - 5 = 9 + 8 - 5$ Use the exponent.
$= 12$ Add, then subtract. ■

WORK PROBLEM 3 AT THE SIDE.

Note Parentheses and fraction bars are used as grouping symbols to indicate an expression that represents a single number. That is why we must first simplify within parentheses and above and below fraction bars.

Calculators follow the order of operations given in this section. You may want to try some of the examples to see that your calculator gives the same answers. Be sure to use the parentheses keys to insert parentheses where they are needed. To work Example 3(c) with a calculator, you must put parentheses around the numerator and the denominator.

3 An expression with double parentheses, such as $2(8 + 3(6 + 5))$, can be confusing. We can avoid confusion by using square brackets, [ ], in place of one pair of parentheses.

■ EXAMPLE 4 *Using Brackets*

Simplify $2[8 + 3(6 + 5)]$.

Begin inside the parentheses. Then follow the order of operations.

$2[8 + 3(6 + 5)] = 2[8 + 3(11)]$ Add.
$= 2[8 + 33]$ Multiply.
$= 2[41]$ Add.
$= 82$ Multiply. ■

WORK PROBLEM 4 AT THE SIDE.

4 So far we have used only the symbols of arithmetic, such as +, −, × (or ·), and ÷. Another common symbol is the one for equality, =, which

3. Find the value of each expression.

(a) $2 \cdot 9 + 7 \cdot 3$

(b) $7 \cdot 6 - 3(8 + 1)$

(c) $\dfrac{2(7 + 8) + 2}{3 \cdot 5 + 1}$

(d) $2 + 3^2 - 5$

4. Find the value of each expression.

(a) $4[7 + 3(6 + 1)]$

(b) $9[(4 + 8) - 3]$

ANSWERS
3. (a) 39 **(b)** 15 **(c)** 2 **(d)** 6
4. (a) 112 **(b)** 81

says that two numbers are equal. This symbol with a slash through it, $\neq$, means "is *not* equal to." For example,

$$7 \neq 8$$

indicates that 7 is not equal to 8.

If two numbers are not equal, then one of the numbers must be less than the other. The symbol $<$ represents "is less than," so "7 is less than 8" is written

$$7 < 8.$$

Also, we write "6 is less than 9" as $6 < 9$.

The symbol $>$ means "is greater than." We write "8 is greater than 2" as

$$8 > 2.$$

The statement "17 is greater than 11" becomes $17 > 11$.

Keep the meanings of the symbols $<$ and $>$ clear by remembering that the symbol always points to the smaller number.

smaller number $\rightarrow$ **8** < 15

$15 >$ **8** $\leftarrow$ **smaller number**

WORK PROBLEM 5 AT THE SIDE.

Two other symbols, $\leq$ and $\geq$, also represent the idea of inequality. The symbol $\leq$ means "is less than or equal to," so

$$5 \leq 9$$

means "5 is less than or equal to 9." If either the $<$ part or the $=$ part is true, then the inequality $\leq$ is true. The statement $5 \leq 9$ is true because $5 < 9$ is true.

The symbol $\geq$ means "is greater than or equal to";

$$9 \geq 5$$

is true because $9 > 5$ is true. Also, $8 \leq 8$ is true because $8 = 8$ is true. But $13 \leq 9$ is not true because neither $13 < 9$ nor $13 = 9$ is true.

EXAMPLE 5 *Using the Symbols $\leq$ and $\geq$*

Tell whether each statement is true or false.

(a) $15 \leq 20$ The statement $15 \leq 20$ is true because $15 < 20$.

(b) $25 \geq 30$ Both $25 > 30$ and $25 = 30$ are false; therefore, $25 \geq 30$ is false.

(c) $12 \geq 12$ Since $12 = 12$, this statement is true. ■

WORK PROBLEM 6 AT THE SIDE.

5 Word phrases or statements often must be converted to symbols in algebra. The next example shows how to do this.

EXAMPLE 6 *Converting Words to Symbols*

Write each word statement in symbols.

(a) Twelve **equals** ten **plus** two. $12 = 10 + 2$

(b) Nine **is less than** ten. $9 < 10$

---

5. Write each statement in words, then decide whether it is true or false.

(a) $7 < 5$

(b) $12 > 6$

(c) $4 \neq 10$

(d) $28 \neq 4 \cdot 7$

6. Tell whether each statement is true or false.

(a) $30 \leq 40$

(b) $25 \geq 10$

(c) $40 \leq 10$

(d) $21 \leq 21$

(e) $3 \geq 3$

ANSWERS

5. (a) Seven is less than five. False
(b) Twelve is greater than six. True
(c) Four is not equal to ten. True
(d) Twenty-eight is not equal to four times seven. False

6. (a) true (b) true (c) false (d) true (e) true

(c) Fifteen **is not equal to** eighteen. $15 \neq 18$

(d) Seven **is greater than** four. $7 > 4$

(e) Thirteen **is less than or equal to** forty. $13 \leq 40$

(f) Six **is greater than or equal to** six. $6 \geq 6$ ■

WORK PROBLEM 7 AT THE SIDE. ▶▶

6 Any statement with $<$ can be converted to one with $>$, and any statement with $>$ can be converted to one with $<$. We do this by reversing both the order of the numbers and the direction of the symbol. For example, the statement $6 < 10$ can be written as $10 > 6$.

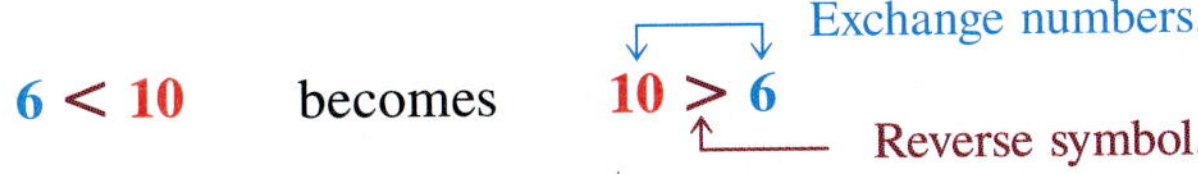

■ **EXAMPLE 7** *Converting Between < and >*

The following list shows the same statements written in two equally correct ways.

(a) $9 < 16$ $16 > 9$

(b) $5 > 2$ $2 < 5$

(c) $3 \leq 8$ $8 \geq 3$

(d) $12 \geq 5$ $5 \leq 12$ ■

WORK PROBLEM 8 AT THE SIDE. ▶▶

Here is a summary of the symbols of equality and inequality.

SYMBOLS OF EQUALITY AND INEQUALITY

Symbol	*Meaning*	*Symbol*	*Meaning*
$=$	is equal to	$>$	is greater than
$\neq$	is not equal to	$\leq$	is less than or equal to
$<$	is less than	$\geq$	is greater than or equal to

Caution The symbols of equality and inequality are used to write mathematical *sentences.* They differ from the symbols for operations ($+, -, \cdot$, and $\div$), discussed earlier, which are used to write mathematical *expressions* that represent a number. For example, compare the sentence $4 < 10$, which gives the relationship between 4 and 10, with the expression $4 + 10$, which tells how to operate on 4 and 10 to get the number 14.

7. Write in symbols.

(a) Nine equals eleven minus two.

(b) Seventeen is less than thirty.

(c) Eight is not equal to ten.

(d) Fourteen is greater than twelve.

(e) Thirty is less than or equal to fifty.

(f) Two is greater than or equal to two.

8. Write each statement with the inequality symbol reversed.

(a) $8 < 10$

(b) $3 > 1$

(c) $9 \leq 15$

(d) $6 \geq 2$

ANSWERS

7. (a) $9 = 11 - 2$ (b) $17 < 30$ (c) $8 \neq 10$ (d) $14 > 12$ (e) $30 \leq 50$ (f) $2 \geq 2$

8. (a) $10 > 8$ (b) $1 < 3$ (c) $15 \geq 9$ (d) $2 \leq 6$

QUEST FOR NUMERACY

Pattern Recognition: An Application of Inductive Reasoning

Two types of reasoning are used in the study of mathematics. They are deductive reasoning, which is characterized by applying general principles to specific examples, and inductive reasoning, which is characterized by making educated guesses from repeated observations of specific outcomes. One of the most beautiful parts of mathematics lies in the countless patterns we can observe. For example, look at the following list of multiplication problems.

$$37 \times 3 = 111$$
$$37 \times 6 = 222$$
$$37 \times 9 = 333$$
$$37 \times 12 = 444$$

You would probably expect the product 37×15 to equal 555 after observing the four results above. We can easily verify that this is a correct conjecture (educated guess). However, you must always be aware that inductive reasoning does not *always* lead to valid conjectures. Mathematicians require proof that results are valid, and we cannot be sure that a particular outcome will always result simply by observing a finite number of examples. Mathematicians use deductive reasoning in their proofs.

FOR GROUP DISCUSSION

1. Use inductive reasoning to make a conjecture about the next equation in each list, and then use your calculator to determine whether your conjecture is correct.

(a)
$(1 \times 9) + 2 = 11$
$(12 \times 9) + 3 = 111$
$(123 \times 9) + 4 = 1111$
$(1234 \times 9) + 5 = 11{,}111$

(b)
$(9 \times 9) + 7 = 88$
$(98 \times 9) + 6 = 888$
$(987 \times 9) + 5 = 8888$
$(9876 \times 9) + 4 = 88{,}888$

(c)
$15{,}873 \times 7 = 111{,}111$
$15{,}873 \times 14 = 222{,}222$
$15{,}873 \times 21 = 333{,}333$
$15{,}873 \times 28 = 444{,}444$

(d)
$3367 \times 3 = 10{,}101$
$3367 \times 6 = 20{,}202$
$3367 \times 9 = 30{,}303$
$3367 \times 12 = 40{,}404$

(e)
$11 \times 11 = 121$
$111 \times 111 = 12{,}321$
$1111 \times 1111 = 1{,}234{,}321$

(f)
$34 \times 34 = 1156$
$334 \times 334 = 111{,}556$
$3334 \times 3334 = 11{,}115{,}556$

2. Complete the following using your calculator.

$12{,}345{,}679 \times 9 =$ ____
$12{,}345{,}679 \times 18 =$ ____
$12{,}345{,}679 \times 27 =$ ____

By what number would you have to multiply 12,345,679 in order to get an answer of 888,888,888?

3. Complete the following using your calculator.

$142{,}857 \times 1 =$ ____ $142{,}857 \times 4 =$ ____
$142{,}857 \times 2 =$ ____ $142{,}857 \times 5 =$ ____
$142{,}857 \times 3 =$ ____ $142{,}857 \times 6 =$ ____

What pattern exists in the successive answers? Now multiply 142,857 by 7 to obtain an interesting result.

NAME DATE HOUR

1.1 EXERCISES

Decide whether each of the following is true or false. If it is false, explain why.

1. The exponential expression 3^5 means $3 \cdot 3 \cdot 3 \cdot 3 \cdot 3 \cdot 3$.

2. $7 + 3 \cdot 2$ and $(7 + 3) \cdot 2$ have the same meaning.

3. In an inequality using $<$ or $>$, the inequality symbol should point toward the smaller number for the inequality to be true.

4. $4 + 9 = 13$ is a mathematical sentence, while $4 + 9 - 13$ is a mathematical expression.

Find the value of each exponential expression. See Example 1.

5. 7^2 **6.** 4^2 **7.** 12^2 **8.** 14^2

9. 4^3 **10.** 5^3 **11.** 10^3 **12.** 11^3

13. 3^4 **14.** 6^4 **15.** 4^5 **16.** 3^5

17. $\left(\frac{2}{3}\right)^4$ **18.** $\left(\frac{3}{4}\right)^3$ **19.** $(.04)^3$ **20.** $(.05)^4$

21. Explain in your own words how to evaluate a power of a number, such as 6^3.

22. Explain why 1 to any power must be equal to 1.

Find the value of each of the following expressions. See Examples 2–4.

23. $9 \cdot 5 - 13$ **24.** $7 \cdot 6 - 11$ **25.** $\frac{1}{4} \cdot \frac{2}{3} + \frac{2}{5} \cdot \frac{11}{3}$ **26.** $\frac{9}{4} \cdot \frac{2}{3} + \frac{4}{5} \cdot \frac{5}{3}$

27. $9 \cdot 4 - 8 \cdot 3$ **28.** $11 \cdot 4 + 10 \cdot 3$ **29.** $(4.3)(1.2) + (2.1)(8.5)$

30. $(2.5)(1.9) + (4.3)(7.3)$ **31.** $5[3 + 4(2^2)]$ **32.** $6[2 + 8(3^3)]$

33. $3^2[(11 + 3) - 4]$ **34.** $4^2[(13 + 4) - 8]$ **35.** $\frac{6(3^2 - 1) + 8}{3 \cdot 2 - 2}$

36. $\frac{2(8^2 - 4) + 8}{4 \cdot 3 - 10}$ **37.** $\frac{4(6 + 2) + 8(8 - 3)}{6(4 - 2) - 2^2}$ **38.** $\frac{6(5 + 1) - 9(1 + 1)}{5(8 - 6) - 2^3}$

39. Explain why, in the expression $3 + 4 \cdot 6$, the product $4 \cdot 6$ should be found *before* the addition is performed.

40. When evaluating $(4^2 + 3^3)^4$, what is the *last* exponent that would be applied?

Tell whether each statement is true or false. In Exercises 45–54, first simplify each expression involving an operation. See Example 5.

41. $5 < 6$

42. $3 < 7$

43. $8 \geq 17$

44. $10 \geq 41$

45. $17 \leq 18 - 1$

46. $12 \geq 10 + 2$

47. $6 \cdot 8 + 6 \cdot 6 \geq 0$

48. $4 \cdot 20 - 16 \cdot 5 \geq 0$

49. $6[5 + 3(4 + 2)] \leq 70$

50. $6[2 + 3(2 + 5)] \leq 135$

51. $\dfrac{9(7 - 1) - 8 \cdot 2}{4(6 - 1)} > 3$

52. $\dfrac{2(5 + 3) + 2 \cdot 2}{2(4 - 1)} > 1$

53. $8 \leq 4^2 - 2^2$

54. $10^2 - 8^2 > 6^2$

Write each word statement in symbols. See Example 6.

55. Fifteen is equal to five plus ten.

56. Twelve is equal to twenty minus eight.

57. Nine is greater than five minus four.

58. Ten is greater than six plus one.

59. Sixteen is not equal to nineteen.

60. Three is not equal to four.

61. Two is less than or equal to three.

62. Five is less than or equal to nine.

Write each statement in words and decide whether it is true or false.

63. $7 < 19$

64. $9 < 10$

65. $3 \neq 6$

66. $9 \neq 13$

67. $8 \geq 11$

68. $4 \leq 2$

69. Construct a true statement that involves an addition on the left side, the symbol $\geq$, and a multiplication on the right side.

70. Construct a false statement that involves subtraction on the left side, the symbol $\leq$, and a division on the right side. Then tell why the statement is false and how it could be changed to become true.

Write each statement with the inequality symbol reversed. See Example 7.

71. $5 < 30$

72. $8 > 4$

73. $12 \geq 3$

74. $25 \leq 41$

PREVIEW EXERCISES

Most of the exercise sets in the book end with brief sets of "Preview Exercises." These exercises *review* concepts previously introduced, to provide a *preview* of the ideas that will be used in the next section or next few sections. If you need help with these, refer to the indicated section(s).

Perform the indicated operations. See Section R.2.

75. $(.53)^2$

76. $5.43 - 2.78$

77. $3.12 + 9.34 - 7.33$

78. $\dfrac{1.331}{.011}$

1.2 VARIABLES, EXPRESSIONS, AND EQUATIONS

A **variable** is a symbol, usually a letter, such as x, y, or z, used to represent any unknown number. An **algebraic expression** is a collection of numbers, variables, symbols for operations, and symbols for grouping, such as parentheses, square brackets, or division bars. For example,

$$x + 5, \quad 2m - 9, \quad \text{and} \quad 8p^2 + 6(p - 2)$$

are all algebraic expressions. In the algebraic expression $2m - 9$, the expression $2m$ means $2 \cdot m$, the product of 2 and m, and $8p^2$ shows the product of 8 and p^2. Also, $6(p - 2)$ means the product of 6 and $p - 2$.

1 An algebraic expression has different numerical values for different values of the variables.

EXAMPLE 1 *Finding the Value of an Expression Given a Value of the Variable*

Find the values of the following algebraic expressions if $m = 5$ and if $m = 9$.

(a) $8m$

Replace m with 5, to get

$$\begin{aligned} 8m &= 8 \cdot 5 && \text{Let } m = 5. \\ &= 40. && \text{Multiply.} \end{aligned}$$

If $m = 9$,

$$\begin{aligned} 8m &= 8 \cdot 9 && \text{Let } m = 9. \\ &= 72. && \text{Multiply.} \end{aligned}$$

(b) $3m^2$

If $m = 5$,

$$\begin{aligned} 3m^2 &= 3 \cdot 5^2 && \text{Let } m = 5. \\ &= 3 \cdot 25 && \text{Square.} \\ &= 75. && \text{Multiply.} \end{aligned}$$

If $m = 9$,

$$\begin{aligned} 3m^2 &= 3 \cdot 9^2 \\ &= 3 \cdot 81 = 243. \end{aligned}$$ ■

Caution In Example 1(b), it is important to notice that $3m^2$ means $3 \cdot m^2$; it *does not* mean $3m \cdot 3m$. Unless parentheses are used, the exponent refers only to the variable or number just before it.

WORK PROBLEM 1 AT THE SIDE. ▶

EXAMPLE 2 *Finding the Value of an Expression with More Than One Variable*

Find the value of each expression if $x = 5$ and $y = 3$.

(a) $2x + 5y$

$$\begin{aligned} 2x + 5y &= 2 \cdot 5 + 5 \cdot 3 && \text{Replace } x \text{ with 5 and } y \text{ with 3.} \\ &= 10 + 15 && \text{Multiply.} \\ &= 25 && \text{Add.} \end{aligned}$$

OBJECTIVES

1. Find the value of algebraic expressions, given values for the variables.
2. Convert phrases from words to algebraic expressions.
3. Identify solutions of equations.
4. Identify solutions of equations from a set of numbers.
5. Distinguish between expressions and equations.

FOR EXTRA HELP

Tape 1 | SSM pp. 12–16 | MAC: A IBM: A

1. Find the value of each expression if $p = 3$.

(a) $6p$

(b) $p + 12$

(c) $5p^2$

ANSWERS
1. (a) 18 **(b)** 15 **(c)** 45

2. Find the value of each expression if $x = 6$ and $y = 9$.

(a) $4x + 7y$

(b) $\dfrac{4x - 2y}{x + 1}$

(c) $2x^2 + y^2$

ANSWERS

2. (a) 87 (b) $\frac{6}{7}$ (c) 153

(b) $\dfrac{9x - 8y}{2x - y}$

$$\frac{9x - 8y}{2x - y} = \frac{9 \cdot 5 - 8 \cdot 3}{2 \cdot 5 - 3} \quad \text{Replace } x \text{ with 5 and } y \text{ with 3.}$$

$$= \frac{45 - 24}{10 - 3} \quad \text{Multiply.}$$

$$= \frac{21}{7} \quad \text{Subtract.}$$

$$= 3 \quad \text{Divide.}$$

(c)
$$x^2 - 2y^2 = 5^2 - 2 \cdot 3^2 \quad \text{Replace } x \text{ with 5 and } y \text{ with 3.}$$

$$= 25 - 2 \cdot 9 \quad \text{Use the exponents.}$$

$$= 25 - 18 \quad \text{Multiply.}$$

$$= 7 \quad \text{Subtract.}$$ ■

WORK PROBLEM 2 AT THE SIDE.

An Introduction to Scientific Calculators in the front of this book explains how to evaluate exponentials with a calculator.

2 In the previous section, we wrote word phrases in symbols. The next example shows how variables are used to translate words to symbols. This process will be very important later for solving applied problems.

■ **EXAMPLE 3** *Using Variables to Change Word Phrases into Algebraic Expressions*

Change the following word phrases to algebraic expressions. Use x as the variable to represent the number.

(a) The **sum** of a number and 9

"Sum" is the answer to an addition problem. This phrase translates as

$$x + 9 \quad \text{or} \quad 9 + x.$$

(b) 7 **minus** a number

"Minus" indicates subtraction, so the answer is

$$7 - x.$$

Note that $x - 7$ would *not* be correct because we cannot do a subtraction in either order and get the same results.

(c) A number **subtracted from 12**

Since a number is subtracted *from* 12, write this as

$$12 - x.$$

Compare this result with "12 is subtracted from 25," which is $25 - 12$.

(d) The **product** of 11 and a number

$$11 \cdot x \quad \text{or} \quad 11x$$

(e) 5 divided by a number

$$\frac{5}{x}$$

(f) The product of 2 and the difference between a number and 8

$$2(x - 8)$$ ■

Caution Notice that in translating the words "the difference between a number and 8" the order is kept the same: $x - 8$. "The difference between 8 and a number" would be written $8 - x$.

WORK PROBLEM 3 AT THE SIDE.

3 An **equation** states that two expressions are equal. Examples of equations are

$$x + 4 = 11, \qquad 2y = 16, \qquad \text{and} \qquad 4p + 1 = 25 - p.$$

To **solve** an equation, we must find all values of the variable that make the equation true. The values of the variable that make the equation true are called the **solutions** of the equation.

■ **EXAMPLE 4** *Deciding Whether a Number is a Solution of an Equation*

Decide whether the given number is a solution of the equation.

(a) $5p + 1 = 36$; 7

$$5p + 1 = 36$$
$$5 \cdot 7 + 1 = 36 \quad \text{Replace } p \text{ with 7.}$$
$$35 + 1 = 36 \quad \text{Multiply.}$$
$$36 = 36 \quad \text{True}$$

The number 7 is a solution of the equation.

(b) $9m - 6 = 32$; 4

$$9m - 6 = 32$$
$$9 \cdot 4 - 6 = 32 \quad \text{Replace } m \text{ with 4.}$$
$$36 - 6 = 32 \quad \text{Multiply.}$$
$$30 = 32 \quad \text{False}$$

The number 4 is not a solution of the equation. ■

WORK PROBLEM 4 AT THE SIDE.

4 Sometimes the solutions of an equation must come from a certain list of numbers. This list of numbers is often written as a **set,** a collection of objects. For example, the set containing the numbers 1, 2, 3, 4, and 5 is written with **set braces,** { }, as

$$\{1, 2, 3, 4, 5\}.$$

The set of numbers from which the solutions of an equation must be chosen is called the **domain** of the equation.

■ **EXAMPLE 5** *Finding the Solution of an Equation from the Domain*

Change each word statement to an equation. Use x as the variable. Then find all solutions for the equation from the domain

$$\{0, 2, 4, 6, 8, 10\}.$$

3. Write as an algebraic expression. Use x as the variable.

(a) The sum of 5 and a number

(b) A number minus 4

(c) A number subtracted from 48

(d) The product of 6 and a number

(e) 9 multiplied by the sum of a number and 5

4. Decide whether the given number is a solution of the equation.

(a) $p - 1 = 3$; 2

(b) $2k + 3 = 15$; 7

(c) $8p - 11 = 5$; 2

ANSWERS

3. (a) $5 + x$ (b) $x - 4$ (c) $48 - x$ (d) $6x$ (e) $9(x + 5)$

4. (a) no (b) no (c) yes

5. Change each statement to an equation. Find all solutions from the domain {0, 2, 4, 6, 8, 10}.

(a) The sum of a number and 13 is 19.

(b) Three times a number is subtracted from 21, giving 15.

6. Decide whether each of the following is an equation or an expression.

(a) $2x + 5y - 7$

(b) $\dfrac{3x - 1}{5}$

(c) $2x + 5 = 7$

(d) $\dfrac{x}{y - 3} = 4x$

ANSWERS
5. (a) $x + 13 = 19$; 6
(b) $21 - 3x = 15$; 2
6. (a) expression (b) expression
(c) equation (d) equation

(a) The sum of a number and four is six.

The word "is" suggests "equals." Let x represent the unknown number and translate as follows.

The sum of a number and four	is	six.
↓	↓	↓
$x + 4$	$=$	6

Try each number from the given domain, {0, 2, 4, 6, 8, 10}, in turn.

$x + 4 = 6$	Given equation
$0 + 4 = 6$	False
$2 + 4 = 6$	True
$4 + 4 = 6$	False
$6 + 4 = 6$	False
$8 + 4 = 6$	False
$10 + 4 = 6$	False

The only solution of $x + 4 = 6$ is 2.

(b) Nine more than five times a number is 49.

"Nine more than" means "nine is added to." Use x to represent the unknown number.

Nine	more than	five times a number	is	49.
↓	↓	↓	↓	↓
9	$+$	$5x$	$=$	49

Try each number from the given domain, {0, 2, 4, 6, 8, 10}. The solution is 8 because $9 + 5 \cdot 8 = 49$. ■

◀◀ WORK PROBLEM 5 AT THE SIDE.

5 Students often have trouble distinguishing between equations and expressions. Remember that an equation is a sentence; an expression is a phrase.

$$4x + 5 = 9 \qquad 4x + 5$$

↑ equation ↑ expression

■ **EXAMPLE 6** *Distinguishing Between Equations and Expressions*

Decide whether each of the following is an equation or an expression.

(a) $2x - 5y$

There is no equals sign, so this is an expression.

(b) $2x = 5y$

Because of the equals sign, this is an equation. ■

◀◀ WORK PROBLEM 6 AT THE SIDE.

1.2 EXERCISES

NAME DATE HOUR

Identify each of the following as an expression or an equation. See Example 6.

1. $3x + 2(x - 4)$

2. $5y - (3y + 6)$

3. $7t + 2(t + 1) = 4$

4. $9r + 3(r - 4) = 2$

5. $x + y = 3$

6. $x + y - 3$

7. Why is $2x^3$ not the same as $2x \cdot 2x \cdot 2x$?

8. Why are "5 less than a number" and "5 is less than a number" translated differently?

*Find the numerical values of the following if **(a)** $x = 4$ and **(b)** $x = 6$. See Example 1.*

9. $x + 9$
(a) **(b)**

10. $x - 1$
(a) **(b)**

11. $5x$
(a) **(b)**

12. $7x$
(a) **(b)**

13. $4x^2$
(a) **(b)**

14. $5x^2$
(a) **(b)**

15. $\frac{x + 1}{3}$
(a) **(b)**

16. $\frac{x - 2}{5}$
(a) **(b)**

17. $\frac{3x - 5}{2x}$
(a) **(b)**

18. $\frac{4x - 1}{3x}$
(a) **(b)**

19. $3x^2 + x$
(a) **(b)**

20. $2x + x^2$
(a) **(b)**

21. $6.459x$
(a) **(b)**

22. $.74x^2$
(a) **(b)**

*Find the numerical values of the following if **(a)** $x = 2$ and $y = 1$ and **(b)** $x = 1$ and $y = 5$. See Example 2.*

23. $8x + 3y + 5$
(a) **(b)**

24. $4x + 2y + 7$
(a) **(b)**

25. $3(x + 2y)$
(a) **(b)**

26. $2(2x + y)$
(a) **(b)**

27. $x + \frac{4}{y}$
(a) **(b)**

28. $y + \frac{8}{x}$
(a) **(b)**

29. $\frac{x}{2} + \frac{y}{3}$
(a) **(b)**

30. $\frac{x}{5} + \frac{y}{4}$
(a) **(b)**

31. $\frac{2x + 4y - 6}{5y + 2}$
(a) **(b)**

32. $\frac{4x + 3y - 1}{x}$
(a) **(b)**

33. $2y^2 + 5x$
(a) **(b)**

34. $6x^2 + 4y$
(a) **(b)**

35. $\dfrac{3x + y^2}{2x + 3y}$
(a) **(b)**

36. $\dfrac{x^2 + 1}{4x + 5y}$
(a) **(b)**

37. $.841x^2 + .32y^2$
(a) **(b)**

38. $.941x^2 + .2y^2$
(a) **(b)**

Change the word phrases to algebraic expressions. Use x as the variable to represent the number. See Example 3.

39. Twelve times a number

40. Nine times a number

41. Seven added to a number

42. Thirteen added to a number

43. Two subtracted from a number

44. Eight subtracted from a number

45. A number subtracted from seven

46. A number subtracted from fourteen

47. The difference between a number and 6

48. The difference between 6 and a number

49. 12 divided by a number

50. A number divided by 12

51. The product of 6 and four less than a number

52. The product of 9 and five more than a number

53. In the phrase "four more than the product of a number and 6," does the word *and* signify the operation of addition? Explain.

54. What value of x would cause the expression $2x + 3$ to equal 9?

55. There are many pairs of values of x and y for which $2x + y$ will equal 6. Name two such pairs.

56. Suppose that the directions on a test read "Solve the following expressions." How would you politely correct the person who wrote these directions?

Decide whether the given number is a solution of the equation. See Example 4.

57. $p - 5 = 12; 7$

58. $x + 6 = 15; 10$

59. $5m + 2 = 7; 1$

60. $3r + 5 = 8; 1$

61. $2y + 3(y - 2) = 14; 3$

62. $6a + 2(a + 3) = 14; 2$

63. $6p + 4p + 9 = 11; \frac{1}{5}$

64. $2x + 3x + 8 = 20; \frac{12}{5}$

65. $3r^2 - 2 = 53.47; 4.3$

66. $2x^2 + 1 = 28.38; 3.7$

67. $\frac{z + 4}{2 - z} = \frac{13}{5}; \frac{1}{3}$

68. $\frac{x + 6}{x - 2} = \frac{37}{5}; \frac{13}{4}$

Change the word statements to equations. Use x as the variable. Find the solutions from the domain {0, 2, 4, 6, 8, 10}. See Example 5.

69. The sum of a number and 8 is 18.

70. A number minus three equals 1.

71. Sixteen minus three-fourths of a number is 13.

72. The sum of six-fifths of a number and 2 is 14.

73. Five more than twice a number is 5.

74. The product of a number and 3 is 6.

75. Three times a number is equal to 8 more than twice the number.

76. Twelve divided by a number equals $\frac{1}{3}$ times that number.

1.3 REAL NUMBERS AND THE NUMBER LINE

OBJECTIVES

1. Use integers to express numbers in applications.
2. Graph rational numbers on the number line.
3. Tell which of two real numbers is smaller.
4. Find the opposite of a number.
5. Find the absolute values of real numbers.

FOR EXTRA HELP

Tape 1

SSM pp. 16–18

MAC: A
IBM: A

In Chapter R we introduced the set of whole numbers.

Whole numbers {0, 1, 2, 3, 4, 5, . . . }

The numbers used for counting are called the **natural numbers.**

Natural numbers {1, 2, 3, 4, 5, . . . }

These numbers, along with many others, can be represented on **number lines** like the one in Figure 1. We draw a number line by choosing any point on the line and labeling it 0. Choose any point to the right of 0 and label it 1. The distance between 0 and 1 gives a unit of measure used to locate other points, as shown in Figure 1. The points labeled in Figure 1 correspond to the first few whole numbers.

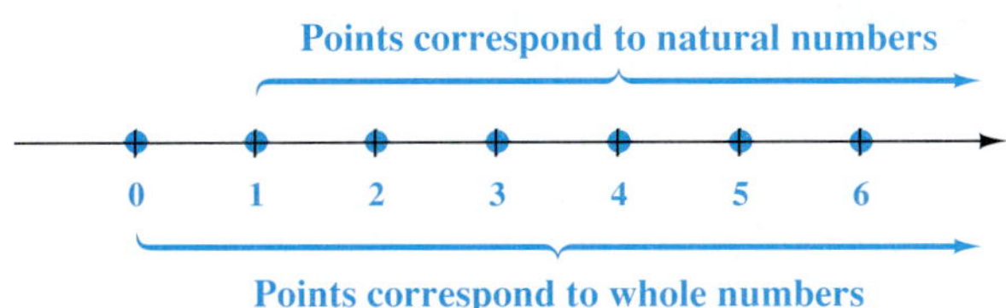

FIGURE 1

1 The natural numbers are located to the right of 0 on the number line. But numbers may also be placed to the left of 0. For each natural number we can place a corresponding number to the left of 0. These numbers, written -1, -2, -3, -4, and so on, are shown in Figure 2. Each is the **opposite** or **negative** of a natural number. The natural numbers, their opposites, and zero form a new set of numbers, called the **integers.**

Integers { . . . -3, -2, -1, 0, 1, 2, 3 . . . }

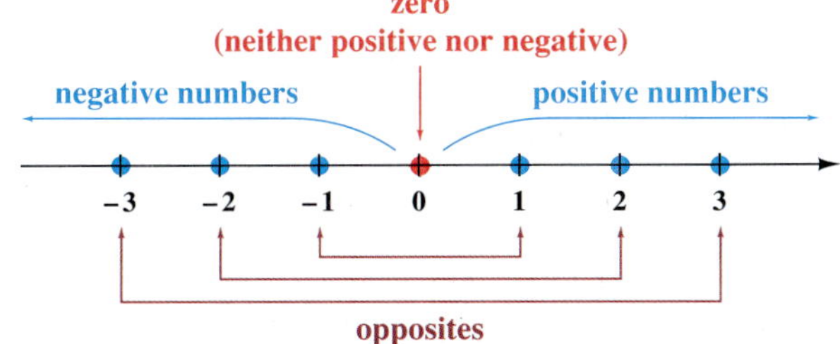

FIGURE 2 Points that correspond to integers

There are many practical applications of negative numbers. For example, a Fahrenheit temperature on a cold January day might be $-10°$, and a business that spends more than it takes in has a negative "profit."

■ **EXAMPLE 1** *Using Negative Numbers in Applications*

Use integers to express the numbers in the following applications.

(a) The lowest Fahrenheit temperature ever recorded in meteorological records was 128.6° below zero at Vostok, Antarctica, on July 22, 1983.

Use -128.6 because "below zero" indicates a negative number.

(b) The shore surrounding the Dead Sea is 1312 feet below sea level.

Again, "below sea level" indicates a negative number, -1312. ■

WORK PROBLEM 1 AT THE SIDE.

2 Not all numbers are integers. For example, $\frac{1}{2}$ is not; it is a number halfway between the integers 0 and 1. Also, $3\frac{1}{4}$ is not an integer. These numbers and others that are quotients of integers are *rational numbers.* (The name comes from the word *ratio,* which indicates a quotient.)

Rational numbers {numbers that can be written as quotients of integers, with denominator not 0}

Since any integer can be written as the quotient of itself and 1, all integers are rational numbers. A decimal number that comes to an end (terminates), such as .23, is a rational number: $.23 = \frac{23}{100}$. Decimal numbers that repeat in a fixed block of digits, such as .3333 . . . and .454545 . . . , are also rational numbers. For example, $.3333\ldots = \frac{1}{3}$.

To **graph** a number we place a dot on the number line at the point that corresponds to the number.

■ **EXAMPLE 2** *Graphing Rational Numbers*

Graph the following numbers on the number line.

$$-\frac{3}{2},\ -\frac{2}{3},\ \frac{1}{2},\ 1\frac{1}{3},\ \frac{23}{8},\ 3\frac{1}{4}$$

To locate the improper fractions on the number line, write them in the form of mixed numbers. The graph is shown in Figure 3. ■

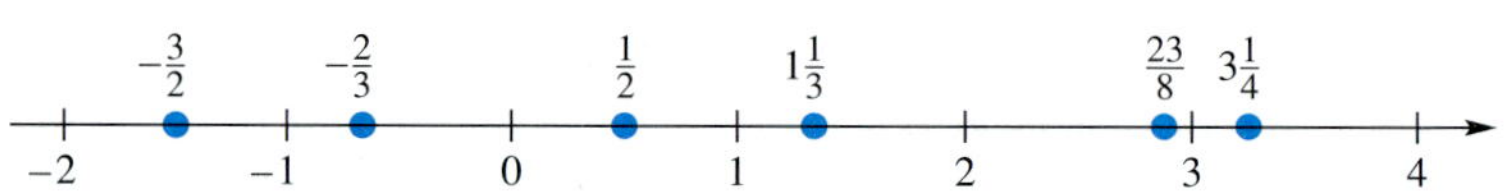

FIGURE 3

WORK PROBLEM 2 AT THE SIDE.

Although a great many numbers are rational, there are also numbers that are not. For example, a floor tile 1 foot on a side has a diagonal whose length is the square root of 2 (written $\sqrt{2}$). See Figure 4. It can be shown that $\sqrt{2}$ cannot be written as a quotient of integers, so it is an example of a number that is not rational. It is **irrational.**

1. Use an integer to express the number(s) in each application.

(a) Erin discovers that she has spent $53 more than she has in her checking account.

(b) The record Fahrenheit high temperature in the U.S. was 134° in Death Valley, California, July 10, 1913.

(c) A football team gained 5 yards, then lost 10 yards on the next play.

2. Graph the following numbers on the number line.

$$-3,\ -2.75,\ -\frac{3}{4},\ 1\frac{1}{2},\ \frac{17}{8}$$

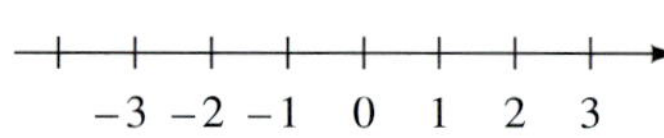

ANSWERS

1. (a) -53 (b) 134 (c) 5, -10

2. -2.75 $-\frac{3}{4}$ $1\frac{1}{2}$ $\frac{17}{8}$

-3 -2 -1 0 1 2 3

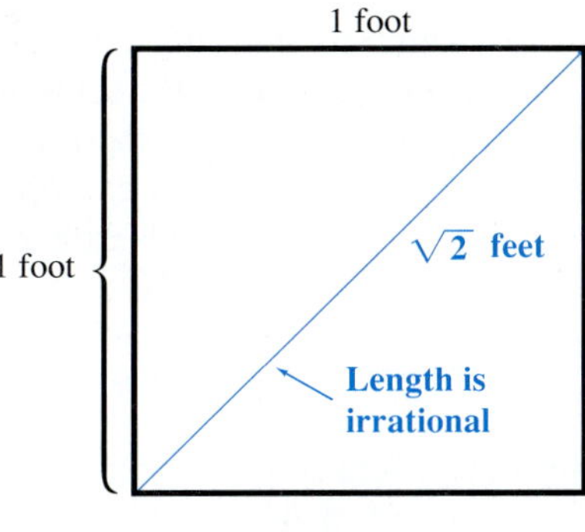

FIGURE 4

Irrational numbers {nonrational numbers represented by points on the number line}

The decimal form of an irrational number never terminates and never repeats. Examples of irrational numbers include $\sqrt{3}$, $\sqrt{7}$, .10110111011110 . . . , $-\sqrt{10}$, and π, which is the ratio of the distance around a circle to the distance across it. These numbers lie between the rational numbers on the number line. Irrational numbers are discussed in Chapter 8.

Finally, *all* numbers that can be represented by points on the number line are called **real numbers.**

Real numbers {all numbers that can be represented by points on the number line} or {all rational and irrational numbers}.

All the numbers mentioned above are real numbers. The relationships between the various types of numbers are shown in two different ways in Figure 5. Notice that any real number is either a rational number or an irrational number.

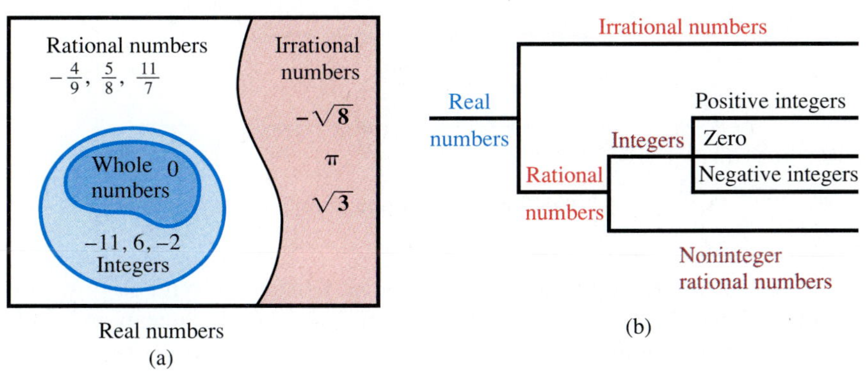

FIGURE 5

3. Tell whether each statement is true or false.

(a) $-2 < 4$

(b) $6 > -3$

(c) $-9 < -12$

(d) $-4 \geq -1$

(e) $-6 \leq 0$

ANSWERS
3. (a) true (b) true (c) false (d) false (e) true

3 Given any two whole numbers, we can tell which number is smaller. But what about two negative numbers, as in the set of integers? Moving from zero to the right along a number line, the positive numbers corresponding to the points on the number line *increase*. For example, $8 < 12$, and 8 is to the left of 12 on a number line. This ordering is extended to all real numbers by definition.

THE ORDERING OF THE REAL NUMBERS

For any two real numbers a and b, ***a* is less than *b*** if a is to the left of b on the number line.

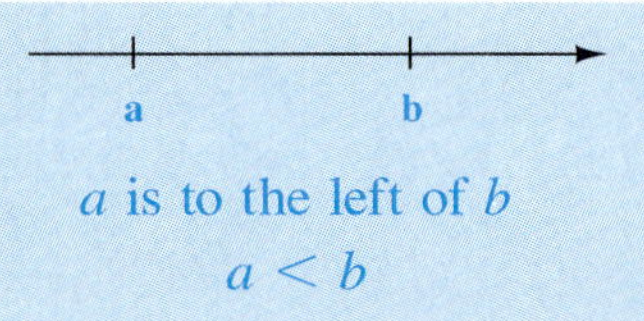

This means that any negative number is smaller than 0, and any negative number is smaller than any positive number. Also, 0 is smaller than any positive number.

EXAMPLE 3 *Determining the Order of Real Numbers*

Is it true that $-3 < -1$?

To find out, locate -3 and -1 on a number line, as shown in Figure 6. Because -3 is to the left of -1 on the number line, -3 is smaller than -1. The statement $-3 < -1$ is true. ■

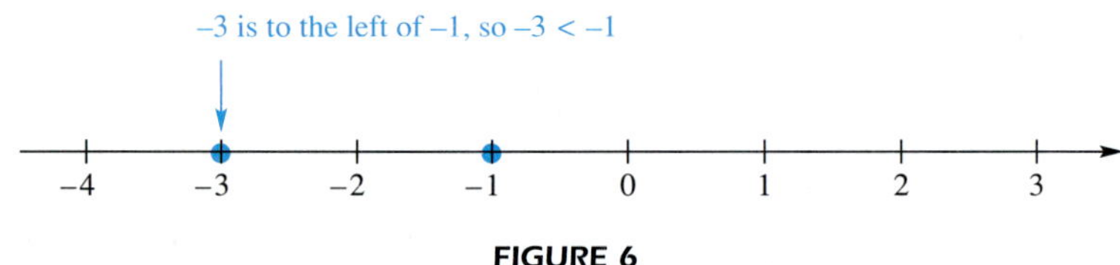

FIGURE 6

WORK PROBLEM 3 AT THE SIDE.

4 Earlier, we saw that every positive integer has a negative integer that is its opposite or negative. This is true for every real number except 0, which is its own opposite. A characteristic of pairs of opposites is that they are the same distance from 0 on the number line. See Figure 7.

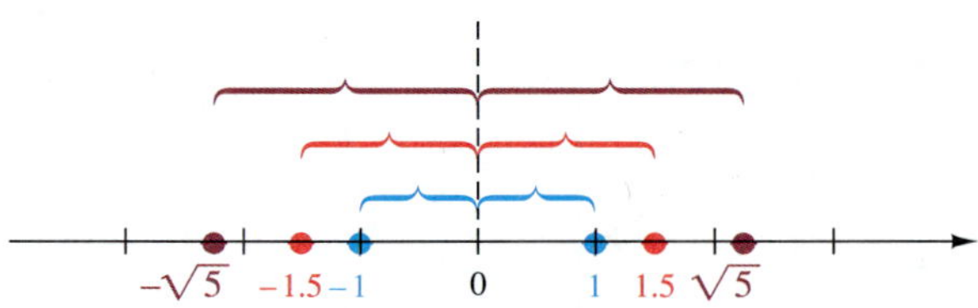

FIGURE 7 Pairs of opposites

We indicate the opposite of a number by writing the symbol $-$ in front of the number. For example, the opposite of 7 is -7 (read "negative 7"). We could write the opposite of -4 as $-(-4)$, but we know that 4 is the opposite of -4. Since a number can have only one opposite, 4 and $-(-4)$ must represent the same number, so that

$$-(-4) = 4.$$

A generalization of this idea is given below.

DOUBLE NEGATIVE RULE

For any real number a.

$$-(-a) = a.$$

EXAMPLE 4 *Finding the Opposite of a Number*

The following chart shows several numbers and their opposites.

Number	*Opposite*
-4	$-(-4)$, or 4
-3	3
0	0
5	-5
19	-19 ■

Example 4 suggests the following rule.

The opposite of a number is found by changing the sign of the number.

WORK PROBLEM 4 AT THE SIDE. ▶▶

4. Find the opposite of each number.

(a) 6

(b) 15

(c) -9

(d) -12

(e) 0

5 As mentioned above, opposites are numbers the same distance from 0 on the number line but on opposite sides of 0. Another way to say this is to say that opposites have the same absolute value. The **absolute value** of a number is the undirected distance between 0 and the number on the number line. The symbol for the absolute value of the number a is $|a|$, read "the absolute value of a." For example, the distance between 2 and 0 on the number line is 2 units, so

$$|2| = 2.$$

Also, the distance between -2 and 0 on the number line is 2, so

$$|-2| = 2.$$

Since distance is a physical measurement, which is never negative, we can make the following statement.

The absolute value of a number can never be negative.

For example,

$$|12| = 12 \quad \text{and} \quad |-12| = 12$$

because both 12 and -12 lie at a distance of 12 units from 0 on the number line. Since 0 is a distance 0 units from 0, we have

$$|0| = 0.$$

ANSWERS

4. (a) -6 (b) -15 (c) 9 (d) 12 (e) 0

5. Simplify by removing absolute value symbols.

(a) $|-6|$

(b) $|9|$

(c) $-|15|$

(d) $-|-9|$

(e) $-|32 - 2|$

ANSWERS
5. (a) 6 (b) 9 (c) −15 (d) −9 (e) −30

■ **EXAMPLE 5** *Evaluating Absolute Value*

Simplify by removing absolute value symbols.

(a) $|5| = 5$

(b) $|-5| = 5$

(c) $-|-5| = -(5) = -5$ Replace $|-5|$ with 5.

(d) $-|-13| = -13$

(e) $|8 - 5|$

Simplify within the absolute value bars first.

$$|8 - 5| = |3| = 3$$ ■

Part (e) in Example 5 suggests that absolute value bars also act as grouping symbols.

◀◀ WORK PROBLEM 5 AT THE SIDE.

EXERCISES

Decide whether each statement is true or false.

1. Every whole number is an integer.

2. Every natural number is a whole number.

3. Every rational number is a real number.

4. No number can be both rational and irrational.

5. Every whole number is positive.

6. No natural number is negative.

7. Some real numbers are not rational.

8. Not every rational number is positive.

9. Some whole numbers are not integers.

10. The number 0 is irrational.

Use an integer to express each number in the following applications of numbers. See Example 1.

11. Between the years 1980 and 1990, the population of Marshalltown, Iowa, decreased by 1760 people.

12. Between 1970 and 1980, the population of the state of New York decreased by 683,226.

13. The city of New Orleans lies 8 feet below sea level.

14. Death Valley lies 282 feet below sea level.

15. The height of Mt. St. Helen's, an active volcano in the state of Washington, is about 8300 feet.

16. Alexader Fedotov, in 1977, flew a jet airplane at an altitude of 123,524 feet.

17. In 1990, New Zealand exported \$66,000,000 less than it imported, thus accounting for a negative trade balance.

18. In 1990, Portugal exported \$90,000,000 more than it imported, thus accounting for a positive trade balance.

Graph each group of numbers on a number line. See Example 2.

19. $0, 3, -5, -6$

20. $2, 6, -2, -1$

21. $-2, -6, -4, 3, 4$

22. $-5, -3, -2, 0, 4$

23. $\frac{1}{4}, 2\frac{1}{2}, -3\frac{4}{5}, -4, -1\frac{5}{8}$

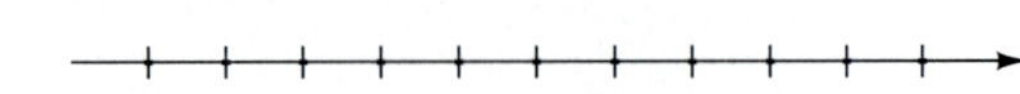

24. $5\frac{1}{4}, 4\frac{5}{9}, -2\frac{1}{3}, 0, -3\frac{2}{5}$

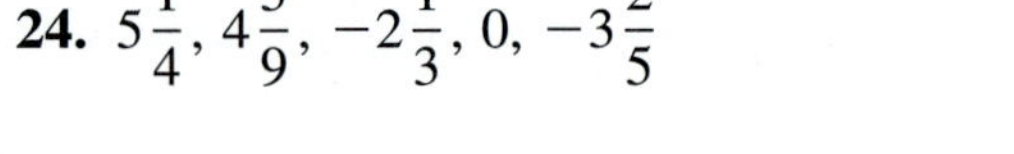

Select the smaller number in each pair. See Example 3.

25. $-11, -4$ **26.** $-9, -16$ **27.** $-21, 1$ **28.** $-57, 3$

29. $0, -100$ **30.** $-215, 0$ **31.** $-\frac{2}{3}, -\frac{1}{4}$ **32.** $-\frac{3}{8}, -\frac{9}{16}$

Decide whether each statement is true or false. See Example 3.

33. $8 < -16$ **34.** $12 < -24$ **35.** $-3 < -2$ **36.** $-10 < -9$

For each of the following, (a) find the opposite of the number and (b) find the absolute value of the number. See Examples 4 and 5.

37. -2 **38.** -8 **39.** 6

40. 11 **41.** $-\frac{3}{4}$ **42.** $-\frac{1}{3}$

Simplify by removing absolute value symbols. See Example 5.

43. $|-7|$ **44.** $|-3|$ **45.** $|4|$ **46.** $|10|$

47. $-|12|$ **48.** $-|23|$ **49.** $-|-14|$ **50.** $-|-19|$

51. $|13 - 4|$ **52.** $|8 - 7|$

53. Students often say "The absolute value of a number is always positive." Is this true? If not, explain.

54. If the absolute value of a number is equal to the number itself, what must be true about the number?

PREVIEW EXERCISES

Add or subtract. Write answers in lowest terms. See Section R.1.

55. $\frac{3}{4} + \frac{9}{10}$ **56.** $\frac{4}{5} - \frac{2}{7}$ **57.** $\frac{3}{4} + \frac{1}{6} - \frac{1}{2}$ **58.** $5\frac{1}{8} - 2\frac{2}{3}$

1.4 ADDITION OF REAL NUMBERS

1 We can use the number line to explain the addition of real numbers. Later, we will give the rules for addition.

EXAMPLE 1 *Adding with the Number Line*

Use the number line to find the sum 2 + 3.

Add the positive numbers 2 and 3 by starting at 0 and drawing an arrow two units to the *right,* as shown in Figure 8. This arrow represents the number 2 in the sum 2 + 3. Next, from the right end of this arrow draw another three units to the right. The number below the end of this second arrow is 5, so 2 + 3 = 5. ■

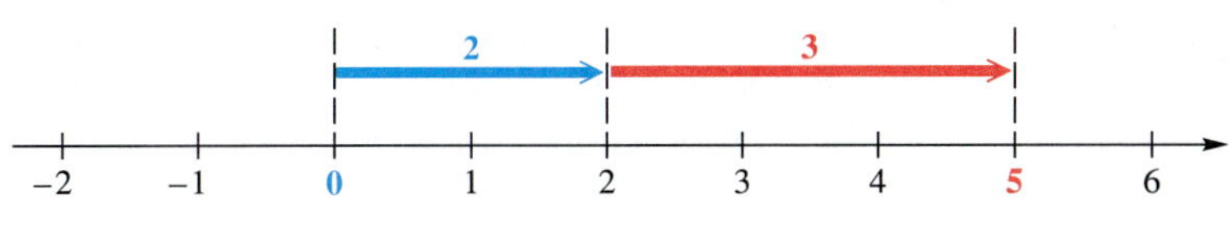

FIGURE 8

EXAMPLE 2 *Adding with the Number Line*

Use the number line to find the sum −2 + (−4). (Parentheses are placed around the −4 to avoid the confusing use of + and − next to each other.)

To add the negative numbers −2 and −4 on the number line, we start at 0 and draw an arrow two units to the *left,* as shown in Figure 9. We draw the arrow to the left to represent the addition of a *negative* number. From the left end of this first arrow, we draw a second arrow four units to the left. The number below the end of this second arrow is −6, so −2 + (−4) = −6. ■

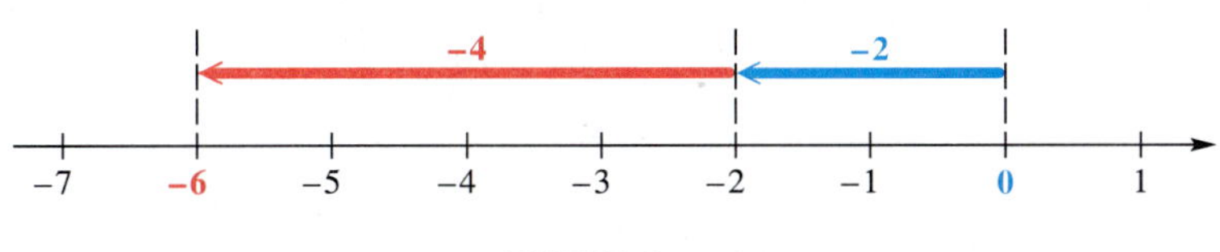

FIGURE 9

WORK PROBLEM 1 AT THE SIDE. ▶

In Example 2, we found that the sum of the two negative numbers −2 and −4 is a negative number whose distance from 0 is the sum of the distance of −2 from 0 and the distance of −4 from 0. That is, *the sum of two negative numbers is the negative of the sum of their absolute values.*

$$-2 + (-4) = -(|-2| + |-4|) = -(2 + 4) = -6$$

To add two numbers having the same signs, add the absolute values of the numbers. Give the result the same sign as the numbers being added. Example: −4 + (−3) = −7.

EXAMPLE 3 *Adding Two Negative Numbers*

Find the sums.

(a) −2 + (−9) = −11 The sum of two negative numbers is negative.

OBJECTIVES

1. Add two numbers with the same sign on a number line.
2. Add positive and negative numbers.
3. Add mentally.
4. Use the order of operations with real numbers.
5. Interpret words and phrases that indicate addition.

FOR EXTRA HELP

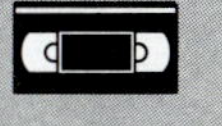

Tape 2 | SSM pp. 18–21 | MAC: A IBM: A

1. Use the number lines to find the sums.

(a) 1 + 4

0 1 2 3 4 5

(b) −2 + (−5)

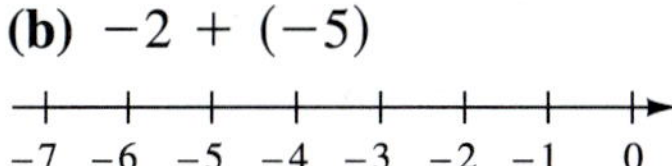

ANSWERS

1. (a) **1 + 4 = 5**

1 4

0 1 2 3 4 5

(b) **−2 + (−5) = −7**

−5 −2

−7 −6 −4 −2 0

2. Find the sums.

(a) $-7 + (-3)$

(b) $-12 + (-18)$

(c) $-15 + (-4)$

3. Use the number lines to find the sums.

(a) $6 + (-3)$

0 1 2 3 4 5 6

(b) $-5 + 1$

−5 −4 −3 −2 −1 0

ANSWERS

2. (a) -10 **(b)** -30 **(c)** -19

3. (a) $6 + (-3) = 3$

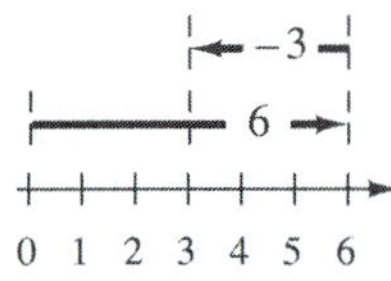

(b) $-5 + 1 = -4$

1
−5
−5 −4 −3 −2 −1 0

(b) $-8 + (-12) = -20$

(c) $-15 + (-3) = -18$ ■

WORK PROBLEM 2 AT THE SIDE.

2 We use the number line again to illustrate the sum of a positive number and a negative number.

■ EXAMPLE 4 *Adding Numbers with Different Signs*

Use the number line to find the sum $-2 + 5$.

We find the sum $-2 + 5$ on the number line by starting at 0 and drawing an arrow two units to the left. From the left end of this arrow, we draw a second arrow five units to the right, as shown in Figure 10. The number below the end of this second arrow is 3, so $-2 + 5 = 3$. ■

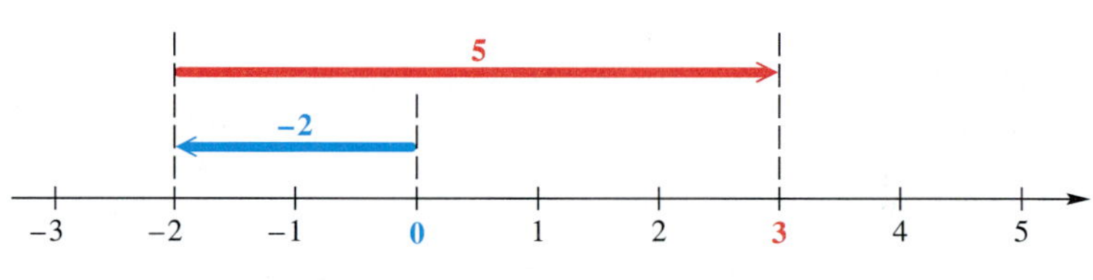

FIGURE 10

WORK PROBLEM 3 AT THE SIDE.

Addition of numbers with different signs also can be defined using absolute value.

> To add numbers with different signs, first find the difference between the absolute values of the numbers. Give the answer the same sign as the number with the larger absolute value. Example: $-12 + 6 = -6$.

For example, to add -12 and 5, we find their absolute values: $|-12| = 12$ and $|5| = 5$; then we find the difference between these absolute values: $12 - 5 = 7$. Since $|-12| > |5|$, the sum will be negative, so the final answer is $-12 + 5 = -7$.

The − or +/− key is used to input a negative number in the calculator. Try using your calculator to add negative numbers.

3 While a number line is useful in showing the rules for addition, it is important to be able to find sums mentally.

■ EXAMPLE 5 *Adding a Positive Number and a Negative Number*

Check each answer, trying to work the addition mentally. If you get stuck, use a number line.

(a) $7 + (-4) = 3$

(b) $-8 + 12 = 4$

(c) $-\frac{1}{2} + \frac{1}{8} = -\frac{4}{8} + \frac{1}{8} = -\frac{3}{8}$ Remember to find a common denominator first.

(d) $\frac{5}{6} + \left(-\frac{4}{3}\right) = \frac{5}{6} + \left(-\frac{8}{6}\right) = -\frac{3}{6} = -\frac{1}{2}$

(e) $-4.6 + 8.1 = 3.5$ ■

WORK PROBLEM 4 AT THE SIDE. ▶▶

The rules for adding signed numbers are summarized below.

ADDING SIGNED NUMBERS

Like signs Add the absolute values of the numbers. Give the sum the same sign as the numbers being added.
Unlike signs Find the difference between the larger absolute value and the smaller. Give the answer the sign of the number having the larger absolute value.

4 Sometimes a problem involves square brackets, [ ]. As we mentioned earlier, brackets are treated just like parentheses. We do the calculations inside the brackets until a single number is obtained. Remember to use the order of operations given in Section 1.1 for adding more than two numbers.

■ EXAMPLE 6 *Adding with Brackets*
Find the sums.

(a) $-3 + [4 + (-8)]$

First work inside the brackets. Follow the rules for the order of operations given in Section 1.1.

$$-3 + [4 + (-8)] = -3 + (-4) = -7$$

(b) $8 + [(-2 + 6) + (-3)] = 8 + [4 + (-3)] = 8 + 1 = 9$ ■

WORK PROBLEM 5 AT THE SIDE. ▶▶

5 Let's now look at the interpretation of words and phrases that involve addition. Problem solving often requires translating words and phrases into symbols. We began this process with translating simple phrases in Section 1.1.

The word *sum* indicates addition. There are other key words and phrases that also indicate addition. Some of these are given in the chart below.

Word or Phrase	*Example*	*Numerical Expression and Simplification*
Sum	The *sum of* −3 and 4	$-3 + 4 = 1$
Added to	5 *added to* −8	$-8 + 5 = -3$
More than	12 *more than* −5	$12 + (-5) = 7$
Increased by	−6 *increased by* 13	$-6 + 13 = 7$

4. Check each answer, trying to work the addition in your head. If you get stuck, use a number line.

(a) $-8 + 2 = -6$

(b) $-15 + 4 = -11$

(c) $17 + (-10) = 7$

(d) $\frac{3}{4} + \left(-\frac{11}{8}\right) = -\frac{5}{8}$

(e) $-9.5 + 3.8 = -5.7$

5. Find the sums.

(a) $2 + [7 + (-3)]$

(b) $6 + [(-2 + 5) + 7]$

(c) $-9 + [-4 + (-8 + 6)]$

ANSWERS
4. All are correct.
5. (a) 6 (b) 16 (c) −15

6. Write a numerical expression for each phrase, and simplify the expression.

(a) 4 more than -12

(b) The sum of 6 and -7

(c) -12 added to -31

(d) 7 increased by the sum of 8 and -3

7. A football team lost 8 yards on the first play from scrimmage, lost 5 yards on the second play, and then gained 7 yards on the third play. How many yards did the team gain or lose altogether?

ANSWERS

6. (a) $-12 + 4$; -8 **(b)** $6 + (-7)$; -1 **(c)** $-31 + (-12)$; -43 **(d)** $7 + [8 + (-3)]$; 12

7. The team lost 6 yards.

■ EXAMPLE 7 *Interpreting Words and Phrases*

Write a numerical expression for each phrase, and simplify the expression.

(a) The sum of -8 and 4 and 6

$$-8 + 4 + 6 = [-8 + 4] + 6 = -4 + 6 = 2$$

Notice that brackets were placed around $-8 + 4$, and this addition was done first, using the order of operations given earlier. In fact, the same result would be obtained if the brackets were placed around $4 + 6$. This idea will be discussed further in Section 1.8.

$$-8 + 4 + 6 = -8 + [4 + 6] = -8 + 10 = 2$$

(b) 3 more than -5, increased by 12

$$-5 + 3 + 12 = [-5 + 3] + 12 = -2 + 12 = 10$$ ■

WORK PROBLEM 6 AT THE SIDE.

Gains (or increases) and losses (or decreases) sometimes appear in applied problems. When they do, the gains may be interpreted as positive numbers and the losses as negative numbers.

■ EXAMPLE 8 *Interpreting Gains and Losses*

A football team gained 3 yards on the first play from scrimmage, lost 12 yards on the second play, and then gained 13 yards on the third play. How many yards did the team gain or lose altogether?

The gains are represented by positive numbers and the loss by a negative number.

$$3 + (-12) + 13$$

Add from left to right.

$$3 + (-12) + 13 = [3 + (-12)] + 13 = (-9) + 13 = 4$$

The team gained 4 yards altogether. ■

WORK PROBLEM 7 AT THE SIDE.

1.4 EXERCISES

Fill in the blank with the correct response.

1. The sum of two negative numbers will always be a ______________ number.
(positive/negative)

2. The sum of a number and its opposite will always be ________.

3. To simplify the expression $8 + [-2 + (-3 + 5)]$, I should begin by adding ____ and ____, according to the rules for order of operations.

4. If I am adding a positive number and a negative number, and the negative number has the larger absolute value, the sum will be a ______________ number.
(positive/negative)

Find the following sums. See Examples 1–6.

5. $6 + (-4)$

6. $12 + (-9)$

7. $7 + (-10)$

8. $4 + (-8)$

9. $-7 + (-3)$

10. $-11 + (-4)$

11. $-10 + (-3)$

12. $-16 + (-7)$

13. $-12.4 + (-3.5)$

14. $-21.3 + (-2.5)$

15. $-8 + 7$

16. $-12 + 10$

17. $5 + [14 + (-6)]$

18. $7 + [3 + (-14)]$

19. $10 + [-3 + (-2)]$

20. $13 + [-4 + (-5)]$

21. $-3 + [5 + (-2)]$

22. $-7 + [10 + (-3)]$

23. $-8 + [3 + (-1) + (-2)]$

24. $-7 + [5 + (-8) + 3]$

25. $-\frac{1}{6} + \frac{2}{3}$

26. $\frac{9}{10} + \left(-\frac{3}{5}\right)$

27. $\frac{5}{8} + \left(-\frac{17}{12}\right)$

28. $-\frac{6}{25} + \frac{19}{20}$

29. $2\frac{1}{2} + \left(-3\frac{1}{4}\right)$

30. $-4\frac{3}{8} + 6\frac{1}{2}$

31. $7.8 + (-9.4)$

32. $14.7 + (-10.1)$

33. $-7.1 + [3.3 + (-4.9)]$

34. $-9.5 + [-6.8 + (-1.3)]$

35. $[-8 + (-3)] + [-7 + (-7)]$

36. $[-5 + (-4)] + [9 + (-2)]$

37. $[-5 + (-7)] + [-4 + (-9)] + [13 + (-12)]$

38. $[-3 + (-11)] + [13 + (-3)] + [19 + (-7)]$

Perform each operation, and then determine whether the statement is true or false. Try to do all work in your head. See Examples 5 and 6.

39. $-5 + 0 = -5$

40. $0 + (-9) = -9$

41. $-11 + 13 = 13 + (-11)$

42. $16 + (-9) = -9 + 16$

43. $-10 + 6 + 7 = -3$

44. $-12 + 8 + 5 = -1$

45. $18 + (-6) + (-12) = 0$

46. $-5 + 21 + (-16) = 0$

47. $|-8 + 10| = -8 + (-10)$

48. $|-4 + 6| = -4 + (-6)$

49. $\frac{11}{5} + \left(-\frac{6}{11}\right) = -\frac{6}{11} + \frac{11}{5}$

50. $-\frac{3}{2} + \frac{5}{8} = \frac{5}{8} + \left(-\frac{3}{2}\right)$

51. $-7 + [-5 + (-3)] = [(-7) + (-5)] + 3$

52. $6 + [-2 + (-5)] = [(-4) + (-2)] + 5$

Write a numerical expression for each phrase, and simplify the expression. See Example 7.

53. The sum of -5 and 12 and 6

54. The sum of -3 and 5 and -12

55. 14 added to the sum of -19 and -4

56. -2 added to the sum of -18 and 11

57. The sum of -4 and -10, increased by 12

58. The sum of -7 and -13, increased by 14

59. 4 more than the sum of 8 and -18

60. 10 more than the sum of -4 and -6

Solve each problem. See Example 8.

61. A college student received a \$100 check in the mail from her parents. She then spent \$53 for a sociology textbook. How much money does she have left?

62. Shalita's checking account balance is \$54.00. She then takes a gamble by writing a check for \$89.00. What is her new balance? (Write the balance as a signed number.)

63. The surface, or rim, of a canyon is at altitude 0. On a hike down into the canyon, a party of hikers stops for a rest at 130 meters below the surface. They then descend another 54 meters. What is their new altitude? (Write the altitude as a signed number.)

64. A pilot announces to the passengers that the current altitude of their plane is 34,000 feet. Because of some unexpected turbulence, the pilot is forced to descend 2100 feet. What is the new altitude of the plane? (Write the altitude as a signed number.)

65. The lowest temperature ever recorded in Little Rock, Arkansas, was -5°F. The highest temperature ever recorded there was 117°F more than the lowest. What was this highest temperature?

66. On a series of three consecutive running plays, Herschel Walker of the Philadelphia Eagles gained 4 yards, lost 3 yards, and lost 2 yards. What positive or negative number represents his total net yardage for the series of plays?

67. On three consecutive passes, Troy Aikman of the Dallas Cowboys passed for a gain of 6 yards, was sacked for a loss of 12 yards, and passed for a gain of 43 yards. What positive or negative number represents the total net yardage for the plays?

68. On January 23, 1943, the temperature rose 49°F in two minutes in Spearfish, South Dakota. If the starting temperature was −4°F, what was the temperature two minutes later?

69. Jennifer owes $153 to a credit card company. She makes a $14 purchase with the card, and then pays $60 on the account. What amount does she still owe?

70. Jack owes his brother $4. He pays him $3 and then borrows another $6. What amount does he still owe?

*71. Kim Falgout owes $870.00 on her Master Card account. She returns two items costing $35.90 and $150.00 and receives credits for these on the account. Next, she makes a purchase of $82.50, and then two more purchases of $10.00 each. She finally makes a payment of $500.00. How much does she still owe?

72. A welder working with stainless steel must use precise measurements. Suppose a welder attaches two pieces of steel that are each 3.60 inches in length, and then attaches an additional three pieces that are each 9.10 inches long. She finally cuts off a piece that is 7.60 inches long. Find the length of the welded piece of steel.

Find the solution of each equation from the domain {−3, −2, −1, 0, 1, 2, 3} by guessing or using trial and error.

73. $x + 3 = 1$

74. $x + 1 = 0$

75. $p + 4 = 1$

76. $r + 9 = 7$

77. $13 + y = 12$

78. $8 + t = 5$

79. $b + (-6) = -6$

80. $k + (-3) = -3$

81. $r + (-4.6) = -3.6$

82. Make up your own equation similar to the ones in Exercises 73–81 that has −2 as a solution.

PREVIEW EXERCISES

Simplify. See Section. 1.1.

83. $8[(9 + 2) - 4]$

84. $4 + 9(7 - 3)$

85. $(3 - 1) + (17 - 14)$

86. $9(4 - 2) - 2(8 - 7)$

* Color exercise numbers are used to indicate exercises designed for calculator use.

1.5 SUBTRACTION OF REAL NUMBERS

1 As we mentioned earlier, the answer to a subtraction problem is called a **difference.** Differences between signed numbers can be found by using a number line. Addition and subtraction are opposite operations. Thus, because *addition* of a positive number on the number line is shown by drawing an arrow to the *right, subtraction* of a positive number is shown by drawing an arrow to the *left.*

■ EXAMPLE 1 *Subtracting with the Number Line*

Use the number line to find the difference $7 - 4$.

To find the difference $7 - 4$ on the number line, begin at 0 and draw an arrow 7 units to the right. From the right end of this arrow, draw an arrow 4 units to the left, as shown in Figure 11. The number at the end of the second arrow shows that $7 - 4 = 3$. ■

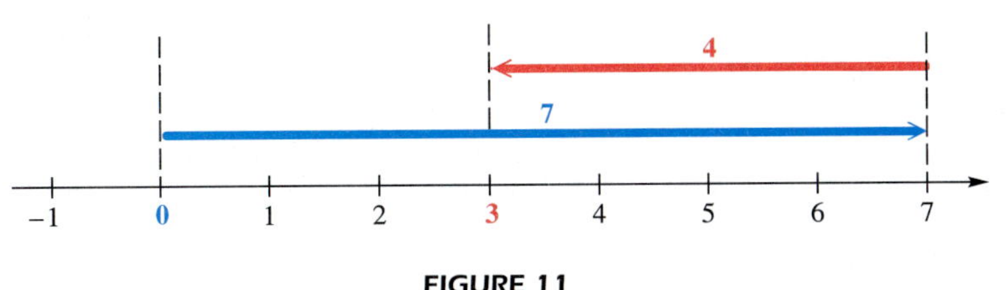

FIGURE 11

WORK PROBLEM 1 AT THE SIDE.

2 The procedure used in Example 1 to find $7 - 4$ is exactly the same procedure that would be used to find $7 + (-4)$, so that

$$7 - 4 = 7 + (-4).$$

This shows that *subtraction* of a positive number from a larger positive number is the same as *adding* the opposite of the smaller number to the larger. This result is extended as the definition of subtraction for all real numbers.

DEFINITION OF SUBTRACTION

For any real numbers a and b,

$$a - b = a + (-b).$$

Example: $4 - 9 = 4 + (-9) = -5$.

That is, to **subtract** b from a, *add the* opposite (or *negative*) of b to a. This definition leads to the following procedure for subtracting signed numbers.

SUBTRACTING SIGNED NUMBERS

Step 1 Change the subtraction symbol to addition.
Step 2 Change the sign of the number being subtracted.
Step 3 Add, as in the previous section.

OBJECTIVES

1. Find a difference on the number line.
2. Use the definition of subtraction.
3. Work subtraction problems that involve brackets.
4. Interpret words and phrases that indicate subtraction.

FOR EXTRA HELP

Tape 2

SSM pp. 22–24

MAC: A
IBM: A

1. Use the number line to find the differences.

(a) $5 - 1$

0 1 2 3 4 5

(b) $6 - 2$

0 1 2 3 4 5 6

ANSWERS

1. (a) $5 - 1 = 4$

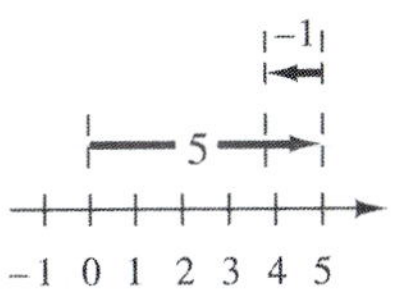

(b) $6 - 2 = 4$

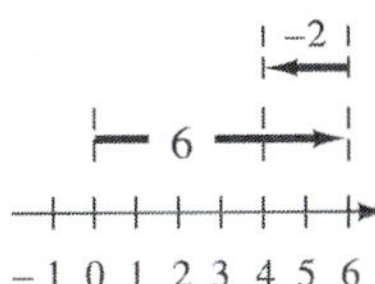

2. Subtract.

(a) $6 - 10$

(b) $-2 - 4$

(c) $3 - (-5)$

(d) $-8 - (-12)$

3. Work each problem.

(a) $2 - [(-3) - (4 + 6)]$

(b) $[(5 - 7) + 3] - 8$

(c) $6 - [(-1 - 4) - 2]$

ANSWERS
2. (a) -4 (b) -6 (c) 8 (d) 4
3. (a) 15 (b) -7 (c) 13

■ EXAMPLE 2 *Using the Definition of Subtraction*

Subtract.

No change; Change − to +.; Opposite of 3

(a) $12 - 3 = 12 + (-3) = 9$

(b) $5 - 7 = 5 + (-7) = -2$

(c) $8 - 15 = 8 + (-15) = -7$

No change; Change − to +.; Opposite of −5

(d) $-3 - (-5) = -3 + (5) = 2$

(e) $-6 - (-9) = -6 + (9) = 3$

(f) $8 - (-5) = 8 + (5) = 13$ ■

WORK PROBLEM 2 AT THE SIDE.

Subtraction can be used to reverse the result of an addition problem. For example, if 4 is added to a number and then subtracted from the sum, the original number is the result.

$$12 + 4 = 16 \quad \text{and} \quad 16 - 4 = 12$$

The symbol − has now been used for three purposes:

1. to represent subtraction, as in $9 - 5 = 4$;
2. to represent negative numbers, such as -10, -2, and -3;
3. to represent the opposite (or negative) of a number, as in "the opposite (or negative) of 8 is -8."

We may see more than one use in the same problem, such as $-6 - (-9)$, where -9 is subtracted from -6. The meaning of the symbol depends on its position in the algebraic expression.

3 As before, with problems that have both parentheses and brackets, first do any operations inside the parentheses and brackets. Work from the inside out. Because subtraction is defined in terms of addition, the order of operations rules from Section 1.4 can still be used.

■ EXAMPLE 3 *Subtracting with Grouping Symbols*

Work each problem.

(a)
$$\begin{aligned} -6 - [2 - (8 + 3)] &= -6 - [2 - 11] \\ &= -6 - [2 + (-11)] \quad \text{Change } - \text{ to } +. \\ &= -6 - (-9) \\ &= -6 + (9) = 3 \end{aligned}$$

(b)
$$\begin{aligned} 5 - [(-3 - 2) - (4 - 1)] &= 5 - [(-3 + (-2)) - 3] \\ &= 5 - [(-5) - 3] \\ &= 5 - [(-5) + (-3)] \\ &= 5 - (-8) \\ &= 5 + 8 = 13 \end{aligned}$$ ■

WORK PROBLEM 3 AT THE SIDE.

4 Now we translate words and phrases that involve subtraction of real numbers. *Difference* is one of them. Some others are given in the chart below.

Word or Phrase	*Example*	*Numerical Expression and Simplification*
Difference	The ***difference between*** −3 and −8	$-3 - (-8) = -3 + 8 = 5$
Subtracted from	12 ***subtracted from*** 18	$18 - 12 = 6$
Less than	6 ***less than*** 5	$5 - 6 = 5 + (-6) = -1$
Decreased by	9 ***decreased by*** −4	$9 - (-4) = 9 + 4 = 13$

Caution When you are subtracting two numbers, it is important that you write them in the correct order, because, in general, $a - b \neq b - a$. For example, $5 - 3 \neq 3 - 5$. For this reason, it is important to *think carefully before interpreting an expression involving subtraction!* (This problem did not arise for addition.)

EXAMPLE 4 *Interpreting Words and Phrases*

Write a numerical expression for each phrase, and simplify the expression.

(a) The **difference between** −8 and 5

It is conventional to write the numbers in the order they are given when "difference between" is used.

$$-8 - 5 = -8 + (-5) = -13$$

(b) 4 **subtracted from** the sum of 8 and −3

Here the operation of addition is also used, as indicated by the word *sum.* First, add 8 and −3. Next, subtract 4 from this sum.

$$[8 + (-3)] - 4 = 5 - 4 = 1$$

(c) 4 **less than** −6

Be careful with order here. 4 must be taken *from* −6.

$$-6 - 4 = -6 + (-4) = -10$$

Notice that "4 less than −6" differs from "4 *is less than* −6." "4 is less than −6" is symbolized as $4 < -6$ (which is a false statement).

(d) 8, **decreased by** 5 **less than** 12

First, write "5 less than 12" as $12 - 5$. Next, subtract $12 - 5$ from 8.

$$8 - (12 - 5) = 8 - 7 = 1$$

WORK PROBLEM 4 AT THE SIDE.

We have seen a few applications of signed numbers in earlier sections. The next example involves subtraction of signed numbers.

4. Write a numerical expression for each phrase, and simplify the expression.

(a) The difference between −5 and −12

(b) −2 subtracted from the sum of 4 and −4

(c) 7 less than −2

(d) 9, decreased by 10 less than 7

ANSWERS

4. (a) $-5 - (-12)$; 7
(b) $[4 + (-4)] - (-2)$; 2
(c) $-2 - 7$; -9
(d) $9 - (7 - 10)$; 12

5. The highest elevation in Argentina is Mt. Aconcagua, which is 6960 meters above sea level. The lowest point in Argentina is the Valdes Peninsula, 40 meters below sea level. Find the difference between the highest and lowest elevations.

EXAMPLE 5 *Solving a Problem Involving Subtraction*

The record high temperature of 134° Fahrenheit in the United States was recorded at Death Valley, California, in 1913. The record low was −80°F, at Prospect Creek, Alaska, in 1971. See Figure 12. What is the difference between these highest and lowest temperatures?

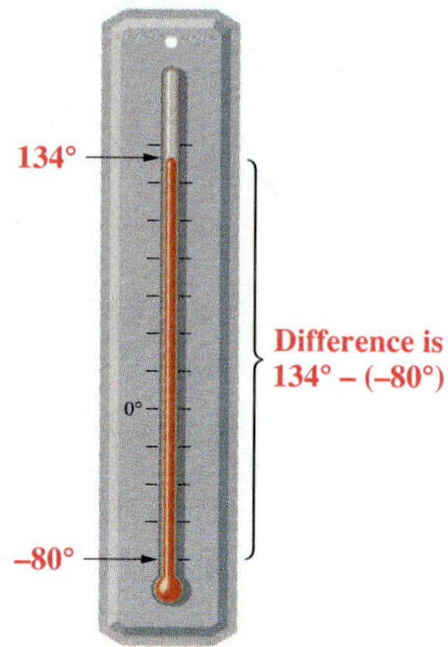

FIGURE 12

We must subtract the lowest temperature from the highest temperature.

$$134 - (-80) = 134 + 80 \quad \text{Use the definition of subtraction.}$$
$$= 214 \quad \text{Add.}$$

The difference between the two temperatures is 214°F. ■

◀◀ WORK PROBLEM 5 AT THE SIDE.

ANSWER
5. 7000 meters

1.5 EXERCISES

NAME DATE HOUR

Fill in the blank with the correct response.

1. By the definition of subtraction, in order to perform the subtraction problem $-6 - (-8)$, we must add the opposite of _____ to _____.

2. "The difference between 7 and 12" translates as _____, while "the difference between 12 and 7" translates as _____.

3. In order to simplify $6 - [(7 - 8) - (8 - 12)]$ according to the rules for order of operations, we should begin by subtracting 8 from _____.

4. $-6 - (-3) = -6 +$ _____

5. $-8 - 4 = -8 +$ _____

6. $0 - 41 = 0 +$ _____

Find the differences. See Examples 1–3.

7. $4 - 7$

8. $8 - 13$

9. $6 - 10$

10. $9 - 14$

11. $-7 - 3$

12. $-12 - 5$

13. $-10 - 6$

14. $-13 - 16$

15. $7 - (-4)$

16. $9 - (-6)$

17. $6 - (-13)$

18. $13 - (-3)$

19. $-7 - (-3)$

20. $-8 - (-6)$

21. $3 - (4 - 6)$

22. $6 - (7 - 14)$

23. $-3 - (6 - 9)$

24. $-4 - (5 - 12)$

25. $\frac{1}{2} - \left(-\frac{1}{4}\right)$

26. $\frac{1}{3} - \left(-\frac{4}{3}\right)$

27. $-\frac{3}{4} - \frac{5}{8}$

28. $-\frac{5}{6} - \frac{1}{2}$

29. $\frac{5}{8} - \left(-\frac{1}{2} - \frac{3}{4}\right)$

30. $\frac{9}{10} - \left(\frac{1}{8} - \frac{3}{10}\right)$

31. $4.4 - (-9.2)$

32. $6.7 - (-12.6)$

33. $-7.4 - 4.5$

34. $-5.4 - 9.6$

35. $-5.2 - (8.4 - 10.8)$

36. $-9.6 - (3.5 - 12.6)$

37. $[(-3.1) - 4.5] - (.8 - 2.1)$

38. $[(-7.8) - 9.3] - (.6 - 3.5)$

39. Explain in your own words how to subtract signed numbers.

40. We know that in general, $a - b \neq b - a$. Can you give values for a and b so that $a - b$ *is equal to* $b - a$? If so, give two such pairs.

Work each problem. See Example 3.

41. $(-3 - 8) - (7 - 4)$

42. $(-5 - 9) - (7 - 2)$

43. $-10 - [(5 - 4) - (-5 - 8)]$

44. $-12 - [(9 - 2) - (-6 - 3)]$

45. $-4 + [(-6 - 9) - (-7 + 4)]$

46. $-8 + [(-3 - 10) - (-4 + 1)]$

47. $\left(-\frac{3}{4} - \frac{5}{2}\right) - \left(-\frac{1}{8} - 1\right)$

48. $\left(-\frac{3}{8} - \frac{2}{3}\right) - \left(-\frac{9}{8} - 3\right)$

49. $[-34.99 + (6.59 - 12.25)] - 8.33$

50. $[-12.25 - (8.34 + 3.57)] - 17.88$

51. Make up a subtraction problem so that the difference between two negative numbers is a negative number.

52. Make up a subtraction problem so that the difference between two negative numbers is a positive number.

Write a numerical expression for each phrase and simplify. See Example 4.

53. The difference between 4 and -8

54. The difference between 7 and -14

55. 8 less than -2

56. 9 less than -13

57. The sum of 9 and -4, decreased by 7

58. The sum of 12 and -7, decreased by 14

59. 12 less than the difference between 8 and -5

60. 19 less than the difference between 9 and -2

Solve each problem. See Example 5.

61. On a cold winter day, the temperature reached $-5°F$ in Glenview, Illinois. The following day it was even colder, and the temperature dropped $10°F$ below $-5°F$. What was the temperature on the following day?

62. A chemist is running an experiment under precise conditions. At first, she runs it at $-174.6°F$. She then lowers it by $2.3°F$. What is the new temperature for the experiment?

63. The top of Mount Whitney, visible from Death Valley, has an altitude of 14,494 feet above sea level. The bottom of Death Valley is 282 feet below sea level. Using zero as sea level, find the difference between these two elevations.

64. The highest point in Louisiana is Driskill Mountain, at an altitude of 535 feet. The lowest point is at Spanish Fort, 8 feet below sea level. Using zero as sea level, find the difference between these two elevations.

65. Chris owed his brother \$10. He later borrowed \$70. What positive or negative number represents his present financial status?

66. Francesca has \$15 in her purse, and Emilio has a debt of \$12. Find the difference between these amounts.

67. For the year of 1993, one health club showed a profit of \$76,000, while another showed a loss of \$29,000. Find the difference between these.

68. At 1:00 A.M., a plant worker found that a dial reading was 7.904. At 2:00 A.M., she found the reading to be -3.291. By how much had the reading declined?

Find the solution of each equation from the domain $\{-3, -2, -1, 0, 1, 2, 3\}$ by guessing or by trial and error.

69. $x - 1 = -3$

70. $x - 2 = -1$

71. $3 - x = 6$

72. $2 - x = 3$

73. $3 - (-x) = 0$

74. $1 - (-x) = 0$

In Exercises 75–78, suppose that a represents a positive number and b represents a negative number. Determine whether the given expression must represent a positive number or a negative number.

75. $a - b$

76. $b - a$

77. $a + |b|$

78. $b - |a|$

PREVIEW EXERCISES

Find the value of each expression, given $x = 5$ and $y = 2$. See Section 1.2.

79. $4x - 2y$

80. $9(2x + y)$

81. $(x + y)(2x - y)$

82. $(3x - y)(2x + 4y)$

1.6 MULTIPLICATION OF REAL NUMBERS

Multiplication of real numbers is defined to preserve the properties of multiplication of positive numbers that we are familiar with. The result of multiplication is called the **product.** In this section we learn how to multiply with positive and negative numbers. We already know the rule for multiplying two positive numbers. The product of two positive numbers is positive. We also know that the product of 0 and any positive number is 0, so we extend that property to all real numbers.

For any number a,

$$a \cdot 0 = 0.$$

1 In order to define the product of a positive and a negative number so that the result is consistent with the multiplication of two positive numbers, look at the following pattern.

$$\begin{aligned} 3 \cdot 5 &= 15 \\ 3 \cdot 4 &= 12 \\ 3 \cdot 3 &= 9 \\ 3 \cdot 2 &= 6 \\ 3 \cdot 1 &= 3 \\ 3 \cdot 0 &= 0 \\ 3 \cdot (-1) &= ? \end{aligned}$$

The numbers decrease by 3.

What should $3(-1)$ equal? The product $3(-1)$ represents the sum

$$-1 + (-1) + (-1) = -3$$

so the product should be -3. Also,

$$3(-2) = -2 + (-2) + (-2) = -6.$$

WORK PROBLEM 1 AT THE SIDE.

The results from Problem 1 maintain the pattern in the list above, which suggests the following rule.

The product of a positive number and a negative number is negative. Example: $6 \cdot -3 = -18$.

■ EXAMPLE 1 *Multiplying a Positive Number and a Negative Number*

Find the following products using the multiplication rule given above.

(a) $8(-5) = -(8 \cdot 5) = -40$

(b) $(-7)(2) = -(7 \cdot 2) = -14$

(c) $(-9)\left(\frac{1}{3}\right) = -3$

(d) $(-6.2)(4.1) = -25.42$ ■

WORK PROBLEM 2 AT THE SIDE.

OBJECTIVES

1. Find the product of a positive number and a negative number.
2. Find the product of two negative numbers.
3. Use the order of operations.
4. Evaluate expressions that use variables.
5. Interpret words and phrases that indicate multiplication.

FOR EXTRA HELP

Tape 2

SSM pp. 24–27

MAC: A IBM: A

1. Find each product by finding the sum of three numbers.

(a) $3(-3)$

(b) $3(-4)$

(c) $3(-5)$

2. Find the products.

(a) $2(-6)$

(b) $7(-8)$

(c) $(-9)(2)$

(d) $(-16)\left(\frac{5}{32}\right)$

(e) $(4.56)(-10)$

ANSWERS

1. (a) -9 **(b)** -12 **(c)** -15

2. (a) -12 **(b)** -56 **(c)** -18 **(d)** $-\frac{5}{2}$ **(e)** -45.6

3. Find the products.

(a) $(-5)(-6)$

(b) $(-7)(-3)$

(c) $(-8)(-5)$

(d) $(-11)(-2)$

(e) $(-17)(-21)$

(f) $(-82)(-13)$

ANSWERS
3. (a) 30 (b) 21 (c) 40 (d) 22 (e) 357 (f) 1066

2 The product of two positive numbers is positive, and the product of a positive number and a negative number is negative. What about the product of two negative numbers? Let's look at another pattern.

$$\begin{aligned}(-5)(4) &= -20\\(-5)(3) &= -15\\(-5)(2) &= -10\\(-5)(1) &= -5\\(-5)(0) &= 0\\(-5)(-1) &= ?\end{aligned}$$

The numbers increase by 5.

The numbers on the left of the equals sign (in color) decrease by 1 for each step down the list. The products on the right increase by 5 for each step down the list. To maintain this pattern, $(-5)(-1)$ should be 5 more than $(-5)(0)$, or 5 more than 0, so

$$(-5)(-1) = 5.$$

The pattern continues with

$$\begin{aligned}(-5)(-2) &= 10\\(-5)(-3) &= 15\\(-5)(-4) &= 20\\(-5)(-5) &= 25,\end{aligned}$$

and so on. This pattern suggests that we should multiply two negative numbers as follows.

> The product of two negative numbers is positive.
> Example: $-5 \cdot -4 = 20$.

EXAMPLE 2 *Multiplying Two Negative Numbers*

Find the products using the multiplication rule given above.

(a) $(-9)(-2) = 18$

(b) $(-6)(-12) = 72$

(c) $(-8)(-1) = 8$

(d) $(-15)(-2) = 30$ ■

WORK PROBLEM 3 AT THE SIDE.

Here is a summary of the results for multiplying signed numbers.

MULTIPLYING SIGNED NUMBERS

> The product of two numbers having the *same* signs is *positive,* and the product of two numbers having *different* signs is *negative.*

3 In the next example, we use the order of operations (discussed in Section 1.1) with positive and negative numbers.

■ **EXAMPLE 3** *Using the Order of Operations*
Simplify.

(a) $(-9)(2) - (-3)(2)$

First find all products, working from left to right.

$$(-9)(2) - (-3)(2) = -18 - (-6)$$

Now perform the subtraction. $-18 - (-6) = -18 + 6 = -12$

(b) $-5(-2 - 3) = -5(-5) = 25$ ■

WORK PROBLEM 4 AT THE SIDE. ▶▶

4 The next examples show numbers substituted for variables where the rules for multiplying signed numbers must be used.

■ **EXAMPLE 4** *Evaluating an Expression for Numerical Values*
Evaluate $(3x + 4y)(-2m)$ if $x = -1$, $y = -2$, and $m = -3$.

First substitute the given values for the variables. Then find the value of the expression. Put parentheses around the number for each variable.

$(3x + 4y)(-2m)$	
$= [3(-1) + 4(-2)][-2(-3)]$	Use parentheses around given values.
$= [-3 + (-8)][6]$	Find the products.
$= (-11)(6)$	Use order of operations.
$= -66$	Multiply.

■ **EXAMPLE 5** *Evaluating an Expression for Numerical Values*
Evaluate $2x^2 - 3y^2$ if $x = -3$ and $y = -4$.

Use parentheses as shown.

$2(-3)^2 - 3(-4)^2 = 2(9) - 3(16)$	Square -3 and -4.
$= 18 - 48$	Multiply.
$= -30$	Subtract. ■

WORK PROBLEM 5 AT THE SIDE. ▶▶

5 The word *product* refers to multiplication. In the following chart, we give some other key words and phrases that indicate multiplication.

Word or Phrase	*Example*	*Numerical Expression and Simplification*
Product	The ***product of*** −5 and −2	$(-5)(-2) = 10$
Times	13 ***times*** −4	$13(-4) = -52$
Twice (meaning "2 times")	***Twice*** 6	$2(6) = 12$
Of (used with fractions)	$\frac{1}{2}$ ***of*** 10	$\frac{1}{2}(10) = 5$
Percent of	12***% of*** −16	$.12(-16) = -1.92$

4. Perform the indicated operations.

(a) $(-3)(4) - (2)(6)$

(b) $-7(-2 - 5)$

(c) $-8[-1 - (-4)(-5)]$

5. Evaluate the following expressions.

(a) $2x - 7(y + 1)$ if $x = -4$ and $y = 3$

(b) $(-3x)(4x - 2y)$ if $x = 2$ and $y = -1$

(c) $2x^2 - 4y^2$ if $x = -2$ and $y = -3$

ANSWERS
4. (a) −24 (b) 49 (c) 168
5. (a) −36 (b) −60 (c) −28

6. Write a numerical expression for each phrase and simplify.

(a) The product of 6 and the sum of -5 and -4

(b) Twice the difference between 8 and -4

(c) Three-fifths of the sum of 2 and -7

(d) 20% of the sum of 9 and -4

ANSWERS

6. (a) $6[(-5) + (-4)]$; -54
(b) $2[8 - (-4)]$; 24
(c) $\frac{3}{5}[2 + (-7)]$; -3
(d) $.20[9 + (-4)]$; 1

■ EXAMPLE 6 *Interpreting Words and Phrases*

Write a numerical expression for each phrase and simplify. Use the order of operations.

(a) The **product of** 12 and the sum of 3 and -6
Here 12 is multiplied by "the sum of 3 and -6."

$$12[3 + (-6)] = 12(-3) = -36$$

(b) **Twice** the difference between 8 and -4

$$2[8 - (-4)] = 2[8 + 4] = 2(12) = 24$$

(c) Two-thirds **of** the sum of -5 and -3

$$\frac{2}{3}[-5 + (-3)] = \frac{2}{3}[-8] = -\frac{16}{3}$$

(d) 15 **% of** the difference between 14 and -2
Remember that 15% = .15.

$$.15[14 - (-2)] = .15(14 + 2) = .15(16) = 2.4 \quad ■$$

◀ WORK PROBLEM 6 AT THE SIDE.

NAME DATE HOUR

1.6 EXERCISES

Answer true or false to each of the following statements.

1. The product of two negative numbers is a positive number.

2. The product of a positive number and a negative number is a negative number.

3. When the sum of two negative numbers is multiplied by a positive number, the product is a positive number.

4. When the sum of two positive numbers is multiplied by a negative number, the product is a negative number.

5. When a negative number is squared, the result is a positive number.

6. When a negative number is raised to the fifth power, the result is a negative number.

Find the products. See Examples 1 and 2.

7. $(-4)(-5)$

8. $(-4)(-6)$

9. $5(-6)$

10. $3(-7)$

11. $(-7)(4)$

12. $(-8)(5)$

13. $(-4)(-20)$

14. $(-8)(-30)$

15. $(-8)(0)$

16. $(-12)(0)$

17. $\left(-\frac{3}{8}\right)\left(-\frac{20}{9}\right)$

18. $\left(-\frac{5}{4}\right)\left(-\frac{6}{25}\right)$

19. $(-6.8)(.35)$

20. $(-4.6)(.24)$

21. $(-6)\left(-\frac{1}{4}\right)$

22. $(-8)\left(-\frac{1}{2}\right)$

Perform the indicated operations. See Example 3.

23. $7 - 3 \cdot 6$

24. $8 - 2 \cdot 5$

25. $-10 - (-4)(2)$

26. $-11 - (-3)(6)$

27. $15(8 - 12)$

28. $3(4 - 16)$

29. $-7(3 - 8)$

30. $-5(4 - 7)$

31. $(12 - 14)(1 - 4)$

32. $(8 - 9)(4 - 12)$

33. $(7 - 10)(10 - 4)$

34. $(5 - 12)(19 - 4)$

35. $(-2 - 8)(-6) + 7$

36. $(-9 - 4)(-2) + 10$

37. $3(-5) - (-7)$

38. $4(-8) - (-11)$

39. $(-9 - 3)(-5) - (-4)$

40. $(-7 - 4)(-9) - (-2)$

41. Explain the method you would use to evaluate $3x + 2y$ if $x = -3$ and $y = 4$.

42. If x and y are both replaced by negative numbers, is the value of $4x + 8y$ positive or negative?

Evaluate the following expressions if $x = 6$, $y = -4$, and $a = 3$. See Examples 4 and 5.

43. $5x - 2y + 3a$

44. $6x - 5y + 4a$

45. $(2x + y)(3a)$

46. $(5x - 2y)(-2a)$

47. $\left(\frac{1}{3}x - \frac{4}{5}y\right)\left(-\frac{1}{5}a\right)$

48. $\left(\frac{5}{6}x + \frac{3}{2}y\right)\left(-\frac{1}{3}a\right)$

49. $(-5 + x)(-3 + y)(3 - a)$

50. $(6 - x)(5 + y)(3 + a)$

51. $-2y^2 + 3a$

52. $5x - 4a^2$

53. $3a^2 - x^2$

54. $4y^2 - 2x^2$

Write a numerical expression for each phrase and simplify. See Example 6.

55. The product of -9 and 2, added to 9

56. The product of 4 and -7, added to -12

57. Twice the product of -1 and 6, subtracted from -4

58. Twice the product of -8 and 2, subtracted from -1

59. Nine subtracted from the product of 7 and -12

60. Three subtracted from the product of -2 and 5

61. The product of 12 and the difference between 9 and -8

62. The product of -3 and the difference between 3 and -7

63. Four-fifths of the sum of -8 and -2

64. Three-tenths of the sum of -2 and -28

Find the solution for each equation from the domain $\{-3, -2, -1, 0, 1, 2, 3\}$ by guessing or by using trial and error.

65. $-2x = -6$

66. $-3x = -9$

67. $-4y = 0$

68. $-6t = 0$

69. $-5x = 10$

70. $-2x = 2$

71. $7x = -14$

72. $6x = -12$

73. $\frac{1}{5}w = -\frac{1}{5}$

74. $\frac{2}{3}t = -\frac{2}{3}$

75. $6x + 10 = -8$

76. $2x + 6 = 4$

PREVIEW EXERCISES

Find the value of each expression, given $x = 2$ and $y = 10$. See Section 1.2.

77. $\frac{3x + 5y}{2}$

78. $\frac{10(x + y)}{y - 2x}$

79. $\frac{2(x + 4y)}{2x + y}$

80. $\frac{x^2 + y^2}{x^2}$

81. $\frac{3x^2 + 2y^2}{10y + 3}$

82. $\frac{x}{y} + \frac{y}{x}$

1.7 DIVISION OF REAL NUMBERS

1 The difference between two numbers is found by adding the opposite of the second number to the first. Division is related to multiplication in a similar way. The *quotient* of two numbers is found by *multiplying* by the *reciprocal.* By definition, since

$$8 \cdot \frac{1}{8} = \frac{8}{8} = 1 \quad \text{and} \quad \frac{5}{4} \cdot \frac{4}{5} = \frac{20}{20} = 1,$$

the reciprocal of 8 is $\frac{1}{8}$, and that of $\frac{5}{4}$ is $\frac{4}{5}$.

Pairs of numbers whose product is 1 are called **reciprocals** of each other.

EXAMPLE 1 *Finding the Reciprocal*

The following chart shows several numbers and the reciprocal (if it exists) of each number.

Number	*Reciprocal*
4	$\frac{1}{4}$
−5	$\frac{1}{-5}$ or $-\frac{1}{5}$
$\frac{3}{4}$	$\frac{4}{3}$
$-\frac{5}{8}$	$-\frac{8}{5}$
0	None ■

Why is there no reciprocal for the number 0? Suppose that k is to be the reciprocal of 0. Then $k \cdot 0$ should equal 1. But $k \cdot 0 = 0$ for any number k. Because there is no value of k that is a solution of the equation $k \cdot 0 = 1$, the following statement can be made.

0 has no reciprocal.

WORK PROBLEM 1 AT THE SIDE.

2 By definition, the *quotient* of a and b is the product of a and the reciprocal of b.

OBJECTIVES

1. Find the reciprocal of a number.
2. Divide with signed numbers.
3. Simplify numerical expressions.
4. Interpret words and phrases that indicate division.
5. Translate simple sentences into equations.

FOR EXTRA HELP

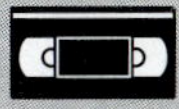
Tape 2

SSM pp. 27–29

MAC: A IBM: A

1. Complete the chart.

Number	*Reciprocal*
(a) 6	
(b) −2	
(c) $\frac{2}{3}$	
(d) $-\frac{1}{4}$	
(e) 0	

ANSWERS

1. (a) $\frac{1}{6}$ **(b)** $-\frac{1}{2}$ or $\frac{1}{-2}$ **(c)** $\frac{3}{2}$ **(d)** −4 **(e)** none

DEFINITION OF DIVISION

For any real numbers a and b, with $b \neq 0$,

$$\frac{a}{b} = a \cdot \frac{1}{b}.$$

The definition above indicates that b, the number to divide by, cannot be 0. The reason is that 0 has no reciprocal, so $\frac{1}{0}$ is not a number.

Division by 0 is undefined and is never permitted.

Note If a division problem turns out to involve division by 0, write "undefined."

Because division is defined in terms of multiplication, all the rules for multiplication of signed numbers also apply to division.

EXAMPLE 2 *Using the Definition of Division*

Write each quotient as a product and evaluate.

(a) $\frac{12}{3} = 12 \cdot \frac{1}{3}$
$= 4$

(b) $\frac{-10}{2} = -10 \cdot \frac{1}{2}$
$= -5$

(c) $\frac{8}{-4} = 8 \cdot \left(\frac{1}{-4}\right)$
$= -2$

(d) $\frac{-1.47}{-7} = -1.47\left(-\frac{1}{7}\right)$
$= .21$

(e) $-\frac{2}{3} \div -\frac{5}{4} = -\frac{2}{3} \cdot -\frac{4}{5}$
$= \frac{8}{15}$

(f) $\frac{-10}{0}$ Undefined ■

WORK PROBLEM 2 AT THE SIDE.

Multiplying by the reciprocal to divide integers is awkward. It is easier to divide in the usual way. On the other hand, to divide by a fraction, it is easier to multiply by the reciprocal. The following rule for division can be used instead of multiplying by the reciprocal.

2. Find the quotients.

(a) $\frac{42}{7}$

(b) $\frac{-36}{6}$

(c) $\frac{-12.56}{-.4}$

(d) $\frac{10}{7} \div -\frac{24}{5}$

(e) $\frac{-3}{0}$

ANSWERS

2. (a) 6 **(b)** −6 **(c)** 31.4 **(d)** $-\frac{25}{84}$ **(e)** undefined

DIVIDING SIGNED NUMBERS

The quotient of two numbers having the *same* sign is *positive;* the quotient of two numbers having *different* signs is *negative.*
Examples: $\frac{-15}{-5} = 3$ and $\frac{-15}{5} = -3$.

EXAMPLE 3 *Dividing Signed Numbers*
Find the quotients.

(a) $\frac{8}{-2} = -4$

(b) $\frac{-4.5}{-.09} = 50$

(c) $-\frac{1}{8} \div \left(-\frac{3}{4}\right) = -\frac{1}{8} \cdot \left(-\frac{4}{3}\right) = \frac{1}{6}$ ■

WORK PROBLEM 3 AT THE SIDE.

From the definitions of multiplication and division of real numbers,

$$\frac{-40}{8} = -40 \cdot \frac{1}{8} = -5$$

and

$$\frac{40}{-8} = 40\left(\frac{1}{-8}\right) = -5$$

so

$$\frac{-40}{8} = \frac{40}{-8}.$$

Based on this example, the quotient of a positive and a negative number can be written in any of the following three forms.

For any positive real numbers a and b,

$$\frac{-a}{b} = \frac{a}{-b} = -\frac{a}{b}.$$

The form $\frac{a}{-b}$ is seldom used.

Similarly, the quotient of two negative numbers can be expressed as the quotient of two positive numbers.

For any positive real numbers a and b,

$$\frac{-a}{-b} = \frac{a}{b}.$$

3. Find the quotients.

(a) $\frac{-8}{-2}$

(b) $\frac{-16.4}{2.05}$

(c) $\frac{1}{4} \div \left(-\frac{2}{3}\right)$

ANSWERS

3. (a) 4 (b) -8 (c) $-\frac{3}{8}$

4. Perform the indicated operations.

(a) $\dfrac{6(-4) - 2(5)}{3(2 - 7)}$

(b) $\dfrac{-6(-8) + (-3)9}{(-2)[4 - (-3)]}$

(c) $\dfrac{5^2 + 3^2}{3(-4) - 5}$

ANSWERS

4. (a) $\dfrac{34}{15}$ (b) $-\dfrac{3}{2}$ (c) -2

The rules for operations with signed numbers are summarized here.

OPERATIONS WITH SIGNED NUMBERS

Addition

Like signs Add the absolute values of the numbers. The result is given the same sign as the numbers.

Unlike signs Subtract the smaller absolute value from the larger absolute value. Give the result the sign of the number having the larger absolute value.

Subtraction

Add the opposite of the second number to the first number.

Multiplication and Division

Like signs The product or quotient of two numbers with like signs is positive.

Unlike signs The product or quotient of two numbers with unlike signs is negative.

Division by 0 is undefined.

3 In the next example, we simplify numerical expressions involving quotients.

EXAMPLE 4 *Simplifying Expressions Involving Division*

Simplify each expression.

(a) $\dfrac{5(-2) - (3)(4)}{2(1 - 6)}$

Follow the order of operations. Simplify the numerator and denominator separately. Then divide or write in lowest terms.

$$\frac{5(-2) - (3)(4)}{2(1 - 6)} = \frac{-10 - 12}{2(-5)}$$ Multiply in numerator. Subtract in denominator.

$$= \frac{-22}{-10}$$ Subtract in numerator. Multiply in denominator.

$$= \frac{11}{5}$$ Express in lowest terms.

(b) $\dfrac{-3^2 - 6^2}{5(-3 + 2)}$

$$\frac{-3^2 - 6^2}{5(-3 + 2)} = \frac{-9 - 36}{5(-1)}$$ Square 3 and 6. Add −3 and 2.

$$= \frac{-45}{-5}$$ Subtract in numerator. Multiply in denominator.

$$= 9$$ Divide. ■

WORK PROBLEM 4 AT THE SIDE.

4 The word *quotient* refers to the result obtained in a division problem. In algebra, quotients are usually represented with a fraction bar. The symbol $\div$ is seldom used.

The following chart gives some key words and phrases associated with division.

Word or Phrase	*Example*	*Numerical Expression and Simplification*
Quotient	The ***quotient of*** -24 and 3	$\frac{-24}{3} = -8$
Divided by	-16 ***divided by*** -4	$\frac{-16}{-4} = 4$

It is customary to write the first number named as the numerator and the second as the denominator when interpreting a phrase involving division.

EXAMPLE 5 *Interpreting Words and Phrases*

Write a numerical expression for each phrase, and simplify the expression.

(a) The **quotient** of 14 and the sum of -9 and 2

"Quotient" indicates division. The number 14 is the numerator and "the sum of -9 and 2" is the denominator.

$$\frac{14}{-9+2} = \frac{14}{-7} = -2$$

(b) The product of 5 and -6, **divided by** the difference between -7 and 8

The numerator of the fraction representing the division is obtained by multiplying 5 and -6. The denominator is found by subtracting -7 and 8.

$$\frac{5(-6)}{-7-8} = \frac{-30}{-15} = 2$$ ■

WORK PROBLEM 5 AT THE SIDE.

5 In this section and the preceding three sections, important words and phrases involving the four operations of arithmetic have been introduced. We can use these words and phrases to interpret sentences that translate into equations. The ability to do this will help us to solve the types of applied problems found in Section 2.4.

EXAMPLE 6 *Translating Words into an Equation*

Write the following in symbols, using x as the variable, and use guessing or trial and error to find the solution. All solutions come from the list of integers between -12 and 12, inclusive.

(a) Three **times** a number **is** -18.

The word *times* indicates multiplication, and the word *is* translates as the equals sign $(=)$.

$$3x = -18$$

Since the integer between -12 and 12, inclusive, that makes this statement true is -6, the solution of the equation is -6.

5. Write a numerical expression for each phrase, and simplify the expression.

(a) The quotient of 20 and the sum of 8 and -3

(b) The product of -9 and 2, divided by the difference between 5 and -1

ANSWERS

5. (a) $\frac{20}{8+(-3)}$; 4
(b) $\frac{(-9)(2)}{5-(-1)}$; -3

6. Write the following in symbols, using x as the variable, and find the solution by guessing or by using trial and error. All solutions come from the list of integers between -12 and 12, inclusive.

(a) Twice a number is -6.

(b) The difference between -8 and a number is -11.

(c) The sum of 5 and a number is 8.

(d) The quotient of a number and -2 is 6.

ANSWERS

6. (a) $2x = -6$; -3
(b) $-8 - x = -11$; 3
(c) $5 + x = 8$; 3
(d) $\frac{x}{-2} = 6$; -12

(b) The **sum** of a number and 9 **is** 12.

$$x + 9 = 12$$

Since $3 + 9 = 12$, the solution of this equation is 3.

(c) The **difference between** a number and 5 **is** 0.

$$x - 5 = 0$$

Since $5 - 5 = 0$, the solution of this equation is 5.

(d) The **quotient of** 24 and a number **is** -2.

$$\frac{24}{x} = -2$$

Here, x must be a negative number because the numerator is positive and the quotient is negative. Since $\frac{24}{-12} = -2$, the solution is -12. ■

WORK PROBLEM 6 AT THE SIDE.

Caution It is important to recognize the distinction between the types of problems found in Example 5 and Example 6. In Example 5, the phrases translate as *expressions,* while in Example 6, the sentences translate as *equations.* Remember that an equation is a sentence, while an expression is a phrase. For example,

$$\frac{5(-6)}{-7-8} \text{ is an expression,}$$

$$3x = -18 \text{ is an equation.}$$

1.7 EXERCISES

NAME DATE HOUR

Fill in the blanks with the correct responses.

1. The quotient of two numbers with the same sign is a ______________ (positive/negative) number.

2. The quotient of two numbers with unlike signs is a ______________ (positive/negative) number.

3. If two negative numbers are multiplied and their product is then divided by a negative number, the result is ______________ (greater than/less than) zero.

4. If a positive number is multiplied by a negative number and their product is then divided by a positive number, the result is ______________ (greater than/less than) zero.

5. The reciprocal of a positive number is a ______________ (positive/negative) number.

6. The reciprocal of a negative number is a ______________ (positive/negative) number.

Find the reciprocal, if one exists, for each number. See Example 1.

7. 11 **8.** 12 **9.** -5 **10.** -3

11. $\frac{5}{6}$ **12.** $\frac{9}{10}$ **13.** $3 - 3$ **14.** $5 + (-5)$

15. $-\frac{8}{7}$ **16.** $-\frac{9}{7}$ **17.** .4 **18.** .50

19. Which one of the following expressions is undefined?

(a) $\frac{5 - 5}{5 + 5}$ **(b)** $\frac{5 + 5}{5 + 5}$ **(c)** $\frac{5 - 5}{5 - 5}$ **(d)** $\frac{5 - 5}{5}$

20. Explain why 0 has no reciprocal.

Find the quotients. See Examples 2 and 3.

21. $\frac{-15}{5}$ **22.** $\frac{-18}{6}$ **23.** $\frac{20}{-10}$ **24.** $\frac{28}{-4}$

25. $\frac{-160}{-10}$ **26.** $\frac{-260}{-20}$ **27.** $\frac{0}{-3}$ **28.** $\frac{0}{-6}$

29. $\frac{-10.252}{-.4}$ **30.** $\frac{-29.584}{-.8}$ **31.** $\left(-\frac{3}{4}\right) \div \left(-\frac{1}{2}\right)$ **32.** $\left(-\frac{3}{16}\right) \div \left(-\frac{5}{8}\right)$

33. $(-6.8) \div (-2)$ **34.** $(-23.5) \div (-5)$ **35.** $\frac{18}{3 - 9}$ **36.** $\frac{21}{4 - 7}$

37. $\frac{-50}{7 - (-3)}$ **38.** $\frac{-36}{4 - (-5)}$ **39.** $\frac{-12 - 36}{-12}$ **40.** $\frac{-18 - 2}{-5}$

Simplify the numerators and denominators separately. Then find the quotients. See Example 4.

41. $\frac{-5(-6)}{9 - (-1)}$ **42.** $\frac{-12(-5)}{7 - (-5)}$ **43.** $\frac{-21(3)}{-3 - 6}$

44. $\frac{-40(3)}{-2 - 3}$ **45.** $\frac{-10(2) + 6(2)}{-3 - (-1)}$ **46.** $\frac{8(-1) + 6(-2)}{-6 - (-1)}$

47. $\dfrac{-27(-2) - (-12)(-2)}{-2(3) - 2(2)}$

48. $\dfrac{-13(-4) - (-8)(-2)}{(-10)(2) - 4(-2)}$

49. $\dfrac{1^2 + 4^2}{3^2 + 5^2}$

50. $\dfrac{3^2 + 4^2}{6^2 + 8^2}$

51. $\dfrac{2^2 - 8^2}{6(-4 + 3)}$

52. $\dfrac{3^2 - 4^2}{7(-8 + 9)}$

Write a numerical expression for each phrase, and simplify the expression. See Example 5.

53. The quotient of -36 and -9

54. The quotient of -48 and -6

55. The quotient of -12 and the sum of -5 and -1

56. The quotient of -20 and the sum of -8 and -2

57. The sum of 15 and -3, divided by the product of 4 and -3

58. The sum of -18 and -6, divided by the product of 2 and -4

59. The product of -34 and 7, divided by -14

60. The product of -25 and 4, divided by -10

Find the solution of each equation from the domain $\{-8, -6, -4, -2, 0, 2, 4, 6, 8\}$ by guessing or by using trial and error.

61. $\frac{x}{-4} = 2$

62. $\frac{x}{-2} = 3$

63. $\frac{n}{-2} = -2$

64. $\frac{t}{-2} = -3$

65. $\frac{2x}{2} = -4$

66. $\frac{2y}{4} = -3$

Write the following in symbols, using x as the variable, and find the solution by guessing or by using trial and error. All solutions come from the list of integers between −12 and 12, inclusive. See Example 6.

67. Six times a number is −42.

68. Four times a number is −36.

69. The quotient of a number and 3 is −3.

70. The quotient of a number and 4 is −1.

71. 6 less than a number is 4.

72. 7 less than a number is 2.

73. When 5 is added to a number, the result is −5.

74. When 6 is added to a number, the result is −3.

Preview Exercises

Perform the indicated operations. See Sections 1.4 and 1.6.

75. $19 + (-19)$

76. $5 + (-6 + 6)$

77. $-\frac{1}{3}(-3)$

78. **(a)** $4 + [5 + (-13)]$
(b) $(4 + 5) + (-13)$

1.8 PROPERTIES OF ADDITION AND MULTIPLICATION

If you are asked to find the sum

$$3 + 89 + 97$$

you might mentally add $3 + 97$ to get 100, and then add $100 + 89$ to get 189. While the rule for order of operations says to add from left to right, it is a fact that we may change the order of the terms and group them in any way we choose without affecting the sum. These are examples of shortcuts that we use in everyday mathematics. These shortcuts are justified by the basic properties of addition and multiplication, which are discussed in this section. In the following statements, a, b, and c represent real numbers.

OBJECTIVES

1. Use the commutative properties.
2. Use the associative properties.
3. Use the identity properties.
4. Use the inverse properties.
5. Use the distributive property.

FOR EXTRA HELP

Tape 2

SSM pp. 29–32

MAC: A IBM: A

1 **Commutative properties** The word *commute* means to go back and forth. Many people commute to work or to school. If you travel from home to work and follow the same route from work to home, you travel the same distance each time. The commutative properties say that if two numbers are added or multiplied in any order, they give the same result.

$$a + b = b + a$$
$$ab = ba$$

EXAMPLE 1 *Using the Commutative Properties*

Use a commutative property to complete each statement.

(a) $-8 + 5 = 5 +$ ____

By the commutative property for addition, the missing number is -8 because $-8 + 5 = 5 + (-8)$.

(b) $(-2)(7) =$ ____ (-2)

By the commutative property for multiplication, the missing number is 7, since $(-2)(7) = (7)(-2)$. ■

WORK PROBLEM 1 AT THE SIDE. ▶

1. Complete each statement. Use a commutative property.

(a) $x + 9 = 9 +$ ____

(b) $(-12)(4) =$ ____ (-12)

(c) $5x = x \cdot$ ____

2 **Associative properties** When we *associate* one object with another, we tend to think of those objects as being grouped together. The associative properties say that when we add or multiply three numbers, we can group them in any order and get the same answer.

$$(a + b) + c = a + (b + c)$$
$$(ab)c = a(bc)$$

EXAMPLE 2 *Using the Associative Properties*

Use an associative property to complete each statement.

(a) $8 + (-1 + 4) = (8 +$ ____ $) + 4$

The missing number is -1.

(b) $[2 \cdot (-7)] \cdot 6 = 2 \cdot$ ____

The completed expression on the right should be $2 \cdot [(-7) \cdot 6]$. ■

WORK PROBLEM 2 AT THE SIDE. ▶

2. Complete each statement. Use an associative property.

(a) $(9 + 10) + (-3)$
$= 9 + [$ ____ $+ (-3)]$

(b) $-5 + (2 + 8)$
$= ($ ____ $) + 8$

(c) $10 \cdot [(-8) \cdot (-3)]$
$=$ ____

ANSWERS
1. (a) x (b) 4 (c) 5
2. (a) 10 (b) $-5 + 2$
(c) $[10 \cdot (-8)] \cdot (-3)$

By the associative property of addition, the sum of three numbers will be the same no matter which way the numbers are "associated" in groups. For this reason, parentheses can be left out in many addition problems. For example, both

$$(-1 + 2) + 3 \quad \text{and} \quad -1 + (2 + 3)$$

can be written as

$$-1 + 2 + 3.$$

In the same way, parentheses also can be left out of many multiplication problems.

EXAMPLE 3 *Distinguishing Between the Associative and Commutative Properties*

(a) Is $(2 + 4) + 5 = 2 + (4 + 5)$ an example of the associative property?

The order of three numbers is the same on both sides of the equals sign. The only change is in the grouping, or association, of the numbers. Therefore, this is an example of the associative property.

(b) Is $6(3 \cdot 10) = 6(10 \cdot 3)$ an example of the associative property or the commutative property?

The same numbers, 3 and 10, are grouped on each side. On the left, however, the 3 appears first in $(3 \cdot 10)$. On the right, the 10 appears first. Since the only change involves the order of the numbers, this statement is an example of the commutative property.

(c) Is $(8 + 1) + 7 = 8 + (7 + 1)$ an example of the associative property or the commutative property?

In the statement, both the order and the grouping are changed. On the left the order of the three numbers is 8, 1, and 7. On the right it is 8, 7, and 1. On the left the 8 and 1 are grouped, and on the right the 7 and 1 are grouped. Therefore, both the associative and the commutative properties are used. ■

WORK PROBLEM 3 AT THE SIDE.

We can use the commutative and associative properties to simplify expressions.

EXAMPLE 4 *Using the Commutative and Associative Properties to Simplify Expressions*

Simplify $16 + 2x + 5$.

By the order of operations, we should add from left to right, but we can't combine 16 and $2x$ into a single term. The commutative and associative properties are used to group 16 and 5, so we can add them.

$(16 + 2x) + 5$	Order of operations
$(2x + 16) + 5$	Commutative property
$2x + (16 + 5)$	Associative property
$2x + 21$	Add.

We don't usually show all these steps, we just add. But this example shows how the properties make it possible. ■

WORK PROBLEM 4 AT THE SIDE.

3. Decide whether each statement is an example of a commutative property, an associative property, or both.

(a) $2(4 \cdot 6) = (2 \cdot 4)6$

(b) $(2 \cdot 4)6 = (4 \cdot 2)6$

(c) $(2 + 4) + 6 = 4 + (2 + 6)$

4. Simplify $8 + 4y + 10$.

ANSWERS

3. (a) associative (b) commutative (c) both

4. $4y + 18$ or $18 + 4y$

3 **Identity properties** If a child wears a costume to masquerade on Halloween, the child's appearance is changed, but his or her *identity* is unchanged. The identity of a real number is left unchanged when identity properties are applied. The identity properties say that the sum of 0 and any number equals that number, and the product of 1 and any number equals that number.

$$a + 0 = a \quad \text{and} \quad 0 + a = a$$
$$a \cdot 1 = a \quad \text{and} \quad 1 \cdot a = a$$

The number 0 leaves the identity, or value, of any real number unchanged by addition. For this reason, 0 is called the **identity element for addition.** Since multiplication by 1 leaves any real number unchanged, 1 is the **identity element for multiplication.**

EXAMPLE 5 *Using the Identity Properties*

These statements are examples of the identity properties.

(a) $-3 + 0 = -3$ **(b)** $1 \cdot 25 = 25$ ■

WORK PROBLEM 5 AT THE SIDE.

We use the identity property for multiplication to write fractions in lowest terms and to get common denominators.

EXAMPLE 6 *Using the Identity Property of 1 to Simplify Expressions*

Simplify the following expressions.

(a) $\frac{49}{35}$

$$\frac{49}{35} = \frac{7 \cdot 7}{5 \cdot 7} \quad \text{Factor.}$$
$$= \frac{7}{5} \cdot \frac{7}{7} \quad \text{Write as a product.}$$
$$= \frac{7}{5} \cdot 1 \quad \text{Property of 1}$$
$$= \frac{7}{5} \quad \text{Identity Property}$$

(b) $\frac{3}{4} + \frac{5}{24}$

$$\frac{3}{4} + \frac{5}{24} = \frac{3}{4} \cdot 1 + \frac{5}{24} \quad \text{Identity property}$$
$$= \frac{3}{4} \cdot \frac{6}{6} + \frac{5}{24} \quad \text{Get a common denominator.}$$
$$= \frac{18}{24} + \frac{5}{24} \quad \text{Multiply.}$$
$$= \frac{23}{24} \quad \text{Add.}$$ ■

WORK PROBLEM 6 AT THE SIDE.

5. Use an identity property to complete each statement.

(a) $9 + 0 =$ ____

(b) ____ $+ (-7) = -7$

(c) $\frac{1}{4} \cdot$ ____ $= \frac{3}{12}$

(d) ____ $\cdot 1 = 5$

6. Use the identity property to simplify each expression.

(a) $\frac{85}{105}$

(b) $\frac{9}{10} - \frac{53}{50}$

ANSWERS

5. (a) 9 (b) 0 (c) $\frac{3}{3}$ (d) 5

6. (a) $\frac{17}{21}$ (b) $-\frac{4}{25}$

4 **Inverse Properties** Each day before you go to work or school, you probably put on your shoes before you leave. Before you go to sleep at night, you probably take them off, and this leads to the same situation that existed before you put them on. These operations from everyday life are examples of inverse operations. The inverse properties of addition and multiplication lead to the additive and multiplicative identities, respectively. The opposite of a, $-a$, is the **additive inverse** of a and the reciprocal of a, $\frac{1}{a}$, is the **multiplicative inverse** of the nonzero number a. The sum of the numbers a and $-a$ is 0, and the product of the nonzero numbers a and $\frac{1}{a}$ is 1.

$$a + (-a) = 0 \quad \text{and} \quad -a + a = 0$$

$$a \cdot \frac{1}{a} = 1 \quad \text{and} \quad \frac{1}{a} \cdot a = 1 \quad (a \neq 0)$$

EXAMPLE 7 *Using the Inverse Properties*

The following statements are examples of the inverse properties.

(a) $\frac{2}{3} \cdot \frac{3}{2} = 1$

(b) $(-5)\left(-\frac{1}{5}\right) = 1$

(c) $-\frac{1}{2} + \frac{1}{2} = 0$

(d) $4 + (-4) = 0$ ■

WORK PROBLEM 7 AT THE SIDE.

EXAMPLE 8 *Using the Additive Inverse Property to Simplify an Expression*

Simplify $-2x + 10 + 2x$.

$$\begin{aligned} -2x + 10 + 2x &= (-2x + 10) + 2x && \text{Order of operations} \\ &= [10 + (-2x)] + 2x && \text{Commutative property} \\ &= 10 + [(-2x) + 2x] && \text{Associative property} \\ &= 10 + 0 && \text{Inverse property} \\ &= 10 && \text{Identity property} \end{aligned}$$ ■

Again, although this example shows how we can use the properties to change the order of operations, in practice we would not show all the steps.

WORK PROBLEM 8 AT THE SIDE.

5 **Distributive property** The everyday meaning of the word *distribute* is "to give out from one to several." An important property of real number operations involves this idea.

Look at the following statements.

$$2(5 + 8) = 2(13) = 26$$

$$2(5) + 2(8) = 10 + 16 = 26$$

Since both expressions equal 26,

$$2(5 + 8) = 2(5) + 2(8).$$

7. Complete the statements so that they are examples of either an identity property or an inverse property. Tell which property is used.

(a) $-6 + \_\_\_\_ = 0$

(b) $\frac{4}{3} \cdot \_\_\_\_ = 1$

(c) $-\frac{1}{9} \cdot \_\_\_\_ = 1$

(d) $275 + \_\_\_\_ = 275$

8. Simplify.

(a) $5m - 3 - 5m$

(b) $\left(\frac{4}{3}\right)(-7)\left(\frac{3}{4}\right)$

ANSWERS

7. (a) 6; inverse (b) $\frac{3}{4}$; inverse (c) -9; inverse (d) 0; identity

8. (a) -3 (b) -7

This result is an example of the *distributive property,* the only property involving *both* addition and multiplication. With this property, a product can be changed to a sum or difference. This idea is illustrated by the divided rectangle in Figure 13.

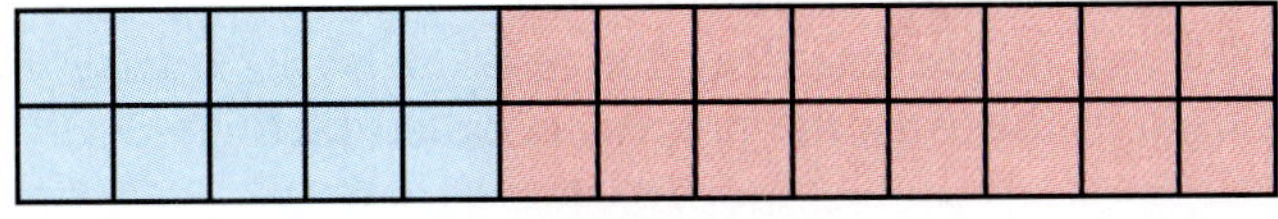

Area of the left part is 2(5) = 10
Area of the right part is 2(8) = 16
Area of total is 2(13) = 2(5 + 8) = 26
2(5 + 8) = 2(5) + 2(8)

FIGURE 13

The distributive property says that multiplying a number a by a sum of numbers $b + c$ gives the same result as multiplying a by b and a by c and then adding the two products.

$$a(b + c) = ab + ac \quad \text{and} \quad (b + c)a = ba + ca$$

As the arrows show, the a outside the parentheses is "distributed" over the b and c inside. Another form of the distributive property is valid for subtraction.

$$a(b - c) = ab - ac \quad \text{and} \quad (b - c)a = ba - ca$$

The distributive property also can be extended to more than two numbers.

$$a(b + c + d) = ab + ac + ad$$

■ EXAMPLE 9 *Using the Distributive Property*

Use the distributive property to rewrite each expression.

(a) $5(9 + 6) = 5 \cdot 9 + 5 \cdot 6$ Multiply both terms by 5.
$= 45 + 30$ Multiply.
$= 75$ Add.

(b) $4(x + 5 + y) = 4x + 4 \cdot 5 + 4y$ Distributive property
$= 4x + 20 + 4y$ Multiply.

(c) $-2(x + 3) = -2x + (-2)(3)$ Distributive property
$= -2x - 6$ Multiply.

(d) $3(k - 9) = 3k - 3 \cdot 9$ Distributive property
$= 3k - 27$ Multiply.

(e) $6 \cdot 8 + 6 \cdot 2 = 6(8 + 2)$ Distributive property
$= 6(10) = 60$ Add, then multiply.

(f) $4x - 4m = 4(x - m)$ Distributive property

(g) $8(3r + 11t + 5z) = 8(3r) + 8(11t) + 8(5z)$ Distributive property

$= (8 \cdot 3)r + (8 \cdot 11)t + (8 \cdot 5)z$ Associative property

$= 24r + 88t + 40z$ ■

Examples 9(e) and (f) use the distributive property in reverse. When the property is used in this way, the process is called **factoring.**

◀◀ WORK PROBLEM 9 AT THE SIDE.

The distributive property is used to remove the parentheses from expressions such as $-(2y + 3)$. We do this by first writing $-(2y + 3)$ as $-1 \cdot (2y + 3)$.

$-(2y + 3) = -1 \cdot (2y + 3)$ Identity Property

$= -1 \cdot (2y) + (-1) \cdot (3)$ Distributive property

$= -2y - 3$ Multiply.

■ **EXAMPLE 10** *Using the Distributive Property to Remove Parentheses*

Write without parentheses.

(a) $-(7r - 8) = -1(7r) + (-1)(-8)$ Distributive property

$= -7r + 8$ Multiply.

(b) $-(-9w + 2) = 9w - 2$ ■

◀◀ WORK PROBLEM 10 AT THE SIDE.

The identity property is often used with the distributive property to simplify expressions.

■ **EXAMPLE 11** *Using the Identity Property With the Distributive Property*

Simply each expression.

(a) $12p - p$

$12p - p = 12p - 1p$ Identity property

$= (12 - 1)p$ Distributive property

$= 11p$ Subtract.

(b) $y + y$

$y + y = 1y + 1y$ Identity property

$= (1 + 1)y$ Distributive property

$= 2y$ Add. ■

◀◀ WORK PROBLEM 11 AT THE SIDE.

9. Use the distributive property to rewrite each expression.

(a) $2(p + 5)$

(b) $-4(y + 7)$

(c) $5(m - 4)$

(d) $9 \cdot k + 9 \cdot 5$

(e) $3a - 3b$

(f) $7(2y + 7k - 9m)$

10. Write without parentheses.

(a) $-(3k - 5)$

(b) $-(2 - r)$

(c) $-(-5y + 8)$

(d) $-(-z + 4)$

11. Simplify.

(a) $4x + x$

(b) $a + a + a$

ANSWERS

9. (a) $2p + 10$ (b) $-4y - 28$ (c) $5m - 20$ (d) $9(k + 5)$ (e) $3(a - b)$ (f) $14y + 49k - 63m$

10. (a) $-3k + 5$ (b) $-2 + r$ (c) $5y - 8$ (d) $z - 4$

11. (a) $5x$ (b) $3a$

The properties of addition and multiplication of real numbers are summarized below.

PROPERTIES OF ADDITION AND MULTIPLICATION

For any real numbers a, b, and c, the following properties hold.

Commutative properties $\quad a + b = b + a \qquad ab = ba$

Associative properties $\quad (a + b) + c = a + (b + c)$

$$(ab)c = a(bc)$$

Identity properties There is a real number 0 such that

$$a + 0 = a \quad \text{and} \quad 0 + a = a.$$

There is a real number 1 such that

$$a \cdot 1 = a \quad \text{and} \quad 1 \cdot a = a.$$

Inverse properties For each real number a, there is a single real number $-a$ such that

$$a + (-a) = 0 \quad \text{and} \quad (-a) + a = 0.$$

For each nonzero real number a, there is a single real number $\frac{1}{a}$ such that

$$a \cdot \frac{1}{a} = 1 \quad \text{and} \quad \frac{1}{a} \cdot a = 1.$$

Distributive property $\quad a(b + c) = ab + ac$

$$(b + c)a = ba + ca$$

QUEST FOR NUMERACY

Are You a Good "Guesstimator"?

While calculators can make life easier when it comes to computations, keep in mind that many times a good guess, or estimate (hence, the term *guesstimate*) is sufficient. In such cases, a calculator may not be necessary or even appropriate. For example, consider the following problem.

A birdhouse for purple martins can accommodate up to 6 nests. How many birdhouses would be necessary to accomodate 92 nests?

If we divide 92 by 6 either by hand or with a calculator, we get an answer of 15.333333 (rounded). Can this possibly be the desired number? Of course not, because we cannot consider fractions of birdhouses. Now we must ask ourselves should we decide on 15 or 16 birdhouses? Because we need to provide nesting space for the nests left over after the 15 birdhouses (as indicated by the decimal fraction), we should plan to use 16 birdhouses. Notice that, in this problem, we must round our answer *up* to the next counting number.

The ability to estimate results from data provided in an everyday situation is important. We are bombarded by information (much of it useless to many of us) in this day and age, and much of it is numerical. By improving our ability to estimate, we can often obtain approximate results when precise results are not absolutely necessary.

FOR GROUP DISCUSSION Use estimation techniques to answer the following. Once you answer correctly, share the technique you used with a class member. (Remember that when it comes to guesstimating, not all people are created equal!)

1. A certain type of carrying case will hold a maximum of 48 audiocassettes. If you need to store 490 audiocassettes, how many carrying cases will you require?
2. A plastic page designed to store baseball cards will hold up to 16 cards. If you must store your collection of 484 cards, how many pages will you need?
3. Each room available for administering a placement test will hold up to 40 students. Two hundred fifty students have signed up for the test. How many rooms will be used?
4. A gardener wants to fertilize 2000 tomato plants. Each bag of fertilizer will supply up to 150 plants. How many bags does she need to do the job?
5. The planet Mercury takes 88.0 Earth days to revolve around the sun. Pluto takes 90,824.2 days to do the same. When Pluto has revolved around the sun once, about how many times will Mercury have revolved around the sun?
   (a) 100 **(b)** 1,000 **(c)** 10,000 **(d)** 100,000
6. Hale County, Texas, has a population of 34,671 and covers 1,005 square miles. About how many inhabitants per square mile does the county have?
   (a) 35 **(b)** 350 **(c)** 3,500 **(d)** 35,000
7. The 1990 United States census showed that the total population of the country was 248,709,873. Of these, 25,223,086 were in the 45—50-year age bracket. On the average, about one in every _____ citizens is in that age bracket.
   (a) 5 **(b)** 10 **(c)** 15 **(d)** 20
8. In voting for which age Elvis was to be in his portrayal on a U.S. postage stamp, voters cast 851,000 votes for the younger Elvis against 277,723 votes for the more mature Elvis. A newspaper article read "Younger Elvis is Winner by _____ to 1 Margin." What number belongs in the blank?
   (a) 5 **(b)** 4 **(c)** 3 **(d)** 2

NAME DATE HOUR

1.8 EXERCISES

Decide whether each of the following statements is true or false.

1. The identity element for addition is 0.

2. The identity element for multiplication is 1.

3. The additive inverse of a is $\frac{1}{a}$.

4. The multiplicative inverse of a is $-a$.

5. The sum of a number and its additive inverse is 0.

6. The product of a number and its multiplicative inverse is 1.

7. Every number has a multiplicative inverse.

8. Every number has an additive inverse.

Decide whether each statement is an example of the commutative, associative, identity, inverse, or distributive property. See Examples 1, 2, 3, 5, 7, and 9.

9. $7 + 18 = 18 + 7$

10. $13 + 12 = 12 + 13$

11. $5(13 \cdot 7) = (5 \cdot 13) \cdot 7$

12. $-4(2 \cdot 6) = (-4 \cdot 2) \cdot 6$

13. $\frac{2}{3} \cdot (-4) = -4 \cdot \left(\frac{2}{3}\right)$

14. $6\left(-\frac{5}{6}\right) = \left(-\frac{5}{6}\right) \cdot 6$

15. $-6 + (12 + 7) = (-6 + 12) + 7$

16. $(-8 + 13) + 2 = -8 + (13 + 2)$

17. $-6 + 6 = 0$

18. $12 + (-12) = 0$

19. $\left(\frac{2}{3}\right)\left(\frac{3}{2}\right) = 1$

20. $\left(\frac{5}{8}\right)\left(\frac{8}{5}\right) = 1$

21. $2.34 \cdot 1 = 2.34$

22. $-8.456 \cdot 1 = -8.456$

23. $(4 + 17) + 3 = 3 + (4 + 17)$

24. $(-8 + 4) + (-12) = -12 + (-8 + 4)$

25. $6(x + y) = 6x + 6y$

26. $14(t + s) = 14t + 14s$

27. $-\frac{5}{9} = -\frac{5}{9} \cdot \frac{3}{3} = -\frac{15}{27}$

28. $\frac{13}{12} = \frac{13}{12} \cdot \frac{7}{7} = \frac{91}{84}$

29. $5(2x) + 5(3y) = 5(2x + 3y)$

30. $3(5t) - 3(7r) = 3(5t - 7r)$

31. The following conversation actually took place between one of the authors of this book and his son, Jack, when Jack was four years old:

Daddy: "Jack, what is 3 + 0?"
Jack: "3."
Daddy: "Jack, what is 4 + 0?"
Jack: "4. And Daddy, *string* plus zero equals *string*!"

What property of addition did Jack recognize?

32. The distributive property holds for multiplication with respect to addition. Is there a distributive property for addition with respect to multiplication? That is, does $a + b \cdot c = (a + b) \cdot (a + c)$? If not, give an example to show why.

33. Using an example, explain why subtraction is not associative.

34. Suppose that a student shows you the following work.

$$-2(5 - 6) = -2(5) - 2(6) = -10 - 12 = -22$$

The student has made a very common error. Write a short paragraph explaining what the error is, and then work the problem correctly.

Use the indicated property to write a new expression that is equal to the given expression. Then simplify the new expression if possible. See Examples 1, 2, 5, 7, and 9.

35. $r + 7$; commutative

36. $t + 9$; commutative

37. $s + 0$; identity

38. $w + 0$; identity

NAME ______________________________

39. $-6(x + 7)$; distributive

40. $-5(y + 2)$; distributive

41. $(w + 5) + (-3)$; associative

42. $(b + 8) + (-10)$; associative

Use the properties of this section to simplify the following expressions. See Examples 4 and 8.

43. $9 + 3x + 7$

44. $-8 + 5t + 10$

45. $6t + 8 - 6t + 3$

46. $9r + 12 - 9r + 1$

47. $-3w + 7 + 3w$

48. $-5t + 3 + 5t$

49. $\frac{2}{3}x - 11 + 11 - \frac{2}{3}x$

50. $\frac{1}{5}y + 4 - 4 - \frac{1}{5}y$

51. $\left(\frac{9}{7}\right)(-.38)\left(\frac{7}{9}\right)$

52. $\left(\frac{4}{5}\right)(-.73)\left(\frac{5}{4}\right)$

53. $t + (-t) + \frac{1}{2}(2)$

54. $w + (-w) + \frac{1}{4}(4)$

Use the distributive property to rewrite each expression. Simplify if possible. See Examples 9 and 11.

55. $5x + x$

56. $6q + q$

57. $4(t + 3)$

58. $5(w + 4)$

59. $-8(r + 3)$

60. $-11(x + 4)$

61. $-5(y - 4)$

62. $-9(g - 4)$

63. $-\frac{4}{3}(12y + 15z)$

64. $-\frac{2}{5}(10b + 20a)$

65. $8 \cdot z + 8 \cdot w$

66. $4 \cdot s + 4 \cdot r$

67. $7(2v) + 7(5r)$

68. $13(5w) + 13(4p)$

69. $8(3r + 4s - 5y)$

70. $2(5u - 3v + 7w)$

71. $q + q + q$

72. $m + m + m + m$

73. $-5x + x$

74. $-9p + p$

Use the distributive property to write each of the following without parentheses. See Example 10.

75. $-(4t + 5m)$

76. $-(9x + 12y)$

77. $-(-5c - 4d)$

78. $-(-13x - 15y)$

79. $-(-3q + 5r - 8s)$

80. $-(-4z + 5w - 9y)$

81. The operations of "getting out of bed" and "taking a shower" are not commutative. Give an example of another pair of everyday operations that are not commutative.

82. The phrase "dog biting man" has two different meanings, depending on how the words are associated:

(dog biting) man dog (biting man)

Give another example of a three-word phrase that has different meanings depending on how the words are associated.

PREVIEW EXERCISES

Perform the indicated operations. See Sections 1.4 and 1.5.

83. $(-12) + 26 + 19 + (-2)$

84. $-5 + (-6) + (-7)$

85. $10 - [-2 - (4 - 6)]$

86. $13 - [(2 - 5) - (3 - 7)]$

1.9 SIMPLIFYING EXPRESSIONS

1 As we saw in the previous section, we can use the properties of addition and multiplication to simplify algebraic expressions.

EXAMPLE 1 *Simplifying Expressions*
Simplify the following expressions.

(a) $4x + 8 + 9$

Since $8 + 9 = 17$, $4x + 8 + 9 = 4x + 17$.

(b) $4(3m - 2n)$

Use the distributive property.

$$4(3m - 2n) = 4(3m) - 4(2n)$$
$$= 12m - 8n$$

(c) $6 + 3(4k + 5) = 6 + 3(4k) + 3(5)$ Distributive property
$= 6 + 12k + 15$ Multiply.
$= 21 + 12k$ Add.

(d) $5 - (2y - 8) = 5 - 1(2y - 8)$ Identity property
$= 5 - 2y + 8$ Distributive property
$= 13 - 2y$ Add. ■

Note Although the steps were not shown, in Examples 1(c) and 1(d) we mentally used the commutative and associative properties to add in the last step. In practice, these steps are usually left out, but we should realize that they are used whenever the ordering in a sum is rearranged.

WORK PROBLEM 1 AT THE SIDE.

2 A **term** is a single number, or a product of a number and one or more variables raised to powers. Examples of terms include

$$-9x^2, \quad 15y, \quad -3, \quad 8m^2n, \quad \text{and} \quad k.$$

The **numerical coefficient** of the term $9m$ is 9; the numerical coefficient of $-15x^3y^2$ is -15; the numerical coefficient of x is 1; and the numerical coefficient of 8 is 8. A coefficient is a multiplier.

Caution It is important to be able to distinguish between *terms* and *factors*. For example, in the expression $8x^3 + 12x^2$, there are two terms. They are $8x^3$ and $12x^2$. Terms are separated by a + or − sign. On the other hand, in the expression $(8x^3)(12x^2)$, $8x^3$ and $12x^2$ are *factors*. Factors are multiplied.

OBJECTIVES

1. Simplify expressions.
2. Identify terms and numerical coefficients.
3. Identify like terms.
4. Combine like terms.
5. Simplify expressions from word phrases.

FOR EXTRA HELP

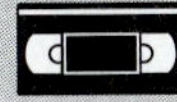
Tape 3

SSM pp. 33–35

MAC: A
IBM: A

1. Simplify each expression.

(a) $9k + 12 - 5$

(b) $7(3p + 2q)$

(c) $2 + 5(3z - 1)$

(d) $-3 - (2 + 5y)$

ANSWERS
1. (a) $9k + 7$ (b) $21p + 14q$ (c) $15z - 3$ (d) $-5 - 5y$

2. Give the numerical coefficient of each term.

(a) $15q$

(b) $-2m^3$

(c) $-18m^7q^4$

(d) $-r$

3. Identify each pair of terms as *like* or *unlike*.

(a) $9x$, $4x$

(b) $-8y^3$, $12y^2$

(c) $7x^2y^4$, $-7x^2y^4$

(d) $13kt$, $4tk$

ANSWERS
2. (a) 15 (b) −2 (c) −18 (d) −1
3. (a) like (b) unlike (c) like (d) like

EXAMPLE 2 *Identifying the Numerical Coefficient of a Term*

Give the numerical coefficient of each term.

Term	*Numerical Coefficient*
$-7y$	-7
$34r^3$	34
$-26x^5yz^4$	-26
$-k$	-1
r	1

WORK PROBLEM 2 AT THE SIDE.

3 Terms with exactly the same variables (including the same exponents) are called **like terms.** For example, $9m$ and $4m$ have the same variables and are like terms. Also, $6x^3$ and $-5x^3$ are like terms. The terms $-4y^3$ and $4y^2$ have different exponents and are **unlike terms.**

Here are some examples of like terms.

$$5x \text{ and } -12x \qquad 3x^2y \text{ and } 5x^2y$$

Here are some examples of unlike terms.

$$4xy^2 \text{ and } 5xy \qquad -7w^3z^3 \text{ and } 2xz^3$$

WORK PROBLEM 3 AT THE SIDE.

4 We add or subtract like terms by using the distributive property. For example.

$$3x + 5x = (3 + 5)x = 8x.$$

This process is called **combining terms.**

Caution Remember that *only like terms* may be combined.

EXAMPLE 3 *Combining Like Terms*

Combine terms in the following expressions.

(a) $6r + 3r + 2r$

Use the distributive property to combine like terms.

$$6r + 3r + 2r = (6 + 3 + 2)r = 11r$$

(b) $4x + x = 4x + 1x = 5x$ Identity property

(c) $16y - 9y = (16 - 9)y = 7y$

(d) $32y + 10y^2$ cannot be simplified because $32y$ and $10y^2$ are unlike terms. ■

WORK PROBLEM 4 AT THE SIDE.

■ **EXAMPLE 4** *Simplifying Expressions Involving Like Terms*

Simplify the following expressions.

(a)
$$\begin{aligned} 14y + 2(6 + 3y) &= 14y + 2(6) + 2(3y) && \text{Distributive property} \\ &= 14y + 12 + 6y && \text{Multiply.} \\ &= 20y + 12 && \text{Combine like terms.} \end{aligned}$$

(b)
$$\begin{aligned} 9k - 6 - 3(2 - 5k) &= 9k - 6 - 3(2) - 3(-5k) && \text{Distributive property} \\ &= 9k - 6 - 6 + 15k && \text{Multiply.} \\ &= 24k - 12 && \text{Combine like terms.} \end{aligned}$$

(c)
$$\begin{aligned} -(2 - r) + 10r &= -1(2 - r) + 10r && -(2 - r) = -1(2 - r) \\ &= -1(2) - 1(-r) + 10r && \text{Distributive property} \\ &= -2 + r + 10r && \text{Multiply.} \\ &= -2 + 11r && \text{Combine like terms.} \end{aligned}$$

(d)
$$\begin{aligned} 5(2a - 6) - 3(4a - 9) &= 10a - 30 - 12a + 27 && \text{Distributive property} \\ &= -2a - 3 && \text{Combine like terms.} \end{aligned}$$
■

WORK PROBLEM 5 AT THE SIDE.

4. Combine terms.

(a) $4k + 7k$

(b) $4r - r$

(c) $5z + 9z - 4z$

(d) $8p + 8p^2$

5. Simplify.

(a) $10p + 3(5 + 2p)$

(b) $7z - 2 - 4(1 + z)$

(c) $-(3 + 5k) + 7k$

ANSWERS

4. (a) $11k$ (b) $3r$ (c) $10z$ (d) cannot be combined

5. (a) $16p + 15$ (b) $3z - 6$ (c) $2k - 3$

5 In the next example, we translate a word phrase to a mathematical expression and then simplify it. We will need to do this in the next chapter when we solve applied problems.

EXAMPLE 5 *Converting Words to a Mathematical Expression*

Write the following phrase as a mathematical expression and simplify: four times a number, subtracted from the sum of twice the number and 4.

Let x represent the number.

the sum of twice the number and 4 ↓	four times the number ↓	
$(2x + 4)$	$- 4x$	Write with symbols.
$-2x + 4$		Combine terms. ■

WORK PROBLEM 6 AT THE SIDE.

6. Write the following phrases as a mathematical expression, and simplify by combining terms.

(a) Three times a number, subtracted from the sum of the number and 8

(b) Twice a number added to the sum of 6 and the number

ANSWERS

6. (a) $(x + 8) - 3x$; $-2x + 8$
(b) $2x + (6 + x)$; $3x + 6$

NAME DATE HOUR

1.9 EXERCISES

Choose the letter of the correct response in each of the following.

1. Which one of the following is correct?
(a) $6 + 2x = 8x$ **(b)** $6 - 2x = 4x$ **(c)** $6x - 2x = 4x$ **(d)** $3 + 8(4t - 6) = 11(4t - 6)$

2. Which one of the following is an example of a pair of like terms?
(a) $6t, 6w$ **(b)** $-8x^2y, 9xy^2$ **(c)** $5ry, 6yr$ **(d)** $-5x^2, 2x^3$

3. Which one of the following is an example of a term with numerical coefficient 5?
(a) $5x^3y^7$ **(b)** x^5 **(c)** $\frac{x}{5}$ **(d)** 5^2xy^3

4. Which one of the following is a correct translation for "six times a number, subtracted from the product of eleven and the number" (if x represents the number)?
(a) $6x - 11x$ **(b)** $11x - 6x$ **(c)** $(11 + x) - 6x$ **(d)** $6x - (11 + x)$

Simplify each expression. See Example 1.

5. $4r + 19 - 8$ **6.** $7t + 18 - 4$ **7.** $8(4q - 3t)$ **8.** $12(9m - 7n)$

9. $5 + 2(x - 3y)$ **10.** $8 + 3(s - 6t)$ **11.** $-2 - (5 - 3p)$ **12.** $-10 - (7 - 14r)$

Give the numerical coefficient of each of the following terms. See Example 2.

13. $14x$ **14.** $9x$ **15.** $-12k$ **16.** $-23y$

17. $5m^2$ **18.** $-3n^6$ **19.** xw **20.** pq

21. $-x$ **22.** $-t$ **23.** 74 **24.** 98

25. Give an example of a pair of like terms with the variable x, such that one of them has a negative numerical coefficient, one has a positive numerical coefficient, and their sum has a positive numerical coefficient.

26. Give an example of a pair of unlike terms such that each term has x as the only variable.

Identify each group of terms as like *or* unlike.

27. $8r, -13r$ **28.** $-7a, 12a$ **29.** $5z^4, 9z^3$ **30.** $8x^5, -10x^3$

31. $4, 9, -24$ **32.** $7, 17, -83$ **33.** x, y **34.** t, s

35. There is an old saying "You can't add apples and oranges." Explain how this saying can be applied to the goal of Objective 3 in this section.

36. Explain how the distributive property is used in combining $6t + 5t$ to get $11t$.

Simplify each expression, and combine like terms. See Examples 3 and 4.

37. $4k + 3 - 2k + 8 + 7k - 16$

38. $9x + 7 - 13x + 12 + 8x - 15$

39. $-\frac{4}{3} + 2t + \frac{1}{3}t - 8 - \frac{8}{3}t$

40. $-\frac{5}{6} + 8x + \frac{1}{6}x - 7 - \frac{7}{6}$

41. $-5.3r + 4.9 - 2r + .7 + 3.2r$

42. $2.7b + 5.8 - 3b + .5 - 4.4b$

43. $2y^2 - 7y^3 - 4y^2 + 10y^3$

44. $9x^4 - 7x^6 + 12x^4 + 14x^6$

45. $13p + 4(4 - 8p)$

46. $5x + 3(7 - 2x)$

47. $-4(y - 7) - 6$

48. $-5(t - 13) - 4$

49. $-5(5y - 9) + 3(3y + 6)$

50. $-3(2t + 4) + 8(2t - 4)$

51. $-4(-3k + 3) - (6k - 4) - 2k + 1$

52. $-5(8j + 2) - (5j - 3) - 3j + 17$

53. $-7.5(2y + 4) - 2.9(3y - 6)$

54. $8.4(6t - 6) + 2.4(9 - 3t)$

Convert the following phrases into mathematical expressions. Use x as the variable. Combine like terms when possible. See Example 5.

55. Five times a number, added to the sum of the number and three

56. Six times a number, added to the sum of the number and six

57. A number multiplied by -7, subtracted from the sum of 13 and six times the number

58. A number multiplied by 5, subtracted from the sum of 14 and eight times the number

59. Six times a number added to -4, subtracted from twice the sum of three times the number and 4

60. Nine times a number added to 6, subtracted from triple the sum of 12 and 8 times the number

61. Write the expression $9x - (x + 2)$ using words, as in Exercises 55–60.

62. Write the expression $2(3x + 5) - 2(x + 4)$ using words, as in Exercises 55–60.

PREVIEW EXERCISES

Find the opposite or additive inverse of each number. See Section 1.8.

63. 5

64. 3.4

65. -15

66. $-\frac{3}{4}$

Add a number to each expression so that the final sum is just x.

67. $x - 2$

68. $x - \frac{1}{2}$

69. $x + 14$

70. $x + 9.6$

CHAPTER 1 SUMMARY

KEY TERMS

Section	Term	Definition
1.1	**exponent**	An exponent, or power, is a number that indicates how many times a factor is repeated.
	base	The base is the number that is a repeated factor when written with an exponent.
	exponential expression	A number written with an exponent is an exponential expression.
1.2	**variable**	A variable is a symbol, usually a letter, used to represent an unknown number.
	algebraic expression	An algebraic expression is a collection of numbers, variables, symbols for operations, and symbols for grouping.
	equation	An equation is a statement that says two expressions are equal.
	solution	A solution of an equation is any replacement for the variable that makes the equation true.
	domain	The domain of an equation is the set of numbers from which the solution is chosen.
1.3	**whole numbers**	The set of whole numbers is $\{0, 1, 2, 3, 4, 5, \ldots\}$.
	natural numbers	The set of natural numbers includes the numbers used for counting: $\{1, 2, 3, 4, \ldots\}$
	negative number	A negative number is located to the *left* of 0 on the number line.
	positive number	A positive number is located to the *right* of 0 on the number line.
	signed numbers	Signed numbers are either positive or negative.
	integers	The set of integers is $\{\ldots, -3, -2, -1, 0, 1, 2, 3, \ldots\}$.
	rational numbers	Rational numbers can be written as quotients of two integers, with denominator not 0.
	irrational numbers	Irrational numbers are nonrational numbers represented by points on the number line.
	real numbers	Real numbers include all numbers that can be represented by points on the number line, or all rational and irrational numbers.
	opposite	The opposite of a number a is the number that is the same distance from 0 on the number line as a, but on the opposite side of 0. This number is also called the **negative** of a or the **additive inverse** of a.
	absolute value	The absolute value of a number is the distance between 0 and the number on the number line.
1.4	**sum**	The answer to an addition problem is called the sum.
1.5	**difference**	The answer to a subtraction problem is called the difference.
1.6	**product**	The result of multiplication is called the product.
1.7	**reciprocal**	Pairs of numbers whose product is 1 are called reciprocals or **multiplicative inverses** of each other.
	quotient	The answer to a division problem is called the quotient.

1.8	**identity element for addition**	When the identity element for addition, which is 0, is added to a number, the number is unchanged.
	identity element for multiplication	When a number is multiplied by the identity element for multiplication, which is 1, the number is unchanged.
	additive inverse	The opposite of a number is also called the additive inverse of the number.
	multiplicative inverse	The reciprocal of a number is also called the multiplicative inverse of the number.
1.9	**term**	A term is a single number, or a product of a number and one or more variables raised to powers.
	numerical coefficient	The numerical factor in a term is its numerical coefficient.
	like terms	Terms with exactly the same variables (including the same exponents) are called like terms.

NEW SYMBOLS

a^n	n factors of a
$=$	is equal to
$\neq$	is not equal to
$<$	is less than
$\leq$	is less than or equal to
$>$	is greater than
$\geq$	is greater than or equal to
$a(b)$, $(a)(b)$, $a \cdot b$, or ab	a times b
$\frac{a}{b}$, a/b or $a \div b$	a divided by b
$\{\ \}$	set braces
$\lvert x \rvert$	absolute value of x

QUICK REVIEW

Concepts	Examples
1.1 Exponents, Order of Operations, and Inequality	
Order of Operations If possible, simplify within parentheses and above and below fraction bars. *Step 1* Apply all exponents. *Step 2* Do any multiplications or divisions in the order in which they occur, working from left to right. *Step 3* Do any additions or subtractions in the order in which they occur, working from left to right.rule	$36 - 4(2^2 + 3) = 36 - 4(4 + 3)$ $= 36 - 4(7)$ $= 36 - 28$ $= 8$

Concepts	Examples
1.2 Variables, Expressions, and Equations	
Evaluate an expression with a variable by substituting a given number for the variable.	Evaluate $2x + y^2$ if $x = 3$ and $y = -4$. $2x + y^2 = 2(3) + (-4)^2$ $= 6 + 16$ $= 22$
Values of a variable that make an equation true are solutions of the equation.	Is 2 a solution of $5x + 3 = 18$? $5(2) + 3 = 18$ $13 = 18$ False 2 is not a solution.
1.3 Real Numbers and the Number Line	
Ordering Real Numbers a is less than b if a is to the left of b on the number line.	−3 −2 −1 0 1 2 3 4 $-2 < 3$ $3 > 0$ $0 < 3$
The opposite or additive inverse of a is $-a$.	$-(5) = -5$ $-(-7) = 7$ $-0 = 0$
The absolute value of a, $\|a\|$, is the distance between a and 0 on the number line.	$\|13\| = 13$ $\|0\| = 0$ $\|-5\| = 5$
1.4 Addition of Real Numbers	
To add two numbers with the same sign, add their absolute values. The sum has that same sign.	$9 + 4 = 13$ $-8 + (-5) = -13$
To add two numbers with different signs, subtract their absolute values. The sum has the sign of the number with larger absolute value.	$7 + (-12) = -5$ $-5 + 13 = 8$
1.5 Subtraction of Real Numbers	
Definition of Subtraction $a - b = a + (-b)$	$5 - (-2) = 5 + 2 = 7$ $-3 - 4 = -3 + (-4) = -7$
Subtracting signed numbers: **1.** Change the subtraction symbol to addition. **2.** Change the sign of the number being subtracted. **3.** Add, as in the previous section.	$-2 - (-6) = -2 + 6$ $= 4$ $13 - (-8) = 13 + 8$ $= 21$

Concepts	Examples
1.6, 1.7 Multiplication and Division of Real Numbers	
Multiplying and Dividing Signed Numbers The product (or quotient) of two numbers having the *same sign* is *positive;* the product (or quotient) of two numbers having *different signs* is *negative.*	$6 \cdot 5 = 30 \quad (-7)(-8) = 56$ $\frac{10}{2} = 5 \quad \frac{-24}{-6} = 4$ $(-6)(5) = -30 \quad (6)(-5) = -30$ $\frac{-18}{9} = -2 \quad \frac{49}{-7} = -7$
Division *by* 0 is undefined.	$\frac{0}{5} = 0 \quad \frac{5}{0}$ is undefined
1.8 Properties of Addition and Multiplication	
Commutative	
$a + b = b + a$ $ab = ba$	$7 + (-1) = -1 + 7$ $5(-3) = (-3)5$
Associative	
$(a + b) + c = a + (b + c)$ $(ab)c = a(bc)$	$(3 + 4) + 8 = 3 + (4 + 8)$ $[(-2)(6)](4) = (-2)[(6)(4)]$
Identity	
$a + 0 = a \quad 0 + a = a$ $a \cdot 1 = a \quad 1 \cdot a = a$	$-7 + 0 = -7 \quad 0 + (-7) = -7$ $9 \cdot 1 = 9 \quad 1 \cdot 9 = 9$
Inverse	
$a + (-a) = 0 \quad -a + a = 0$ $a \cdot \frac{1}{a} = 1 \quad \frac{1}{a} \cdot a = 1 \ (a \neq 0)$	$7 + (-7) = 0 \quad -7 + 7 = 0$ $-2\left(-\frac{1}{2}\right) = 1 \quad -\frac{1}{2}(-2) = 1$
Distributive	
$a(b + c) = ab + ac$ $(b + c)a = ba + ca$	$5(4 + 2) = 5(4) + 5(2)$ $(4 + 2) \cdot 5 = 4(5) + 2(5)$
1.9 Simplifying Expressions	
Only like terms may be combined.	$-3y^2 + 6y^2 + 14y^2 = 17y^2$ $4(3 + 2x) - 6(5 - x) = 12 + 8x - 30 + 6x$ $= 14x - 18$

NAME DATE HOUR

CHAPTER 1 REVIEW EXERCISES

[1.1] *Find the value of each exponential expression.**

1. 5^4 **2.** $\left(\frac{3}{5}\right)^3$ **3.** $(.02)^5$ **4.** $(.001)^3$

Find the value of each of the following expressions.

5. $8 \cdot 5 - 13$ **6.** $7[3 + 6(3^2)]$ **7.** $\frac{9(4^2 - 3)}{4 \cdot 5 - 17}$ **8.** $\frac{6(5 - 4) + 2(4 - 2)}{3^2 - (4 + 3)}$

Tell whether each statement is true or false.

9. $12 \cdot 3 - 6 \cdot 6 \leq 0$ **10.** $3[5(2) - 3] > 20$ **11.** $9 \leq 4^2 - 8$ **12.** $9 \cdot 2 - 6 \cdot 3 \geq 0$

Write each word statement in symbols.

13. Thirteen is less than seventeen. **14.** Five plus two is not equal to ten.

[1.2] *Find the numerical values of the following if* **(a)** $x = 3$ *and* **(b)** $x = 15$.

15. $4x$ **16.** $\frac{x + 2}{x + 1}$ **17.** $3x + x^2$ **18.** $7.45(x + 1)$

Find the numerical values of the following if $x = 6$ *and* $y = 3$.

19. $2x + 6y$ **20.** $4(3x - y)$ **21.** $\frac{x}{3} + 4y$ **22.** $\frac{x^2 + 3}{3y - x}$

Change the word phrases to algebraic expressions. Use x *as the variable to represent the number.*

23. Six added to a number **24.** A number subtracted from eight

25. Nine subtracted from six times a number **26.** Three-fifths of a number added to 12

Decide whether the given number is a solution of the equation.

27. $5x + 3(x + 2) = 22;\ 2$ **28.** $\frac{t + 5}{3t} = 1;\ 6$

* For help with any of these exercises, refer to the section given in brackets.

Change the word statements to equations. Use x as the variable. Then find the solutions from the domain {0, 2, 4, 6, 8, 10}.

29. Six less than twice a number is 10.

30. The product of a number and 4 is 8.

[1.3] *Graph each group of numbers on a number line.*

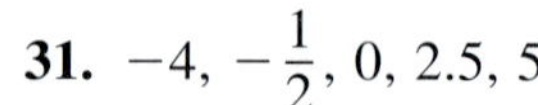

31. $-4, -\frac{1}{2}, 0, 2.5, 5$

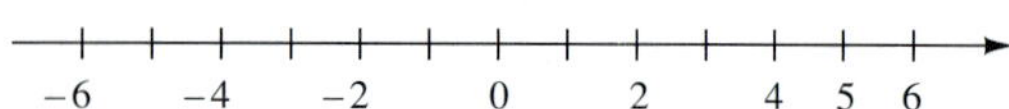

32. $-2, -3, |-3|, |-1|$

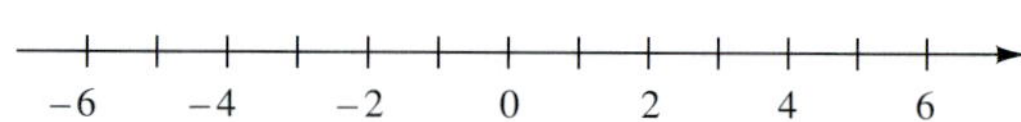

33. $-3\frac{1}{4}, 2\frac{4}{5}, -1\frac{1}{8}, \frac{5}{6}$

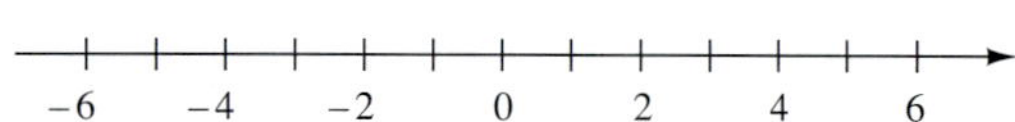

34. $|-4|, -|-3|, -|-5|, -6$

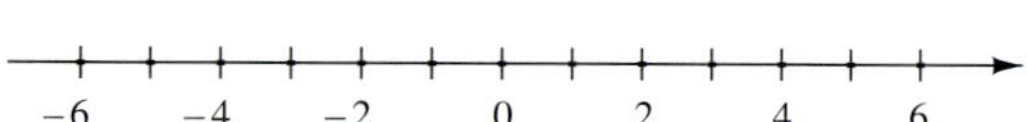

Select the smaller number in each pair.

35. $-10, 5$

36. $-8, -9$

37. $-\frac{2}{3}, -\frac{3}{4}$

38. $0, -|23|$

Decide whether each statement is true or false.

39. $12 > -13$

40. $0 > -5$

41. $-9 < -7$

42. $-13 > -13$

For each of the following, (a) find the opposite of the number and (b) find the absolute value of the number.

43. -9

44. 0

45. 6

46. $-\frac{5}{7}$

Simplify by removing absolute value symbols.

47. $|-12|$

48. $-|3|$

49. $-|-19|$

50. $-|9 - 2|$

[1.4] *Find the following sums.*

51. $-10 + 4$

52. $14 + (-18)$

53. $-8 + (-9)$

54. $\frac{4}{9} + \left(-\frac{5}{4}\right)$

55. $-13.5 + (-8.3)$

56. $(-10 + 7) + (-11)$

57. $[-6 + (-8) + 8] + [9 + (-13)]$

58. $(-4 + 7) + (-11 + 3) + (-15 + 1)$

Write a numerical expression for each phrase, and simplify the expression.

59. 19 added to the sum of -31 and 12

60. 13 more than the sum of -4 and -8

Solve each problem.

61. Tri Nguyen has \$18 in his checking account. He then writes a check for \$26. What negative number represents his balance?

62. The temperature at noon on an August day in Houston was 93°F. After a thunderstorm, it dropped 6°. What was the new temperature?

Find the solution of each equation from the domain $\{-3, -2, -1, 0, 1, 2, 3\}$ by guessing or using trial and error.

63. $x + (-2) = -4$

64. $12 + x = 11$

[1.5] *Find the following differences.*

65. $-7 - 4$

66. $-12 - (-11)$

67. $5 - (-2)$

68. $-\frac{3}{7} - \frac{4}{5}$

69. $2.56 - (-7.75)$

70. $(-10 - 4) - (-2)$

71. $(-3 + 4) - (-1)$

72. $-(-5 + 6) - 2$

Write a numerical expression for each phrase, and simplify the expression.

73. The difference between -4 and -6

74. Five less than the sum of 4 and -8

Solve each problem.

75. Eric owed his brother \$28. He repaid \$13 but then borrowed another \$14. What positive or negative amount represents his present financial status?

76. If the temperature drops 7° below its previous level of $-3°$, what is the new temperature?

77. Explain in your own words how the subtraction problem $-8 - (-6)$ is performed.

78. Can the difference of two negative numbers be positive? Explain with an example.

[1.6] *Perform the indicated operations.*

79. $(-12)(-3)$

80. $15(-7)$

81. $\left(-\frac{4}{3}\right)\left(-\frac{3}{8}\right)$

82. $(-4.8)(-2.1)$

83. $5(8 - 12)$

84. $(5 - 7)(8 - 3)$

85. $2(-6) - (-4)(-3)$

86. $3(-10) - 5$

Evaluate the following expressions if $x = -5$, $y = 4$, and $z = -3$.

87. $6x - 4z$

88. $5x + y - z$

89. $5x^2$

90. $z^2(3x - 8y)$

Write a numerical expression for each phrase, and simplify the expression.

91. Nine less than the product of -4 and 5

92. Five-sixths of the sum of 12 and -6

[1.7] *Find the following quotients.*

93. $\dfrac{-36}{-9}$

94. $\dfrac{220}{-11}$

95. $-\dfrac{1}{2} \div \dfrac{2}{3}$

96. $-33.9 \div (-3)$

97. $\dfrac{-5(3) - 1}{8 - 4(-2)}$

98. $\dfrac{5(-2) - 3(4)}{-2[3 - (-2)] - 1}$

99. $\dfrac{10^2 - 5^2}{8^2 + 3^2 - (-2)}$

100. $\dfrac{(.6)^2 + (.8)^2}{(-1.2)^2 - (-.56)}$

Write a numerical expression for each phrase, and simplify the expression.

101. The quotient of 12 and the sum of 8 and -4

102. The product of -20 and 12, divided by the difference between 15 and -15

Write the following in symbols, using x as the variable, and find the solution by guessing or by using trial and error. All solutions come from the list of integers between -12 and 12.

103. 8 times a number is -24.

104. The quotient of a number and 3 is -2.

105. 3 less than a number is -7.

106. The sum of a number and 5 is -6.

[1.8] *Decide whether each statement is an example of the commutative, associative, identity, inverse, or distributive property.*

107. $6 + 0 = 6$

108. $5 \cdot 1 = 5$

109. $-\frac{2}{3}\left(-\frac{3}{2}\right) = 1$

110. $17 + (-17) = 0$

111. $5 + (-9 + 2) = [5 + (-9)] + 2$

112. $w(xy) = (wx)y$

113. $3x + 3y = 3(x + y)$

114. $(1 + 2) + 3 = 3 + (1 + 2)$

Use the distributive property to rewrite each expression. Simplify if possible.

115. $7y + y$

116. $-12(4 - t)$

117. $3(2s) + 3(4y)$

118. $-(-4r + 5s)$

119. Evaluate $25 - (5 - 2)$ and $(25 - 5) - 2$. Use this example to explain why subtraction is not associative.

120. Evaluate $180 \div (15 \div 5)$ and $(180 \div 15) \div 5$. Use this example to explain why division is not associative.

[1.9] *Use the distributive property as necessary and combine like terms.*

121. $16p^2 - 8p^2 + 9p^2$

122. $4r^2 - 3r + 10r + 12r^2$

123. $-8(5k - 6) + 3(7k + 2)$

124. $2s - (-3s + 6)$

125. $-7(2t - 4) - 4(3t + 8) - 19(t + 1)$

126. $3.6t^2 + 9t - 8.1(6t^2 + 4t)$

Convert the following phrases into mathematical expressions. Use x as the variable, and combine like terms when possible.

127. Seven times a number subtracted from the product of -2 and three times the number

128. The quotient of 9 more than a number and 6 less than the number

129. In Exercise 127, does the word *and* signify addition? Explain.

130. Write the expression $3(4x - 6)$ using words, as in Exercises 127–128.

MIXED REVIEW EXERCISES*

Perform the indicated operations.

131. $[(-2) + 7 - (-5)] + [-4 - (-10)]$

132. $\left(-\frac{5}{6}\right)^2$

133. $-|(-7)(-4)| - (-2)$

134. $\dfrac{6(-4) + 2(-12)}{5(-3) + (-3)}$

135. $\dfrac{3}{8} - \dfrac{5}{12}$

136. $\dfrac{12^2 + 2^2 - 8}{10^2 - (-4)(-15)}$

137. $\dfrac{8^2 + 6^2}{7^2 + 1^2}$

138. $-16(-3.5) - 7.2(-3)$

139. $2\frac{5}{6} - 4\frac{1}{3}$

140. $-8 + [(-4 + 17) - (-3 - 3)]$

141. $-\dfrac{12}{5} \div \dfrac{9}{7}$

142. $(-8 - 3) - 5(2 - 9)$

143. $[-7 + (-2) - (-3)] + [8 + (-13)]$

144. $\dfrac{15}{2} \cdot \left(-\dfrac{4}{5}\right)$

Write a numerical expression for each, and simplify it if possible. Use x as the variable if one is needed.

145. In 1993, a company spent $13,600 on advertising. In 1994, the amount spent on advertising was reduced by $1400. How much was spent on advertising in 1994?

146. The quotient of a number and 14 less than three times the number

*The order of exercises in this final group does not correspond to the order in which topics occur in the chapter. This random ordering should help you prepare for the chapter test in yet another way.

NAME DATE HOUR

CHAPTER 1 TEST

Decide whether the statement is true or false.

1. $4[-20 + 7(-2)] \leq 135$

1. ________________

2. $(-3)^2 + 2^2 = 5^2$

2. ________________

3. Graph the group of numbers $-1, -3, |-4|, |-1|$ on the number line.

3. 

Select the smaller number from each pair.

4. $6, -|-8|$

4. ________________

5. $-.742, -1.277$

5. ________________

6. Write in symbols: The quotient of -6 and the sum of 2 and -8. Simplify the expression.

6. ________________

7. If a and b are both negative, is $\dfrac{a + b}{a \cdot b}$ positive or negative?

7. ________________

Perform the indicated operations whenever possible.

8. $-2 - (5 - 17) + (-6)$

8. ________________

9. $-5\frac{1}{2} + 2\frac{2}{3}$

9. ________________

10. $-6 - [-7 + (2 - 3)]$

10. ________________

11. $4^2 + (-8) - (2^3 - 6)$

11. ________________

12. $(-5)(-12) + 4(-4) + (-8)^2$

12. ________________

13. $\dfrac{-7 - (-6 + 2)}{-5 - (-4)}$

13. ________________

14. $\dfrac{30(-1 - 2)}{-9[3 - (-2)] - 12(-2)}$

14. ________________

Find the solution for each equation from the domain $\{-6, -4, -2, 0, 2, 4, 6\}$ by guessing or by trial and error.

15. ______

15. $-3x = -12$

16. ______

16. $\frac{x}{-4} = -1$

Evaluate the following expressions, given $x = -2$ and $y = 4$.

17. ______

17. $3x - 4y^2$

18. ______

18. $\frac{5x + 7y}{3(x + y)}$

Solve the following problem.

19. ______

19. The highest Fahrenheit temperature ever recorded in Idaho was 118°, while the lowest was −60°. What is the difference between these highest and lowest temperatures?

Match the property in Column I with the example of it in Column II.

20. ______

21. ______

22. ______

23. ______

24. ______

I	II
20. Commutative	A. $3x + 0 = 3x$
21. Associative	B. $(5 + 2) + 8 = 8 + (5 + 2)$
22. Inverse	C. $-3(x + y) = -3x + (-3y)$
23. Identity	D. $-5 + (3 + 2) = (-5 + 3) + 2$
24. Distributive	E. $-\frac{5}{3}\left(-\frac{3}{5}\right) = 1$

25. ______

25. Simplify by using the distributive property and combining like terms: $-2(3x^2 + 4) - 3(x^2 + 2x)$

2 Solving Equations and Inequalities

2.1 THE ADDITION PROPERTY OF EQUALITY

OBJECTIVES

1. Identify linear equations.
2. Use the addition property of equality.
3. Simplify equations, and then use the addition property of equality.
4. Solve equations that have no solution or infinitely many solutions.

FOR EXTRA HELP

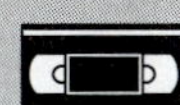
Tape 3

SSM pp. 48–50

MAC: A IBM: A

To solve applied problems, we must be able to solve equations. The simplest type of equation is a *linear equation.* Methods for solving linear equations will be introduced in this section. We will be using the definitions and properties of real numbers that we learned in Chapter 1.

1 Before we can solve a linear equation, we must be able to recognize one.

A **linear equation** can be written in the form

$$Ax + B = C$$

for real numbers A, B, and C, with $A \neq 0$.

As we saw in Chapter 1, a solution of an equation is a number that makes the equation a true statement when it replaces the variable. Equations that have exactly the same solutions are **equivalent equations.** Linear equations are solved by using a series of steps to produce a simpler equivalent equation of the form

$$x = \text{a number.}$$

2 In the equation $x - 5 = 2$, both $x - 5$ and 2 represent the same number because this is the meaning of the equals sign. To solve the equation, we change the left side from $x - 5$ to just x. This is done by adding 5 to $x - 5$. To keep the two sides equal, we must also add 5 to the right side.

$x - 5 = 2$	Given equation
$x - 5 + 5 = 2 + 5$	Add 5 to each side.
$x + 0 = 7$	Additive inverse property
$x = 7$	Identity property

The solution of the given equation is 7. Check by replacing x with 7 in the given equation.

$$x - 5 = 2 \quad \text{Given equation}$$
$$7 - 5 = 2 \quad ? \quad \text{Let } x = 7.$$
$$2 = 2 \quad \text{True}$$

Since the final statement is true, 7 checks as the solution.

To solve the equation above, we added the same number to each side. The **addition property of equality** justifies this step.

ADDITION PROPERTY OF EQUALITY

If A, B, and C are real numbers, then the equations

$$A = B \quad \text{and} \quad A + C = B + C$$

are equivalent equations. That is, we can add the same number to each side of an equation without changing the solution.

In the addition property, C represents a real number. This means that any quantity that represents a real number can be added to both sides of an equation to change it to an equivalent equation.

EXAMPLE 1 *Using the Addition Property of Equality*

Solve $x - 16.2 = 7.5$.

If the left side of this equation were just x, the solution would be known. Get x alone by using the addition property of equality, adding 16.2 to each side.

$$x - 16.2 = 7.5$$
$$x - 16.2 + 16.2 = 7.5 + 16.2 \quad \text{Add 16.2 on each side.}$$
$$x = 23.7 \quad \text{Combine terms.}$$

Here we combined the steps that change $x - 16.2 + 16.2$ to $x + 0$ and $x + 0$ to x. We will combine these steps from now on. Check by substituting 23.7 for x in the original equation.

$$x - 16.2 = 7.5 \quad \text{Given equation}$$
$$23.7 - 16.2 = 7.5 \quad ? \quad \text{Let } x = 23.7.$$
$$7.5 = 7.5 \quad \text{True}$$

Since the check results in a true statement, 23.7 is the solution. ■

WORK PROBLEM 1 AT THE SIDE.

The addition property of equality says that the same number may be *added* to each side of an equation. In Chapter 1, subtraction was defined as addition of the opposite. Thus, we can also use the following rule when solving an equation.

The same number may be subtracted from each side of an equation without changing the solution.

WORK PROBLEM 2 AT THE SIDE.

1. Solve.

(a) $m - 2.9 = -6.4$

(b) $y - 4.1 = 6.3$

2. Solve.

(a) $a + 2 = -3$

(b) $r + 16 = 22$

ANSWERS

1. (a) −3.5 (b) 10.4

2. (a) −5 (b) 6

EXAMPLE 2 *Subtracting a Variable Expression*

Solve $\frac{3}{5}k + 17 = \frac{8}{5}k$.

Get all terms with variables on the same side of the equation. One way to do this is to subtract $\frac{3}{5}k$ from each side.

$$\frac{3}{5}k + 17 = \frac{8}{5}k$$

$$\frac{3}{5}k + 17 - \frac{3}{5}k = \frac{8}{5}k - \frac{3}{5}k \qquad \text{Subtract } \tfrac{3}{5}k.$$

$$17 = 1k \qquad \text{Combine terms.}$$

$$17 = k \qquad \text{Identity property}$$

The solution is 17. From now on we will skip the step that changes $1k$ to k. Check the solution by replacing k with 17 in the original equation. ■

Another way to solve the equation in Example 2 is to first subtract $\frac{8}{5}k$ from each side.

$$\frac{3}{5}k + 17 = \frac{8}{5}k$$

$$\frac{3}{5}k + 17 - \frac{8}{5}k = \frac{8}{5}k - \frac{8}{5}k \qquad \text{Subtract } \tfrac{8}{5}k.$$

$$17 - k = 0 \qquad \text{Combine terms. Additive identity}$$

$$17 - k - 17 = 0 - 17 \qquad \text{Subtract 17.}$$

$$-k = -17 \qquad \text{Combine terms.}$$

This result gives the value of $-k$, but not of k itself. However, this result does say that the additive inverse of k is -17, which means that k must be 17, the same result we obtained in Example 2.

$$-k = -17$$

$$k = 17$$

This situation may be generalized as shown below.

> If a is a number and $-x = a$, then $x = -a$.

WORK PROBLEM 3 AT THE SIDE.

3 Sometimes an equation must be simplified as a first step in its solution.

EXAMPLE 3 *Using the Distributive Property to Simplify an Equation*

Solve $3(2 + 5x) - (1 + 14x) = 6$.

$$3(2 + 5x) - (1 + 14x) = 6$$

$$3(2) + 3(5x) - 1(1) - 1(14x) = 6 \qquad \text{Distributive property}$$

$$6 + 15x - 1 - 14x = 6 \qquad \text{Multiply.}$$

$$x + 5 = 6 \qquad \text{Combine terms.}$$

$$x = 1 \qquad \text{Subtract 5 from each side.}$$

Check by substituting 1 for x in the original equation. ■

WORK PROBLEM 4 AT THE SIDE.

3. Solve $\frac{7}{2}m + 1 = \frac{9}{2}m$.

4. Solve.

(a) $-(5 - 3r) + 4(-r + 1) = 1$

(b) $-3(m - 4) + 2(5 + 2m) = 29$

ANSWERS
3. 1
4. (a) -2 (b) 7

5. Solve each equation.

(a) $2(x - 6) = 2x - 12$

(b) $3x + 6(x + 1) = 9x - 4$

ANSWERS
5. (a) all real numbers
(b) no solution

4 The equations solved so far each have had exactly one solution. Sometimes this is not the case, as shown in the next examples.

EXAMPLE 4 *Solving an Equation That Has Infinitely Many Solutions*

Solve $5x - 15 = 5(x - 3)$.

$5x - 15 = 5(x - 3)$	
$5x - 15 = 5x - 15$	Distributive property
$5x - 15 + 15 = 5x - 15 + 15$	Add 15 to each side.
$5x = 5x$	Combine terms.
$5x - 5x = 5x - 5x$	Subtract $5x$ from each side.
$0 = 0$	

The final step leads us to an equation that contains no variables ($0 = 0$ in this case). Whenever such a statement is true, as it is in this example, *any* real number is a solution. (Try several replacements for x in the given equation to see that they all satisfy the equation.) We indicate the solution as "all real numbers." ■

Caution When you are solving an equation like the one in Example 4, do not write "0" as the solution. While 0 is a solution, there are infinitely many other solutions.

EXAMPLE 5 *Solving an Equation That Has No Solution*

Solve $2x + 3(x + 1) = 5x + 4$.

$2x + 3(x + 1) = 5x + 4$	
$2x + 3x + 3 = 5x + 4$	Distributive property
$5x + 3 = 5x + 4$	Combine terms.
$5x + 3 - 5x = 5x + 4 - 5x$	Subtract $5x$ from each side.
$3 = 4$	Combine terms.

Again, the variable has disappeared, but this time a *false* statement ($3 = 4$) results. Whenever this happens in solving an equation, it is a signal that the equation has no solution, and we write "no solution." ■

WORK PROBLEM 5 AT THE SIDE.

2.1 EXERCISES

NAME DATE HOUR

Solve each equation by using the addition property of equality. Check each solution. See Examples 1, 2, 4, and 5.

1. $x - 4 = 8$

2. $x - 8 = 9$

3. $7 + r = -3$

4. $8 + k = -4$

5. $t + 2.3 = 8.9$

6. $s + 4.1 = 12.6$

7. $x - 6.5 = -2.3$

8. $y - 5.5 = -1.2$

9. $\frac{9}{7}r - 3 = \frac{2}{7}r$

10. $\frac{8}{5}w - 6 = \frac{3}{5}w$

11. $5.6x + 2 = 4.6x$

12. $9.1x - 5 = 8.1x$

13. $3p + 6 = 10 + 2p$

14. $8b - 4 = -6 + 7b$

15. $1.2y - 4 = .2y - 4$

16. $7.7r + 6 = 6.7r + 6$

17. $3x + 9 = 3x + 8$

18. $-2x + 5 = -2x$

19. $8x + 1 = 1 + 8x$

20. $4w - 5 = -5 + 4w$

21. Refer to the definition of *linear equation* given in this section. Why is the restriction $A \neq 0$ necessary?

22. Which of the following equations are not linear equations?
(a) $x^2 - 5x + 6 = 0$ **(b)** $x^3 = x$ **(c)** $3x - 4 = 0$ **(d)** $7x - 6x = 3 + 9x$

Solve the following equations. First simplify each side of the equation as much as possible. Check each solution. See Examples 3, 4, and 5.

23. $5t + 3 + 2t - 6t = 4 + 12$

24. $4x + 3x - 6 - 6x = 10 + 3$

25. $10x + 5x + 7 - 8 = 12x + 3 + 2x$

26. $7p + 4p + 13 - 7 = 7p + 9 + 3p$

27. $6x + 5 - 7x + 3 = 5x - 6x - 4$

28. $4x - 3 - 8x + 1 = 5x - 9x + 7$

29. $5.2q - 4.6 - 7.1q = -2.1 - 1.9q - 2.5$

30. $-4.0x + 2.7 - 1.6x = 1.3 - 5.6x + 1.4$

31. $\frac{5}{7}x + \frac{1}{3} = \frac{2}{5} - \frac{2}{7}x + \frac{2}{5}$

32. $\frac{6}{7}s - \frac{3}{4} = \frac{4}{5} - \frac{1}{7}s + \frac{1}{6}$

33. $(5y + 6) - (3 + 4y) = 10$

34. $(8r - 3) - (7r + 1) = -6$

35. $2(p + 5) - (9 + p) = -3$

36. $4(k - 6) - (3k + 2) = -5$

37. $-6(2b + 1) + (13b - 7) = 0$

38. $-5(3w - 3) + (1 + 16w) = 0$

39. $10(-2x + 1) = -14(x + 2) + 38 - 6x$

40. $2(2 - 3r) = 5(1 - r) - r - 1$

41. $-2(8p + 2) - 3(2 - 7p) = 2(4 + 2p)$

42. $-5(1 - 2z) + 4(3 - z) = 7(3 + z)$

43. $4(7x - 1) + 3(2 - 5x) = 4(3x + 5) - 6$

44. $9(2m - 3) - 4(5 + 3m) = 5(4 + m) - 3$

45. In your own words, state how you would find the solution of a linear equation if your next-to-last step reads "$-x = 5$."

46. If the final step in solving a linear equation leads to the statement $0 = 0$, why is it incorrect to say that 0 is the solution of the equation? What are the solutions of the equation?

PREVIEW EXERCISES

Use the associative, commutative, inverse, and identity properties to simplify. See Section 1.8.

47. $4\left(\frac{1}{4}m\right)$

48. $-6\left(-\frac{1}{6}y\right)$

49. $\frac{2}{3}\left(\frac{3}{2}z\right)$

50. $\frac{.6x}{.6}$

By what number must the expression be multiplied to give a result of just x?

51. $7x$

52. $-9x$

53. $-\frac{9}{7}x$

54. $.2x$

2.2 THE MULTIPLICATION PROPERTY OF EQUALITY

The addition property of equality by itself is not enough to solve some equations, such as $3x + 2 = 17$.

$$3x + 2 = 17$$
$$3x + 2 - 2 = 17 - 2 \quad \text{Subtract 2 from each side.}$$
$$3x = 15 \quad \text{Simplify.}$$

Instead of just x on the left side, the equation has $3x$. Another property is needed to change $3x = 15$ to $x =$ a number.

1 If $3x = 15$, then $3x$ and 15 both represent the same number. Multiplying both $3x$ and 15 by the same number will also result in an equality. The **multiplication property of equality** states that we can multiply each side of an equation by the same number without changing the solution.

MULTIPLICATION PROPERTY OF EQUALITY

If A, B, and C ($C \neq 0$) represent real numbers, the equations

$$A = B \quad \text{and} \quad AC = BC$$

have exactly the same solution.

In other words, we can multiply each side of an equation by the same nonzero number without changing the solution.

This property can be used to solve $3x = 15$. The $3x$ on the left must be changed to $1x$, or x, instead of $3x$. Get x by multiplying each side of the equation by $\frac{1}{3}$. We use $\frac{1}{3}$ because $\frac{1}{3} \cdot 3 = \frac{3}{3} = 1$, since $\frac{1}{3}$ is the reciprocal of 3.

$$3x = 15$$
$$\frac{1}{3}(3x) = \frac{1}{3} \cdot 15 \quad \text{Multiply each side by } \tfrac{1}{3}.$$
$$\left(\frac{1}{3} \cdot 3\right)x = \frac{1}{3} \cdot 15 \quad \text{Associative property}$$
$$1x = 5 \quad \text{Multiplicative inverse property}$$
$$x = 5 \quad \text{Multiplicative identity property}$$

The solution of the equation is 5. We can check this result in the original equation. As in the previous section, we shall combine the last two steps shown in the example above.

WORK PROBLEM 1 AT THE SIDE.

Just as the addition property of equality permits *subtracting* the same number from each side of an equation, the multiplication property of equality permits *dividing* each side of an equation by the same nonzero number. For example, the equation $3x = 15$, which we just solved by multiplication, also could be solved by dividing each side by 3, as follows.

$$3x = 15$$
$$\frac{3x}{3} = \frac{15}{3} \quad \text{Divide by 3.}$$
$$x = 5$$

We can divide each side of an equation by the same nonzero number without changing the solution.

OBJECTIVES

1. Use the multiplication property of equality.
2. Use the multiplication property of equality to solve equations with decimals.
3. Simplify equations, and then use the multiplication property of equality.
4. Solve equations such as $-r = 4$.

FOR EXTRA HELP

Tape 3 | SSM pp.50–53 | MAC: A IBM: A

1. Check that 5 is the solution of $3x = 15$.

ANSWERS

1. Since $3(5) = 15$, the solution of $3x = 15$ is 5.

2. Solve.

(a) $-6p = -14$

(b) $3r = -12$

(c) $-2m = 16$

ANSWERS

2. (a) $\frac{7}{3}$ (b) -4 (c) -8

Note In practice, it is usually easier to multiply on each side if the coefficient of the variable is a fraction, and divide on each side if the coefficient is an integer. For example, to solve

$$-\frac{3}{4}x = 12$$

it is easier to multiply by $-\frac{4}{3}$ than to divide by $-\frac{3}{4}$. On the other hand, to solve

$$-5x = -20$$

it is easier to divide by -5 than to multiply by $-\frac{1}{5}$.

EXAMPLE 1 *Dividing Each Side of an Equation by a Nonzero Number*

Solve $25p = 30$.

Get p (instead of $25p$) on the left by using the multiplication property of equality. Divide each side of the equation by 25, the coefficient of p.

$$25p = 30$$

$$\frac{25p}{25} = \frac{30}{25} \qquad \text{Divide by 25.}$$

$$p = \frac{30}{25} = \frac{6}{5} \qquad \text{Reduce to lowest terms.}$$

To check, substitute $\frac{6}{5}$ for p in the given equation.

$$25p = 30$$

$$\frac{25}{1}\left(\frac{6}{5}\right) = 30 \qquad ? \quad \text{Let } p = \tfrac{6}{5}.$$

$$30 = 30 \qquad \text{True}$$

The solution is $\frac{6}{5}$. ■

WORK PROBLEM 2 AT THE SIDE.

In the next two examples, multiplication produces the solution more quickly than division would.

EXAMPLE 2 *Using the Multiplication Property of Equality*

Solve $\frac{a}{4} = 3$.

Replace $\frac{a}{4}$ by $\frac{1}{4}a$, since division by 4 is the same as multiplication by $\frac{1}{4}$. To get a alone on the left, multiply each side by 4, the reciprocal of the coefficient of a.

$$\frac{a}{4} = 3$$

$$\frac{1}{4}a = 3 \qquad \text{Change } \tfrac{a}{4} \text{ to } \tfrac{1}{4}a.$$

$$4 \cdot \frac{1}{4}a = 4 \cdot 3 \qquad \text{Multiply by 4.}$$

$$1a = 12 \qquad \text{Multiplicative inverse property}$$

$$a = 12 \qquad \text{Multiplicative identity property}$$

Check the answer.

$$\frac{a}{4} = 3 \qquad \text{Given equation}$$

$$\frac{12}{4} \stackrel{?}{=} 3 \qquad \text{Let } a = 12.$$

$$3 = 3 \qquad \text{True}$$

12 is the correct solution. ■

WORK PROBLEM 3 AT THE SIDE. ▶▶

3. Solve.

(a) $\frac{y}{5} = 5$

(b) $\frac{p}{4} = -6$

■ **EXAMPLE 3** *Using the Multiplication Property of Equality*

Solve $\frac{3}{4}h = 6$.

Get h alone on the left by multiplying each side of the equation by $\frac{4}{3}$. Use $\frac{4}{3}$ because $\frac{4}{3} \cdot \frac{3}{4}h = 1 \cdot h = h$.

$$\frac{3}{4}h = 6$$

$$\frac{4}{3}\left(\frac{3}{4}h\right) = \frac{4}{3} \cdot 6 \qquad \text{Multiply by } \tfrac{4}{3}.$$

$$1 \cdot h = \frac{4}{3} \cdot \frac{6}{1} \qquad \text{Multiplicative inverse property}$$

$$h = 8 \qquad \text{Multiplicative identity property}$$

The solution is 8. Check the answer by substitution in the given equation. ■

WORK PROBLEM 4 AT THE SIDE. ▶▶

4. Solve.

(a) $-\frac{5}{6}t = -15$

(b) $\frac{3}{4}k = -21$

2 In the next example, we solve an equation with decimals.

■ **EXAMPLE 4** *Solving an Equation with Decimals*

Solve $2.1x = 6.09$.

Divide both sides by 2.1.

$$\frac{2.1x}{2.1} = \frac{6.09}{2.1}$$

You may use a calculator to simplify the work at this point.

$$1x = 2.9 \qquad \text{Divide.}$$

$$x = 2.9 \qquad \text{Multiplicative identity property}$$

Check that the solution is 2.9. ■

WORK PROBLEM 5 AT THE SIDE. ▶▶

5. Solve.

(a) $-.7m = -5.04$

(b) $12.5k = -63.75$

3 In the next example, it is necessary to simplify the equation before using the multiplication property of equality.

ANSWERS

3. (a) 25 (b) −24

4. (a) 18 (b) −28

5. (a) 7.2 (b) −5.1

6. Solve.

(a) $4r - 9r = 20$

(b) $7m - 5m = -12$

7. Solve.

(a) $-m = 2$

(b) $-p = -7$

ANSWERS
6. (a) −4 (b) −6
7. (a) −2 (b) 7

■ EXAMPLE 5 *Simplifying Terms in an Equation*

Solve $5m + 6m = 33$.

$$5m + 6m = 33$$

$11m = 33$	Combine terms.
$\dfrac{11m}{11} = \dfrac{33}{11}$	Divide by 11.
$1m = 3$	Divide.
$m = 3$	Multiplicative identity property

The solution is 3. Check this solution. ■

WORK PROBLEM 6 AT THE SIDE.

4 The following example shows how to use the multiplication property of equality to solve equations such as $-r = 4$.

■ EXAMPLE 6 *Using −1 with the Multiplication Property of Equality*

Solve $-r = 4$.

On the left side, change $-r$ to r by first writing $-r$ as $-1 \cdot r$.

$-r = 4$	
$-1 \cdot r = 4$	$-r = -1 \cdot r$
$-1(-1 \cdot r) = -1 \cdot 4$	Multiply by -1, since $-1 \cdot -1 = 1$.
$[(-1)(-1)] \cdot r = -4$	Associative property
$1 \cdot r = -4$	Multiplicative inverse property
$r = -4$	Multiplicative identity property

Check this solution.

$-r = 4$	Given equation
$-(-4) = 4$?	Let $r = -4$.
$4 = 4$	True

The solution, −4, checks. ■

WORK PROBLEM 7 AT THE SIDE.

2.2 EXERCISES

NAME DATE HOUR

By what number is it necessary to multiply both sides of the equation in order to obtain just x on the left side? Do not actually solve these equations.

1. $\frac{2}{3}x = 9$

2. $\frac{4}{5}x = 7$

3. $.1x = 2$

4. $.01x = 7$

5. $-\frac{9}{2}x = -3$

6. $-\frac{8}{3}x = -10$

7. $-x = .45$

8. $-x = .23$

By what number is it necessary to divide both sides of the equation in order to obtain just x on the left side? Do not actually solve these equations.

9. $6x = 4$

10. $7x = 9$

11. $-4x = 10$

12. $-13x = 5$

13. $.12x = 36$

14. $.21x = 42$

15. $-x = 32$

16. $-x = 94$

Solve each equation and check your solution. See Examples 1–6.

17. $5x = 30$

18. $7x = 56$

19. $2m = 15$

20. $3m = 10$

21. $3a = -15$

22. $5k = -70$

23. $10t = -36$

24. $4s = -34$

25. $-6x = -72$

26. $-8x = -64$

27. $2r = 0$

28. $5x = 0$

29. $\frac{1}{4}y = -12$

30. $\frac{1}{5}p = -3$

31. $-y = 12$

32. $-t = 14$

33. $-x = -\frac{4}{7}$

34. $-m = -\frac{9}{5}$

35. $.2t = 8$

36. $.9x = 18$

37. $4x + 3x = 21$

38. $9x + 2x = 121$

39. $5m + 6m - 2m = 63$

40. $11r - 5r + 6r = 168$

41. $3r - 5r = 10$

42. $9p - 13p = 24$

43. $\frac{x}{7} = -5$

44. $\frac{k}{8} = -3$

45. $\frac{2}{3}t = 6$

46. $\frac{4}{3}m = 24$

47. $-\frac{2}{7}p = -5$

48. $-\frac{3}{8}y = -2$

49. $-\frac{7}{9}c = \frac{3}{5}$

50. $-\frac{5}{6}d = \frac{4}{9}$

51. $-2.1m = 25.62$

52. $-3.9a = -31.2$

53. Write an equation that requires the use of the multiplication property of equality, where both sides must be multiplied by $\frac{2}{3}$, and the solution is a negative number.

54. Write an equation that requires the use of the multiplication property of equality, where both sides must be divided by 100, and the solution is not an integer.

PREVIEW EXERCISES

Simplify each expression. See Section 1.9.

55. $8(3q + 4)$

56. $6(2m - 6)$

57. $-7(4p - 3) + 8$

58. $-(3 - 9r) + 12r$

59. $6 - 7(2 - 8p)$

60. $9(6 + 4y) - 3(3 - 2y)$

2.3 MORE ON SOLVING LINEAR EQUATIONS

OBJECTIVES

1 Learn the four steps for solving a linear equation and how to use them.

2 Solve equations by clearing fractions and decimals.

FOR EXTRA HELP

Tape 3

SSM pp. 53–57

MAC: A IBM: A

1 In this section we use the addition and multiplication properties together to solve more complicated equations. We will use the following four-step method.

SOLVING LINEAR EQUATIONS

Step 1 Clear parentheses using the distributive property, if needed; combine terms.

Step 2 Use the addition property to simplify further if necessary, so that the variable term is on one side of the equation and a number is on the other.

Step 3 Use the multiplication property if necessary to get the equation in the form $x =$ a number.

Step 4 Check your answer by substituting into the *original* equation.

The check is used only to catch errors in carrying out the steps.

EXAMPLE 1 *Using the Four Steps to Solve an Equation*

Solve the equation $3r + 4 - 2r - 7 = 4r + 3$.

We use the four steps described above.

Step 1

$$3r + 4 - 2r - 7 = 4r + 3$$

$$r - 3 = 4r + 3 \qquad \text{Combine like terms.}$$

Step 2

$$r - 3 + 3 = 4r + 3 + 3 \qquad \text{Add 3.}$$

$$r = 4r + 6$$

$$r - 4r = 4r + 6 - 4r \qquad \text{Subtract } 4r.$$

$$-3r = 6 \qquad \text{Combine terms.}$$

Step 3

$$\frac{-3r}{-3} = \frac{6}{-3} \qquad \text{Divide by } -3.$$

$$r = -2 \qquad \text{Reduce.}$$

Step 4 Substitute -2 for r in the original equation.

$$3r + 4 - 2r - 7 = 4r + 3$$

$$3(-2) + 4 - 2(-2) - 7 = 4(-2) + 3 \qquad ? \quad \text{Let } r = -2.$$

$$-6 + 4 + 4 - 7 = -8 + 3 \qquad ? \quad \text{Multiply.}$$

$$-5 = -5 \qquad \text{True}$$

The solution of the given equation is -2. ■

In Step 2 of Example 1, we added and subtracted the terms in such a way that the variable term ended up on the left side of the equation. Choosing differently would have put the variable term on the right side of the equation. Usually there is no real advantage either way.

WORK PROBLEM 1 AT THE SIDE. ▶▶

1. Solve.

(a) $5y - 7y + 6y - 9 = 3 + 2y$

(b) $-3k - 5k - 6 + 11 = 2k - 5$

ANSWERS

1. (a) 6 (b) 1

2. Solve.

(a) $7(p - 2) + p = 2p + 4$

(b) $3(m + 5) - 1 + 2m$
$= 5(m + 2)$

EXAMPLE 2 *Using the Four Steps to Solve an Equation*

Solve the equation $4(k - 3) - k = k - 6$.

Step 1

$$4(k - 3) - k = k - 6$$

$$4k - 12 - k = k - 6 \quad \text{Distributive property}$$

$$3k - 12 = k - 6 \quad \text{Combine terms.}$$

Step 2

$$3k - 12 + 12 = k - 6 + 12 \quad \text{Add 12.}$$

$$3k = k + 6 \quad \text{Combine terms.}$$

$$3k - k = k + 6 - k \quad \text{Subtract } k.$$

$$2k = 6 \quad \text{Combine terms.}$$

Step 3

$$\frac{2k}{2} = \frac{6}{2} \quad \text{Divide by 2.}$$

$$k = 3 \quad \text{Reduce.}$$

Step 4 Check this answer by substituting 3 for k in the given equation. Remember to do all the work inside the parentheses first.

$$4(k - 3) - k = k - 6$$

$$4(3 - 3) - 3 = 3 - 6 \quad ? \quad \text{Let } k = 3.$$

$$4(0) - 3 = 3 - 6 \quad ? \quad 3 - 3 = 0$$

$$0 - 3 = 3 - 6 \quad ? \quad 4(0) = 0$$

$$-3 = -3 \quad \text{True}$$

The solution of the equation is 3. ■

WORK PROBLEM 2 AT THE SIDE.

EXAMPLE 3 *Using the Four Steps to Solve an Equation*

Solve the equation $8a - (3 + 2a) = 3a + 1$.

Step 1 Simplify.

$$8a - (3 + 2a) = 3a + 1$$

$$8a - 1 \cdot (3 + 2a) = 3a + 1 \quad \text{Multiplicative identity property}$$

$$8a - 3 - 2a = 3a + 1 \quad \text{Distributive property}$$

$$6a - 3 = 3a + 1 \quad \text{Combine terms.}$$

Step 2 First, add 3 to each side; then subtract $3a$.

$$6a - 3 + 3 = 3a + 1 + 3 \quad \text{Add 3.}$$

$$6a = 3a + 4 \quad \text{Combine terms.}$$

$$6a - 3a = 3a + 4 - 3a \quad \text{Subtract } 3a.$$

$$3a = 4 \quad \text{Combine terms.}$$

Step 3

$$\frac{3a}{3} = \frac{4}{3} \quad \text{Divide by 3.}$$

$$a = \frac{4}{3} \quad \tfrac{3}{3} = 1;\ 1a = a$$

Step 4 Check that the solution is $\frac{4}{3}$. ■

ANSWERS
2. (a) 3 (b) no solution

Caution Be very careful with signs when solving an equation like the one in Example 3. When clearing parentheses in the expression

$$8a - (3 + 2a),$$

remember that the $-$ sign acts like a factor of -1, changing the sign of *every* term in the parentheses. Thus,

$$8 - (3 + 2a) = 8 - 3 - 2a.$$

Change to $-$ in both terms.

WORK PROBLEM 3 AT THE SIDE.

■ EXAMPLE 4 *Using the Four Steps to Solve an Equation*

Solve the equation $4(8 - 3t) = 32 - 8(t + 2)$.

Step 1

$$4(8 - 3t) = 32 - 8(t + 2)$$

$$32 - 12t = 32 - 8t - 16 \quad \text{Distributive property}$$

$$32 - 12t = 16 - 8t \quad \text{Combine terms.}$$

Step 2

$$32 - 12t - 32 = 16 - 8t - 32 \quad \text{Subtract 32.}$$

$$-12t = -16 - 8t \quad \text{Combine terms.}$$

$$-12t + 8t = -16 - 8t + 8t \quad \text{Add } 8t.$$

$$-4t = -16 \quad \text{Combine terms.}$$

Step 3

$$\frac{-4t}{-4} = \frac{-16}{-4} \quad \text{Divide by } -4.$$

$$t = 4 \quad \text{Reduce.}$$

Step 4 Check this solution in the given equation.

$$4(8 - 3t) = 32 - 8(t + 2)$$

$$4(8 - 3 \cdot 4) = 32 - 8(4 + 2) \quad ? \quad \text{Let } t = 4.$$

$$4(8 - 12) = 32 - 8(6) \quad ? \quad \text{Combine terms.}$$

$$4(-4) = 32 - 48 \quad ? \quad \text{Combine terms.}$$

$$-16 = -16 \quad \text{True}$$

The solution, 4, checks. ■

WORK PROBLEM 4 AT THE SIDE.

2 We can clear an equation of fractions by multiplying both sides by the LCD of all denominators in the equation. It is a good idea to do this before starting the four-step method to avoid working with fractions.

■ EXAMPLE 5 *Clearing an Equation of Fractions*

Solve $\frac{2}{3}x - \frac{1}{2}x = -\frac{1}{6}x - 2$.

The least common denominator of all the fractions in the equation is 6. Start by multiplying both sides of the equation by 6.

$$\frac{2}{3}x - \frac{1}{2}x = -\frac{1}{6}x - 2$$

$$6\left(\frac{2}{3}x - \frac{1}{2}x\right) = 6\left(-\frac{1}{6}x - 2\right) \quad \text{Multiply by 6.}$$

$$6\left(\frac{2}{3}x\right) + 6\left(-\frac{1}{2}x\right) = 6\left(-\frac{1}{6}x\right) + 6(-2) \quad \text{Distributive property}$$

$$4x - 3x = -x - 12$$

3. Solve.

(a) $7m - (2m - 9) = 39$

(b) $4x + 2(3 - 2x) = 6$

4. Solve.

(a) $2(4 + 3r) = 3(r + 1) + 11$

(b) $2 - 3(2 + 6z) = 4(z + 1) + 18$

ANSWERS

3. (a) 6 **(b)** all real numbers

4. (a) 2 **(b)** $-\frac{13}{11}$

5. Solve $\frac{1}{4}x - 4 = \frac{3}{2}x + \frac{3}{4}x$.

Now use the four steps to solve this equivalent equation.

Step 1 $\quad x = -x - 12 \quad$ Combine like terms.

Step 2 $\quad x + x = x - x - 12 \quad$ Add x.

$$2x = -12$$

Step 3 $\quad \frac{2x}{2} = \frac{-12}{2} \quad$ Divide by 2.

$$x = -6$$

Step 4 Check by substituting -6 for x in the original equation.

$$\frac{2}{3}(-6) - \frac{1}{2}(-6) = -\frac{1}{6}(-6) - 2 \quad ? \quad \text{Let } x = -6.$$

$$-4 + 3 = 1 - 2 \quad ?$$

$$-1 = -1 \quad \text{True}$$

The solution of the equation is -6. ■

Caution When clearing equations of fractions, be sure you multiply *every* term on both sides of the equation by the LCD.

WORK PROBLEM 5 AT THE SIDE.

The multiplication property can also be used to clear an equation of decimals.

6. Solve $.06(100 - y) + .04y = .05(92)$.

■ **EXAMPLE 6** *Clearing an Equation of Decimals*

Solve $.10t + .05(20 - t) = .09(20)$.

Since the decimals are all hundredths, start the solution by multiplying both sides of the equation by 100. A number can be multiplied by 100 by moving the decimal point two places to the right.

$$.10t + .05(20 - t) = .09(20)$$

$$10t + 5(20 - t) = 9(20)$$

Now use the four steps.

Step 1 $\quad 10t + 5(20) + 5(-t) = 180 \quad$ Distributive property

$$10t + 100 - 5t = 180$$

$$5t + 100 = 180 \quad \text{Combine terms.}$$

Step 2 $\quad 5t + 100 - 100 = 180 - 100 \quad$ Subtract 100.

$$5t = 80 \quad \text{Combine terms.}$$

Step 3 $\quad \frac{5t}{5} = \frac{80}{5} \quad$ Divide by 5.

$$t = 16$$

Step 4 Check to see that 16 is the solution by substituting into the original equation. ■

WORK PROBLEM 6 AT THE SIDE.

ANSWERS
5. -2
6. 70

2.3 EXERCISES

NAME DATE HOUR

Solve each equation, and check your solution. See Examples 1–4.

1. $2h + 4 = 8$

2. $3x + 5 = 17$

3. $5m + 8 = 7 + 4m$

4. $4r + 2 = 3r - 6$

5. $10p + 6 = 12p - 4$

6. $-5x + 8 = -3x + 10$

7. $x + 3 = -(2x + 2)$

8. $2x + 1 = -(x + 3)$

9. $4(2x - 1) = -6(x + 3)$

10. $6(3w + 5) = 2(10w + 10)$

11. $6(4x - 1) = 12(2x + 3)$

12. $6(2x + 8) = 4(3x - 6)$

13. $3(2x - 4) = 6(x - 2)$

14. $3(6 - 4x) = 2(-6x + 9)$

15. $7r - 5r + 2 = 5r - r$

16. $9p - 4p + 6 = 7p - 3p$

17. After working correctly through several steps of the solution of a linear equation, a student obtains the equation $7x = 3x$. Then the student divides both sides by x to get $7 = 3$ and gives "no solution" as the answer. Is this correct? If not, explain why.

18. Which one of the following linear equations does *not* have all real numbers as solutions?

(a) $5x = 4x + x$ **(b)** $2(x + 6) = 2x + 12$ **(c)** $\frac{1}{2}x = .5x$ **(d)** $3x = 2x$

Solve each equation by first clearing it of fractions or decimals. See Examples 5 and 6.

19. $\frac{3}{5}t - \frac{1}{10}t = t - \frac{5}{2}$

20. $-\frac{2}{7}r + 2r = \frac{1}{2}r + \frac{17}{2}$

21. $-\frac{1}{4}(x - 12) + \frac{1}{2}(x + 2) = x + 4$

22. $\frac{1}{9}(y + 18) + \frac{1}{3}(2y + 3) = y + 3$

23. $\frac{2}{3}k - \left(k + \frac{1}{4}\right) = \frac{1}{12}(k + 4)$

24. $-\frac{5}{6}q - \left(q - \frac{1}{2}\right) = \frac{1}{4}(q + 1)$

25. $.20(60) + .05x = .10(60 + x)$

26. $.30(30) + .15x = .20(30 + x)$

27. $1.00x + .05(12 - x) = .10(63)$

28. $.92x + .98(12 - x) = .96(12)$

29. $.06(10{,}000) + .08x = .072(10{,}000 + x)$

30. $.02(5000) + .03x = .025(5000 + x)$

Solve each equation, and check your solution. See Examples 1–6.

31. $10(2x - 1) = 8(2x + 1) + 14$

32. $9(3k - 5) = 12(3k - 1) - 51$

33. $-2(2s - 4) - 8 = -3(4s + 4) - 1$

34. $-3(5z + 24) + 2 = 2(3 - 2z) - 4$

35. $-(4y + 2) - (-3y - 5) = 3$

36. $-(6k - 5) - (-5k + 8) = -3$

37. $\frac{1}{2}(x + 2) + \frac{3}{4}(x + 4) = x + 5$

38. $\frac{1}{3}(x + 3) + \frac{1}{6}(x - 6) = x + 3$

39. $.10(x + 80) + .20x = 14$

40. $.30(x + 15) + .40(x + 25) = 25$

41. $4(x + 8) = 2(2x + 6) + 20$

42. $4(x + 3) = 2(2x + 8) - 4$

43. $9(v + 1) - 3v = 2(3v + 1) - 8$

44. $8(t - 3) + 4t = 6(2t + 1) - 10$

45. Explain in your own words the major steps used in solving a linear equation that does not contain fractions or decimals as coefficients.

46. Explain in your own words the major steps used in solving a linear equation that contains fractions or decimals as coefficients.

PREVIEW EXERCISES

Write each of the following as a mathematical expression, using x as the variable. See Sections 1.4, 1.5, 1.6, and 1.7.

47. A number added to -6

48. A number decreased by 9

49. The sum of a number and twice the number

50. The difference between -5 and a number

51. The quotient of -6 and a nonzero number

52. A number divided by 17

53. The product of 12 and the difference between a number and 9

54. The quotient of 9 more than a number and 6 less than the number

2.4 AN INTRODUCTION TO APPLIED PROBLEMS

OBJECTIVE

1 Translate sentences of an applied problem into an equation, and then solve the problem.

FOR EXTRA HELP

Tape 3 | SSM pp. 57–61 | MAC: A IBM: A

Earlier, we practiced translating words, phrases, and sentences into mathematical expressions and equations. Now we will begin to use these translations in solving applied problems using algebra. Some of the problems will seem contrived, and to some extent they are. But the skills you develop in solving simple problems will help you in solving more realistic problems in chemistry, biology, business, and other fields. We suggest the following general procedure for solving applied problems.

SOLVING APPLIED PROBLEMS

Step 1 Read the problem carefully, and choose a variable to represent the numerical value that you are asked to find—the unknown number. *Write down* what the variable represents.

Step 2 *Write down* a mathematical expression using the variable for any other unknown quantities. Draw figures or diagrams if they apply.

Step 3 Translate the problem into an equation.

Step 4 Solve the equation.

Step 5 Answer the question asked in the problem.

Step 6 Check your solution by using the words of the original problem. Be sure that the answer is appropriate and makes sense.

1 The third step in solving an applied problem is often the hardest. Begin to translate the problem into an equation by writing the given phrases as mathematical expressions. Because equal mathematical expressions are names for the same number, replace any words that mean *equal* or *same* with an = sign. Other forms of the verb "to be," such as *is, are, was,* and *were,* also translate this way. The = sign leads to an equation to be solved.

EXAMPLE 1 *Finding an Unknown Number*

If three times the sum of a number and 4 is decreased by twice the number, the result is -6. Find the number.

Step 1 Read the problem carefully. We are asked to find a number, so we write

Let x represent the number.

Step 2 $3(x + 4)$ = three times the sum of the number and 4

$2x$ = twice the number.

Step 3 Translate the information given in the problem.

Three times the sum of a number and 4	decreased by	twice the number	is	-6.
↓	↓	↓	↓	↓
$3(x + 4)$	$-$	$2x$	$=$	-6

Step 4 Solve the equation.

$$3(x + 4) - 2x = -6$$

$$3x + 12 - 2x = -6 \quad \text{Distributive property}$$

$$x + 12 = -6 \quad \text{Combine terms.}$$

$$x = -18 \quad \text{Subtract 12.}$$

1. Write an equation, and then solve the problem. Use x as the variable. If 5 is added to the product of 9 and a number, the result is 19 less than the number. Find the number.

2. Solve the problem. On one day of their vacation, Annie drove three times as far as Jim. Altogether they drove 84 miles that day. Find the number of miles driven by each.

ANSWERS
1. $9x + 5 = x - 19$; -3
2. 21 miles for Jim; 63 miles for Annie

Step 5 The number is -18.

Step 6 Check that -18 is the correct answer by substituting this result into the words of the original problem. Three times the sum of -18 and 4 is $3(-18 + 4) = 3(-14) = -42$. Twice -18 is -36; subtract -36 from -42 to get $-42 - (-36) = -6$, as required. ■

WORK PROBLEM 1 AT THE SIDE.

See if you can tell how the six steps are used in the rest of the examples.

■ **EXAMPLE 2** *Finding the Number of Coffee Orders*

The owner of a small café found one day that the number of orders for tea was $\frac{1}{3}$ the number of orders for coffee. If the total number of orders for the two drinks was 76, how many orders were placed for coffee?

$$\text{Let} \quad x = \text{the number of orders for coffee}$$
$$\frac{1}{3}x = \text{the number of orders for tea.}$$

To set up an equation, use the fact that the total number of orders was 76.

The total	is	orders for coffee	plus	orders for tea.
↓	↓	↓	↓	↓
76	$=$	x	$+$	$\frac{1}{3}x$

Now solve the equation.

$$76 = \frac{4}{3}x \qquad \text{Combine terms.}$$
$$\frac{3}{4}(76) = \frac{3}{4}\left(\frac{4}{3}x\right) \qquad \text{Multiply by } \tfrac{3}{4}.$$
$$57 = x \qquad \text{Combine terms.}$$

There were 57 orders for coffee. The 57 coffee orders and the $(\frac{1}{3})(57) = 19$ orders for tea give a total of $57 + 19 = 76$ orders, as required. ■

WORK PROBLEM 2 AT THE SIDE.

Sometimes it is necessary to find three unknown quantities in an applied problem. Frequently the three unknowns are compared in *pairs*. When this happens, it is usually easiest to let the variable represent the unknown found in both pairs. The next example illustrates this idea.

■ **EXAMPLE 3** *Dividing a Board into Pieces*

Maria Gonzales is building a cabinet. The instructions require three pieces of shelving. The longest piece must be twice the length of the middle-sized piece, and the shortest piece must be 10 inches shorter than the middle-sized piece. She has a piece of shelving 70 inches long. How long will each of the three pieces be if she cuts them from this 70-inch piece?

Since the middle-sized piece appears in both pairs of comparisons, let x represent the length of the middle-sized piece. We have

$$x = \text{the length of the middle-sized piece}$$
$$2x = \text{the length of the longest piece}$$
$$x - 10 = \text{the length of the shortest piece.}$$

A sketch is helpful here.

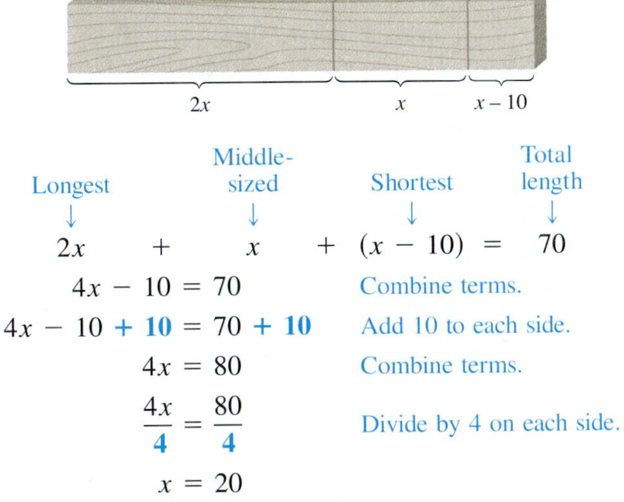

$$\underset{\text{Longest}}{2x} + \underset{\text{Middle-sized}}{x} + \underset{\text{Shortest}}{(x - 10)} = \underset{\text{Total length}}{70}$$

$$4x - 10 = 70 \qquad \text{Combine terms.}$$
$$4x - 10 + 10 = 70 + 10 \qquad \text{Add 10 to each side.}$$
$$4x = 80 \qquad \text{Combine terms.}$$
$$\frac{4x}{4} = \frac{80}{4} \qquad \text{Divide by 4 on each side.}$$
$$x = 20$$

The length of the middle-sized piece is 20 inches, the length of the longest piece is $2(20) = 40$ inches, and the length of the shortest piece is $20 - 10 = 10$ inches. Check to see that the sum of the three lengths is 70 inches. ■

WORK PROBLEM 3 AT THE SIDE.

3. Solve the problem. A piece of pipe is 50 inches long. It is cut into three pieces. The longest piece is 10 inches more than the middle-sized piece, and the shortest piece measures 5 inches less than the middle-sized piece. Find the lengths of the three pieces.

■ EXAMPLE 4 *Analyzing a Gasoline/Oil Mixture*

A lawn trimmer uses a mixture of gasoline and oil. For each ounce of oil, the mixture contains 16 ounces of gasoline. If the tank holds 68 ounces of the mixture, how many ounces of oil and how many ounces of gasoline does it require when it is full?

Let x = the number of ounces of oil required when full

$16x$ = the number of ounces of gasoline required when full.

ANSWERS
3. 25 inches, 15 inches, 10 inches

4. Solve the problem. At a meeting of the local stamp club, each member brought two non-members. If a total of 27 people attended, how many were members and how many were non-members?

Amount of gasoline		Amount of oil		Total amount in tank
↓		↓		↓
$16x$	$+$	x	$=$	68

$$17x = 68 \quad \text{Combine terms.}$$

$$\frac{17x}{17} = \frac{68}{17} \quad \text{Divide by 17.}$$

$$x = 4 \quad \text{Reduce.}$$

When the tank is full, it holds 4 ounces of oil and $16(4) = 64$ ounces of gasoline. This checks, since $4 + 64 = 68$. ■

WORK PROBLEM 4 AT THE SIDE.

The final example of problem solving in this section deals with concepts from geometry. An angle can be measured by a unit called the **degree** (°). See Figure 1. Two angles whose sum is 90° are said to be **complementary,** or complements of each other. Two angles whose sum is 180° are said to be **supplementary,** or supplements of each other. If x represents the degree measure of an angle, then

$90 - x$ represents the degree measure of its complement, and

$180 - x$ represents the degree measure of its supplement.

FIGURE 1

5. Twice the complement of an angle is 30° less than its supplement. Find the measure of the angle.

■ **EXAMPLE 5** *Finding the Measure of an Angle*

Find the measure of an angle whose supplement is 10 degrees more than twice its complement.

Let x = the degree measure of the angle;

$90 - x$ = the degree measure of its complement;

$180 - x$ = the degree measure of its supplement.

Supplement	is	10	more than	twice		its complement.
↓	↓	↓	↓	↓		↓
$180 - x$	$=$	10	$+$	2	$\cdot$	$(90 - x)$

Solve the equation.

$$180 - x = 10 + 180 - 2x \quad \text{Distributive property}$$

$$180 - x = 190 - 2x \quad \text{Combine terms.}$$

$$180 - x + 2x = 190 - 2x + 2x \quad \text{Add } 2x.$$

$$180 + x = 190 \quad \text{Combine terms.}$$

$$180 + x - 180 = 190 - 180 \quad \text{Subtract 180.}$$

$$x = 10$$

The measure of the angle is 10 degrees. The complement of 10° is 80° and the supplement of 10° is 170°. 170° is equal to 10° more than twice 80° ($170 = 10 + 2(80)$ is true); therefore, the answer is correct. ■

WORK PROBLEM 5 AT THE SIDE.

ANSWERS

4. Nine were members and 18 were nonmembers.

5. 30°

2.4 EXERCISES

NAME DATE HOUR

1. Which one of the following would not be a reasonable answer in an applied problem that requires finding the number of coins in a jar?

   (a) 7 **(b)** 0 **(c)** $6\frac{2}{3}$ **(d)** 80

2. Which one of the following would not be a reasonable answer in an applied problem that requires finding someone's age (in years)?

   (a) $5\frac{1}{2}$ **(b)** 7 **(c)** 12.25 **(d)** -4

3. Explain in your own words the general procedure for solving applied problems described in this section.

4. List some words that will translate as "=" in an applied problem.

Solve the following problems. See Example 1.

5. If 1 is added to a number and this sum is doubled, the result is 5 more than the number. Find the number.

6. If 2 is subtracted from a number and this difference is tripled, the result is 4 more than the number. Find the number.

7. If 3 is added to twice a number and this sum is multiplied by 4, the result is the same as if the number is multiplied by 7 and 8 is added to the product. What is the number?

8. The sum of three times a number and 12 more than the number is the same as the difference between -6 and twice the number. What is the number?

Solve the following problems. See Examples 2 and 3.

9. The U.S. Senate has 100 members. After the 1990 election, there were 14 more Democrats than Republicans, with no other parties represented. How many members of each party were there in the Senate?

10. The total number of Democrats and Republicans in the U.S. House of Representatives in 1990 was 434. There were 100 fewer Republicans than Democrats. How many members of each party were there in the House of Representatives?

11. In his coaching career with the Boston Celtics, Red Auerbach had 558 more wins than losses. His total number of games coached was 1516. How many wins did Auerbach have?

12. In the first Super Bowl, played in 1966, Green Bay and Kansas City scored a total of 45 points. Green Bay won by 25 points. What was the score of the first Super Bowl?

13. Nagaraj Nanjappa has a strip of paper 39 inches long. He wants to cut it into two pieces so that one piece will be 9 inches shorter than the other. How long should each of the two pieces be?

14. On Professor Brandsma's algebra test, the highest grade was 34 points higher than the lowest grade. The sum of the two grades was 160. What were the highest and lowest grades?

15. In one day, Gwen Boyle received 13 packages. Federal Express delivered three times as many as Airborne Express, while Airborne Express delivered two more than United Parcel Service. How many packages did each service deliver to Gwen?

16. In her job at the post office, Janie Quintana works a $6\frac{1}{2}$-hour day. She sorts mail, sells stamps, and does supervisory work. On one day, she sold stamps twice as long as she sorted mail, and sold stamps 1 hour longer than the time she spent doing supervisory work. How many hours did she spend at each task?

17. Venus is 31.2 million miles farther from the sun than Mercury, while Earth is 25.7 million miles farther from the sun than Venus. If the total of the distances for these three planets from the sun is 196.1 million miles, how far away from the sun is Mercury? (All distances given here are *mean* (*average*) distances.)

18. It is believed that Saturn has 5 more satellites (moons) than the known number of satellites for Jupiter, and 20 more satellites than the known number for Mars. If the total of these numbers is 41, how many satellites does Mars have?

19. During their National League championship year of 1992, Atlanta Braves hitters Ron Gant, David Justice, and Francisco Cabrera combined for a total of 268 base hits. Gant had 47 times as many hits as Cabrera, while Justice had 17 fewer hits than Gant. How many hits did each player have?

20. During their World Series championship year of 1992, Toronto Blue Jays pitchers Duane Ward, David Wells, and Jack Morris combined to pitch 462 innings. Together, Morris and Ward pitched 342 innings. Wells pitched $18\frac{2}{3}$ more innings than Ward. How many innings did each player pitch?

21. The sum of the measures of the angles of any triangle is 180 degrees. In triangle *ABC*, angles *A* and *B* have the same measure, while the measure of angle *C* is 60 degrees larger than each of *A* and *B*. What are the measures of the three angles?

22. (See Exercise 21.) In triangle *ABC*, the measure of angle *A* is 141 degrees more than the measure of angle *B*. The measure of angle *B* is the same as the measure of angle *C*. Find the measure of each angle.

Solve the following problems. See Example 4.

23. On the 1992 Eagle Premier, the suggested list price for the antilock brake system is $\frac{10}{3}$ of the suggested list price of power door locks. Together, these two options cost \$1040. What is the suggested list price for each of these options?

24. The 1993 edition of *A Guide Book of United States Coins* lists the value of a Mint State-65 (uncirculated) 1950 Jefferson nickel minted at Denver as $\frac{6}{5}$ the value of a similar condition 1944 nickel minted at Denver. Together the total value of the two coins is \$22.00. What is the value of each coin?

25. An insecticide contains 95 centigrams of inert ingredient for every 1 centigram of active ingredient. If a quantity of the insecticide weighs 672 centigrams, how much of each type of ingredient does it contain?

26. In a mixture of concrete, there are 3 pounds of cement mix for every 1 pound of gravel. If the mixture contains a total of 840 pounds of these two ingredients, how many pounds of gravel are there?

Solve the following problems. See Example 5.

27. Find the measure of an angle whose supplement measures 10 times the measure of its complement.

28. Find the measure of an angle whose supplement measures 4 times the measure of its complement.

29. Find the measure of an angle whose supplement measures 38° less than three times its complement.

30. Find the measure of an angle whose supplement measures 39° more than twice its complement.

31. Find the measure of an angle such that the sum of the measures of its complement and its supplement is 160°.

32. Find the measure of an angle such that the difference between the measures of its supplement and three times its complement is 10°.

The following problems are a bit different from those described in the examples of this section. Solve these using the general method described. (In Exercises 33–36, consecutive integers *and* consecutive even integers *are mentioned. Some examples of consecutive integers are 5, 6, 7, and 8; some examples of consecutive even integers are 4, 6, 8, and 10.)*

33. If x represents an integer, then $x + 1$ represents the next larger consecutive integer. If the sum of two consecutive integers is 243, find the integers.

34. (See Exercise 33.) If the sum of two consecutive integers is -29, find the integers.

35. If x represents an even integer, then $x + 2$ represents the next larger even integer. Find two consecutive even integers such that the smaller added to three times the larger equals 86.

36. (See Exercise 35.) Find two consecutive even integers such that six times the smaller added to the larger gives a sum of 58.

PREVIEW EXERCISES

Use the given values to evaluate each expression. See Section 1.2.

37. LW; $L = 6$, $W = 4$

38. rt; $r = 25$, $t = 4.5$

39. prt; $p = 4000$, $r = .04$, $t = 2$

40. $\frac{1}{2}Bh$; $B = 27$, $h = 6$

2.5 FORMULAS AND GEOMETRY APPLICATIONS

Many applied problems can be solved with a formula. For example, formulas exist for geometric figures such as squares and circles, for distance, for money earned on bank savings, and for converting English measurements to metric measurements. The formulas used in this book are given in the endsheets.

OBJECTIVES

1. Solve a formula for one variable, given the values of the other variables.
2. Use a formula to solve an applied problem.
3. Solve problems about vertical angles and straight angles.
4. Solve a formula for a specified variable.

FOR EXTRA HELP

Tape 4	SSM pp. 62–65	MAC: A IBM: A

1 Given the values of all but one of the variables in a formula, we can find the value of the remaining variable by using the methods introduced in this chapter.

EXAMPLE 1 *Using a Formula to Evaluate a Variable*
Find the value of the remaining variable in each of the following.

(a) $A = LW$; $A = 64$, $L = 10$

As shown in Figure 2, this formula gives the area of a rectangle with length L and width W. Substitute the given values into the formula and then solve for W.

$$A = LW$$

$$64 = 10W \qquad \text{Let } A = 64 \text{ and } L = 10.$$

$$6.4 = W \qquad \text{Divide by 10.}$$

Check that the width of the rectangle is 6.4.

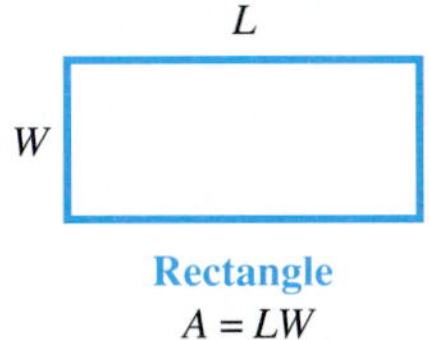

Rectangle
$A = LW$

FIGURE 2

(b) $A = \frac{1}{2}h\,(b + B)$; $A = 210$, $B = 27$, $h = 10$

This formula gives the area of a trapezoid with parallel sides of length b and B and distance h between the parallel sides. See Figure 3.

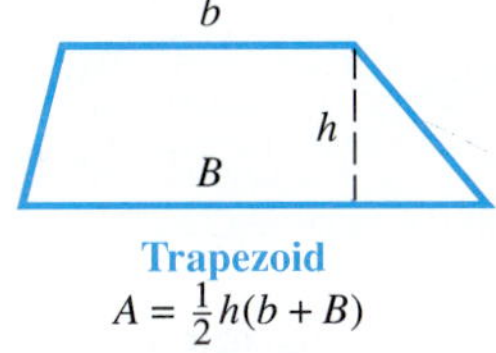

Trapezoid
$A = \frac{1}{2}h(b + B)$

FIGURE 3

Again, begin by substituting the given values into the formula.

$$A = \frac{1}{2}h(b + B)$$

$$210 = \frac{1}{2}(10)(b + 27) \qquad A = 210,\ B = 27,\ h = 10$$

1. Find the value of the remaining variable in each of the following.

(a) $I = prt$; $I = \$246$, $r = .06$, $t = 2$

(b) $P = 2L + 2W$; $P = 126$, $W = 25$

2. A farmer has 800 meters of fencing material to enclose a rectangular field. The width of the field is 175 meters. Find the length of the field.

ANSWERS
1. (a) $p = \$2050$ **(b)** $L = 38$
2. 225 meters

Now solve for b.

$$210 = 5(b + 27) \quad \text{Multiply.}$$
$$210 = 5b + 135 \quad \text{Distributive property}$$
$$210 - 135 = 5b + 135 - 135 \quad \text{Subtract 135.}$$
$$75 = 5b \quad \text{Combine terms.}$$
$$\frac{75}{5} = \frac{5b}{5} \quad \text{Divide by 5.}$$
$$15 = b \quad \text{Reduce.}$$

Check that the length of the shorter parallel side, b, is 15. ■

WORK PROBLEM 1 AT THE SIDE.

2 As the next examples show, formulas are often used to solve applied problems. *It is a good idea to draw a sketch when a geometric figure is involved.* Example 2 uses the idea of *perimeter*. The **perimeter** of a figure is the sum of the lengths of its sides.

■ **EXAMPLE 2** *Finding the Width of a Rectangular Lot*

A rectangular lot has perimeter 80 meters and length 25 meters. Find the width of the lot. See Figure 4.

We are told to find the width of the lot, so

let W = the width of the lot in meters.

The formula for the perimeter of a rectangle is

$$P = 2L + 2W.$$

Find the width by substituting 80 for P and 25 for L in the formula.

$$80 = 2(25) + 2W \quad P = 80, L = 25$$

FIGURE 4

Solve the equation.

$$80 = 50 + 2W \quad \text{Multiply.}$$
$$80 - 50 = 50 + 2W - 50 \quad \text{Subtract 50.}$$
$$30 = 2W$$
$$15 = W \quad \text{Divide by 2.}$$

The width is 15 meters. Check this result. If the width is 15 meters and the length is 25 meters, the distance around the rectangular lot (perimeter) is $2(25) + 2(15) = 50 + 30 = 80$ meters, as required. ■

WORK PROBLEM 2 AT THE SIDE.

The **area** of a geometric figure is a measure of the surface covered by the figure. Example 3 shows an application of area.

EXAMPLE 3 *Finding the Height of a Triangular Sail*

The area of a triangular sail of a sailboat is 126 square meters. The base of the sail is 21 meters. Find the height of the sail. See Figure 5.

Since we must find the height of the triangular sail,

let h = the height of the sail in meters.

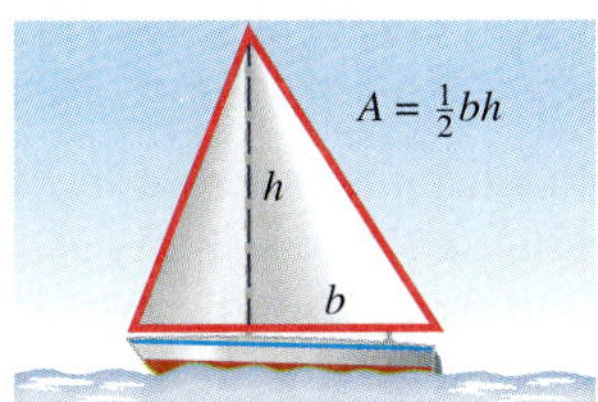

FIGURE 5

The formula for the area of a triangle is $A = \frac{1}{2}bh$, where A is the area, b is the base, and h is the height. Using the information given in the problem, substitute 126 for A and 21 for b in the formula.

$$A = \frac{1}{2}bh$$

$$126 = \frac{1}{2}(21)h \qquad A = 126, b = 21$$

Solve the equation.

$$126 = \frac{21}{2}h$$

$$\frac{2}{21}(126) = \frac{2}{21} \cdot \frac{21}{2}h \qquad \text{Multiply by } \frac{2}{21}.$$

$$12 = h \quad \text{or} \quad h = 12$$

The height of the sail is 12 meters. Check to see that the values $A = 126$, $b = 21$, and $h = 12$ satisfy the formula for the area of a triangle. ■

WORK PROBLEM 3 AT THE SIDE.

3 Figure 6 shows two intersecting lines forming angles that are numbered ①, ②, ③, and ④. Angles ① and ③ lie "opposite" each other. They are called **vertical angles.** Another pair of vertical angles are ② and ④. In geometry, it is shown that the following property holds.

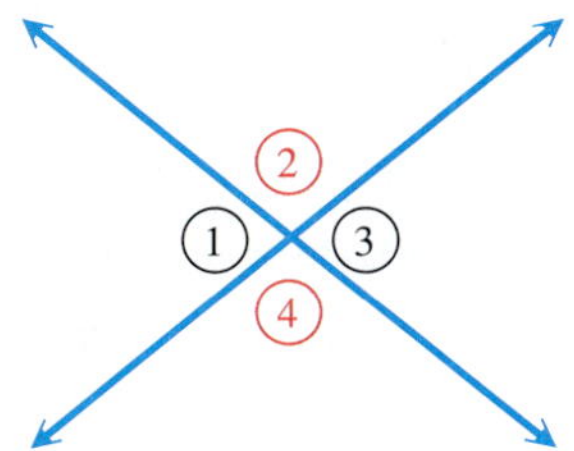

FIGURE 6

3. The area of a triangle is 120 square meters. The height is 24 meters. Find the length of the base of the triangle.

ANSWERS
3. 10 meters

4. Find the measures of each marked angle.

(a)

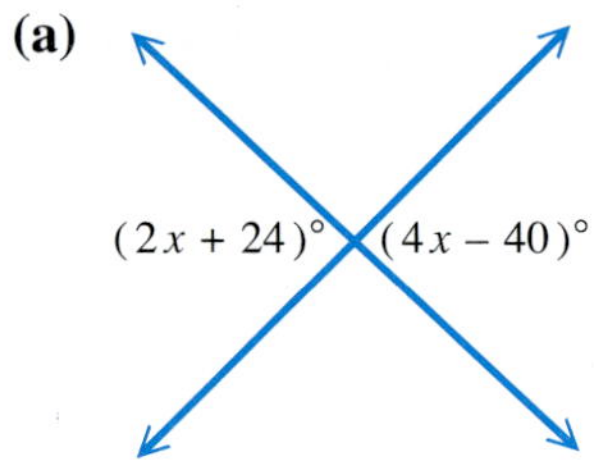

(b)

(5x + 12)° (3x)°

4. (a) Both measure 88°.
(b) 117° and 63°

VERTICAL ANGLES

> Vertical angles have equal measures.

Now look at angles ① and ②. When their measures are added, we get the measure of a **straight angle,** which is 180°. There are three other such pairs of angles: ② and ③, ③ and ④, ① and ④.

The next example uses these ideas.

■ EXAMPLE 4 *Finding Angle Measures*

Refer to the appropriate figures in each part.

(a) Find the measure of each marked angle in Figure 7.

Since the marked angles are vertical angles, they have equal measures. Set $4x + 19$ equal to $6x - 5$ and solve.

$$\begin{aligned} 4x + 19 &= 6x - 5 \\ -4x + 4x + 19 &= -4x + 6x - 5 && \text{Add } -4x. \\ 19 &= 2x - 5 \\ 19 + 5 &= 2x - 5 + 5 && \text{Add 5.} \\ 24 &= 2x \\ 12 &= x && \text{Divide by 2.} \end{aligned}$$

Since $x = 12$, one angle has measure $4(12) + 19 = 67$ degrees. The other has the same measure, since $6(12) - 5 = 67$ as well. Each angle measures 67°.

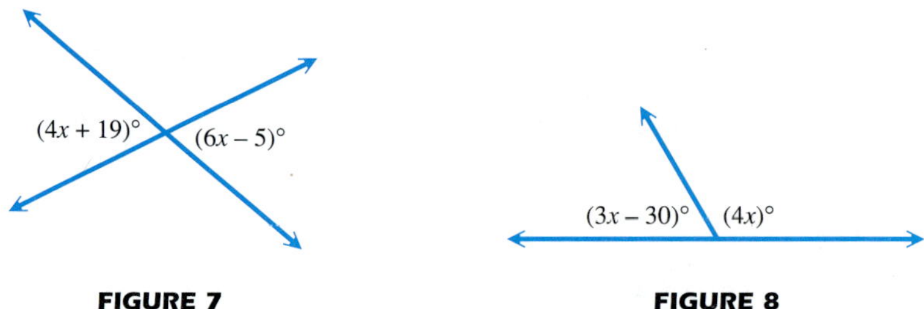

FIGURE 7 **FIGURE 8**

(b) Find the measure of each marked angle in Figure 8.

The measures of the marked angles must add to 180° because together they form a straight angle. The equation to solve is

$$\begin{aligned} (3x - 30) + 4x &= 180. \\ 7x - 30 &= 180 && \text{Combine like terms.} \\ 7x - 30 + 30 &= 180 + 30 && \text{Add 30.} \\ 7x &= 210 \\ x &= 30 && \text{Divide by 7.} \end{aligned}$$

To find the measures of the angles, replace x with 30 in the two expressions.

$$\begin{aligned} 3x - 30 &= 3(30) - 30 = 90 - 30 = 60 \\ 4x &= 4(30) = 120 \end{aligned}$$

The two angle measures are 60° and 120°. ■

◀◀ WORK PROBLEM 4 AT THE SIDE.

4 Sometimes it is necessary to solve a large number of problems that use the same formula. For example, a surveying class might need to solve several problems that involve the formula for the area of a rectangle, $A = LW$. Suppose that in each problem the area (A) and the length (L) of a rectangle are given, and the width (W) must be found. Rather than solving for W each time the formula is used, it would be simpler to rewrite the *formula* so that it is solved for W. This process is called **solving for a specified variable.** As the following examples will show, solving a formula for a specified variable requires the same steps used earlier to solve equations with just one variable.

In solving a formula for a specified variable, we treat the specified variable as if it were the *only* variable in the equation, and treat the other variables as if they were numbers.

EXAMPLE 5 *Solving for a Specified Variable*

Solve $A = LW$ for W.

Think of undoing what has been done to W. Since W is multiplied by L, undo the multiplication by dividing both sides of $A = LW$ by L.

$$A = LW$$

$$\frac{A}{L} = \frac{LW}{L} \quad \text{Divide by } L.$$

$$\frac{A}{L} = W \quad \frac{L}{L} = 1;\ 1W = W$$

The formula is now solved for W. ■

WORK PROBLEM 5 AT THE SIDE.

EXAMPLE 6 *Solving for a Specified Variable*

Solve $P = 2L + 2W$ for L.

Begin by subtracting $2W$ on both sides.

$$P = 2L + 2W$$

$$P - 2W = 2L + 2W - 2W \quad \text{Subtract } 2W.$$

$$P - 2W = 2L \quad \text{Combine terms.}$$

$$\frac{P - 2W}{2} = \frac{2L}{2} \quad \text{Divide by 2.}$$

$$\frac{P - 2W}{2} = L \quad \frac{2}{2} = 1;\ 1L = L$$

The last step gives the formula solved for L, as required. ■

WORK PROBLEM 6 AT THE SIDE.

5. (a) Solve $I = prt$ for t.

(b) Solve $P = a + b + c$ for a.

6. (a) Solve $A = p + prt$ for t.

(b) Solve $y = mx + b$ for x.

ANSWERS

5. (a) $t = \frac{I}{pr}$ (b) $a = P - b - c$

6. (a) $t = \frac{A - p}{pr}$ (b) $x = \frac{y - b}{m}$

QUEST FOR NUMERACY

Regular Polyhedra: The Platonic Solids

An interesting occurrence of geometry in the world around us involves three-dimensional figures known as regular polyhedra (singular: polyhedron). Because they were studied by Plato and his followers, they are also known as the platonic solids.

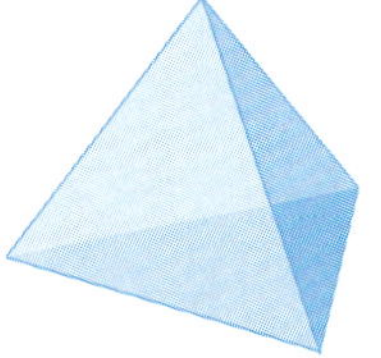
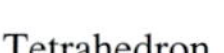

Tetrahedron

Hexahedron (Cube)

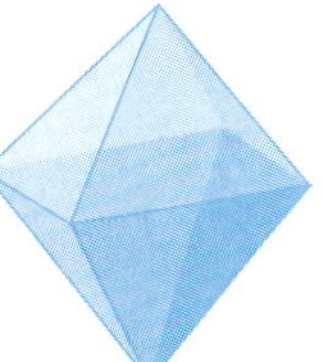

Octahedron

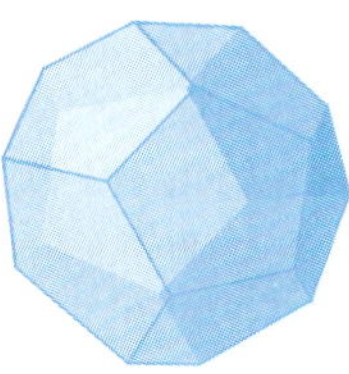

Dodecahedron

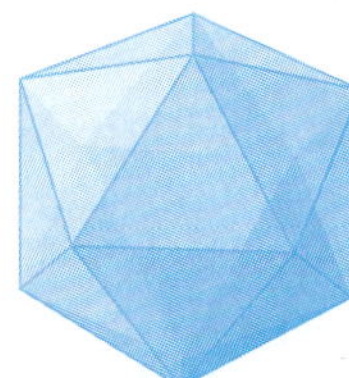

Icosahedron

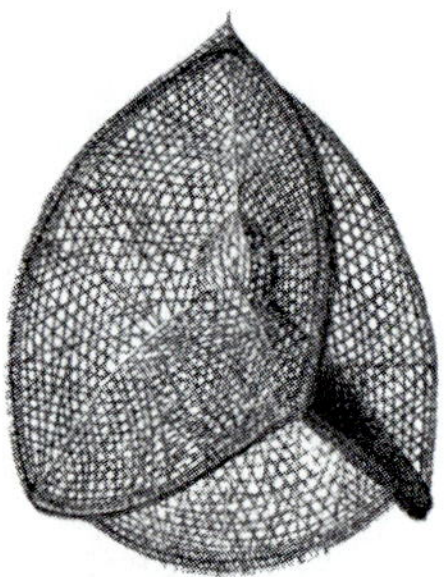

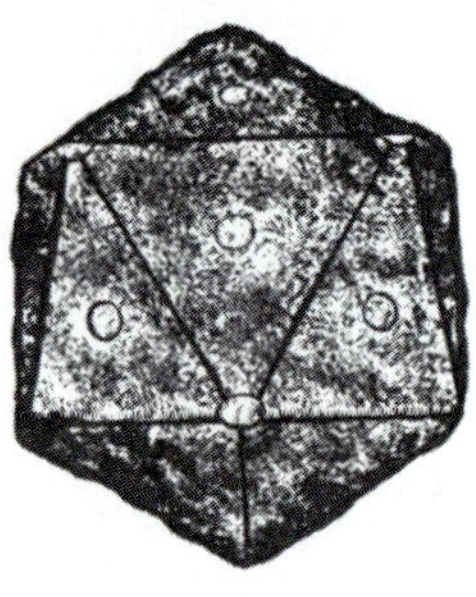

Each regular polyhedron is formed by regular polygons (two-dimensional figures having equal sides and equal angles). The polygons are called faces. They meet at edges, and the "corners" formed are called vertices. There are only five regular polyhedra, pictured above.

Many crystals and some viruses are constructed in the shapes of regular polyhedra. Many viruses that infect animals were once thought to have spherical shapes, until X-ray analysis revealed that their shells are icosahedral in nature. The sketch on the left (bottom) shows a polio virus.

Radiolaria, shown on the top left, is a group of microorganisms. The skeletons of these single-celled animals take on beautiful geometric shapes. The one pictured is based on the tetrahedron.

There are patterns that may be used to actually construct three-dimensional models of regular polyhedra. The Swiss mathematician Leonhard Euler (1707–1783) investigated a remarkable relationship among these numbers. The group discussion exercise below will allow you to discover it as well.

FOR GROUP DISCUSSION As a class, use the figures in the text, or models brought to class by your instructor or class members, to complete the following table.

Polyhedron	*Faces* (F)	*Vertices* (V)	*Edges* (E)	*Value of* $F + V - E$
Tetrahedron				
Hexahedron (Cube)				
Octahedron				
Dodecahedron				
Icosahedron				

Now, state your conjecture. Verify it by looking up **Euler's formula** in a more advanced book on geometry or the history of mathematics.

* Polio virus and radiolaria from: *On Growth and Form* by Arch Wentworth Thompson, Cambridge University Press. Abridged edition edited by John Tyler Bonner.

2.5 EXERCISES

NAME DATE HOUR

1. In your own words, explain what is meant by the *perimeter* of a geometric figure.

2. In your own words, explain what is meant by the *area* of a geometric figure.

Decide whether perimeter or area would be used to solve a problem concerning the measure of the quantity.

3. carpeting for a bedroom

4. sod for a lawn

5. fencing for a yard

6. baseboards for a living room

7. tile for a bathroom

8. fertilizer for a garden

9. determining the cost for replacing a linoleum floor with a wood floor

10. determining the cost for planting rye grass in a lawn for the winter

In the following exercises a formula is given, along with the values of all but one of the variables in the formula. Find the value of the variable that is not given. See Example 1.

11. $P = 2L + 2W$ (perimeter of a rectangle); $L = 6$, $W = 4$

12. $P = 2L + 2W$; $L = 8$, $W = 5$

13. $P = 4s$ (perimeter of a square); $s = 6$

14. $P = 4s$; $s = 12$

15. $A = \frac{1}{2}bh$ (area of a triangle); $b = 10$, $h = 14$

16. $A = \frac{1}{2}bh$; $b = 8$, $h = 16$

17. $P = a + b + c$ (perimeter of a triangle); $P = 15$, $a = 3$, $b = 7$

18. $P = a + b + c$; $P = 12$, $a = 3$, $c = 5$

19. $d = rt$ (distance formula); $d = 100$, $t = 2.5$

20. $d = rt$; $d = 252$, $r = 45$

21. $I = prt$ (simple interest); $p = 5000$, $r = .025$, $t = 7$

22. $I = prt$; $p = 7500$, $r = .035$, $t = 6$

23. $A = \frac{1}{2}h(b + B)$ (area of a trapezoid); $h = 7$, $b = 12$, $B = 14$

24. $A = \frac{1}{2}h(b + B)$; $h = 3$, $b = 19$, $B = 31$

25. $C = 2\pi r$ (circumference of a circle); $C = 8.164$, $\pi = 3.14$*

26. $C = 2\pi r$; $C = 16.328$, $\pi = 3.14$

27. $A = \pi r^2$ (area of a circle); $r = 12$, $\pi = 3.14$

28. $A = \pi r^2$; $r = 4$, $\pi = 3.14$

*The **volume** of a three-dimensional object is a measure of the space occupied by the object. For example, we would need to know the volume of a gasoline tank in order to know how many gallons of gasoline it would take to completely fill the tank. In each of the following exercises, a formula for the volume (V) of a three-dimensional object is given, along with values for the other variables. Evaluate V. See Example 1.*

29. $V = LWH$ (volume of a rectangular-sided box); $L = 12$, $W = 8$, $H = 4$

30. $V = LWH$; $L = 10$, $W = 5$, $H = 3$

31. $V = \frac{1}{3}Bh$ (volume of a pyramid); $B = 36$, $h = 4$

32. $V = \frac{1}{3}Bh$; $B = 12$, $h = 13$

33. $V = \frac{4}{3}\pi r^3$ (volume of a sphere); $r = 6$, $\pi = 3.14$

34. $V = \frac{4}{3}\pi r^3$; $r = 12$, $\pi = 3.14$

35. Give several examples of how volume is used in everyday measurement.

36. Explain the distinction among perimeter, area, and volume.

---

*Actually, π is approximately equal to 3.14, not *exactly* equal to 3.14.

Use a formula to write an equation for each of the following applications, and use the problem-solving method of Section 2.4 to solve them. Formulas may be found in the endsheets of this book. See Examples 2 and 3.

37. A radio telescope at the Max Planck Institute in West Germany has a 328-foot diameter. What is the circumference of this telescope? (Use $\pi = 3.14$.)

38. The largest drum ever constructed was played at the Royal Festival Hall in London in 1987. It had a diameter of 13 feet. What was the area of the circular face of the drum? (Use $\pi = 3.14$.)

39. The newspaper *The Constellation,* printed in 1859 in New York City as part of the Fourth of July celebration, had length 51 inches and width 35 inches. What was the perimeter? What was the area?

40. The *Daily Banner,* published in Roseberg, Oregon, in the nineteenth century, had page size 3 inches by 3.5 inches. What was the perimeter? What was the area?

41. A color television set with a liquid crystal display was manufactured by Epson in 1985 and had dimensions 3 inches by $6\frac{3}{4}$ inches by $1\frac{1}{8}$ inches. What was the volume of this set?

42. A sea lock at Zeebrugge, Belgium, measures 1640 feet by 187 feet by 75.4 feet. What is the volume of the lock?

43. The survey plat shown in the figure shows two lots that form a figure called a trapezoid. The measures of the parallel sides are 115.80 feet and 171.00 feet. The height of the trapezoid is 165.97 feet. Find the combined area of the two lots. Round your answer to the nearest hundredth of a square foot.

44. Lot *A* in the figure is in the shape of a trapezoid. The parallel sides measure 26.84 feet and 82.05 feet. The height of the trapezoid is 165.97 feet. Find the area of Lot *A*. Round your answer to the nearest hundredth of a square foot.

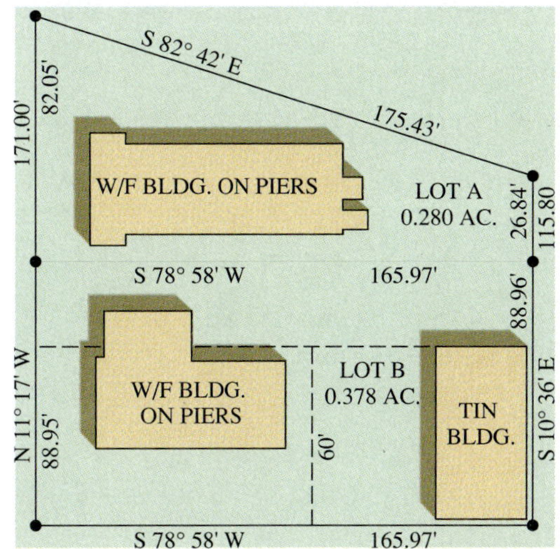

Find the measure of each marked angle. See Example 4.

45.

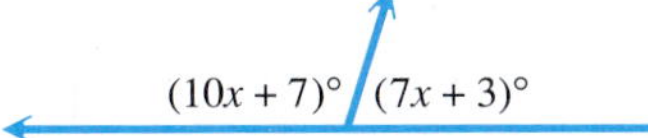

46.

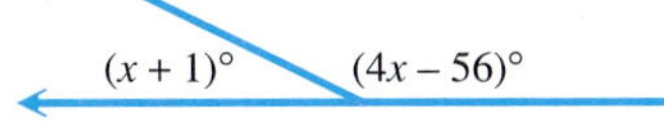

47.

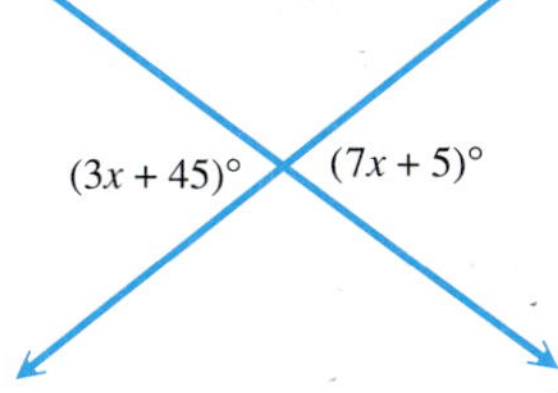

48.

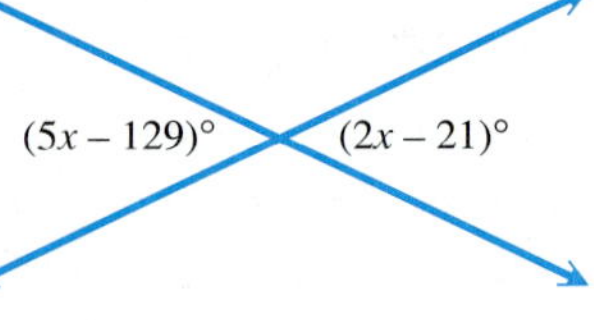

49.

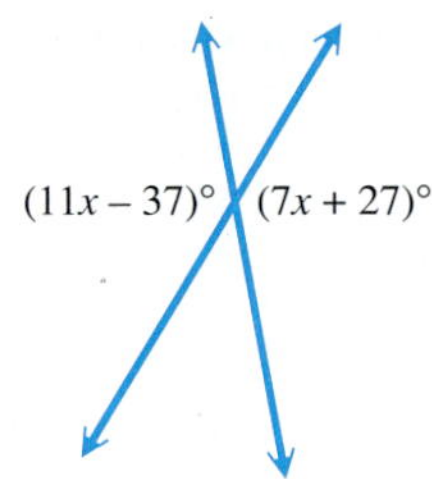

50.

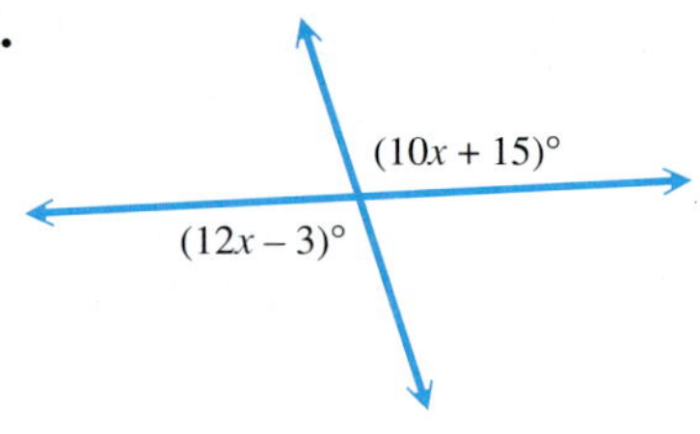

Solve the formula for the specified variable. See Examples 5 and 6.

51. $A = LW$ for L

52. $d = rt$ for t

53. $d = rt$ for r

54. $I = prt$ for r

55. $I = prt$ for p

56. $V = LWH$ for H

57. $P = a + b + c$ for a

58. $P = a + b + c$ for b

59. $A = \frac{1}{2}bh$ for b

60. $A = \frac{1}{2}bh$ for h

61. $A = p + prt$ for r

62. $P = 2L + 2W$ for W

63. $V = \pi r^2 h$ for h

64. $V = \frac{1}{3}\pi r^2 h$ for h

65. $y = mx + b$ for m

66. $F = \frac{9}{5}C + 32$ for C

PREVIEW EXERCISES

Change each decimal to a percent. See Section R.2.

67. .05

68. .23

69. .005

Solve each equation. See Section 2.2.

70. $4x = 12$

71. $\frac{3}{4}y = 21$

72. $.06x = 300$

2.6 PERCENT; RATIO AND PROPORTION

We all see examples of percent almost every day. We can use the techniques for solving equations to solve percent problems. Recall, we move the decimal point two places to the left to change a percent to a decimal.

Many calculators have a percent key that allows the user to work problems involving percent.

OBJECTIVES

1. Find percentages and percents.
2. Write ratios.
3. Solve proportions.
4. Solve applied problems using proportions.

FOR EXTRA HELP

Tape 4 | SSM pp. 65–69 | MAC: A IBM: A

1 A **percentage** is part of a whole. For example, since 50% represents $\frac{50}{100} = \frac{1}{2}$ of a whole, 50% of 800 is one-half of 800, or 400.

EXAMPLE 1 *Finding Percentages*

Solve each problem.

(a) Find 15% of 600.

In examples like this, the word *of* indicates multiplication, so 15% of 600 is found by multiplying .15 and 600.

$$15\%\,(600) = .15\,(600) = 90$$

(b) A video with a regular price of \$18 is on sale this week at 22% off. Find the amount of the discount and the sale price of the video.

The discount is 22% of \$18. Translate this as

$$22\%\,(\$18) = .22\,(\$18) = \$3.96$$

The discount is \$3.96. The video is on sale for \$18 − \$3.96 = \$14.04. ■

WORK PROBLEM 1 AT THE SIDE.

1. Solve each problem.

(a) Find 20% of 70.

(b) Find the amount of discount on a television set with a regular price of \$270 if the set is on sale at 25% off. Find the sale price of the set.

EXAMPLE 2 *Finding Percent Given the Percentage and the Amount*

Find the following percents.

(a) 30 is what percent of 50?

Translate the statement into symbols, and then solve the equation. Let x represent the percent in decimal form.

30 is what percent of 50?
↓ ↓ ↓ ↓ ↓
$30 = x \cdot 50$

$$\frac{30}{50} = \frac{x \cdot 50}{50} \quad \text{Divide by 50.}$$

$$.60 = x \quad \text{Write the left side in decimal form.}$$

$$x = 60\% \quad \text{Change .60 to percent.}$$

(b) A newspaper ad offered a mountain bike at a sale price of \$258. The regular price was \$300. What percent of the regular price was the savings?

ANSWERS
1. (a) 14 (b) \$67.50; \$202.50

The savings amounted to $300 − $258 = $42. We can now restate the problem: what percent of 300 is 42?

Let x represent the percent in decimal form.

$$x \cdot 300 = 42 \qquad \text{Translate into symbols.}$$
$$\frac{x \cdot 300}{300} = \frac{42}{300} \qquad \text{Divide by 300.}$$
$$x = .14 = 14\% \qquad \text{Change to percent.}$$

The sale price represented a 14% savings. ■

While not absolutely necessary, you could use a calculator to save time in performing the division of 42 by 300 in Example 2.

WORK PROBLEM 2 AT THE SIDE.

2 Ratios provide a way of comparing two numbers or quantities using division. A **ratio** is a quotient of two quantities with the same units.

> The ratio of the number a to the number b is written
>
> $$a \text{ to } b, \quad a : b, \quad \text{or} \quad \frac{a}{b}.$$

This last way of writing a ratio is most common in algebra.

Percents are ratios where the second number is always 100. For example 50% represents the ratio of 50 to 100, 27% represents the ratio of 27 to 100, and so on.

■ EXAMPLE 3 ***Writing a Word Phrase as a Ratio***

Write a ratio for each word phrase.

(a) The ratio of 5 hours to 3 hours is

$$\frac{5}{3}.$$

(b) To find the ratio of 6 hours to 3 days, first convert 3 days to hours.

$$3 \text{ days} = 3 \cdot 24$$
$$= 72 \text{ hours}$$

The ratio of 6 hours to 3 days is thus

$$\frac{6}{72} = \frac{1}{12}. \quad ■$$

WORK PROBLEM 3 AT THE SIDE.

3 A ratio is used to compare two numbers or amounts. A **proportion** says that two ratios are equal. For example,

$$\frac{3}{4} = \frac{15}{20}$$

---

2. Answer each of the following.

(a) 90 is what percent of 270?

(b) The interest in one year on deposits of $11,000 was $682. What percent interest was paid?

3. Write each ratio.

(a) 9 women to 5 women

(b) 4 inches to 1 foot

ANSWERS

2. (a) $33\frac{1}{3}\%$ (b) 6.2%

3. (a) $\frac{9}{5}$ (b) $\frac{4}{12} = \frac{1}{3}$

is a proportion that says that the ratios $\frac{3}{4}$ and $\frac{15}{20}$ are equal. In the proportion

$$\frac{a}{b} = \frac{c}{d}$$

a, b, c, and d are the **terms** of the proportion. Beginning with the proportion

$$\frac{a}{b} = \frac{c}{d}$$

and multiplying both sides by the common denominator, bd, gives

$$bd \cdot \frac{a}{b} = bd \cdot \frac{c}{d}$$

$$\frac{b}{b}(d \cdot a) = \frac{d}{d}(b \cdot c) \qquad \text{Associative and commutative properties}$$

$$ad = bc. \qquad \text{Commutative and identity properties}$$

The products ad and bc are found by multiplying diagonally, as shown below.

$$\frac{a}{b} = \frac{c}{d}$$

bc

ad

For this reason, ad and bc are called **cross products.**

In our discussion below, we assume that no denominators are zero.

> If $\frac{a}{b} = \frac{c}{d}$, then the cross products ad and bc are equal.
> Also, if $ad = bc$, then $\frac{a}{b} = \frac{c}{d}$.

From the rule given above, if $\frac{a}{b} = \frac{c}{d}$ then $ad = bc$. However, if $\frac{a}{c} = \frac{b}{d}$, then $ad = cb$, or $ad = bc$. This means that the two proportions are equivalent, and

the proportion $\frac{a}{b} = \frac{c}{d}$ can always be written as $\frac{a}{c} = \frac{b}{d}$.

Sometimes one form is more convenient to work with than the other.

Four numbers are used in a proportion. If any three of these numbers are known, the fourth can be found.

■ EXAMPLE 4 ***Finding an Unknown in a Proportion***

Find x in the proportion

$$\frac{5}{9} = \frac{x}{63}.$$

The cross products must be equal.

$$5 \cdot 63 = 9 \cdot x \qquad \text{Cross products}$$

$$315 = 9x \qquad \text{Multiply.}$$

$$35 = x \qquad \text{Divide by 9.} \quad ■$$

WORK PROBLEM 4 AT THE SIDE. ▶▶

4. Solve each equation.

(a) $\frac{y}{6} = \frac{35}{42}$

(b) $\frac{a}{24} = \frac{15}{16}$

ANSWERS

4. (a) 5 **(b)** $\frac{45}{2}$

Caution The cross product method cannot be used directly if there is more than one term on either side.

EXAMPLE 5 *Solving an Equation Using Cross Products*

Solve the equation

$$\frac{m-2}{5} = \frac{m+1}{3}.$$

Find the cross products.

$$3(m-2) = 5(m+1) \quad \text{Be sure to use parentheses.}$$
$$3m - 6 = 5m + 5 \quad \text{Distributive property}$$
$$3m = 5m + 11 \quad \text{Add 6.}$$
$$-2m = 11 \quad \text{Subtract } 5m.$$
$$m = -\frac{11}{2} \quad \text{Divide by } -2.$$

WORK PROBLEM 5 AT THE SIDE.

Note Remember, when you set cross products equal, you are really multiplying each ratio in the proportion by a common denominator.

4 Proportions are useful in many practical applications.

EXAMPLE 6 *Applying Proportions*

A local drugstore is offering 3 packs of toothpicks for \$.87. How much would be charged for 10 packs?

Let x represent the cost of 10 packs of toothpicks.

Set up a proportion. One ratio in the proportion can involve the number of packs, and the other can involve the costs. Make sure that the corresponding numbers appear in the numerator and denominator.

$$\frac{\text{Cost of 3}}{\text{Cost of 10}} = \frac{3}{10}$$
$$\frac{.87}{x} = \frac{3}{10}$$
$$3x = .87(10) \quad \text{Cross products}$$
$$3x = 8.7$$
$$x = 2.90 \quad \text{Divide by 3.}$$

The 10 packs should cost \$2.90. As we saw earlier, the proportion could also be written as

$$\frac{3}{.87} = \frac{10}{x},$$

which would give the same cross products.

WORK PROBLEM 6 AT THE SIDE.

5. Solve each equation.

(a) $\frac{z}{2} = \frac{z+1}{3}$

(b) $\frac{p+3}{3} = \frac{p-5}{4}$

6. On a map, 12 inches represents 500 miles. How many miles would be represented by 30 inches?

ANSWERS
5. (a) 2 (b) −27
6. 1250 miles

An example of the use of a ratio is in unit pricing, to see which size of an item offered in different sizes produces the best price per unit. To do this, set up the ratio of the price of the item to the number of units on the label. Then divide to obtain a decimal.

A calculator will be helpful in such problems.

EXAMPLE 7 *Finding the Price per Unit*
The local supermarket charges the following prices for a popular brand of pancake syrup:

Size	*Price*
36-ounce	\$3.89
24-ounce	\$2.79
12-ounce	\$1.89

Which size is the best buy? That is, which size has the lowest unit price?

To find the best buy, divide the price by the number of units (ounces here) to get the price per unit. The results in the following table were rounded to the nearest thousandth.

Size	*Unit Cost (dollars per ounce)*
36-ounce	$\frac{\$3.89}{36} = \$.108$ ←the best buy
24-ounce	$\frac{\$2.79}{24} = \$.116$
12-ounce	$\frac{\$1.89}{12} = \$.158$

Since the 36-ounce size produces the lowest price per unit, it would be the best buy. (Be careful: Sometimes the largest container is not the best buy.) ■

WORK PROBLEM 7 AT THE SIDE.

7. Paper towels sell for \$.85 for 1 roll of 63 square feet and \$2.89 for a package of 3 rolls of the same size. Find the best buy by finding the price per square foot in each case.

ANSWER
7. \$.85 for 1 roll

QUEST FOR NUMERACY

Inquiring Minds Want to Know . . . What's So Curious and Interesting About Numbers?

The theory of numbers is the branch of mathematics that deals with the study of the properties of the natural numbers. One of the most remarkable books on number theory is *The Penguin Dictionary of Curious and Interesting Numbers* (1986) by David Wells. This book contains fascinating numbers and their properties, some of which are given here.

- 113 and 199 and 337 are the only three-digit numbers that are prime and all rearrangements of the digits are prime.
- Find the sum of the cubes of the digits of 136: $1^3 + 3^3 + 6^3 = 244$. Repeat the process with the digits of 244: $2^3 + 4^3 + 4^3 = 136$. We're back to where we started.
- 635,318,657 is the smallest number that can be expressed as the sum of two fourth powers in two ways: $635{,}318{,}657 = 59^4 + 158^4 = 133^4 + 134^4$.
- The number 24,678,050 has an interesting property: $24{,}678{,}050 = 2^8 + 4^8 + 6^8 + 7^8 + 8^8 + 0^8 + 5^8 + 0^8$.
- The number 54,748 has a similar interesting property: $54{,}748 = 5^5 + 4^5 + 7^5 + 4^5 + 8^5$.
- The number 3435 has this property: $3435 = 3^5 + 4^4 + 3^3 + 5^5$.

If your curiosity is piqued by such facts, this book is for you!

FOR GROUP DISCUSSION Discover some curious and interesting properties of numbers by going through the following processes.

1. Choose a three-digit number that is a multiple of 3. Add the cubes of the digits. Repeat the process until the same number is obtained over and over. Compare your results with another student.
2. Take any three-digit number whose digits are not all the same. Arrange the digits in decreasing order, and then arrange them in increasing order. Now subtract. Repeat the process, using a 0 if necessary in the event that the difference consists of only two digits. Continue until the same number repeats over and over. What is the number?
3. Repeat Item 2 for a four-digit number. What result do you get?
4. When told that the number 1729 was rather dull, the Indian mathematician Srinivasa Ramanujan commented that, on the contrary, it was interesting because it is the smallest number that can be expressed as the sum of two cubes in exactly two ways. What are these two ways? Can you express 85 as the sum of two squares in two ways? (For more on Ramanujan, read *The Man Who Knew Infinity,* by Robert Kanigel.)

NAME DATE HOUR

2.6 EXERCISES

Answer each of the following. See Examples 1(a) and 2(a).

1. Find 38% of 12.

2. Find 10.5% of 28.

3. What is 26% of 480?

4. What is 48.6% of 19?

5. What percent of 30 is 36?

6. What percent of 48 is 20?

7. 25% of what number is 150?

8. 12% of what number is 3600?

9. .392 is what percent of 28?

10. 78.84 is what percent of 292?

Work each of the following problems. Round all money amounts to the nearest dollar. See Examples 1(b) and 2(b).

11. According to a Knight-Ridder Newspapers report, as of May 31, 1992, the nation's "consumer-debt burden" was 16.4%. This means that the average American had consumer debts, such as credit card bills and auto loans, totaling 16.4% of his or her take-home pay. Suppose that Paul Eldersveld has a take-home pay of $3250 per month. What is 16.4% of his monthly take-home pay?

12. In 1992 General Motors announced that it would raise prices on its 1993 vehicles by an average of 1.6%. If a certain vehicle had a 1992 price of $10,526 and this price was raised 1.6%, what would the 1993 price be?

13. At the 1992 Olympic Games in Barcelona, 25% of the medals won by Great Britain were gold medals. The team won a total of 20 medals. How many of the medals were *not* gold?

14. Forty percent of Florence Griffith Joyner's Olympic medals are not gold medals. She has 2 medals that are not gold. How many gold medals does she have?

15. An advertisement for a dot matrix printer gives a sale price of $150.00. The regular price is $180.00. What is the percent discount on this printer?

16. Jane Gunton bought a boat five years ago for $5000 and sold it this year for $2000. What percent of her original purchase price did she lose on the sale?

17. Quinhon Dac Ho earns $3200 per month. He wants to save 12% of this amount. How much will he save?

18. A family of four with a monthly income of $3800 plans to spend 8% of this amount on entertainment. How much will be spent on entertainment?

19. The 1916 dime minted in Denver is quite rare. The 1979 edition of *A Guide Book of United States Coins* listed its value in extremely fine condition as $625.00. The 1991 value had increased to $1750. What was the percent increase in the value of this coin?

20. In 1963, the value of a 1903 Morgan dollar minted in New Orleans in uncirculated condition was $1500. Due to a discovery of a large hoard of these dollars late that year, the value plummeted. Its value as listed in the 1991 edition of *A Guide Book of United States Coins* was $200. What percent of its 1963 value was its 1991 value?

At the start of play on August 16, 1992, the standings of the Eastern division of the National League were as follows:

	Won	*Lost*
Pittsburgh	65	51
Montreal	64	53
Chicago	56	60
St. Louis	54	61
New York	52	63
Philadelphia	49	67

"Winning percentage" is commonly expressed as a decimal rounded to the nearest thousandth. To find the winning percentage of a team, divide the number of wins by the total number of games played. Find the winning percentage of each of the following teams.

21. Pittsburgh

22. Chicago

23. New York

24. Philadelphia

25. Explain the difference between .5 and .5%.

26. Suppose that an item is discounted 10% and then marked up 10%. Is the final price the same as the starting price? If not, how do they compare?

Write a ratio for each word phrase. Write fractions in lowest terms. See Example 3.

27. 40 miles to 30 miles

28. 60 feet to 70 feet

29. 120 people to 90 people

30. 72 dollars to 220 dollars

31. 20 yards to 8 feet

32. 30 inches to 8 feet

33. 24 minutes to 2 hours

34. 16 minutes to 1 hour

35. 8 days to 40 hours

36. 50 hours to 5 days

37. Explain how percent and ratio are related.

38. Explain the distinction between *ratio* and *proportion.*

Solve each of the following. See Examples 4 and 5.

39. $\frac{k}{4} = \frac{175}{20}$

40. $\frac{49}{56} = \frac{z}{8}$

41. $\frac{x}{6} = \frac{18}{4}$

42. $\frac{z}{80} = \frac{20}{100}$

43. $\frac{3y - 2}{5} = \frac{6y - 5}{11}$

44. $\frac{2p + 7}{3} = \frac{p - 1}{4}$

45. $\frac{2r + 8}{4} = \frac{3r - 9}{3}$

46. $\frac{5k + 1}{6} = \frac{3k - 2}{3}$

Solve each of the following problems involving proportion. See Example 6.

47. Two slices of bacon contain 85 calories. How many calories are there in twelve slices of bacon?

48. Three ounces of liver contain 22 grams of protein. How many ounces of liver provide 121 grams of protein?

49. If 6 gallons of premium unleaded gasoline cost \$9.72, how much would 9 gallons cost?

50. If sales tax on a \$24 headset radio is \$1.68, how much would the sales tax be on a pair of binoculars that costs \$36?

51. The distance between Singapore and Tokyo is 3300 miles. On a certain wall map, this distance is represented by 11 inches. The actual distance between Mexico City and Cairo is 7700 miles. How far apart are they on the same map?

52. The distance between Kanasas City, Missouri, and Denver is 600 miles. On a certain wall map, this is represented by a length of 2.4 feet. On the map, how many feet would there be between Memphis and Philadelphia, two cities that are actually 1000 miles apart?

53. Biologists tagged 250 fish in Willow Lake on October 5. On a later date they found 7 tagged fish in a sample of 350. Estimate the total number of fish in Willow Lake to the nearest hundred.

54. On May 13, researchers at Argyle Lake tagged 420 fish. When they returned a few weeks later, their sample of 500 fish contained 9 that were tagged. Give an approximation of the fish population in Argyle Lake to the nearest hundred.

A supermarket was surveyed to find the prices charged for items in various sizes. Find the best buy (based on price per unit) for each of the following. See Example 7.

55. Trash bags
20-count: \$2.49
30-count: \$4.29

56. Black pepper
4-ounce size: \$1.57
8-ounce size: \$2.27

57. Spaghetti sauce
$15\frac{1}{2}$-ounce size: \$1.19
32-ounce size: \$1.69
48-ounce size: \$2.69

58. Breakfast cereal
10-ounce size: \$1.49
13-ounce size: \$1.85
19-ounce size: \$2.81

59. Tomato ketchup
14-ounce size: \$.93
32-ounce size: \$1.19
44-ounce size: \$2.19

60. Extra crunchy peanut butter
12-ounce size: \$1.49
28-ounce size: \$1.99
40-ounce size: \$3.99

PREVIEW EXERCISES

Place > or < in each blank to make the statement true. See Section 1.1.

61. -9 _____ 6

62. -5.3 _____ 2.4

63. -13 _____ -6

64. -12 _____ -14

65. $-\frac{2}{3}$ _____ $-\frac{7}{8}$

66. $-\frac{14}{3}$ _____ $-\frac{15}{4}$

2.7 THE ADDITION AND MULTIPLICATION PROPERTIES OF INEQUALITY

The addition and multiplication properties can be extended to inequalities. **Inequalities** are statements with algebraic expressions related in the following ways:

$<$ "is less than"
$\leq$ "is less than or equal to"
$>$ "is greater than"
$\geq$ "is greater than or equal to."

We solve an inequality by finding all real number solutions for it. For example, the solutions of $x \leq 2$ include all *real numbers* that are less than or equal to 2, and not just the *integers* less than or equal to 2.

1 Graphing is a good way to show the solutions of an inequality. To graph all real numbers satisfying $x \leq 2$, we place a dot at 2 on a number line and draw an arrow extending from the dot to the left (to represent the fact that all numbers less than 2 are also part of the graph). The graph is shown in Figure 9.

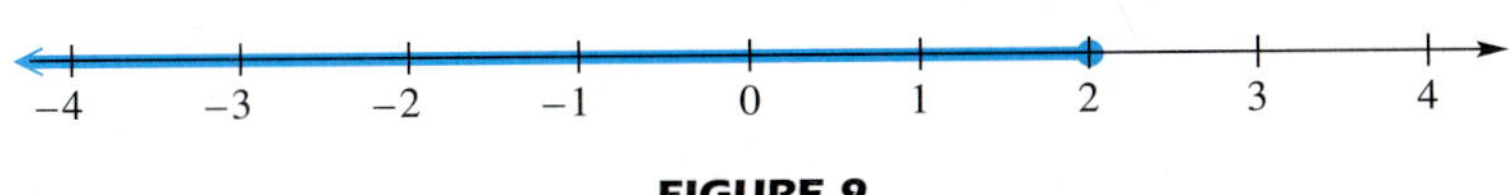

FIGURE 9

■ EXAMPLE 1 *Graphing the Solutions of an Inequality*
Graph $x > -5$.

The statement $x > -5$ says that x can take any value greater than -5, but x cannot equal -5 itself. We show this on a graph by placing an open circle at -5 and drawing an arrow to the right, as in Figure 10. The open circle at -5 shows that -5 is not part of the graph. ■

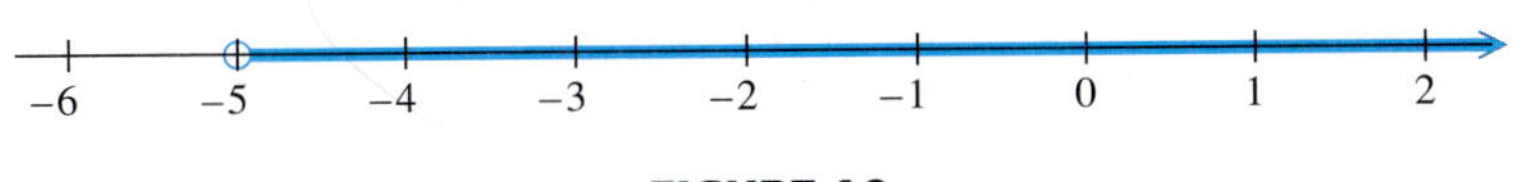

FIGURE 10

■ EXAMPLE 2 *Graphing the Solutions of an Inequality*
Graph $3 > x$.

The statement $3 > x$ means the same as $x < 3$. The graph of $x < 3$ is shown in Figure 11. ■

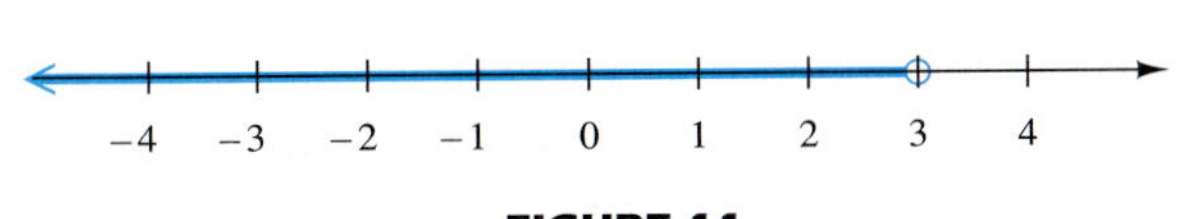

FIGURE 11

WORK PROBLEM 1 AT THE SIDE.

OBJECTIVES

1. Graph inequalities on a number line.
2. Use the addition property of inequality.
3. Use the multiplication property of inequality.
4. Solve inequalities using both properties of inequality.
5. Use inequalities to solve applied problems.

FOR EXTRA HELP

Tape 4

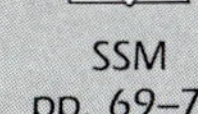
SSM pp. 69–74

MAC: A
IBM: A

1. Graph each of the following.

(a) $x \leq 3$

(b) $x > -4$

(c) $-4 \geq x$

(d) $0 < x$

ANSWERS

1. (a) −4 −2 0 2 4

(b) −8 −6 −4 −2

(c) −8 −6 −4 −2 0

(d) −4 −2 0 2 4

2. Graph the solutions of each inequality.

(a) $-7 < x < -2$

(b) $-6 < x \leq -4$

3. Solve each inequality, and graph the solutions.

(a) $-1 + 8r < 7r + 2$

(b) $4m \geq 5m - \frac{4}{3}$

■ **EXAMPLE 3** *Graphing the Solutions of an Inequality*

Graph $-3 \leq x < 2$.

The statement $-3 \leq x < 2$ is read "-3 is less than or equal to x *and* x is less than 2." We graph the solutions of this inequality by placing a solid dot at -3 (because -3 is part of the graph) and an open circle at 2 (because 2 is not part of the graph), then drawing a line segment between the two circles. Notice that the graph includes all points *between* -3 and 2, and includes -3 as well. See Figure 12. ■

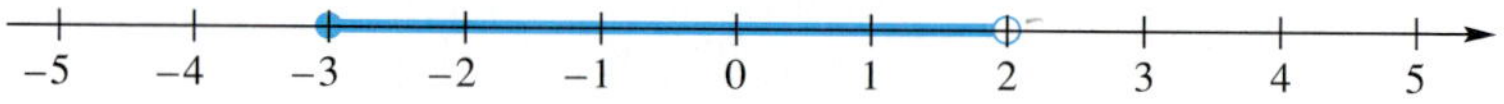

FIGURE 12

◀ WORK PROBLEM 2 AT THE SIDE.

2 We can solve inequalities such as $x + 4 \leq 9$ in much the same way as equations. The **addition property of inequality** is very similar to the addition property of equality.

ADDITION PROPERTY OF INEQUALITY

For any real numbers A, B, and C, the inequalities

$$A < B \quad \text{and} \quad A + C < B + C$$

have exactly the same solutions. In other words, the same number may be added to each side of an inequality without changing the solutions.

We can replace $<$ in the addition property of inequality with $>$, $\leq$, or $\geq$. Also, as with the addition property of equality, the same number may be *subtracted* on each side of an inequality.

The following examples show how the addition property is used to solve inequalities.

■ **EXAMPLE 4** *Using the Addition Property of Inequality*

Solve $7 + 3k > 2k - 5$.

$$7 + 3k > 2k - 5$$

$$7 + 3k - 2k > 2k - 5 - 2k \qquad \text{Subtract } 2k.$$

$$7 + k > -5 \qquad \text{Combine terms.}$$

$$7 + k - 7 > -5 - 7 \qquad \text{Subtract 7.}$$

$$k > -12 \qquad \text{Combine terms.}$$

A graph of the solutions, $k > -12$, is shown in Figure 13. ■

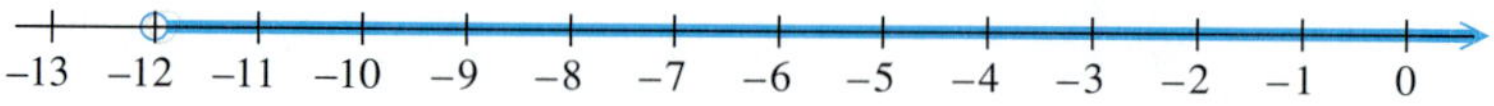

FIGURE 13

◀ WORK PROBLEM 3 AT THE SIDE.

ANSWERS

2. (a)
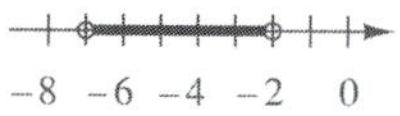

(b)
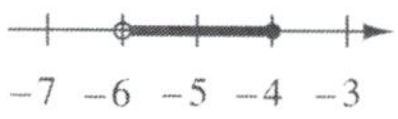

3. (a) $r < 3$
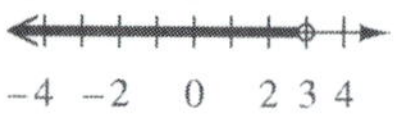

(b) $m \leq \frac{4}{3}$
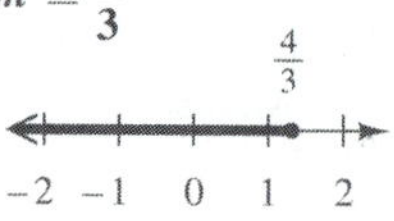

3 The addition property of inequality alone cannot be used to solve inequalities such as $4y \geq 28$. These inequalities require the *multiplication property of inequality*. To see how this property works, it will be helpful to look at some examples.

First, write the inequality $3 < 7$ and then multiply each side by the positive number 2.

$$3 < 7$$
$$2(3) < 2(7) \quad \text{Multiply each side by 2.}$$
$$6 < 14 \quad \text{True}$$

Now multiply each side of $3 < 7$ by the negative number -5.

$$3 < 7$$
$$-5(3) < -5(7) \quad \text{Multiply each side by } -5.$$
$$-15 < -35 \quad \text{False}$$

Getting a true statement when multiplying each side by -5 requires reversing the direction of the inequality symbol.

$$3 < 7$$
$$-5(3) > -5(7) \quad \text{Multiply by } -5\text{; reverse the symbol.}$$
$$-15 > -35 \quad \text{True}$$

Take the inequality $-6 < 2$ as another example. Multiply each side by the positive number 4.

$$-6 < 2$$
$$4(-6) < 4(2) \quad \text{Multiply by 4.}$$
$$-24 < 8 \quad \text{True}$$

Multiplying each side of $-6 < 2$ by -5 *and at the same time reversing the direction of the inequality symbol* gives

$$-6 < 2$$
$$(-5)(-6) > (-5)(2) \quad \text{Multiply by } -5\text{; reverse the symbol.}$$
$$30 > -10. \quad \text{True}$$

WORK PROBLEM 4 AT THE SIDE.

In summary, the multiplication property of inequality has two parts.

MULTIPLICATION PROPERTY OF INEQUALITY

For any real numbers A, B, and C $(C \neq 0)$,

1. if C is *positive,* then the inequalities

$$A < B \quad \text{and} \quad AC < BC$$

have exactly the same solutions;

2. if C is *negative,* then the inequalities

$$A < B \quad \text{and} \quad AC > BC$$

have exactly the same solutions.

In other words, each side of an inequality may be multiplied by the same positive number without changing the solutions. If the number is negative, we must reverse the direction of the inequality symbol.

4. (a) Multiply each side of $-2 < 8$ by 6 and then by -5. Reverse the direction of the inequality symbol if necessary.

(b) Multiply each side of $-4 > -9$ by 2 and then by -8. Reverse the direction of the inequality symbol if necessary.

ANSWERS
4. (a) $-12 < 48$; $10 > -40$
(b) $-8 > -18$; $32 < 72$

5. Solve each inequality. Graph the solutions.

(a) $9y < -18$

(b) $-2r > -12$

(c) $-5p \leq 0$

We can replace $<$ in the multiplication property of inequality with $>$, $\leq$, or $\geq$. As with the multiplication property of equality, the same nonzero number may be divided into each side.

It is important to remember the differences in the multiplication property for positive and negative numbers.

1. When each side of an inequality is multiplied or divided by a positive number, the direction of the inequality symbol *does not change.* Adding or subtracting terms on each side also does not change the symbol.
2. When each side of an inequality is multiplied or divided by a negative number, the direction of the symbol *does change. Reverse the direction of the symbol of inequality only when multiplying or dividing by a negative number.*

The next examples show how to use the multiplication property to solve inequalities.

EXAMPLE 5 *Using the Multiplication Property of Inequality*

Solve $3r < -18$.

Using the multiplication property of inequality, we divide each side by 3. Since 3 is a positive number, the direction of the inequality symbol *does not* change. *It does not matter that the number on the right side of the inequality is negative.*

$$3r < -18$$

$$\frac{3r}{3} < \frac{-18}{3} \quad \text{Divide by 3.}$$

$$r < -6 \quad \text{Reduce.}$$

The graph of the solutions is shown in Figure 14. ■

FIGURE 14

EXAMPLE 6 *Using the Multiplication Property of Inequality*

Solve the inequality $-4t \geq 8$.

Here each side of the inequality must be divided by -4, a negative number, which *does* change the direction of the inequality symbol.

$$-4t \geq 8$$

$$\frac{-4t}{-4} \leq \frac{8}{-4} \quad \text{Divide by } -4\text{; symbol reversed.}$$

$$t \leq -2 \quad \text{Reduce.}$$

The solutions are graphed in Figure 15. ■

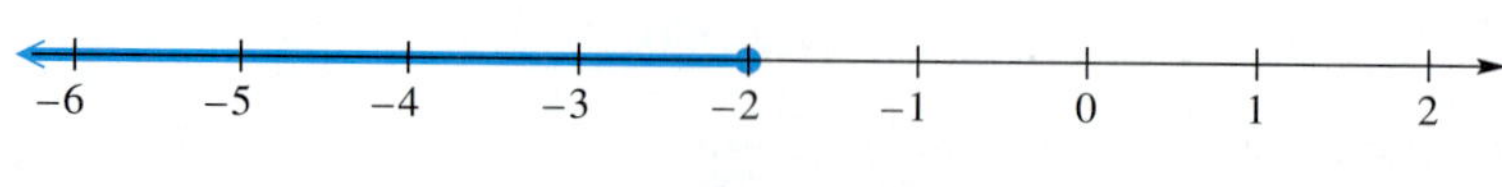

FIGURE 15

WORK PROBLEM 5 AT THE SIDE.

ANSWERS

5. (a) $y < -2$

−5 −4 −3 −2 −1 0 1

(b) $r < 6$

3 4 5 6 7 8 9

(c) $p \geq 0$

−3 −2 −1 0 1 2 3

4 The steps in solving an inequality are summarized below. (Remember that $<$ can be replaced with $>$, $\leq$, or $\geq$ in this summary.)

SOLVING INEQUALITIES

Step 1 Clear parentheses and combine terms on each side.

Step 2 Use the addition property to simplify the inequality to the form $ax < b$.

Step 3 Use the multiplication property to simplify further to the form $x < c$ or $x > c$.

Notice how these steps are used in the next example.

EXAMPLE 7 *Solving an Inequality*

Solve $5(k - 3) - 7k \geq 4(k - 3) + 9$.

Step 1 Use the distributive property to clear parentheses; then combine like terms.

$$5(k - 3) - 7k \geq 4(k - 3) + 9$$
$$5k - 15 - 7k \geq 4k - 12 + 9 \quad \text{Distributive property}$$
$$-2k - 15 \geq 4k - 3 \quad \text{Combine like terms.}$$

Step 2 Use the addition property.

$$-2k - 15 - 4k \geq 4k - 3 - 4k \quad \text{Subtract } 4k.$$
$$-6k - 15 \geq -3 \quad \text{Combine terms.}$$
$$-6k - 15 + 15 \geq -3 + 15 \quad \text{Add 15.}$$
$$-6k \geq 12 \quad \text{Combine terms.}$$

Step 3 Divide each side by -6, a negative number. Change the direction of the inequality symbol.

$$\frac{-6k}{-6} \leq \frac{12}{-6} \quad \text{Divide by } -6\text{; reverse the symbol.}$$
$$k \leq -2$$

A graph of the solutions is shown in Figure 16. ■

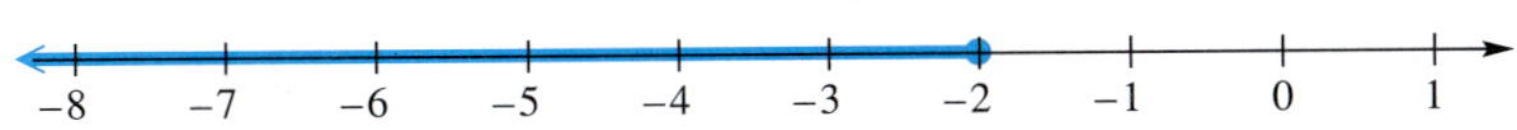

FIGURE 16

WORK PROBLEMS 6 AND 7 AT THE SIDE. ▶

5 Inequalities can be used to solve applied problems involving phrases that suggest inequality. The chart below gives some of the more common such phrases, along with examples and translations.

Phrase	*Example*	*Inequality*
Is more than	A number *is more than* 4	$x > 4$
Is less than	A number *is less than* −12	$x < -12$
Is at least	A number *is at least* 6	$x \geq 6$
Is at most	A number *is at most* 8	$x \leq 8$

6. Solve, and graph the solutions.
$5r - r + 2 < 7r - 5$

7. Solve, and graph the solutions.
$4(y - 1) - 3y > -15 - (2y + 1)$

ANSWERS

6. $r > \frac{7}{3}$

$\frac{7}{3}$

−1 0 1 2 3 4 5

7. $y > -4$

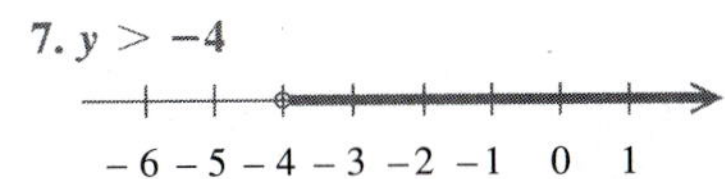

8. Maggie has scores of 98, 86, and 88 on her first three tests in algebra. If she wants an average of at least 90 after her fourth test, what score must she make on her fourth test?

Caution Do not confuse phrases like "5 more than a number" and statements like "5 *is* more than a number." The first of these is expressed as "$x + 5$" while the second is expressed as "$5 > x$."

EXAMPLE 8 *Finding an Average Test Score*

Brent has test grades of 86, 88, and 78 on his first three tests in geometry. If he wants an average of at least 80 after his fourth test, what score must he make on his fourth test?

Let x represent Brent's score on his fourth test. To find the average of the four scores, add them and find $\frac{1}{4}$ of the sum.

Average ↓ ; is at least 80. ↓ ↓

$$\frac{1}{4}(86 + 88 + 78 + x) \geq 80$$

$$4 \cdot \frac{1}{4}(252 + x) \geq 4 \cdot 80 \qquad \text{Multiply by 4.}$$

$$252 + x \geq 320 \qquad \text{Multiply.}$$

$$252 - 252 + x \geq 320 - 252 \qquad \text{Subtract 252.}$$

$$x \geq 68 \qquad \text{Combine terms.}$$

He must score 68 or more on the fourth test to have an average of *at least* 80. ■

WORK PROBLEM 8 AT THE SIDE.

ANSWER
8. 88 or more

NAME DATE HOUR

2.7 EXERCISES

Graph each inequality on the given number line. See Examples 1–3.

1. $k \leq 4$

2. $r \leq -11$

3. $x < -3$

4. $y < 3$

5. $t > 4$

6. $m > 5$

7. $8 \leq x \leq 10$

8. $3 \leq x \leq 5$

9. $0 < y \leq 10$

10. $-3 \leq x < 5$

11. How can you determine whether to use an open circle or a closed circle at the endpoint when graphing an inequality on a number line?

12. How does the graph of $t \geq -7$ differ from the graph of $t > -7$?

13. Why is it *wrong* to write $3 < x < -2$ to indicate that x is between -2 and 3?

14. If $p < q$ and $r < 0$, which one of the following statements is *false?*
(a) $pr < qr$ **(b)** $pr > qr$ **(c)** $p + r < q + r$ **(d)** $p - r < q - r$

Solve each inequality and graph the solutions. See Example 4.

15. $z - 8 \geq -7$

16. $p - 3 \geq -11$

17. $2k + 3 \geq k + 8$

18. $3x + 7 \geq 2x + 11$

19. $3n + 5 < 2n - 6$

20. $5x - 2 < 4x - 5$

21. Under what conditions must the inequality symbol be reversed when solving an inequality?

22. Explain the steps you would use to solve the inequality $-5x > 20$.

23. Your friend tells you that when solving the inequality $6x < -42$, he reversed the direction of the inequality because of the presence of -42. How would you respond?

24. By what number must you *multiply* both sides of $.2x > 6$ to get just x on the left side?

Solve each inequality and graph the solutions. See Examples 5 and 6.

25. $3x < 18$

26. $5x < 35$

27. $2y \geq -20$

28. $6m \geq -24$

29. $-8t > 24$

30. $-7x > 49$

31. $-x \geq 0$

32. $-k < 0$

33. $-\frac{3}{4}r < -15$

34. $-\frac{7}{8}t < -14$

35. $-.02x \leq .06$

36. $-.03v \geq -.12$

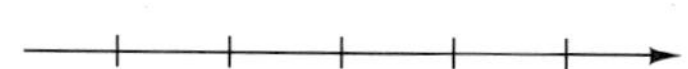

Solve each inequality, and graph the solutions. See Example 7.

37. $5r + 1 \geq 3r - 9$

38. $6t + 3 < 3t + 12$

39. $6x + 3 + x < 2 + 4x + 4$

40. $-4w + 12 + 9w \geq w + 9 + w$

41. $-x + 4 + 7x \leq -2 + 3x + 6$

42. $14y - 6 + 7y > 4 + 10y - 10$

43. $5(x + 3) - 6x \leq 3(2x + 1) - 4x$

44. $2(x - 5) + 3x < 4(x - 6) + 1$

45. $\frac{2}{3}(p + 3) > \frac{5}{6}(p - 4)$

46. $\frac{7}{9}(y - 4) \leq \frac{4}{3}(y + 5)$

47. $4x - (6x + 1) \leq 8x + 2(x - 3)$

48. $2y - (4y + 3) < 6y + 3(y + 4)$

49. $5(2k + 3) - 2(k - 8) > 3(2k + 4) + k - 2$

50. $2(3z - 5) + 4(z + 6) \geq 2(3z + 2) + 3z - 15$

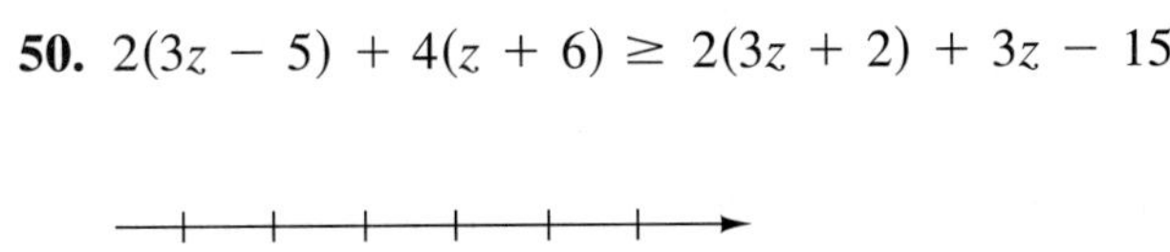

Solve each application of inequalities. See Example 8.

51. When 8 is subtracted from the sum of three times a number and 6, the result is less than 4 more than the number. Find all such numbers.

52. When 2 is added to the difference between six times a number and 5, the result is greater than 13 added to 5 times the number. Find all such numbers.

53. Twylene Johnson has grades of 76 and 81 on her first two algebra tests. If she wants an average of at least 80 after her third test, what score must she make on her third test?

54. Mabimi Pampo has grades of 96 and 86 on his first two geometry tests. What must he score on his third test so that his average is at least 90?

55. The formula for converting Fahrenheit temperature to Celsius is

$$C = \frac{5}{9}(F - 32).$$

If the Celsius temperature on a certain summer day in Houston is never more than 30 degrees, how would you describe the corresponding Fahrenheit temperatures?

56. The formula for converting Celsius temperature to Fahrenheit is

$$F = \frac{9}{5}C + 32.$$

The Fahrenheit temperature of Key West, Florida, has never exceeded 95 degrees. How would you describe this using Celsius temperature?

57. A company will break even or produce a profit from the sales of a product if the revenue R from selling the product is at least equal to the cost C of producing it. Suppose that the cost C (in dollars) to produce x units of bicycle helmets is $C = 50x + 5000$, while the revenue R (in dollars) collected from the sale of x units is $R = 60x$. For what values of x does the company break even or produce a profit?

58. See Exercise 57. The cost to produce x units of basketball cards is $C = 100x + 6000$ (in dollars), and the revenue collected from selling x units is $R = 500x$ (in dollars). For what values of x does the company break even or produce a profit?

59. Audrey earned \$200 at odd jobs during July, \$300 during August, and \$225 during September. If her average salary for the four months from July through October is to be at least \$250, how much must she earn during October?

60. In order to qualify for a company pension plan, an employee must average at least \$1000 per month in earnings. During the first four months of the year, an employee made \$900, \$1200, \$1040, and \$760. What amount of earnings during the fifth month will qualify the employee for the pension plan?

PREVIEW EXERCISES

Evaluate each expression. See Section 1.1.

61. 2^6 **62.** $(-2)^5$ **63.** $(-2)^6$ **64.** $\left(\frac{2}{3}\right)^3$

Perform the indicated operations. See Sections 1.4, 1.5, and 1.6.

65. $-3 + 8 + (-2)$ **66.** $5 - (-3) - (-8)$ **67.** $1 - [2 - (-4)]$

68. $(-3)(2)(-5)$

Evaluate each expression for $x = 4$. See Section 1.2.

69. $3x^2 - 4x + 7$ **70.** $-2x^3 - 5x^2 + x - 3$

QUEST FOR NUMERACY

Figurate Numbers

The Greek mathematician Pythagoras (c. 540 B.C.) was the founder of the Pythagorean brotherhood. This group studied, among other things, numbers of geometric arrangements of points, such as *triangular numbers, square numbers,* and *pentagonal numbers.* The figure shown here illustrates the first few of these types of numbers.

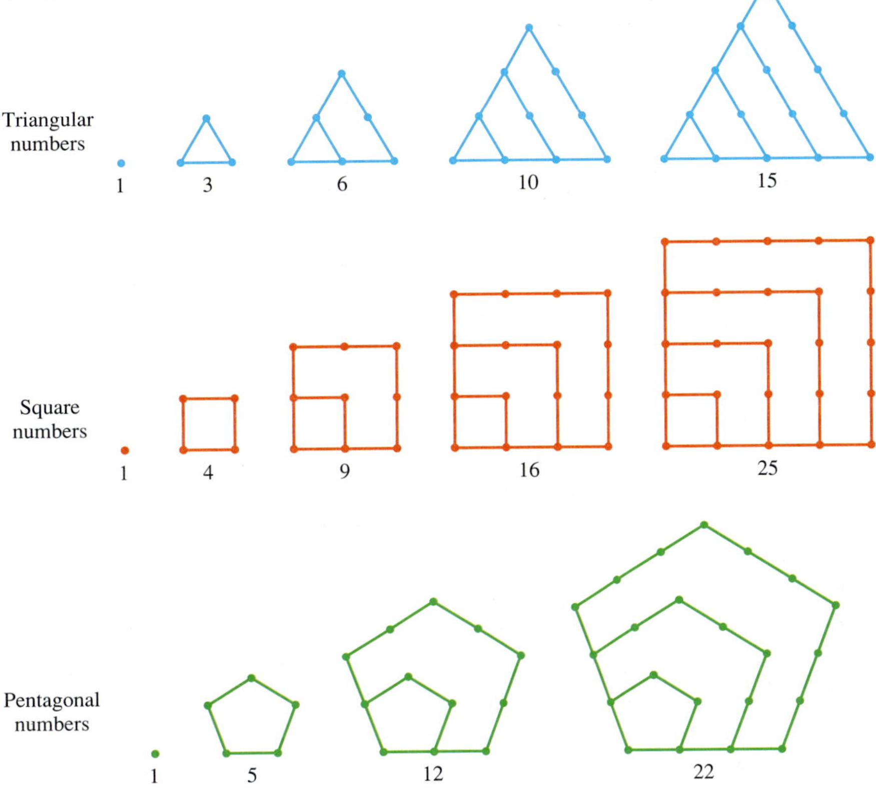

These arrangements continue to include hexagonal numbers (six-sided figures), heptagonal, octagonal, and so on. The patterns among these numbers are virtually endless.

FOR GROUP DISCUSSION

1. Use patterns to complete the table of figurate numbers. Then discuss some of the patterns you notice.

Figurate Number	*1st*	*2nd*	*3rd*	*4th*	*5th*	*6th*	*7th*	*8th*
Triangular	1	3	6	10	15	21		
Square	1	4	9	16	25			
Pentagonal	1	5	12	22				
Hexagonal	1	6	15					
Heptagonal	1	7						
Octagonal	1							

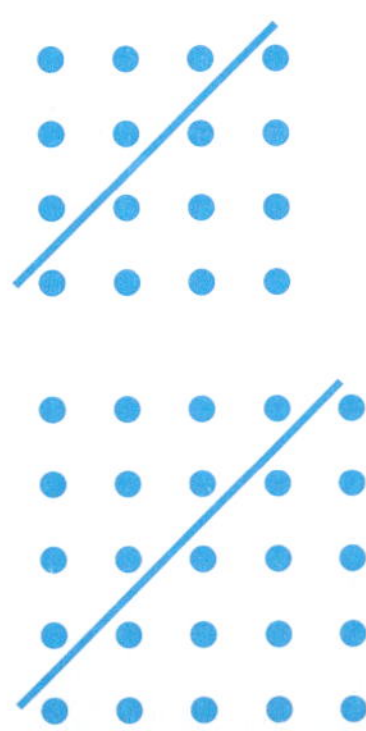

2. Observe the figures shown in the margin. They suggest a relationship between two types of figurate numbers. Now complete the following statement: Any _____ number can be written as the sum of two _____ numbers. (Verify this for 36.)

CHAPTER 2 SUMMARY

KEY TERMS

2.1	**linear equation**	A linear equation is an equation that can be written in the form $Ax + B = C$, for real numbers A, B, and C, with $A \neq 0$.
	equivalent equations	Equations that have the same solution set are equivalent equations.
2.5	**vertical angles**	Vertical angles are angles formed by intersecting lines. They have the same measure.
	straight angle	A straight angle measures 180°.
2.6	**percentage**	A percentage is a part of the whole amount.
	ratio	A ratio is a quotient of two quantities with the same units.
	proportion	A proportion is a statement that two ratios are equal.
	cross products	The method of cross products provides a way of determining whether a proportion is true.
2.7	**inequality**	An inequality is a statement with algebraic expressions related by $<$, $\leq$, $>$, or $\geq$.

NEW SYMBOLS

***a* to *b*, *a* : *b*, or $\frac{a}{b}$** the ratio of a to b

QUICK REVIEW

Concepts	Examples
2.1 The Addition Property of Equality The same number may be added to (or subtracted from) each side of an equation without changing the solution.	$x - 6 = 12$ $x - 6 + 6 = 12 + 6$ Add 6. $x = 18$ Combine terms.
2.2 The Multiplication Property of Equality Each side of an equation may be multiplied (or divided) by the same nonzero number without changing the solution.	$\frac{3}{4}x = -9$ $\frac{4}{3} \cdot \frac{3}{4}x = \frac{4}{3}(-9)$ Multiply by $\frac{4}{3}$. $x = -12$

Concepts	Examples
2.3 More on Solving Linear Equations	Solve the equation $2x + 3(x + 1) = 38$.
1. Combine like terms to simplify each side.	**1.** $2x + 3x + 3 = 38$ Distributive property $5x + 3 = 38$ Combine like terms.
2. Get the variable term on one side, a number on the other.	**2.** $5x + 3 - 3 = 38 - 3$ Subtract 3. $5x = 35$ Combine terms.
3. Get the equation into the form $x =$ a number.	**3.** $\frac{5x}{5} = \frac{35}{5}$ Divide by 5. $x = 7$ Reduce.
4. Check by substituting the result into the original equation.	**4.** $2x + 3(x + 1) = 38$ Check. $2(7) + 3(7 + 1) = 38$? Let $x = 7$. $14 + 24 = 38$? Add; then multiply. $38 = 38$ True
2.4 An Introduction to Applied Problems	One number is 5 more than another. Their sum is 21. Find both numbers.
1. Choose a variable to represent the unknown.	**1.** Let x be the smaller number.
2. Determine expressions for any other unknown quantities, using the variable. Draw figures or diagrams if they apply.	**2.** Let $x + 5$ be the larger number.
3. Translate the problem into an equation.	**3.** $x + (x + 5) = 21$
4. Solve the equation.	**4.** $2x + 5 = 21$ Combine terms. $2x + 5 - 5 = 21 - 5$ Subtract 5. $2x = 16$ Combine terms. $\frac{2x}{2} = \frac{16}{2}$ Divide by 2. $x = 8$ Reduce.
5. Answer the question asked in the problem.	**5.** The numbers are 8 and 13.
6. Check your solution by using the original words of the problem. Be sure that the answer is appropriate and makes sense.	**6.** 13 is 5 more than 8, and $8 + 13 = 21$. It checks.

Concepts	Examples
2.5 Formulas and Geometry Applications	
To find the values of one of the variables in a formula, given values for the others, substitute the known values into the formula.	Find L if $A = LW$, given that $A = 24$ and $W = 3$. $24 = L \cdot 3$ $A = 24, W = 3$ $\frac{24}{3} = \frac{L \cdot 3}{3}$ Divide by 3. $8 = L$ Reduce.
To solve a formula for one of the variables, isolate that variable by treating the other variables as numbers and using the steps for solving equations.	Solve $P = 2L + 2W$ for W. $P - 2L = 2L + 2W - 2L$ Subtract $2L$. $P - 2L = 2W$ Combine terms. $\frac{P - 2L}{2} = \frac{2W}{2}$ Divide by 2. $\frac{P - 2L}{2} = W$ or $W = \frac{P - 2L}{2}$
2.6 Percent; Ratio and Proportion	
To write a ratio, express quantities in the same units.	4 feet to 8 inches $= 48$ inches to 8 inches $= \frac{48}{8} = \frac{6}{1}$
To solve a proportion, use the method of cross products.	Solve $\frac{x}{12} = \frac{35}{60}$. $60x = 12 \cdot 35$ Cross products $60x = 420$ Multiply. $\frac{60x}{60} = \frac{420}{60}$ Divide by 60. $x = 7$ Reduce.
2.7 The Addition and Multiplication Properties of Inequality	
To solve an inequality: **1.** Clear parentheses and combine like terms. **2.** Add or subtract the same number on each side to get the variable term on one side and a number on the other side. **3.** Multiply or divide by the same number on each side to get the form $x > a$ or $x < a$. (When multiplying or dividing by a negative number, reverse the direction of the inequality symbol.)	Solve $3(1 - x) + 5 - 2x > 9 - 6$. $3 - 3x + 5 - 2x > 9 - 6$ Clear parentheses. $8 - 5x > 3$ Combine terms. $8 - 5x - 8 > 3 - 8$ Subtract 8. $-5x > -5$ Combine terms. $\frac{-5x}{-5} < \frac{-5}{-5}$ Divide by -5; change $>$ to $<$. $x < 1$ Reduce. −2 −1 0 1 2

NAME DATE HOUR

CHAPTER 2 REVIEW EXERCISES

[2.1–2.3] *Solve each equation. Check the solution.*

1. $x - 7 = 2$

2. $4r - 6 = 10$

3. $5x + 8 = 4x + 2$

4. $8t = 7t + \frac{3}{2}$

5. $(4r - 8) - (3r + 12) = 0$

6. $7(2x + 1) = 6(2x - 9)$

7. $-\frac{6}{5}y = -18$

8. $\frac{1}{2}r - \frac{1}{6}r + 3 = 2 + \frac{1}{6}r + 1$

9. $3x - (-2x + 6) = 4(x - 4) + x$

10. $.10(x + 80) + .20x = 14$

[2.4] *Solve each problem.*

11. If 7 is added to 5 times a number, the result is equal to 3 times the number. Find the number.

12. If 4 is subtracted from twice a number, the result is 36. Find the number.

13. The land area of Hawaii is 5213 square miles greater than that of Rhode Island. Together, the areas total 7637 square miles. What is the area of each state?

14. The height of Seven Falls in Colorado is $\frac{5}{2}$ the height (in feet) of Twin Falls in Idaho. The sum of the heights is 420 feet. Find the height of each.

15. The supplement of an angle measures 10 times the measure of its complement. What is the measure of the angle (in degrees)?

16. During a recent baseball season, Bret Saberhagen pitched 9 fewer innings than Jack Morris. Mark Langston pitched 6 more innings than Morris. How many innings did each pitch if their total number of innings pitched was 795?

[2.5] *A formula is given in each exercise, along with the values for all but one of the variables. Find the value of the variable that is not given.*

17. $A = \frac{1}{2}bh$; $A = 44$, $b = 8$

18. $A = \frac{1}{2}h(b + B)$; $b = 3$, $B = 4$, $h = 8$

19. $C = 2\pi r$; $C = 29.83$, $\pi = 3.14$

20. $V = \frac{4}{3}\pi r^3$; $r = 6$, $\pi = 3.14$

Solve the formula for the specified variable.

21. $A = LW$ for L

22. $A = \frac{1}{2}h(b + B)$ for h

Find the measure of each marked angle.

23.

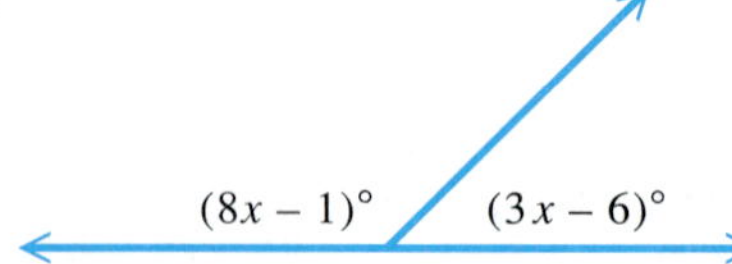

24.

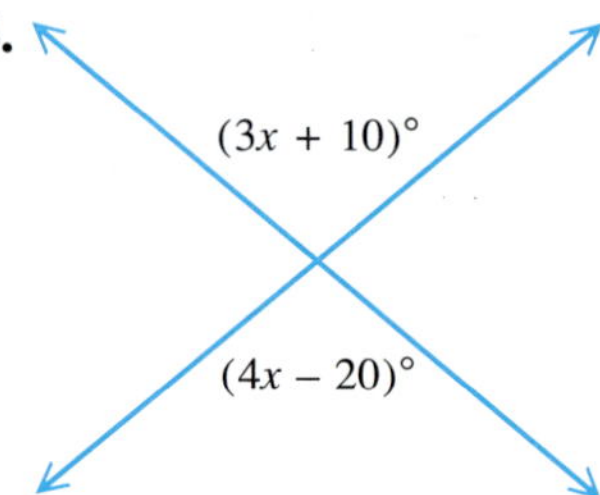

Solve each application of geometry.

25. A cinema screen in Indonesia has length 92.75 feet and width 70.5 feet. What is the perimeter? What is the area?

26. The Ziegfield Room in Reno, Nevada, has a circular turntable on which its show girls dance. The circumference of the turntable is 62.5 feet. What is the diameter of the turntable? What is the radius? What is its area? (Use $\pi = 3.14$.)

[2.6]

27. What is 23% of 76? 17.48

28. What percent of 12 is 21?

29. 6 is what percent of 18?

30. 36% of what number is 900?

31. Vinh must pay 6.5% sales tax on a new car. The cost of the car is $17,200. Find the amount of the tax.

32. An ad for steel-belted radials promises 15% better mileage when using them. Alexandra's import now gets 380 miles on a tank of gas. If she is to believe the ad, how many miles should she get per tank if she uses these tires?

Write a ratio for each word phrase. Write fractions in lowest terms.

33. 60 centimeters to 40 centimeters

34. 5 days to 2 weeks

35. 90 inches to 10 feet

36. 3 months to 3 years

Solve each proportion.

37. $\frac{p}{21} = \frac{5}{30}$

38. $\frac{5 + x}{3} = \frac{2 - x}{6}$

39. $\frac{y}{5} = \frac{6y - 5}{11}$

40. Explain how 40% can be expressed as a ratio of two whole numbers.

Solve each of the following problems involving proportion.

41. If 2 pounds of fertilizer will cover 150 sqaure feet of lawn, how many pounds would be needed to cover 500 square feet?

42. If 8 ounces of medicine must be mixed with 20 ounces of water, how many ounces of medicine should be mixed with 90 ounces of water?

43. The tax on a \$24.00 item is \$2.04. How much tax would be paid on a \$36.00 item?

44. The distance between two cities on a road map is 32 centimeters. The two cities are actually 150 kilometers apart. The distance on the map between two other cities is 80 centimeters. How far apart are these cities?

[2.7] *Graph each inequality on the number line provided.*

45. $p \geq -4$

46. $x < 7$

47. $-5 \leq y < 6$

48. $r \geq \frac{1}{2}$

Solve each inequality. Graph the solutions.

49. $y + 6 \geq 3$

50. $5t < 4t + 2$

51. $-6x \leq -18$

52. $8(k - 5) - (2 + 7k) \geq 4$

53. $4x - 3x > 10 - 4x + 7x$

54. $3(2w + 5) + 4(8 + 3w) < 5(3w + 2) + 2w$

55. Carlotta Valdez has grades of 94 and 88 on her first two calculus tests. What possible scores on a third test will give her an average of at least 90?

56. If nine times a number is added to 6, the result is at most 3. Find all such numbers.

Mixed Review Exercises

Solve each of the following.

57. $\frac{y}{7} = \frac{y - 5}{2}$

58. $I = prt$ for r

59. $-2x > -4$

60. $2k - 5 = 4k + 13$

61. $.05x + .02x = 4.9$

62. $2 - 3(y - 5) = 4 + y$

63. $9x - (7x + 2) = 3x + (2 - x)$

64. $\frac{1}{3}s + \frac{1}{2}s + 7 = \frac{5}{6}s + 5 + 2$

Solve each problem.

65. Two-thirds of a number added to the number is 10. What is the number?

66. If three-fourths of a number is subtracted from twice the number, the result is 15. Find the number.

67. Buddy defeated Bob in an election. Buddy had twice as many votes as Bob, and together they had 1800 votes. How many votes did each of the candidates receive?

68. John and Gwen commute to work. John travels three times as far as Gwen each day, and together they travel 112 miles. How far does each travel?

69. On a recent diet, Duc lost 18 pounds more than Hoa. Their total weight loss was 42 pounds. How much did Hoa lose?

70. Rick and Steve drove from different towns to a family reunion. Rick drove 43 miles farther than Steve. The two men drove a total of 293 miles. How far did Steve drive to the reunion?

71. A teacher noted that one week he had graded 32 more algebra tests than geometry tests. He had graded 102 tests altogether. How many geometry tests did he grade that week?

72. A parking lot attendant parked 63 cars one day. He parked 11 fewer large cars than small cars. How many small cars did he park?

73. The perimeter of a rectangle is 288 feet. The length is 4 feet longer than the width. Find the width.

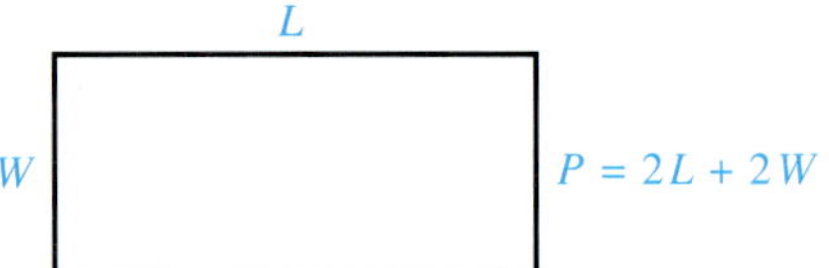

74. The perimeter of a triangle is 96 meters. One side is twice as long as another, and the third side is 30 meters long. What is the length of the longest side?

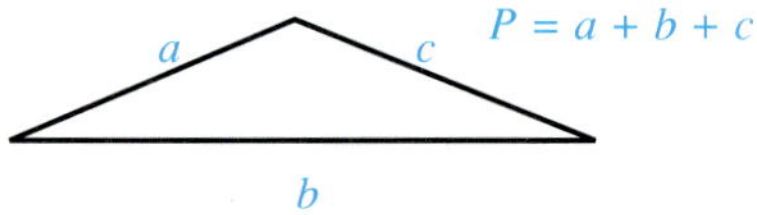

75. The area of a triangle is 182 square inches. The height is 14 inches. Find the length of the base.

76. The perimeter of a rectangle is 75 inches. The width is 17 inches. What is the length?

77. Find the measure of each marked angle.

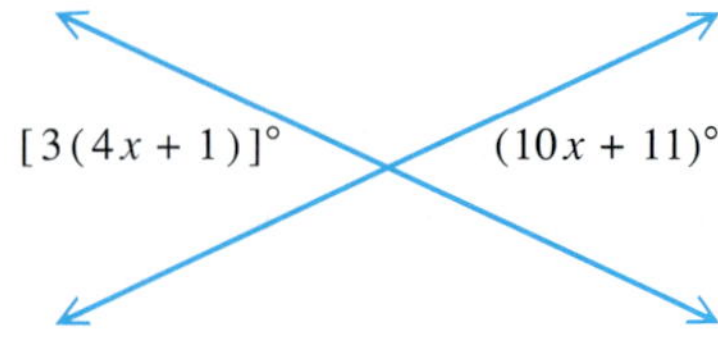

78. Nalima has grades of 82 and 96 on her first two English tests. What must she make on her third test so that her average will be at least 90?

NAME DATE HOUR

CHAPTER 2 TEST

Solve each equation and check the solution.

1. $3x - 7 = 11$ **1.** ____

2. $5x + 9 = 7x + 21$ **2.** ____

3. $2 - 3(y - 5) = 3 + (y + 1)$ **3.** ____

4. $2.3x + 13.7 = 1.3x + 2.9$ **4.** ____

5. $7 - (m - 4) = -3m + 2(m + 1)$ **5.** ____

6. $-\frac{4}{7}x = -12$ **6.** ____

7. $.06(x + 20) + .08(x - 10) = 4.6$ **7.** ____

8. $-8(2x + 4) = -4(4x + 8)$ **8.** ____

Solve each problem.

9. If 3 is subtracted from four times a number, the result is 10 less than five times the number. What is the number? **9.** ____

10. The Golden Gate Bridge in San Francisco is 2605 feet longer than the Brooklyn Bridge. Together, their spans total 5795 feet. How long is each bridge? **10.** ____

11. A piece of string is 40 centimeters long. It is cut into three pieces. The longest piece is 3 times as long as the middle-sized piece, and the shortest piece is 23 centimeters shorter than the longest piece. Find the lengths of the three pieces. **11.** ____

12. The formula for the perimeter of a rectangle is $P = 2L + 2W$. If $P = 116$ and $L = 40$, find W. **12.** ____

13. ______________

13. Solve the formula $P = 2L + 2W$ for W.

Find the measure of each marked angle.

14. ______________

15. ______________

14.

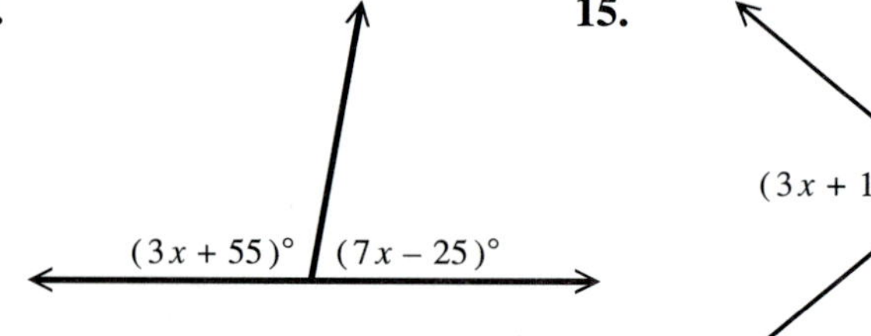

15.

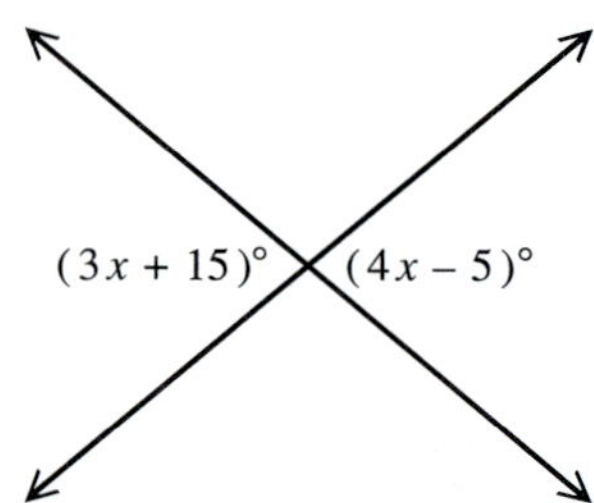

16. ______________

16. Find the measure of an angle if its supplement measures 10° more than three times its complement.

Solve each proportion.

17. ______________

17. $\frac{z}{8} = \frac{12}{16}$

18. ______________

18. $\frac{y + 5}{3} = \frac{y - 3}{4}$

19. ______________

19. Which is the better buy for processed cheese slices: 8 slices for \$2.19 or 12 slices for \$3.30?

20. ______________

20. The distance between Milwaukee and Boston is 1050 miles. On a certain map, this distance is represented by 42 inches. On the same map, Seattle and Cincinnati are 92 inches apart. What is the actual distance between Seattle and Cincinnati?

21. ______________

21. Of the 56 people who signed the Declaration of Independence, 25% were judges. How many of the signers were judges?

Solve each inequality and graph the solutions.

22. ——+——+——+——+——+——+——➤

22. $-3x > -33$

23. ——+——+——+——+——+——+——➤

23. $-4x + 2(x - 3) \geq 4x - (3 + 5x) - 7$

24. ______________

24. Julie Tarr has grades of 84 and 100 on her first two college algebra tests. What must she score on her third test so that her average is at least 90?

25. ______________

25. Write a short explanation of the additional (extra) rule that must be remembered when solving an inequality (as opposed to solving an equation).

NAME DATE HOUR

CUMULATIVE REVIEW EXERCISES CHAPTERS R-2

Beginning with this chapter, each chapter in the text will conclude with a set of cumulative review exercises designed to cover the major topics from the beginning of the course. This feature will allow the student to constantly review topics that have been introduced up to that point.

Write each fraction in lowest terms.

1. $\frac{15}{40}$

2. $\frac{108}{144}$

Work the following problems.

3. $\frac{5}{6} + \frac{1}{4} + \frac{7}{15}$

4. $16\frac{7}{8} - 3\frac{1}{10}$

5. $\frac{9}{8} \cdot \frac{16}{3}$

6. $\frac{3}{4} \div \frac{5}{8}$

7. $4.8 + 12.5 + 16.73$

8. $56.3 - 28.99$

9. $(67.8)(.45)$

10. $236.46 \div 4.2$

11. In making dresses, Earth Works uses $\frac{5}{8}$ yard of trim per dress. How many yards of trim would be used to make 56 dresses?

12. A cook wants to increase a recipe that serves 6 to make enough for 20 people. The recipe calls for $1\frac{1}{4}$ cups of cheese. How much cheese will be needed to serve 20?

13. One dog weighs $8\frac{1}{3}$ pounds, and another dog weighs $12\frac{5}{8}$ pounds. Find the total weight of both dogs.

14. A purchasing agent bought 3 desks at $211.40 each and 3 chairs for $195, $189.95, and $168.50. What was the final bill (without tax)?

Tell whether each of the following is true or false.

15. $\frac{8(7) - 5(6 + 2)}{3 \cdot 5 + 1} \geq 1$

16. $\frac{4(9 + 3) - 8(4)}{2 + 3 - 3} \geq 2$

Perform the indicated operations.

17. $-11 + 20 + (-2)$

18. $13 + (-19) + 7$

19. $9 - (-4)$

20. $-2(-5)(-4)$

21. $\frac{4 \cdot 9}{-3}$

22. $\frac{8}{7 - 7}$

23. $(-5 + 8) + (-2 - 7)$

24. $(-7 - 1)(-4) + (-4)$

25. $\frac{-3 - (-5)}{1 - (-1)}$

26. $\frac{6(-4) - (-2)(12)}{3^2 + 7^2}$

27. $\frac{(-3)^2 - (-4)(2^4)}{5 \cdot 2 - (-2)^3}$

28. $\frac{-2(5^3) - 6}{4^2 + 2(-5) + (-2)}$

Find the value of each expression when $x = -2$, $y = -4$, and $z = 3$.

29. $xz^3 - 5y^2$

30. $\frac{3x - y^3}{-4z}$

Name the property illustrated by each of the following examples.

31. $7(k + m) = 7k + 7m$

32. $3 + (5 + 2) = 3 + (2 + 5)$

33. $7 + (-7) = 0$

34. $3.5(1) = 3.5$

Simplify the following expressions by combining terms.

35. $4p - 6 + 3p - 8$

36. $-4(k + 2) + 3(2k - 1)$

Solve the following equations and check each solution.

37. $2r - 6 = 8$

38. $2(p - 1) = 3p + 2$

39. $4 - 5(a + 2) = 3(a + 1) - 1$

40. $2 - 6(z + 1) = 4(z - 2) + 10$

41. $-(m - 1) = 3 - 2m$

42. $\frac{y - 2}{3} = \frac{2y + 1}{5}$

43. $\frac{2x + 3}{5} = \frac{x - 4}{2}$

44. $\frac{2}{3}y + \frac{3}{4}y = -17$

Solve the formula for the indicated variable.

45. $P = a + b + c$ for c

46. $P = 4s$ for s

Solve the following inequalities. Graph the solutions.

47. $-5z \geq 4z - 18$

48. $6(r - 1) + 2(3r - 5) \leq -4$

Solve the following problems.

49. The purchasing agent in Exercise 14 paid a sales tax of $6\frac{1}{4}\%$ on his purchase. What was the final bill, including tax?

50. A car has a price of \$5000. For trading in her old car, Rosalie will get 25% off. Find the price of the car with the trade-in.

51. Margaret Boyle bought textbooks at the college bookstore for \$52.47, including 6% sales tax. What did the books cost?

52. Louise Palla received a bill from her credit card company for \$104.93. The bill included interest at $1\frac{1}{2}\%$ per month for one month and a \$5.00 late charge. How much did her purchases amount to?

53. The perimeter of a rectangle is 98 centimeters. The width is 19 centimeters. Find the length.

54. The area of a triangle is 104 square inches. The base is 13 inches. Find the height.

55. On the first two days of their vacation to Florida, Wally and Joanna drove 430 miles and 470 miles. If they must average at least 450 miles per day, what possible distances must they drive on the third day?

56. Vern paid \$57 more to tune up his Bronco than his Oldsmobile. He paid \$257 in all. How much did he pay for the tune-up on the Oldsmobile?

$.05x \quad x^{-n} \quad 17^0 \quad 5.1 \times 10^{-3}$

3 Exponents and Polynomials

3.1 THE PRODUCT RULE AND POWER RULES FOR EXPONENTS

OBJECTIVES

1. Use exponents.
2. Use the product rule for exponents.
3. Use the rule $(a^m)^n = a^{mn}$.
4. Use the rule $(ab)^m = a^m b^m$.
5. Use the rule $\left(\frac{a}{b}\right)^m = \frac{a^m}{b^m}$.
6. Use combinations of the rules for exponents.

FOR EXTRA HELP

Tape 4

SSM pp. 96–98

MAC: A IBM: A

1 In Chapter 1 we used exponents to write repeated products. Recall that in the expression 5^2, the number 5 is called the *base* and 2 is called the *exponent* or *power*. The expression 5^2 is called an *exponential expression*. Usually we do not write a quantity with an exponent of 1; however, sometimes it is convenient to do so. In general, for any quantity a, $a = a^1$.

EXAMPLE 1 *Using Exponents*

Write $3 \cdot 3 \cdot 3 \cdot 3 \cdot 3$ in exponential form, and evaluate the exponential expression.

Since 3 occurs as a factor five times, the base is 3 and the exponent is 5. The exponential expression is 3^5, read "3 to the fifth power" or simply "3 to the fifth." The value is

$$3^5 = 3 \cdot 3 \cdot 3 \cdot 3 \cdot 3 = 243. \blacksquare$$

WORK PROBLEM 1 AT THE SIDE.

EXAMPLE 2 *Evaluating an Exponential Expression*

Evaluate each exponential expression. Name the base and the exponent.

	Base	Exponent
(a) $5^4 = 5 \cdot 5 \cdot 5 \cdot 5 = 625$	5	4
(b) $-5^4 = -1 \cdot 5^4 = -1 \cdot (5 \cdot 5 \cdot 5 \cdot 5) = -625$	5	4
(c) $(-5)^4 = (-5)(-5)(-5)(-5) = 625$	-5	4 ■

Caution It is important to understand the difference between parts (b) and (c) of Example 2. In -5^4 the lack of parentheses shows that the exponent 4 refers only to the base 5, and not -5; in $(-5)^4$ the parentheses show that the exponent 4 refers to the base -5. In summary, $-a^n$ and $(-a)^n$ are not always the same.

Expression	*Base*	*Exponent*	*Example*
$-a^n$	a	n	$-3^2 = -(3 \cdot 3) = -9$
$(-a)^n$	$-a$	n	$(-3)^2 = (-3)(-3) = 9$

WORK PROBLEM 2 AT THE SIDE.

1. Write $2 \cdot 2 \cdot 2 \cdot 2$ in exponential form, and evaluate.

2. Evaluate each exponential expression. Name the base and the exponent.

(a) $(-2)^5$ **(b)** -2^5

(c) -4^2 **(d)** $(-4)^2$

ANSWERS

1. $2^4 = 16$
2. (a) -32; -2; 5 (b) -32; 2; 5 (c) -16; 4; 2 (d) 16; -4; 2

3. Find each product by the product rule, if possible.

(a) $8^2 \cdot 8^5$

(b) $(-7)^5 \cdot (-7)^3$

(c) $y^3 \cdot y$

(d) $4^2 \cdot 3^5$

(e) $6^4 + 6^2$

2 By the definition of exponential expressions,

$$\begin{aligned} 2^4 \cdot 2^3 &= \underbrace{(2 \cdot 2 \cdot 2 \cdot 2)}_{\text{4 factors}}\underbrace{(2 \cdot 2 \cdot 2)}_{\text{3 factors}} \\ &= \underbrace{2 \cdot 2 \cdot 2 \cdot 2 \cdot 2 \cdot 2 \cdot 2}_{4 + 3 = 7 \text{ factors}} \\ &= 2^7. \end{aligned}$$

Also,

$$\begin{aligned} 6^2 \cdot 6^3 &= (6 \cdot 6)(6 \cdot 6 \cdot 6) \\ &= 6 \cdot 6 \cdot 6 \cdot 6 \cdot 6 \\ &= 6^5. \end{aligned}$$

Generalizing from these examples, $2^4 \cdot 2^3 = 2^{4+3} = 2^7$ and $6^2 \cdot 6^3 = 6^5$. In each case, adding the exponents gives the exponent of the product, suggesting the **product rule for exponents.**

PRODUCT RULE FOR EXPONENTS

For any positive integers m and n, $a^m \cdot a^n = a^{m+n}$.
(Keep the same base and add the exponents.)
Example: $6^2 \cdot 6^5 = 6^{2+5} = 6^7$

EXAMPLE 3 *Using the Product Rule*

Use the product rule for exponents to find each product, if possible.

(a) $6^3 \cdot 6^5 = 6^{3+5} = 6^8$ by the product rule.

(b) $(-4)^7(-4)^2 = (-4)^{7+2} = (-4)^9$ by the product rule.

(c) $x^2 \cdot x = x^2 \cdot x^1 = x^{2+1} = x^3$

(d) $m^4 \cdot m^3 = m^{4+3} = m^7$

(e) The product rule does not apply to the product $2^3 \cdot 3^2$ because the bases are different.

$$2^3 \cdot 3^2 = 8 \cdot 9 = 72.$$

(f) The product rule does not apply to $2^3 + 2^4$ because it is a *sum,* not a *product.*

$$2^3 + 2^4 = 8 + 16 = 24$$ ■

Caution The bases must be the same before we can apply the product rule for exponents.

WORK PROBLEM 3 AT THE SIDE.

EXAMPLE 4 *Using the Product Rule*

Multiply $2x^3$ and $3x^7$.

Since $2x^3$ means $2 \cdot x^3$ and $3x^7$ means $3 \cdot x^7$, we use the associative and commutative properties and the product rule to get

$$2x^3 \cdot 3x^7 = 2 \cdot 3 \cdot x^3 \cdot x^7 = 6x^{10}.$$ ■

ANSWERS
3. (a) 8^7 (b) $(-7)^8$ (c) y^4
(d) cannot use the product rule
(e) cannot use the product rule

Caution Be sure you understand the difference between *adding* and *multiplying* exponential expressions. For example,

$$8x^3 + 5x^3 = 13x^3$$

but $$(8x^3)(5x^3) = 8 \cdot 5 \cdot x^{3+3} = 40x^6.$$

WORK PROBLEM 4 AT THE SIDE.

3 We can simplify an expression such as $(8^3)^2$ with the product rule for exponents, as follows.

$$(8^3)^2 = (8^3)(8^3) = 8^{3+3} = 8^6$$

The product of the exponents, 3 and 2, in $(8^3)^2$ gives the exponent in 8^6. As another example,

$$\begin{aligned}(5^2)^3 &= 5^2 \cdot 5^2 \cdot 5^2 \\ &= 5^{2+2+2} \\ &= 5^6,\end{aligned}$$

and $2 \cdot 3 = 6$. These examples suggest power rule (a) for exponents.

POWER RULE (a) FOR EXPONENTS

For any positive integers m and n, $(a^m)^n = a^{mn}$.
(Raise a power to a power by multiplying exponents.)
Example: $(3^2)^4 = 3^{2 \cdot 4} = 3^8$

EXAMPLE 5 *Using Power Rule (a)*

Use power rule (a) for exponents to simplify each expression.

(a) $(2^5)^3 = 2^{5 \cdot 3} = 2^{15}$

(b) $(5^7)^2 = 5^{7 \cdot 2} = 5^{14}$

(c) $(x^2)^5 = x^{2 \cdot 5} = x^{10}$

(d) $(n^3)^2 = n^{3 \cdot 2} = n^6$ ■

WORK PROBLEM 5 AT THE SIDE.

4 The properties studied in Chapter 1 can be used to develop two more rules for exponents. Using the definition of an exponential expression and the commutative and associative properties, we can evaluate the expression $(4 \cdot 8)^3$ as shown below.

$$\begin{aligned}(4 \cdot 8)^3 &= (4 \cdot 8)(4 \cdot 8)(4 \cdot 8) && \text{Definition of exponent} \\ &= 4 \cdot 4 \cdot 4 \cdot 8 \cdot 8 \cdot 8 && \text{Commutative and associative properties} \\ &= 4^3 \cdot 8^3 && \text{Definition of exponent}\end{aligned}$$

This example suggests power rule (b) for exponents.

POWER RULE (b) FOR EXPONENTS

For any positive integer m, $(ab)^m = a^m b^m$.
(Raise a product to a power by raising each factor to the power.)
Example: $(2p)^5 = 2^5 p^5$

4. Multiply.

(a) $5m^2 \cdot 2m^6$

(b) $3p^5 \cdot 9p^4$

(c) $-7p^5 \cdot (3p^8)$

5. Simplify each expression.

(a) $(5^3)^4$

(b) $(6^2)^5$

(c) $(3^2)^4$

(d) $(a^6)^5$

ANSWERS
4. (a) $10m^8$ (b) $27p^9$ (c) $-21p^{13}$
5. (a) 5^{12} (b) 6^{10} (c) 3^8 (d) a^{30}

EXAMPLE 6 *Using Power Rule (b)*

Use power rule (b) to simplify each expression.

(a) $(3xy)^2 = 3^2x^2y^2$
$= 9x^2y^2$

(b) $9(pq)^2 = 9(p^2q^2)$ Power rule (b)
$= 9p^2q^2$ Multiply.

(c) $5(2m^2p^3)^4 = 5[2^4(m^2)^4(p^3)^4]$ Power rule (b)
$= 5(2^4m^8p^{12})$ Power rule (a)
$= 5 \cdot 2^4m^8p^{12}$
$= 80m^8p^{12}$ $5 \cdot 2^4 = 5 \cdot 16 = 80$ ■

WORK PROBLEM 6 AT THE SIDE.

5 Since the quotient $\frac{a}{b}$ can be written as $a \cdot \frac{1}{b}$, we can use power rule (b), together with some of the properties of real numbers, to get power rule (c) for exponents.

POWER RULE (c) FOR EXPONENTS

For any positive integer m, $\left(\frac{a}{b}\right)^m = \frac{a^m}{b^m}$ $(b \neq 0)$.

(Raise a quotient to a power by raising both the numerator and the denominator to the power.)

Example: $\left(\frac{5}{3}\right)^2 = \frac{5^2}{3^2}$

EXAMPLE 7 *Using Power Rule (c)*

Simplify each expression.

(a) $\left(\frac{2}{3}\right)^5 = \frac{2^5}{3^5}$

(b) $\left(\frac{m}{n}\right)^4 = \frac{m^4}{n^4}$, $n \neq 0$ ■

WORK PROBLEM 7 AT THE SIDE.

The rules for exponents discussed in this section are summarized below. These rules are basic to the study of algebra and should be *memorized.*

RULES FOR EXPONENTS

For positive integers m and n:

			Examples
Product rule		$a^m \cdot a^n = a^{m+n}$	$6^2 \cdot 6^5 = 6^{2+5} = 6^7$
Power rules	(a)	$(a^m)^n = a^{mn}$	$(3^2)^4 = 3^{2 \cdot 4} = 3^8$
	(b)	$(ab)^m = a^mb^m$	$(2p)^5 = 2^5p^5$
	(c)	$\left(\frac{a}{b}\right)^m = \frac{a^m}{b^m}$ $(b \neq 0)$	$\left(\frac{5}{3}\right)^2 = \frac{5^2}{3^2}$

6. Simplify.

(a) $5(mn)^3$

(b) $(3a^2b^4)^5$

(c) $(5m^2)^3$

7. Simplify. Assume all variables represent nonzero real numbers.

(a) $\left(\frac{5}{2}\right)^4$

(b) $\left(\frac{p}{q}\right)^2$

(c) $\left(\frac{r}{t}\right)^3$

ANSWERS

6. (a) $5m^3n^3$ (b) $3^5a^{10}b^{20}$ (c) 5^3m^6

7. (a) $\frac{5^4}{2^4}$ (b) $\frac{p^2}{q^2}$ (c) $\frac{r^3}{t^3}$

6 As shown in the next example, more than one rule may be needed to simplify an expression.

EXAMPLE 8 *Using Combinations of Rules*

Simplify each expression.

(a) $\left(\frac{2}{3}\right)^2 \cdot 2^3$

$$\begin{aligned}\left(\frac{2}{3}\right)^2 \cdot 2^3 &= \frac{2^2}{3^2} \cdot \frac{2^3}{1} && \text{Power rule (c)}\\ &= \frac{2^2 \cdot 2^3}{3^2 \cdot 1} && \text{Multiply fractions.}\\ &= \frac{2^5}{3^2} && \text{Product rule}\end{aligned}$$

(b) $(5x)^3(5x)^4$

$$\begin{aligned}(5x)^3(5x)^4 &= (5x)^7 && \text{Product rule}\\ &= 5^7x^7 && \text{Power rule (b)}\end{aligned}$$

(c) $(2x^2y^3)^4(3xy^2)^3$

$$\begin{aligned}(2x^2y^3)^4(3xy^2)^3 &= 2^4(x^2)^4(y^3)^4 \cdot 3^3x^3(y^2)^3 && \text{Power rule (b)}\\ &= 2^4 \cdot 3^3x^8y^{12}x^3y^6 && \text{Power rule (a)}\\ &= 16 \cdot 27x^{11}y^{18} && \text{Product rule}\\ &= 432x^{11}y^{18}\end{aligned}$$

Caution Refer to Example 8(c). Notice that

$$(2x^2y^3)^4 \neq (2 \cdot 4)x^{2 \cdot 4}y^{3 \cdot 4}.$$

Do not multiply the coefficient 2 and the exponent 4.

WORK PROBLEM 8 AT THE SIDE.

8. Simplify.

(a) $(2m)^3(2m)^4$

(b) $\left(\frac{5k^3}{3}\right)^2$

(c) $\left(\frac{1}{5}\right)^4(2x)^2$

(d) $(3xy^2)^3(x^2y)^4$

ANSWERS

8. (a) 2^7m^7 (b) $\frac{5^2k^6}{3^2}$ (c) $\frac{2^2x^2}{5^4}$ (d) $3^3x^{11}y^{10}$

QUEST FOR NUMERACY

Mathematics of the Egyptians

At least five thousand years go, the Egyptians used a method of writing numerals much different from our own. Our Hindu-Arabic system involves place value based on powers of ten (hence the name *decimal system*). The Egyptians used a system based on simple grouping, using the symbols shown below for writing numbers through 9,999,999.

Number	*Symbol*	*Description*
1	𓏺	Stroke
10	𓎆	Heel bone
100	𓍢	Scroll
1000	𓆼	Lotus flower
10,000	𓂭	Pointing finger
100,000	𓆐	Burbot fish
1,000,000	𓁨	Astonished person

A typical Egyptian numeral is shown below:

𓆐𓆐𓆼𓆼𓆼𓆼𓆼𓍢𓍢𓍢𓍢𓎆𓎆𓎆𓎆𓎆𓎆𓎆𓎆𓎆𓏺𓏺𓏺𓏺𓏺𓏺𓏺

There are 2 symbols for 100,000, 5 symbols for 1000, 4 symbols for 100, 9 symbols for 10, and 7 symbols for 1. Thus, the Egyptian numeral above represents

$$2(100{,}000) + 5(1000) + 4(100) + 9(10) + 7(1) = 205{,}497$$

in our system of numeration.

Much of our knowledge of Egyptian mathematics comes from the Rhind papyrus, which dates back some 3800 years. A small portion of the papyrus is shown here.

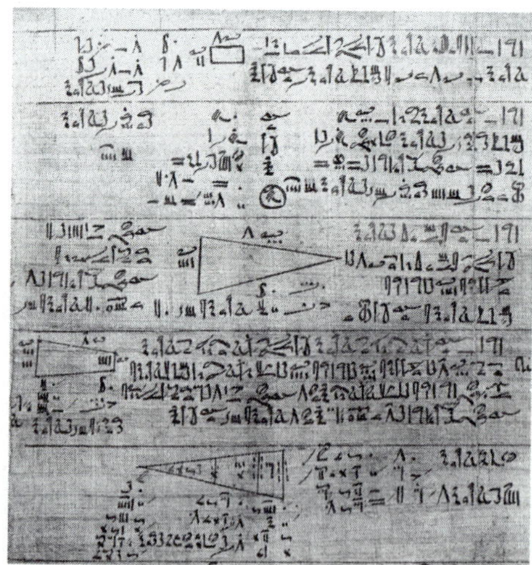

FOR GROUP DISCUSSION What are some of the other ancient numeration systems that you have heard of?

Art: Courtesy of the Trustees of the British Museum

NAME DATE HOUR

3.1 EXERCISES

Write each expression using exponents. See Example 1.

1. $3 \cdot 3 \cdot 3 \cdot 3 \cdot 3 \cdot 3 \cdot 3$

2. $8 \cdot 8 \cdot 8 \cdot 8 \cdot 8$

3. $(-6)(-6)(-6)(-6)$

4. $(-2)(-2)(-2)(-2)(-2)$

5. $w \cdot w \cdot w \cdot w \cdot w \cdot w$

6. $t \cdot t \cdot t \cdot t \cdot t \cdot t$

7. $\dfrac{1}{4 \cdot 4 \cdot 4 \cdot 4}$

8. $\dfrac{1}{3 \cdot 3 \cdot 3}$

9. $(-7x)(-7x)(-7x)(-7x)$

10. $(-8p)(-8p)$

11. $\left(\frac{1}{2}\right)\left(\frac{1}{2}\right)\left(\frac{1}{2}\right)\left(\frac{1}{2}\right)\left(\frac{1}{2}\right)\left(\frac{1}{2}\right)$

12. $\left(-\frac{1}{4}\right)\left(-\frac{1}{4}\right)\left(-\frac{1}{4}\right)\left(-\frac{1}{4}\right)\left(-\frac{1}{4}\right)$

13. Explain how the expressions $(-3)^4$ and -3^4 are different.

14. Explain how the expressions $(5x)^3$ and $5x^3$ are different.

Identify the base and the exponent for each exponential expression. In Exercises 15–18, also evaluate the expression. See Example 2.

15. 3^5

16. 2^7

17. $(-3)^5$

18. $(-2)^7$

19. $(-6x)^4$

20. $(-8x)^4$

21. $-6x^4$

22. $-8x^4$

23. Explain why the product rule does not apply to the expression $5^2 + 5^3$. Then evaluate the expression by finding the individual powers and then adding the results.

24. Repeat Exercise 23 for the expression $(-4)^3 + (-4)^4$.

Use the product rule to simplify each expression. Write each answer in exponential form. See Examples 3 and 4.

25. $5^2 \cdot 5^6$

26. $3^6 \cdot 3^7$

27. $4^2 \cdot 4^7 \cdot 4^3$

28. $5^3 \cdot 5^8 \cdot 5^2$

29. $(-7)^3(-7)^6$

30. $(-9)^8(-9)^5$

31. $t^3 \cdot t^8 \cdot t^{13}$

32. $n^5 \cdot n^6 \cdot n^9$

33. $(-8r^4)(7r^3)$

34. $(10a^7)(-4a^3)$

35. $(-6p^5)(-7p^5)$

36. $(-5w^8)(-9w^8)$

37. Explain why the product rule does not apply to the expression $3^2 \cdot 4^3$. Then evaluate the expression by finding the individual powers and multiplying the results.

38. Repeat Exercise 37 for the expression $(-3)^3 \cdot (-2)^5$.

In each of the following exercises, first add the given terms. Then start over and multiply them.

39. $5x^4, 9x^4$

40. $8t^5, 3t^5$

41. $-7a^2, 2a^2, 10a^2$

42. $6x^3, 9x^3, -2x^3$

Use the power rules for exponents to simplify each expression. Write each answer in exponential form. See Examples 5–7.

43. $(4^3)^2$ **44.** $(8^3)^6$ **45.** $(t^4)^5$ **46.** $(y^6)^5$

47. $(7r)^3$ **48.** $(11x)^4$ **49.** $(5xy)^5$ **50.** $(9pq)^6$

51. $(-5^2)^6$ **52.** $(-9^4)^8$ **53.** $(-8^3)^5$ **54.** $(-7^5)^7$

55. $8(qr)^3$ **56.** $4(vw)^5$ **57.** $\left(\frac{1}{2}\right)^3$ **58.** $\left(\frac{1}{3}\right)^5$

59. $\left(\frac{a}{b}\right)^3 \quad (b \neq 0)$ **60.** $\left(\frac{r}{t}\right)^4 \quad (t \neq 0)$ **61.** $\left(\frac{9}{5}\right)^8$ **62.** $\left(\frac{12}{7}\right)^3$

Use a combination of the rules of exponents introduced in this section to simplify each expression. See Example 8.

63. $\left(\frac{5}{2}\right)^3 \cdot \left(\frac{5}{2}\right)^2$ **64.** $\left(\frac{3}{4}\right)^5 \cdot \left(\frac{3}{4}\right)^6$ **65.** $\left(\frac{9}{8}\right)^3 \cdot 9^2$ **66.** $\left(\frac{8}{5}\right)^4 \cdot 8^3$

67. $(2x)^9(2x)^3$ **68.** $(6y)^5(6y)^8$ **69.** $(-6p)^4(-6p)$ **70.** $(-13q)^3(-13q)$

71. $(6x^2y^3)^5$ **72.** $(5r^5t^6)^7$ **73.** $(x^2)^3(x^3)^5$ **74.** $(y^4)^5(y^3)^5$

75. $(2w^2x^3y)^2(x^4y)^5$ **76.** $(3x^4y^2z)^3(yz^4)^5$ **77.** $(-r^4s)^2(-r^2s^3)^5$

78. $(-ts^6)^4(-t^3s^5)^3$ **79.** $\left(\frac{5a^2b^5}{c^6}\right)^3 \quad (c \neq 0)$ **80.** $\left(\frac{6x^3y^9}{z^5}\right)^4 \quad (z \neq 0)$

81. A student tried to simplify $(10^2)^3$ as 1000^6. Is this correct? If not, how is it simplified using the product rule for exponents?

82. Explain why $(3x^2y^3)^4$ is *not* equivalent to $(3 \cdot 4)x^8y^{12}$.

PREVIEW EXERCISES

Perform each operation. See Section 1.5.

83. $8 - (-2)$ **84.** $-7 - (-3)$ **85.** $9 - 12$ **86.** $0 - (-5)$

3.2 INTEGER EXPONENTS AND THE QUOTIENT RULE

OBJECTIVES

1. Use zero as an exponent.
2. Use negative numbers as exponents.
3. Use the quotient rule for exponents.
4. Use combinations of rules.

FOR EXTRA HELP

Tape 4 | SSM pp. 99–101 | MAC: A IBM: A

1 In the previous section we studied the product rule for exponents. In all of our work, exponents were positive integers. To develop meanings for exponents other than positive integers (such as 0 and negative integers), we want to define them in such a way that rules for exponents are the same, regardless of the kind of number used for the exponents.

Suppose we want to find a meaning for an expression such as

$$6^0,$$

where 0 is used as an exponent. If we were to multiply this by 6^2, for example, we would want the product rule to still be valid. Therefore, we would have

$$6^0 \cdot 6^2 = 6^{0+2} = 6^2.$$

So multiplying 6^2 by 6^0 should give 6^2. Because 6^0 is acting as if it were 1 here, we should define 6^0 to equal 1. This is the definition for 0 used as an exponent with any nonzero base.

DEFINITION OF ZERO EXPONENT

For any nonzero real number a, $\quad a^0 = 1 \quad (a \neq 0)$.
Example: $17^0 = 1$

EXAMPLE 1 *Using Zero Exponents*

Evaluate each exponential expression.

(a) $60^0 = 1$

(b) $(-60)^0 = 1$

(c) $-60^0 = -(1) = -1$

(d) $y^0 = 1$, if $y \neq 0$

(e) $-r^0 = -1$, if $r \neq 0$ ■

Caution Notice the difference between parts (b) and (c) of Example 1. In Example 1(b) the base is -60 and the exponent is 0. Any nonzero base raised to a zero exponent is 1. But in Example 1(c), the base is 60. Then $60^0 = 1$, and $-60^0 = -1$.

WORK PROBLEM 1 AT THE SIDE.

1. Evaluate.

(a) 28^0

(b) $(-16)^0$

(c) -7^0

(d) m^0, $m \neq 0$

(e) $-p^0$, $p \neq 0$

2 Now let us consider how we can define negative integers as exponents. Suppose that we want to give a meaning to

$$6^{-2}$$

so that the product rule is still valid. If we multiply 6^{-2} by 6^2, we get

$$6^{-2} \cdot 6^2 = 6^{-2+2} = 6^0 = 1.$$

The expression 6^{-2} is acting as if it were the reciprocal of 6^2, because their product is 1. The reciprocal of 6^2 may be written $\frac{1}{6^2}$, leading us to define 6^{-2} as $\frac{1}{6^2}$. This is a particular case of the definition of negative exponents.

ANSWERS
1. (a) 1 (b) 1 (c) −1 (d) 1 (e) −1

DEFINITION OF NEGATIVE EXPONENTS

For any nonzero real number a and any integer n, $a^{-n} = \frac{1}{a^n}$ $(a \neq 0)$.

Example: $3^{-2} = \frac{1}{3^2}$

By definition, a^{-n} and a^n are reciprocals, since

$$a^n \cdot a^{-n} = a^n \cdot \frac{1}{a^n} = 1.$$

Since $1^n = 1$, the definition of a^{-n} also can be written

$$a^{-n} = \frac{1}{a^n} = \frac{1^n}{a^n} = \left(\frac{1}{a}\right)^n.$$

For example,

$$6^{-3} = \left(\frac{1}{6}\right)^3 \quad \text{and} \quad \left(\frac{1}{3}\right)^{-2} = 3^2.$$

EXAMPLE 2 *Using Negative Exponents*

Simplify by writing with positive exponents.

(a) $3^{-2} = \frac{1}{3^2} = \frac{1}{9}$

(b) $5^{-3} = \frac{1}{5^3} = \frac{1}{125}$

(c) $\left(\frac{1}{2}\right)^{-3} = 2^3 = 8$ $\quad$ $\frac{1}{2}$ and 2 are reciprocals.

As shown above, we can change the base to its reciprocal if we also change the sign of the exponent.

(d) $\left(\frac{2}{5}\right)^{-4} = \left(\frac{5}{2}\right)^4$ $\quad$ $\frac{2}{5}$ and $\frac{5}{2}$ are reciprocals.

(e) $4^{-1} - 2^{-1} = \frac{1}{4} - \frac{1}{2} = \frac{1}{4} - \frac{2}{4} = -\frac{1}{4}$

Apply the exponents first, then subtract.

(f) $p^{-2} = \frac{1}{p^2}, p \neq 0$

(g) $\frac{1}{x^{-4}}, x \neq 0$

$$\begin{aligned} \frac{1}{x^{-4}} &= \frac{1^{-4}}{x^{-4}} && 1^{-4} = 1 \\ &= \left(\frac{1}{x}\right)^{-4} && \text{Power rule (c)} \\ &= x^4 && \tfrac{1}{x} \text{ and } x \text{ are reciprocals.} \end{aligned}$$

Caution A negative exponent does not indicate a negative number; negative exponents lead to reciprocals.

Expression	*Example*	
a^{-n}	$3^{-2} = \frac{1}{3^2} = \frac{1}{9}$	Not negative
$-a^{-n}$	$-3^{-2} = -\frac{1}{3^2} = -\frac{1}{9}$	Negative

WORK PROBLEM 2 AT THE SIDE.

The definition of negative exponents allows us to move factors in a fraction if we also change the signs of the exponents. For example,

$$\frac{2^{-3}}{3^{-4}} = \frac{\frac{1}{2^3}}{\frac{1}{3^4}} = \frac{1}{2^3} \cdot \frac{3^4}{1} = \frac{3^4}{2^3}$$

so that

$$\frac{2^{-3}}{3^{-4}} = \frac{3^4}{2^3}.$$

CHANGING FROM NEGATIVE TO POSITIVE EXPONENTS

For any nonzero numbers a and b, and any integers m and n,

$$\frac{a^{-m}}{b^{-n}} = \frac{b^n}{a^m}.$$

Example: $\frac{3^{-5}}{2^{-4}} = \frac{2^4}{3^5}$

EXAMPLE 3 *Changing from Negative to Positive Exponents*

Write with only positive exponents. Assume all variables represent nonzero real numbers.

(a) $\frac{4^{-2}}{5^{-3}} = \frac{5^3}{4^2}$

(b) $\frac{m^{-5}}{p^{-1}} = \frac{p^1}{m^5} = \frac{p}{m^5}$

(c) $\frac{a^{-2}b}{3d^{-3}} = \frac{bd^3}{3a^2}$

(d) $x^3y^{-4} = \frac{x^3y^{-4}}{1} = \frac{x^3}{y^4}$ ■

WORK PROBLEM 3 AT THE SIDE.

2. Write with positive exponents.

(a) 4^{-3}

(b) 6^{-2}

(c) $\left(\frac{2}{3}\right)^{-2}$

(d) $2^{-1} + 5^{-1}$

(e) m^{-5}, $m \neq 0$

(f) $\frac{1}{z^{-4}}$, $z \neq 0$

3. Write with only positive exponents. Assume all variables represent nonzero real numbers.

(a) $\frac{7^{-1}}{5^{-4}}$

(b) $\frac{x^{-3}}{y^{-2}}$

(c) $\frac{4h^{-5}}{m^{-2}k}$

(d) p^2q^{-5}

ANSWERS

2. (a) $\frac{1}{4^3}$ (b) $\frac{1}{6^2}$ (c) $\left(\frac{3}{2}\right)^2$ (d) $\frac{1}{2} + \frac{1}{5} = \frac{7}{10}$ (e) $\frac{1}{m^5}$ (f) z^4

3. (a) $\frac{5^4}{7}$ (b) $\frac{y^2}{x^3}$ (c) $\frac{4m^2}{h^5k}$ (d) $\frac{p^2}{q^5}$

4. Simplify. Write answers with positive exponents.

(a) $\frac{5^{11}}{5^8}$

(b) $\frac{4^7}{4^{10}}$

(c) $\frac{6^{-5}}{6^{-2}}$

(d) $\frac{8^4 \cdot m^9}{8^5 \cdot m^{10}}$

3 What about the quotient of two exponential expressions with the same base? We know that

$$\frac{6^5}{6^3} = \frac{6 \cdot 6 \cdot 6 \cdot 6 \cdot 6}{6 \cdot 6 \cdot 6} = 6^2.$$

Notice that $5 - 3 = 2$. Also,

$$\frac{6^2}{6^4} = \frac{6 \cdot 6}{6 \cdot 6 \cdot 6 \cdot 6} = \frac{1}{6^2} = 6^{-2}$$

Here, $2 - 4 = -2$. These examples suggest the quotient rule for exponents.

QUOTIENT RULE FOR EXPONENTS

For any nonzero real number a and any integers m and n,

$\frac{a^m}{a^n} = a^{m-n}$ $(a \neq 0)$.

(Keep the base and subtract the exponents.)

Example: $\frac{5^8}{5^4} = 5^{8-4} = 5^4$

EXAMPLE 4 *Using the Quotient Rule for Exponents*

Simplify, using the quotient rule for exponents. Write answers with positive exponents.

(a) $\frac{5^8}{5^6} = 5^{8-6} = 5^2$

(b) $\frac{4^2}{4^9} = 4^{2-9} = 4^{-7} = \frac{1}{4^7}$

(c) $\frac{5^{-3}}{5^{-7}} = 5^{-3-(-7)} = 5^4$

(d) $\frac{q^5}{q^{-3}} = q^{5-(-3)} = q^8, q \neq 0$

(e) $\frac{3^2x^5}{3^4x^3} = \frac{3^2}{3^4} \cdot \frac{x^5}{x^3} = 3^{2-4} \cdot x^{5-3}$

$= 3^{-2}x^2 = \frac{x^2}{3^2}, x \neq 0$

Sometimes numerical expressions with small exponents, such as 3^2, are evaluated. Doing that would give the result as $\frac{x^2}{9}$. ■

WORK PROBLEM 4 AT THE SIDE.

Since exponential expressions with negative exponents can be written with positive exponents, the rules for exponents also are true for negative exponents.

ANSWERS

4. (a) 5^3 (b) $\frac{1}{4^3}$ (c) $\frac{1}{6^3}$ (d) $\frac{1}{8m}$

The definitions and rules for exponents given in this section and the previous one are summarized below.

DEFINITIONS AND RULES FOR EXPONENTS

For any integers m and n:			Examples
Product rule	$a^m \cdot a^n = a^{m+n}$		$7^4 \cdot 7^5 = 7^9$
Zero exponent	$a^0 = 1$	$(a \neq 0)$	$(-3)^0 = 1$
Negative exponent	$a^{-n} = \dfrac{1}{a^n}$	$(a \neq 0)$	$5^{-3} = \dfrac{1}{5^3}$
Quotient rule	$\dfrac{a^m}{a^n} = a^{m-n}$	$(a \neq 0)$	$\dfrac{2^2}{2^5} = 2^{-3} = \dfrac{1}{2^3}$
Power rules (a)	$(a^m)^n = a^{mn}$		$(4^2)^3 = 4^6$
(b)	$(ab)^m = a^m b^m$		$(3k)^4 = 3^4 k^4$
(c)	$\left(\dfrac{a}{b}\right)^m = \dfrac{a^m}{b^m}$	$(b \neq 0)$	$\left(\dfrac{2}{3}\right)^{-2} = \dfrac{2^{-2}}{3^{-2}} = \dfrac{3^2}{2^2}$
Negative to positive exponent	$\dfrac{a^{-m}}{b^{-n}} = \dfrac{b^n}{a^m}$		$\dfrac{2^{-4}}{5^{-3}} = \dfrac{5^3}{2^4}$

4 As shown in the next example, we may sometimes need to use more than one rule to simplify an expression.

EXAMPLE 5 *Using a Combination of Rules*

Use a combination of the rules for exponents to simplify each expression. Assume all variables represent nonzero real numbers.

(a) $\dfrac{(4^2)^3}{4^5}$

$$\frac{(4^2)^3}{4^5} = \frac{4^6}{4^5} \qquad \text{Power rule (a)}$$

$$= 4^{6-5} \qquad \text{Quotient rule}$$

$$= 4^1 = 4$$

(b) $(2x)^3(2x)^2$

$$(2x)^3(2x)^2 = (2x)^5 \qquad \text{Product rule}$$

$$= 2^5x^5 \text{ or } 32x^5 \qquad \text{Power rule (b)}$$

(c) $\left(\dfrac{2x^3}{5}\right)^{-4}$

By the definition of a negative exponent and the power rules,

$$\left(\frac{2x^3}{5}\right)^{-4} = \left(\frac{5}{2x^3}\right)^4 \qquad \text{Change the base to its reciprocal, and change the sign of the exponent.}$$

$$= \frac{5^4}{2^4x^{12}}. \qquad \text{Power rules}$$

(d) $\left(\dfrac{3x^{-2}}{4^{-1}y^{3}}\right)^{-3} = \dfrac{3^{-3}x^{6}}{4^{3}y^{-9}}$ Power rules

$= \dfrac{x^{6}y^{9}}{4^{3}\cdot 3^{3}}$ Negative to positive exponent rule

(e) $\dfrac{(4m)^{-3}}{(3m)^{-4}} = \dfrac{4^{-3}m^{-3}}{3^{-4}m^{-4}}$ Power rule (b)

$= \dfrac{3^{4}m^{4}}{4^{3}m^{3}}$ Negative to positive exponent rule

$= \dfrac{3^{4}m^{4-3}}{4^{3}}$ Quotient rule

$= \dfrac{3^{4}m}{4^{3}}$ ■

Note Since the steps can be done in several different orders, there are many equally good ways to simplify problems like Examples 5(d) and 5(e).

◀ **WORK PROBLEM 5 AT THE SIDE.**

5. Simplify. Assume all variables represent nonzero real numbers.

(a) $12^{5} \cdot 12^{-7} \cdot 12^{6}$

(b) $y^{-2} \cdot y^{5} \cdot y^{-8}$

(c) $\dfrac{(6x)^{-1}}{(3x^{2})^{-2}}$

(d) $\dfrac{3^{9} \cdot (x^{2}y)^{-2}}{3^{3} \cdot x^{-4}y}$

ANSWERS

5. (a) 12^{4} (b) $\dfrac{1}{y^{5}}$ (c) $\dfrac{3x^{3}}{2}$ (d) $\dfrac{3^{6}}{y^{3}}$

3.2 EXERCISES

NAME DATE HOUR

Each of the following expressions is equal to either 0, 1, or −1. Decide which is correct. See Example 1.

1. 9^0

2. 5^0

3. $(-4)^0$

4. $(-10)^0$

5. -9^0

6. -5^0

7. $(-2)^0 - 2^0$

8. $(-8)^0 - 8^0$

9. $\dfrac{0^{10}}{10^0}$

10. $\dfrac{0^5}{5^0}$

11. $x^0\ (x \neq 0)$

12. $y^0\ (y \neq 0)$

Evaluate each expression. See Examples 1 and 2.

13. $7^0 + 9^0$

14. $8^0 + 6^0$

15. 4^{-3}

16. 5^{-4}

17. $\left(\dfrac{1}{2}\right)^{-4}$

18. $\left(\dfrac{1}{3}\right)^{-3}$

19. $\left(\dfrac{6}{7}\right)^{-2}$

20. $\left(\dfrac{2}{3}\right)^{-3}$

21. $(-3)^{-4}$

22. $(-4)^{-3}$

23. $5^{-1} + 3^{-1}$

24. $6^{-1} + 2^{-1}$

Decide whether each of the following is positive, negative, or zero.

25. $(-2)^{-3}$

26. $(-3)^{-2}$

27. -2^4

28. -3^6

29. $\left(\dfrac{1}{4}\right)^{-2}$

30. $\left(\dfrac{1}{5}\right)^{-2}$

31. $1 - 5^0$

32. $1 - 7^0$

Use the quotient rule to simplify each expression. Write each expression with positive exponents. Assume that all variables represent nonzero real numbers. See Examples 2, 3, and 4.

33. $\dfrac{5^8}{5^5}$

34. $\dfrac{11^6}{11^3}$

35. $\dfrac{9^4}{9^5}$

36. $\dfrac{7^3}{7^4}$

37. $\frac{6^{-3}}{6^2}$

38. $\frac{4^{-2}}{4^3}$

39. $\frac{x^{12}}{x^{-3}}$

40. $\frac{y^4}{y^{-6}}$

41. $\frac{1}{6^{-3}}$

42. $\frac{1}{5^{-2}}$

43. $\frac{2}{r^{-4}}$

44. $\frac{3}{s^{-8}}$

45. $\frac{4^{-3}}{5^{-2}}$

46. $\frac{6^{-2}}{5^{-4}}$

47. p^5q^{-8}

48. $x^{-8}y^4$

49. $\frac{r^5}{r^{-4}}$

50. $\frac{a^6}{a^{-4}}$

51. $\frac{6^4x^8}{6^5x^3}$

52. $\frac{3^8y^5}{3^{10}y^2}$

Use a combination of the rules for exponents to simplify each expression. Write answers with only positive exponents. Assume that all variables represent nonzero real numbers. See Example 5.

53. $\frac{(7^4)^3}{7^9}$

54. $\frac{(5^3)^2}{5^2}$

55. $x^{-3} \cdot x^5 \cdot x^{-4}$

56. $y^{-8} \cdot y^5 \cdot y^{-2}$

57. $\frac{(3x)^{-2}}{(4x)^{-3}}$

58. $\frac{(2y)^{-3}}{(5y)^{-4}}$

59. $\left(\frac{x^{-1}y}{z^2}\right)^{-2}$

60. $\left(\frac{p^{-4}q}{r^{-3}}\right)^{-3}$

61. $(6x)^4(6x)^{-3}$

62. $(10y)^9(10y)^{-8}$

63. $\frac{(m^7n)^{-2}}{m^{-4}n^3}$

64. $\frac{(m^8n^{-4})^2}{m^{-2}n^5}$

PREVIEW EXERCISES

Perform each operation.

65. 10,000(.5687)

66. 1000(.0004)

67. $\frac{12}{10,000}$

68. $\frac{5.8}{1,000,000}$

3.3 AN APPLICATION OF EXPONENTS: SCIENTIFIC NOTATION

OBJECTIVES

1. Express numbers in scientific notation.
2. Convert numbers in scientific notation to numbers without exponents.
3. Use scientific notation in calculations.

FOR EXTRA HELP

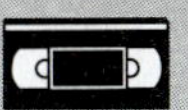
Tape 5

SSM pp. 101–102

MAC: A IBM: A

1 One example of the use of exponents comes from science. The numbers occurring in science are often extremely large (such as the distance from the earth to the sun, 93,000,000 miles) or extremely small (the wavelength of yellow-green light is approximately .0000006 meter). Because of the difficulty of working with many zeros, scientists often express such numbers with exponents. Each number is written as $a \times 10^n$, where $1 \leq |a| < 10$ and n is an integer. This form is called **scientific notation.** There is always one nonzero digit before the decimal point. For example, 35 is written 3.5×10^1, or 3.5×10; 56,200 is written 5.62×10^4, since

$$56{,}200 = 5.62 \times 10{,}000 = 5.62 \times 10^4,$$

and .09 is written as 9×10^{-2}.

The steps involved in writing a number in scientific notation are given below. For negative numbers, follow these steps using the absolute value of the number; then make the result negative.

WRITING A NUMBER IN SCIENTIFIC NOTATION

Step 1 Move the decimal point to the right of the first nonzero digit.

Step 2 Count the number of places you moved the decimal point.

Step 3 The number of places in Step 2 is the absolute value of the exponent on 10.

Step 4 The exponent on 10 is positive if you made the number smaller in Step 1; the exponent is negative if you made the number larger in Step 1.

EXAMPLE 1 *Using Scientific Notation*

Write each number in scientific notation.

(a) 93,000,000

The number will be written in scientific notation as 9.3×10^n. To find the value of n, first compare 9.3 with the original number, 93,000,000. Did the number get larger or smaller? Here the number 9.3 is *smaller than* 93,000,000. Therefore, we must multiply by a *positive* power of 10 so the product 9.3×10^n will equal the larger number.

Move the decimal point to follow the first nonzero digit. Count the number of places the decimal point was moved.

9.3 000 000   7 places

Since the decimal point was moved 7 places, and since n is positive, $93{,}000{,}000 = 9.3 \times 10^7$.

(b) $463{,}000{,}000{,}000{,}000 = 4.63\ 000\ 000\ 000\ 000$ 14 places
$= 4.63 \times 10^{14}$

(c) $-302{,}100 = -3.021 \times 10^5$

(d) .00462

Move the decimal point to the right of the first nonzero digit and count the number of places the decimal point was moved.

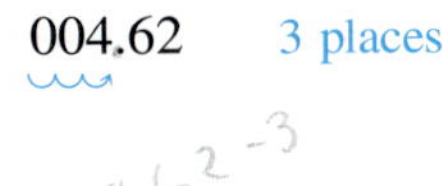
004.62   3 places

Because 4.62 is *larger* than .00462, the exponent must be *negative.*

$$.00462 = 4.62 \times 10^{-3}$$

(e) $.0000762 = 7.62 \times 10^{-5}$ ■

WORK PROBLEM 1 AT THE SIDE.

2 To convert a number written in scientific notation to a number without exponents, work in reverse. Multiplying a number by a positive power of 10 will make the number larger; multiplying by a negative power of 10 will make the number smaller.

■ **EXAMPLE 2** *Writing Numbers Without Exponents*

Write each number without exponents.

(a) 6.2×10^3

Since the exponent is positive, make 6.2 larger by moving the decimal point 3 places to the right.

$$6.2 \times 10^3 = 6.200 = 6200$$

(b) $4.283 \times 10^5 = 4.28300 = 428{,}300$ Move 5 places to the right.

(c) $-9.73 \times 10^{-2} = -09.73 = -.0973$ Move 2 places to the left. ■

As these examples show, the exponent tells the number of places and the direction that the decimal point is moved.

WORK PROBLEM 2 AT THE SIDE.

3 The next example shows how scientific notation can be used with products and quotients.

■ **EXAMPLE 3** *Multiplying and Dividing with Scientific Notation*

Write each product or quotient without exponents.

(a) $(6 \times 10^3)(5 \times 10^{-4})$

$(6 \times 10^3)(5 \times 10^{-4})$

$= (6 \times 5)(10^3 \times 10^{-4})$ Commutative and associative properties

$= 30 \times 10^{-1}$ Product rule for exponents

$= 30. = 3$ Write without exponents.

(b) $\dfrac{6 \times 10^{-5}}{2 \times 10^3} = \dfrac{6}{2} \times \dfrac{10^{-5}}{10^3} = 3 \times 10^{-8} = .00000003$ ■

WORK PROBLEM 3 AT THE SIDE.

Scientific calculators usually have a key labeled EE or EXP for scientific notation. See An Introduction to Scientific Calculators at the front of this book for more information.

---

1. Write each number in scientific notation.

(a) 63,000

(b) 5,870,000

(c) .0571

(d) .000062

2. Write without exponents.

(a) 4.2×10^3

(b) 8.7×10^5

(c) 6.42×10^{-3}

3. Simplify, and write without exponents.

(a) $(2.6 \times 10^4)(2 \times 10^{-6})$

(b) $\dfrac{4.8 \times 10^2}{2.4 \times 10^{-3}}$

ANSWERS

1. (a) 6.3×10^4 (b) 5.87×10^6 (c) 5.71×10^{-2} (d) 6.2×10^{-5}
2. (a) 4200 (b) 870,000 (c) .00642
3. (a) .052 (b) 200,000

3.3 NAME DATE HOUR

EXERCISES

Determine whether or not the given number is written in scientific notation as defined in Objective 1. If it is not, write it as such.

1. 4.56×10^3

2. 7.34×10^5

3. 5,600,000

4. 34,000

5. $.8 \times 10^2$

6. $.9 \times 10^3$

7. .004

8. .0007

9. Explain in your own words what it means for a number to be written in scientific notation.

10. Explain how to multiply a number by a positive power of ten. Then explain how to multiply a number by a negative power of ten.

Write each number in scientific notation. See Example 1.

11. 5,876,000,000

12. 9,994,000,000

13. 82,350

14. 78,330

15. .000007

16. .0000004

17. $-.00203$

18. $-.0000578$

Write each number without exponents. See Example 2.

19. 7.5×10^5

20. 8.8×10^6

21. 5.677×10^{12}

22. 8.766×10^9

23. -6.21×10^0

24. -8.56×10^0

25. 7.8×10^{-4}

26. 8.9×10^{-5}

27. 5.134×10^{-9}

28. 7.123×10^{-10}

Perform the indicated operations, and write the answers without exponents. See Example 3.

29. $(2 \times 10^8) \times (3 \times 10^3)$

30. $(4 \times 10^7) \times (3 \times 10^3)$

31. $(5 \times 10^4) \times (3 \times 10^2)$

32. $(8 \times 10^5) \times (2 \times 10^3)$

33. $(3 \times 10^{-4}) \times (2 \times 10^8)$

34. $(4 \times 10^{-3}) \times (2 \times 10^7)$

Perform the indicated operations, and write the answers in sientific notation. See Example 3.

35. $\dfrac{9 \times 10^{-5}}{3 \times 10^{-1}}$

36. $\dfrac{12 \times 10^{-4}}{4 \times 10^{-3}}$

37. $\dfrac{8 \times 10^3}{2 \times 10^2}$

38. $\dfrac{5 \times 10^4}{1 \times 10^3}$

39. $\dfrac{2.6 \times 10^{-3}}{2 \times 10^2}$

40. $\dfrac{9.5 \times 10^{-1}}{5 \times 10^3}$

If the number in the statement is written in scientific notation, write it without exponents. If it is written without exponents, write it in scientific notation. See Examples 1 and 2.

41. The number of possible hands in contract bridge is about 6.35×10^{11}.

42. If there are forty numbers to choose from in a lottery, and a player must choose six different ones, the player has about 3.84×10^6 ways to make a choice.

43. In 1990, ESPN's regular season baseball telecasts averaged just under 1,150,000 viewers.

44. In 1990, ESPN's estimated losses were at least 36,000,000 dollars.

45. The body of a 150-pound person contains about 2.3×10^{-4} pounds of copper and about 6×10^{-3} pounds of iron.

46. The mean distance from Venus to the sun is about 6.7×10^6 miles.

PREVIEW EXERCISES

Simplify. See Section 1.9.

47. $8x + 4x$

48. $5.6y - 2.3y + 8.1y$

49. $6(2x - 3) + 5$

50. $4 - 8(x + 2)$

3.4 ADDITION AND SUBTRACTION OF POLYNOMIALS

1 Recall that in an expression such as

$$4x^3 + 6x^2 + 5x + 8,$$

the quantities that are added, $4x^3$, $6x^2$, $5x$, and 8 are called *terms.* In the term $4x^3$, the number 4 is called the *numerical coefficient,* or simply the *coefficient,* of x^3. In the same way, 6 is the coefficient of x^2 in the term $6x^2$, 5 is the coefficient of x in the term $5x$, and 8 is the coefficient in the term 8. A constant term, like 8 in the polynomial above, can be thought of as $8x^0$, where x^0 is defined to equal 1, as shown in Section 3.2.

EXAMPLE 1 *Identifying Coefficients*
Name the coefficient of each term in these expressions.

(a) $4x^3$
The coefficient is 4.

(b) $x - 6x^4$
The coefficient of x is 1 because $x = 1 \cdot x$. The coefficient of x^4 is -6, since $x - 6x^4$ can be written as the sum $x + (-6x^4)$.

(c) $5 - v^3$

The coefficient of the term 5 is 5 since $5 = 5v^0$. By writing $5 - v^3$ as a sum, $5 + (-v^3)$, or $5 + (-1v^3)$, the coefficient of v^3 can be identified as -1. ■

WORK PROBLEM 1 AT THE SIDE. ▶

2 Recall from Section 1.9 that *like terms* have exactly the same combination of variables, with the same exponents on the variables. Only the coefficients may be different. Examples of like terms are

$$19m^5 \quad \text{and} \quad 14m^5;$$
$$6y^9, \quad -37y^9, \quad \text{and} \quad y^9;$$
$$3pq, \quad -2pq, \quad \text{and} \quad 4pq.$$

By the distributive property, we add like terms by adding their coefficients.

EXAMPLE 2 *Adding Like Terms*
Simplify each expression by adding like terms.

(a) $-4x^3 + 6x^3 = (-4 + 6)x^3$ Distributive property
$= 2x^3$

(b) $9x^6 - 14x^6 + x^6 = (9 - 14 + 1)x^6 = -4x^6$

OBJECTIVES

1. Identify terms and coefficients.
2. Add like terms.
3. Know the vocabulary for polynomials.
4. Add polynomials.
5. Subtract polynomials.
6. Apply the rules and definitions for polynomials to multivariable polynomials.

FOR EXTRA HELP

Tape 5

SSM pp. 103–106

MAC: A IBM: A

1. Name the coefficient of each term in these expressions.

(a) $3m^2$

(b) $2x^2 - x$

(c) $x + 8$

ANSWERS
1. (a) 3 (b) 2; −1 (c) 1; 8

(c) $12m^2 + 5m + 4m^2 = (12 + 4)m^2 + 5m$
$= 16m^2 + 5m$

(d) $3x^2y + 4x^2y - x^2y = (3 + 4 - 1)x^2y = 6x^2y$ ■

In Example 2(c), it is not possible to add $16m^2$ and $5m$. These two terms are unlike because the exponents on the variables are different. *Unlike terms* have different variables or different exponents on the same variables.

WORK PROBLEM 2 AT THE SIDE.

3 Polynomials are basic to algebra. A **polynomial in x** is a term or the sum of a finite number of terms of the form ax^n, for any real number a and any whole number n. For example,

$$16x^8 - 7x^6 + 5x^4 - 3x^2 + 4$$

is a polynomial in x (here 4 can be written as $4x^0$). This polynomial is written in **descending powers** of the variable, because the exponents on x decrease from left to right. On the other hand,

$$2x^3 - x^2 + \frac{4}{x}$$

is not a polynomial in x, since $\frac{4}{x} = 4x^{-1}$ is not a *product*, ax^n, for a whole number n. (Of course, a polynomial could be defined using any variable, or variables, and not just x.)

WORK PROBLEM 3 AT THE SIDE.

The **degree of a term with one variable** is the exponent on the variable. For example, $3x^4$ has degree 4, $6x^{17}$ has degree 17, $5x$ has degree 1, and -7 has degree 0 (since -7 can be written as $-7x^0$). The **degree of a polynomial** is the highest degree in any nonzero term of the polynomial. For example, $3x^4 - 5x^2 + 6$ is degree 4, the polynomial $5x + 7$ is degree 1, and 3 (or $3x^0$) is degree 0.

Three types of polynomials are very common and are given special names. A polynomial with exactly three terms is called a **trinomial.** (*Tri*- means "three," as in *tri*angle.) Examples are

$$9m^3 - 4m^2 + 6, \quad \frac{19}{3}y^2 + \frac{8}{3}y + 5, \quad \text{and} \quad -3m^5 - 9m^2 + 2.$$

A polynomial with exactly two terms is called a **binomial.** (*Bi*- means "two," as in *bi*cycle.) Examples are

$$-9x^4 + 9x^3, \quad 8m^2 + 6m, \quad \text{and} \quad 3m^5 - 9m^2.$$

A polynomial with only one term is called a **monomial.** (*Mon(o)*- means "one," as in *mono*rail.) Examples are

$$9m, \quad -6y^5, \quad a^2, \quad \text{and} \quad 6.$$

2. Add like terms.

(a) $5x^4 + 7x^4$

(b) $9pq + 3pq - 2pq$

(c) $r^2 + 3r + 5r^2$

(d) $8t + 6w$

3. Choose all descriptions that apply for each of the expressions in parts (a)–(d).
(1) Polynomial
(2) Polynomial written in descending order
(3) Not a polynomial

(a) $3m^5 + 5m^2 - 2m + 1$

(b) $2p^4 + p^6$

(c) $\frac{1}{x} + 2x^2 + 3$

(d) $x - 3$

ANSWERS
2. (a) $12x^4$ (b) $10pq$ (c) $6r^2 + 3r$ (d) cannot be added—unlike terms
3. (a) 1 and 2 (b) 1 (c) 3 (d) 1 and 2

EXAMPLE 3 *Classifying Polynomials*

For each polynomial, first simplify if possible by combining like terms. Then give the degree and tell whether the polynomial is a monomial, a binomial, a trinomial, or none of these.

(a) $2x^3 + 5$

The polynomial cannot be simplified. The degree is 3. The polynomial is a binomial.

(b) $4x - 5x + 2x$

Add like terms to simplify: $4x - 5x + 2x = x$. The degree is 1. The simplified polynomial is a monomial. ■

WORK PROBLEM 4 AT THE SIDE.

4 Polynomials may be added, subtracted, multiplied, and divided. Polynomial addition and subtraction are explained in the rest of this section.

ADDING POLYNOMIALS

To add two polynomials, add like terms.

EXAMPLE 4 *Adding Polynomials Horizontally*

(a) Add $6x^3 - 4x^2 + 3$ and $-2x^3 + 7x^2 - 5$.
Write the sum.

$$(6x^3 - 4x^2 + 3) + (-2x^3 + 7x^2 - 5)$$

Rewrite this sum with the parentheses removed.

$$6x^3 + (-4x^2) + 3 + (-2x^3) + 7x^2 + (-5)$$

Place like terms together.

$$6x^3 + (-2x^3) + (-4x^2) + 7x^2 + 3 + (-5)$$

Combine like terms to get

$$4x^3 + 3x^2 + (-2) \quad \text{or} \quad 4x^3 + 3x^2 - 2.$$

(b) Add $(2x^2 - 4x + 3)$ and $(x^3 + 5x)$.

$$\begin{aligned}(2x^2 - 4x + 3) + (x^3 + 5x) &= 2x^2 - 4x + 3 + x^3 + 5x \\ &= x^3 + 2x^2 + x + 3 \quad \text{Combine like terms.}\end{aligned}$$

■

WORK PROBLEM 5 AT THE SIDE.

The polynomials in Example 4 also could be added vertically, as shown in the next example. This form of addition is used in multiplication of polynomials.

4. For each polynomial, first simplify if possible. Then give the degree and tell whether the polynomial is a monomial, binomial, trinomial, or none of these.

(a) $3x^2 + 2x - 4$

(b) $x^3 + 4x^3$

(c) $x^8 - x^7 + 2x^8$

5. Find each sum.

(a) $(2x^4 - 6x^2 + 7)$
$+ (-3x^4 + 5x^2 + 2)$

(b) $(3x^2 + 4x + 2)$
$+ (6x^3 - 5x - 7)$

ANSWERS

4. (a) degree 2; trinomial **(b)** degree 3; monomial (simplify to $5x^3$) **(c)** degree 8; binomial (simplify to $3x^8 - x^7$)

5. (a) $-x^4 - x^2 + 9$
(b) $6x^3 + 3x^2 - x - 5$

6. Add each pair of polynomials.

(a)
$$\begin{array}{r} 4x^3 - 3x^2 + 2x \\ \underline{6x^3 + 2x^2 - 3x} \end{array}$$

(b)
$$\begin{array}{r} x^2 - 2x + 5 \\ \underline{4x^2 + 3x - 2} \end{array}$$

EXAMPLE 5 *Adding Polynomials Vertically*

Add $6x^3 - 4x^2 + 3$ and $-2x^3 + 7x^2 - 5$.

Write like terms in columns.

$$\begin{array}{r} 6x^3 - 4x^2 + 3 \\ \underline{-2x^3 + 7x^2 - 5} \end{array}$$

Now add, column by column.

$$\begin{array}{rrr} 6x^3 & -4x^2 & 3 \\ \underline{-2x^3} & \underline{7x^2} & \underline{-5} \\ 4x^3 & 3x^2 & -2 \end{array}$$

Add the three sums together.

$$4x^3 + 3x^2 + (-2) = 4x^3 + 3x^2 - 2$$ ■

This is the same answer found in Example 4(a).

WORK PROBLEM 6 AT THE SIDE.

5 Earlier, the difference $x - y$ was defined as $x + (-y)$. (We find the difference $x - y$ by adding x and the opposite of y.) For example,

$$7 - 2 = 7 + (-2) = 5$$

and

$$-8 - (-2) = -8 + 2 = -6.$$

A similar method is used to subtract polynomials.

SUBTRACTING POLYNOMIALS

To subtract polynomials, change all the signs of the second polynomial and add the result to the first polynomial.

EXAMPLE 6 *Subtracting Polynomials*

Subtract: $(5x - 2) - (3x - 8)$.

By the definition of subtraction,

$$(5x - 2) - (3x - 8) = (5x - 2) + [-(3x - 8)].$$

From Chapter 1,

$$\begin{aligned} -(3x - 8) &= -1(3x - 8) \\ &= -3x + 8 \end{aligned}$$

so

$$\begin{aligned} (5x - 2) - (3x - 8) &= (5x - 2) + (-3x + 8) \\ &= 2x + 6. \end{aligned}$$ ■

EXAMPLE 7 *Subtracting Polynomials*

Subtract $6x^3 - 4x^2 + 2$ from $11x^3 + 2x^2 - 8$.

Start with

$$(11x^3 + 2x^2 - 8) - (6x^3 - 4x^2 + 2).$$

ANSWERS

6. (a) $10x^3 - x^2 - x$ (b) $5x^2 + x + 3$

Change all the signs on the second polynomial and add like terms.

$$(11x^3 + 2x^2 - 8) + (-6x^3 + 4x^2 - 2)$$
$$= 11x^3 + 2x^2 - 8 - 6x^3 + 4x^2 - 2$$
$$= 5x^3 + 6x^2 - 10$$

We can check a subtraction problem by using the fact that if $a - b = c$, then $a = b + c$. For example, $6 - 2 = 4$. Check by writing $6 = 2 + 4$, which is correct. Check the polynomial subtraction above by adding $6x^3 - 4x^2 + 2$ and $5x^3 + 6x^2 - 10$. Since the sum is $11x^3 + 2x^2 - 8$, the subtraction was performed correctly. ■

WORK PROBLEM 7 AT THE SIDE. ▶▶

Subtraction also can be done in columns. We will use vertical subtraction in polynomial division in Section 3.8.

EXAMPLE 8 *Subtracting Polynomials Vertically*

Use the method of subtracting by columns to find $(14y^3 - 6y^2 + 2y - 5) - (2y^3 - 7y^2 - 4y + 6)$.

Step 1 Arrange like terms in columns.

$$\begin{array}{r} 14y^3 - 6y^2 + 2y - 5 \\ \underline{2y^3 - 7y^2 - 4y + 6} \end{array}$$

Step 2 Change all signs in the second row, and then add (*Step 3*).

$$\begin{array}{rl} 14y^3 - 6y^2 + 2y - 5 & \\ \underline{-2y^3 + 7y^2 + 4y - 6} & \text{All signs changed} \\ 12y^3 + y^2 + 6y - 11 & \text{Add.} \end{array}$$ ■

Either the horizontal or the vertical method may be used for adding and subtracting polynomials.

WORK PROBLEM 8 AT THE SIDE. ▶▶

6 A **multivariable polynomial** has more than one variable. Some examples are $a + ab + b$ and $2x^2y + xy - 5x$. Most of the definitions and rules in this section also apply to multivariable polynomials. The **degree of a term with more than one variable** is the sum of the exponents on the variables. Thus, the term $2x^2y$ has degree $2 + 1 = 3$, and the polynomial $2x^2y + xy - 5$ has degree 3.

Multivariable polynomials are added and subtracted by combining like terms, just as we did with single variable polynomials.

7. Subtract, and check your answers by addition.

(a) $(14y^3 - 6y^2 + 2y - 5) - (2y^3 - 7y^2 - 4y + 6)$

(b) $\left(\frac{7}{2}y^2 - \frac{11}{3}y + 8\right) - \left(-\frac{3}{2}y^2 + \frac{4}{3}y + 6\right)$

8. Subtract, using the method of subtracting by columns.

(a) $(14y^3 - 6y^2 + 2y) - (2y^3 - 7y^2 + 6)$

(b) $(6p^4 - 8p^3 + 2p - 1) - (-7p^4 + 6p^2 - 12)$

ANSWERS

7. (a) $12y^3 + y^2 + 6y - 11$
(b) $5y^2 - 5y + 2$

8. (a) $12y^3 + y^2 + 2y - 6$
(b) $13p^4 - 8p^3 - 6p^2 + 2p + 11$

9. Add or subtract. Give the degree of the answer.

(a) $(3mn + 2m - 4n) + (-mn + 4m + n)$

(b) $(5p^2q^2 - 4p^2 + 2q) - (2p^2q^2 - p^2 - 3q)$

ANSWERS

9. (a) $2mn + 6m - 3n$; degree 2
(b) $3p^2q^2 - 3p^2 + 5q$; degree 4

■ **EXAMPLE 9** *Adding and Subtracting Multivariable Polynomials*

Add or subtract as indicated, and give the degree of the answer.

(a) $(4a + 2ab - b) + (3a - ab + b)$

$$\begin{aligned}(4a + 2ab - b) &+ (3a - ab + b)\\ &= 4a + 2ab - b + 3a - ab + b\\ &= 7a + ab\end{aligned}$$

The polynomial $7a + ab$ has degree 2.

(b) $(2x^2y + 3xy + y^2) - (3x^2y - xy - 2y^2)$

$$\begin{aligned}(2x^2y &+ 3xy + y^2) - (3x^2y - xy - 2y^2)\\ &= 2x^2y + 3xy + y^2 - 3x^2y + xy + 2y^2\\ &= -x^2y + 4xy + 3y^2\end{aligned}$$

The polynomial $-x^2y + 4xy + 3y^2$ has degree 3. ■

WORK PROBLEM 9 AT THE SIDE.

NAME DATE HOUR

3.4 EXERCISES

For each of the following, determine the number of terms in the polynomial, and name the coefficients of the terms. See Example 1.

1. $6x^4$

2. $-9y^5$

3. t^4

4. s^7

5. $-19r^2 - r$

6. $2y^3 - y$

7. $x + 8x^2$

8. $v - 2v^3$

In each polynomial, add like terms whenever possible. Write the result in descending powers of the variable. See Example 2.

9. $-3m^5 + 5m^5$

10. $-4y^3 + 3y^3$

11. $2r^5 + (-3r^5)$

12. $-19y^2 + 9y^2$

13. $.2m^5 - .5m^2$

14. $-.9y + .9y^2$

15. $-3x^5 + 2x^5 - 4x^5$

16. $6x^3 - 8x^3 + 9x^3$

17. $-4p^7 + 8p^7 + 5p^9$

18. $-3a^8 + 4a^8 - 3a^2$

19. $-4y^2 + 3y^2 - 2y^2 + y^2$

20. $3r^5 - 8r^5 + r^5 + 2r^5$

Tell whether each statement is true always, sometimes, or never.

21. A polynomial is a binomial.

22. A polynomial is a trinomial.

23. A trinomial is a polynomial.

24. A binomial is a polynomial.

25. A trinomial is a binomial.

26. A binomial is a trinomial.

For each polynomial, first simplify, if possible, and write it in descending powers of the variable. Then give the degree of the resulting polynomial, and tell whether it is a monomial, a binomial, a trinomial, or none of these. See Example 3.

	Simplified form	*Degree*	*Kind of polynomial*
27. $6x^4 - 9x$			
28. $7t^3 - 3t$			

	Simplified form	*Degree*	*Kind of polynomial*
29. $5m^4 - 3m^2 + 6m^5 - 7m^3$			
30. $6p^5 + 4p^3 - 8p^4 + 10p^2$			
31. $\frac{5}{3}x^4 - \frac{2}{3}x^4 + \frac{1}{3}x^2 - 4$			
32. $\frac{4}{5}r^6 + \frac{1}{5}r^6 - r^4 + \frac{2}{5}r$			
33. $.8x^4 - .3x^4 - .5x^4 + 7$			
34. $1.2t^3 - .9t^3 - .3t^3 + 9$			

Add or subtract as indicated. See Examples 5 and 8.

35. Add.

$$\begin{array}{r} 3m^2 + 5m \\ \underline{2m^2 - 2m} \end{array}$$

36. Add.

$$\begin{array}{r} 4a^3 - 4a^2 \\ \underline{6a^3 + 5a^2} \end{array}$$

37. Subtract.

$$\begin{array}{r} 12x^4 - x^2 \\ \underline{8x^4 + 3x^2} \end{array}$$

38. Subtract.

$$\begin{array}{r} 13y^5 - y^3 \\ \underline{7y^5 + 5y^3} \end{array}$$

39. Add.

$$\begin{array}{r} \frac{2}{3}x^2 + \frac{1}{5}x + \frac{1}{6} \\ \underline{\frac{1}{2}x^2 - \frac{1}{3}x + \frac{2}{3}} \end{array}$$

40. Add.

$$\begin{array}{r} \frac{4}{7}y^2 - \frac{1}{5}y + \frac{7}{9} \\ \underline{\frac{1}{3}y^2 - \frac{1}{3}y + \frac{2}{5}} \end{array}$$

41. Subtract.

$$\begin{array}{r} 12m^3 - 8m^2 + 6m + 7 \\ \underline{-3m^3 + 5m^2 - 2m - 4} \end{array}$$

42. Subtract.

$$\begin{array}{r} 5a^4 - 3a^3 + 2a^2 - a + 6 \\ \underline{-6a^4 + a^3 - a^2 + a - 1} \end{array}$$

43. After reading Examples 5–8, explain whether you have a preference regarding horizontal or vertical addition of polynomials and why.

44. Repeat Exercise 47, but for subtraction of polynomials.

Perform the indicated operations. See Examples 4, 6, and 7.

45. $(2r^2 + 3r - 12) + (6r^2 + 2r)$

46. $(3r^2 + 5r - 6) + (2r - 5r^2)$

47. $(8m^2 - 7m) - (3m^2 + 7m - 6)$

48. $(x^2 + x) - (3x^2 + 2x - 1)$

49. $(16x^3 - x^2 + 3x) + (-12x^3 + 3x^2 + 2x)$

50. $(-2b^6 + 3b^4 - b^2) + (b^6 + 2b^4 + 2b^2)$

51. $(7y^4 + 3y^2 + 2y) - (18y^5 - 5y^3 + y)$

52. $(8t^5 + 3t^3 + 5t) - (19t^4 - 6t^2 + t)$

53. $[(8m^2 + 4m - 7) - (2m^3 - 5m + 2)] - (m^2 + m + 1)$

54. $[(9b^3 - 4b^2 + 3b + 2) - (-2b^3 - 3b^2 + b)] - (8b^3 + 6b + 4)$

Find the perimeter of each of the following geometric figures.

55. A rectangle with length $4x^2 + 3x + 1$ and width $x + 2$

56. A rectangle with length $5y^2 + 3y + 8$ and width $y + 4$

57. A triangle with sides $3t^2 + 2t + 7$, $5t^2 + 2$, $6t + 4$

58. A triangle with sides $9p^3 + 2p^2 + 1$, $6p^2 + p$, $2p + 5$

59. Subtract $9x^2 - 3x + 7$ from $-2x^2 - 6x + 4$.

60. Subtract $-5w^3 + 5w^2 - 7$ from $6w^3 + 8w + 5$.

61. Explain why the degree of the term 3^4 is not 4. What is its degree?

62. Can the sum of two polynomials in x, both of degree 3, be of degree 2? If so, give an example.

Add or subtract as indicated. Give the degree of the answer. See Example 9.

63. $(9a^2b - 3a^2 + 2b) + (4a^2b - 4a^2 - 3b)$

64. $(4xy^3 - 3x + y) + (5xy^3 + 13x - 4y)$

65. $(2c^4d + 3c^2d^2 - 4d^2) - (c^4d + 8c^2d^2 - 5d^2)$

66. $(3k^2h^3 + 5kh + 6k^3h^2) - (2k^2h^3 - 9kh + k^3h^2)$

67. Subtract. $\begin{array}{r} 9m^3n - 5m^2n^2 + 4mn^2 \\ \underline{-3m^3n + 6m^2n^2 + 8mn^2} \end{array}$

68. Subtract. $\begin{array}{r} 12r^5t + 11r^4t^2 - 7r^3t^3 \\ \underline{-\ 8r^5t + 10r^4t^2 + 3r^3t^3} \end{array}$

PREVIEW EXERCISES

Multiply. See Section 3.1.

69. $-8n^3(3n^4)$

70. $(5x^2)(10x^4)$

71. $(-4y^3)(-8y)$

72. $(.2x^8)(.8x^4)$

3.5 MULTIPLICATION OF POLYNOMIALS

1 As shown earlier, we find the product of two monomials by using the rules for exponents and the commutative and associative properties. For example,

$$(6x^3)(4x^4) = 6 \cdot 4 \cdot x^3 \cdot x^4 = 24x^7.$$

Also,

$$(-8m^6)(-9n^6) = (-8)(-9)(m^6)(n^6) = 72m^6n^6.$$

To find the product of a monomial and a polynomial with more than one term, we use the distributive property and then the method shown above.

EXAMPLE 1 *Multiplying a Monomial and a Polynomial*
Use the distributive property to find each product.

(a) $4x^2(3x + 5)$

$$4x^2(3x + 5) = (4x^2)(3x) + (4x^2)(5) \quad \text{Distributive property}$$
$$= 12x^3 + 20x^2 \quad \text{Multiply monomials.}$$

(b) $-8m^3(4m^3 + 3m^2 + 2m - 1)$

$$= (-8m^3)(4m^3) + (-8m^3)(3m^2) \quad \text{Distributive property}$$
$$+ (-8m^3)(2m) + (-8m^3)(-1)$$
$$= -32m^6 - 24m^5 - 16m^4 + 8m^3 \quad \text{Multiply monomials.}$$ ■

WORK PROBLEM 1 AT THE SIDE.

2 The distributive property is used also to find the product of any two polynomials. For example, to find the product of the polynomials $x + 1$ and $x - 4$, think of $x + 1$ as a single quantity and use the distributive property as follows.

$$(x + 1)(x - 4) = (x + 1)x + (x + 1)(-4)$$

Now use the distributive property to find $(x + 1)x$ and $(x + 1)(-4)$.

$$(x + 1)x + (x + 1)(-4) = x(x) + 1(x) + x(-4) + 1(-4)$$
$$= x^2 + x + (-4x) + (-4)$$
$$= x^2 - 3x - 4$$

EXAMPLE 2 *Multiplying Two Binomials*
Find the product $(2x + 1)(3x + 5)$.

$$(2x + 1)(3x + 5) = (2x + 1)(3x) + (2x + 1)(5)$$
$$= (2x)(3x) + (1)(3x) + (2x)(5) + (1)(5)$$
$$= 6x^2 + 3x + 10x + 5$$
$$= 6x^2 + 13x + 5$$ ■

WORK PROBLEM 2 AT THE SIDE.

OBJECTIVES

1. Multiply a monomial and a polynomial.
2. Multiply two polynomials.
3. Multiply vertically.
4. Find higher powers of binomials.

FOR EXTRA HELP

Tape 5

SSM pp. 106–110

MAC: A
IBM: A

1. Find each product.

(a) $5m^3(2m + 7)$

(b) $2x^4(3x^2 + 2x - 5)$

(c) $-4y^2(3y^3 + 2y^2 - 4y + 8)$

2. Find each product.

(a) $(4x + 3)(2x + 1)$

(b) $(3k - 2)(2k + 1)$

(c) $(m + 5)(3m - 4)$

ANSWERS
1. (a) $10m^4 + 35m^3$
(b) $6x^6 + 4x^5 - 10x^4$
(c) $-12y^5 - 8y^4 + 16y^3 - 32y^2$
2. (a) $8x^2 + 10x + 3$
(b) $6k^2 - k - 2$
(c) $3m^2 + 11m - 20$

A rule for multiplying any two polynomials is given below.

MULTIPLYING POLYNOMIALS

Multiply two polynomials by multiplying each term of the second polynomial by each term of the first polynomial and adding the products.

■ EXAMPLE 3 *Multiplying Any Two Polynomials*

Find the product $(3y^2 + 2y - 1)(y^3 + y^2 - 5)$.

Multiply each term in the first polynomial by each term in the second polynomial, then add any like terms.

$$\begin{aligned}(3y^2 + 2y - 1)(y^3 + y^2 - 5) &= 3y^2(y^3) + 3y^2(y^2) + 3y^2(-5) + 2y(y^3) + 2y(y^2) \\ &\quad + 2y(-5) - 1(y^3) - 1(y^2) - 1(-5) \\ &= 3y^5 + 3y^4 - 15y^2 + 2y^4 + 2y^3 - 10y \\ &\quad - y^3 - y^2 + 5 \\ &= 3y^5 + 5y^4 + y^3 - 16y^2 - 10y + 5 \quad ■\end{aligned}$$

WORK PROBLEM 3 AT THE SIDE.

3 We can also multiply two polynomials by writing one polynomial above the other.

■ EXAMPLE 4 *Multiplying Polynomials Vertically*

Multiply $2x^2 + 4x + 1$ by $3x + 5$.

Start with

$$\begin{array}{r} 2x^2 + 4x + 1 \\ 3x + 5. \\ \hline \end{array}$$

It is not necessary to line up like terms in columns, because any terms may be multiplied (not just like terms). Begin by multiplying each of the terms in the top row by 5.

Step 1

$$\begin{array}{r} 2x^2 + 4x + 1 \\ 3x + 5 \\ \hline 10x^2 + 20x + 5 \end{array} \quad \leftarrow \text{Product of 5 and } 2x^2 + 4x + 1$$

Notice how this process is similar to multiplication of whole numbers. Now multiply each term in the top row by $3x$. Be careful to place the like terms in columns, since the final step will involve addition (as in multiplying two whole numbers).

3. Multiply.

(a) $(m^3 - 2m + 1) \cdot (2m^2 + 4m + 3)$

(b) $(6p^2 + 2p - 4)(3p^2 - 5)$

ANSWERS

3. (a) $2m^5 + 4m^4 - m^3 - 6m^2 - 2m + 3$
(b) $18p^4 + 6p^3 - 42p^2 - 10p + 20$

Step 2

$$\begin{array}{r} 2x^2 + 4x + 1 \\ 3x + 5 \\ \hline 10x^2 + 20x + 5 \\ 6x^3 + 12x^2 + 3x \end{array} \leftarrow \text{Product of } 3x \text{ and } 2x^2 + 4x + 1$$

Step 3 Add like terms.

$$\begin{array}{r} 2x^2 + 4x + 1 \\ 3x + 5 \\ \hline 10x^2 + 20x + 5 \\ 6x^3 + 12x^2 + 3x \quad\quad \\ \hline 6x^3 + 22x^2 + 23x + 5 \end{array}$$

The product is $6x^3 + 22x^2 + 23x + 5$. ■

WORK PROBLEM 4 AT THE SIDE.

■ **EXAMPLE 5** *Multiplying Polynomials Vertically*

Find the product of $3p - 5q$ and $2p + 7q$.

$$\begin{array}{r} 3p - 5q \\ 2p + 7q \\ \hline 21pq - 35q^2 \\ 6p^2 - 10pq \quad\quad\quad \\ \hline 6p^2 + 11pq - 35q^2 \end{array} \quad \begin{array}{l} \\ \\ \leftarrow 7q(3p - 5q) \\ \leftarrow 2p(3p - 5q) \\ \\ \end{array}$$ ■

■ **EXAMPLE 6** *Multiplying Polynomials Vertically*

Find the product of $4m^3 - 2m^2 + 4m$ and $m^2 + 5$.

$$\begin{array}{r} 4m^3 - 2m^2 + 4m \\ m^2 + 5 \\ \hline 20m^3 - 10m^2 + 20m \\ 4m^5 - 2m^4 + 4m^3 \quad\quad\quad\quad\quad \\ \hline 4m^5 - 2m^4 + 24m^3 - 10m^2 + 20m \end{array}$$ ■

Terms of top row multiplied by 5

Terms of top row multiplied by m^2

WORK PROBLEM 5 AT THE SIDE.

■ **EXAMPLE 7** *Squaring a Polynomial*

Find $(3n^2 + 5n - 1)^2$.

By definition, $(3n^2 + 5n - 1)^2 = (3n^2 + 5n - 1)(3n^2 + 5n - 1)$. Use the vertical method of multiplication.

$$\begin{array}{r} 3n^2 + 5n - 1 \\ 3n^2 + 5n - 1 \\ \hline -3n^2 - 5n + 1 \\ 15n^3 + 25n^2 - 5n \quad\quad \\ 9n^4 + 15n^3 - 3n^2 \quad\quad\quad\quad \\ \hline 9n^4 + 30n^3 + 19n^2 - 10n + 1 \end{array}$$ ■

4. Find each product.

(a) $\begin{array}{r} 4k - 6 \\ 2k + 5 \\ \hline \end{array}$

(b) $\begin{array}{r} 3x^2 + 4x - 5 \\ x + 4 \\ \hline \end{array}$

5. Find each product.

(a) $\begin{array}{r} 2m + 3p \\ 5m - 4p \\ \hline \end{array}$

(b) $\begin{array}{r} k^3 - k^2 + k + 1 \\ k + 1 \\ \hline \end{array}$

(c) $\begin{array}{r} a^3 + 3a - 4 \\ 2a^2 + 6a + 5 \\ \hline \end{array}$

ANSWERS

4. (a) $8k^2 + 8k - 30$
(b) $3x^3 + 16x^2 + 11x - 20$

5. (a) $10m^2 + 7mp - 12p^2$
(b) $k^4 + 2k + 1$
(c) $2a^5 + 6a^4 + 11a^3 + 10a^2 - 9a - 20$

4 To find higher powers of binomials, such as $(x + 5)^3$, we use the definition of exponent and perform repeated multiplication.

EXAMPLE 8 *Cubing a Binomial*

Find $(x + 5)^3$.

Since $(x + 5)^3 = (x + 5)(x + 5)(x + 5)$, the first step is to find the product $(x + 5)(x + 5)$.

$$\begin{aligned}(x + 5)(x + 5) &= x^2 + 5x + 5x + 25 \\ &= x^2 + 10x + 25\end{aligned}$$

Now multiply this result by $x + 5$.

$$\begin{aligned}(x + 5)(x^2 + 10x + 25) &= x^3 + 10x^2 + 25x + 5x^2 + 50x + 125 \\ &= x^3 + 15x^2 + 75x + 125\end{aligned}$$ ■

WORK PROBLEM 6 AT THE SIDE.

6. Find each product.

(a) $(3k - 2)^2$

(b) $(2x^2 - 3x + 4)^2$

(c) $(m + 1)^3$

ANSWERS

6. (a) $9k^2 - 12k + 4$
(b) $4x^4 - 12x^3 + 25x^2 - 24x + 16$
(c) $m^3 + 3m^2 + 3m + 1$

NAME DATE HOUR

3.5 EXERCISES

1. In multiplying two monomials, we use (but do not always show) the commutative and associative properties. In each item below, name the property that has been used.

$$\begin{aligned}(5x^3)(6x^5) &= (5x^3 \cdot 6)x^5 \quad \_\_\_\_\_\_\_\_\_\_\_\_\_\_\_\_ \\ &= (5 \cdot 6x^3)x^5 \quad \_\_\_\_\_\_\_\_\_\_\_\_\_\_\_\_ \\ &= (5 \cdot 6)(x^3 \cdot x^5) \quad \_\_\_\_\_\_\_\_\_\_\_\_\_\_\_\_ \\ &= 30x^8\end{aligned}$$

2. In multiplying a monomial by a polynomial, such as in $4x(3x^2 + 7x^3) = 4x(3x^2) + 4x(7x^3)$, the first property that is used is the ____________ property.

Find each product. See Example 1.

3. $(-5a^9)(-8a^5)$

4. $(-3m^6)(-5m^4)$

5. $-2m(3m + 2)$

6. $-5p(6 + 3p)$

7. $3p(8 - 6p + 12p^3)$

8. $4x(3 + 2x + 5x^3)$

9. $2y^5(3 + 2y + 5y^4)$

10. $2m^4(3m^2 + 5m + 6)$

Find each binomial product. See Examples 2 and 5.

11. $(n - 2)(n + 3)$

12. $(r - 6)(r + 8)$

13. $(4r + 1)(2r - 3)$

14. $(5x + 2)(2x - 7)$

15. $(3x + 2)(3x - 2)$

16. $(7x + 3)(7x - 3)$

17. $(3q + 1)(3q + 1)$

18. $(4w + 7)(4w + 7)$

19. $(3t + 4s)(2t + 5s)$

20. $(8v + 5w)(2v + 3w)$

21. $(-3t + 4)(t + 6)$

22. $(-5x + 9)(x - 2)$

23. Find a polynomial that represents the area of this square.

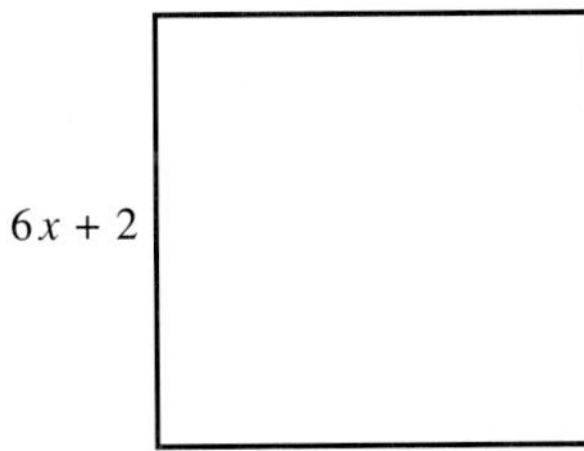

24. Find a polynomial that represents the area of this rectangle.

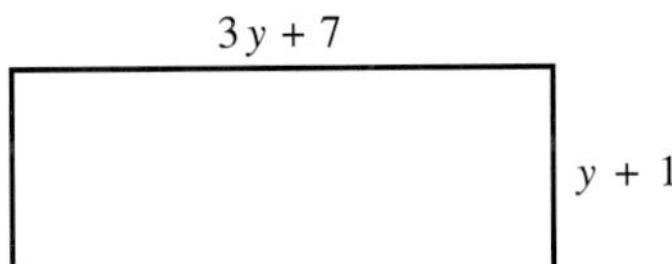

25. Perform the following multiplications: $(x + 4)(x - 4)$; $(y + 2)(y - 2)$; $(r + 7)(r - 7)$. Observe your answers, and explain the pattern that can be found in the answers.

26. Repeat Exercise 25 for the following: $(x + 4)(x + 4)$; $(y - 2)(y - 2)$; $(r + 7)(r + 7)$.

Find each product. See Examples 3, 4, and 6.

27. $(6x + 1)(2x^2 + 4x + 1)$

28. $(9y - 2)(8y^2 - 6y + 1)$

29. $(4m + 3)(5m^3 - 4m^2 + m - 5)$

30. $(y + 4)(3y^3 - 2y^2 + y + 3)$

31. $(5x^2 + 2x + 1)(x^2 - 3x + 5)$

32. $(2m^2 + m - 3)(m^2 - 4m + 5)$

Find each product. See Examples 7 and 8.

33. $(x + 12)^2$

34. $(y + 11)^2$

35. $(3t + 1)^2$

36. $(4y + 3)^2$

37. $(5x - 2y)^2$

38. $(6p - 5s)^2$

39. $(h - 5)^3$

40. $(n - 3)^3$

41. $(r + 1)^4$

42. $(k - 1)^4$

43. $(3x^2 + x - 4)^2$

44. $(4y^2 - y + 3)^2$

PREVIEW EXERCISES

Find two numbers having the given product and sum. See Sections 1.4 and 1.6.

	Product	Sum
45.	8	6
46.	−35	2
47.	−56	−1
48.	25	−10

3.6 PRODUCTS OF BINOMIALS

OBJECTIVES

1. Multiply binomials by the FOIL method.
2. Square binomials.
3. Find the product of the sum and difference of two terms.

FOR EXTRA HELP

Tape 5 | SSM pp. 110–112 | MAC: A IBM: A

1 We can use the methods introduced in the last section to find the product of any two polynomials. They are the only practical methods for multiplying polynomials with three or more terms. However, many of the polynomials to be multiplied are binomials, with only two terms, so in this section we discuss a shortcut that eliminates the need to write all the steps. To develop this shortcut, let us first multiply $x + 3$ and $x + 5$ using the distributive property.

$$\begin{aligned}(x + 3)(x + 5) &= (x + 3)x + (x + 3)5\\ &= (x)(x) + (3)(x) + (x)(5) + (3)(5)\\ &= x^2 + 3x + 5x + 15\\ &= x^2 + 8x + 15\end{aligned}$$

The first term in the second line, $(x)(x)$, is the product of the first terms of the two binomials.

$(x + 3)(x + 5)$ Multiply the **first** terms: $(x)(x)$.

The term $(x)(5)$ is the product of the first term of the first binomial and the last term of the second binomial. This is the **outer product.**

$(x + 3)(x + 5)$ Multiply the **outer** terms: $(x)(5)$.

The term $(3)(x)$ is the product of the last term of the first binomial and the first term of the second binomial. The product of these middle terms is called the **inner product.**

$(x + 3)(x + 5)$ Multiply the **inner** terms: $(3)(x)$.

Finally, $(3)(5)$ is the product of the last terms of the two binomials.

$(x + 3)(x + 5)$ Multiply the **last** terms: $(3)(5)$.

In the third step of the multiplication above, the inner product and the outer product are added. This step should be performed mentally, so that the three terms of the answer can be written without extra steps as

$$(x + 3)(x + 5) = x^2 + 8x + 15.$$

WORK PROBLEM 1 AT THE SIDE.

A summary of these steps is given below. This procedure is sometimes called the **FOIL method,** which comes from the initial letters of the words *First, Outer, Inner,* and *Last.*

1. For the product $(2p - 5)(3p + 7)$, find the following.

(a) Product of first terms

(b) Outer product

(c) Inner product

(d) Product of last terms

(e) Complete product in simplified form

ANSWERS

1. (a) $2p(3p) = 6p^2$
(b) $2p(7) = 14p$
(c) $-5(3p) = -15p$
(d) $-5(7) = -35$
(e) $6p^2 - p - 35$

2. Use the FOIL method to find each product.

(a) $(m + 4)(m - 3)$

(b) $(y + 7)(y + 2)$

(c) $(r - 8)(r - 5)$

ANSWERS
2. (a) $m^2 + m - 12$
(b) $y^2 + 9y + 14$
(c) $r^2 - 13r + 40$

MULTIPLYING BINOMIALS BY THE FOIL METHOD

Step 1 Multiply the two **F**irst terms of the binomials to get the first term of the answer.

Step 2 Find the **O**uter product and the **I**nner product and add them (mentally, if possible) to get the middle term of the answer.

Step 3 Multiply the two **L**ast terms of the binomials to get the last term of the answer.

$$\mathbf{F} = x^2 \quad \mathbf{L} = 15$$
$$(x + 3)(x + 5)$$
$$\mathbf{I} = 3x$$
$$\underline{\mathbf{O} = 5x}$$
$$8x \quad \text{Add.}$$

EXAMPLE 1 *Using the FOIL Method*

Use the FOIL method to find the product $(x + 8)(x - 6)$.

Step 1 **F** Multiply the **first** terms.

$$x(x) = x^2$$

Step 2 **O** Find the product of the **outer** terms.

$$x(-6) = -6x$$

I Find the product of the **inner** terms.

$$8(x) = 8x$$

Add the outer and inner products mentally.

$$-6x + 8x = 2x$$

Step 3 **L** Multiply the **last** terms.

$$8(-6) = -48$$

The product of $x + 8$ and $x - 6$ is found by adding the terms found in the three steps above, so

$$(x + 8)(x - 6) = x^2 + 2x - 48. \blacksquare$$

As a shortcut, this product can be found in the following manner.

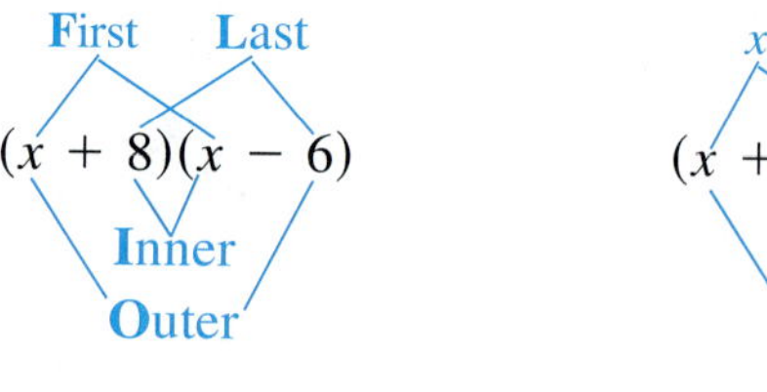

$$x^2 \quad -48$$
$$(x + 8)(x - 6)$$
$$8x$$
$$\underline{-6x}$$
$$2x \quad \text{Add.}$$

WORK PROBLEM 2 AT THE SIDE.

EXAMPLE 2 *Using the FOIL Method*

Multiply $9x - 2$ and $3x + 1$.

First	$(9x - 2)(3x + 1)$	$27x^2$
Outer	$(9x - 2)(3x + 1)$	$9x$

Inner	$(9x - 2)(3x + 1)$	$-6x$
Last	$(9x - 2)(3x + 1)$	-2

$$\begin{aligned}(9x - 2)(3x + 1) &= \overset{F}{27x^2} + \overset{O}{9x} - \overset{I}{6x} - \overset{L}{2}\\ &= 27x^2 + 3x - 2\end{aligned}$$ ■

■ **EXAMPLE 3** *Using the FOIL Method*

Find the following products.

(a) $$\begin{aligned}(2k + 5y)(k + 3y) &= \overset{F}{(2k)(k)} + \overset{O}{(2k)(3y)} + \overset{I}{(5y)(k)} + \overset{L}{(5y)(3y)}\\ &= 2k^2 + 6ky + 5ky + 15y^2\\ &= 2k^2 + 11ky + 15y^2\end{aligned}$$

(b) $(7p + 2q)(3p - q) = 21p^2 - pq - 2q^2$ ■

WORK PROBLEM 3 AT THE SIDE.

2 Certain special types of binomial products occur so often that the form of the answers should be memorized. For example, to find the square of a binomial quickly, use the method shown in Example 4.

■ **EXAMPLE 4** *Squaring a Binomial*

Find $(2m + 3)^2$.

Squaring $2m + 3$ by the FOIL method gives

$$(2m + 3)(2m + 3) = 4m^2 + 6m + 6m + 9 = 4m^2 + 12m + 9.$$

The result has the square of both the first and the last terms of the binomial:

$$4m^2 = (2m)^2 \quad \text{and} \quad 9 = 3^2.$$

The middle term is twice the product of the two terms of the binomial, that is,

$$12m = 2(6m) = 2(2m)(3).$$ ■

This example suggests the following rule.

SQUARE OF A BINOMIAL

The square of a binomial is a trinomial made up of the square of the first term, plus twice the product of the two terms, plus the square of the last term of the binomial. For a and b,

$$(a + b)^2 = a^2 + 2ab + b^2.$$

Also, $$(a - b)^2 = a^2 - 2ab + b^2.$$

■ **EXAMPLE 5** *Squaring Binomials*

Use the formula to square each binomial.

$$(a - b)^2 = a^2 - 2 \cdot a \cdot b + b^2$$

(a) $$\begin{aligned}(5z - 1)^2 &= (5z)^2 - 2(5z)(1) + (1)^2\\ &= 25z^2 - 10z + 1\end{aligned}$$

Recall that $(5z)^2 = 5^2z^2 = 25z^2$.

3. Find each product.

(a) $(4k - 1)(2k + 3)$

(b) $(6m + 5)(m - 4)$

(c) $(8y + 3)(2y + 1)$

(d) $(3r + 2t)(3r + 4t)$

ANSWERS

3. (a) $8k^2 + 10k - 3$
(b) $6m^2 - 19m - 20$
(c) $16y^2 + 14y + 3$
(d) $9r^2 + 18rt + 8t^2$

(b) $(3b + 5r)^2 = (3b)^2 + 2(3b)(5r) + (5r)^2$
$= 9b^2 + 30br + 25r^2$

(c) $(2a - 9x)^2 = 4a^2 - 36ax + 81x^2$

(d) $\left(4m + \frac{1}{2}\right)^2 = (4m)^2 + 2(4m)\left(\frac{1}{2}\right) + \left(\frac{1}{2}\right)^2$
$= 16m^2 + 4m + \frac{1}{4}$ ■

Caution A common error in squaring a binomial is forgetting the middle term of the product. In general,

$$(a + b)^2 \neq a^2 + b^2.$$

◀ WORK PROBLEM 4 AT THE SIDE.

3 Binomial products of the form $(a + b)(a - b)$ also occur frequently. In these products, one binomial is the sum of two terms, and the other is the difference of the same two terms. As an example, the product of $x + 2$ and $x - 2$ is

$$(x + 2)(x - 2) = x^2 - 2x + 2x - 4$$
$$= x^2 - 4.$$

As this example suggests, the product of $a + b$ and $a - b$ is the difference between two squares.

PRODUCT OF THE SUM AND DIFFERENCE OF TWO TERMS

$$(a + b)(a - b) = a^2 - b^2$$

■ **EXAMPLE 6** *Finding the Product of the Sum and Difference of Two Terms*

Find each product.

$(a \quad + b)(a \quad - b)$
↓ ↓ ↓ ↓

(a) $(5m + 3)(5m - 3)$

Use the pattern for the sum and difference of two terms.

$$(5m + 3)(5m - 3) = (5m)^2 - 3^2$$
$$= 25m^2 - 9$$

(b) $(4x + y)(4x - y) = (4x)^2 - y^2$
$= 16x^2 - y^2$

(c) $\left(z - \frac{1}{4}\right)\left(z + \frac{1}{4}\right) = z^2 - \frac{1}{16}$ ■

◀ WORK PROBLEM 5 AT THE SIDE.

The product formulas of this section will be very useful in later work, particularly in Chapter 5. Therefore, it is important to memorize these formulas and practice using them.

4. Find each square by using the pattern for the square of a binomial.

(a) $(t + u)^2$

(b) $(2m - p)^2$

(c) $(4p + 3q)^2$

(d) $(5r - 6s)^2$

(e) $\left(3k - \frac{1}{2}\right)^2$

5. Find each product by using the pattern for the sum and difference of two terms.

(a) $(6a + 3)(6a - 3)$

(b) $(10m + 7)(10m - 7)$

(c) $(7p + 2q)(7p - 2q)$

(d) $\left(3r - \frac{1}{2}\right)\left(3r + \frac{1}{2}\right)$

ANSWERS
4. (a) $t^2 + 2tu + u^2$
(b) $4m^2 - 4mp + p^2$
(c) $16p^2 + 24pq + 9q^2$
(d) $25r^2 - 60rs + 36s^2$
(e) $9k^2 - 3k + \frac{1}{4}$
5. (a) $36a^2 - 9$ (b) $100m^2 - 49$
(c) $49p^2 - 4q^2$ (d) $9r^2 - \frac{1}{4}$

NAME DATE HOUR

3.6 EXERCISES

1. Consider the product $(2x + 3)(x - 5)$.
   (a) What is the product of the first terms, $2x(x)$?_____
   (b) What is the product of the outer terms, $2x(-5)$?_____
   (c) What is the product of the inner terms, $3(x)$?_____
   (d) What is the product of the last terms, $3(-5)$?_____
   (e) What is the sum of the outer and inner products found in parts b and c?_____
   (f) Write the complete product, which is a trinomial, using your results in parts a, e, and d._____

2. Repeat Exercise 1 for the product $(3y - 8)(2y + 5)$.

Find each product using the FOIL method. See Examples 1–3.

3. $(r + 1)(r + 3)$
4. $(p + 2)(p + 3)$
5. $(w - 4)(w - 6)$
6. $(y - 3)(y - 7)$
7. $(s - 12)(s + 4)$
8. $(t - 6)(t + 1)$
9. $(2x - 1)(5x + 3)$
10. $(5y - 3)(6y + 5)$
11. $(9x + 2)(3x + 7)$
12. $(6x + 7)(2x + 3)$
13. $(3m + 7)(3m + 5)$
14. $(7x - 1)(7x - 8)$
15. $(3 - 2x)(5 - 3x)$
16. $(4 - 3r)(5 - 2r)$
17. $(-5 + 6z)(3 - z)$
18. $(-4 + 5w)(5 - w)$
19. $(-8 + 3k)(-2 - k)$
20. $(-5 + 2r)(-3 - r)$
21. $(3w + 2z)(9w - z)$
22. $(4x + 5y)(7x - y)$
23. $(-8p + 3s)(2p + s)$
24. $(7h + k)(-4h + 3k)$
25. $(x - .3)(x + .8)$
26. $(y + .7)(y + .4)$
27. $\left(x - \frac{2}{3}\right)\left(x + \frac{1}{4}\right)$
28. $\left(z - \frac{5}{6}\right)\left(z + \frac{3}{4}\right)$

29. Consider the square $(2x + 3)^2$.
   (a) What is the square of the first term, $(2x)^2$?_____
   (b) What is twice the product of the two terms, $2(2x)(3)$?_____
   (c) What is the square of the last term, 3^2?_____
   (d) Write the final product, which is a trinomial, using your results in parts a–c._____

30. Repeat Exercise 29 for the square $(3x - 2)^2$.

Find each square. See Examples 4 and 5.

31. $(p + 2)^2$ **32.** $(r + 5)^2$ **33.** $(a - c)^2$ **34.** $(p - y)^2$

35. $(4x - 3)^2$ **36.** $(5y + 2)^2$ **37.** $(8t + 7s)^2$ **38.** $(7z - 3w)^2$

39. $\left(5x + \frac{2}{5}y\right)^2$ **40.** $\left(6m - \frac{4}{5}n\right)^2$

41. Consider the product $(7x + 3y)(7x - 3y)$.
- **(a)** What is the product of the first terms, $(7x)(7x)$?_____
- **(b)** Multiply the outer terms, $(7x)(-3y)$. Then multipy the inner terms, $(3y)(7x)$. Add the results. What is this sum?_____
- **(c)** What is the product of the last terms, $(3y)(-3y)$?_____
- **(d)** Write the complete product using your answers in parts a and c. ______________ Why is the sum found in part b omitted here?

42. Repeat Exercise 41 for the product $(5x + 7y)(5x - 7y)$.

Find the following products. See Example 6.

43. $(q + 2)(q - 2)$ **44.** $(x + 8)(x - 8)$ **45.** $(2w + 5)(2w - 5)$

46. $(3z + 8)(3z - 8)$ **47.** $(10x + 3y)(10x - 3y)$ **48.** $(13r + 2z)(13r - 2z)$

49. $(2x^2 - 5)(2x^2 + 5)$ **50.** $(9y^2 - 2)(9y^2 + 2)$ **51.** $\left(7x + \frac{3}{7}\right)\left(7x - \frac{3}{7}\right)$

52. $\left(9y + \frac{2}{3}\right)\left(9y - \frac{2}{3}\right)$

PREVIEW EXERCISES

Simplify. Write answers with positive exponents. See Section 3.2.

53. $\dfrac{6x^5}{2x}$ **54.** $\dfrac{-8x^4y^2}{2xy}$ **55.** $\dfrac{64m^9}{8m^2}$ **56.** $\dfrac{-18p^7q^4}{27p^5q^9}$

3.7 DIVISION OF A POLYNOMIAL BY A MONOMIAL

1 We add two fractions with a common denominator as follows.

$$\frac{a}{c} + \frac{b}{c} = \frac{a+b}{c}.$$

Looking at this statement in reverse gives us a rule for dividing a polynomial by a monomial.

DIVIDING A POLYNOMIAL BY A MONOMIAL

To divide a polynomial by a monomial, divide each term of the polynomial by the monomial:

$$\frac{a+b}{c} = \frac{a}{c} + \frac{b}{c} \quad (c \neq 0).$$

The quotient rule for exponents is used to reduce each fraction as shown in the following examples.

EXAMPLE 1 *Dividing a Polynomial by a Monomial*

Divide $5m^5 - 10m^3$ by $5m^2$.

Use the rule above, with + replaced by −.

$$\frac{5m^5 - 10m^3}{5m^2} = \frac{5m^5}{5m^2} - \frac{10m^3}{5m^2} = m^3 - 2m$$

Check by multiplication.

$$5m^2(m^3 - 2m) = 5m^5 - 10m^3$$

Because division by 0 is undefined, the quotient

$$\frac{5m^5 - 10m^3}{5m^2}$$

is undefined if $m = 0$. In the rest of the chapter, we assume that no denominators are 0. ■

WORK PROBLEM 1 AT THE SIDE. ▶▶

EXAMPLE 2 *Dividing a Polynomial by a Monomial*

Divide: $\dfrac{16a^5 - 12a^4 + 8a^2}{4a^3}$.

Divide each term of $16a^5 - 12a^4 + 8a^2$ by $4a^3$.

$$\frac{16a^5 - 12a^4 + 8a^2}{4a^3} = \frac{16a^5}{4a^3} - \frac{12a^4}{4a^3} + \frac{8a^2}{4a^3}$$

$$= 4a^2 - 3a + \frac{2}{a}$$

The quotient is not a polynomial because of the expression $\frac{2}{a}$, which has a variable in the denominator. While the sum, difference, and product of two polynomials are always polynomials, the quotient of two polynomials may not be.

Again, check by multiplying.

$$4a^3\left(4a^2 - 3a + \frac{2}{a}\right) = 4a^3(4a^2) - 4a^3(3a) + 4a^3\left(\frac{2}{a}\right)$$

$$= 16a^5 - 12a^4 + 8a^2 \quad ■$$

OBJECTIVE

1 Divide a polynomial by a monomial.

FOR EXTRA HELP

Tape 5

SSM pp. 112–113

MAC: A
IBM: A

1. Divide.

(a) $\dfrac{6p^4 + 18p^7}{3p^2}$

(b) $\dfrac{12m^6 + 18m^5 + 30m^4}{6m^2}$

(c) $(18r^7 - 9r^2) \div (3r)$

ANSWERS

1. (a) $2p^2 + 6p^5$
(b) $2m^4 + 3m^3 + 5m^2$
(c) $6r^6 - 3r$

2. Divide.

(a) $\dfrac{20x^4 - 25x^3 + 5x}{5x^2}$

(b) $\dfrac{50m^4 - 30m^3 + 20m}{10m^3}$

3. Divide.

(a) $\dfrac{8y^7 - 9y^6 - 11y - 4}{y^2}$

(b) $\dfrac{12p^5 + 8p^4 + 3p^3 - 5p^2}{3p^3}$

(c) $\dfrac{45x^4 + 30x^3 - 60x^2}{-15x^2}$

ANSWERS

2. (a) $4x^2 - 5x + \dfrac{1}{x}$ (b) $5m - 3 + \dfrac{2}{m^2}$

3. (a) $8y^5 - 9y^4 - \dfrac{11}{y} - \dfrac{4}{y^2}$

(b) $4p^2 + \dfrac{8p}{3} + 1 - \dfrac{5}{3p}$

(c) $-3x^2 - 2x + 4$

WORK PROBLEM 2 AT THE SIDE.

EXAMPLE 3 *Dividing a Polynomial by a Monomial*

Divide.

$$\frac{12x^4 - 7x^3 + 4x}{4x} = \frac{12x^4}{4x} - \frac{7x^3}{4x} + \frac{4x}{4x}$$
$$= 3x^3 - \frac{7x^2}{4} + 1$$

Check by multiplication. ■

Caution In Example 3, notice that the quotient $\frac{4x}{4x} = 1$. It is a common error to leave that term out of the answer. Checking by multiplication will show that the answer $3x^3 - \frac{7}{4}x^2$ is not correct.

EXAMPLE 4 *Dividing a Polynomial by a Monomial*

Divide the polynomial

$$180y^{10} - 150y^8 + 120y^6 - 90y^4 + 100y$$

by the monomial $-30y^2$.

Using the methods of this section,

$$\frac{180y^{10} - 150y^8 + 120y^6 - 90y^4 + 100y}{-30y^2}$$
$$= \frac{180y^{10}}{-30y^2} - \frac{150y^8}{-30y^2} + \frac{120y^6}{-30y^2} - \frac{90y^4}{-30y^2} + \frac{100y}{-30y^2}$$
$$= -6y^8 + 5y^6 - 4y^4 + 3y^2 - \frac{10}{3y}.$$

To check, multiply this answer and $-30y^2$. ■

WORK PROBLEM 3 AT THE SIDE.

NAME DATE HOUR

3.7 EXERCISES

Perform each division.

1. $\dfrac{12x^5}{-2x}$

2. $\dfrac{16y^8}{-4y^2}$

3. $\dfrac{15x^9y^4}{3x^2y^3}$

4. $\dfrac{26w^4z^5}{13w^3z}$

5. Explain why the division problem $\dfrac{16m^3 - 12m^2}{4m}$ can be performed using the methods of this section, while the division problem $\dfrac{4m}{16m^3 - 12m^2}$ cannot.

6. If the area of a rectangle is $24x^6$ and the width is $4x^2$, what is the length?

Divide each polynomial by 2m. See Examples 1–4.

7. $60m^4 - 20m^2 + 10m$

8. $120m^6 - 60m^3 + 80m^2$

9. $10m^5 - 16m^4 + 8m^3$

10. $6m^5 - 4m^3 + 2m^2$

11. $8m^5 - 4m^3 + 4m^2$

12. $8m^4 - 4m^3 + 6m^2$

13. $m^5 - 4m^2 + 8$

14. $m^3 + m^2 + 6$

Divide each polynomial by $3x^2$. See Examples 1–4.

15. $12x^5 - 9x^4 + 6x^3$

16. $24x^6 - 12x^5 + 30x^4$

17. $3x^2 + 15x^3 - 27x^4$

18. $3x^2 - 18x^4 + 30x^5$

19. $36x + 24x^2 + 6x^3$

20. $9x - 12x^2 + 9x^3$

21. $4x^4 + 3x^3 + 2x$

22. $5x^4 - 6x^3 + 8x$

Perform each division. See Examples 1–4.

23. $\dfrac{27r^4 - 36r^3 - 6r^2 + 26r - 2}{3r}$

24. $\dfrac{8k^4 - 12k^3 - 2k^2 + 7k - 3}{2k}$

25. $\dfrac{2m^5 - 6m^4 + 8m^2}{-2m^3}$

26. $\dfrac{6r^5 - 8r^4 + 10r^2}{-2r^4}$

27. $(20a^4 - 15a^5 + 25a^3) \div (5a^4)$

28. $(16y^5 - 8y^2 + 12y) \div (4y^2)$

29. $(120x^{11} - 60x^{10} + 140x^9 - 100x^8) \div (10x^{12})$

30. $(120x^{12} - 84x^9 + 60x^8 - 36x^7) \div (12x^9)$

31. The quotient in Exercise 21 is $\frac{4x^2}{3} + x + \frac{2}{3x}$. Notice how the third term is written with x in the denominator. Would $\frac{2}{3}x$ be an acceptable form for this term? Explain why or why not. Is $\frac{4}{3}x^2$ an acceptable form for the first term? Why or why not?

32. If the area of a rectangle is represented by $12x^2 - 4x + 2$ and the width is 2x, what expression represents the length?

33. What polynomial, when divided by $5x^3$, yields $3x^2 - 7x + 7$ as a quotient?

34. The quotient of a certain polynomial and $-12y^3$ is $6y^3 - 5y^2 + 2y - 3 + \frac{7}{y}$. Find the polynomial.

PREVIEW EXERCISES

Find each product. See Sections 3.5 and 3.6.

35. $-3k(8k^2 - 12k + 2)$

36. $(3r + 5)(2r + 1)$

37. $(-2k + 1)(8k^2 + 9k + 3)$

Subtract. See Section 3.4.

38.
$$\begin{array}{r} 5t^2 + 2t - 6 \\ \underline{5t^2 - 3t - 9} \end{array}$$

39.
$$\begin{array}{r} -4x^3 + 2x^2 - 3x + 7 \\ \underline{-4x^3 - 8x^2 + x - 4} \end{array}$$

40.
$$\begin{array}{r} x^4 + 0x^3 - 4x^2 + 3x - 8 \\ \underline{x^4 - 2x^3 + 4x^2 + 6x - 7} \end{array}$$

3.8 THE QUOTIENT OF TWO POLYNOMIALS

OBJECTIVE

1. Divide a polynomial by a polynomial.

FOR EXTRA HELP

Tape 6

SSM pp. 114–116

MAC: A IBM: A

1 A method of "long division" is used to divide a polynomial by a polynomial (other than a monomial). This method is similar to the method of long division used for two whole numbers. For comparison, the division of whole numbers is shown alongside the division of polynomials. The polynomial must be in descending order.

Division of Whole Numbers	Division of Polynomials
Step 1	
Divide 27 into 6696.	Divide $2x + 3$ into $8x^3 - 4x^2 - 14x + 15$.
$27\overline{)6696}$	$2x + 3\overline{)8x^3 - 4x^2 - 14x + 15}$
Step 2	
27 divides into 66 **2** times; $2 \cdot 27 =$ **54**.	$2x$ divides into $8x^3$ $\mathbf{4x^2}$ times; $4x^2(2x + 3) = \mathbf{8x^3 + 12x^2}$.
$\begin{array}{r} 2 \\ 27\overline{)6696} \\ \underline{54} \end{array}$	$\begin{array}{r} 4x^2 \\ 2x + 3\overline{)8x^3 - 4x^2 - 14x + 15} \\ \underline{8x^3 + 12x^2} \end{array}$
Step 3	
Subtract; then bring down the next digit.	Subtract; then bring down the next term.
$\begin{array}{r} 2 \\ 27\overline{)6696} \\ \underline{54}\downarrow \\ 129 \end{array}$	$\begin{array}{r} 4x^2 \\ 2x + 3\overline{)8x^3 - 4x^2 - 14x + 15} \\ \underline{8x^3 + 12x^2} \quad\downarrow \\ -16x^2 - 14x \end{array}$
	(To subtract two polynomials, change the sign of the second and then add.)
Step 4	
27 divides into 129 **4** times; $4 \cdot 27 =$ **108**.	$2x$ divides into $-16x^2$ $\mathbf{-8x}$ times; $-8x(2x + 3) = \mathbf{-16x^2 - 24x}$.
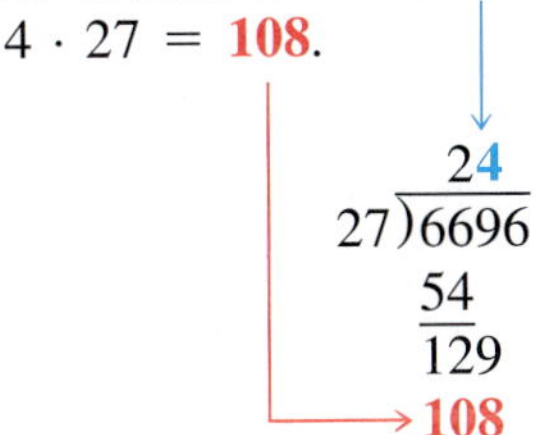	$\begin{array}{r} 4x^2 - 8x \\ 2x + 3\overline{)8x^3 - 4x^2 - 14x + 15} \\ \underline{8x^3 + 12x^2} \\ -16x^2 - 14x \\ \underline{-16x^2 - 24x} \end{array}$

Step 5

Subtract; then bring down the next digit.

$$\begin{array}{r} 24 \\ 27\overline{)6696} \\ \underline{54} \\ 129 \\ \underline{108} \\ 216 \end{array}$$

Subtract; then bring down the next term.

$$\begin{array}{r} 4x^2 - 8x \\ 2x + 3\overline{)8x^3 - 4x^2 - 14x + 15} \\ \underline{8x^3 + 12x^2} \\ -16x^2 - 14x \\ \underline{-16x^2 - 24x} \\ 10x + 15 \end{array}$$

Step 6

27 divides into 216 **8** times; $8 \cdot 27 =$ **216**.

$$\begin{array}{r} 248 \\ 27\overline{)6696} \\ \underline{54} \\ 129 \\ \underline{108} \\ 216 \\ \underline{216} \end{array}$$

6696 divided by 27 is 248. There is no remainder.

$2x$ divides into $10x$ **5** times; $5(2x + 3) =$ **$10x + 15$**.

$$\begin{array}{r} 4x^2 - 8x + 5 \\ 2x + 3\overline{)8x^3 - 4x^2 - 14x + 15} \\ \underline{8x^3 + 12x^2} \\ -16x^2 - 14x \\ \underline{-16x^2 - 24x} \\ 10x + 15 \\ \underline{10x + 15} \end{array}$$

$8x^3 - 4x^2 - 14x + 15$ divided by $2x + 3$ is $4x^2 - 8x + 5$. There is no remainder.

Step 7

Check by multiplication.

$$27 \cdot 248 = 6696$$

Check by multiplication.

$$(2x + 3)(4x^2 - 8x + 5) = 8x^3 - 4x^2 - 14x + 15$$

EXAMPLE 1 *Dividing a Polynomial by a Polynomial*

Divide $4x^3 - 4x^2 + 5x - 8$ by $2x - 1$.

$$\begin{array}{r} 2x^2 - x + 2 \\ 2x - 1\overline{)4x^3 - 4x^2 + 5x - 8} \\ \underline{4x^3 - 2x^2} \\ -2x^2 + 5x \\ \underline{-2x^2 + x} \\ 4x - 8 \\ \underline{4x - 2} \\ -6 \end{array} \quad \leftarrow \text{remainder}$$

Step 1 $2x$ divides into $4x^3$ $2x^2$ times; $2x^2(2x - 1) = 4x^3 - 2x^2$.

Step 2 Subtract; bring down the next term.

Step 3 $2x$ divides into $-2x^2$ $-x$ times; $-x(2x - 1) = -2x^2 + x$.

Step 4 Subtract; bring down the next term.

Step 5 $2x$ divides into $4x$ 2 times; $2(2x - 1) = 4x - 2$.

Step 6 Subtract. The remainder is -6. Thus $2x - 1$ divides into $4x^3 - 4x^2 + 5x - 8$ with a quotient of $2x^2 - x + 2$ and a remainder of -6. Write the remainder as the numerator of a fraction that has $2x - 1$ as its denominator. The answer is not a polynomial because of the remainder.

$$\frac{4x^3 - 4x^2 + 5x - 8}{2x - 1} = 2x^2 - x + 2 + \frac{-6}{2x - 1}$$

Step 7 Check by multiplication.

$$\begin{aligned}
&(2x - 1)\left(2x^2 - x + 2 + \frac{-6}{2x - 1}\right)\\
&= (2x - 1)(2x^2) + (2x - 1)(-x) + (2x - 1)(2)\\
&\quad + (2x - 1)\left(\frac{-6}{2x - 1}\right)\\
&= 4x^3 - 2x^2 - 2x^2 + x + 4x - 2 - 6\\
&= 4x^3 - 4x^2 + 5x - 8
\end{aligned}$$ ■

WORK PROBLEM 1 AT THE SIDE.

1. Divide.

(a) $(2y^2 - y - 21) \div (y + 3)$

(b) $(x^3 + x^2 + 4x - 6) \div (x - 1)$

(c) $\dfrac{p^3 - 2p^2 - 5p + 9}{p + 2}$

■ **EXAMPLE 2** *Dividing into a Polynomial with Missing Terms*

Divide $x^3 - 1$ by $x - 1$.

Here the polynomial $x^3 - 1$ is missing the x^2 term and the x term. When terms are missing, use 0 as the coefficient for the missing terms. (Zero acts as a placeholder here, just as it does in our number system.)

$$x^3 - 1 = x^3 + 0x^2 + 0x - 1$$

Now divide.

$$\begin{array}{r}
x^2 + x + 1\\
x - 1\overline{)\,x^3 + 0x^2 + 0x - 1}\\
\underline{x^3 - x^2}\\
x^2 + 0x\\
\underline{x^2 - x}\\
x - 1\\
\underline{x - 1}\\
0
\end{array}$$

The remainder is 0. The quotient is $x^2 + x + 1$. Check by multiplication.

$$(x^2 + x + 1)(x - 1) = x^3 - 1$$ ■

ANSWERS

1. (a) $2y - 7$ (b) $x^2 + 2x + 6$ (c) $p^2 - 4p + 3 + \dfrac{3}{p + 2}$

2. Divide.

(a) $\dfrac{r^2 - 5}{r + 4}$

(b) $(x^3 - 8) \div (x - 2)$

3. Divide.

(a)
$(2x^4 + 3x^3 - x^2 + 6x + 5)$
$\div (x^2 - 1)$

(b)
$\dfrac{2m^5 + m^4 + 6m^3 - 3m^2 - 18}{m^2 + 3}$

ANSWERS

2. (a) $r - 4 + \dfrac{11}{r + 4}$

(b) $x^2 + 2x + 4$

3. (a) $2x^2 + 3x + 1 + \dfrac{9x + 6}{x^2 - 1}$

(b) $2m^3 + m^2 - 6$

WORK PROBLEM 2 AT THE SIDE.

EXAMPLE 3 *Dividing by a Polynomial with Missing Terms*

Divide $x^4 + 2x^3 + 2x^2 - x - 1$ by $x^2 + 1$.

Since $x^2 + 1$ has a missing x term, write it as $x^2 + 0x + 1$. Then go through the division process as follows.

$$
\begin{array}{r@{}l}
 & x^2 + 2x + 1 \\
x^2 + 0x + 1 & \overline{)\,x^4 + 2x^3 + 2x^2 - x - 1} \\
 & \underline{x^4 + 0x^3 + x^2} \\
 & 2x^3 + x^2 - x \\
 & \underline{2x^3 + 0x^2 + 2x} \\
 & x^2 - 3x - 1 \\
 & \underline{x^2 + 0x + 1} \\
 & -3x - 2
\end{array}
$$

When the result of subtracting ($-3x - 2$, in this case) is a polynomial of smaller degree than the divisor ($x^2 + 0x + 1$), that polynomial is the remainder. Write the answer as

$$x^2 + 2x + 1 + \frac{-3x - 2}{x^2 + 1}.$$

WORK PROBLEM 3 AT THE SIDE.

3.8 NAME DATE HOUR

EXERCISES

Perform each division. See Example 1.

1. $\dfrac{x^2 - x - 6}{x - 3}$

2. $\dfrac{m^2 - 2m - 24}{m - 6}$

3. $\dfrac{2y^2 + 9y - 35}{y + 7}$

4. $\dfrac{2y^2 + 9y + 7}{y + 1}$

5. $\dfrac{p^2 + 2p + 20}{p + 6}$

6. $\dfrac{x^2 + 11x + 16}{x + 8}$

7. $(r^2 - 8r + 15) \div (r - 3)$

8. $(t^2 + 2t - 35) \div (t - 5)$

9. $\dfrac{12m^2 - 20m + 3}{2m - 3}$

10. $\dfrac{12y^2 + 20y + 7}{2y + 1}$

11. $\dfrac{4a^2 - 22a + 32}{2a + 3}$

12. $\dfrac{9w^2 + 6w + 10}{3w - 2}$

13. $\dfrac{8x^3 - 10x^2 - x + 3}{2x + 1}$

14. $\dfrac{12t^3 - 11t^2 + 9t + 18}{4t + 3}$

15. In the division problem $(4x^4 + 2x^3 - 14x^2 + 19x + 10) \div (2x + 5) = 2x^3 - 4x^2 + 3x + 2$, which polynomial is the divisor? Which is the quotient?

16. When dividing one polynomial by another, how do you know when to stop dividing?

Perform each division. See Examples 2 and 3.

17. $\dfrac{3y^3 + y^2 + 2}{y + 1}$

18. $\dfrac{2r^3 - 6r - 36}{r - 3}$

19. $\dfrac{3k^3 - 4k^2 - 6k + 10}{k^2 - 2}$

20. $\dfrac{5z^3 - z^2 + 10z + 2}{z^2 + 2}$

21. $(x^4 - x^2 - 2) \div (x^2 - 2)$

22. $(r^4 + 2r^2 - 3) \div (r^2 - 1)$

23. $\dfrac{6p^4 - 15p^3 + 14p^2 - 5p + 10}{3p^2 + 1}$

24. $\dfrac{6r^4 - 10r^3 - r^2 + 15r - 8}{2r^2 - 3}$

25. $\dfrac{2x^5 + 9x^4 + 8x^3 + 10x^2 + 14x + 5}{2x^2 + 3x + 1}$

26. $\dfrac{4t^5 - 11t^4 - 6t^3 + 5t^2 - t + 3}{4t^2 + t - 3}$

27. $\dfrac{x^4 - 1}{x^2 - 1}$

28. $\dfrac{y^3 + 1}{y + 1}$

PREVIEW EXERCISES

List all positive integer factors of each number. See Section R.1.

29. 18

30. 36

31. 48

32. 23

CHAPTER 3 SUMMARY

KEY TERMS

3.3	**scientific notation**	A number written as $a \times 10^n$, where $1 \leq \|a\| < 10$ and n is an integer, is in scientific notation.
3.4	**polynomial**	A polynomial is a term or the sum of a finite number of terms.
	descending powers	A polynomial is written in descending powers if the degrees of its terms are in decreasing order.
	degree of a term	The degree of a term with one variable is the exponent on the variable.
	degree of a polynomial	The degree of a polynomial in one variable is the highest exponent found in any term of the polynomial.
	trinomial	A trinomial is a polynomial with three terms.
	binomial	A binomial is a polynomial with two terms.
	monomial	A monomial is a polynomial with one term.

NEW SYMBOLS

x^{-n} x to the negative n

QUICK REVIEW

Concepts	Examples
3.1 The Product Rule and Power Rules for Exponents	
For any integers m and n:	
Product rule $a^m \cdot a^n = a^{m+n}$	$2^4 \cdot 2^5 = 2^9$
Power rules (a) $(a^m)^n = a^{mn}$	$(3^4)^2 = 3^8$
(b) $(ab)^m = a^m b^m$	$(6a)^5 = 6^5 a^5$
(c) $\left(\frac{a}{b}\right)^m = \frac{a^m}{b^m}$ $(b \neq 0)$	$\left(\frac{2}{3}\right)^4 = \frac{2^4}{3^4}$
3.2 Integer Exponents and the Quotient Rule	
If $a \neq 0$, for integers m and n:	
Zero exponent $a^0 = 1$	$15^0 = 1$
Negative exponent $a^{-n} = \frac{1}{a^n}$	$5^{-2} = \frac{1}{5^2} = \frac{1}{25}$
Quotient rule $\frac{a^m}{a^n} = a^{m-n}$	$\frac{4^8}{4^3} = 4^5$
$\frac{a^{-m}}{b^{-n}} = \frac{b^n}{a^m}$ $\left(\frac{a}{b}\right)^{-m} = \left(\frac{b}{a}\right)^m$	$\frac{6^{-2}}{7^{-3}} = \frac{7^3}{6^2}$ $\left(\frac{5}{3}\right)^{-4} = \left(\frac{3}{5}\right)^4$
3.3 An Application of Exponents: Scientific Notation	
To write a number in scientific notation (as $a \times 10^n$), move the decimal point n places to follow the first nonzero digit. If this makes the number smaller, n is positive; otherwise, n is negative. If the decimal point is not moved, n is 0.	$247 = 2.47 \times 10^2$ $0.0051 = 5.1 \times 10^{-3}$ $4.8 = 4.8 \times 10^0$

Concepts	Examples
3.4 Addition and Subtraction of Polynomials	
Addition: Add like terms.	Add: $\begin{array}{r} 2x^2 + 5x - 3 \\ 5x^2 - 2x + 7 \\ \hline 7x^2 + 3x + 4 \end{array}$
Subtraction: Change the signs of the terms in the second polynomial and add to the first polynomial.	$(2x^2 + 5x - 3) - (5x^2 - 2x + 7)$ $= (2x^2 + 5x - 3) + (-5x^2 + 2x - 7)$ $= -3x^2 + 7x - 10$
3.5 Multiplication of Polynomials	
Multiply each term of the first polynomial by each term of the second polynomial. Then add like terms.	Multiply: $\begin{array}{rrrrr} & 3x^3 & - 4x^2 & + 2x & - 7 \\ & & & 4x & + 3 \\ \hline & 9x^3 & - 12x^2 & + 6x & - 21 \\ 12x^4 & - 16x^3 & + 8x^2 & - 28x & \\ \hline 12x^4 & - 7x^3 & - 4x^2 & - 22x & - 21 \end{array}$
3.6 Products of Binomials	
FOIL Method	Find $(2x + 3)(5x - 4)$.
Step 1 Multiply the two first terms to get the first term of the answer.	$2x(5x) = 10x^2$
Step 2 Find the outer product and the inner product and mentally add them, when possible, to get the middle term of the answer.	$2x(-4) + 3(5x) = 7x$
Step 3 Multiply the two last terms to get the last term of the answer.	$3(-4) = -12$ The product of $(2x + 3)$ and $(5x - 4)$ is $10x^2 + 7x - 12$.
Square of a Binomial $(a + b)^2 = a^2 + 2ab + b^2$ $(a - b)^2 = a^2 - 2ab + b^2$	$(3x + 1)^2 = 9x^2 + 6x + 1$ $(2m - 5n)^2 = 4m^2 - 20mn + 25n^2$
Product of the Sum and Difference of Two Terms $(a + b)(a - b) = a^2 - b^2$	$(4a + 3)(4a - 3) = 16a^2 - 9$
3.7 Division of a Polynomial by a Monomial	
Divide each term of the polynomial by the monomial: $\frac{a + b}{c} = \frac{a}{c} + \frac{b}{c}$	Divide: $\frac{4x^3 - 2x^2 + 6x - 8}{2x}$ $= 2x^2 - x + 3 - \frac{4}{x}$
3.8 The Quotient of Two Polynomials	
Use "long division."	Divide: $\begin{array}{r} 2x - 5 + \frac{-1}{3x + 4} \\ 3x + 4 \overline{)6x^2 - 7x - 21} \\ \underline{6x^2 + 8x} \quad\quad \\ -15x - 21 \\ \underline{-15x - 20} \\ -1 \end{array}$

NAME DATE HOUR

CHAPTER 3 REVIEW EXERCISES

[3.1] *Use the product rule to simply each expression. Write each answer in exponential form.*

1. $4^3 \cdot 4^8$ **2.** $(-5)^6(-5)^5$ **3.** $(-8x^4)(9x^3)$ **4.** $(2x^2)(5x^3)(x^9)$

Use the power rules to simplify each expression. Write each answer in exponential form.

5. $(19x)^5$ **6.** $(-4y)^7$ **7.** $5(pt)^4$ **8.** $\left(\frac{7}{5}\right)^6$

Use a combination of rules to simplify each expression. Leave answers in exponential form.

9. $(3x^2y^3)^3$ **10.** $(t^4)^8(t^2)^5$ **11.** $(6x^2z^4)^2(x^3yz^2)^4$

12. Explain why the product rule for exponents does not apply to the expression $7^2 + 7^4$.

[3.2] *Evaluate each expression.*

13. $5^0 + 8^0$ **14.** 2^{-5} **15.** $\left(\frac{6}{5}\right)^{-2}$ **16.** $4^{-2} - 4^{-1}$

Simplfy. Write each answer in exponential form, using only positive exponents. Assume all variables are nonzero.

17. $\frac{6^{-3}}{6^{-5}}$ **18.** $\frac{x^{-7}}{x^{-9}}$ **19.** $\frac{p^{-8}}{p^4}$ **20.** $\frac{r^{-2}}{r^{-6}}$

21. $(2^4)^2$ **22.** $(9^3)^{-2}$ **23.** $(5^{-2})^{-4}$

24. $(8^{-3})^4$ **25.** $\frac{(m^2)^3}{(m^4)^2}$ **26.** $\frac{y^4 \cdot y^{-2}}{y^{-5}}$

27. $\frac{r^9 \cdot r^{-5}}{r^{-2} \cdot r^{-7}}$ **28.** $(-5m^3)^2$ **29.** $(2y^{-4})^{-3}$

30. $\frac{ab^{-3}}{a^4b^2}$ **31.** $\frac{(6r^{-1})^2 \cdot (2r^{-4})}{r^{-5}(r^2)^{-3}}$ **32.** $\frac{(2m^{-5}n^2)^3(3m^2)^{-1}}{m^{-2}n^{-4}(m^{-1})^2}$

[3.3] *Write each number in scientific notation.*

33. 48,000,000 **34.** 28,988,000,000 **35.** .000065 **36.** .0000000824

Write each number without exponents.

37. 2.4×10^4 **38.** 7.83×10^7 **39.** 8.97×10^{-7} **40.** 9.95×10^{-12}

Perform the indicated operation and write the answer without exponents.

41. $(2 \times 10^{-3}) \times (4 \times 10^5)$ **42.** $\dfrac{8 \times 10^4}{2 \times 10^{-2}}$ **43.** $\dfrac{12 \times 10^{-5} \times 5 \times 10^4}{4 \times 10^3 \times 6 \times 10^{-2}}$ **44.** $\dfrac{2.5 \times 10^5 \times 4.8 \times 10^{-4}}{7.5 \times 10^8 \times 1.6 \times 10^{-5}}$

[3.4] *For each polynomial, first simplify, if possible, and write it in descending powers of the variable. Then give the degree of the resulting polynomial, and tell whether it is a monomial, a binomial, a trinomial, or none of these.*

	Simplified form	*Degree*	*Kind of polynomial*
45. $9m^2 + 11m^2 + 2m^2$			
46. $-4p + p^3 - p^2 + 8p + 2$			
47. $2r^4 - r^3 + 8r^4 + r^3 - 6r^4 + 8r^5$			
48. $12a^5 + 19a^4 + 8a^3 + 2a^3 - 9a^4 + 3a^5$			

Add or subtract as indicated.

49. Add.
$$\begin{array}{r} -2a^3 + 5a^2 \\ \underline{-3a^3 - a^2} \end{array}$$

50. Add.
$$\begin{array}{r} 4r^3 - 8r^2 + 6r \\ \underline{-2r^3 + 5r^2 + 3r} \end{array}$$

51. Subtract.
$$\begin{array}{r} 6y^2 - 8y + 2 \\ \underline{-5y^2 + 2y - 7} \end{array}$$

52. Subtract.
$$\begin{array}{r} -12k^4 - 8k^2 + 7k - 5 \\ \underline{k^4 + 7k^2 + 11k + 1} \end{array}$$

53. $(2m^3 - 8m^2 + 4) + (8m^3 + 2m^2 - 7)$ **54.** $(-5y^2 + 3y + 11) + (4y^2 - 7y + 15)$

55. $(6p^2 - p - 8) - (-4p^2 + 2p + 3)$ **56.** $(12r^4 - 7r^3 + 2r^2) - (5r^4 - 3r^3 + 2r^2 + 1)$

[3.5] *Find each product.*

57. $5x(2x + 14)$

58. $-3p^3(2p^2 - 5p)$

59. $(m - 9)(m + 2)$

60. $(3k - 6)(2k + 1)$

61. $(3r - 2)(2r^2 + 4r - 3)$

62. $(2y + 3)(4y^2 - 6y + 9)$

63. $(r + 2)^3$

64. Explain why $(a + b)^2$ is not equal to $a^2 + b^2$.

[3.6] *Find each product.*

65. $(3k + 1)(2k + 3)$

66. $(a + 3b)(2a - b)$

67. $(6k - 3q)(2k - 7q)$

68. $(a + 4)^2$

69. $(3p - 2)^2$

70. $(2r + 5s)^2$

71. $(6m - 5)(6m + 5)$

72. $(2z + 7)(2z - 7)$

73. $(5a + 6b)(5a - 6b)$

74. $(2x^2 + 5)(2x^2 - 5)$

[3.7] *Perform each division.*

75. $\dfrac{-15y^4}{-9y^2}$

76. $\dfrac{-12x^3y^2}{6xy}$

77. $\dfrac{6y^4 - 12y^2 + 18y}{-6y}$

78. $\dfrac{2p^3 - 6p^2 + 5p}{2p^2}$

79. $(5x^{13} - 10x^{12} + 20x^7 - 35x^5) \div (-5x^4)$

80. $(-10m^4n^2 + 5m^3n^3 + 6m^2n^4) \div (5m^2n)$

[3.8] *Perform each division.*

81. $(2r^2 + 3r - 14) \div (r - 2)$

82. $\dfrac{12m^2 - 11m - 10}{3m - 5}$

83. $\dfrac{10a^3 + 5a^2 - 14a + 9}{5a^2 - 3}$

84. $\dfrac{2k^4 + 4k^3 + 9k^2 - 8}{2k^2 + 1}$

MIXED REVIEW EXERCISES

Simplify. Perform the indicated operations. Write with positive exponents. Assume that no denominators are equal to zero.

85. $19^0 - 3^0$

86. $(3p)^4(3p^{-7})$

87. 7^{-2}

88. $(-7 + 2k)^2$

89. $\dfrac{2y^3 + 17y^2 + 37y + 7}{2y + 7}$

90. $\left(\dfrac{6r^2s}{5}\right)^4$

91. $-m^5(8m^2 + 10m + 6)$

92. $\left(\dfrac{1}{2}\right)^{-5}$

93. $(25x^2y^3 - 8xy^2 + 15x^3y) \div (5x)$

94. $(6r^{-2})^{-1}$

95. $(2x + y)^3$

96. $2^{-1} + 4^{-1}$

97. $(a + 2)(a^2 - 4a + 1)$

98. $(5y^3 - 8y^2 + 7) - (-3y^3 + y^2 + 2)$

99. $(2r + 5)(5r - 2)$

100. $(12a + 1)(12a - 1)$

NAME DATE HOUR

CHAPTER 3 TEST

Evaluate each expression.

1. 5^{-4}

2. $(-3)^0 + 4^0$

3. $4^{-1} + 3^{-1}$

Simplify, and write each answer using only positive exponents. Assume that variables represent nonzero numbers.

4. $\dfrac{8^{-1} \cdot 8^4}{8^{-2}}$

5. $\dfrac{(x^{-3})^{-2}(x^{-1}y)^2}{(xy^{-2})^2}$

Write each number in scientific notation.

6. **(a)** 344,000,000,000
(b) .00000557

Write each number without exponents.

7. **(a)** 2.96×10^7
(b) 6.07×10^{-8}

For each polynomial, combine like terms when possible, and write the polynomial in descending powers of the variable. Give the degree of the simplified polynomial. Decide whether the simplified polynomial is a monomial, binomial, trinomial, or none of these.

8. $5x^2 + 8x - 12x^2$

9. $13n^3 - n^2 + n^4 + 3n^4 - 9n^2$

Perform the indicated operations.

10. $(5t^4 - 3t^2 + 7t + 3) - (t^4 - t^3 + 3t^2 + 8t + 3)$

11. $(2y^2 - 8y + 8) + (-3y^2 + 2y + 3) - (y^2 + 3y - 6)$

12. Subtract.

$$\begin{array}{r} 9t^3 - 4t^2 + 2t + 2 \\ \underline{9t^3 + 8t^2 - 3t - 6} \end{array}$$

1. ______________
2. ______________
3. ______________
4. ______________
5. ______________
6. (a) ______________
   (b) ______________
7. (a) ______________
   (b) ______________
8. ______________
9. ______________
10. ______________
11. ______________
12. ______________

13. ______________________

14. ______________________

15. ______________________

16. ______________________

17. ______________________

18. ______________________

19. ______________________

20. ______________________

21. ______________________

22. ______________________

23. ______________________

24. ______________________

25. ______________________

13. $3x^2(-9x^3 + 6x^2 - 2x + 1)$

14. $(t - 8)(t + 3)$

15. $(4x + 3y)(2x - y)$

16. $(5x - 2y)^2$

17. $(10v + 3w)(10v - 3w)$

18. $(2r - 3)(r^2 + 2r - 5)$

19. $(x + 1)^3$

20. $\dfrac{8y^3 - 6y^2 + 4y + 10}{2y}$

21. $(-9x^2y^3 + 6x^4y^3 + 12xy^3) \div (3xy)$

22. $\dfrac{2x^2 + x - 36}{x - 4}$

23. $(3x^3 - x + 4) \div (x - 2)$

24. What polynomial expression represents the area of this square?

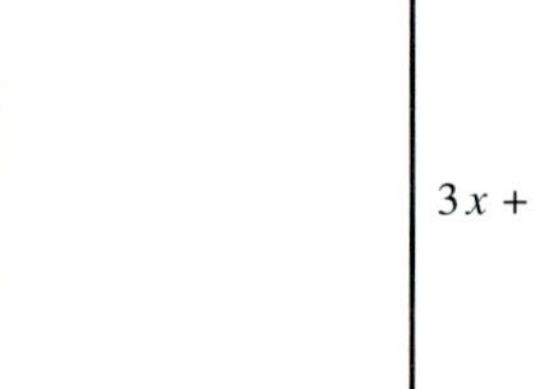

25. Give an example of this situation: the sum of two fourth degree polynomials in x is a third degree polynomial in x.

NAME DATE HOUR

CUMULATIVE REVIEW EXERCISES CHAPTERS R–3

Write each fraction in lowest terms.

1. $\frac{28}{16}$

2. $\frac{55}{11}$

Work the following problems.

3. $\frac{2}{3} + \frac{1}{8}$

4. $\frac{7}{4} - \frac{9}{5}$

5. $8.32 - 4.6$

6. 7.21×8.6

7. A contractor installs toolsheds. Each requires $1\frac{1}{4}$ cubic yards of concrete. How much concrete would be needed for 25 sheds?

8. A retailer has $34,000 invested in her business. She finds that last year she earned 5.4% on this investment. How much did she earn?

Find the value of the expression if $x = -2$ and $y = 4$.

9. $\frac{4x - 2y}{x + y}$

10. $x^3 - 4xy$

Perform the indicated operations.

11. $\frac{(-13 + 15) - (3 + 2)}{6 - 12}$

12. $-7 - 3[2 + (5 - 8)]$

Decide what property justifies each statement.

13. $(9 + 2) + 3 = 9 + (2 + 3)$

14. $-7 + 7 = 0$

15. $6(4 + 2) = 6(4) + 6(2)$

Solve each equation.

16. $2x - 7x + 8x = 30$

17. $2 - 3(t - 5) = 4 + t$

18. $2(5h + 1) = 10h + 4$

19. $d = rt$ for r

20. $\frac{x}{5} = \frac{x - 2}{7}$

21. $\frac{1}{3}p - \frac{1}{6}p = -2$

22. $.05x + .15(50 - x) = 5.50$

23. $4 - (3x + 12) = (2x - 9) - (5x - 1)$

Solve each problem.

24. Each month, Janet's allowance is six times as much as Louis's allowance. Together their allowances total \$56. What is the monthly allowance for each?

25. If 8 is subtracted from a number and this difference is tripled, the result is -3 times the number. Find the number.

26. A farmer has 28 more hens than roosters, with 190 chickens in all. Find the number of hens and the number of roosters on this farm.

27. One side of a triangle is twice as long as a second side. The third side of the triangle is 17 feet long. The perimeter of the triangle cannot be more than 50 feet. Find the longest possible values for the other two sides of the triangle.

Solve each inequality.

28. $-8x \leq -80$

29. $-2(x + 4) > 3x + 6$

30. $-3 \leq 2x + 5 < 9$

Evaluate each expression.

31. $4^{-1} + 3^0$

32. $2^{-4} \cdot 2^5$

33. $\dfrac{8^{-5} \cdot 8^7}{8^2}$

34. Write with positive exponents only: $\dfrac{(a^{-3}b^2)^2}{(2a^{-4}b^{-3})^{-1}}$

35. Write in scientific notation: 34,500

Perform the indicated operations.

36. $(7x^3 - 12x^2 - 3x + 8) + (6x^2 + 4) - (-4x^3 + 8x^2 - 2x - 2)$

37. $6x^5(3x^2 - 9x + 10)$

38. $(7x + 4)(9x + 3)$

39. $(5x + 8)^2$

40. $\dfrac{y^3 - 3y^2 + 8y - 6}{y - 1}$

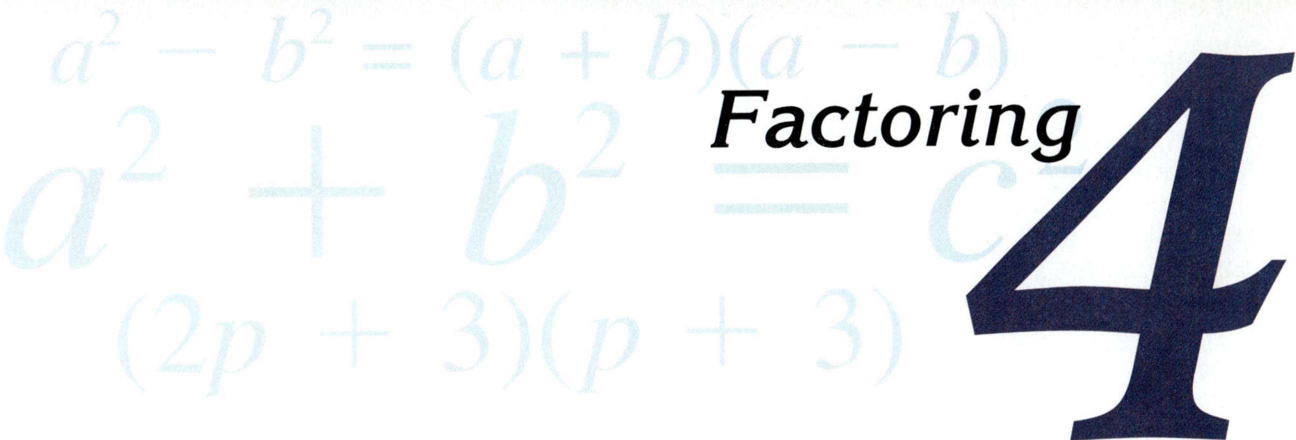

4 Factoring

4.1 FACTORS; THE GREATEST COMMON FACTOR

We discussed prime factors of whole numbers in Section R.1. Factoring is the opposite of multiplication; to **factor** a number means to write it as the product of two or more numbers. The product is called the **factored form** of the number. Now we extend the idea of factoring to polynomials.

OBJECTIVES

1. Find the greatest common factor of a list of numbers.
2. Find the greatest common factor of a list of variable terms.
3. Factor out the greatest common factor.
4. Factor by grouping.

FOR EXTRA HELP

Tape 6

SSM pp. 129–132

MAC: A
IBM: A

1 An integer that is a factor of two or more integers is a **common factor** of those integers. For example, 6 is a common factor of 18 and 24 because 6 is a factor of both 18 and 24. Other common factors of 18 and 24 are 1, 2, and 3. The **greatest common factor** of a list of integers is the largest common factor of those integers. This means 6 is the greatest common factor of 18 and 24, since it is the largest of their common factors.

EXAMPLE 1 *Finding the Greatest Common Factor for Numbers*

Find the greatest common factor for each group of numbers.

(a) 30, 45

First write each number in prime factored form.

$$30 = 2 \cdot 3 \cdot 5 \qquad 45 = 3^2 \cdot 5$$

Use each prime the *least* number of times it appears in all the factored forms. There is no 2 in the prime factored form of 45, so there will be no 2 in the greatest common factor. The least number of times 3 appears in all the factored forms is 1; the least number of times 5 appears is also 1. From this, the greatest common factor is

$$3^1 \cdot 5^1 = 3 \cdot 5 = 15.$$

(b) 72, 120, 432

Find the prime factored form of each number.

$$72 = 2^3 \cdot 3^2 \quad 120 = 2^3 \cdot 3 \cdot 5 \quad 432 = 2^4 \cdot 3^3$$

The least number of times 2 appears in all the factored forms is 3, and the least number of times 3 appears is 1. There is no 5 in the prime factored form of either 72 or 432, so the greatest common factor is

$$2^3 \cdot 3 = 24.$$

1. Find the greatest common factor for each group of numbers.

(a) 30, 20, 15

(b) 42, 28, 35

(c) 12, 18, 26, 32

(d) 10, 15, 21

(c) 10, 11, 14

Write the prime factored form of each number.

$$10 = 2 \cdot 5 \quad 11 = 11 \quad 14 = 2 \cdot 7$$

There are no primes common to all three numbers, so the greatest common factor is 1. ■

◀ WORK PROBLEM 1 AT THE SIDE.

2 The greatest common factor can also be found for a list of variable terms. For example, the terms x^4, x^5, x^6, and x^7 have x^4 as the greatest common factor because each of these terms can be written with x^4 as a factor.

$$x^4 = 1 \cdot x^4, \quad x^5 = x \cdot x^4, \quad x^6 = x^2 \cdot x^4, \quad x^7 = x^3 \cdot x^4$$

Note The exponent on a variable in the greatest common factor is the *smallest* exponent that appears on that variable.

■ EXAMPLE 2 *Finding Greatest Common Factors for Variable Terms*

Find the greatest common factor for each list of terms.

(a) $21m^7, -18m^6, 45m^8, -24m^5$

$$\begin{aligned} 21m^7 &= 3 \cdot 7 \cdot m^7 \\ -18m^6 &= -1 \cdot 2 \cdot 3^2 \cdot m^6 \\ 45m^8 &= 3^2 \cdot 5 \cdot m^8 \\ -24m^5 &= -1 \cdot 2^3 \cdot 3 \cdot m^5 \end{aligned}$$

First, 3 is the greatest common factor of the coefficients 21, −18, 45, and −24. The smallest exponent on m is 5, so the greatest common factor of the terms is $3m^5$.

(b) $x^4y^2, x^7y^5, x^3y^7, y^{15}$

$$\begin{aligned} x^4y^2 &= x^4 \cdot y^2 \\ x^7y^5 &= x^7 \cdot y^5 \\ x^3y^7 &= x^3 \cdot y^7 \\ y^{15} &= y^{15} \end{aligned}$$

There is no x in the last term, y^{15}, so x will not appear in the greatest common factor. There is a y in each term, however, and 2 is the smallest exponent on y. The greatest common factor is y^2.

(c) $-a^2b, -ab^2$

Write $-a^2b$ as $-1a^2b$ and $-ab^2$ as $-1ab^2$. The factors of -1 are -1 and 1. Since $1 > -1$, the greatest common factor is $1ab$ or ab. ■

Note In a group of terms with negatives, many times a negative common factor is preferable (even though it is not the greatest common factor). In Example 2(c), for instance, we might prefer $-ab$ as the common factor. In factoring exercises, either answer will be acceptable.

ANSWERS
1. (a) 5 (b) 7 (c) 2 (d) 1

In summary, we find the greatest common factor of a list of terms as follows.

FINDING THE GREATEST COMMON FACTOR

Step 1 Write each number in prime factored form.

Step 2 List each prime number that is a factor of every number in the list.

Step 3 Use as exponents on the prime factors the *smallest* exponent from the prime factored forms. (If a prime does not appear in one of the prime factored forms, it cannot appear in the greatest common factor.)

Step 4 Multiply together the primes from Step 3. If there are no primes left after Step 3, the greatest common factor is 1.

WORK PROBLEM 2 AT THE SIDE.

3 The idea of a greatest common factor can be used to write a polynomial in factored form. For example, the polynomial

$$3m + 12$$

consists of the two terms $3m$ and 12. The greatest common factor for these two terms is 3. Write $3m + 12$ so that each term is a product with 3 as one factor.

$$3m + 12 = 3 \cdot m + 3 \cdot 4$$

Now use the distributive property.

$$3m + 12 = 3 \cdot m + 3 \cdot 4 = 3(m + 4)$$

The factored form of $3m + 12$ is $3(m + 4)$. This process is called **factoring out the greatest common factor.**

Caution Notice that the polynomial $3m + 12$ is *not* in factored form when written as

$$3 \cdot m + 3 \cdot 4.$$

The *terms* are factored, but the polynomial is not. The factored form of $3m + 12$ is the *product*

$$3(m + 4).$$

EXAMPLE 3 *Factoring Out the Greatest Common Factor*

Factor out the greatest common factor.

(a) $20m^5 + 10m^4 + 15m^3$

The greatest common factor for the terms of this polynomial is $5m^3$.

$$\begin{aligned} &20m^5 + 10m^4 + 15m^3 \\ &= (5m^3)(4m^2) + (5m^3)(2m) + (5m^3)3 && \text{Factor each term.} \\ &= 5m^3(4m^2 + 2m + 3) && \text{Factor out } 5m^3. \end{aligned}$$

Check this work by multiplying $5m^3$ and $4m^2 + 2m + 3$. You should get the original polynomial as your answer.

2. Find the greatest common factor for each list of terms.

(a) $6m^4, 9m^2, 12m^5$

(b) $-12p^5, -18q^4$

(c) y^4z^2, y^6z^8, z^9

(d) $12p^{11}, 17q^5$

ANSWERS

2. (a) $3m^2$ (b) 6 (c) z^2 (d) 1

3. Factor out the greatest common factor.

(a) $32p^2 + 16p + 48$

(b) $10y^5 - 8y^4 + 6y^2$

(c) $m^7 + m^9$

(d) $8p^5q^2 + 16p^6q^3 - 12p^4q^7$

(e) $13x^2 - 27$

(f) $r(t - 4) + 5(t - 4)$

ANSWERS
3. (a) $16(2p^2 + p + 3)$
(b) $2y^2(5y^3 - 4y^2 + 3)$
(c) $m^7(1 + m^2)$
(d) $4p^4q^2(2p + 4p^2q - 3q^5)$
(e) no common factor (except 1)
(f) $(t - 4)(r + 5)$

(b) $x^5 + x^3 = (x^3)x^2 + (x^3)1 = x^3(x^2 + 1)$

(c) $20m^7p^2 - 36m^3p^4 = 4m^3p^2(5m^4 - 9p^2)$

(d) $a(a + 3) + 4(a + 3)$

The binomial $a + 3$ is the greatest common factor here.

$$a(a + 3) + 4(a + 3) = (a + 3)(a + 4)$$ ■

Caution Be careful to avoid the common error of leaving out the 1 in a problem like Example 3(b). Always be sure that the factored form can be multiplied out to give the original polynomial.

WORK PROBLEM 3 AT THE SIDE.

4 Common factors are used in **factoring by grouping,** explained in the next example.

■ EXAMPLE 4 *Factoring by Grouping*

Factor by grouping.

(a) $2x + 6 + ax + 3a$

The first two terms have a common factor of 2, and the last two terms have a common factor of a.

$$2x + 6 + ax + 3a = 2(x + 3) + a(x + 3)$$

The expression is still not in factored form because it is the *sum* of two terms. Now, however, $x + 3$ is a common factor and can be factored out.

$$2x + 6 + ax + 3a = 2(x + 3) + a(x + 3) = (x + 3)(2 + a)$$

The final result is in factored form because it is a *product*. Note that the goal in factoring by grouping is to get a common factor, $x + 3$ here, so that the last step is possible.

(b) $m^2 + 6m + 2m + 12 = m(m + 6) + 2(m + 6)$ Factor each group.
$= (m + 6)(m + 2)$ Factor out the common binomial factor.

(c) $2xy + 12 - 3y - 8x$

We need to rearrange these terms to get two groups that each have a common factor. Trial and error suggests the following grouping.

$2xy + 12 - 3y - 8x = (2xy - 3y) + (-8x + 12)$
$= y(2x - 3) - 4(2x - 3)$ Factor each group. (Must be same)
$= (2x - 3)(y - 4)$ Factor out the common binomial factor.

Since the quantities in parentheses in the second step must be the same, in part (c) we needed to factor out -4 rather than 4. ■

Caution Note the careful use of signs in Example 4(c). Sign errors often occur when grouping with negative signs.

WORK PROBLEM 4 AT THE SIDE.

Use these steps when factoring by grouping.

FACTORING BY GROUPING

Step 1 Group the terms so that each group has a common factor.
Step 2 Use the distributive property to factor each group of terms.
Step 3 If possible, factor a common binomial factor from the results of Step 2.
Step 4 If Step 2 does not result in a common binomial factor, try grouping the terms of the original polynomial in a different way.

4. Factor by grouping.

(a) $pq + 5q + 2p + 10$

(b) $2mn - 8n + 3m - 12$

(c) $6x - yx + 2y - 3x^2$

(d) $2a + 3ax - 2b^2 - 3b^2x$

ANSWERS
4. (a) $(p + 5)(q + 2)$
(b) $(m - 4)(2n + 3)$
(c) $(3x + y)(2 - x)$
(d) $(2 + 3x)(a - b^2)$

QUEST FOR NUMERACY

Big Numbers

$10^{100000\cdots0}$

The term **googol,** meaning 10^{100}, was coined by Professor Edward Kasner of Columbia University. A googol is made up of a 1 with 100 zeros following it. This number exceeds the estimated number of electrons in the universe, which is 10^{79}.

From *Mathematical Circles Revisited* by Howard Eves comes this list of estimates of large numbers:

1. the boiling point of iron is 5.4×10^3 degrees Fahrenheit;
2. the temperature at the center of an atomic bomb explosion is 2×10^8 degrees Fahrenheit;
3. the total number of bridge hands is 6.35×10^{11};
4. the total number of words spoken since the beginning of the world is about 10^{16};
5. the total number of printed words since the Gutenberg Bible appeared is somewhat larger than 10^{16};
6. the age of the earth is set at about 3,350 million years, or about 10^{17} seconds;
7. the half-life of uranium 238 is 1.42×10^{17} seconds;
8. the number of grains of sand on the beach at Coney Island, New York, is about 10^{20};
9. the total age of the expanding universe is probably less than 2,000 million million years, or some 10^{22} seconds;
10. the mass of the earth is about 1.2×10^{25} pounds;
11. the number of atoms of oxygen in an average thimble is perhaps about 10^{27};
12. the diameter of the universe, as assigned by relativity theory, is about 10^{29} centimeters;
13. the number of snow crystals necessary to form the ice age would be about 10^{30};
14. the total number of ways of arranging 52 cards is of the order 8×10^{67};
15. the total number of electrons in the universe is, by an estimate made by Sir Arthur Eddington, about 10^{79}.

FOR GROUP DISCUSSION Suppose that you have just applied for and been offered a new job with "salary negotiable." You and your new employer meet to discuss the salary, and you present her with this offer: For 30 days, you will work and be paid 1¢ on Day 1, 2¢ on Day 2, 4¢ on Day 3, 8¢ on Day 4, and so on, doubling your pay each day. Your employer then accepts your offer.

1. When will you have received half of your month's wages?
2. When will you have received one-fourth of your total month's wages?
3. What do you think your approximate monthly salary will be? Use your calculator to find out.

4.1 EXERCISES

NAME DATE HOUR

Find the greatest common factor for each group of numbers. See Example 1.

1. 12, 16

2. 18, 24

3. 40, 20, 4

4. 50, 30, 5

5. 18, 24, 36, 48

6. 15, 30, 45, 75

7. 4, 9, 12

8. 9, 16, 24

9. How would you respond to the statement that "25 and 36 have no greatest common factor"?

10. Give an example of three numbers whose greatest common factor is 5.

Find the greatest common factor for each list of terms. See Example 2.

11. $16y$, 24

12. $18w$, 27

13. $30x^3$, $40x^6$, $50x^7$

14. $60z^4$, $70z^8$, $90z^9$

15. $12m^3n^2$, $18m^5n^4$, $36m^8n^3$

16. $25p^5r^7$, $30p^7r^8$, $50p^5r^3$

17. $-x^4y^3$, $-xy^2$

18. $-a^4b^5$, $-a^3b$

19. $42ab^3$, $-36a$, $90b$, $-48ab$

20. $45c^3d$, $75c$, $90d$, $-105cd$

21. Is $-xy$ a common factor of $-x^4y^3$ and $-xy^2$? If so, what is the other factor that when multiplied by $-xy$ gives $-x^4y^3$?

22. Is $-a^5b^2$ a common factor of $-a^4b^5$ and $-a^3b$? Explain why or why not.

Complete the factoring.

23. $12 = 6(\quad)$

24. $18 = 9(\quad)$

25. $3x^2 = 3x(\quad)$

26. $8x^3 = 8x(\quad)$

27. $9m^4 = 3m^2(\quad)$

28. $12p^5 = 6p^3(\quad)$

29. $-8z^9 = -4z^5(\quad)$

30. $-15k^{11} = -5k^8(\quad)$

31. $6m^4n^5 = 3m^3n(\quad)$

32. $27a^3b^2 = 9a^2b(\quad)$

33. $-14x^4y^3 = 2xy(\quad)$

34. $-16m^3n^3 = 4mn^2(\quad)$

Factor out the greatest common factor for each expression. See Example 3.

35. $12y - 24$

36. $18p + 36$

37. $10a^2 - 20a$

38. $15x^3 - 30x^2$

39. $65y^{10} + 35y^6$

40. $100a^5 + 16a^3$

41. $11w^3 - 100$

42. $13z^5 - 80$

43. $8m^2n^3 + 24m^2n^2$

44. $19p^2y - 38p^2y^3$

45. $13y^8 + 26y^4 - 39y^2$

46. $5x^5 + 25x^4 - 20x^3$

47. $45q^4p^5 + 36qp^6 + 81q^2p^3$

48. $125a^3z^5 + 60a^4z^4 - 85a^5z^2$

49. $a^5 + 2a^3b^2 - 3a^5b^2 + 4a^4b^3$

50. $x^6 + 5x^4y^3 - 6xy^4 + 10xy$

Factor by grouping. See Example 4.

51. $p^2 + 4p + 3p + 12$

52. $m^2 + 2m + 5m + 10$

53. $a^2 - 2a + 5a - 10$

54. $y^2 - 6y + 4y - 24$

55. $7z^2 + 14z - az - 2a$

56. $5m^2 + 15mp - 2mp - 6p^2$

57. $18r^2 + 12ry - 3xr - 2xy$

58. $8s^2 - 4st + 6sy - 3yt$

59. $3a^3 + 3ab^2 + 2a^2b + 2b^3$

60. $4x^3 + 3x^2y + 4xy^2 + 3y^3$

61. $1 - a + ab - b$

62. $6 - 3x - 2y + xy$

63. $16m^3 - 4m^2p^2 - 4mp + p^3$

64. $10t^3 - 2t^2s^2 - 5ts + s^3$

65. Refer to Exercise 61. The answer given in the back of the book is $(1 - a)(1 - b)$. A student factored this same polynomial and got the result $(a - 1)(b - 1)$.
(a) Is this student's answer correct?
(b) If your answer to part a is *yes,* explain why these two seemingly different answers are both acceptable.

66. Explain how you, while tutoring someone in introductory algebra, could make up a polynomial like those in Exercises 51–64 that could be factored by grouping.

PREVIEW EXERCISES

Find each product. See Section 3.6.

67. $(x + 6)(x - 9)$

68. $(x - 3)(x - 6)$

69. $(x + 2)(x + 7)$

70. $2x(x + 5)(x - 1)$

4.2 FACTORING TRINOMIALS

Using FOIL, we find the product of the binomials $k - 3$ and $k + 1$ is

$$(k - 3)(k + 1) = k^2 - 2k - 3.$$

In this section, we show how to *factor* the polynomial $k^2 - 2k - 3$, that is, write it as the product $(k - 3)(k + 1)$. This product is called the *factored form* of $k^2 - 2k - 3$. The discussion of factoring in this section is limited to trinomials where the coefficient of the squared term is 1.

OBJECTIVES

1. Factor trinomials with a coefficient of 1 for the squared term.
2. Factor such polynomials after factoring out the greatest common factor.

FOR EXTRA HELP

Tape 6 | SSM pp. 132–136 | MAC: A IBM: A

1 When factoring polynomials with only integer coefficients, we use only integers for the numerical coefficients. For example, we can factor $x^2 + 5x + 6$ by finding integers m and n such that

$$x^2 + 5x + 6 = (x + m)(x + n).$$

To find these integers m and n, we first use FOIL to multiply the two binomials:

$$(x + m)(x + n) = x^2 + nx + mx + mn.$$

By the distributive property,

$$x^2 + nx + mx + mn = x^2 + (n + m)x + mn.$$

Comparing this result with $x^2 + 5x + 6$ shows that we must find integers m and n having a sum of 5 and a product of 6.

Product of m and n is 6.

$$x^2 + 5x + 6 = x^2 + (n + m)x + mn$$

Sum of m and n is 5.

Because many pairs of integers have a sum of 5, it is best to begin by listing those pairs of integers whose product is 6. Both 5 and 6 are positive, so we need to consider only pairs in which both integers are positive.

WORK PROBLEM 1 AT THE SIDE.

1. (a) List all pairs of positive integers whose product is 6.

From Problem 1 at the side, we see that the numbers 1 and 6 and the numbers 2 and 3 both have a product of 6, but only the pair 2 and 3 has a sum of 5. So 2 and 3 are the required integers, and

$$x^2 + 5x + 6 = (x + 2)(x + 3).$$

(b) Find the pair from part (a) whose sum is 5.

We can check by multiplying the binomials, using FOIL. Make sure that the sum of the outer and inner products produces the correct middle term.

$$(x + 2)(x + 3) = x^2 + 5x + 6$$

$$\begin{array}{r} 2x \\ 3x \\ \hline 5x \end{array}$$

This method of factoring can be used only for trinomials having the coefficient of the squared term equal to 1. Methods for factoring other trinomials will be given in the next section.

EXAMPLE 1 *Factoring a Trinomial with Only Positive Terms*

Factor $m^2 + 9m + 14$.

Look for two integers whose product is 14 and whose sum is 9. List the pairs of integers whose products are 14. Then examine the sums.

ANSWERS
1. (a) 1, 6; 2, 3 (b) 2, 3

2. Complete the given lists of numbers; then factor $y^2 + 12y + 20$.

Factors of 20	*Sum of factors*
20, 1	$20 + 1 = 21$
10, ___	$10 +$ ___ $=$ ___
5, ___	$5 +$ ___ $=$ ___

Again, only positive integers are needed because all signs in $m^2 + 9m + 14$ are positive.

Factors of 14	*Sum of factors*	
14, 1	$14 + 1 = 15$	
7, 2	$7 + 2 = 9$	Sum is 9.

From the list, 7 and 2 are the required integers, since $7 \cdot 2 = 14$ and $7 + 2 = 9$. Thus

$$m^2 + 9m + 14 = (m + 2)(m + 7). \blacksquare$$

Note In Example 1, the answer also could have been written $(m + 7)(m + 2)$. Because of the commutative property of multiplication, the order of the factors does not matter.

WORK PROBLEM 2 AT THE SIDE.

3. Factor each trinomial.

(a) $a^2 - 9a - 22$

(b) $r^2 - 6r - 16$

■ EXAMPLE 2 *Factoring a Trinomial with Two Negative Terms*

Factor $p^2 - 2p - 15$.

Find two integers whose product is -15 and whose sum is -2. If these numbers do not come to mind right away, find them (if they exist) by listing all the pairs of integers whose product is -15. Because the last term, -15, is negative, we need pairs of integers with different signs.

Factors of −15	*Sum of factors*	
15, −1	$15 + (-1) = 14$	
5, −3	$5 + (-3) = 2$	
−15, 1	$-15 + 1 = -14$	
−5, 3	$-5 + 3 = -2$	Sum is −2.

The required integers are -5 and 3, and

$$p^2 - 2p - 15 = (p - 5)(p + 3). \blacksquare$$

WORK PROBLEM 3 AT THE SIDE.

As shown in the next example, some trinomials cannot be factored using only integer coefficients. We call such trinomials **prime polynomials.**

■ EXAMPLE 3 *Factoring a Trinomial with One Negative Term*

Factor each trinomial.

(a) $x^2 - 5x + 12$

Since the middle term is negative, the sum must be negative. The last term is positive, so both factors must be negative to give a positive product and a negative sum. First, list all pairs of negative integers whose product is 12. Then examine the sums.

Factors of 12	*Sum of factors*
−12, −1	$-12 + (-1) = -13$
−6, −2	$-6 + (-2) = -8$
−3, −4	$-3 + (-4) = -7$

ANSWERS
2. 2; 2; 12; 4; 4; 9; $(y + 10)(y + 2)$
3. (a) $(a - 11)(a + 2)$
(b) $(r - 8)(r + 2)$

None of the pairs of integers has a sum of -5. Because of this, the trinomial $x^2 - 5x + 12$ *cannot be factored using only integer coefficients,* showing that it is a *prime polynomial.*

(b) $k^2 - 8k + 11$

There is no pair of integers whose product is 11 and whose sum is -8, so $k^2 - 8k + 11$ is a prime polynomial. ■

WORK PROBLEM 4 AT THE SIDE.

We can now summarize the procedure for factoring a trinomial of the form $x^2 + bx + c$.

FACTORING $x^2 + bx + c$

Find two integers whose product is c and whose sum is b.

1. Both integers must be positive if b and c are positive.
2. Both integers must be negative if c is positive and b is negative.
3. One integer must be positive and one must be negative if c is negative.

■ EXAMPLE 4 *Factoring a Trinomial with Two Variables*

Factor $z^2 - 2bz - 3b^2$.

Look for two expressions whose product is $-3b^2$ and whose sum is $-2b$. The expressions are $-3b$ and b, so

$$z^2 - 2bz - 3b^2 = (z - 3b)(z + b).$$ ■

WORK PROBLEM 5 AT THE SIDE.

2 The trinomial in the next example does not have a coefficient of 1 for the squared term. (In fact, there is no squared term.) A preliminary step must be taken before using the steps discussed above.

■ EXAMPLE 5 *Factoring a Trinomial with a Common Factor*

Factor $4x^5 - 28x^4 + 40x^3$.

First, factor out the greatest common factor, $4x^3$.

$$4x^5 - 28x^4 + 40x^3 = 4x^3(x^2 - 7x + 10)$$

Now factor $x^2 - 7x + 10$. The integers -5 and -2 have a product of 10 and a sum of -7. The complete factored form is

$$4x^5 - 28x^4 + 40x^3 = 4x^3(x - 5)(x - 2).$$ ■

Caution When factoring, always remember to look for a common factor first. Do not forget to include the common factor as part of the answer. Multiplying out the factored form should always give the original polynomial. (This is a good way to check your answer.)

WORK PROBLEM 6 AT THE SIDE.

4. Factor each trinomial, where possible.

(a) $r^2 - 3r - 4$

(b) $m^2 - 2m + 5$

5. Factor each trinomial.

(a) $b^2 - 3ab - 4a^2$

(b) $r^2 - 6rs + 8s^2$

6. Factor each trinomial as completely as possible.

(a) $2p^3 + 6p^2 - 8p$

(b) $3x^4 - 15x^3 + 18x^2$

ANSWERS

4. (a) $(r - 4)(r + 1)$
   (b) prime
5. (a) $(b - 4a)(b + a)$
   (b) $(r - 4s)(r - 2s)$
6. (a) $2p(p + 4)(p - 1)$
   (b) $3x^2(x - 3)(x - 2)$

QUEST FOR NUMERACY

The Magic of Number Squares

Consider the square arrangements of numbers shown below.

6	1	8
7	5	3
2	9	4

16	3	2	13
5	10	11	8
9	6	7	12
4	15	14	1

10	18	1	14	22
11	24	7	20	3
17	5	13	21	9
23	6	19	2	15
4	12	25	8	16

If you add along any row, column, or diagonal within any arrangement, you will find that the sum is always the same. These arrangements are known as **magic squares**, and they have been studied since ancient times. The number of rows and columns is called the *order* of the magic square. Notice that the magic square of order three has a magic sum of 15. The order four magic square has magic sum 34, while the order five magic square has magic sum 65.

Magic squares have fascinated both lay people and mathematicians over the years. The emperor Charlemagne, the artist Albrecht Dürer, and the statesman Benjamin Franklin have all been linked in some way to magic squares.

FOR GROUP DISCUSSION

1. Consider the following square (which is not, by definition, a magic square).

1	2	3	4	5
6	7	8	9	10
11	12	13	14	15
16	17	18	19	20
21	22	23	24	25

Have each class member individually perform the following. Choose any number in the first row. Circle it, and cross out all entries in the column below it. (For example, if you circle 4, cross out 9, 14, 19, and 24.) Now circle any remaining number in the second row, and cross out all entries in the column below it. Repeat this procedure for the third and fourth rows, and then circle the final remaining number in the fifth row. Add the numbers you circled. Now compare with others in the class. How does the sum compare in each case?

2. Repeat the procedure, starting with a different number in the first row. What is the sum of all the circled entries? Make a conjecture about this procedure.

3. How does the sum obtained in Items 1 and 2 compare with the magic sum for the order five magic square shown earlier?

NAME DATE HOUR

4.2 EXERCISES

In Exercises 1–8, list all pairs of integers with the given product. Then find the pair whose sum is given.

1. Product: 12 Sum: 7

2. Product: 18 Sum: 9

3. Product: -24 Sum: -5

4. Product: -36 Sum: -16

5. Product: 27 Sum: 28

6. Product: 32 Sum: 33

7. Product: -48 Sum: 2

8. Product: -48 Sum: 8

9. Which one of the following is the correct factored form of $x^2 - 12x + 32$?
(a) $(x - 8)(x + 4)$ **(b)** $(x + 8)(x - 4)$ **(c)** $(x - 8)(x - 4)$ **(d)** $(x + 8)(x + 4)$

10. What would be the first step in factoring $2x^3 + 8x^2 - 10x$?

Complete the factoring.

11. $p^2 + 11p + 30 = (p + 5)(\quad)$

12. $x^2 + 10x + 21 = (x + 7)(\quad)$

13. $x^2 + 15x + 44 = (x + 4)(\quad)$

14. $r^2 + 15r + 56 = (r + 7)(\quad)$

15. $x^2 - 9x + 8 = (x - 1)(\quad)$

16. $t^2 - 14t + 24 = (t - 2)(\quad)$

17. $y^2 - 2y - 15 = (y + 3)(\quad)$

18. $t^2 - t - 42 = (t + 6)(\quad)$

19. $x^2 + 9x - 22 = (x - 2)(\quad)$

20. $x^2 + 6x - 27 = (x - 3)(\quad)$

21. $y^2 - 7y - 18 = (y + 2)(\quad)$

22. $y^2 - 2y - 24 = (y + 4)(\quad)$

Factor completely. If a polynomial cannot be factored, write prime. *See Examples 1–3.*

23. $y^2 + 9y + 8$

24. $a^2 + 9a + 20$

25. $b^2 + 8b + 15$

26. $x^2 + 6x + 8$

27. $m^2 + m - 20$

28. $p^2 + 4p - 5$

29. $y^2 - 8y + 15$

30. $y^2 - 6y + 8$

31. $r^2 - r - 30$

32. $q^2 - q - 42$

33. $t^2 - 8t + 16$

34. $s^2 - 10s + 25$

Factor completely. See Examples 4 and 5.

35. $r^2 + 3ra + 2a^2$

36. $x^2 + 5xa + 4a^2$

37. $t^2 - tz - 6z^2$

38. $a^2 - ab - 12b^2$

39. $x^2 + 4xy + 3y^2$

40. $p^2 + 9pq + 8q^2$

41. $v^2 - 11vw + 30w^2$

42. $v^2 - 11vx + 24x^2$

43. $4x^2 + 12x - 40$

44. $5y^2 - 5y - 30$

45. $2t^3 + 8t^2 + 6t$

46. $3t^3 + 27t^2 + 24t$

47. $2x^6 + 8x^5 - 42x^4$

48. $4y^5 + 12y^4 - 40y^3$

49. $m^3n - 10m^2n^2 + 24mn^3$

50. $y^3z + 3y^2z^2 - 54yz^3$

51. Use the FOIL method from Section 3.6 to show that $(2x + 4)(x - 3) = 2x^2 - 2x - 12$. Why, then, is it incorrect to completely factor $2x^2 - 2x - 12$ as $(2x + 4)(x - 3)$?

52. Why is it incorrect to completely factor $3x^2 + 9x - 12$ as the product $(x - 1)(3x + 12)$?

53. What polynomial can be factored to give $(a + 9)(a + 4)$?

54. What polynomial can be factored to give $(y - 7)(y + 3)$?

PREVIEW EXERCISES

Find each product. See Section 3.6.

55. $(2y - 7)(y + 4)$

56. $(3a + 2)(2a + 1)$

57. $(5z + 2)(3z - 2)$

58. $(4m - 3)(2m + 5)$

59. $(4p + 1)(2p - 3)$

60. $(6r - 5)(3r + 2)$

4.3 MORE ON FACTORING TRINOMIALS

Trinomials such as $2x^2 + 7x + 6$, in which the coefficient of the squared term is *not* 1, are factored with an extension of the method we used in the previous section.

OBJECTIVES

1. Use trial and error to factor trinomials with the coefficient of the squared term not 1.
2. Factor trinomials using factoring by grouping.

FOR EXTRA HELP

Tape 6 | SSM pp. 136–142 | MAC: A IBM: A

1 To factor $2x^2 + 7x + 6$ by trial and error, we must use FOIL backwards. We want to write $2x^2 + 7x + 6$ as the product of two binomials.

$$2x^2 + 7x + 6 = (\quad)(\quad)$$

The product of the two first terms of the binomials is $2x^2$. The possible factors of $2x^2$ are $2x$ and x or $-2x$ and $-x$. Since all terms of the trinomial are positive, only positive factors should be considered. Thus, we have

$$2x^2 + 7x + 6 = (2x \quad)(x \quad).$$

The product of the two last terms, 6, can be factored as $6 \cdot 1$, $1 \cdot 6$, $2 \cdot 3$, or $3 \cdot 2$. Try each pair to find the pair that gives the correct middle term.

WORK PROBLEM 1 AT THE SIDE. ▶

1. Decide whether each factored form is correct or incorrect for $2x^2 + 7x + 6$.

(a) $(2x + 1)(x + 6)$

(b) $(2x + 6)(x + 1)$

(c) $(2x + 2)(x + 3)$

In part (b) of Problem 1 at the side, the factor $2x + 6 = 2(x + 3)$, so the binomial $2x + 6$ has a common factor of 2. However, $2x^2 + 7x + 6$ does not have a common factor other than 1. The product $(2x + 6)(x + 1)$ cannot be correct.

> *Note* If the original polynomial has no common factor, then none of its binomial factors will either.

Now try 2 and 3 as factors of 6. Because of the common factor of 2 in $2x + 2$, the factored form $(2x + 2)(x + 3)$ will not work. Try $(2x + 3)(x + 2)$.

$$(2x + 3)(x + 2) = 2x^2 + 7x + 6 \qquad \text{Correct}$$

$$\begin{array}{l} 3x \\ \underline{4x} \\ 7x \quad \text{Add.} \end{array}$$

Finally, $2x^2 + 7x + 6$ factors as

$$2x^2 + 7x + 6 = (2x + 3)(x + 2).$$

Check by multiplying $2x + 3$ and $x + 2$.

EXAMPLE 1 *Factoring a Trinomial by Trial and Error*

Factor $8p^2 + 14p + 5$.

The number 8 has several possible pairs of factors, but 5 has only 1 and 5 or -1 and -5. For this reason, it is easier to begin by considering the factors of 5. We ignore the negative factors since all coefficients in the trinomial are positive. If $8p^2 + 14p + 5$ can be factored, it will be factored as

$$(\quad + 5)(\quad + 1).$$

The possible pairs of factors of $8p^2$ are $8p$ and p, or $4p$ and $2p$. Try various combinations, checking the middle term in each case.

ANSWERS
1. (a) incorrect (b) incorrect (c) incorrect

$$(8p + 5)(p + 1) = 8p^2 + 13p + 5 \quad \text{Incorrect}$$

$5p + 8p = 13p$

$$(p + 5)(8p + 1) = 8p^2 + 41p + 5 \quad \text{Incorrect}$$

$40p + p = 41p$

$$(4p + 5)(2p + 1) = 8p^2 + 14p + 5 \quad \text{Correct}$$

$10p + 4p = 14p$

Because $14p$ is the correct middle term, $8p^2 + 14p + 5$ factors as $(4p + 5)(2p + 1)$. ■

◀◀ WORK PROBLEM 2 AT THE SIDE.

■ EXAMPLE 2 *Factoring a Trinomial by Trial and Error*

Factor $6x^2 - 11x + 3$.

There are several possible pairs of factors for 6, but 3 has only 1 and 3 or -1 and -3, so it is better to begin by factoring 3. The middle term of $6x^2 - 11x + 3$ has a negative coefficient, so negative factors must be considered. Try -3 and -1 as factors of 3:

$$(\quad - 3)(\quad - 1).$$

The factors of $6x^2$ are either $6x$ and x, or $2x$ and $3x$. Let us try $2x$ and $3x$.

$$(2x - 3)(3x - 1) = 6x^2 - 11x + 3 \quad \text{Correct}$$

$-9x + (-2x) = -11x$

Finally, $6x^2 - 11x + 3 = (2x - 3)(3x - 1)$. ■

■ EXAMPLE 3 *Factoring a Trinomial by Trial and Error*

Factor $8x^2 + 6x - 9$.

The integer 8 has several possible pairs of factors, as does -9. Since the coefficient of the middle term is small, it is wise to avoid large factors such as 8 or 9. The last term is negative, so we need one positive factor and one negative factor. Let us try 4 and 2 as factors of 8, and 3 and -3 as factors of -9.

$$(4x + 3)(2x - 3) = 8x^2 - 6x - 9 \quad \text{Incorrect}$$

$6x + (-12x) = -6x$

When the opposite sign is obtained, simply exchange the constants, 3 and -3.

$$(4x - 3)(2x + 3) = 8x^2 + 6x - 9 \quad \text{Correct} \quad ■$$

$-6x + 12x = 6x$

◀◀ WORK PROBLEM 3 AT THE SIDE.

2. Factor each trinomial.

(a) $2p^2 + 9p + 9$

(b) $6p^2 + 19p + 10$

(c) $8x^2 + 14x + 3$

3. Factor each trinomial.

(a) $6x^2 + 5x - 4$

(b) $6m^2 - 11m - 10$

(c) $4x^2 - 3x - 7$

ANSWERS

2. (a) $(2p + 3)(p + 3)$
   (b) $(3p + 2)(2p + 5)$
   (c) $(4x + 1)(2x + 3)$
3. (a) $(3x + 4)(2x - 1)$
   (b) $(2m - 5)(3m + 2)$
   (c) $(4x - 7)(x + 1)$

EXAMPLE 4 *Factoring a Trinomial with Two Variables*

Factor $12a^2 - ab - 20b^2$.

There are several possible pairs of factors of $12a^2$, including $12a$ and a, $6a$ and $2a$, and $3a$ and $4a$, just as there are many possible pairs of factors of $-20b^2$, including $-20b$ and b, $10b$ and $-2b$, $-10b$ and $2b$, $4b$ and $-5b$, and $-4b$ and $5b$. Once again, because the desired middle term is small, it is better to avoid the larger factors. Let us try as factors $6a$ and $2a$ and $4b$ and $-5b$.

$$(6a + 4b)(2a - 5b)$$

This cannot be correct since $6a + 4b$ has a common factor of 2 while the given trinomial does not. Let us try $3a$ and $4a$ with $4b$ and $-5b$.

$$(3a + 4b)(4a - 5b) = 12a^2 + ab - 20b^2 \quad \text{Incorrect}$$

Here the middle term has the opposite sign, so we change the signs in the factors.

$$(3a - 4b)(4a + 5b) = 12a^2 - ab - 20b^2 \quad \text{Correct}$$ ■

WORK PROBLEM 4 AT THE SIDE.

EXAMPLE 5 *Factoring a Trinomial with a Common Factor*

Factor $28x^5 - 58x^4 - 30x^3$.

First factor out the greatest common factor, $2x^3$.

$$28x^5 - 58x^4 - 30x^3 = 2x^3(14x^2 - 29x - 15)$$

Now factor $14x^2 - 29x - 15$. Let us try $7x$ and $2x$ as factors of $14x^2$ and -3 and 5 as factors of -15.

$$(7x - 3)(2x + 5) = 14x^2 + 29x - 15 \quad \text{Incorrect}$$

The middle term differs only in sign, so we change the signs in the two factors.

$$(7x + 3)(2x - 5) = 14x^2 - 29x - 15 \quad \text{Correct}$$

Finally, the factored form of $28x^5 - 58x^4 - 30x^3$ is

$$28x^5 - 58x^4 - 30x^3 = 2x^3(7x + 3)(2x - 5).$$ ■

Caution Do not forget to include the common factor in the final result.

WORK PROBLEM 5 AT THE SIDE.

2 The rest of this section shows an alternative method of factoring trinomials in which the coefficient of the squared term is not 1. This method uses factoring by grouping, introduced in Section 4.1. In the next example, we use the alternative method to factor $2x^2 + 7x + 6$, the same trinomial factored at the beginning of this section.

Recall that we factor a trinomial such as $m^2 + 3m + 2$ by finding two numbers whose product is 2 and whose sum is 3. To factor $2x^2 + 7x + 6$ by the alternative method, look for two integers whose product is 12 (that is, $2 \cdot 6$) and whose sum is 7.

4. Factor each trinomial.

(a) $2x^2 - 5xy - 3y^2$

(b) $8a^2 + 2ab - 3b^2$

(c) $12x^2 - 16xy - 3y^2$

5. Factor each polynomial as completely as possible.

(a) $4x^2 - 2x - 30$

(b) $18p^4 + 63p^3 + 27p^2$

(c) $6a^2 + 3ab - 18b^2$

ANSWERS

4. (a) $(2x + y)(x - 3y)$
(b) $(4a + 3b)(2a - b)$
(c) $(6x + y)(2x - 3y)$

5. (a) $2(2x + 5)(x - 3)$
(b) $9p^2(2p + 1)(p + 3)$
(c) $3(2a - 3b)(a + 2b)$

6. Find two integers whose product is 12 and whose sum is 7.

$$2x^2 + 7x + 6$$

Sum is 7. Product is $2 \cdot 6 = 12$.

WORK PROBLEM 6 AT THE SIDE.

As the problem at the side shows, the necessary integers are 3 and 4. Use these integers to write the middle term, $7x$, as $7x = 3x + 4x$. With this, the trinomial $2x^2 + 7x + 6$ becomes

$$2x^2 + 7x + 6 = 2x^2 + \underbrace{3x + 4x}_{7x = 3x + 4x} + 6.$$

$$= x(2x + 3) + 2(2x + 3) \quad \text{Factor each group.}$$

must be same

$$2x^2 + 7x + 6 = (2x + 3)(x + 2) \quad \text{Factor out } 2x + 3.$$

In the example above, we could have written $7x$ as $4x + 3x$. Check that the resulting factoring by grouping would give the same answer.

7. Factor each trinomial by grouping.

(a) $2m^2 + 7m + 3$

(b) $5p^2 - 2p - 3$

(c) $15k^2 - k - 2$

■ EXAMPLE 6 *Factoring Trinomials by Grouping*

Factor each trinomial.

(a) $6r^2 + r - 1$

Find two integers whose product is $6(-1) = -6$ and whose sum is 1.

$$6r^2 + r - 1 = 6r^2 + 1r - 1$$

Sum is 1. Product is -6.

The integers are -2 and 3.

$$\begin{aligned} 6r^2 + r - 1 &= 6r^2 - 2r + 3r - 1 && r = -2r + 3r \\ &= 2r(3r - 1) + 1(3r - 1) && \text{Factor each group.} \\ &= (3r - 1)(2r + 1) && \text{Factor out } 3r - 1. \end{aligned}$$

(b) $12z^2 - 5z - 2$

Look for two integers whose product is $12(-2) = -24$ and whose sum is -5. The required integers are 3 and -8, and

$$\begin{aligned} 12z^2 - 5z - 2 &= 12z^2 + 3z - 8z - 2 && -5z = 3z - 8z \\ &= 3z(4z + 1) - 2(4z + 1) && \text{Factor each group.} \\ &= (4z + 1)(3z - 2). && \text{Factor out } 4z + 1. \end{aligned}$$ ■

WORK PROBLEM 7 AT THE SIDE.

ANSWERS

6. 3 and 4

7. (a) $(2m + 1)(m + 3)$
(b) $(5p + 3)(p - 1)$
(c) $(5k - 2)(3k + 1)$

EXERCISES

Decide which is the correct factored form of the given polynomial.

1. $2x^2 - x - 1$
(a) $(2x - 1)(x + 1)$ **(b)** $(2x + 1)(x - 1)$

2. $3a^2 - 5a - 2$
(a) $(3a + 1)(a - 2)$ **(b)** $(3a - 1)(a + 2)$

3. $4y^2 + 17y - 15$
(a) $(y + 5)(4y - 3)$ **(b)** $(2y - 5)(2y + 3)$

4. $12c^2 - 7c - 12$
(a) $(6c - 2)(2c + 6)$ **(b)** $(4c + 3)(3c - 4)$

5. $4k^2 + 13mk + 3m^2$
(a) $(4k + m)(k + 3m)$ **(b)** $(4k + 3m)(k + m)$

6. $2x^2 + 11x + 12$
(a) $(2x + 3)(x + 4)$ **(b)** $(2x + 4)(x + 3)$

Complete the factoring.

7. $6a^2 + 7ab - 20b^2 = (3a - 4b)(\qquad)$

8. $9m^2 - 3mn - 2n^2 = (3m + n)(\qquad)$

9. $2x^2 + 6x - 8 = 2(\qquad) = 2(\qquad)(\qquad)$

10. $3x^2 - 9x - 30 = 3(\qquad) = 3(\qquad)(\qquad)$

11. $4z^3 - 10z^2 - 6z = 2z(\qquad) = 2z(\qquad)(\qquad)$

12. $15r^3 - 39r^2 - 18r = 3r(\qquad) = 3r(\qquad)(\qquad)$

13. For the polynomial $12x^2 + 7x - 12$, 2 is not a common factor. Explain why the binomial $2x - 6$, then, cannot be a factor of the polynomial.

14. Explain how the signs of the last terms of the two binomial factors of a trinomial are determined.

Factor completely. Use either method described in this section. See Examples 1–6.

15. $2x^2 + 7x + 3$

16. $3y^2 + 13y + 4$

17. $3a^2 + 10a + 7$

18. $7r^2 + 8r + 1$

19. $4r^2 + r - 3$

20. $4r^2 + 3r - 10$

21. $15m^2 + m - 2$

22. $6x^2 + x - 1$

23. $8m^2 - 10m - 3$

24. $12s^2 + 11s - 5$

25. $20x^2 + 11x - 3$

26. $20x^2 - 28x - 3$

27. $21m^2 + 13m + 2$

28. $38x^2 + 23x + 2$

29. $20y^2 + 39y - 11$

30. $10x^2 + 11x - 6$

31. $6b^2 + 7b + 2$

32. $6w^2 + 19w + 10$

33. $24x^2 - 42x + 9$

34. $48b^2 - 74b - 10$

35. $40m^2q + mq - 6q$

36. $15a^2b + 22ab + 8b$

37. $2m^3 + 2m^2 - 40m$

38. $3x^3 + 12x^2 - 36x$

39. $15n^4 - 39n^3 + 18n^2$

40. $24a^4 + 10a^3 - 4a^2$

41. $18x^5 + 15x^4 - 75x^3$

42. $32z^5 - 20z^4 - 12z^3$

43. $15x^2y^2 - 7xy^2 - 4y^2$

44. $14a^2b^3 + 15ab^3 - 9b^3$

45. $12p^2 + 7pq - 12q^2$

46. $6m^2 - 5mn - 6n^2$

47. $25a^2 + 25ab + 6b^2$

48. $6x^2 - 5xy - y^2$

49. $6a^2 - 7ab - 5b^2$

50. $25g^2 - 5gh - 2h^2$

51. $6m^6n + 7m^5n^2 + 2m^4n^3$

52. $12k^3q^4 - 4k^2q^5 - kq^6$

53. $5 - 6x + x^2$

54. $7 + 8x + x^2$

55. $16 + 16x + 3x^2$

56. $18 + 65x + 7x^2$

If a trinomial has a negative coefficient for the squared term, such as $-2x^2 + 11x - 12$, it may be easier to factor by first factoring out the common factor -1:

$$\begin{aligned} -2x^2 + 11x - 12 &= -1(2x^2 - 11x + 12) \\ &= -1(2x - 3)(x - 4). \end{aligned}$$

Use this method to factor the trinomials in Exercises 57–62.

57. $-x^2 - 4x + 21$

58. $-x^2 + x + 72$

59. $-3x^2 - x + 4$

60. $-5x^2 + 2x + 16$

61. $-2a^2 - 5ab - 2b^2$

62. $-3p^2 + 13pq - 4q^2$

63. The answer given in the back of the book for Exercise 57 is $-1(x + 7)(x - 3)$. Is $(x + 7)(3 - x)$ also a correct answer? Explain why or why not.

64. If $(2x - 3)(x - 5)$ is a correctly factored form of a polynomial, is $(3 - 2x)(5 - x)$ also a factored form of the polynomial? Explain why or why not.

Preview Exercises

Find each product. See Section 3.6.

65. $(7p + 3)(7p - 3)$

66. $(3h + 5k)(3h - 5k)$

67. $\left(r^2 + \frac{1}{2}\right)\left(r^2 - \frac{1}{2}\right)$

68. $(x + 6)^2$

69. $(3t + 4)^2$

70. $\left(c - \frac{2}{3}\right)^2$

4.4 SPECIAL FACTORIZATIONS

By reversing the rules for multiplication of binomials that we learned in the last chapter, we get three rules for factoring polynomials in certain forms.

1 The formula for the product of the sum and difference of the same two terms is

$$(a + b)(a - b) = a^2 - b^2.$$

Reversing this rule leads to the following factorization.

FACTORING A DIFFERENCE OF TWO SQUARES

$$a^2 - b^2 = (a + b)(a - b)$$

For example,

$$m^2 - 16 = m^2 - 4^2 = (m + 4)(m - 4).$$

EXAMPLE 1 *Factoring a Difference of Squares*
Factor each binomial that is the difference of two squares.

$$a^2 - b^2 = (a + b)(a - b)$$

(a) $x^2 - 49 = x^2 - 7^2 = (x + 7)(x - 7)$

(b) $z^2 - \frac{9}{16} = z^2 - \left(\frac{3}{4}\right)^2 = \left(z + \frac{3}{4}\right)\left(z - \frac{3}{4}\right)$

(c) $y^2 - m^2 = (y + m)(y - m)$

(d) $p^2 + 16$

Since $p^2 + 16$ is the *sum* of two squares, it is not equal to $(p + 4)(p - 4)$. Also, using FOIL,

$$(p - 4)(p - 4) = p^2 - 8p + 16 \neq p^2 + 16$$

and

$$(p + 4)(p + 4) = p^2 + 8p + 16 \neq p^2 + 16$$

so $p^2 + 16$ is a prime polynomial. ■

Caution As Example 1(d) suggests, after any common factor is removed, the sum of two squares cannot be factored.

EXAMPLE 2 *Factoring a Difference of Squares*
Factor each binomial that is the difference of two squares.

$$a^2 - b^2 = (a + b)(a - b)$$

(a) $25m^2 - 16 = (5m)^2 - 4^2 = (5m + 4)(5m - 4)$

(b) $49z^2 - 64 = (7z)^2 - (8)^2 = (7z + 8)(7z - 8)$ ■

WORK PROBLEM 1 AT THE SIDE.

OBJECTIVES

1. Factor the difference of two squares.
2. Factor a perfect square trinomial.

FOR EXTRA HELP

Tape 6	SSM pp. 142–144	MAC: A IBM: A

1. Factor if possible.

(a) $p^2 - 100$

(b) $9m^2 - 49$

(c) $64a^2 - 25$

(d) $x^2 + y^2$

ANSWERS
1. (a) $(p + 10)(p - 10)$
(b) $(3m + 7)(3m - 7)$
(c) $(8a + 5)(8a - 5)$
(d) prime

2. Factor.

(a) $50r^2 - 32$

(b) $27y^2 - 75$

(c) $k^4 - 49$

(d) $81r^4 - 16$

ANSWERS

2. (a) $2(5r + 4)(5r - 4)$
(b) $3(3y + 5)(3y - 5)$
(c) $(k^2 + 7)(k^2 - 7)$
(d) $(9r^2 + 4)(3r + 2)(3r - 2)$

EXAMPLE 3 *Factoring More Involved Differences of Squares*

Factor completely.

$$a^2 - b^2 = (a + b)(a - b)$$

(a) $9x^2 - 4z^2 = (3x)^2 - (2z)^2 = (3x + 2z)(3x - 2z)$

(b) $81y^2 - 36$

$$81y^2 - 36 = 9(9y^2 - 4) \quad \text{Factor out 9.}$$
$$= 9(3y + 2)(3y - 2) \quad \text{Difference of squares}$$

(c) $p^4 - 36 = (p^2)^2 - 6^2 = (p^2 + 6)(p^2 - 6)$

Neither $p^2 + 6$ nor $p^2 - 6$ can be factored further.

(d) $m^4 - 16 = (m^2)^2 - 4^2$

$$= (m^2 + 4)(m^2 - 4) \quad \text{Difference of squares}$$
$$= (m^2 + 4)(m + 2)(m - 2) \quad \text{Difference of squares}$$

Caution A common error is to forget to factor the difference of two squares a second time when several steps are required, as in Example 3(d).

WORK PROBLEM 2 AT THE SIDE.

2 The expressions 144, $4x^2$, and $81m^6$ are called perfect squares because

$$144 = 12^2, \quad 4x^2 = (2x)^2, \quad \text{and} \quad 81m^6 = (9m^3)^2.$$

A **perfect square trinomial** is a trinomial that is the square of a binomial. As an example, $x^2 + 8x + 16$ is a perfect square trinomial because it is the square of the binomial $x + 4$:

$$x^2 + 8x + 16 = (x + 4)^2.$$

For a trinomial to be a perfect square, two of its terms must be perfect squares. For this reason, $16x^2 + 4x + 15$ is not a perfect square trinomial because only the term $16x^2$ is a perfect square.

On the other hand, even if two of the terms are perfect squares, the trinomial may not be a perfect square trinomial. For example, $x^2 + 6x + 36$ has two perfect square terms, but it is not a perfect square trinomial. (Try to find a binomial that can be squared to give $x^2 + 6x + 36$.)

We can multiply to see that the square of a binomial gives the following perfect square trinomials.

FACTORING PERFECT SQUARE TRINOMIALS

$$a^2 + 2ab + b^2 = (a + b)^2$$
$$a^2 - 2ab + b^2 = (a - b)^2$$

The middle term of a perfect square trinomial is always twice the product of the two terms in the squared binomial. (This was shown in Section 3.6.) Use this to check any attempt to factor a trinomial that appears to be a perfect square.

■ EXAMPLE 4 *Factoring a Perfect Square Trinomial*

Factor $x^2 + 10x + 25$.

The term x^2 is a perfect square, and so is 25. We can try to factor the trinomial as

$$x^2 + 10x + 25 = (x + 5)^2.$$

To check this we take twice the product of the two terms in the squared binomial.

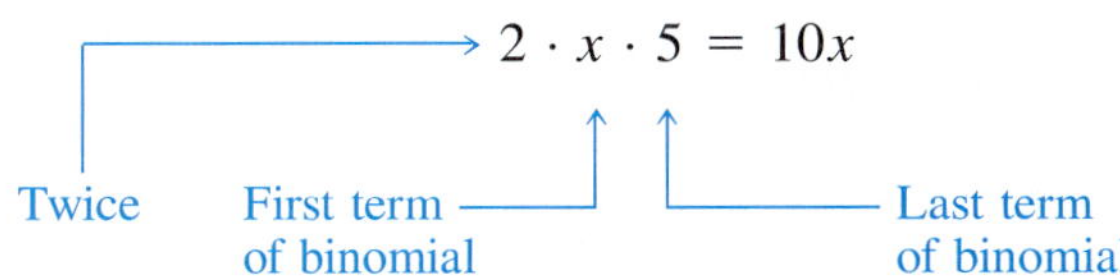

Since $10x$ is the middle term of the trinomial, the trinomial is a perfect square and can be factored as $(x + 5)^2$. ■

WORK PROBLEM 3 AT THE SIDE.

■ EXAMPLE 5 *Factoring Perfect Square Trinomials*

Factor each perfect square trinomial.

(a) $x^2 - 22x + 121$

The first and last terms are perfect squares ($121 = 11^2$). Check to see whether the middle term of $x^2 - 22x + 121$ is twice the product of the first and last terms of the binomial $(x - 11)$.

$$2 \cdot x \cdot 11 = 22x$$

Twice ↑, First term ↑, Last term ↑

Since twice the product of the first and last terms of the binomial is the middle term, $x^2 - 22x + 121$ is a perfect square trinomial and

$$x^2 - 22x + 121 = (x - 11)^2.$$

(b) $9m^2 - 24m + 16 = (3m)^2 - 2(3m)(4) + 4^2 = (3m - 4)^2$

Twice ↑, First term ↑, Last term ↑

(c) $25y^2 + 20y + 16$

The first and last terms are perfect squares.

$$25y^2 = (5y)^2 \quad \text{and} \quad 16 = 4^2$$

Twice the product of the first and last terms of the binomial $5y + 4$ is

$$2 \cdot 5y \cdot 4 = 40y$$

which is not the middle term of $25y^2 + 20y + 16$. This polynomial is not a perfect square. In fact, the polynomial cannot be factored even with the methods of Section 4.3; it is a prime polynomial. ■

3. Factor each trinomial that is a perfect square.

(a) $p^2 + 14p + 49$

(b) $m^2 + 8m + 16$

(c) $x^2 + 2x + 1$

ANSWERS

3. (a) $(p + 7)^2$ (b) $(m + 4)^2$ (c) $(x + 1)^2$

4. Factor each trinomial that is a perfect square.

(a) $p^2 - 18p + 81$

(b) $16a^2 + 56a + 49$

(c) $121p^2 + 110p + 100$

(d) $64x^2 - 48x + 9$

ANSWERS
4. (a) $(p - 9)^2$ (b) $(4a + 7)^2$
(c) not a perfect square trinomial
(d) $(8x - 3)^2$

Note The sign of the second term in the squared binomial is always the same as the sign of the middle term in the trinomial. Also, the first and last terms of a perfect square trinomial must be *positive,* because they are squares. For example, the polynomial $x^2 - 2x - 1$ cannot be a perfect square because the last term is negative.

WORK PROBLEM 4 AT THE SIDE.

The methods of factoring discussed in this section are summarized here. These rules should be memorized.

SPECIAL FACTORIZATIONS

Difference of two squares	$a^2 - b^2 = (a + b)(a - b)$
Perfect square trinomials	$a^2 + 2ab + b^2 = (a + b)^2$
	$a^2 - 2ab + b^2 = (a - b)^2$

NAME DATE HOUR

4.4 EXERCISES

1. In order to develop the factoring techniques described in this section, complete the following list of squares.

$1^2 =$ ____ $2^2 =$ ____ $3^2 =$ ____ $4^2 =$ ____ $5^2 =$ ____

$6^2 =$ ____ $7^2 =$ ____ $8^2 =$ ____ $9^2 =$ ____ $10^2 =$ ____

$11^2 =$ ____ $12^2 =$ ____ $13^2 =$ ____ $14^2 =$ ____ $15^2 =$ ____

$16^2 =$ ____ $17^2 =$ ____ $18^2 =$ ____ $19^2 =$ ____ $20^2 =$ ____

2. In order to use the factoring techniques descibed in this section, you will sometimes need to recognize fourth powers of integers. Complete the following list of fourth powers.

$1^4 =$ ______ $2^4 =$ ______ $3^4 =$ ______ $4^4 =$ ______ $5^4 =$ ______

3. The following powers of x are all perfect squares: x^2, x^4, x^6, x^8, x^{10}. Based on this observation, we may make a conjecture (an educated guess) that if the power of a variable is divisible by ____ (with 0 remainder) then we have a perfect square.

4. Is the conjecture of Exercise 3 true in all cases? Why or why not?

Factor each binomial completely. Use your answers in Exercises 1 and 2 as necessary. See Examples 1–3.

5. $y^2 - 25$ **6.** $t^2 - 16$ **7.** $9r^2 - 4$ **8.** $4x^2 - 9$

9. $36m^2 - \frac{16}{25}$ **10.** $100b^2 - \frac{4}{49}$ **11.** $36x^2 - 16$ **12.** $32a^2 - 8$

13. $196p^2 - 225$ **14.** $361q^2 - 400$ **15.** $16r^2 - 25a^2$ **16.** $49m^2 - 100p^2$

17. $100x^2 + 49$ **18.** $81w^2 + 16$ **19.** $p^4 - 49$ **20.** $r^4 - 25$

21. $x^4 - 1$ **22.** $y^4 - 16$ **23.** $p^4 - 256$ **24.** $16k^4 - 1$

25. When a student was directed to factor $x^4 - 81$ completely, his teacher did not give him full credit when the answered $(x^2 + 9)(x^2 - 9)$. The student argued that because his answer does indeed give $x^4 - 81$ when multiplied out, he should be given full credit. Was the teacher justified in her grading of this item? Why or why not?

26. The binomial $4x^2 + 16$ is a sum of two squares that *can* be factored. How is this binomial factored? When can the sum of two squares be factored?

Factor each trinomial completely. It may be necessary to factor out the greatest common factor first. See Examples 4 and 5.

27. $w^2 + 2w + 1$

28. $p^2 + 4p + 4$

29. $x^2 - 8x + 16$

30. $x^2 - 10x + 25$

31. $t^2 + t + \frac{1}{4}$

32. $m^2 + \frac{2}{3}m + \frac{1}{9}$

33. $x^2 - 1.0x + .25$

34. $y^2 - 1.4y + .49$

35. $2x^2 + 24x + 72$

36. $3y^2 - 48y + 192$

37. $16x^2 - 40x + 25$

38. $36y^2 - 60y + 25$

39. $49x^2 - 28xy + 4y^2$

40. $4z^2 - 12zw + 9w^2$

41. $64x^2 + 48xy + 9y^2$

42. $9t^2 + 24tr + 16r^2$

43. $-50h^2 + 40hy - 8y^2$

44. $-18x^2 - 48xy - 32y^2$

45. In the polynomial $9y^2 + 14y + 25$, the first and last terms are perfect squares. Can the polynomial be factored? If it can, factor it. If it cannot, explain why it is not a perfect square trinomial.

46. Repeat Exercise 45 for $16m^2 + 42m + 49$.

47. Find the value of a so that $ay^2 - 12y + 4 = (3y - 2)^2$ is true.

48. Find the value of b so that $100a^2 + ba + 9 = (10a + 3)^2$ is true.

PREVIEW EXERCISES

Solve each equation. See Section 2.3.

49. $m - 4 = 0$

50. $3t + 2 = 0$

51. $4z - 9 = 0$

52. $2t + 10 = 0$

53. $9x - 6 = 0$

54. $7x = 0$

NAME DATE HOUR

SUMMARY EXERCISES ON FACTORING*

As you factor a polynomial, these questions will help you decide on a suitable factoring technique.

FACTORING A POLYNOMIAL

1. Is there a common factor? If so, factor it out.
2. How many terms are in the polynomial?
   Two terms: Check to see whether it is the difference of two squares.
   Three terms: Is it a perfect square trinomial? If the trinomial is not a perfect square, check to see whether the coefficient of the squared term is 1. If so, use the method of Section 4.2. If the coefficient of the squared term of the trinomial is not 1, use the general factoring methods of Section 4.3.
   Four terms: Try to factor the polynomial by grouping.
3. Can any factors be factored further? If so, factor them.

Factor as completely as possible.

1. $32m^9 + 16m^5 + 24m^3$

2. $2m^2 - 10m - 48$

3. $14k^3 + 7k^2 - 70k$

4. $9z^2 + 64$

5. $6z^2 + 31z + 5$

6. $m^2 - 3mn - 4n^2$

7. $49z^2 - 16y^2$

8. $100n^2r^2 + 30nr^3 - 50n^2r$

9. $16x + 20$

10. $2m^2 + 5m - 3$

11. $10y^2 - 7yz - 6z^2$

12. $y^4 - 16$

13. $m^2 + 2m - 15$

14. $6y^2 - 5y - 4$

15. $32z^3 + 56z^2 - 16z$

16. $15y + 5$

17. $z^2 - 12z + 36$

18. $9m^2 - 64$

19. $y^2 - 4yk - 12k^2$

20. $16z^2 - 8z + 1$

21. $6y^2 - 6y - 12$

22. $72y^3z^2 + 12y^2 - 24y^4z^2$

23. $p^2 - 17p + 66$

24. $a^2 + 17a + 72$

---

*This exercise set includes all kinds of factoring methods. The exercises are randomly mixed to give you practice at deciding which method should be used.

25. $k^2 + 9$

26. $108m^2 - 36m + 3$

27. $z^2 - 3za - 10a^2$

28. $45a^3b^5 - 60a^4b^2 + 75a^6b^4$

29. $4k^2 - 12k + 9$

30. $a^2 - 3ab - 28b^2$

31. $16r^2 + 24rm + 9m^2$

32. $3k^2 + 4k - 4$

33. $3k^3 - 12k^2 - 15k$

34. $a^4 - 625$

35. $16k^2 - 48k + 36$

36. $8k^2 - 10k - 3$

37. $36y^6 - 42y^5 - 120y^4$

38. $8p^2 + 23p - 3$

39. $5z^3 - 45z^2 + 70z$

40. $8k^2 - 2kh - 3h^2$

41. $54m^2 - 24z^2$

42. $4k^2 - 20kz + 25z^2$

43. $6a^2 + 10a - 4$

44. $15h^2 + 11hg - 14g^2$

45. $m^2 - 81$

46. $10z^2 - 7z - 6$

47. $125m^4 - 400m^3n + 195m^2n^2$

48. $9y^2 + 12y - 5$

49. $m^2 - 4m + 4$

50. $27p^{10} - 45p^9 - 252p^8$

51. $24k^4p + 60k^3p^2 + 150k^2p^3$

52. $10m^2 + 25m - 60$

53. $12p^2 + pq - 6q^2$

54. $k^2 - 64$

55. $64p^2 - 100m^2$

56. $2m^2 + 7mn - 15n^2$

57. $100a^2 - 81y^2$

58. $8a^2 + 23ab - 3b^2$

59. $a^2 + 8a + 16$

60. $4y^2 - 25$

4.5 SOLVING QUADRATIC EQUATIONS BY FACTORING

OBJECTIVES

1. Solve quadratic equations by factoring.
2. Solve other equations by factoring.

FOR EXTRA HELP

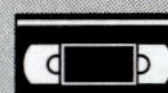

Tape 7 | SSM pp. 145–150 | MAC: A IBM: A

In this section we introduce **quadratic equations,** which are equations that contain a squared term and no terms of higher degree.

QUADRATIC EQUATIONS

Quadratic equations can be written in the form

$$ax^2 + bx + c = 0$$

where a, b, and c are real numbers, with $a \neq 0$.

The form $ax^2 + bx + c = 0$ is the **standard form** of a quadratic equation. For example,

$$x^2 + 5x + 6 = 0, \quad 2a^2 - 5a = 3, \quad \text{and} \quad y^2 = 4$$

are all quadratic equations but only $x^2 + 5x + 6 = 0$ is in standard form.

1 Some quadratic equations can be solved by factoring. A more general method for solving those equations that cannot be solved by factoring is given in Chapter 9. We use the **zero-factor property** to solve a quadratic equation by factoring.

ZERO-FACTOR PROPERTY

If a and b are real numbers and if $ab = 0$, then $a = 0$ or $b = 0$.

In other words, if the product of two numbers is zero, then at least one of the numbers must be zero. One number *must* be 0, but both *may* be 0.

EXAMPLE 1 *Using the Zero-Factor Property*

Solve the equation $(x + 3)(2x - 1) = 0$.

The product $(x + 3)(2x - 1)$ is equal to zero. By the zero-factor property, the only way that the product of these two factors can be zero is if at least one of the factors is zero. Therefore, either $x + 3 = 0$ or $2x - 1 = 0$. Solve each of these two linear equations as in Chapter 2.

$$x + 3 = 0 \qquad 2x - 1 = 0$$

$$x = -3 \qquad 2x = 1 \qquad \text{Add to both sides.}$$

$$x = \frac{1}{2} \qquad \text{Divide by 2.}$$

The given equation $(x + 3)(2x - 1) = 0$ has two solutions, $x = -3$ and $x = \frac{1}{2}$. Check these answers by substituting -3 for x in the original equation, $(x + 3)(2x - 1) = 0$. Then start over and substitute $\frac{1}{2}$ for x.

If $x = -3$, then

$$(-3 + 3)[2(-3) - 1] = 0 \quad ?$$

$$0(-7) = 0. \quad \text{True}$$

If $x = \frac{1}{2}$, then

$$\left(\frac{1}{2} + 3\right)\left(2 \cdot \frac{1}{2} - 1\right) = 0 \quad ?$$

$$\frac{7}{2}(1 - 1) = 0 \quad ?$$

$$\frac{7}{2} \cdot 0 = 0. \quad \text{True}$$

1. Solve each equation. Check your answers.

(a) $(x - 5)(x + 2) = 0$

(b) $(3x - 2)(x + 6) = 0$

(c) $x(5x + 3) = 0$

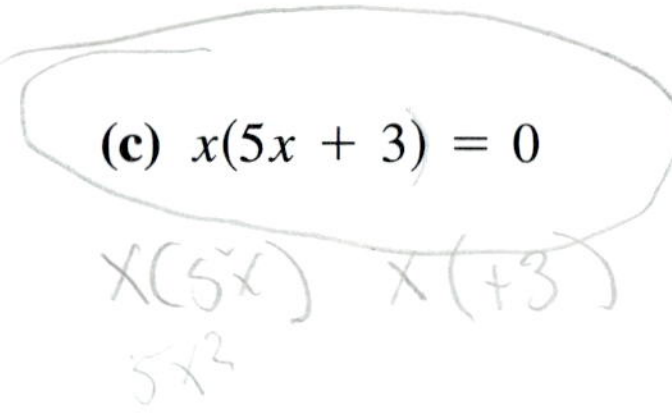

2. Solve each equation.

(a) $m^2 - 3m - 10 = 0$

(b) $x^2 - 7x = 0$

(c) $r^2 + 2r = 8$

ANSWERS

1. (a) 5, −2 (b) $\frac{2}{3}$, −6 (c) 0, $-\frac{3}{5}$
2. (a) −2, 5 (b) 0, 7 (c) −4, 2

Both −3 and $\frac{1}{2}$ result in true equations, so they are solutions to the original equation. ■

Note The word "or" as used in Example 2 means "one or the other or both."

WORK PROBLEM 1 AT THE SIDE.

In Example 1 the equation to be solved was presented with the polynomial in factored form. If the polynomial in an equation is not already factored, first make sure that the equation is in standard form. Then factor.

■ **EXAMPLE 2** *Solving a Quadratic Equation*

Solve each equation.

(a) $x^2 - 5x = -6$

First, rewrite the equation in standard form by adding 6 to both sides.

$$x^2 - 5x + 6 = -6 + 6 \qquad \text{Add 6.}$$
$$x^2 - 5x + 6 = 0$$

Now factor $x^2 - 5x + 6$. Find two numbers whose product is 6 and whose sum is −5. These two numbers are −2 and −3, so the equation becomes

$$(x - 2)(x - 3) = 0. \qquad \text{Factor.}$$
$$x - 2 = 0 \quad \text{or} \quad x - 3 = 0 \qquad \text{Zero-factor property}$$
$$x = 2 \quad \text{or} \quad x = 3 \qquad \text{Solve each equation.}$$

Check both solutions by substituting first 2 and then 3 for x in the original equation.

(b) $y^2 = y + 20$

$$y^2 = y + 20$$
$$y^2 - y - 20 = 0 \qquad \text{Subtract } y \text{ and 20.}$$
$$(y - 5)(y + 4) = 0 \qquad \text{Factor.}$$
$$y - 5 = 0 \quad \text{or} \quad y + 4 = 0 \qquad \text{Zero-factor property}$$
$$y = 5 \quad \text{or} \quad y = -4 \qquad \text{Solve each equation.}$$

Check these solutions by substituting in the original equation. ■

WORK PROBLEM 2 AT THE SIDE.

In summary, we go through the following steps to solve quadratic equations by factoring.

SOLVING A QUADRATIC EQUATION BY FACTORING

Step 1 Write the equation in standard form: all terms on one side of the equals sign, with 0 on the other side.

Step 2 Factor completely.

Step 3 Set each factor with a variable equal to 0, and solve the resulting equations.

Step 4 Check each solution in the original equation.

EXAMPLE 3 *Solving a Quadratic Equation with a Common Factor*

Solve the equation $4p^2 + 40 = 26p$.

Subtract $26p$ from each side and write in descending powers to get

$$4p^2 - 26p + 40 = 0.$$

$$2(2p^2 - 13p + 20) = 0 \quad \text{Factor out 2.}$$

$$2(2p - 5)(p - 4) = 0 \quad \text{Factor the trinomial.}$$

$$2p - 5 = 0 \quad \text{or} \quad p - 4 = 0 \quad \text{Zero-factor property}$$

$$2p = 5 \quad \text{or} \quad p = 4$$

$$p = \frac{5}{2}$$

The solutions of $4p^2 + 40 = 26p$ are $\frac{5}{2}$ and 4; check them by substituting in the original equation. ■

Caution A common error is to include the common factor 2 as a solution in Example 3.

WORK PROBLEM 3 AT THE SIDE.

EXAMPLE 4 *Solving Quadratic Equations*

Solve each equation.

(a) $16m^2 - 25 = 0$

$$16m^2 - 25 = 0$$

$$(4m + 5)(4m - 5) = 0 \quad \text{Factor.}$$

$$4m + 5 = 0 \quad \text{or} \quad 4m - 5 = 0 \quad \text{Zero-factor property}$$

$$4m = -5 \quad \text{or} \quad 4m = 5$$

$$m = -\frac{5}{4} \quad \text{or} \quad m = \frac{5}{4}$$

You should check the two solutions, $-\frac{5}{4}$ and $\frac{5}{4}$, using the original equation.

(b) $k(2k + 5) = 3$

$$k(2k + 5) = 3$$

$$2k^2 + 5k = 3 \quad \text{Multiply.}$$

$$2k^2 + 5k - 3 = 0 \quad \text{Subtract 3.}$$

$$(2k - 1)(k + 3) = 0 \quad \text{Factor.}$$

$$2k - 1 = 0 \quad \text{or} \quad k + 3 = 0 \quad \text{Zero-factor property}$$

$$2k = 1$$

$$k = \frac{1}{2} \quad \text{or} \quad k = -3$$

Check that the two solutions are $\frac{1}{2}$ and -3.

3. Solve each equation.

(a) $10a^2 - 5a - 15 = 0$

(b) $4x^2 - 2x = 42$

ANSWERS

3. (a) $-1, \frac{3}{2}$ **(b)** $-3, \frac{7}{2}$

(c) $y^2 = 2y$

First write the equation in standard form.

$$y^2 - 2y = 0 \quad \text{Standard form}$$
$$y(y - 2) = 0 \quad \text{Factor.}$$
$$y = 0 \quad \text{or} \quad y - 2 = 0 \quad \text{Zero-factor property}$$
$$y = 2$$

The solutions are 0 and 2. ■

Caution In Example 4(b) the zero-factor property could not be used to solve the equation as given because of the 3 on the right. Remember that the zero-factor property applies only to a product that equals 0.

In Example 4(c) it is tempting to begin by dividing both sides of the equation by y to get $y = 2$. Note that we do not get the other solution, 0, by this method.

We *may* divide both sides of an equation by a *nonzero* real number, however. For instance, in Example 3 we could have divided both sides by 2 to begin. This would not have affected the solutions.

WORK PROBLEM 4 AT THE SIDE.

2 The zero-factor property can also be used to solve equations that result in more than two factors, as shown in Example 5. (These equations are not quadratic equations. Why not?)

■ **EXAMPLE 5** *Solving Equations with More Than Two Factors*

Solve the equation $6z^3 - 6z = 0$.

$$6z^3 - 6z = 0$$
$$6z(z^2 - 1) = 0 \quad \text{Factor out } 6z.$$
$$6z(z + 1)(z - 1) = 0 \quad \text{Factor } z^2 - 1.$$

By an extension of the zero-factor property, this product can equal 0 only if at least one of the factors is 0. Write and solve three equations, one for each factor with a variable.

$$6z = 0 \quad \text{or} \quad z + 1 = 0 \quad \text{or} \quad z - 1 = 0$$
$$z = 0 \quad \text{or} \quad z = -1 \quad \text{or} \quad z = 1$$

Check by substituting, in turn, 0, -1, and 1 in the original equation. ■

WORK PROBLEM 5 AT THE SIDE.

4. Solve each equation.

(a) $49m^2 - 9 = 0$

(b) $p(4p + 7) = 2$

(c) $m^2 = 3m$

5. Solve each equation.

(a) $r^3 - 16r = 0$

(b) $x^3 - 3x^2 - 18x = 0$

ANSWERS

4. (a) $-\frac{3}{7}, \frac{3}{7}$ (b) $-2, \frac{1}{4}$ (c) 0, 3

5. (a) $-4, 0, 4$ (b) $-3, 0, 6$

EXAMPLE 6 *Solving Equations with More Than Two Factors*

Solve the equation $(2x - 1)(x^2 - 9x + 20) = 0$.

$$(2x - 1)(x^2 - 9x + 20) = 0$$

$$(2x - 1)(x - 5)(x - 4) = 0 \qquad \text{Factor } x^2 - 9x + 20.$$

$$2x - 1 = 0 \quad \text{or} \quad x - 5 = 0 \quad \text{or} \quad x - 4 = 0 \qquad \text{Zero-factor property}$$

$$x = \frac{1}{2} \quad \text{or} \quad x = 5 \quad \text{or} \quad x = 4$$

The solutions are $\frac{1}{2}$, 5, and 4. Check each solution. ■

WORK PROBLEM 6 AT THE SIDE.

Caution In Example 6, it would be unproductive to begin by multiplying the two factors together. Keep in mind the zero-factor property requires the product of two or more factors equal to zero. Always consider first whether an equation is given in the appropriate form for the zero-factor property.

6. Solve each equation.

(a) $(m + 3)(m^2 - 11m + 10) = 0$

(b) $(2x + 5)(4x^2 - 9) = 0$

ANSWERS

6. (a) $-3, 1, 10$ **(b)** $-\frac{5}{2}, -\frac{3}{2}, \frac{3}{2}$

QUEST FOR NUMERACY

A Look at Pi

π

The March 22, 1992, issue of *The New Yorker* contains the article "The Mountains of Pi," profiling Gregory and David Chudnovsky. With their own computer, housed in their Manhattan apartment, the Chudnovskys have become two of today's foremost researchers on the investigation of the decimal digits of π. Their research deals with observing the digits for possible patterns. For example, one conjecture about π deals with what mathematicians term "normalcy." The normality conjecture says that all digits appear with the same average frequency. According to Gregory "There is absolutely no doubt that π is a 'normal' number. Yet we can't prove it. We don't even know how to *try* to prove it."

In mid-1991, the Chudnovskys stopped their calculation of the decimal digits of π at 2,260,321,336 digits. If printed in ordinary type, they would stretch from New York to Southern California.

The computation of π has fascinated mathematicians and lay people for centuries. In the nineteenth century the British mathematician William Shanks spent many years of his life calculating π to 707 decimal places. It turned out that only the first 527 were correct. The advent of the computer greatly revolutionized the quest of calculating π. In fact, the accuracy of computers and computer programs is sometimes tested by performing the computation of π. In 1767, J. H. Lambert proved that π is irrational (and thus its decimal will never terminate and never repeat).

This poem, dedicated to Archimedes ("the immortal Syracusan"), allows us to learn the first thirty digits of the decimal representation of π. By replacing each word with the number of letters it contains, with a decimal point following the initial 3, the decimal is found. The poem was written by A. C. Orr and appeared in the *Literary Digest* in 1906.

Now I, even I, would celebrate
In rhymes unapt, the great
Immortal Syracusan, rivaled nevermore,
Who in his wondrous lore
Passed on before,
Left men his guidance
How to circles mensurate.

From this poem, we can determine these digits of π: 3.14159265358979323846264338327 9.

A fascinating history of π has been chronicled by Petr Beckman in the book *A History of Pi.*

FOR GROUP DISCUSSION

1. Ask someone outside of class "What is π?" Then as a class, discuss the various responses obtained.
2. As with Mount Everest, some people enjoy climbing the mountain of π simply because it is there. Have you ever tackled a project for no reason other than to simply say "I did it"? Share any such experiences with the class.

NAME DATE HOUR

4.5 EXERCISES

Solve each equation, and check your answer. See Example 1.

1. $(x - 4)(x + 5) = 0$

2. $(y + 3)(y - 8) = 0$

3. $(3k + 8)(k + 7) = 0$

4. $(5r + 9)(r + 6) = 0$

5. $t(t + 4) = 0$

6. $x(x - 10) = 0$

7. $2x(3x - 4) = 0$

8. $6y(4y + 9) = 0$

9. Students often become confused as to how to handle a constant, such as 2 in the equation $2x(3x - 4) = 0$ of Exercise 7. How would you explain to someone how to solve this equation, and how to handle the constant 2?

10. The zero-factor property can be extended to more than two factors. For example, to solve $(x - 4)(x + 3)(2x - 7) = 0$, we would set each factor equal to zero and solve three equations. Find the solutions to this equation.

11. Why do you think that 9 is called a *double solution* of the equation $(x - 9)^2 = 0$?

12. Write an equation with the two solutions 5 and $-\frac{4}{3}$.

Solve each equation, and check your answer. See Examples 2–4.

13. $y^2 + 3y + 2 = 0$

14. $p^2 + 8p + 7 = 0$

15. $y^2 - 3y + 2 = 0$

16. $r^2 - 4r + 3 = 0$

17. $x^2 = 24 - 5x$

18. $t^2 = 2t + 15$

19. $x^2 = 3 + 2x$

20. $m^2 = 4 + 3m$

21. $z^2 = -2 - 3z$

22. $p^2 = 2p + 3$

23. $m^2 + 8m + 16 = 0$

24. $b^2 - 6b + 9 = 0$

25. $3x^2 + 5x - 2 = 0$

26. $6r^2 - r - 2 = 0$

27. $6p^2 = 4 - 5p$

28. $6x^2 = 4 + 5x$

29. $9s^2 + 12s = -4$

30. $36x^2 + 60x = -25$

31. $y^2 - 9 = 0$

32. $m^2 - 100 = 0$

33. $16k^2 - 49 = 0$

34. $4w^2 - 9 = 0$

35. $n^2 = 121$

36. $x^2 = 400$

37. What is wrong with this reasoning in solving the equation in Exercise 35? "To solve $n^2 = 121$, I must find a number that, when multiplied by itself, gives 121. Because $11^2 = 121$, the solution of the equation is 11."

38. What is wrong with this reasoning in solving $x^2 = 7x$? "To solve $x^2 = 7x$, first divide both sides by x to get $x = 7$. Therefore, the solution is 7."

Solve each equation, and check your answer. See Examples 4, 5, and 6.

39. $x^2 = 7x$

40. $t^2 = 9t$

41. $6r^2 = 3r$

42. $10y^2 = -5y$

43. $g(g - 7) = -10$

44. $r(r - 5) = -6$

45. $z(2z + 7) = 4$

46. $b(2b + 3) = 9$

47. $2(y^2 - 66) = -13y$

48. $3(t^2 + 4) = 20t$

49. $3x(x + 1) = (2x + 3)(x + 1)$

50. $2k(k + 3) = (3k + 1)(k + 3)$

51. $(2r + 5)(3r^2 - 16r + 5) = 0$

52. $(3m + 4)(6m^2 + m - 2) = 0$

53. $(2x + 7)(x^2 + 2x - 3) = 0$

54. $(x + 1)(6x^2 + x - 12) = 0$

55. $9y^3 - 49y = 0$

56. $16r^3 - 9r = 0$

PREVIEW EXERCISES

Solve each problem. See Section 2.4.

57. Florida has 9 more counties than California. Together the two states have 125 counties. How many counties does each state have?

58. If a number is doubled and 6 is subtracted from this result, the answer is 3684. The unknown number is the year that Texas was admitted to the Union. What year was Texas admitted?

4.6 APPLICATIONS OF QUADRATIC EQUATIONS

OBJECTIVES

1. Solve applied problems about geometric figures.
2. Solve applied problems about consecutive integers.
3. Solve applied problems using the Pythagorean formula.

FOR EXTRA HELP

Tape 7 | SSM pp. 150–154 | MAC: A IBM: A

We are now ready to use factoring to solve quadratic equations that arise from applied problems. Many problems in this section will require a formula from the endsheets. The general approach is the same as in Chapter 2. We still follow the six steps listed in Section 2.4.

1 We begin with a geometry problem.

EXAMPLE 1 *Solving an Area Problem*

The Goldsteins are planning to add a rectangular porch to their house. The width of the porch will be 4 feet less than its length, and they want it to have an area of 96 square feet. Find the length and width of the porch.

Let x = the length of the porch,

$x - 4$ = the width (the width is 4 less than the length).

See Figure 1. The area of a rectangle is given by the formula

$$\text{Area} = LW = \text{Length} \times \text{Width}.$$

FIGURE 1

Substitute 96 for the area, x for the length, and $x - 4$ for the width in the formula.

$$A = LW$$

$96 = x(x - 4)$	Let $A = 96$, $L = x$, $W = x - 4$.
$96 = x^2 - 4x$	Distributive property
$0 = x^2 - 4x - 96$	Subtract 96 from both sides.
$0 = (x - 12)(x + 8)$	Factor.
$x - 12 = 0$ or $x + 8 = 0$	Zero-factor property
$x = 12$ or $x = -8$	

The solutions are $x = 12$ and $x = -8$. Because a rectangle cannot have a side of negative length, discard the solution -8. Then the length of the porch will be 12 feet, and the width will be $12 - 4 = 8$ feet. As a check, the width is 4 feet less than the length and the area is $8 \cdot 12 = 96$ square feet, as required. ■

Caution In an applied problem, always be careful to check solutions against physical facts.

WORK PROBLEM 1 AT THE SIDE.

1. The length of a room is 2 meters more than the width. The area of the floor is 48 square meters. Find the length and width of the room.

2 The next problem involves **perimeter,** the distance around a figure, as well as area.

ANSWERS
1. length: 8 meters; width: 6 meters

2. The width of a rectangular yard is 4 meters less than the length. The area is numerically 92 more than the perimeter. Find the length and width of the yard.

■ **EXAMPLE 2** *Solving an Area and Perimeter Problem*

The length of a rectangular rug is 4 feet more than the width. The area of the rug is numerically 1 more than the perimeter. See Figure 2. Find the length and width of the rug.

Let x = the width of the rug

$x + 4$ = the length of the rug.

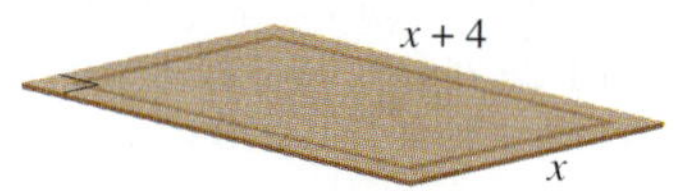

FIGURE 2

The area is the product of the length and width, so

$$A = LW.$$

Substituting $x + 4$ for the length and x for the width gives

$$A = (x + 4)x.$$

Now substitute into the formula for perimeter.

$$P = 2L + 2W$$
$$P = 2(x + 4) + 2x.$$

According to the information given in the problem, the area is numerically 1 more than the perimeter.

The area	is	1	more than	the perimeter.
↓	↓	↓	↓	↓
$x(x + 4)$	$=$	1	$+$	$2(x + 4) + 2x$

Simplify and solve this equation.

$x^2 + 4x = 1 + 2x + 8 + 2x$	Distributive property
$x^2 + 4x = 9 + 4x$	Combine terms.
$x^2 = 9$	Subtract $4x$ from both sides.
$x^2 - 9 = 0$	Subtract 9 from both sides.
$(x + 3)(x - 3) = 0$	Factor.
$x + 3 = 0$ or $x - 3 = 0$	Zero-factor property
$x = -3$ or $x = 3$	

A rectangle cannot have a negative width, so ignore -3. The only valid solution is 3, so the width is 3 feet and the length is $3 + 4 = 7$ feet. Check to see that the area is numerically 1 more than the perimeter. The rug is 3 feet wide and 7 feet long. ■

WORK PROBLEM 2 AT THE SIDE.

2 **Consecutive integers** are integers that are next to each other on a number line, such as 5 and 6, or -11 and -10. **Consecutive odd integers** are odd integers that are next to each other, such as 21 and 23, or -17 and -15. **Consecutive even integers** are defined similarly.

ANSWERS

2. length: 14 meters; width: 10 meters

The following list may be helpful in working with consecutive integers. Here x represents the first of the integers.

CONSECUTIVE INTEGERS

Two consecutive integers	$x, x + 1$
Three consecutive integers	$x, x + 1, x + 2$
Two consecutive odd integers	$x, x + 2$
Two consecutive even integers	$x, x + 2$

In the next example we show how quadratic equations can occur in work with consecutive integers.

EXAMPLE 3 *Solving a Consecutive Integer Problem*

The product of two consecutive odd integers is 1 less than five times their sum. Find the integers.

Let s = the small integer

$s + 2$ = the next larger odd integer.

Because the problem mentions consecutive *odd* integers, use $s + 2$ for the larger of the two integers. According to the problem, the product is 1 less than five times the sum.

The product	is	five times the sum	less 1.
↓	↓	↓	↓
$s(s + 2)$	$=$	$5(s + s + 2)$	$- 1$

Simplify this equation and solve it.

$$s^2 + 2s = 5s + 5s + 10 - 1 \quad \text{Distributive property}$$
$$s^2 + 2s = 10s + 9 \quad \text{Combine terms.}$$
$$s^2 - 8s - 9 = 0 \quad \text{Subtract } 10s \text{ and } 9.$$
$$(s - 9)(s + 1) = 0 \quad \text{Factor.}$$
$$s - 9 = 0 \quad \text{or} \quad s + 1 = 0 \quad \text{Zero-factor property}$$
$$s = 9 \quad \text{or} \quad s = -1$$

We need to find two consecutive odd integers.

If $s = 9$ is the first, then $s + 2 = 11$ is the second.

If $s = -1$ is the first, then $s + 2 = 1$ is the second.

Check that two pairs of integers satisfy the problem: 9 and 11 or -1 and 1. ■

WORK PROBLEM 3 AT THE SIDE.

3 The next example requires the **Pythagorean formula** from geometry.

3. The product of two consecutive even integers is 4 more than two times their sum. Find the integers.

ANSWERS

3. 4 and 6 or −2 and 0

4. The hypotenuse of a right triangle is 3 inches longer than the longer leg. The shorter leg is 3 inches shorter than the longer leg. Find the lengths of the sides of the triangle.

PYTHAGOREAN FORMULA

If a right triangle (a triangle with a 90° angle) has longest side of length c and two other sides of lengths a and b, then

$$a^2 + b^2 = c^2.$$

hypotenuse c

leg a

90°

leg b

The longest side is called the **hypotenuse,** and the two shorter sides are the **legs** of the triangle.

EXAMPLE 4 *Using the Pythagorean Formula*

Ed and Mark leave their office, with Ed traveling north and Mark traveling east. When Mark is 1 mile farther than Ed from the office, the distance between them is 2 miles more than Ed's distance from the office. Find their distances from the office and the distance between them.

Let x represent Ed's distance from the office, $x + 1$ represent Mark's distance from the office, and $x + 2$ represent the distance between them. Place these on a right triangle, as in Figure 3.

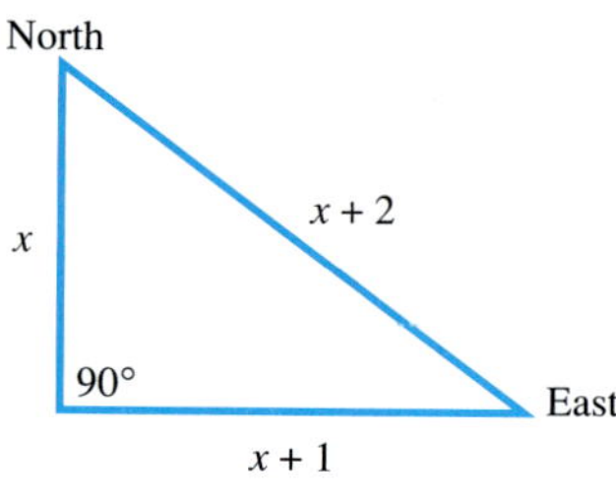

FIGURE 3

Substitute into the Pythagorean formula.

$$a^2 + b^2 = c^2$$

$$x^2 + (x + 1)^2 = (x + 2)^2$$

Because $(x + 1)^2 = x^2 + 2x + 1$, and $(x + 2)^2 = x^2 + 4x + 4$, the equation becomes

$$x^2 + x^2 + 2x + 1 = x^2 + 4x + 4.$$

$$x^2 - 2x - 3 = 0 \qquad \text{Standard form}$$

$$(x - 3)(x + 1) = 0 \qquad \text{Factor.}$$

$$x - 3 = 0 \quad \text{or} \quad x + 1 = 0 \qquad \text{Zero-factor property}$$

$$x = 3 \quad \text{or} \quad x = -1$$

Since -1 cannot represent a distance, 3 is the only possible answer. Ed's distance is 3 miles, Mark's distance is 4 miles, and the distance between them is 5 miles. Check that $3^2 + 4^2 = 5^2$. ■

Caution When solving a problem involving the Pythagorean formula, be sure that the expressions for the sides are properly placed.

$$\text{leg}^2 + \text{leg}^2 = \text{hypotenuse}^2$$

WORK PROBLEM 4 AT THE SIDE.

ANSWERS
4. 9 inches, 12 inches, 15 inches

4.6 EXERCISES

NAME DATE HOUR

In Exercises 1–6, a figure and a correspondng geometric formula are given. (These and other geometric formulas may be found in the endsheets of this book.)

(a) *Write an equation using the formula and the given information.*
(b) *Solve the equation, giving only solutions that make sense in the problem.*
(c) *Use the solution to find the indicated dimensions of the figure.*

1.

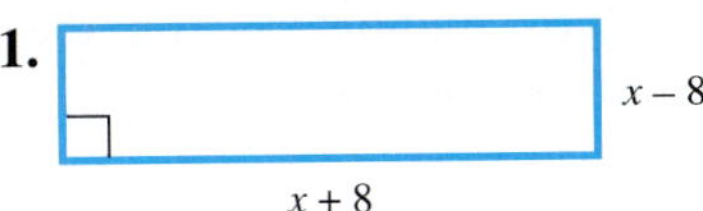

Area of a rectangle: $A = LW$

The area of this rectangle is 80 square units. Find its length and its width.

2.

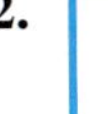

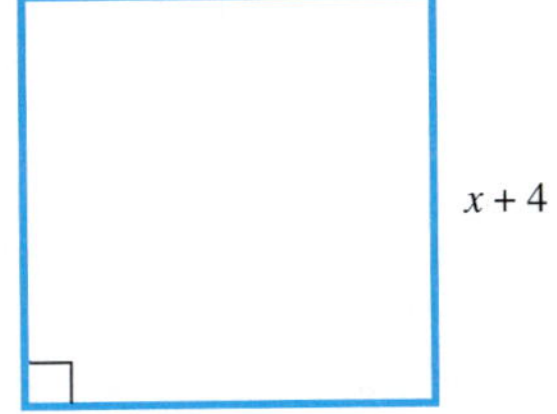

Area of a square: $A = s^2$

The area of this square is 196 square units. Find the length of each side.

3.

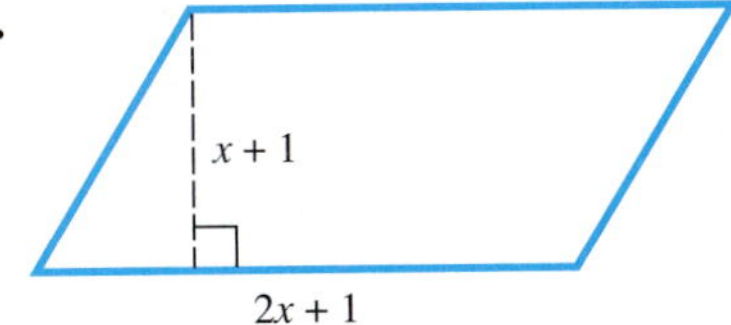

Area of a parallelogram: $A = bh$

The area of this parallelogram is 45 square units. Find its base and its height.

4.

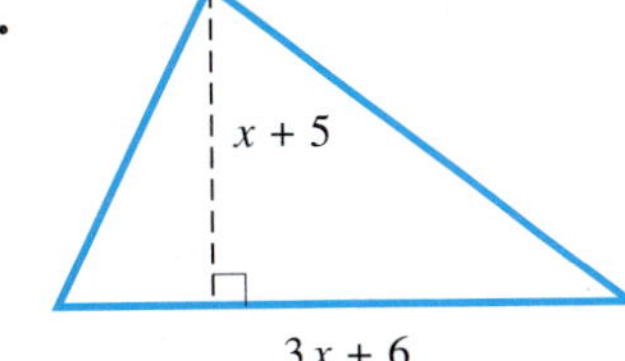

Area of a triangle: $A = \frac{1}{2}bh$

The area of this triangle is 60 square units. Find its base and its height.

5.

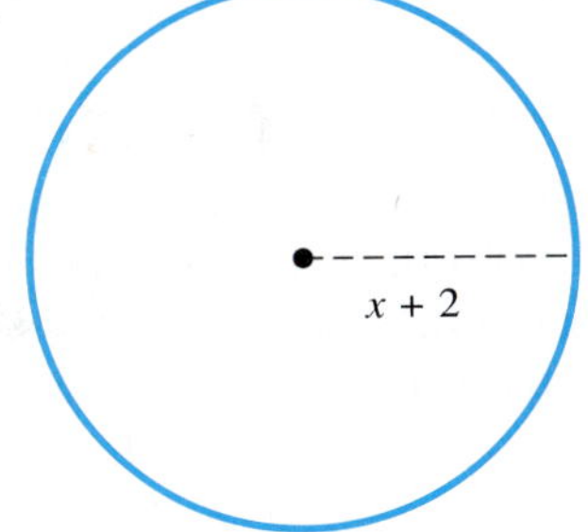

Area of a circle: $A = \pi r^2$

The area of this circle is 36π square units. Find its radius.

6.

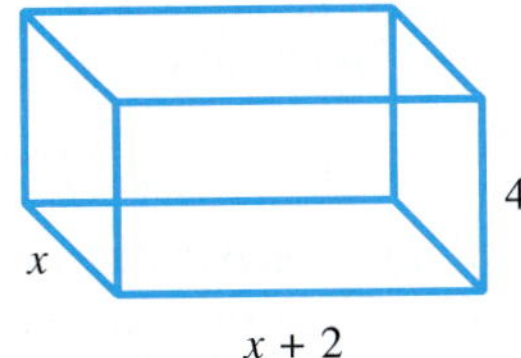

Volume of a rectangular box: $V = LWH$

The volume of this box is 192 cubic units. Find its length and its width.

Solve each problem. Check your answer to be sure that it is reasonable. Refer to the formulas in Exercises 1–6 above, or the endsheets. See Examples 1 and 2.

7. The length of a VHS videocassette shell is 3 inches more than its width. The area of the rectangular top side of the shell is 28 square inches. Find the length and the width of the videocassette shell.

8. A plastic box that holds a standard audiocassette has a length 4 centimeters longer than its width. The area of the rectangular top of the box is 77 square centimeters. Find the length and the width of the box.

9. The dimensions of a certain IBM computer monitor screen are such that its length is 3 inches more than its width. If the length were increased by 1 inch while the width remained the same, the area would increase by 8 square inches. What are the dimensions of the screen?

10. The keyboard of the computer mentioned in Exercise 9 is 11 inches longer than it is wide. If both its length and width were increased by 2 inches, the area of the top of the keyboard would increase by 58 square inches. What are the length and the width of the keyboard?

11. A square poster has sides measuring 2 feet less than the sides of a square sign. If the difference between their areas is 32 square feet, find the lengths of the sides of the poster and the sign.

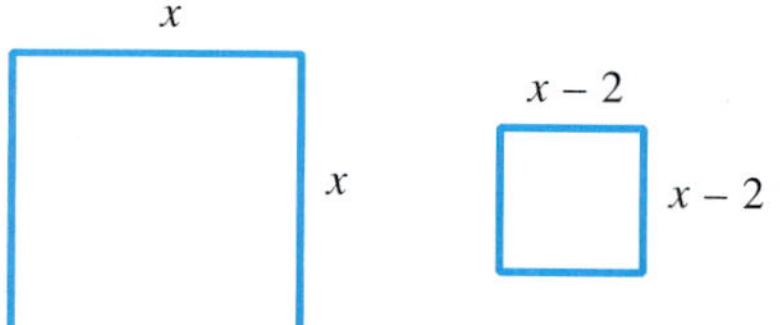

12. The sides of one square have a length 3 meters more than the sides of a second square. If the area of the larger square is subtracted from 4 times the area of the smaller square, the result is 36 square meters. What are the lengths of the sides of each square?

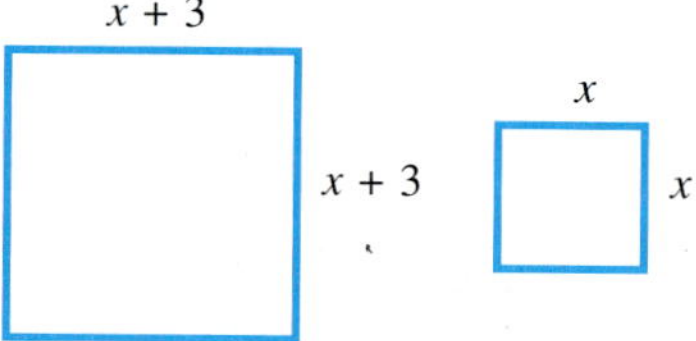

13. The area of a triangle is 30 square inches. The base of the triangle measures 2 inches more than twice the height of the triangle. Find the measures of the base and the height.

14. A certain triangle has its base equal in measure to its height. The area of the triangle is 72 square meters. Find the equal base and height measures.

15. A ten gallon aquarium holding African cichlids has height 3 inches more than its width. Its length is 21 inches, and its volume is 2730 cubic inches. What are the height and width of the aquarium?

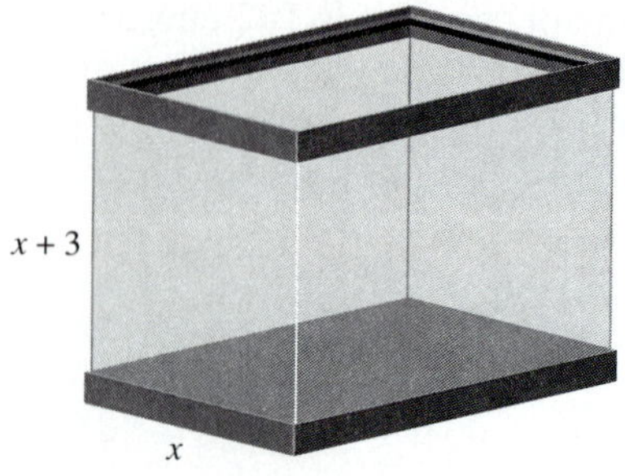

16. Nana wishes to build a box to hold her tools. It is to be 2 feet high, and the width is to be 3 feet less than its length. If its volume will be 80 cubic feet, find the length and the width of the box.

Solve each problem. See Example 3.

17. The product of two consecutive integers is 11 more than their sum. Find the integers.

18. The product of two consecutive integers is 4 less than 4 times their sum. Find the integers.

19. Find three consecutive odd integers such that 3 times the sum of all three is 18 more than the product of the smaller two.

20. Find three consecutive odd integers such that the sum of all three is 42 less than the product of the larger two.

21. Find three consecutive even integers such that the sum of the squares of the smaller two is equal to the square of the largest.

22. Find three consecutive even integers such that the square of the sum of the smaller two is equal to twice the largest.

Use the Pythagorean formula to solve the following problems. See Example 4.

23. Wei-Jen works due north of home. Her husband Alan works due east. They leave for work at the same time. By the time Wei-Jen is 5 miles from home, the distance between them is one mile more than Alan's distance from home. How far from home is Alan?

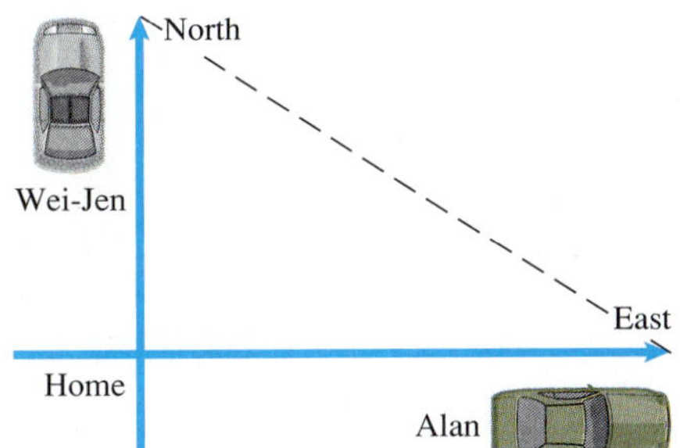

24. Two cars left an intersection at the same time. One traveled north. The other traveled 14 miles farther, but to the east. How far apart were they then, if the distance between them was 4 miles more than the distance traveled east?

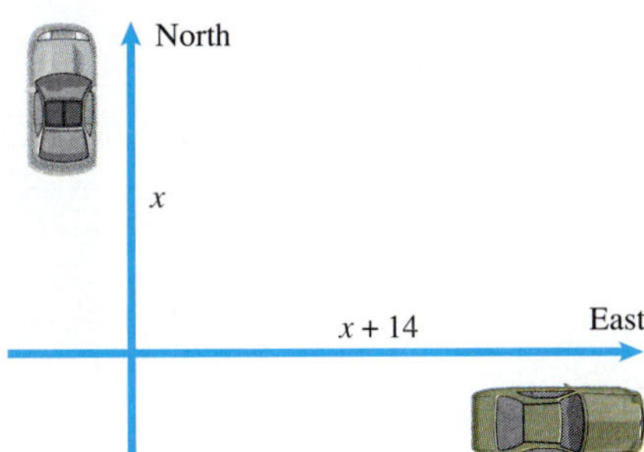

25. A ladder is leaning against a building. The distance from the bottom of the ladder to the building is 4 feet less than the length of the ladder. How high up the side of the building is the top of the ladder if that distance is 2 feet less than the length of the ladder?

26. A lot has the shape of a right triangle with one leg 2 meters longer than the other. The hypotenuse is two meters less than twice the length of the shorter leg. Find the length of the shorter leg.

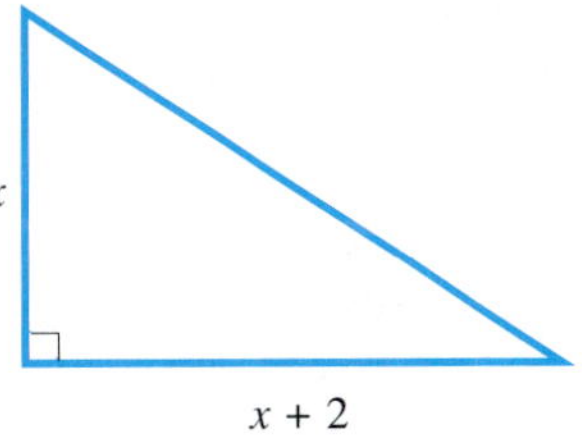

Solve Problems 27 and 28 using the formula for the volume of a pyramid, $V = \frac{1}{3}Bh$, where B is the area of the base.

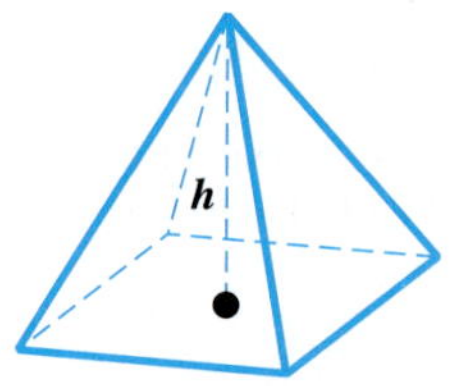

27. The volume of a pyramid is 32 cubic meters. Suppose the numerical value of the height is 10 meters less than the numerical value of the area of the base. Find the height and the area of the base.

28. Suppose a pyramid has a rectangular base whose width is 3 centimeters less than the length. If the height is 8 centimeters and the volume is 144 cubic centimeters, find the dimensions of the base.

29. If an object is dropped from a height as shown in the figure, the distance d (in feet) that it falls in t seconds (disregarding air resistance) is given by the formula

$$d = 16t^2.$$

(a) How far would the object fall in 3 seconds?

(b) How long would it take for the object to fall 1600 feet?

30. If an object is thrown straight upward with an initial velocity of 64 feet per second, its height h after t seconds is given by the formula

$$h = -16t^2 + 64t.$$

(a) Find the number of seconds it would take for the object to reach a height of 48 feet.

(b) How long will it take for the object to return to the ground? (Hint: When it hits the ground, $h = 0$.)

31. Explain why two answers are possible in Exercise 30(a).

32. In Exercise 30, can h be a negative number? Why or why not?

Find a polynomial representing the area of each shaded region.

33.

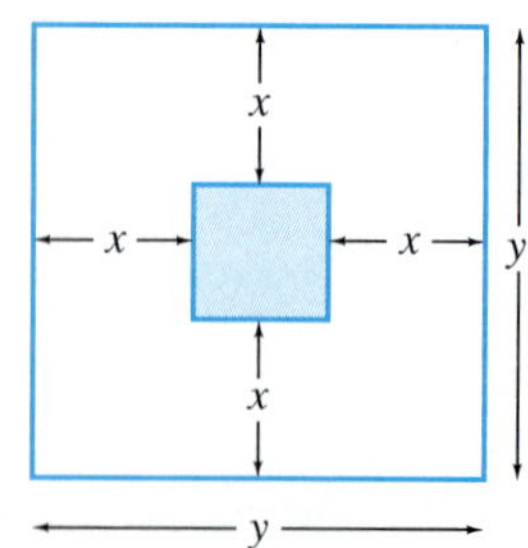

34.

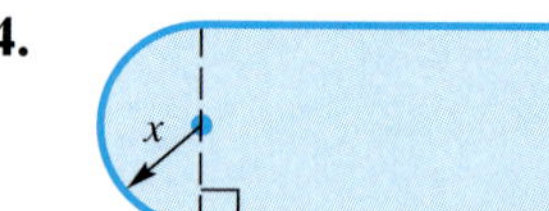

PREVIEW EXERCISES

Write each fraction in lowest terms. See Section R.1.

35. $\frac{26}{156}$ **36.** $\frac{60}{25}$ **37.** $-\frac{42}{14}$ **38.** $\frac{25}{36}$

QUEST FOR NUMERACY

The Man Who Loved Numbers: Srinivasa Ramanujan (1887–1920)

The Indian mathematician Srinivasa Ramanujan (1887–1920) developed many ideas in the branch of mathematics known as number theory. His friend and collaborator on occasion was G. H. Hardy, also a number theorist and professor at Cambridge University in England.

Ramanujan introduced himself to Hardy in 1913 by a letter in which he stated without proof a number of complicated formulas. Hardy at first thought it was a crank letter, but he soon realized that the formulas had to have been the work of a genius. Hardy arranged for Ramanujan to receive a modest stipend, enough to live fairly well at Cambridge.

A story has been told about Ramanujan that illustrates his genius. Hardy once mentioned to Ramanujan that he had just taken a taxicab with a rather dull number: 1729. Ramanujan countered by saying that this number isn't dull at all: it is the smallest natural number that can be expressed as the sum of two cubes in two different ways:

$$1^3 + 12^3 = 1729 \qquad \text{and} \qquad 9^3 + 10^3 = 1729.$$

Ramanujan was the subject of a 1988 installment of the Nova series aired on the Public Broadcasting System. It was aptly titled "The Man Who Loved Numbers."

FOR GROUP DISCUSSION See who can be the first in your group to discover the smallest number that can be written as the sum of two squares in two different ways. (To see if you are correct, look up the ZIP code for Surprise, Arizona. The number you're looking for is formed by the first two digits of the ZIP code.)

CHAPTER 4 SUMMARY

KEY TERMS

4.1	**factor**	An expression A is a factor of an expression B if B can be divided by A with zero remainder.
	factored form	An expression is in factored form when it is written as a product.
	greatest common factor	The greatest common factor is the largest quantity that is a factor of each of a group of quantities.
4.2	**prime polynomial**	A prime polynomial is a polynomial that cannot be factored.
4.4	**perfect square trinomial**	A perfect square trinomial is a trinomial that can be factored as the square of a binomial.
4.5	**quadratic equation**	A quadratic equation is an equation that can be written in the form $ax^2 + bx + c = 0$, with $a \neq 0$.
	standard form	The form $ax^2 + bx + c = 0$ is the standard form of a quadratic equation.
4.6	**perimeter**	The perimeter of a geometric figure is the distance around the figure.

QUICK REVIEW

Concepts	Examples
4.1 Factors; the Greatest Common Factor	
Finding the Greatest Common Factor	Find the greatest common factor of $4x^2y$, $-6x^2y^3$, $2xy^2$.
1. Include the largest numerical factor of every term. **2.** Include each variable that is a factor of every term raised to the smallest exponent that appears in a term.	$4x^2y = 2^2 \cdot x^2 \cdot y$ $-6x^2y^3 = -1 \cdot 2 \cdot 3 \cdot x^2 \cdot y^3$ $2xy^2 = 2 \cdot x \cdot y^2$ The greatest common factor is $2xy$.
Factoring by Grouping	Factor $2a^2 + 2ab + a + b$.
1. Group the terms so that each group has a common factor.	$= (2a^2 + 2ab) + (a + b)$
2. Factor out the greatest common factor in each group.	$= 2a(a + b) + 1(a + b)$
3. Factor a common binomial factor from the result of Step 2.	$= (a + b)(2a + 1)$
4. If Step 3 cannot be performed, try a different grouping.	

Concepts	Examples
4.2 Factoring Trinomials	
To factor $x^2 + bx + c$, find m and n such that $mn = c$ and $m + n = b$. $mn = c$ $x^2 + bx + c$ $m + n = b$ Then $x^2 + bx + c = (x + m)(x + n)$.	Factor $x^2 + 6x + 8$. $mn = 8$ $x^2 + 6x + 8$ $m + n = 6$ $m = 2$ and $n = 4$ $x^2 + 6x + 8 = (x + 2)(x + 4)$
4.3 More on Factoring Trinomials	
To factor $ax^2 + bx + c$: **By Trial and Error** Use FOIL backwards. **By Grouping** Find m and n: $mn = ac$ $ax^2 + bx + c$ $m + n = b$	Factor $3x^2 + 14x - 5$. $mn = -15$ $m + n = 14$ $m = -1, n = 15$ By trial and error or by grouping, $3x^2 + 14x - 5 = (3x - 1)(x + 5)$.
4.4 Special Factorizations	
$a^2 - b^2 = (a + b)(a - b)$ $a^2 + 2ab + b^2 = (a + b)^2$ $a^2 - 2ab + b^2 = (a - b)^2$	$4x^2 - 9 = (2x + 3)(2x - 3)$ $9x^2 + 6x + 1 = (3x + 1)^2$ $4x^2 - 20x + 25 = (2x - 5)^2$
4.5 Solving Quadratic Equations by Factoring	
Zero-Factor Property If a and b are real numbers and if $ab = 0$, then $a = 0$ or $b = 0$.	If $(x - 2)(x + 3) = 0$, then $x - 2 = 0$ or $x + 3 = 0$.
Solving a Quadratic Equation by Factoring	Solve $2x^2 = 7x + 15$.
1. Write in standard form.	**1.** $2x^2 - 7x - 15 = 0$
2. Factor.	**2.** $(2x + 3)(x - 5) = 0$
3. Use the zero-factor property.	**3.** $2x + 3 = 0$ or $x - 5 = 0$ $2x = -3$ $x = 5$ $x = -\frac{3}{2}$
4. Check.	**4.** Both solutions satisfy the original equation.

Concepts	Examples
4.6 Applications of Quadratic Equations	
Pythagorean Formula In a right triangle, the square of the hypotenuse equals the sum of the squares of the legs. $$a^2 + b^2 = c^2$$ 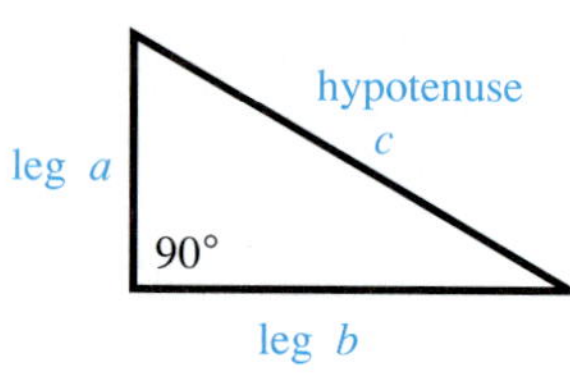	In a right triangle with legs measuring 8 meters and 15 meters, the square of the hypotenuse is equal to $$8^2 + 15^2 = 289$$ Therefore, the hypotenuse measures 17 meters, since $289 = 17^2$.

NAME DATE HOUR

CHAPTER 4 REVIEW EXERCISES

[4.1] *Factor out the greatest common factor or factor by grouping.*

1. $7t + 14$

2. $60z^3 + 30z$

3. $35x^3 + 70x^2$

4. $2xy - 8y + 3x - 12$

5. $100m^2n^3 - 50m^3n^4 + 150m^2n^2$

6. $6y^2 + 9y + 4y + 6$

[4.2] *Factor completely.*

7. $x^2 + 5x + 6$

8. $y^2 - 13y + 40$

9. $q^2 + 6q - 27$

10. $r^2 - r - 56$

11. $r^2 - 4rs - 96s^2$

12. $p^2 + 2pq - 120q^2$

13. $8p^3 - 24p^2 - 80p$

14. $3x^4 + 30x^3 + 48x^2$

15. $m^2 - 3mn - 18n^2$

16. $y^2 - 8yz + 15z^2$

17. $p^7 - p^6q - 2p^5q^2$

18. $3r^5 - 6r^4s - 45r^3s^2$

19. $x^2 + x + 1$

20. $3x^2 + 6x + 6$

[4.3]

21. In order to begin factoring $6r^2 - 5r - 6$, what are the possible first terms of the two binomial factors, if we consider only positive integer coefficients?

22. What is the first step you would use to factor $2z^3 + 9z^2 - 5z$?

Factor completely.

23. $2k^2 - 5k + 2$

24. $3r^2 + 11r - 4$

25. $6r^2 - 5r - 6$

26. $10z^2 - 3z - 1$

27. $8v^2 + 17v - 21$

28. $24x^5 - 20x^4 + 4x^3$

29. $-6x^2 + 3x + 30$

30. $10r^3s + 17r^2s^2 + 6rs^3$

[4.4]

31. Which one of the following is the difference of two squares?
(a) $32x^2 - 1$ **(b)** $4x^2y^2 - 25z^2$ **(c)** $x^2 + 36$ **(d)** $25y^3 - 1$

32. Which one of the following is a perfect square trinomial?
(a) $x^2 + x + 1$ **(b)** $y^2 - 4y + 9$ **(c)** $4x^2 + 10x + 25$ **(d)** $x^2 - 20x + 100$

Factor completely.

33. $n^2 - 49$

34. $25b^2 - 121$

35. $49y^2 - 25w^2$

36. $144p^2 - 36q^2$

37. $x^2 + 100$

38. $z^2 + 10z + 25$

39. $r^2 - 12r + 36$

40. $9t^2 - 42t + 49$

41. $16m^2 + 40mn + 25n^2$

42. $54x^3 - 72x^2 + 24x$

[4.5] *Solve each equation and check the solutions.*

43. $(4t + 3)(t - 1) = 0$

44. $(x + 7)(x - 4)(x + 3) = 0$

45. $x(2x - 5) = 0$

46. $z^2 + 4z + 3 = 0$

47. $m^2 - 5m + 4 = 0$

48. $x^2 = -15 + 8x$

49. $3z^2 - 11z - 20 = 0$

50. $81t^2 - 64 = 0$

51. $y^2 = 8y$

52. $n(n - 5) = 6$

53. $t^2 - 14t + 49 = 0$

54. $t^2 = 12(t - 3)$

55. $(5z + 2)(z^2 + 3z + 2) = 0$

56. $x^2 = 9$

[4.6] *Solve each problem.*

57. The length of a rectangle is 6 meters more than the width. The area is 40 square meters. Find the length and width of the rectangle.

58. The length of a rectangle is three times the width. If the width were increased by 3 meters while the length remained the same, the new rectangle would have an area of 30 square meters. Find the length and width of the original rectangle.

59. The length of a rectangle is 2 centimeters more than the width. The area is numerically 44 more than the perimeter. Find the length and width of the rectangle.

60. The volume of a box is to be 120 cubic meters. The width of the box is to be 4 meters, and the height 1 meter less than the length. Find the length and height of the box.

61. The length of a rectangle is 4 feet more than the width. The area is numerically 1 more than the perimeter. Find the dimensions of the rectangle.

62. The width of a rectangle is 5 inches less than the length. The area is numerically 10 more than the perimeter. Find the dimensions of the rectangle.

63. The product of two consecutive integers is 29 more than their sum. What are the integers?

64. The sides of a right triangle have lengths (in inches) that are consecutive even integers. What are the lengths of the sides?

If an object is thrown straight up with an initial velocity of 128 feet per second, its height h after t seconds is

$$h = 128t - 16t^2.$$

Find the height of the object after the following periods of time.

65. 1 second

66. 2 seconds

67. 4 seconds

68. For the object described above, when does it return to the ground?

69. A 9-inch by 12-inch picture is to be placed on a cardboard mat so that there is an equal border around the picture. The area of the finished mat and picture is to be 208 square inches. How wide will the border be?

70. A box is made from a 12-centimeter by 10-centimeter piece of cardboard by cutting a square from each corner and folding up the sides. The area of the bottom of the box is to be 48 square centimeters. Find the length of a side of the cutout squares.

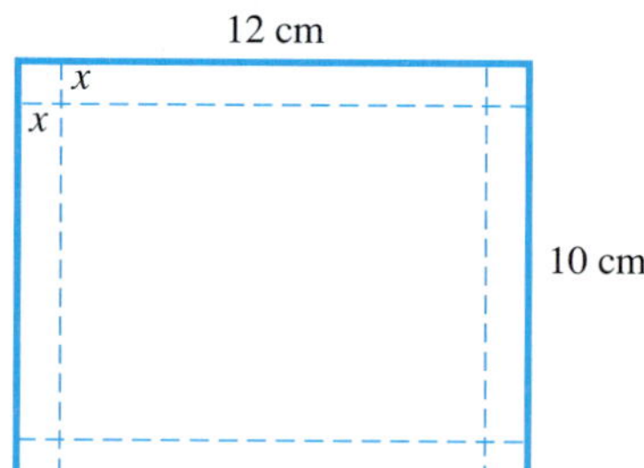

Mixed Review Exercises

Factor completely.

71. $z^2 - 11zx + 10x^2$

72. $3k^2 + 11k + 10$

73. $15m^2 + 20mp - 12mp - 16p^2$

74. $y^4 - 625$

75. $6m^3 - 21m^2 - 45m$

76. $24ab^3c^2 - 56a^2bc^3 + 72a^2b^2c$

77. $25a^2 + 15ab + 9b^2$

78. $12x^2yz^3 + 12xy^2z - 30x^3y^2z^4$

79. $2a^5 - 8a^4 - 24a^3$

80. $12r^2 + 18rq - 10rq - 15q^2$

81. $100a^2 - 9$

82. $49t^2 + 56t + 16$

Solve.

83. $t(t - 7) = 0$

84. $x(x + 3) = 10$

85. $25x^2 + 20x + 4 = 0$

86. A lot is shaped like a right triangle. The hypotenuse is 3 meters longer than the longer leg. The longer leg is 6 meters longer than twice the length of the shorter leg. Find the lengths of the sides of the lot.

87. A pyramid has a rectangular base with a length that is 2 meters more than the width. The height of the pyramid is 6 meters, and its volume is 48 cubic meters. Find the length and width of the base.

88. The product of the smaller two of three consecutive integers is equal to 23 plus the largest. Find the integers.

89. The sum of two consecutive even integers is 34 less than their product. Find the integers.

90. The floor plan for a house is a rectangle with length 7 meters more than its width. The area is 170 square meters. Find the width and length of the house.

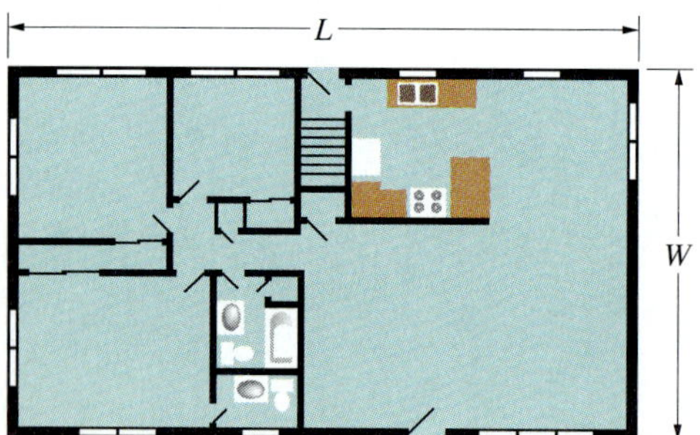

91. The triangular sail of a schooner has an area of 30 square meters. The height of the sail is 4 meters more than the base. Find the base of the sail.

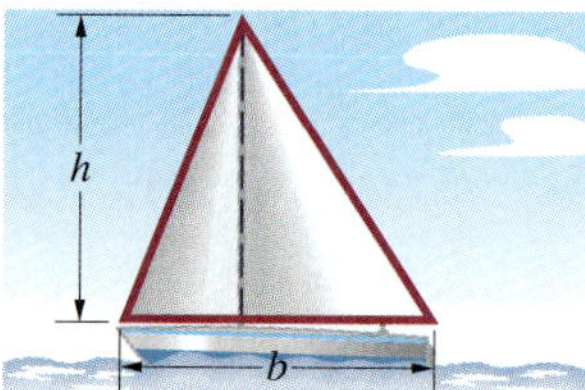

NAME DATE HOUR

CHAPTER 4 TEST

1. Which one of the following is the correct completely factored form of $2x^2 - 2x - 24$?
   (a) $(2x + 6)(x - 4)$ **(b)** $(x + 3)(2x - 8)$
   (c) $2(x + 4)(x - 3)$ **(d)** $2(x + 3)(x - 4)$

Factor each polynomial as completely as possible.

2. $12x^2 - 30x$
3. $2m^3n^2 + 3m^3n - 5m^2n^2$
4. $x^2 - 5x - 24$
5. $x^2 - 9x + 14$
6. $2x^2 + x - 3$
7. $6x^2 - 19x - 7$
8. $3x^2 - 12x - 15$
9. $10z^2 - 17z + 3$
10. $t^2 + 2t + 3$
11. $x^2 + 36$
12. $y^2 - 49$
13. $9y^2 - 64$
14. $x^2 + 16x + 64$
15. $4x^2 - 28xy + 49y^2$
16. $-2x^2 - 4x - 2$
17. $6t^4 + 3t^3 - 108t^2$
18. $4r^2 + 10rt + 25t^2$

1. ______
2. ______
3. ______
4. ______
5. ______
6. ______
7. ______
8. ______
9. ______
10. ______
11. ______
12. ______
13. ______
14. ______
15. ______
16. ______
17. ______
18. ______

19. ______

19. $4t^3 + 32t^2 + 64t$

20. ______

20. $x^4 - 81$

21. ______

21. Why is $(p+3)(p+3)$ *not* the correct factorization of $p^2 + 9$?

Solve each equation.

22. ______

22. $(x + 3)(x - 9) = 0$

23. ______

23. $2r^2 - 13r + 6 = 0$

24. ______

24. $25x^2 - 4 = 0$

25. ______

25. $x(x - 20) = -100$

26. ______

26. $t^2 = 3t$

27. ______

27. Why isn't "$x = \frac{2}{3}$" the correct response to "Solve the equation $x^2 = \frac{4}{9}$"?

Solve each problem.

28. ______

28. The length of a rectangular flower bed is 3 feet less than twice its width. The area of the bed is 54 square feet. Find the dimensions of the flower bed.

29. ______

29. A carpenter needs to cut a brace to support a wall stud, as shown in the figure. The brace should be 7 feet less than three times the length of the stud. If the brace will be anchored on the floor 15 feet away from the stud, how long should the brace be?

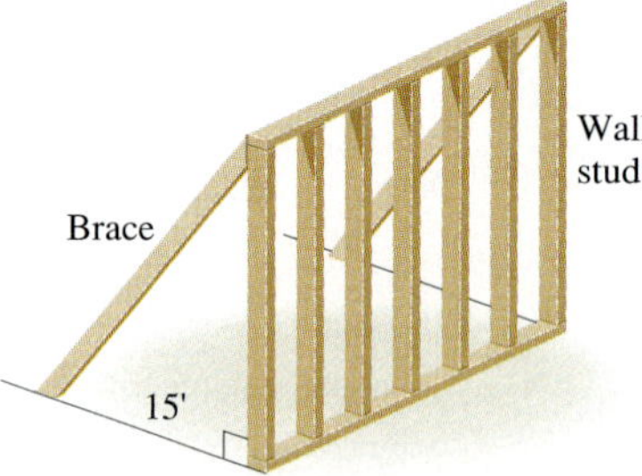

30. ______

30. Find two consecutive integers such that the square of the sum of the two integers is 11 more than the smaller integer.

NAME DATE HOUR

CUMULATIVE REVIEW EXERCISES CHAPTERS R–4

Solve each equation.

1. $3x + 2(x - 4) = 4(x - 2)$

2. $.3x + .9x = .06$

3. $\frac{2}{3}y - \frac{1}{2}(y - 4) = 3$

4. Solve for P: $A = P + Prt$

Solve each problem.

5. Four times a number, added to the sum of the number and twelve, gives a result of 3. Find the number.

6. A pharmacist found that at the end of the day she had $\frac{2}{3}$ as many prescriptions for antibiotics as she had for tranquilizers. If she had 90 prescriptions in all, how many did she have for each type of drug?

7. Louise has 6% of her salary deducted for her daughter's educational fund. After the deduction, her weekly check is \$423. How much does she earn before the deduction?

8. In a mixture of concrete, there are 3 pounds of cement mix for every 1 pound of gravel. If the mixture contains a total of 140 pounds of these two ingredients, how many pounds of gravel are there?

Evaluate each expression.

9. $2^{-3} \cdot 2^5$

10. $\left(\frac{3}{4}\right)^{-2}$

11. $\frac{6^5 \cdot 6^{-2}}{6^3}$

12. $\left(\frac{4^{-3} \cdot 4^4}{4^5}\right)^{-1}$

Simplify each expression and write the answer using only positive exponents. Assume no denominators are zero.

13. $\frac{(p^2)^3 p^{-4}}{(p^{-3})^{-1} p}$

14. $\frac{(m^{-2})^3 m}{m^5 m^{-4}}$

Perform the indicated operations.

15. $(2k^2 + 4k) - (5k^2 - 2) - (k^2 + 8k - 6)$

16. $3m^3(2m^5 - 5m^3 + m)$

17. $(y^2 + 3y + 5)(3y - 1)$

18. $(3p + 2)^2$

19. $(2p + 3q)(2p - 3q)$

20. $(9x + 6)(5x - 3)$

21. $\frac{8x^4 + 12x^3 - 6x^2 + 20x}{2x}$

22. $(12p^3 + 2p^2 - 12p + 4) \div (2p - 2)$

Factor completely.

23. $2a^2 + 7a - 4$

24. $10m^2 + 19m + 6$

25. $15x^2 - xy - 6y^2$

26. $8t^2 + 10tv + 3v^2$

27. $9x^2 + 6x + 1$

28. $4p^2 - 12p + 9$

29. $-32t^2 - 112tz - 98z^2$

30. $25r^2 - 81t^2$

31. $100x^2 + 25$

32. $6a^2m + am - 2m$

33. $2pq + 6p^3q + 8p^2q$

34. $2ax - 2bx + ay - by$

Solve each equation.

35. $(2p - 3)(p + 2)(p - 6) = 0$

36. $6m^2 + m - 2 = 0$

37. $8x^2 = 64x$

Solve each problem.

38. The length of a rectangle is 1 centimeter less than twice the width. The area is 15 square centimeters. Find the width of the rectangle.

39. The sum of the squares of two consecutive even integers is 5 times the larger. What are the two integers?

40. The length of the hypotenuse of a right triangle is twice the length of the shorter leg, plus 3 meters. The longer leg is 7 meters longer than the shorter leg. Find the lengths of the sides.

5 Rational Expressions

5.1 THE FUNDAMENTAL PROPERTY OF RATIONAL EXPRESSIONS

OBJECTIVES

1. Find the values of the variable for which a rational expression is undefined.
2. Find the numerical value of a rational expression.
3. Write rational expressions in lowest terms.

FOR EXTRA HELP

Tape 7

SSM pp. 173–175

MAC: A IBM: A

The quotient of two integers (with denominator not zero) is called a rational number. In the same way, the quotient of two polynomials with denominator not equal to zero is called a **rational expression.** Our work with rational expressions will require much of what we learned in Chapters 3 and 4 on polynomials and factoring, as well as the rules for fractions from Chapter R.

A rational expression is an expression of the form

$$\frac{P}{Q},$$

where P and Q are polynomials, with $Q \neq 0$.

Examples of rational expressions include

$$\frac{-6x}{x^3 + 8}, \quad \frac{9x}{y + 3}, \quad \text{and} \quad \frac{2m^3}{8}.$$

1 A number with a zero denominator is *not* a rational expression because division by zero is undefined. For that reason, be careful when substituting a number in the denominator of a rational expression. For example, in

$$\frac{8x^2}{x - 3}$$

the variable x can take on any value except 3. When $x = 3$ the denominator becomes $3 - 3 = 0$, making the quotient undefined.

EXAMPLE 1 *Finding Values That Make a Rational Expression Undefined*

Find any values of the variable for which the following are undefined.

(a) $\dfrac{p + 5}{3p + 2}$

1. Find all values for which the following rational expressions are undefined.

(a) $\dfrac{x+2}{x-5}$

(b) $\dfrac{3r}{r^2+6r+8}$

(c) $\dfrac{-5m}{m^2+4}$

2. Find the value of each rational expression when $x = 3$.

(a) $\dfrac{x}{2x+1}$

(b) $\dfrac{2x+6}{x-3}$

ANSWERS

1. (a) 5 (b) $-4, -2$ (c) never undefined

2. (a) $\frac{3}{7}$ (b) undefined

This rational expression is undefined for any value of p that makes the denominator equal to zero. We find these values by solving the equation $3p + 2 = 0$ to get

$$3p = -2 \quad \text{or} \quad p = -\frac{2}{3}.$$

Since $p = -\frac{2}{3}$ will make the denominator zero, the given expression is undefined for $-\frac{2}{3}$.

(b) $\dfrac{9m^2}{m^2-5m+6}$

Solve the equation $m^2 - 5m + 6 = 0$.

$$(m-2)(m-3) = 0 \qquad \text{Factor.}$$

$$m - 2 = 0 \quad \text{or} \quad m - 3 = 0 \qquad \text{Set each factor equal to 0.}$$

$$m = 2 \quad \text{or} \quad m = 3$$

The original fraction is undefined for $m = 2$ and for $m = 3$.

(c) $\dfrac{2r}{r^2+1}$

This denominator cannot equal zero for any value of r because r^2 is always greater than or equal to zero, and adding 1 makes the sum greater than zero. Thus, there are no values for which this rational expression is undefined. ■

WORK PROBLEM 1 AT THE SIDE.

2 The next example shows how to find the numerical value of a rational expression for a given value of the variable.

■ EXAMPLE 2 *Evaluating a Rational Expression*

Find the numerical value of $\dfrac{3x+6}{2x-4}$ for each of the following values of x.

(a) $x = 1$

$$\frac{3x+6}{2x-4} = \frac{3(1)+6}{2(1)-4} \qquad \text{Let } x = 1.$$

$$= \frac{9}{-2} = -\frac{9}{2}$$

(b) $x = 2$

$$\frac{3x+6}{2x-4} = \frac{3(2)+6}{2(2)-4} = \frac{6+6}{0} \qquad \text{Let } x = 2.$$

Substituting 2 for x makes the denominator zero, so the quotient is undefined when $x = 2$. ■

WORK PROBLEM 2 AT THE SIDE.

3 A rational expression represents a number for each value of the variable that does not make the denominator zero. For this reason, the properties of rational numbers discussed in Chapter R also apply to rational expressions. For example, the **fundamental property of rational expressions** permits rational expressions to be written in lowest terms. A rational expression is in **lowest terms** when there are no common factors in the numerator and the denominator (except 1).

FUNDAMENTAL PROPERTY OF RATIONAL EXPRESSIONS

If $\frac{P}{Q}$ is a rational expression and if K represents any factor where $K \neq 0$, then

$$\frac{PK}{QK} = \frac{P}{Q}.$$

This property is based on the identity property of multiplication:

$$\frac{PK}{QK} = \frac{P}{Q} \cdot \frac{K}{K} = \frac{P}{Q} \cdot 1 = \frac{P}{Q}.$$

The next example shows how to write both a rational number and a rational expression in lowest terms. Notice the similarity in the procedures.

EXAMPLE 3 *Writing in Lowest Terms*

Write in lowest terms.

(a) $\dfrac{30}{72}$

Begin by factoring.

$$\frac{30}{72} = \frac{2 \cdot 3 \cdot 5}{2 \cdot 2 \cdot 2 \cdot 3 \cdot 3}$$

(b) $\dfrac{14k^2}{2k^3}$

Write k^2 as $k \cdot k$.

$$\frac{14k^2}{2k^3} = \frac{2 \cdot 7 \cdot k \cdot k}{2 \cdot k \cdot k \cdot k}$$

Group any factors common to the numerator and denominator.

$$\frac{30}{72} = \frac{5 \cdot (2 \cdot 3)}{2 \cdot 2 \cdot 3 \cdot (2 \cdot 3)} \qquad \frac{14k^2}{2k^3} = \frac{7(2 \cdot k \cdot k)}{k(2 \cdot k \cdot k)}$$

Use the fundamental property.

$$\frac{30}{72} = \frac{5}{2 \cdot 2 \cdot 3} = \frac{5}{12} \qquad \frac{14k^2}{2k^3} = \frac{7}{k}$$ ■

WORK PROBLEM 3 AT THE SIDE.

EXAMPLE 4 *Writing in Lowest Terms*

Write $\dfrac{3x - 12}{5x - 20}$ in lowest terms.

Begin by factoring both numerator and denominator. Then use the fundamental property.

$$\frac{3x - 12}{5x - 20} = \frac{3(x - 4)}{5(x - 4)} = \frac{3}{5}$$ ■

3. Use the fundamental property to write the following rational expressions in lowest terms.

(a) $\dfrac{5x^4}{15x^2}$

(b) $\dfrac{6p^3}{2p^2}$

ANSWERS

3. (a) $\dfrac{x^2}{3}$ (b) $3p$

4. Write each rational expression in lowest terms.

(a) $\dfrac{4y + 2}{6y + 3}$

(b) $\dfrac{8p + 8q}{5p + 5q}$

5. Write each rational expression in lowest terms.

(a) $\dfrac{x^2 + 4x + 4}{4x + 8}$

(b) $\dfrac{a^2 - b^2}{a^2 + 2ab + b^2}$

ANSWERS

4. (a) $\frac{2}{3}$ (b) $\frac{8}{5}$

5. (a) $\dfrac{x + 2}{4}$ (b) $\dfrac{a - b}{a + b}$

Caution Although x appears in both the numerator and denominator in Example 4, and 12 and 20 have a common factor of 4, the fundamental property cannot be used before factoring because $3x$, $5x$, 12, and 20 are *terms,* not *factors.* Terms are *added* or *subtracted;* factors are *multiplied* or *divided.* For example,

$$\frac{6 + 2}{3 + 2} = \frac{8}{5}, \quad \text{not} \quad \frac{6}{3} + \frac{2}{2} = 2 + 1 = 3.$$

Also, $$\frac{2x + 3}{4x + 6} = \frac{2x + 3}{2(2x + 3)} = \frac{1}{2}, \quad \text{but} \quad \frac{x^2 + 6}{x + 3} \neq x + 2.$$

WORK PROBLEM 4 AT THE SIDE.

EXAMPLE 5 *Writing in Lowest Terms*

Write $\dfrac{m^2 + 2m - 8}{2m^2 - m - 6}$ in lowest terms.

Always begin by factoring both numerator and denominator, if possible. Then use the fundamental property to remove common factors.

$$\frac{m^2 + 2m - 8}{2m^2 - m - 6} = \frac{(m + 4)(m - 2)}{(2m + 3)(m - 2)} = \frac{m + 4}{2m + 3} \quad ■$$

WORK PROBLEM 5 AT THE SIDE.

In Example 4, the rational expression $\dfrac{3x - 12}{5x - 20}$ is restricted to values of x not equal to 4. For this reason

$$\frac{3x - 12}{5x - 20} = \frac{3}{5} \quad \text{for} \quad x \neq 4.$$

Similarly, in Example 5,

$$\frac{m^2 + 2m - 8}{2m^2 - m - 6} = \frac{m + 4}{2m + 3} \quad \text{for} \quad m \neq -\frac{3}{2} \text{ or } 2.$$

From now on we will assume that such restrictions are understood without actually writing them down.

EXAMPLE 6 *Writing in Lowest Terms*

Write $\dfrac{x - y}{y - x}$ in lowest terms.

At first glance, there does not seem to be any way in which $x - y$ and $y - x$ can be factored to get a common factor. However, $y - x$ can be factored as

$$y - x = -1(-y + x) = -1(x - y).$$

With these factors, use the fundamental property to simplify the rational expression.

$$\frac{x - y}{y - x} = \frac{1(x - y)}{-1(x - y)} = \frac{1}{-1} = -1 \quad ■$$

In Example 6, notice that $y - x$ is the negative of $x - y$. A general rule for this situation follows.

A fraction with numerator and denominator that are negatives equals -1.

Caution Although x and y appear in both the numerator and denominator in Example 6, it is not possible to use the fundamental property right away because they are *terms,* not *factors.* Terms are *added,* while factors are *multiplied.*

EXAMPLE 7 *Writing in Lowest Terms*

Write each rational expression in lowest terms.

(a) $\dfrac{2 - m}{m - 2}$

Since $2 - m$ is the negative of $m - 2$ (or $-2 + m$),

$$\frac{2 - m}{m - 2} = -1.$$

(b) $\dfrac{3 + r}{3 - r}$

The quantity $3 - r$ *is not* the negative of $3 + r$. This rational expression cannot be written in simpler form. ■

WORK PROBLEM 6 AT THE SIDE.

6. Write each rational expression in lowest terms.

(a) $\dfrac{5 - y}{y - 5}$

(b) $\dfrac{m - n}{n - m}$

(c) $\dfrac{9 - k}{9 + k}$

ANSWERS

6. (a) -1 **(b)** -1
(c) cannot be written in simpler form

QUEST FOR NUMERACY

The Mathematics of Identification: ISBN Numbers

Identification numbers are used in various ways for many kinds of different products.* Books, for example, are assigned International Standard Book Numbers (ISBNs). Each ISBN is a ten-digit number. It includes a check digit, which is determined on the basis of *modular arithmetic.* The ISBN for *Mathematical Ideas,* seventh edition, is

0-673-46738-4

The first digit, 0, identifies the book as being published in an English-speaking country. The next digits, 673, identify the publisher, while 46738 identifies this particular book. The final digit, 4, is a check digit. To find this check digit, start at the left and multiply the digits of the ISBN number by 10,9,8,7,6,5,4,3, and 2, respectively. Then add these products. For that book we get

$$(10 \times 0) + (9 \times 6) + (8 \times 7) + (7 \times 3) + (6 \times 4)$$
$$+ (5 \times 6) + (4 \times 7) + (3 \times 3) + (2 \times 8) = 238.$$

The check digit is the smallest number that must be added to this result to get a multiple of 11. Because $238 + 4 = 242$, a multiple of 11, the check digit is 4. (It is possible to have a check "digit" of 10; the letter X is used instead of 10.)

When an order for this book is received, the ISBN is entered into a computer, and the check digit evaluated. If this result does not match the check digit on the order, the order will not be processed.

FOR GROUP DISCUSSION

1. Without looking at the cover of this book, determine the correct check digit, knowing that the first nine digits are 0-673-46745. Then, check your work (no pun intended).
2. Find the appropriate check digit for each of the following ISBN numbers.

   The Beauty of Fractals, by H. O. Peitgen and P. H. Richter, 3-540-15851-
   Women in Science, by Vivian Gornick, 0-671-41738-
   Beyond Numeracy, by John Allen Paulos, 0-394-58640-
   Iron John, by Robert Bly, 0-201-51720-

*For an interesting general discussion, see "The Mathematics of Identification Numbers," by Joseph A. Gallian in *The College Mathematics Journal,* May 1991, page 194.

NAME DATE HOUR

5.1 EXERCISES

Find any values for which the following rational expressions are undefined. See Example 1.

1. $\frac{2}{5y}$

2. $\frac{7}{3z}$

3. $\frac{4x^2}{3x - 5}$

4. $\frac{2x^3}{3x - 4}$

5. $\frac{m + 2}{m^2 + m - 6}$

6. $\frac{r - 5}{r^2 - 5r + 4}$

7. $\frac{3x}{x^2 + 2}$

8. $\frac{4q}{q^2 + 9}$

Find the numerical value of each rational expression when (a) $x = 2$ and (b) $x = -3$. See Example 2.

9. $\frac{5x - 2}{4x}$

10. $\frac{3x + 1}{5x}$

11. $\frac{2x^2 - 4x}{3x}$

12. $\frac{4x^2 - 1}{5x}$

13. $\frac{(-3x)^2}{4x + 12}$

14. $\frac{(-2x)^3}{3x + 9}$

15. $\frac{5x + 2}{2x^2 + 11x + 12}$

16. $\frac{7 - 3x}{3x^2 - 7x + 2}$

17. If 2 is substituted for x in the rational expression $\frac{x - 2}{x^2 - 4}$, the result is $\frac{0}{0}$. A commonly heard statement is "Any number divided by itself is 1." Does this mean that this expression is equal to 1 for $x = 2$? If not, explain.

18. For $x \neq 2$, the rational expression $\frac{2(x - 2)}{x - 2}$ is equal to 2. Can the same be said for $\frac{2x - 2}{x - 2}$? Explain.

Write each rational expression in lowest terms. See Examples 3–5.

19. $\frac{18r^3}{6r}$

20. $\frac{27p^2}{3p}$

21. $\frac{4(y - 2)}{10(y - 2)}$

22. $\frac{15(m - 1)}{9(m - 1)}$

23. $\frac{(x + 1)(x - 1)}{(x + 1)^2}$

24. $\frac{(t + 5)(t - 3)}{(t - 1)(t + 5)}$

25. $\frac{7m + 14}{5m + 10}$

26. $\frac{8z - 24}{4z - 12}$

27. $\frac{m^2 - n^2}{m + n}$

28. $\dfrac{a^2 - b^2}{a - b}$

29. $\dfrac{12m^2 - 3}{8m - 4}$

30. $\dfrac{20p^2 - 45}{6p - 9}$

31. $\dfrac{3m^2 - 3m}{5m - 5}$

32. $\dfrac{6t^2 - 6t}{2t - 2}$

33. $\dfrac{9r^2 - 4s^2}{9r + 6s}$

34. $\dfrac{16x^2 - 9y^2}{12x - 9y}$

35. $\dfrac{zw + 4z - 3w - 12}{zw + 4z + 5w + 20}$

36. $\dfrac{km + 4k + 4m + 16}{km + 4k + 5m + 20}$

37. $\dfrac{2x^2 - 3x - 5}{2x^2 - 7x + 5}$

38. $\dfrac{3x^2 + 8x + 4}{3x^2 - 4x - 4}$

Write each rational expression in lowest terms. See Examples 6 and 7.

39. $\dfrac{6 - t}{t - 6}$

40. $\dfrac{2 - k}{k - 2}$

41. $\dfrac{m^2 - 1}{1 - m}$

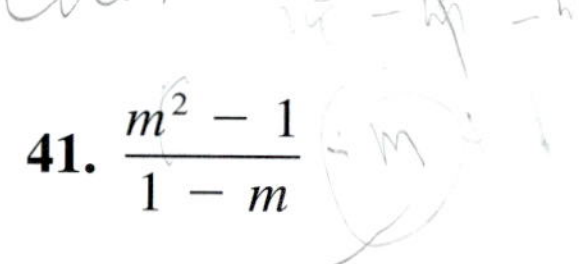

42. $\dfrac{a^2 - b^2}{b - a}$

43. $\dfrac{q^2 - 4q}{4q - q^2}$

44. $\dfrac{z^2 - 5z}{5z - z^2}$

45. Which one of the following rational expressions is *not* equivalent to $\dfrac{4 - 3x}{7}$?

(a) $-\dfrac{-4 + 3x}{7}$ **(b)** $-\dfrac{4 - 3x}{-7}$ **(c)** $\dfrac{-4 + 3x}{-7}$ **(d)** $\dfrac{-(3x + 4)}{7}$

46. One of the following is equal to 1 for *all* real numbers. Which one is it?

(a) $\dfrac{k^2 + 2}{k^2 + 2}$ **(b)** $\dfrac{4 - m}{4 - m}$ **(c)** $\dfrac{2x + 9}{2x + 9}$ **(d)** $\dfrac{x^2 - 1}{x^2 - 1}$

PREVIEW EXERCISES

Multiply or divide as indicated. See Section R.1.

47. $\dfrac{2}{3} \cdot \dfrac{5}{6}$

48. $\dfrac{3}{7} \cdot \dfrac{2}{5}$

49. $\dfrac{6}{15} \cdot \dfrac{25}{3}$

50. $\dfrac{10}{8} \div \dfrac{7}{12}$

51. $\dfrac{10}{3} \div \dfrac{5}{6}$

52. $\dfrac{7}{12} \div \dfrac{15}{4}$

5.2 MULTIPLICATION AND DIVISION OF RATIONAL EXPRESSIONS

1 The product of two fractions is found by multiplying the numerators and multiplying the denominators. Rational expressions are multiplied in the same way.

MULTIPLYING RATIONAL EXPRESSIONS

The product of the rational expressions P/Q and R/S is

$$\frac{P}{Q} \cdot \frac{R}{S} = \frac{PR}{QS}.$$

In words: Multiply the numerators and multiply the denominators.

The next example shows the multiplication of both two rational numbers and two rational expressions. This parallel discussion lets you compare the steps.

EXAMPLE 1 *Multiplying Rational Expressions*

Multiply. Write answers in lowest terms.

(a) $\frac{3}{10} \cdot \frac{5}{9}$ | (b) $\frac{6}{x} \cdot \frac{x^2}{12}$

Find the product of the numerators and the denominators.

$$\frac{3}{10} \cdot \frac{5}{9} = \frac{3 \cdot 5}{10 \cdot 9} \qquad \frac{6}{x} \cdot \frac{x^2}{12} = \frac{6 \cdot x^2}{x \cdot 12}$$

Factor the numerator and denominator to identify any common factors. Then use the fundamental property to write each product in lowest terms.

$$\frac{3}{10} \cdot \frac{5}{9} = \frac{3 \cdot 5}{2 \cdot 5 \cdot 3 \cdot 3} = \frac{1}{6} \qquad \frac{6}{x} \cdot \frac{x^2}{12} = \frac{6 \cdot x \cdot x}{2 \cdot 6 \cdot x} = \frac{x}{2}$$

Notice in the second step above that the products were left in factored form because common factors are needed to write the product in lowest terms. ■

WORK PROBLEM 1 AT THE SIDE.

EXAMPLE 2 *Multiplying Rational Expressions*

Find the product of $\frac{x + y}{2x}$ and $\frac{x^2}{(x + y)^2}$.

Use the definition of multiplication.

$$\frac{x + y}{2x} \cdot \frac{x^2}{(x + y)^2} = \frac{(x + y)x^2}{2x(x + y)^2}$$ Multiply numerators. Multiply denominators.

$$= \frac{(x + y)x \cdot x}{2x(x + y)(x + y)}$$ Factor; identify common factors.

$$= \frac{x}{2(x + y)}$$ Lowest terms ■

WORK PROBLEM 2 AT THE SIDE.

OBJECTIVES

1 Multiply rational expressions.

2 Find reciprocals.

3 Divide rational expressions.

FOR EXTRA HELP

Tape 7

SSM pp. 175–178

MAC: A IBM: A

1. Multiply.

(a) $\frac{3m^2}{2} \cdot \frac{10}{m}$

(b) $\frac{8p^2q}{3} \cdot \frac{9}{q^2p}$

2. Multiply.

(a) $\frac{a + b}{5} \cdot \frac{30}{2(a + b)}$

(b) $\frac{3(p - q)}{p} \cdot \frac{q}{2(p - q)}$

ANSWERS

1. (a) $15m$ (b) $\frac{24p}{q}$

2. (a) 3 (b) $\frac{3q}{2p}$

3. Multiply.

(a)
$$\frac{x^2 + 7x + 10}{3x + 6} \cdot \frac{6x - 6}{x^2 + 2x - 15}$$

(b)
$$\frac{m^2 + 4m - 5}{m + 5} \cdot \frac{m^2 + 8m + 15}{m - 1}$$

4. Find the reciprocal.

(a) $\frac{6b^5}{3r^2b}$

(b) $\frac{t^2 - 4t}{t^2 + 2t - 3}$

ANSWERS

3. (a) $\frac{2(x - 1)}{x - 3}$ (b) $(m + 5)(m + 3)$

4. (a) $\frac{3r^2b}{6b^5}$ (b) $\frac{t^2 + 2t - 3}{t^2 - 4t}$

■ **EXAMPLE 3** *Multiplying Rational Expressions*

Find the product of $\frac{x^2 + 3x}{x^2 - 3x - 4}$ and $\frac{x^2 - 5x + 4}{x^2 + 2x - 3}$.

Use the definition of multiplication. Factor the numerators and denominators wherever possible. Use the fundamental property to write the product in lowest terms.

$$\frac{x^2 + 3x}{x^2 - 3x - 4} \cdot \frac{x^2 - 5x + 4}{x^2 + 2x - 3} = \frac{(x^2 + 3x)(x^2 - 5x + 4)}{(x^2 - 3x - 4)(x^2 + 2x - 3)}$$
$$= \frac{x(x + 3)(x - 4)(x - 1)}{(x - 4)(x + 1)(x + 3)(x - 1)}$$
$$= \frac{x}{x + 1}$$ ■

◀◀ WORK PROBLEM 3 AT THE SIDE.

In Chapter R, we defined the reciprocal of c/d as d/c. To divide by the nonzero fraction c/d, we multiply by the reciprocal of c/d. Division of rational expressions is defined in the same way.

DIVIDING RATIONAL EXPRESSIONS

If P/Q and R/S are any two rational expressions, with $R/S \neq 0$, then

$$\frac{P}{Q} \div \frac{R}{S} = \frac{P}{Q} \cdot \frac{S}{R} = \frac{PS}{QR}.$$

In words: Multiply the first rational expression by the reciprocal of the second rational expression.

2 The reciprocal of a rational expression is found by inverting the fraction. For example, the reciprocal of $\frac{2x - 1}{x - 5}$ is $\frac{x - 5}{2x - 1}$.

■ **EXAMPLE 4** *Find the Reciprocal of Each Rational Expression*

(a) $\frac{4p^3}{9q}$

Invert the rational expression. The reciprocal is $\frac{9q}{4p^3}$.

(b) $\frac{k^2 - 9}{k^2 - k - 20}$

The reciprocal is $\frac{k^2 - k - 20}{k^2 - 9}$. ■

◀◀ WORK PROBLEM 4 AT THE SIDE.

3 The next example shows the division of two rational numbers and the division of two rational expressions.

EXAMPLE 5 *Dividing Rational Expressions*

Divide. Write answers in lowest terms.

(a) $\frac{5}{8} \div \frac{7}{16}$

(b) $\frac{y}{y+3} \div \frac{4y}{y+5}$

Multiply the first expression and the reciprocal of the second.

(a)

$$\frac{5}{8} \div \frac{7}{16} = \frac{5}{8} \cdot \frac{16}{7} \quad \text{Reciprocal of } \frac{7}{16}$$

$$= \frac{5 \cdot 16}{8 \cdot 7}$$

$$= \frac{5 \cdot 8 \cdot 2}{8 \cdot 7}$$

$$= \frac{10}{7}$$

(b)

$$\frac{y}{y+3} \div \frac{4y}{y+5}$$

$$= \frac{y}{y+3} \cdot \frac{y+5}{4y} \quad \text{Reciprocal of } \frac{4y}{y+5}$$

$$= \frac{y(y+5)}{(y+3)(4y)}$$

$$= \frac{y+5}{4(y+3)}$$ ■

WORK PROBLEM 5 AT THE SIDE.

EXAMPLE 6 *Dividing Rational Expressions*

Divide: $\frac{(3m)^2}{(2p)^3} \div \frac{6m^3}{16p^2}$.

$$\frac{(3m)^2}{(2p)^3} \div \frac{6m^3}{16p^2} = \frac{(3m)^2}{(2p)^3} \cdot \frac{16p^2}{6m^3} \quad \text{Multiply by the reciprocal.}$$

$$= \frac{(3m)(3m)}{(2p)(2p)(2p)} \cdot \frac{16p^2}{6m^3} \quad \text{Factor.}$$

$$= \frac{9 \cdot 16m^2p^2}{8 \cdot 6p^3m^3} \quad \text{Multiply numerators. Multiply denominators.}$$

$$= \frac{3}{mp} \quad \text{Lowest terms}$$ ■

WORK PROBLEM 6 AT THE SIDE.

EXAMPLE 7 *Dividing Rational Expressions*

Divide: $\frac{x^2-4}{(x+3)(x-2)} \div \frac{(x+2)(x+3)}{-2x}$.

First, use the definition of division.

$$\frac{x^2-4}{(x+3)(x-2)} \div \frac{(x+2)(x+3)}{-2x}$$

$$= \frac{x^2-4}{(x+3)(x-2)} \cdot \frac{-2x}{(x+2)(x+3)} \quad \text{Multiply by the reciprocal of second expression.}$$

5. Divide.

(a) $\frac{r}{r-1} \div \frac{3r}{r+4}$

(b) $\frac{6x-4}{3} \div \frac{15x-10}{9}$

6. Divide.

(a) $\frac{5a^2b}{2} \div \frac{10ab^2}{8}$

(b) $\frac{(3t)^2}{w} \div \frac{3t^2}{5w^4}$

ANSWERS

5. (a) $\frac{r+4}{3(r-1)}$ (b) $\frac{6}{5}$

6. (a) $\frac{2a}{b}$ (b) $15w^3$

7. Divide.

(a) $$\frac{y^2+4y+3}{y+3} \div \frac{y^2-4y-5}{y-3}$$

(b) $$\frac{4x(x+3)}{2x+1} \div \frac{-x^2(x+3)}{4x^2-1}$$

8. Divide.

(a) $$\frac{ab-a^2}{a^2-1} \div \frac{b-a}{a^2+2a+1}$$

(b) $$\frac{x^2-y^2}{x^2-1} \div \frac{x^2+2xy+y^2}{x^2+x}$$

ANSWERS

7. (a) $\frac{y-3}{y-5}$ (b) $-\frac{4(2x-1)}{x}$

8. (a) $\frac{a(a+1)}{a-1}$ (b) $\frac{x(x-y)}{(x-1)(x+y)}$

Next, be sure all numerators and all denominators are factored.

$$= \frac{(x+2)(x-2)}{(x+3)(x-2)} \cdot \frac{-2x}{(x+2)(x+3)}$$

Now multiply numerators and denominators and simplify.

$$\frac{(x+2)(x-2)}{(x+3)(x-2)} \cdot \frac{-2x}{(x+2)(x+3)}$$
$$= \frac{-2x(x+2)(x-2)}{(x+3)(x-2)(x+2)(x+3)}$$
$$= -\frac{2x}{(x+3)^2} \quad \text{Lowest terms}$$

In the last step, we used the fact that the quotient of a negative number and a positive number is negative to write the negative sign in front of the fraction. ■

WORK PROBLEM 7 AT THE SIDE.

■ **EXAMPLE 8** *Dividing Rational Expressions*

Divide: $\frac{m^2-4}{m^2-1} \div \frac{2m^2+4m}{1-m}$.

$$\frac{m^2-4}{m^2-1} \div \frac{2m^2+4m}{1-m} = \frac{m^2-4}{m^2-1} \cdot \frac{1-m}{2m^2+4m} \quad \text{Definition of division}$$
$$= \frac{(m+2)(m-2)}{(m+1)(m-1)} \cdot \frac{1-m}{2m(m+2)} \quad \text{Factor.}$$

As shown in Section 5.1, $\frac{1-m}{m-1} = -1$, so

$$\frac{(m+2)(m-2)}{(m+1)(m-1)} \cdot \frac{1-m}{2m(m+2)} = \frac{-1(m-2)}{2m(m+1)}$$
$$= \frac{2-m}{2m(m+1)}. \quad -1(m-2) = 2-m$$ ■

WORK PROBLEM 8 AT THE SIDE.

5.2 EXERCISES

NAME DATE HOUR

Multiply. Write answers in lowest terms. See Examples 1 and 2.

1. $\dfrac{15a^2}{14} \cdot \dfrac{7}{5a}$

2. $\dfrac{27k^3}{9k} \cdot \dfrac{24}{9k^2}$

3. $\dfrac{12x^4}{18x^3} \cdot \dfrac{-8x^5}{4x^2}$

4. $\dfrac{12m^5}{-2m^2} \cdot \dfrac{6m^6}{28m^3}$

5. $\dfrac{2(c + d)}{3} \cdot \dfrac{18}{6(c + d)^2}$

6. $\dfrac{4(y - 2)}{x} \cdot \dfrac{3x}{6(y - 2)^2}$

Find the reciprocal. See Example 4.

7. $\dfrac{3p^3}{16q}$

8. $\dfrac{6x^4}{9y^2}$

9. $\dfrac{r^2 + rp}{7}$

10. $\dfrac{16}{9a^2 + 36a}$

11. $\dfrac{z^2 + 7z + 12}{z^2 - 9}$

12. $\dfrac{p^2 - 4p + 3}{p^2 - 3p}$

Divide. Write answers in lowest terms. See Examples 5 and 6.

13. $\dfrac{9z^4}{3z^5} \div \dfrac{3z^2}{5z^3}$

14. $\dfrac{35q^8}{9q^5} \div \dfrac{25q^6}{10q^5}$

15. $\dfrac{4t^4}{2t^5} \div \dfrac{(2t)^3}{-6}$

16. $\dfrac{-12a^6}{3a^2} \div \dfrac{(2a)^3}{27a}$

17. $\dfrac{3}{2y - 6} \div \dfrac{6}{y - 3}$

18. $\dfrac{4m + 16}{10} \div \dfrac{3m + 12}{18}$

19. Explain in your own words how to multiply rational expressions.

20. Explain in your own words how to divide rational expressions.

Multiply or divide. Write answers in lowest terms. See Examples 3, 7, and 8.

21. $\dfrac{5x - 15}{3x + 9} \cdot \dfrac{4x + 12}{6x - 18}$

22. $\dfrac{8r + 16}{24r - 24} \cdot \dfrac{6r - 6}{3r + 6}$

23. $\dfrac{2 - t}{8} \div \dfrac{t - 2}{6}$

24. $\frac{4}{m-2} \div \frac{16}{2-m}$

25. $\frac{27-3z}{4} \cdot \frac{12}{2z-18}$

26. $\frac{5-x}{5+x} \cdot \frac{x+5}{x-5}$

27. $\frac{6(m-2)^2}{5(m+4)^2} \cdot \frac{15(m+4)}{2(2-m)}$

28. $\frac{7(q-1)}{3(q+1)^2} \cdot \frac{6(q+1)}{3(1-q)^2}$

29. $\frac{p^2+4p-5}{p^2+7p+10} \div \frac{p-1}{p+4}$

30. $\frac{z^2-3z+2}{z^2+4z+3} \div \frac{z-1}{z+1}$

31. $\frac{2k^2-k-1}{2k^2+5k+3} \div \frac{4k^2-1}{2k^2+k-3}$

32. $\frac{2m^2-5m-12}{m^2+m-20} \div \frac{4m^2-9}{m^2+4m-5}$

33. $\frac{2k^2+3k-2}{6k^2-7k+2} \cdot \frac{4k^2-5k+1}{k^2+k-2}$

34. $\frac{2m^2-5m-12}{m^2-10m+24} \div \frac{4m^2-9}{m^2-9m+18}$

35. $\frac{m^2+2mp-3p^2}{m^2-3mp+2p^2} \div \frac{m^2+4mp+3p^2}{m^2+2mp-8p^2}$

36. $\frac{r^2+rs-12s^2}{r^2-rs-20s^2} \div \frac{r^2-2rs-3s^2}{r^2+rs-30s^2}$

37. $\left(\frac{x^2+10x+25}{x^2+10x} \cdot \frac{10x}{x^2+15x+50}\right) \div \frac{x+5}{x+10}$

38. $\left(\frac{m^2-12m+32}{8m} \cdot \frac{m^2-8m}{m^2-8m+16}\right) \div \frac{m-8}{m-4}$

PREVIEW EXERCISES

Write the prime factored form of each number. See Section R.1.

39. 18

40. 48

41. 108

42. 60

Find the greatest common factor of each set of terms. See Section 4.1.

43. $24m$, $18m^2$, 6

44. $14x^2$, $28x$, 7

45. $84q^3$, $90q^6$

46. $54b^3$, $36b^4$

5.3 LEAST COMMON DENOMINATORS

1 In this section, we demonstrate a preliminary step needed to add or subtract rational expressions with different denominators. Just as with rational numbers, adding or subtracting rational expressions (to be discussed in the next section) often requires a **least common denominator,** the simplest expression that all denominators divide into without a remainder. For example, the least common denominator for $\frac{2}{9}$ and $\frac{5}{12}$ is 36 because 36 is the smallest number that both 9 and 12 divide into.

Least common denominators often can be found by inspection. For example, the least common denominator for $\frac{1}{6}$ and $\frac{2}{3m}$ is $6m$. In other cases, a least common denominator can be found by a procedure similar to that used in Chapter 4 for finding the greatest common factor.

OBJECTIVES

1. Find least common denominators.
2. Rewrite rational expressions with the least common denominator.

FOR EXTRA HELP

Tape 7	SSM pp. 178–181	MAC: A IBM: A

FINDING A LEAST COMMON DENOMINATOR

1. Factor each denominator into prime factors.
2. List each different denominator factor the *greatest* number of times it appears in any denominator.
3. Multiply the denominators from Step 2 to get the least common denominator.

When each denominator is factored into prime factors, every prime factor must divide evenly into the least common denominator. The least common denominator is often abbreviated LCD.

In Example 1, the least common denominator is found both for numerical denominators and algebraic denominators.

EXAMPLE 1 *Finding the Least Common Denominator*

Find the least common denominator for each pair of fractions.

(a) $\frac{1}{24}, \frac{7}{15}$ **(b)** $\frac{1}{8x}, \frac{3}{10x}$

Write each denominator in factored form, with numerical coefficients in prime factored form.

(a)
$$24 = 2 \cdot 2 \cdot 2 \cdot 3 = 2^3 \cdot 3$$
$$15 = 3 \cdot 5$$

(b)
$$8x = 2 \cdot 2 \cdot 2 \cdot x = 2^3 \cdot x$$
$$10x = 2 \cdot 5 \cdot x$$

The LCD is found by taking each different factor the greatest number of times it appears as a factor in any denominator. That is, each factor must be in the LCD and raised to its highest power.

(a)
$$\text{LCD} = 2 \cdot 2 \cdot 2 \cdot 3 \cdot 5 = 2^3 \cdot 3 \cdot 5 = 120$$

(b)
$$\text{LCD} = 2 \cdot 2 \cdot 2 \cdot 5 \cdot x = 2^3 \cdot 5 \cdot x = 40x$$ ■

WORK PROBLEM 1 AT THE SIDE.

1. Find the least common denominator.

(a) $\frac{7}{20p}, \frac{11}{30p}$

(b) $\frac{9}{8m^4}, \frac{11}{12m^6}$

EXAMPLE 2 *Finding the LCD*

Find the LCD for $\frac{5}{6r^2}$ and $\frac{3}{4r^3}$.

Factor each denominator.

$$6r^2 = 2 \cdot 3 \cdot r^2$$
$$4r^3 = 2^2 \cdot r^3$$

ANSWERS
1. (a) $60p$ (b) $24m^6$

2. Find the LCD.

(a) $\frac{4}{16m^3n}, \frac{5}{9m^5}$

(b) $\frac{3}{25a^2}, \frac{2}{10a^3b}$

3. Find the LCD.

(a) $\frac{7}{3a}, \frac{5}{3a - 10}$

(b) $\frac{1}{12a}, \frac{5}{a^2 - 4a}$

(c) $\frac{2m}{m^2 - 3m + 2}, \frac{5m - 3}{m^2 + 3m - 10}$

(d) $\frac{6}{x - 4}, \frac{3x - 1}{4 - x}$

ANSWERS
2. (a) $144m^5n$ (b) $50a^3b$
3. (a) $3a(3a - 10)$
(b) $12a(a - 4)$
(c) $(m - 1)(m - 2)(m + 5)$
(d) either $x - 4$ or $4 - x$

The highest power of 2 is 2, the highest power of 3 is 1, and the highest power of r is 3; therefore,

$$\text{LCD} = 2^2 \cdot 3^1 \cdot r^3 = 12r^3. \blacksquare$$

WORK PROBLEM 2 AT THE SIDE.

EXAMPLE 3 *Finding the LCD*

Find the LCD.

(a) $\frac{6}{5m}, \frac{4}{m^2 - 3m}$

Factor each denominator.

$$5m = 5 \cdot m$$
$$m^2 - 3m = m(m - 3)$$
$$\text{LCD} = 5 \cdot m \cdot (m - 3) = 5m(m - 3)$$

Because m is not a *factor* of $m - 3$, both factors, m and $m - 3$, must appear in the least common denominator.

(b) $\frac{1}{r^2 - 4r - 5}, \frac{3}{r^2 - r - 20}$

Factor each denominator.

$$r^2 - 4r - 5 = (r - 5)(r + 1)$$
$$r^2 - r - 20 = (r - 5)(r + 4)$$

The LCD is $(r - 5)(r + 1)(r + 4)$.

(c) $\frac{1}{q - 5}, \frac{3}{5 - q}$

The expressions $q - 5$ and $5 - q$ are negatives of each other because

$$-(q - 5) = -q + 5 = 5 - q.$$

Therefore, either $q - 5$ or $5 - q$ can be used as the LCD. ■

WORK PROBLEM 3 AT THE SIDE.

2 Once the least common denominator has been found, the next step in preparing two fractions for addition or subtraction is to use the fundamental property to rewrite each fraction with the least common denominator. The next example shows how to do this with both numerical and algebraic fractions.

EXAMPLE 4 *Writing a Fraction with a Given Denominator*

Rewrite each rational expression with the indicated denominator.

(a) $\frac{3}{8} = \frac{}{40}$

(b) $\frac{9k}{25} = \frac{}{50k}$

For each example, first factor the denominator on the right. Then compare the denominator on the left with the one on the right to decide what factors are missing. (It may be necessary to factor both denominators.)

$$\frac{3}{8} = \frac{}{5 \cdot 8}$$

A factor of 5 is missing. Using the property of 1, multiply $\frac{3}{8}$ by $\frac{5}{5}$.

$$\frac{3}{8} = \frac{3}{8} \cdot \frac{5}{5} = \frac{15}{40}$$

$\frac{5}{5} = 1$

$$\frac{9k}{25} = \frac{}{25 \cdot 2k}$$

Factors of 2 and k are missing. Multiply by $\frac{2k}{2k}$.

$$\frac{9k}{25} = \frac{9k}{25} \cdot \frac{2k}{2k} = \frac{18k^2}{50k}$$

$\frac{2k}{2k} = 1$ ■

■ EXAMPLE 5 *Writing a Fraction with a Given Denominator*

Rewrite the following rational expression with the indicated denominator.

$$\frac{12p}{p^2 + 8p} = \frac{}{p(p + 8)(p - 4)}$$

Factor $p^2 + 8p$ as $p(p + 8)$. Compare with the denominator on the right. The factor $p - 4$ is missing, so multiply $\frac{12p}{p(p + 8)}$ by $\frac{p - 4}{p - 4}$.

$$\frac{12p}{p^2 + 8p} = \frac{12p}{p(p + 8)} \cdot \frac{p - 4}{p - 4} = \frac{12p(p - 4)}{p(p + 8)(p - 4)}$$ ■

WORK PROBLEM 4 AT THE SIDE. ▶▶

4. Rewrite each rational expression with the indicated denominator.

(a) $\frac{7k}{5} = \frac{}{30p}$

(b) $\frac{9}{2a + 5} = \frac{}{6a + 15}$

(c) $\frac{5k + 1}{k^2 + 2k} = \frac{}{k(k + 2)(k - 1)}$

ANSWERS

4. (a) $\frac{42kp}{30p}$ (b) $\frac{27}{6a + 15}$ (c) $\frac{(5k + 1)(k - 1)}{k(k + 2)(k - 1)}$

QUEST FOR NUMERACY

Snap, Crackle, Pop . . . Your Age on a Cereal Box

Recently the Kellogg's Company has included on its Rice Krispies boxes a variation of an old number trick, billed with the title "Age Detector Magic Trick." The following rectangular arrays of numbers are found on the back of the box, with directions to cut out the six cards.

32	37	42	47	52	57
33	38	43	48	53	58
34	39	44	49	54	59
35	40	45	50	55	60
36	41	46	51	56	★

1	11	21	31	41	51
3	13	23	33	43	53
5	15	25	35	45	55
7	17	27	37	47	57
9	19	29	39	49	59

2	11	22	31	42	51
3	14	23	34	43	54
6	15	26	35	46	55
7	18	27	38	47	58
10	19	30	39	50	59

16	21	26	31	52	57
17	22	27	48	53	58
18	23	28	49	54	59
19	24	29	50	55	60
20	25	30	51	56	★

4	13	22	31	44	53
5	14	23	36	45	54
6	15	28	37	46	55
7	20	29	38	47	60
12	21	30	39	52	★★

8	13	26	31	44	57
9	14	27	40	45	58
10	15	28	41	46	59
11	24	29	42	47	60
12	25	30	43	56	★

Cut along dotted lines ✂

You, the performer, are to spread the six cards on a table face up, and someone in the audience is chosen to think of their age (from one to sixty). The person is then to hand you the cards with his or her age on them. Immediately you can guess the person's age.

For example, if you are handed the top left and bottom right cards, you can immediately tell the person his or her age: 40. How is it done? Simply add the numbers in the upper left hand corners of the cards you are handed. In the example given, you would add 32 + 8 to get 40.

This "trick" works because of a property of the binary number system—a system that uses two as its base (rather than ten, as in our decimal system).

* "Age Detector Math Problem" as appeared on Kellogg's RICE KRISPIES. Reprinted by permission of Kellogg's.

5.3 EXERCISES

NAME DATE HOUR

1. What is the least common denominator for $\frac{7}{15}$, $\frac{11}{21}$, and $\frac{5}{24}$?

2. What is the least common denominator for $\dfrac{2}{m^5}$ and $\dfrac{3}{m^8}$?

Find the least common denominator for each list. See Examples 1–3.

3. $\dfrac{2}{15}, \dfrac{3}{10}, \dfrac{7}{30}$

4. $\dfrac{5}{24}, \dfrac{7}{12}, \dfrac{9}{28}$

5. $\dfrac{3}{x^4}, \dfrac{5}{x^7}$

6. $\dfrac{2}{y^5}, \dfrac{3}{y^6}$

7. $\dfrac{5}{36q}, \dfrac{17}{24q}$

8. $\dfrac{4}{30p}, \dfrac{9}{50p}$

9. $\dfrac{6}{21r^3}, \dfrac{8}{12r^5}$

10. $\dfrac{9}{35t^2}, \dfrac{5}{49t^6}$

11. If two denominators have greatest common factor equal to 1, how can you easily find their least common denominator?

12. Suppose two fractions have denominators a^k and a^r, where k and r are natural numbers, with $k > r$. What is their least common denominator?

Find the least common denominator for each pair. See Examples 1–3.

13. $\dfrac{9}{28m^2}, \dfrac{3}{12m - 20}$

14. $\dfrac{15}{27a^3}, \dfrac{8}{9a - 45}$

15. $\dfrac{7}{5b - 10}, \dfrac{11}{6b - 12}$

16. $\dfrac{3}{7x^2 + 21x}, \dfrac{1}{5x^2 + 15x}$

17. $\dfrac{5}{c - d}, \dfrac{8}{d - c}$

18. $\dfrac{4}{y - x}, \dfrac{7}{x - y}$

19. $\dfrac{3}{k^2 + 5k}, \dfrac{2}{k^2 + 3k - 10}$

20. $\dfrac{1}{z^2 - 4z}, \dfrac{4}{z^2 - 3z - 4}$

21. $\dfrac{5}{p^2 + 8p + 15}, \dfrac{3}{p^2 - 3p - 18}$

22. $\dfrac{10}{y^2 - 10y + 21}, \dfrac{2}{y^2 - 2y - 3}$

Rewrite each rational expression with the given denominator. See Examples 4 and 5.

23. $\frac{4}{11} = \frac{}{55}$

24. $\frac{6}{7} = \frac{}{42}$

25. $\frac{-5}{k} = \frac{}{9k}$

26. $\frac{-3}{q} = \frac{}{6q}$

27. $\frac{13}{40y} = \frac{}{80y^3}$

28. $\frac{5}{27p} = \frac{}{108p^4}$

29. $\frac{5t^2}{6r} = \frac{}{42r^4}$

30. $\frac{8y^2}{3x} = \frac{}{30x^3}$

31. $\frac{5}{2(m + 3)} = \frac{}{8(m + 3)}$

32. $\frac{7}{4(y - 1)} = \frac{}{16(y - 1)}$

33. $\frac{-4t}{3t - 6} = \frac{}{6t - 12}$

34. $\frac{-7k}{5k + 20} = \frac{}{15k + 60}$

35. $\frac{14}{z^2 - 3z} = \frac{}{z(z - 3)(z - 2)}$

36. $\frac{12}{x(x + 4)} = \frac{}{x(x + 4)(x - 9)}$

37. $\frac{2(b - 1)}{b^2 + b} = \frac{}{b^3 + 3b^2 + 2b}$

38. $\frac{3(c + 2)}{c(c - 1)} = \frac{}{c^3 - 5c^2 + 4c}$

PREVIEW EXERCISES

Add or subtract as indicated. See Section R.1.

39. $\frac{3}{4} + \frac{7}{4}$

40. $\frac{2}{5} + \frac{9}{5}$

41. $\frac{1}{2} + \frac{7}{8}$

42. $\frac{2}{3} + \frac{8}{27}$

43. $\frac{7}{5} - \frac{3}{4}$

44. $\frac{11}{6} - \frac{2}{5}$

45. $\frac{4}{3} - \frac{1}{4}$

46. $\frac{7}{8} - \frac{10}{3}$

5.4 ADDITION AND SUBTRACTION OF RATIONAL EXPRESSIONS

We are now ready to add and subtract rational expressions. We will need the skills developed in the previous section to find least common denominators and to write fractions with the least common denominator.

1 We find the sum of two rational expressions with a procedure similar to the one that we used for adding two fractions in Chapter R.

ADDING RATIONAL EXPRESSIONS

If P/Q and R/Q are rational expressions, then

$$\frac{P}{Q} + \frac{R}{Q} = \frac{P + R}{Q}.$$

Again, the first example shows how the addition of rational expressions compares with that of rational numbers.

EXAMPLE 1 *Adding Rational Expressions with the Same Denominator*

Add.

(a) $\frac{4}{7} + \frac{2}{7}$ | (b) $\frac{3x}{x+1} + \frac{2x}{x+1}$

The denominators are the same, so the sum is found by adding the two numerators and keeping the same (common) denominator.

$$\frac{4}{7} + \frac{2}{7} = \frac{4+2}{7} = \frac{6}{7}$$

$$\frac{3x}{x+1} + \frac{2x}{x+1} = \frac{3x+2x}{x+1} = \frac{5x}{x+1}$$ ■

WORK PROBLEM 1 AT THE SIDE.

2 We use the steps given below to add two rational expressions with different denominators. These are the same steps that we used to add fractions with different denominators in Chapter R.

ADDING RATIONAL EXPRESSIONS WITH DIFFERENT DENOMINATORS

Step 1 Find the least common denominator (LCD).

Step 2 Rewrite each rational expression as an equivalent fraction with the least common denominator as the denominator.

Step 3 Add the numerators to get the numerator of the sum. The least common denominator is the denominator of the sum.

Step 4 Write the answer in lowest terms.

OBJECTIVES

1. Add rational expressions having the same denominator.
2. Add rational expressions having different denominators.
3. Subtract rational expressions.

FOR EXTRA HELP

Tape 8

SSM pp. 181–186

MAC: A IBM: A

1. Find each sum.

(a) $\frac{3}{y+4} + \frac{2}{y+4}$

(b) $\frac{x}{x+y} + \frac{1}{x+y}$

(c) $\frac{a}{a+b} + \frac{b}{a+b}$

ANSWERS

1. (a) $\frac{5}{y+4}$ (b) $\frac{x+1}{x+y}$ (c) 1

2. Find each sum.

(a) $\frac{6}{5x} + \frac{9}{2x}$

(b) $\frac{m}{3n} + \frac{2}{7n}$

3. Find the sums.

(a) $\frac{2p}{3p+3} + \frac{5p}{2p+2}$

(b) $\frac{4}{y^2-1} + \frac{6}{y+1}$

(c) $\frac{-2}{p+1} + \frac{4p}{p^2-1}$

ANSWERS

2. (a) $\frac{57}{10x}$ (b) $\frac{7m+6}{21n}$

3. (a) $\frac{19p}{6(p+1)}$ (b) $\frac{2(3y-1)}{(y+1)(y-1)}$

(c) $\frac{2}{p-1}$

EXAMPLE 2 *Adding Rational Expressions with Different Denominators*

Add.

(a) $\frac{1}{12} + \frac{7}{15}$

First find the LCD.

$$\text{LCD} = 2^2 \cdot 3 \cdot 5 = 60$$

(b) $\frac{2}{3y} + \frac{1}{4y}$

$$\text{LCD} = 2^2 \cdot 3 \cdot y = 12y$$

Now rewrite each rational expression as a fraction with the LCD, either 60 or $12y$, as the denominator.

$$\frac{1}{12} + \frac{7}{15} = \frac{1}{12} \cdot \frac{5}{5} + \frac{7}{15} \cdot \frac{4}{4} = \frac{5}{60} + \frac{28}{60}$$

$$\frac{2}{3y} + \frac{1}{4y} = \frac{2}{3y} \cdot \frac{4}{4} + \frac{1}{4y} \cdot \frac{3}{3} = \frac{8}{12y} + \frac{3}{12y}$$

Since the fractions now have common denominators, add the numerators, and use the LCD as the denominator of the sum. Write in lowest terms.

$$\frac{5}{60} + \frac{28}{60} = \frac{5+28}{60} = \frac{33}{60} = \frac{11}{20}$$

$$\frac{8}{12y} + \frac{3}{12y} = \frac{8+3}{12y} = \frac{11}{12y}$$ ■

WORK PROBLEM 2 AT THE SIDE.

EXAMPLE 3 *Adding Rational Expressions*

Add $\frac{2x}{x^2-1}$ and $\frac{-1}{x+1}$.

Find the least common denominator by factoring both denominators.

$$x^2 - 1 = (x+1)(x-1); \qquad x + 1 \text{ cannot be factored.}$$

Write the sum with denominators in factored form as

$$\frac{2x}{(x+1)(x-1)} + \frac{-1}{x+1}.$$

The LCD is $(x+1)(x-1)$. Here only the second fraction must be rewritten. Multiply the numerator and denominator of the second fraction by $x - 1$.

$$\frac{2x}{(x+1)(x-1)} + \frac{-1(x-1)}{(x+1)(x-1)} \qquad \text{Multiply by } \frac{x-1}{x-1}.$$

With both denominators now the same, add the numerators and use the LCD as the denominator of the sum.

$$\frac{2x - 1(x-1)}{(x+1)(x-1)} = \frac{2x - x + 1}{(x+1)(x-1)} \qquad \text{Add numerators.}$$

$$= \frac{x+1}{(x+1)(x-1)} \qquad \text{Combine terms.}$$

$$= \frac{1}{x-1} \qquad \text{Lowest terms}$$ ■

WORK PROBLEM 3 AT THE SIDE.

EXAMPLE 4 *Adding Rational Expressions*

Add $\dfrac{2x}{x^2 + 5x + 6}$ and $\dfrac{x + 1}{x^2 + 2x - 3}$.

$$\frac{2x}{(x + 2)(x + 3)} + \frac{x + 1}{(x + 3)(x - 1)} \qquad \text{Factor denominators.}$$

The LCD is $(x + 2)(x + 3)(x - 1)$. By the fundamental property,

$$\frac{2x}{(x + 2)(x + 3)} + \frac{x + 1}{(x + 3)(x - 1)} = \frac{2x(x - 1)}{(x + 2)(x + 3)(x - 1)} + \frac{(x + 1)(x + 2)}{(x + 3)(x - 1)(x + 2)}.$$

Since the two rational expressions now have the same denominator, add their numerators. The LCD is the denominator of the sum.

$$\frac{2x(x - 1)}{(x + 2)(x + 3)(x - 1)} + \frac{(x + 1)(x + 2)}{(x + 3)(x - 1)(x + 2)}$$

$$= \frac{2x(x - 1) + (x + 1)(x + 2)}{(x + 2)(x + 3)(x - 1)} \qquad \text{Add numerators.}$$

$$= \frac{2x^2 - 2x + x^2 + 3x + 2}{(x + 2)(x + 3)(x - 1)} \qquad \text{Distributive property}$$

$$= \frac{3x^2 + x + 2}{(x + 2)(x + 3)(x - 1)} \qquad \text{Combine terms.}$$

It is usually more convenient to leave the denominator in factored form. The numerator cannot be factored here, so the expression is in lowest terms. ■

WORK PROBLEM 4 AT THE SIDE.

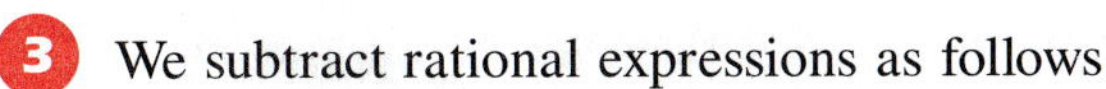

3 We subtract rational expressions as follows.

SUBTRACTING RATIONAL EXPRESSIONS

If P/Q and R/Q are rational expressions, then

$$\frac{P}{Q} - \frac{R}{Q} = \frac{P - R}{Q}.$$

We will not show a parallel subtraction problem from arithmetic because the steps for subtraction are essentially the same as for addition.

EXAMPLE 5 *Subtracting Rational Expressions*

Subtract: $\dfrac{2m}{m - 1} - \dfrac{m + 3}{m - 1}$.

$$\frac{2m}{m - 1} - \frac{m + 3}{m - 1} = \frac{2m - (m + 3)}{m - 1} \qquad \text{Subtract numerators.}$$

$$= \frac{2m - m - 3}{m - 1} \qquad \text{Distributive property}$$

$$= \frac{m - 3}{m - 1} \qquad \text{Combine terms. ■}$$

4. Add.

(a) $\dfrac{2k}{k^2 - 5k + 4} + \dfrac{3}{k^2 - 1}$

(b) $\dfrac{4m}{m^2 + 3m + 2} + \dfrac{2m - 1}{m^2 + 6m + 5}$

ANSWERS

4. (a) $\dfrac{(2k - 3)(k + 4)}{(k - 4)(k - 1)(k + 1)}$
(b) $\dfrac{6m^2 + 23m - 2}{(m + 2)(m + 1)(m + 5)}$

5. Find each difference. Write answers in lowest terms.

(a) $\frac{3}{m^2} - \frac{2}{m^2}$

(b) $\frac{x}{2x + 3} - \frac{3x + 4}{2x + 3}$

6. Subtract.

(a) $\frac{1}{k + 4} - \frac{2}{k}$

(b) $\frac{6}{a + 2} - \frac{1}{a - 3}$

7. Subtract.

(a) $\frac{5}{x - 1} - \frac{3x}{1 - x}$

(b) $\frac{2y}{y - 2} - \frac{1 + y}{2 - y}$

ANSWERS

5. (a) $\frac{1}{m^2}$ (b) $\frac{-2(x + 2)}{2x + 3}$

6. (a) $\frac{-k - 8}{k(k + 4)}$ (b) $\frac{5(a - 4)}{(a + 2)(a - 3)}$

7. (a) $\frac{5 + 3x}{x - 1}$ (b) $\frac{3y + 1}{y - 2}$

WORK PROBLEM 5 AT THE SIDE.

Caution Sign errors often occur in subtraction problems like the one in Example 5. Remember that the numerator of the fraction being subtracted must be treated as a single quantity. Be sure to use parentheses after the subtraction sign to avoid this common error.

EXAMPLE 6 *Subtracting Rational Expressions*

Subtract: $\frac{9}{x - 2} - \frac{3}{x}$.

The LCD is $x(x - 2)$.

$$\frac{9}{x - 2} - \frac{3}{x} = \frac{9x}{x(x - 2)} - \frac{3(x - 2)}{x(x - 2)} \quad \text{Get least common denominator.}$$

$$= \frac{9x - 3(x - 2)}{x(x - 2)} \quad \text{Subtract numerators; keep same denominator.}$$

$$= \frac{9x - 3x + 6}{x(x - 2)} \quad \text{Distributive property}$$

$$= \frac{6x + 6}{x(x - 2)} \quad \text{Combine terms in the numerator.}$$

$$= \frac{6(x + 1)}{x(x - 2)} \quad \text{Factor.}$$

■

WORK PROBLEM 6 AT THE SIDE.

EXAMPLE 7 *Subtracting Rational Expressions*

Subtract: $\frac{3x}{x - 5} - \frac{2x - 25}{5 - x}$.

The denominators are negatives, so either may be used as the common denominator. Let us choose $x - 5$.

$$\frac{3x}{x - 5} - \frac{2x - 25}{5 - x} = \frac{3x}{x - 5} - \frac{2x - 25}{5 - x} \cdot \frac{-1}{-1} \quad \text{Fundamental property}$$

$$= \frac{3x}{x - 5} - \frac{-2x + 25}{x - 5} \quad \text{Multiply.}$$

$$= \frac{3x - (-2x + 25)}{x - 5} \quad \text{Subtract numerators.}$$

$$= \frac{3x + 2x - 25}{x - 5} \quad \text{Distributive property}$$

$$= \frac{5x - 25}{x - 5} \quad \text{Combine terms.}$$

$$= \frac{5(x - 5)}{x - 5} \quad \text{Factor.}$$

$$= 5 \quad \text{Lowest terms}$$

■

WORK PROBLEM 7 AT THE SIDE.

EXAMPLE 8 *Subtracting Rational Expressions*

Find $\dfrac{6x}{x^2 - 2x + 1} - \dfrac{1}{x^2 - 1}$.

Begin by factoring the denominators.

$$x^2 - 2x + 1 = (x - 1)(x - 1) \quad \text{and} \quad x^2 - 1 = (x - 1)(x + 1)$$

From the factored denominators, identify the LCD, $(x - 1)(x - 1)(x + 1)$. Use the factor $x - 1$ twice because it appears twice in the first denominator.

$$\frac{6x}{(x-1)(x-1)} - \frac{1}{(x-1)(x+1)}$$

$$= \frac{6x(x+1)}{(x-1)(x-1)(x+1)} - \frac{1(x-1)}{(x-1)(x-1)(x+1)} \quad \text{Fundamental property}$$

$$= \frac{6x(x+1) - 1(x-1)}{(x-1)(x-1)(x+1)} \quad \text{Subtract numerators.}$$

$$= \frac{6x^2 + 6x - x + 1}{(x-1)(x-1)(x+1)} \quad \text{Distributive property}$$

$$= \frac{6x^2 + 5x + 1}{(x-1)(x-1)(x+1)} \quad \text{Combine terms.}$$

$$= \frac{(3x+1)(2x+1)}{(x-1)^2(x+1)} \quad \text{Factor.}$$

WORK PROBLEM 8 AT THE SIDE.

EXAMPLE 9 *Subtracting Rational Expressions*

Find $\dfrac{q}{q^2 - 4q - 5} - \dfrac{3}{2q^2 - 13q + 15}$.

To find the LCD, factor each denominator.

$$q^2 - 4q - 5 = (q + 1)(q - 5)$$
$$2q^2 - 13q + 15 = (q - 5)(2q - 3)$$

The LCD is $(q + 1)(q - 5)(2q - 3)$. Rewrite each of the two rational expressions with the LCD, using the fundamental property.

$$\frac{q}{(q+1)(q-5)} - \frac{3}{(q-5)(2q-3)}$$

$$= \frac{q(2q-3)}{(q+1)(q-5)(2q-3)} - \frac{3(q+1)}{(q+1)(q-5)(2q-3)}$$

$$= \frac{q(2q-3) - 3(q+1)}{(q+1)(q-5)(2q-3)} \quad \text{Subtract numerators.}$$

$$= \frac{2q^2 - 3q - 3q - 3}{(q+1)(q-5)(2q-3)} \quad \text{Distributive property}$$

$$= \frac{2q^2 - 6q - 3}{(q+1)(q-5)(2q-3)} \quad \text{Combine terms.}$$

WORK PROBLEM 9 AT THE SIDE.

8. Subtract.

(a) $\dfrac{4y}{y^2 - 1} - \dfrac{5}{y^2 + 2y + 1}$

(b) $\dfrac{3r}{r^2 - 5r} - \dfrac{4}{r^2 - 10r + 25}$

9. Subtract.

(a) $\dfrac{2}{p^2 - 5p + 4} - \dfrac{3}{p^2 - 1}$

(b) $\dfrac{q}{2q^2 + 5q - 3} - \dfrac{3q + 4}{3q^2 + 10q + 3}$

ANSWERS

8. (a) $\dfrac{4y^2 - y + 5}{(y+1)^2(y-1)}$ (b) $\dfrac{3r - 19}{(r-5)^2}$

9. (a) $\dfrac{14 - p}{(p-4)(p-1)(p+1)}$

(b) $\dfrac{(-3q + 2)(q + 2)}{(2q - 1)(q + 3)(3q + 1)}$

QUEST FOR NUMERACY

Numbers Don't Lie . . . or Do They?

In baseball statistics, a player's "batting average" gives the average number of hits per time at bat. For example a player who has gotten 84 hits for 250 times at bat has a batting average of 84/250 = .336. This "average" can be interpreted as the empirical probability of that player's getting a hit the next time at bat.

The following are actual comparisons of hits and at-bats for two major league players in the 1989 and 1990 seasons. The numbers illustrate a puzzling statistical occurrence known as **Simpson's paradox.** (This information was reported by Richard J. Friedlander on page 845 of the November 1992 issue of the *MAA Journal.*)

	Dave Justice			*Andy Van Slyke*		
	Hits	*At-bats*	*Batting Average*	*Hits*	*At-bats*	*Batting Average*
1989	12	51	_____	113	476	_____
1990	124	439	_____	140	493	_____
Combined (1989–90)	_____	_____	_____	_____	_____	_____

FOR GROUP DISCUSSION

1. Fill in the ten blanks in the table, giving batting averages to three decimal places.
2. Which player had a better average in 1989?
3. Which player had a better average in 1990?
4. Which player had a better average in 1989 and 1990 combined?
5. Did the results above surprise you? How can it be that one player's batting average leads another's for each of two years, and yet trails the other's for the combined years?

5.4 EXERCISES

NAME DATE HOUR

1. Give the steps used to add rational expressions with the same denominators.

2. Give the steps used to subtract rational expressions with the same denominators.

Add or subtract as indicated. Write each answer in lowest terms. See Examples 1 and 5.

3. $\frac{4}{m} + \frac{7}{m}$

4. $\frac{5}{p} + \frac{11}{p}$

5. $\frac{a + b}{2} - \frac{a - b}{2}$

6. $\frac{x - y}{2} - \frac{x + y}{2}$

7. $\frac{x^2}{x + 5} + \frac{5x}{x + 5}$

8. $\frac{t^2}{t - 3} + \frac{-3t}{t - 3}$

9. $\frac{y^2 - 3y}{y + 3} + \frac{-18}{y + 3}$

10. $\frac{r^2 - 8r}{r - 5} + \frac{15}{r - 5}$

11. Explain how to add rational expressions with different denominators.

12. Explain how to subtract rational expressions with different denominators.

Add or subtract as indicated. Write each answer in lowest terms. See Examples 2–4 and 6–9.

13. $\frac{z}{5} + \frac{1}{3}$

14. $\frac{p}{8} + \frac{3}{5}$

15. $\frac{5}{7} - \frac{r}{2}$

16. $\frac{10}{9} - \frac{z}{3}$

17. $-\frac{3}{4} - \frac{1}{2x}$

18. $-\frac{5}{8} - \frac{3}{2a}$

19. $\frac{5 + 5k}{4} + \frac{1 + k}{8}$

20. $\frac{6 - 5r}{9} - \frac{2 - 3r}{6}$

21. $\frac{b + 3}{b} + \frac{b + 7}{3b}$

$\frac{3b+9+b+7}{3b}$ $\frac{4b+16}{3b}$

22. $\frac{3q - 1}{q} + \frac{q + 2}{4q}$

23. $\frac{7}{3p^2} - \frac{2}{p}$

24. $\frac{12}{5m^2} - \frac{2}{m}$

25. $\dfrac{1}{p-2} - \dfrac{3}{p}$

26. $\dfrac{2}{k-4} - \dfrac{1}{k}$

27. $\dfrac{4}{x-5} + \dfrac{6}{5-x}$

28. $\dfrac{10}{m-2} + \dfrac{5}{2-m}$

29. $\dfrac{-1}{1-y} - \dfrac{3}{y-1}$

30. $\dfrac{-4}{p-3} - \dfrac{7}{3-p}$

31. $\dfrac{6}{c-2} - \dfrac{8}{c+2}$

32. $\dfrac{5}{r-3} - \dfrac{2}{r+3}$

33. $\dfrac{2m}{m-n} - \dfrac{5m+n}{2m-2n}$

34. $\dfrac{5p}{p-q} - \dfrac{3p+1}{4p-4q}$

35. $\dfrac{-2}{x^2-4} + \dfrac{7}{4x+8}$

36. $\dfrac{-4}{z^2-16} + \dfrac{3}{2z+8}$

37. $\dfrac{1}{a^2-1} - \dfrac{a-1}{a^2+3a-4}$

38. $\dfrac{5}{x^2-9} - \dfrac{x+2}{x^2+4x+3}$

39. $\dfrac{8}{m-2} + \dfrac{3}{5m} + \dfrac{7}{5m(m-2)}$

40. $\dfrac{-1}{7z} + \dfrac{3}{z+2} + \dfrac{4}{7z(z+2)}$

41. $\dfrac{4y-1}{2y^2+5y-3} - \dfrac{y+3}{6y^2+y-2}$

42. $\dfrac{2q+1}{3q^2+10q-8} - \dfrac{3q+5}{2q^2+5q-12}$

PREVIEW EXERCISES

Simplify each expression using the order of operations as necessary. See Sections R.1 and 1.1.

43. $\dfrac{\frac{5}{6}+\frac{7}{6}}{\frac{2}{3}-\frac{1}{3}}$

44. $\dfrac{\frac{3}{8}-\frac{5}{8}}{\frac{1}{4}+\frac{7}{4}}$

45. $\dfrac{\frac{3}{2}-\frac{5}{4}}{\frac{7}{4}+\frac{1}{3}}$

46. $\dfrac{\frac{5}{7}-\frac{3}{14}}{\frac{5}{3}+\frac{1}{2}}$

5.5 COMPLEX FRACTIONS

A rational expression with fractions in the numerator, denominator, or both, is called a **complex fraction.** Examples of complex fractions include

$$\frac{3 + \frac{4}{x}}{5}, \quad \frac{\frac{3x^2 - 5x}{6x^2}}{2x - \frac{1}{x}}, \quad \text{and} \quad \frac{3x + x}{5 - \frac{2}{x}}.$$

The parts of a complex fraction are named as follows.

$$\frac{\frac{2}{p} - \frac{1}{q}}{\frac{3}{p} + \frac{5}{q}}$$

← Numerator of complex fraction
← Main fraction bar (indicates division)
← Denominator of complex fraction

Complex fractions can always be simplified to rational expressions without fractions in the numerator and denominator. Two methods are commonly used to do this. We show both methods in this section.

OBJECTIVES

1. Simplify complex fractions by simplifying numerator and denominator (Method 1).
2. Simplify complex fractions by multiplying by the least common denominator (Method 2).

FOR EXTRA HELP

Tape 8 | SSM pp. 186–188 | MAC: A IBM: A

1 **Method 1** Because a fraction represents a quotient, one method of simplifying complex fractions is to rewrite both the numerator and denominator as single fractions, and then to perform the indicated division.

■ **EXAMPLE 1** *Simplifying Complex Fractions by Method 1*

Simplify each complex fraction.

(a) $\dfrac{\frac{2}{3} + \frac{5}{9}}{\frac{1}{4} + \frac{1}{12}}$ **(b)** $\dfrac{6 + \frac{3}{x}}{\frac{x}{4} + \frac{1}{8}}$

First, write each numerator as a single fraction.

(a) $$\frac{2}{3} + \frac{5}{9} = \frac{2(3)}{3(3)} + \frac{5}{9} = \frac{6}{9} + \frac{5}{9} = \frac{11}{9}$$

(b) $$6 + \frac{3}{x} = \frac{6}{1} + \frac{3}{x} = \frac{6x}{x} + \frac{3}{x} = \frac{6x + 3}{x}$$

Do the same thing with each denominator.

(a) $$\frac{1}{4} + \frac{1}{12} = \frac{1(3)}{4(3)} + \frac{1}{12} = \frac{3}{12} + \frac{1}{12} = \frac{4}{12} = \frac{1}{3}$$

(b) $$\frac{x}{4} + \frac{1}{8} = \frac{x(2)}{4(2)} + \frac{1}{8} = \frac{2x}{8} + \frac{1}{8} = \frac{2x + 1}{8}$$

The original complex fraction can now be rewritten as follows.

(a) $$\frac{\frac{11}{9}}{\frac{1}{3}}$$

(b) $$\frac{\frac{6x + 3}{x}}{\frac{2x + 1}{8}}$$

1. Simplify the complex fractions.

(a) $\dfrac{6+\dfrac{1}{x}}{5-\dfrac{2}{x}}$

(b) $\dfrac{9-\dfrac{4}{p}}{\dfrac{2}{p}+1}$

2. Simplify the complex fractions.

(a) $\dfrac{\dfrac{rs^2}{t}}{\dfrac{r^2s}{t^2}}$

(b) $\dfrac{\dfrac{m^2n^3}{p}}{\dfrac{m^4n}{p^2}}$

3. Simplify the complex fraction $\dfrac{\dfrac{2}{x-1}+\dfrac{1}{x+1}}{\dfrac{3}{x-1}-\dfrac{4}{x+1}}$.

ANSWERS

1. (a) $\dfrac{6x+1}{5x-2}$ (b) $\dfrac{9p-4}{2+p}$

2. (a) $\dfrac{st}{r}$ (b) $\dfrac{n^2p}{m^2}$

3. $\dfrac{3x+1}{-x+7}$

Now use the rule for division and the fundamental property.

$$\frac{11}{9}\div\frac{1}{3}=\frac{11}{9}\cdot\frac{3}{1}=\frac{11\cdot 3}{3\cdot 3\cdot 1}=\frac{11}{3}$$

$$\frac{6x+3}{x}\div\frac{2x+1}{8}=\frac{6x+3}{x}\cdot\frac{8}{2x+1}=\frac{3(2x+1)}{x}\cdot\frac{8}{2x+1}=\frac{24}{x}$$ ■

WORK PROBLEM 1 AT THE SIDE.

■ EXAMPLE 2 *Simplifying a Complex Fraction by Method 1*

Simplify the complex fraction $\dfrac{\dfrac{xp}{q^3}}{\dfrac{p^2}{qx^2}}$.

Here the numerator and denominator are already single fractions, so use the rule for division and then the fundamental property.

$$\frac{xp}{q^3}\div\frac{p^2}{qx^2}=\frac{xp}{q^3}\cdot\frac{qx^2}{p^2}=\frac{x^3}{q^2p}$$ ■

WORK PROBLEM 2 AT THE SIDE.

■ EXAMPLE 3 *Simplifying a Complex Fraction by Method 1*

Simplify $\dfrac{\dfrac{3}{x+2}-4}{\dfrac{2}{x+2}+1}$.

$$\frac{\dfrac{3}{x+2}-4}{\dfrac{2}{x+2}+1}=\frac{\dfrac{3}{x+2}-\dfrac{4(x+2)}{x+2}}{\dfrac{2}{x+2}+\dfrac{1(x+2)}{x+2}}$$ Write both second terms with a denominator of $x+2$.

$$=\frac{\dfrac{3-4(x+2)}{x+2}}{\dfrac{2+1(x+2)}{x+2}}$$ Subtract in the numerator. Add in the denominator.

$$=\frac{\dfrac{3-4x-8}{x+2}}{\dfrac{2+x+2}{x+2}}$$ Distributive property

$$=\frac{\dfrac{-5-4x}{x+2}}{\dfrac{4+x}{x+2}}$$ Combine terms.

$$=\frac{-5-4x}{x+2}\cdot\frac{x+2}{4+x}$$ Multiply by the reciprocal.

$$=\frac{-5-4x}{4+x}$$ Lowest terms ■

WORK PROBLEM 3 AT THE SIDE.

2 **Method 2** As an alternative method, complex fractions may be simplified by multiplying both numerator and denominator by the least common denominator of all the denominators appearing in the complex fraction. In the next example, this second method is used to simplify the same complex fractions as in Example 1.

EXAMPLE 4 *Simplifying Complex Fractions by Method 2*

Simplify each complex fraction.

(a) $\dfrac{\frac{2}{3}+\frac{5}{9}}{\frac{1}{4}+\frac{1}{12}}$ **(b)** $\dfrac{6+\frac{3}{x}}{\frac{x}{4}+\frac{1}{8}}$

Find the least common denominator for all the denominators in the complex fraction.

The LCD for 3, 9, 4, and 12 is 36. | The LCD for x, 4, and 8 is $8x$.

Multiply the numerator and denominator of the complex fraction by the LCD. Then use the distributive property to simplify and combine terms. Write the answers in lowest terms.

$$\frac{\frac{2}{3}+\frac{5}{9}}{\frac{1}{4}+\frac{1}{12}} = \frac{36\left(\frac{2}{3}+\frac{5}{9}\right)}{36\left(\frac{1}{4}+\frac{1}{12}\right)} = \frac{36\left(\frac{2}{3}\right)+36\left(\frac{5}{9}\right)}{36\left(\frac{1}{4}\right)+36\left(\frac{1}{12}\right)} = \frac{24+20}{9+3} = \frac{44}{12} = \frac{11}{3}$$

$$\frac{6+\frac{3}{x}}{\frac{x}{4}+\frac{1}{8}} = \frac{8x\left(6+\frac{3}{x}\right)}{8x\left(\frac{x}{4}+\frac{1}{8}\right)} = \frac{8x(6)+8x\left(\frac{3}{x}\right)}{8x\left(\frac{x}{4}\right)+8x\left(\frac{1}{8}\right)} = \frac{48x+24}{2x^2+x} = \frac{24(2x+1)}{x(2x+1)} = \frac{24}{x}$$

EXAMPLE 5 *Simplifying a Complex Fraction by Method 2*

Simplify $\dfrac{\frac{3}{m+5}}{\frac{9}{m+2}}$.

The LCD is $(m+5)(m+2)$. Multiply the numerator and denominator of the complex fraction by the LCD.

$$\frac{\frac{3}{m+5}}{\frac{9}{m+2}} \cdot \frac{(m+5)(m+2)}{(m+5)(m+2)} = \frac{3(m+2)}{9(m+5)} \quad \text{Fundamental property}$$

$$= \frac{m+2}{3(m+5)} \quad \text{Lowest terms}$$

WORK PROBLEM 4 AT THE SIDE.

4. Simplify by the second method.

(a) $\dfrac{2-\frac{6}{a}}{3+\frac{4}{a}}$

(b) $\dfrac{\frac{5}{p}-6}{\frac{2p+1}{p}}$

(c) $\dfrac{\frac{-4}{3+x}}{\frac{5}{2-x}}$

ANSWERS

4. (a) $\dfrac{2a-6}{3a+4}$ (b) $\dfrac{5-6p}{2p+1}$ (c) $\dfrac{4(x-2)}{5(3+x)}$

The two methods for simplifying a complex fraction are summarized below.

SIMPLIFYING COMPLEX FRACTIONS

Method 1 Simplify the numerator and denominator of the complex fraction separately. Then divide the simplified numerator by the simplified denominator.

Method 2 Multiply numerator and denominator of the complex fraction by the least common denominator of all the denominators appearing in the complex fraction.

You may want to choose one method of simplifying complex fractions and stick to it. Although either method can be used correctly to simplify any complex fraction, some students prefer to use Method 1 for problems like Example 2 and Example 5, which are the quotient of two fractions. They find Method 2 works best for problems like Example 1 (or Example 4), which has a sum or difference in the numerator or denominator or both.

5.5 EXERCISES

NAME DATE HOUR

1. In a fraction, what operation does the fraction bar represent?

2. What property of real numbers justifies Method 2 for simplifying complex fractions?

Simplify each complex fraction. Use either method. See Examples 1–5.

3. $\dfrac{-\frac{4}{3}}{\frac{2}{9}}$

4. $\dfrac{-\frac{5}{6}}{\frac{5}{4}}$

5. $\dfrac{\frac{p}{q^2}}{\frac{p^2}{q}}$

6. $\dfrac{\frac{a}{x}}{\frac{a^2}{2x}}$

7. $\dfrac{\frac{x}{y^2}}{\frac{x^2}{y}}$

8. $\dfrac{\frac{p^4}{r}}{\frac{p^2}{r^2}}$

9. $\dfrac{\frac{4a^4b^3}{3a}}{\frac{2ab^4}{b^2}}$

10. $\dfrac{\frac{2r^4t^2}{3t}}{\frac{5r^2t^5}{3r}}$

11. $\dfrac{\frac{m+2}{3}}{\frac{m-4}{m}}$

12. $\dfrac{\frac{q-5}{q}}{\frac{q+5}{3}}$

13. $\dfrac{\frac{2}{x}-3}{\frac{2-3x}{2}}$

14. $\dfrac{6+\frac{2}{r}}{\frac{3r+1}{4}}$

15. $\dfrac{\frac{1}{x}+x}{\frac{x^2+1}{8}}$

16. $\dfrac{\frac{3}{m}-m}{\frac{3-m^2}{4}}$

17. $\dfrac{a-\frac{5}{a}}{a+\frac{1}{a}}$

18. $\dfrac{q+\frac{1}{q}}{q+\frac{4}{q}}$

19. $\dfrac{\dfrac{1}{2}+\dfrac{1}{p}}{\dfrac{2}{3}+\dfrac{1}{p}}$

20. $\dfrac{\dfrac{3}{4}-\dfrac{1}{r}}{\dfrac{1}{5}+\dfrac{1}{r}}$

21. $\dfrac{\dfrac{t}{t+2}}{\dfrac{4}{t^2-4}}$

22. $\dfrac{\dfrac{m}{m+1}}{\dfrac{3}{m^2-1}}$

23. $\dfrac{\dfrac{1}{k+1}-1}{\dfrac{1}{k+1}+1}$

24. $\dfrac{\dfrac{2}{p-1}+2}{\dfrac{3}{p-1}-2}$

25. $\dfrac{\dfrac{1}{m-1}+\dfrac{2}{m+2}}{\dfrac{2}{m+2}-\dfrac{1}{m-3}}$

26. $\dfrac{\dfrac{5}{r+3}-\dfrac{1}{r-1}}{\dfrac{2}{r+2}+\dfrac{3}{r+3}}$

27. $2-\dfrac{2}{2+\dfrac{2}{2+2}}$

28. $3-\dfrac{2}{4+\dfrac{2}{4-2}}$

PREVIEW EXERCISES

Solve each equation. See Section 2.3.

29. $4x-5=8x+3$

30. $6k-1=3k+2$

31. $5(2q+1)-2=8q$

32. $4(p-2)+3=6p$

33. $6-(4-3y)=9$

34. $-(7-r)+3r=-12$

5.6 EQUATIONS INVOLVING RATIONAL EXPRESSIONS

OBJECTIVES

1. Solve equations involving rational expressions.
2. Solve a formula for a specified variable.

FOR EXTRA HELP

Tape	SSM	MAC: A
8	pp. 188–194	IBM: A

1 When an equation involves fractions, the multiplication property of equality can be used first to clear it of fractions. The equation can then be solved in the usual way. The goal is to get an equivalent equation (one with the same solutions) that does not have fractions. Choose as multiplier the least common denominator of all denominators in the fractions of the equation.

EXAMPLE 1 *Solving an Equation Involving Rational Expressions*

Solve $\frac{x}{3} + \frac{x}{4} = 10 + x$.

Because the least common denominator of the two fractions is 12, begin by multiplying each side of the equation by 12.

$$12\left(\frac{x}{3} + \frac{x}{4}\right) = 12(10 + x)$$

$$12\left(\frac{x}{3}\right) + 12\left(\frac{x}{4}\right) = 12(10) + 12x \qquad \text{Distributive property}$$

$$\frac{12x}{3} + \frac{12x}{4} = 120 + 12x$$

$$4x + 3x = 120 + 12x$$

This equation has no fractions. Solve it using the methods given earlier for solving linear equations.

$$7x = 120 + 12x \qquad \text{Combine terms.}$$

$$-5x = 120 \qquad \text{Subtract } 12x.$$

$$x = -24 \qquad \text{Divide by } -5.$$

Check: $\frac{-24}{3} + \frac{-24}{4} = 10 - 24 \quad ? \qquad \text{Let } x = -24.$

$$-8 - 6 = -14 \qquad \text{True}$$

WORK PROBLEM 1 AT THE SIDE.

1. Solve each equation and check your answer.

(a) $\frac{x}{5} + 3 = \frac{3}{5}$

(b) $\frac{x}{2} - \frac{x}{3} = \frac{5}{6}$

Caution Note that use of the LCD here is different from its use in the previous section. Here, we use the Multiplication Property of Equality to multiply each side of an *equation* by the LCD. Earlier, we used the property of 1 to multiply a *fraction* by another fraction that had the LCD as both its numerator and denominator. Be careful not to confuse these two methods.

EXAMPLE 2 *Solving an Equation Involving Rational Expressions*

Solve $\frac{p}{2} - \frac{p-1}{3} = 1$.

$$6\left(\frac{p}{2} - \frac{p-1}{3}\right) = 6 \cdot 1 \qquad \text{Multiply by the LCD, 6.}$$

$$6\left(\frac{p}{2}\right) - 6\left(\frac{p-1}{3}\right) = 6 \qquad \text{Distributive property}$$

$$3p - 2(p - 1) = 6$$

Be very careful to put parentheses around $p - 1$; otherwise you may find an incorrect solution. Continue simplifying and solve.

ANSWERS
1. (a) −12 (b) 5

2. Solve each equation, and check your answer.

(a) $\frac{k}{6} - \frac{k+1}{4} = -\frac{1}{2}$

(b) $\frac{2m-3}{5} - \frac{m}{3} = -\frac{6}{5}$

3. Solve $1 - \frac{2}{x+1} = \frac{2x}{x+1}$ and check your answer.

ANSWERS

2. (a) 3 (b) -9

3. When the equation is solved, -1 is found. However, because $x = -1$ leads to a 0 denominator in the original equation, there is no solution.

$$3p - 2p + 2 = 6 \quad \text{Distributive property}$$
$$p + 2 = 6 \quad \text{Combine terms.}$$
$$p = 4 \quad \text{Subtract 2.}$$

Check to see that 4 is correct by replacing p with 4 in the original equation. ■

WORK PROBLEM 2 AT THE SIDE.

In solving equations that have a variable in the denominator, remember that the number 0 cannot be used as a denominator. Therefore, the solution cannot be a number that will make the denominator equal 0.

EXAMPLE 3 *Solving an Equation Involving Rational Expressions*

Solve $\frac{x}{x-2} = \frac{2}{x-2} + 2$.

The common denominator is $x - 2$. Because $x = 2$ makes a denominator 0, x cannot equal 2. Solve the equation by multiplying each side of the equation by $x - 2$.

$$(x-2)\left(\frac{x}{x-2}\right) = (x-2)\left(\frac{2}{x-2} + 2\right)$$
$$(x-2)\left(\frac{x}{x-2}\right) = (x-2)\left(\frac{2}{x-2}\right) + (x-2)(2)$$
$$x = 2 + 2x - 4$$
$$x = -2 + 2x \quad \text{Combine terms.}$$
$$-x = -2 \quad \text{Subtract } 2x.$$
$$x = 2 \quad \text{Divide by } -1.$$

The proposed solution is 2. However, as shown above, 2 cannot be a solution because 2 makes a denominator equal 0. *The equation has no solution.* (Equations with no solutions are one of the main reasons that it is important to always check proposed solutions.) ■

WORK PROBLEM 3 AT THE SIDE.

In summary, solve equations with rational expressions as follows.

SOLVING EQUATIONS WITH RATIONAL EXPRESSIONS

Step 1 Multiply each side of the equation by the least common denominator. (This clears the equation of fractions.)

Step 2 Solve the resulting equation.

Step 3 Check each proposed solution by substituting it in the original equation.

EXAMPLE 4 *Solving an Equation Involving Rational Expressions*

Solve $\frac{2m}{m^2-4} + \frac{1}{m-2} = \frac{2}{m+2}$.

Multiply by the LCD, $(m+2)(m-2)$.

$$(m+2)(m-2)\left(\frac{2m}{m^2-4} + \frac{1}{m-2}\right) = (m+2)(m-2)\frac{2}{m+2}$$
$$(m+2)(m-2)\frac{2m}{m^2-4} + (m+2)(m-2)\frac{1}{m-2}$$
$$= (m+2)(m-2)\frac{2}{m+2}$$

$$2m + m + 2 = 2(m - 2)$$
$$3m + 2 = 2m - 4 \quad \text{Distributive property; combine terms.}$$
$$m + 2 = -4 \quad \text{Subtract } 2m.$$
$$m = -6 \quad \text{Subtract 2.}$$

Check to see that -6 is indeed a valid solution for the given equation. ■

WORK PROBLEMS 4 AND 5 AT THE SIDE.

■ EXAMPLE 5 *Solving an Equation Involving Rational Expressions*

Solve $\dfrac{2}{x^2 - x} = \dfrac{1}{x^2 - 1}$.

Begin by finding a least common denominator. Since $x^2 - x = x(x - 1)$, and $x^2 - 1 = (x + 1)(x - 1)$, the LCD is $x(x + 1)(x - 1)$. Multiply each side of the equation by the LCD.

$$x(x + 1)(x - 1)\frac{2}{x(x - 1)} = x(x + 1)(x - 1)\frac{1}{(x + 1)(x - 1)}$$
$$2(x + 1) = x$$
$$2x + 2 = x \quad \text{Distributive property}$$
$$x + 2 = 0 \quad \text{Subtract } x.$$
$$x = -2 \quad \text{Subtract 2.}$$

To be sure that $x = -2$ is a solution, substitute -2 for x in the original equation. Since -2 satisfies the equation, the solution is -2. ■

WORK PROBLEM 6 AT THE SIDE.

■ EXAMPLE 6 *Solving an Equation Involving Rational Expressions*

Solve $\dfrac{1}{x - 1} + \dfrac{1}{2} = \dfrac{2}{x^2 - 1}$.

The least common denominator is $2(x + 1)(x - 1)$. Multiply each side of the equation by the LCD, $2(x + 1)(x - 1)$.

$$2(x + 1)(x - 1)\left(\frac{1}{x - 1} + \frac{1}{2}\right) = 2(x + 1)(x - 1)\frac{2}{(x + 1)(x - 1)}$$
$$2(x + 1)(x - 1)\frac{1}{x - 1} + 2(x + 1)(x - 1)\frac{1}{2}$$
$$= 2(x + 1)(x - 1)\frac{2}{(x + 1)(x - 1)}$$
$$2(x + 1) + (x + 1)(x - 1) = 4$$
$$2x + 2 + x^2 - 1 = 4 \quad \text{Distributive property}$$
$$x^2 + 2x + 1 = 4 \quad \text{Combine terms.}$$
$$x^2 + 2x - 3 = 0 \quad \text{Subtract 4.}$$
$$(x + 3)(x - 1) = 0 \quad \text{Factor.}$$

Solving this equation suggests that $x = -3$ or $x = 1$. But 1 makes a denominator of the original equation equal 0, so 1 is not a solution. However, -3 is a solution, as shown by substituting -3 for x in the original equation.

4. Check -6 as a solution to Example 4. Is the solution correct?

5. Solve each equation, and check your answer.

(a) $\dfrac{2p}{p^2 - 1} = \dfrac{2}{p + 1} - \dfrac{1}{p - 1}$

(b) $\dfrac{8r}{4r^2 - 1} = \dfrac{3}{2r + 1} + \dfrac{3}{2r - 1}$

6. Solve each equation, and check your answer.

(a) $\dfrac{4}{3m + 3} = \dfrac{m + 1}{m^2 + m}$

(b) $\dfrac{2}{p^2 - 2p} = \dfrac{3}{p^2 - p}$

ANSWERS

4. yes

5. (a) -3 **(b)** 0

6. (a) 3 **(b)** 4

$$\frac{1}{x-1}+\frac{1}{2}=\frac{2}{x^2-1}$$

$$\frac{1}{-3-1}+\frac{1}{2}\stackrel{?}{=}\frac{2}{(-3)^2-1} \quad \text{Let } x=-3.$$

$$\frac{1}{-4}+\frac{1}{2}\stackrel{?}{=}\frac{2}{9-1} \quad \text{Simplify.}$$

$$\frac{1}{4}=\frac{1}{4} \quad \text{True}$$

The check shows that -3 is a solution. ■

■ **EXAMPLE 7** *Solving an Equation Involving Rational Expressions*

Solve $\dfrac{1}{k^2+4k+3}+\dfrac{1}{2k+2}=\dfrac{3}{4k+12}$.

Factor the three denominators to get the common denominator, $4(k+1)(k+3)$. Multiply each side by this product.

$$4(k+1)(k+3)\left(\frac{1}{(k+1)(k+3)}+\frac{1}{2(k+1)}\right)=4(k+1)(k+3)\frac{3}{4(k+3)}$$

$$4(k+1)(k+3)\frac{1}{(k+1)(k+3)}+2\cdot 2(k+1)(k+3)\frac{1}{2(k+1)}=4(k+1)(k+3)\frac{3}{4(k+3)}$$

$$4+2(k+3)=3(k+1) \quad \text{Simplify.}$$

$$4+2k+6=3k+3 \quad \text{Distributive property}$$

$$2k+10=3k+3 \quad \text{Combine terms.}$$

$$7=k \quad \text{Subtract } 2k \text{ and } 3.$$

Check to see that 7 actually is a solution for the given equation. ■

WORK PROBLEM 7 AT THE SIDE.

2 Solving a formula for a specified variable was discussed in Chapter 2. In the next example this process is applied to formulas with fractions.

■ **EXAMPLE 8** *Solving for a Specified Variable*

Solve the formula $S=\dfrac{a(r^n-1)}{r-1}$ for a.

We need to isolate a on one side of the equation.

$$S=\frac{a(r^n-1)}{r-1}$$

$$(r-1)S=(r-1)\frac{a(r^n-1)}{r-1} \quad \text{Multiply each side by the LCD, } r-1.$$

$$(r-1)S=a(r^n-1)$$

$$\frac{(r-1)S}{r^n-1}=a \quad \text{Divide each side by } r^n-1.$$

The equation is now solved for a. ■

WORK PROBLEM 8 AT THE SIDE.

7. Solve each equation and check your answer.

(a) $\dfrac{1}{x-2}+\dfrac{1}{5}=\dfrac{2}{5(x^2-4)}$

(b) $\dfrac{6}{5a+10}-\dfrac{1}{a-5}=\dfrac{4}{a^2-3a-10}$

8. Solve each equation for the specified variable.

(a) $z=\dfrac{x}{x+y}$ for y

(b) $a=\dfrac{v-w}{t}$ for v

ANSWERS

7. (a) $-4, -1$ **(b)** 60

8. (a) $y=\dfrac{x-zx}{z}$ **(b)** $v=at+w$

5.6 EXERCISES

NAME DATE HOUR

1. Which one of the following numbers satisfies $\frac{y-1}{2} - \frac{y-3}{4} = 1$?

   (a) 9 (b) 0 (c) 6 (d) 3

2. Which one of the following numbers satisfies $\frac{r+5}{3} - \frac{r-1}{4} = \frac{7}{4}$?

   (a) $-\frac{61}{4}$ (b) $-\frac{85}{4}$ (c) -2 (d) 4

Solve each equation, and check your answers. See Examples 1–3.

3. $\frac{5}{m} - \frac{3}{m} = 8$

4. $\frac{4}{y} + \frac{1}{y} = 2$

5. $\frac{5}{y} + 4 = \frac{2}{y}$

6. $\frac{11}{q} = 3 - \frac{1}{q}$

7. $\frac{p}{3} - \frac{p}{6} = 4$

8. $\frac{x}{15} + \frac{x}{5} = 4$

9. $\frac{3x}{5} - 6 = x$

10. $\frac{5t}{4} + t = 9$

11. $\frac{4m}{7} + m = 11$

12. $a - \frac{3a}{2} = 1$

13. $\frac{z-1}{4} = \frac{z+3}{3}$

14. $\frac{r-5}{2} = \frac{r+2}{3}$

15. $\frac{3p+6}{8} = \frac{3p-3}{16}$

16. $\frac{2z+1}{5} = \frac{7z+5}{15}$

17. $\frac{2x+3}{x} = \frac{3}{2}$

18. $\frac{5 - 2y}{y} = \frac{1}{4}$

19. $\frac{k}{k - 4} - 5 = \frac{4}{k - 4}$

20. $\frac{-5}{a + 5} = \frac{a}{a + 5} + 2$

21. $\frac{q + 2}{3} + \frac{q - 5}{5} = \frac{7}{3}$

22. $\frac{b + 7}{8} - \frac{b - 2}{3} = \frac{4}{3}$

23. $\frac{t}{6} + \frac{4}{3} = \frac{t - 2}{3}$

24. $\frac{x}{2} = \frac{5}{4} + \frac{x - 1}{4}$

25. $\frac{3m}{5} - \frac{3m - 2}{4} = \frac{1}{5}$

26. $\frac{8p}{5} = \frac{3p - 4}{2} + \frac{5}{2}$

27. What values of x cannot be solutions of the equation $\frac{1}{x - 4} = \frac{3}{2x}$?

28. What is wrong with the following problem? "Solve $\frac{2}{3x} + \frac{1}{5x}$."

Solve each equation, and check your answers. See Examples 3–7.

29. $\frac{3}{x - 1} + \frac{2}{4x - 4} = \frac{7}{4}$

30. $\frac{2}{p + 3} + \frac{3}{8} = \frac{5}{4p + 12}$

31. $\frac{y}{3y + 3} = \frac{2y - 3}{y + 1} - \frac{2y}{3y + 3}$

32. $\frac{2k + 3}{k + 1} - \frac{3k}{2k + 2} = \frac{-2k}{2k + 2}$

33. $\dfrac{2}{m} = \dfrac{m}{5m + 12}$

34. $\dfrac{x}{4 - x} = \dfrac{2}{x}$

35. $\dfrac{-2}{z + 5} + \dfrac{3}{z - 5} = \dfrac{20}{z^2 - 25}$

36. $\dfrac{3}{r + 3} - \dfrac{2}{r - 3} = \dfrac{-12}{r^2 - 9}$

37. $\dfrac{3y}{y^2 + 5y + 6} = \dfrac{5y}{y^2 + 2y - 3} - \dfrac{2}{y^2 + y - 2}$

38. $\dfrac{x + 4}{x^2 - 3x + 2} - \dfrac{5}{x^2 - 4x + 3} = \dfrac{x - 4}{x^2 - 5x + 6}$

39. $\dfrac{5x}{14x + 3} = \dfrac{1}{x}$

40. $\dfrac{m}{8m + 3} = \dfrac{1}{3m}$

41. $\dfrac{2}{z - 1} - \dfrac{5}{4} = \dfrac{-1}{z + 1}$

42. $\dfrac{5}{p - 2} = 7 - \dfrac{10}{p + 2}$

Solve each formula for the specified variable. See Example 8.

43. $m = \frac{kF}{a}$ for F

44. $I = \frac{kE}{R}$ for E

45. $m = \frac{kF}{a}$ for a

46. $I = \frac{kE}{R}$ for R

47. $I = \frac{E}{R + r}$ for R

48. $I = \frac{E}{R + r}$ for r

49. $h = \frac{2A}{B + b}$ for A

50. $d = \frac{2S}{n(a + L)}$ for S

51. $d = \frac{2S}{n(a + L)}$ for a

52. $h = \frac{2A}{B + b}$ for B

PREVIEW EXERCISES

Write a mathematical expression for each of the following. See Section 2.4 and the list of formulas in the endsheets.

53. Linda drives 288 miles from Philadelphia to Pittsburgh in t hours. Find her rate in miles per hour.

54. Natalie flies her small plane from St. Louis to Chicago, a distance of 289 miles, at z miles per hour. Find her time in hours.

55. Joshua can do a job in x hours. What portion of the job is done in 1 hour?

56. Jennifer needs h hours to tune up her car. How much of the job does she do in 1 hour?

NAME DATE HOUR

SUMMARY EXERCISES ON RATIONAL EXPRESSIONS

A common error when working with rational expressions is to confuse *operations* on rational expressions with the *solution of equations* with rational expressions. For example, the four possible operations on the rational expressions

$$\frac{1}{x} \quad \text{and} \quad \frac{1}{x-2}$$

can be performed as follows.

Add. $\frac{1}{x} + \frac{1}{x-2} = \frac{x-2}{x(x-2)} + \frac{x}{x(x-2)} = \frac{x-2+x}{x(x-2)} = \frac{2x-2}{x(x-2)}$

Subtract. $\frac{1}{x} - \frac{1}{x-2} = \frac{x-2}{x(x-2)} - \frac{x}{x(x-2)} = \frac{x-2-x}{x(x-2)} = \frac{-2}{x(x-2)}$

Multiply. $\frac{1}{x} \cdot \frac{1}{x-2} = \frac{1}{x(x-2)}$

Divide. $\frac{1}{x} \div \frac{1}{x-2} = \frac{1}{x} \cdot \frac{x-2}{1} = \frac{x-2}{x}$

On the other hand, the equation

$$\frac{1}{x} + \frac{1}{x-2} = \frac{3}{4}$$

is solved by multiplying each side by the least common denominator, $4x(x-2)$, giving an equation with no denominators.

$$4x(x-2)\frac{1}{x} + 4x(x-2)\frac{1}{x-2} = 4x(x-2)\frac{3}{4}$$

$$4x - 8 + 4x = 3x^2 - 6x$$

$$0 = 3x^2 - 14x + 8$$

$$0 = (3x-2)(x-4)$$

$$x = \frac{2}{3} \quad \text{or} \quad x = 4$$

In each of the following exercises, first decide whether it is a rational expression to be added, subtracted, multiplied, or divided or an equation to be solved. Then perform the operation, or solve the given equation.

1. $\frac{4}{p} + \frac{6}{p}$

2. $\frac{x^3y^2}{x^2y^4} \cdot \frac{y^5}{x^4}$

3. $\frac{1}{x^2+x-2} \div \frac{4x^2}{2x-2}$

4. $\frac{8}{m-5} = 2$

5. $\dfrac{2y^2 + y - 6}{2y^2 - 9y + 9} \cdot \dfrac{y^2 - 2y - 3}{y^2 - 1}$

6. $\dfrac{2}{k^2 - 4k} + \dfrac{3}{k^2 - 16}$

7. $\dfrac{x - 4}{5} = \dfrac{x + 3}{6}$

8. $\dfrac{3t^2 - t}{6t^2 + 15t} \div \dfrac{6t^2 + t - 1}{2t^2 - 5t - 25}$

9. $\dfrac{4}{p + 2} + \dfrac{1}{3p + 6}$

10. $\dfrac{1}{y} + \dfrac{1}{y - 3} = -\dfrac{5}{4}$

11. $\dfrac{3}{t - 1} + \dfrac{1}{t} = \dfrac{7}{2}$

12. $\dfrac{6}{y} - \dfrac{2}{3y}$

13. $\dfrac{5}{4z} - \dfrac{2}{3z}$

14. $\dfrac{k + 2}{3} = \dfrac{2k - 1}{5}$

15. $\dfrac{1}{m^2 + 5m + 6} + \dfrac{2}{m^2 + 4m + 3}$

16. $\dfrac{2k^2 - 3k}{20k^2 - 5k} \div \dfrac{2k^2 - 5k + 3}{4k^2 + 11k - 3}$

5.7 APPLICATIONS OF RATIONAL EXPRESSIONS

Each time we learn to solve a new type of equation, we are able to apply the knowledge to new applications. We can now solve applications involving rational expressions. The problem-solving techniques of earlier chapters still apply. The main difference between the problems in this section and those of earlier sections is that the equations for these problems involve rational expressions.

OBJECTIVES

1. Solve problems about numbers using rational expressions.
2. Solve applied problems about distance using rational expressions.
3. Solve applied problems about work using rational expressions.
4. Solve applied problems about variation using rational expressions.

FOR EXTRA HELP

Tape 8	SSM pp. 195–200	MAC: A IBM: A

1 In order to prepare for more meaningful applications, we begin with an example about an unknown number.

EXAMPLE 1 *Solving a Problem About Numbers*

If the same number is added to both the numerator and denominator of the fraction $\frac{3}{4}$, the result is $\frac{5}{6}$. Find the number.

Let x = the number added to numerator and denominator. Then

$$\frac{3+x}{4+x} \qquad \text{Same number added to numerator and denominator}$$

represents the result of adding the same number to both the numerator and denominator. Since this result is to equal $\frac{5}{6}$,

$$\frac{3+x}{4+x} = \frac{5}{6}.$$

Solve this equation by multiplying each side by the common denominator, $6(4 + x)$.

$$6(4+x)\frac{3+x}{4+x} = 6(4+x)\frac{5}{6}$$

$$6(3+x) = 5(4+x)$$

$$18 + 6x = 20 + 5x \qquad \text{Distributive property}$$

$$x = 2 \qquad \text{Subtract 18 and } 5x.$$

Check this solution in the words of the original problem: If 2 is added to both the numerator and denominator of $\frac{3}{4}$, the result is $\frac{5}{6}$, as needed. ■

WORK PROBLEM 1 AT THE SIDE.

1. Solve each problem.

(a) A certain number is added to the numerator and subtracted from the denominator of $\frac{5}{8}$. The new fraction equals the reciprocal of $\frac{5}{8}$. Find the number.

(b) The denominator of a fraction is 1 more than the numerator. If 6 is added to the numerator and subtracted from the denominator, the result is $\frac{15}{4}$. Find the original fraction.

2 If an automobile travels at an average rate of 50 miles per hour for two hours, then it travels $50 \times 2 = 100$ miles. This is an example of the basic relationship between distance, rate, and time:

$$\text{distance} = \text{rate} \times \text{time}.$$

This relationship is given by the formula $d = rt$. By solving, in turn, for r and t in the formula, we obtain two other equivalent forms of the formula. The three forms are given below.

DISTANCE, RATE, AND TIME RELATIONSHIP

$$d = rt \qquad r = \frac{d}{t} \qquad t = \frac{d}{r}$$

The next example illustrates the uses of these formulas.

ANSWERS

1. (a) 3 (b) $\frac{9}{10}$

2. Solve each problem.

(a) In 1989, Emerson Fitipaldi won the Indianapolis 500 (mile) race in 2.984 hours. What was his average speed?

(b) The winner of the 1988 Indianapolis 500 (mile) race was Rick Mears, who drove his Penske-Chevy V8 at an average speed of 144.8 miles per hour. What was Mears' driving time?

(c) A small plane flew from Warsaw to Rome averaging 164 miles per hour. The trip took 2 hours. What is the distance between Warsaw and Rome?

ANSWERS

2. (a) 167.560 miles per hour (b) 3.453 hours (c) 328 miles

■ **EXAMPLE 2** *Finding Distance, Rate, or Time*

(a) The speed of sound is 1088 feet per second at sea level at 32°F. In 5 seconds under these conditions, sound travels

$$\underset{\text{Rate}}{\underset{\uparrow}{1088}} \times \underset{\text{Time}}{\underset{\uparrow}{5}} = \underset{\text{Distance}}{\underset{\uparrow}{5440}} \text{ feet.}$$

Here, we found distance given rate and time, using $d = rt$.

(b) Over a short distance, an elephant can travel at a rate of 25 miles per hour. In order to travel $\frac{1}{4}$ mile, it would take an elephant

$$\begin{matrix}\text{Distance} \rightarrow \\ \text{Rate} \rightarrow\end{matrix} \frac{\frac{1}{4}}{25} = \frac{1}{4} \times \frac{1}{25} = \frac{1}{100} \text{ hour.} \quad \leftarrow \text{Time}$$

Here, we find time given rate and distance, using $t = \frac{d}{r}$. To convert $\frac{1}{100}$ hour to minutes, multiply $\frac{1}{100}$ by 60 to get $\frac{60}{100}$ or $\frac{3}{5}$ minute. To convert $\frac{3}{5}$ minute to seconds, multiply $\frac{3}{5}$ by 60 to get 36 seconds.

(c) In the 1988 Olympic Games, the USSR won the 400-meter relay with a time of 38.19 seconds. The rate of the team was

$$\begin{matrix}\text{Distance} \rightarrow \\ \text{Time} \rightarrow\end{matrix} \frac{400}{38.19} = 10.47 \text{ (rounded) meters per second.} \quad \leftarrow \text{Rate}$$

This answer was obtained using a calculator. Here, we found rate given distance and time, using $r = \frac{d}{t}$. ■

◀ WORK PROBLEM 2 AT THE SIDE.

Many applied problems use the formulas just discussed. The next example shows how to solve a typical application of the formula $d = rt$. A strategy for solving such problems involves two major steps:

SOLVING MOTION PROBLEMS

Step 1 Set up a sketch showing what is happening in the problem.

Step 2 Make a chart using the information given in the problem, along with the unknown quantities.

The chart will help you organize the information, and the sketch will help you set up the equation.

■ **EXAMPLE 3** *Solving a Motion Problem*

Two cars leave Baton Rouge, Louisiana, at the same time and travel east on Interstate 12. One travels at a constant speed of 55 miles per hour and the other travels at a constant speed of 63 miles per hour. In how many hours will the distance between them be 24 miles?

Since we are looking for time, let t = the number of hours until the distance between them is 24 miles. The sketch in Figure 1 shows what is happening in the problem. Now, construct a chart like the one that follows. Fill in the information given in the problem, and use t for the time traveled by each car. Multiply rate by time to get the expressions for distances traveled.

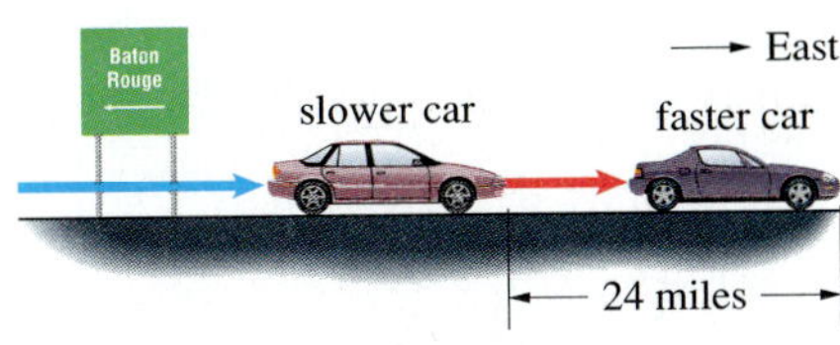

FIGURE 1

	RATE ×	TIME =	DISTANCE	
Faster car	63	t	$63t$	Difference is 24 miles.
Slower car	55	t	$55t$	

The quantities $63t$ and $55t$ represent the different distances. Refer to Figure 1, and notice that the *difference* between the larger distance and the smaller distance is 24 miles. Now write the equation, and solve it.

$$63t - 55t = 24$$

$$8t = 24 \quad \text{Combine terms.}$$

$$t = 3 \quad \text{Divide by 8.}$$

After 3 hours the faster car will have traveled $63 \times 3 = 189$ miles, and the slower car will have traveled $55 \times 3 = 165$ miles. Since $189 - 165 = 24$, the conditions of the problem are satisfied. It will take 3 hours for the distance between them to be 24 miles. ■

Note In motion problems like the one in Example 3, once you have filled in two pieces of information in each row of the chart, you should automatically fill in the third piece of information, using the appropriate form of the formula relating distance, rate, and time. Set up the equation based upon your sketch and the information in the chart.

WORK PROBLEM 3 AT THE SIDE. ▶▶

EXAMPLE 4 *Solving a Problem About Distance, Rate, and Time*

The Big Muddy River has a current of 3 miles per hour. A motorboat takes as long to go 12 miles downstream as to go 8 miles upstream. What is the speed of the boat in still water?

Let $x =$ the speed of the boat in still water. Because the current pushes the boat when the boat is going downstream, the speed of the boat downstream will be the sum of the speed of the boat and the speed of the current, $x + 3$ miles per hour. Also, the boat's speed going upstream is given by $x - 3$ miles per hour. This information is summarized in the following chart.

	d	r	t
Downstream	12	$x + 3$	
Upstream	8	$x - 3$	

3. Work each problem.

(a) From a point on a straight road, Lupe and Maria ride bicycles in opposite directions. Lupe rides 10 miles per hour and Maria rides 12 miles per hour. In how many hours will they be 55 miles apart?

(b) At a given hour, two steamboats leave a city in the same direction on a straight canal. One travels at 18 miles per hour, and the other travels at 25 miles per hour. In how many hours will the boats be 35 miles apart?

ANSWERS

3. (a) $2\frac{1}{2}$ **hours (b) 5 hours**

4. Work each problem.

(a) A boat can go 20 miles against the current in the same time it can go 60 miles with the current. The current is flowing at 4 miles per hour. Find the speed of the boat with no current.

(b) An airplane, maintaining a constant airspeed, takes as long to go 450 miles with the wind as it does to go 375 miles against the wind. If the wind is blowing at 15 miles per hour, what is the speed of the plane?

ANSWERS
4. **(a)** 8 miles per hour
(b) 165 miles per hour

Fill in the column representing time by using the formula $t = \frac{d}{r}$. Then the time upstream is the distance divided by the rate, or

$$t = \frac{d}{r} = \frac{8}{x - 3},$$

and the time downstream is also the distance divided by the rate, or

$$t = \frac{d}{r} = \frac{12}{x + 3}.$$

Now complete the chart.

	d	r	t
Downstream	12	$x + 3$	$\frac{12}{x + 3}$
Upstream	8	$x - 3$	$\frac{8}{x - 3}$

Times are equal.

According to the original problem, the time upstream equals the time downstream. The two times from the chart must therefore be equal, giving the equation

$$\frac{12}{x + 3} = \frac{8}{x - 3}.$$

Solve this equation by multiplying each side by $(x + 3)(x - 3)$.

$$(x + 3)(x - 3)\frac{12}{x + 3} = (x + 3)(x - 3)\frac{8}{x - 3}$$

$$12(x - 3) = 8(x + 3)$$

$$12x - 36 = 8x + 24 \qquad \text{Distributive property}$$

$$4x = 60 \qquad \text{Subtract } 8x\text{; add } 36.$$

$$x = 15 \qquad \text{Divide by 4.}$$

The speed of the boat in still water is 15 miles per hour.

Check the solution by first finding the speed of the boat downstream, which is $15 + 3 = 18$ miles per hour. Traveling 12 miles would take

$$d = rt$$

$$12 = 18t$$

$$t = \frac{2}{3} \text{ hour.}$$

On the other hand, the speed of the boat upstream is $15 - 3 = 12$ miles per hour, and traveling 8 miles would take

$$d = rt$$

$$8 = 12t$$

$$t = \frac{2}{3} \text{ hour.}$$

The time upstream equals the time downstream, as required. ■

WORK PROBLEM 4 AT THE SIDE.

3 Suppose that you can mow your lawn in 4 hours. Then after 1 hour, you will have mowed $\frac{1}{4}$ of the lawn. After 2 hours, you will have mowed $\frac{2}{4}$ or $\frac{1}{2}$ of the lawn, and so on. This idea is generalized as follows.

RATE OF WORK

If a job can be completed in t units of time, then the rate of work is

$$\frac{1}{t} \text{ job per unit of time.}$$

The relationship between problems involving work and problems involving distance is a very close one. Recall that the formula $d = rt$ says that distance traveled is equal to rate of travel multiplied by time traveled. Similarly, the fractional part of a job accomplished is equal to the rate of the work multiplied by the time worked. In the lawn mowing example, after 3 hours, the fractional part of the job done is

$$\underbrace{\frac{1}{4}}_{\text{Rate of work}} \cdot \underbrace{3}_{\text{Time worked}} = \underbrace{\frac{3}{4}}_{\text{Fractional part of job done}}.$$

After 4 hours, $(\frac{1}{4})(4) = 1$ whole job has been done.

These ideas are used in solving problems about the length of time needed to do a job. These problems are often called work problems.

EXAMPLE 5 *Solving a Problem About Work Rates*

With a riding lawn mower, John, the groundskeeper in a large park, can cut the lawn in 8 hours. With a small mower, his assistant Walt needs 14 hours to cut the same lawn. If both John and Walt work on the lawn, how long will it take to cut it?

Let $x =$ the number of hours it will take for John and Walt to mow the lawn, working together.

Certainly, x will be less than 8, since John alone can mow the lawn in 8 hours. Begin by making a chart as shown. Remember that based on the previous discussion, John's rate alone is $\frac{1}{8}$ job per hour, and Walt's rate is $\frac{1}{14}$ job per hour.

	Rate	*Time working together*	*Fractional part of the job done when working together*
John	$\frac{1}{8}$	x	$\frac{1}{8}x$
Walt	$\frac{1}{14}$	x	$\frac{1}{14}x$

Sum is 1 whole job.

Since together John and Walt complete 1 whole job, we must add their individual fractional parts and set the sum equal to 1.

5. (a) Michael can paint a room, working alone, in 8 hours. Lindsay can paint the same room, working alone, in 6 hours. How long will it take them if they work together?

(b) Roberto can tune up his Bronco in 2 hours working alone. His brother Marco can do the job in 3 hours working alone. How long would it take them if they worked together?

6. (a) If z varies directly as t, and $z = 11$ when $t = 4$, find z when $t = 32$.

(b) The circumference of a circle varies directly as the radius. A circle with a radius of 7 centimeters has a circumference of 43.96 centimeters. Find the circumference if the radius is 11 centimeters.

$$\underbrace{\text{Fractonal part done by John}}_{\frac{1}{8}x} + \underbrace{\text{Fractional part done by Walt}}_{\frac{1}{14}x} = \underbrace{\text{1 whole job}}_{1}$$

$$56\left(\frac{1}{8}x + \frac{1}{14}x\right) = 56(1) \qquad \text{Multiply by LCD, 56.}$$

$$56\left(\frac{1}{8}x\right) + 56\left(\frac{1}{14}x\right) = 56(1) \qquad \text{Distributive property}$$

$$7x + 4x = 56$$

$$11x = 56 \qquad \text{Combine like terms.}$$

$$x = \frac{56}{11} \qquad \text{Divide by 11.}$$

Working together, John and Walt can mow the lawn in $\frac{56}{11}$ hours, or $5\frac{1}{11}$ hours. Check to be sure the answer is reasonable. ■

WORK PROBLEM 5 AT THE SIDE.

4 Suppose that gasoline costs \$1.50 per gallon. Then 1 gallon costs \$1.50, 2 gallons cost 2(\$1.50) = \$3.00, 3 gallons cost 3(\$1.50) = \$4.50, and so on. Each time, the total cost is obtained by multiplying the number of gallons by the price per gallon. In general, if k equals the price per gallon and x equals the number of gallons, then the total cost y is equal to kx. Notice that as the number of gallons increases, the total cost increases.

The preceding discussion is an example of variation. Equations with fractions often result when discussing variation. As in the gasoline example, two variables **vary directly** if one is a constant multiple of the other.

DIRECT VARIATION

y* varies directly as *x if there exists a constant k such that

$$y = kx.$$

■ EXAMPLE 6 *Using Direct Variation*

Suppose y varies directly as x, and $y = 20$ when $x = 4$. Find y when $x = 9$.

Since y varies directly as x, there is a constant k such that $y = kx$. We know that $y = 20$ when $x = 4$. Substituting these values into $y = kx$ gives

$$y = kx$$

$$20 = k \cdot 4,$$

from which

$$k = 5.$$

Since $y = kx$ and $k = 5$,

$$y = 5x. \qquad \text{Let } k = 5.$$

When $x = 9$,

$$y = 5x = 5 \cdot 9 = 45. \qquad \text{Let } x = 9.$$

Thus, $y = 45$ when $x = 9$. ■

WORK PROBLEM 6 AT THE SIDE.

ANSWERS

5. (a) $3\frac{3}{7}$ hours (b) $1\frac{1}{5}$ hours

6. (a) 88 (b) 69.08 centimeters

In another common type of variation, the value of one variable increases while the value of another decreases. For example, an increase in the supply of an item causes a decrease in the price of the item.

INVERSE VARIATION

y varies inversely as x if there exists a constant k such that

$$y = \frac{k}{x}.$$

EXAMPLE 7 *Using Inverse Variation*

In a certain manufacturing process, the cost of producing an item varies inversely as the number of items produced. If 10,000 items are produced, the cost is \$2 per item. Find the cost per item to produce 25,000 items.

Let x = the number items produced and
c = the cost per item.

Since c varies inversely as x, there is a constant k such that

$$c = \frac{k}{x}.$$

Find k by replacing c with 2 and x with 10,000.

$$2 = \frac{k}{10{,}000}$$

$$20{,}000 = k \qquad \text{Multiply by 10,000.}$$

$$c = \frac{20{,}000}{25{,}000} = .80 \qquad \text{Let } k = 20{,}000 \text{ and } x = 25{,}000.$$

The cost per item to make 25,000 items is \$.80. ■

WORK PROBLEM 7 AT THE SIDE. ▶▶

7. **(a)** Suppose z varies inversely as t, and $z = 8$ when $t = 2$. Find z when $t = 32$.

(b) The current in a simple electrical circuit varies inversely as the resistance. If the current is 80 amperes when the resistance is 10 ohms, find the current if the resistance is 16 ohms.

ANSWERS

7. (a) $\frac{1}{2}$ **(b)** 50 amperes

QUEST FOR NUMERACY

Can We Average Averages?

Consider the following problem.

Suppose that a plane travels from Boston to New Orleans averaging 200 miles per hour, and makes the return trip averaging 400 miles per hour. What is the average rate for the entire trip?

At first glance, it may seem that the average rate for the entire trip is 300 miles per hour (the average of 200 and 400), but this is not the case. To answer the question, let x represent the distance between Boston and New Orleans. Remembering the relationship between distance, rate, and time, we can find the average rate for the entire trip by simplifying a complex fraction.

$$
\begin{aligned}
\text{Average rate for entire trip} &= \frac{\text{Total distance}}{\text{Total time}} \\
&= \frac{x + x}{\frac{x}{200} + \frac{x}{400}} \\
&= \frac{2x}{\frac{3x}{400}} \\
&= 2x \cdot \frac{400}{3x} \\
&= \frac{800}{3} = 266\frac{2}{3}
\end{aligned}
$$

The answer is $266\frac{2}{3}$ miles per hour.

Therefore, in a problem such as this one, we cannot expect to get the correct answer by "averaging averages," since the times traveled at the two rates were different.

FOR GROUP DISCUSSION

Discuss why a baseball player cannot determine his or her batting average over two seasons by "averaging averages."

5.7 NAME DATE HOUR

EXERCISES

Solve each problem. See Example 1.

1. One-third of a number is two more than one-sixth of the same number. What is the number?

2. The numerator of the fraction $\frac{13}{15}$ is increased by an amount so that the value of the resulting fraction is $\frac{7}{5}$. By what amount was the numerator increased?

$$\frac{13+x}{15}=\frac{7}{5}$$

3. In a certain fraction, the denominator is 4 less than the numerator. If 3 is added to both the numerator and the denominator, the resulting fraction is equal to $\frac{3}{2}$. Find the original fraction.

4. The denominator of a certain fraction is 3 times the numerator. If 2 is added to the numerator and subtracted from the denominator, the resulting fraction is equal to 1. Find the original fraction.

5. Calgene, a company that sells genetically engineered tomatoes, expects these tomatoes to provide about $\frac{1}{2}$ of its total revenue in 1995. Production of oils will contribute about $\frac{1}{4}$ its total revenue in 1995. If these two products produce \$74.8 million that year, what will the total revenue be?

6. The profits from a carnival are to be given to two scholarships so that one scholarship receives $\frac{3}{2}$ as much money as the other. If the total amount given to the two scholarships is \$780, how much goes to the scholarship that receives the lesser amount?

7. A recent news article stated that Germany invests $\frac{1}{10}$ as much as the United States in Mexico. If the total amount invested in the Mexican economy by these countries is \$2 billion, how much does each country invest?

8. Japan invests about $\frac{3}{4}$ as much as Germany in the Mexican economy, according to a recent news article. If Japan and Germany invest a total of \$.32 billion, how much does each of these countries invest?

Solve each problem. See Examples 2–4.

9. In the 1988 Olympic Nordic Skiing, Gunde Svan of Sweden won the 50 kilometer event in 2.08 hours. What was his average speed?

10. Bonnie Blair won the women's 500 meter speed skating in the 1988 Olympics. Her average speed was 12.79 meters per second. What was her time?

11. Suppose Amanda walks D miles at R miles per hour in the same time that Kenneth walks d miles at r miles per hour. Give an equation relating D, R, d, and r.

12. If a boat travels m miles per hour in still water, what is its rate when it travels upstream in a river with a current of 5 miles per hour? What is its rate downstream in the river?

13. Ibrahim Hussein of Kenya won the 1992 Boston marathon (a 26-mile race) in (about) .8 of an hour less time than the winner of the first Boston marathon in 1897, John J. McDermott of New York. Find each runner's speed, if McDermott's speed was (about) .73 times Hussein's speed. (*Hint:* Don't round until the end of the calculations.)

	d	r	t
Hussein			
McDermott			

14. Women first ran in the Boston marathon in 1972, when Nina Kuscsik of New York won the race. In 1992, the winner was Olga Markova of Russia, whose time was .8 of an hour less than Kuscsik's in 1972. If Markova ran $\frac{4}{3}$ as fast as Kuscsik, find each runner's speed. (See Exercise 13 for distance.)

15. Sandi Goldstein flew from Dallas to Indianapolis at 180 miles per hour and then flew back at 150 miles per hour. The trip at the slower speed took 1 hour longer than the trip at the higher speed. Find the distance between the two cities.

16. The distance from Seattle, Washington, to Victoria, British Columbia, is about 148 miles by ferry. It takes about 4 hours less to travel by the same ferry to Vancouver, British Columbia, a distance of about 74 miles. What is the average speed of the ferry?

17. A boat can go 20 miles against a current in the same time that it can go 60 miles with the current. The current is 4 miles per hour. Find the speed of the boat in still water.

	d	r	t
Against current			
With current			

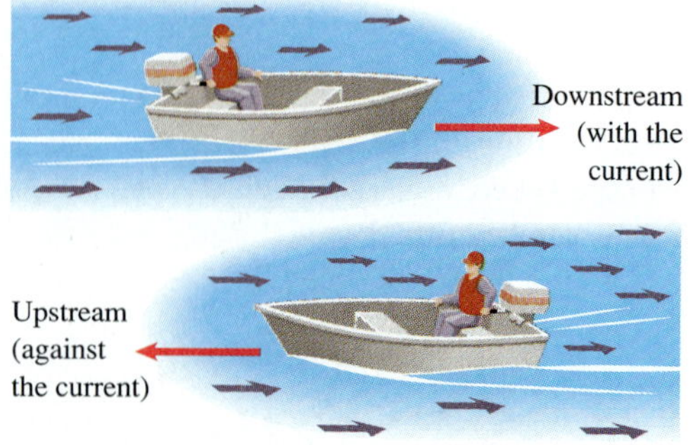

18. A plane flies 350 miles with the wind in the same time that it can fly 310 miles against the wind. The plane has a still-air speed of 165 miles per hour. Find the speed of the wind.

Solve each problem. See Example 5.

19. If it takes Bill Young 10 hours to do a job, what is his work rate?

20. If it takes Karin Sandberg 12 hours to do a job, what part of the job does she do in 8 hours?

21. Geraldo and Luisa Hernandez operate a small cleaners. Luisa, working alone, can clean a day's laundry in 9 hours. Geraldo can clean a day's laundry in 8 hours. How long would it take them if they work together?

22. Lee can clean the house alone in 5 hours, while Tran needs 4 hours. How long will it take them to clean the house if they work together?

23. A pump can pump the water out of a flooded basement in 10 hours. A smaller pump takes 12 hours. How long would it take to pump the water from the basement using both pumps?

24. A copier can do a printing job in 7 hours. A smaller copier can do the same job in 12 hours. How long would it take to do the job using both copiers?

25. An experienced employee can enter tax data into a computer twice as fast as a new employee. Working together, it takes the employees 2 hours. How long would it take the experienced employee working alone?

26. One roofer can put a new roof on a house three times faster than another. Working together they can roof a house in 4 days. How long would it take the faster roofer working alone?

Solve the following variation problems. See Examples 6 and 7.

27. If y varies directly as x, then y increases as x ______?______ .
(decreases/increases)

28. If y varies inversely as x, then y increases as x ______?______ .
(decreases/increases)

29. The interest on an investment varies directly as the rate of interest. If the interest is \$48 when the interest rate is 5%, find the interest when the rate is 4.2%.

30. For a given base, the area of a triangle varies directly as its height. Find the area of a triangle with a height of 6 inches, if the area is 10 square inches when the height is 4 inches.

31. Over a specified distance, speed varies inversely with time. If a car goes a certain distance in one-half hour at 30 miles per hour, what speed is needed to go the same distance in three-fourths of an hour?

32. For a constant area, the length of a rectangle varies inversely as the width. The length of a rectangle is 27 feet when the width is 10 feet. Find the length of a rectangle with the same area if the width is 18 feet.

33. The pressure exerted by a certain liquid at a given point varies directly as the depth of the point beneath the surface of the liquid. The pressure at 10 feet is 50 pounds per square inch. What is the pressure at 20 feet?

34. If the volume is constant, the pressure of a gas in a container varies directly as the temperature. If the pressure is 5 pounds per square inch at a temperature of 200 Kelvin, what is the pressure at a temperature of 300 Kelvin?

35. If the temperature is constant, the pressure of a gas in a container varies inversely as the volume of the container. If the pressure is 10 pounds per square foot in a container with 3 cubic feet, what is the pressure in a container with 1.5 cubic feet?

36. The force required to compress a spring varies directly as the change in the length of the spring. If a force of 12 pounds is required to compress a certain spring 3 inches, how much force is required to compress the spring 5 inches?

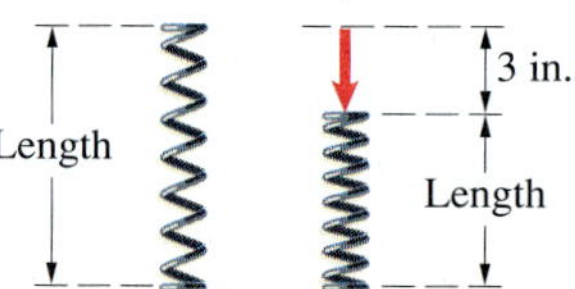

PREVIEW EXERCISES

Find the value of y when (a) $x = -2$ and (b) $x = 4$. See Sections 1.2 and 2.3.

37. $y = 5x + 3$

38. $y = 4 - 3x$

39. $6x - 2 = y$

40. $4x + 7y = 11$

41. $2x - 5y = 10$

42. $y + 3x = 8$

QUEST FOR NUMERACY

Never Be Afraid To Make An Educated Guess

A simple method that allows us to find a rational number between two different rational numbers is to find their average; that is, add them and divide by 2. This will give the number halfway between them. For example, the number halfway between $\frac{1}{3}$ and $\frac{1}{4}$ is $\frac{7}{24}$. There are actually *infinitely many* rational numbers between two different rational numbers. This is a consequence of the *density property of rational numbers.*

In the March 1973 issue of *The Mathematics Teacher* there appeared an article by Laurence Sherzer, an eighth-grade mathematics teacher, that immortalized one of his students, Robert McKay. The class was studying the density property and the teacher was explaining how to find a rational number between two given positive rational numbers by finding the average. McKay pointed out that there was no need to go to all that trouble. To find a number (not necessarily their average) between two positive rational numbers a/b and c/d, he claimed, simply "add the tops and add the bottoms." Much to his teacher's surprise, this method really does work and Sherzer proved this in the article. For example, to find a rational number between $\frac{1}{3}$ and $\frac{1}{4}$, add $1 + 1 = 2$ to get the numerator and $3 + 4 = 7$ to get the denominator. Therefore, by McKay's theorem, $\frac{2}{7}$ is between $\frac{1}{3}$ and $\frac{1}{4}$.

FOR GROUP DISCUSSION

Choose two different positive fractions, and show that McKay's method does indeed work for your choice.

CHAPTER 5 SUMMARY

KEY TERMS

5.1	**rational expression**	The quotient of two polynomials with denominator not zero is called a rational expression.
	lowest terms	A rational expression is written in lowest terms if there are no common factors in the numerator and denominator (except 1).
5.3	**least common denominator**	(LCD) The smallest expression that all denominators divide into without remainder is called the least common denominator.
5.5	**complex fraction**	A rational expression with one or more fractions in the numerator, denominator, or both is a complex fraction.
5.7	**direct variation**	y varies directly as x if there is a constant k such that $y = kx$.
	inverse variation	y varies inversely as x if there is a constant k such that $y = \frac{k}{x}$.

QUICK REVIEW

Concepts	Examples
5.1 The Fundamental Property of Rational Expressions	
To find the values for which a rational expression is not defined, set the denominator equal to zero and solve the equation.	Given: $\frac{x-4}{x^2-16}$. $x^2 - 16 = 0$; $(x-4)(x+4) = 0$; $x - 4 = 0$ or $x + 4 = 0$; $x = 4$ or $x = -4$. The rational expression is not defined for 4 or -4.
To write a rational expression in lowest terms, (1) factor; and (2) use the fundamental property to replace the quotient of common factors with 1.	Given: $\frac{x^2-1}{(x-1)^2}$. $\frac{(x-1)(x+1)}{(x-1)(x-1)} = 1\left(\frac{x+1}{x-1}\right) = \frac{x+1}{x-1}$

Concepts	Examples
5.2 Multiplication and Division of Rational Expressions	
Multiplication	$\dfrac{3x+9}{x-5} \cdot \dfrac{x^2-3x-10}{x^2-9}$
1. Multiply numerators and multiply denominators.	$= \dfrac{(3x+9)(x^2-3x-10)}{(x-5)(x^2-9)}$
2. Factor.	$= \dfrac{3(x+3)(x-5)(x+2)}{(x-5)(x+3)(x-3)}$
3. Write in lowest terms.	$= \dfrac{3(x+2)}{x-3}$
Division	$\dfrac{2x+1}{x+5} \div \dfrac{6x^2-x-2}{x^2-25}$
1. Multiply the first rational expression by the reciprocal of the second rational expression.	$= \dfrac{2x+1}{x+5} \cdot \dfrac{x^2-25}{6x^2-x-2}$
2. Factor.	$= \dfrac{2x+1}{x+5} \cdot \dfrac{(x+5)(x-5)}{(2x+1)(3x-2)}$
3. Write in lowest terms.	$= \dfrac{x-5}{3x-2}$
5.3 Least Common Denominators	
Finding the LCD	Find the LCD for $\dfrac{3}{k^2-16}$ and $\dfrac{1}{4k^2-16k}$.
1. Factor each denominator into prime factors.	$k^2-16 = (k+4)(k-4)$ $4k^2-16k = 4k(k-4)$
2. Choose each different denominator factor the greatest number of times it appears in any denominator.	
3. Multiply the factors from Step 2 to get the LCD.	$\text{LCD} = 4k(k+4)(k-4)$
To write a rational expression with the LCD as denominator:	$\dfrac{5}{2z^2-6z} = \dfrac{}{4z^3-12z^2}$
1. Factor both denominators.	$\dfrac{5}{2z(z-3)} = \dfrac{}{4z^2(z-3)}$
2. Decide what factors the denominator must be multiplied by to equal the LCD.	$2z(z-3)$ must be multiplied by $2z$.
3. Multiply the rational expression by that factor over itself (multiply by 1).	$\dfrac{5}{2z(z-3)} \cdot \dfrac{2z}{2z} = \dfrac{10z}{4z^2(z-3)}$

Concepts	Examples
5.4 Addition and Subtraction of Rational Expressions	
Adding Rational Expressions	Add $\frac{2}{3m+6} + \frac{m}{m^2-4}$.
1. Find the LCD.	$\frac{2}{3(m+2)} + \frac{m}{(m+2)(m-2)}$ Factor. The LCD is $3(m+2)(m-2)$.
2. Rewrite each rational expression as a fraction with the LCD as denominator.	$= \frac{2(m-2)}{3(m+2)(m-2)} + \frac{3m}{3(m+2)(m-2)}$
3. Add the numerators to get the numerator of the sum. The LCD is the denominator of the sum.	$= \frac{2m-4+3m}{3(m+2)(m-2)}$
4. Write in lowest terms.	$= \frac{5m-4}{3(m+2)(m-2)}$
Subtracting Rational Expressions Follow the same steps as for addition, but subtract in Step 3.	Subtract $\frac{6}{k+4} - \frac{2}{k}$. The LCD is $k(k+4)$. $\frac{6k}{(k+4)k} - \frac{2(k+4)}{k(k+4)} = \frac{6k-2(k+4)}{k(k+4)}$ $= \frac{6k-2k-8}{k(k+4)}$ $= \frac{4k-8}{k(k+4)}$
5.5 Complex Fractions	
Simplifying Complex Fractions *Method 1* Simplify the numerator and denominator separately. Then divide the simplified numerator by the simplified denominator.	$\frac{\frac{1}{a}-a}{1-a} = \frac{\frac{1}{a}-\frac{a^2}{a}}{1-a} = \frac{\frac{1-a^2}{a}}{1-a}$ $= \frac{1-a^2}{a} \cdot \frac{1}{1-a}$ $= \frac{(1-a)(1+a)}{a(1-a)} = \frac{1+a}{a}$
Method 2 Multiply numerator and denominator of the complex fraction by the LCD of all the denominators in the complex fraction.	$\frac{\frac{1}{a}-a}{1-a} = \frac{\frac{1}{a}-a}{1-a} \cdot \frac{a}{a} = \frac{\frac{a}{a}-a^2}{(1-a)a}$ $= \frac{1-a^2}{(1-a)a} = \frac{(1+a)(1-a)}{(1-a)a}$ $= \frac{1+a}{a}$

Concepts	Examples
5.6 Equations Involving Rational Expressions	Solve $\frac{2}{x-1} + \frac{3}{4} = \frac{5}{4}$.
1. Find the LCD of all denominators in the equation.	The LCD is $4(x - 1)$.
2. Using the multiplication property of equality, multiply each side of the equation by the LCD.	$4(x-1)\left(\frac{2}{x-1} + \frac{3}{4}\right) = 4(x-1)\left(\frac{5}{4}\right)$ $4(x-1)\left(\frac{2}{x-1}\right) + 4(x-1)\left(\frac{3}{4}\right) = 4(x-1)\left(\frac{5}{4}\right)$
3. Solve the resulting equation, which should have no fractions.	$8 + 3(x-1) = (x-1)5$ $8 + 3x - 3 = 5x - 5$ $-2x = -10$ $x = 5$
4. Check each proposed solution.	The proposed solution, 5, checks.
5.7 Applications of Rational Expressions	
Solving Motion Problems	On a trip from Sacramento to Monterey, Marge traveled at an average speed of 60 miles per hour. The return trip, at an average speed of 64 miles per hour, took $\frac{1}{4}$ hour less. How far did she travel between the two cities?
1. Draw a sketch showing what is happening in the problem.	Sacramento —60 mph→ Monterey ←64 mph—
2. Make a chart showing the information in the problem and assigning the variables. **3.** Use a form of the equation $d = rt$ to complete the chart.	(see chart below)
4. Use the sketch and the chart to write an equation.	Since the time for the return trip was $\frac{1}{4}$ hour less, the time going equals the time returning plus $\frac{1}{4}$. $\frac{x}{60} = \frac{x}{64} + \frac{1}{4}$
5. Solve the equation.	The solution of this equation is $x = 240$. She traveled 240 miles.
6. Check the answer.	Traveling 240 miles at 60 miles per hour takes 4 hours. Traveling 240 miles at 64 miles per hour takes $3\frac{3}{4}$ hours, which is $\frac{1}{4}$ hour less, as required.

	d	r	$t = \frac{d}{r}$
Going	x	60	$\frac{x}{60}$
Returning	x	64	$\frac{x}{64}$

Concepts	Examples
5.7 (Continued)	
Solving Problems About Work Rates	
1. If a job can be completed in t units of time, then the rate of work is $\frac{1}{t}$ job per unit of time.	It takes the regular mail carrier 6 hours to cover the route. A substitute took 8 hours to cover the route. How long would it take them together?
2. Make a chart giving the work rate, time, and part of the job done for each person.	Let t be the number of hours to cover the route together. (see chart below)
3. Add the fractional parts of the job done by each person and set the sum equal to 1 (the whole job).	$$\frac{1}{6}t + \frac{1}{8}t = 1$$
4. Solve the equation.	The solution of this equation is $t = \frac{24}{7} = 3\frac{3}{7}$. It would take them $3\frac{3}{7}$ hours to cover the route together.
5. Check the solution.	
Solving Variation Problems	
1. Write the variation equation using $y = kx$ or $y = \frac{k}{x}$.	If a varies inversely as b, and $a = 5$ when $b = 4$, find a when $b = 6$. The equation is $a = \frac{k}{b}$.
2. Find k by substituting the given values of x and y into the equation.	Subsitute $a = 5$ and $b = 4$. $$5 = \frac{k}{4}$$
3. Write the equation with the value of k from Step 2 and the given value of x or y. Solve for the remaining variable.	The solution is $k = 20$. Let $k = 20$ and $b = 6$ in the variation equation. $$a = \frac{20}{6} = \frac{10}{3}$$

	Rate	*Time Working Together*	*Part of Job Done*
Regular	$\frac{1}{6}$	t	$\left(\frac{1}{6}\right)t$
Sub	$\frac{1}{8}$	t	$\left(\frac{1}{8}\right)t$

NAME DATE HOUR

CHAPTER 5 REVIEW EXERCISES

[5.1] *Find any values of the variables for which the following rational expressions are undefined.*

1. $\frac{4}{x-3}$

2. $\frac{y+3}{2y}$

3. $\frac{m-2}{m^2-2m-3}$

4. $\frac{2k+1}{3k^2+17k+10}$

Find the numerical value of each rational expression when (a) $x = -2$ and (b) $x = 4$.

5. $\frac{x^2}{x-5}$

6. $\frac{4x-3}{5x+2}$

7. $\frac{3x}{x^2-4}$

8. $\frac{x-1}{x+2}$

Write each rational expression in lowest terms.

9. $\frac{5a^3b^3}{15a^4b^2}$

10. $\frac{m-4}{4-m}$

11. $\frac{4x^2-9}{6-4x}$

12. $\frac{4p^2+8pq-5q^2}{10p^2-3pq-q^2}$

[5.2] *Find each product or quotient. Write each answer in lowest terms.*

13. $\frac{18p^3}{6} \cdot \frac{24}{p^4}$

14. $\frac{8x^2}{12x^5} \cdot \frac{6x^4}{2x}$

15. $\frac{9m^2}{(3m)^4} \div \frac{6m^5}{36m}$

16. $\frac{x-3}{4} \cdot \frac{5}{2x-6}$

17. $\frac{3q+3}{5-6q} \div \frac{4q+4}{2(5-6q)}$

18. $\frac{2r+3}{r-4} \cdot \frac{r^2-16}{6r+9}$

19. $\frac{6a^2+7a-3}{2a^2-a-6} \div \frac{a+5}{a-2}$

20. $\frac{y^2-6y+8}{y^2+3y-18} \div \frac{y-4}{y+6}$

21. $\dfrac{2p^2 + 13p + 20}{p^2 + p - 12} \cdot \dfrac{p^2 + 2p - 15}{2p^2 + 7p + 5}$

22. $\dfrac{3z^2 + 5z - 2}{9z^2 - 1} \cdot \dfrac{9z^2 + 6z + 1}{z^2 + 5z + 6}$

[5.3] *Find the least common denominator for each list of fractions.*

23. $\dfrac{1}{8}, \dfrac{5}{12}, \dfrac{7}{32}$

24. $\dfrac{4}{9y}, \dfrac{7}{12y^2}, \dfrac{5}{27y^4}$

25. $\dfrac{1}{m^2 + 2m}, \dfrac{4}{m^2 + 7m + 10}$

26. $\dfrac{3}{x^2 + 4x + 3}, \dfrac{5}{x^2 + 5x + 4}$

Rewrite each rational expression with the given denominator.

27. $\dfrac{5}{8} = \dfrac{}{56}$

28. $\dfrac{10}{k} = \dfrac{}{4k}$

29. $\dfrac{3}{2a^3} = \dfrac{}{10a^4}$

30. $\dfrac{9}{x - 3} = \dfrac{}{18 - 6x}$

31. $\dfrac{-3y}{2y - 10} = \dfrac{}{50 - 10y}$

32. $\dfrac{4b}{b^2 + 2b - 3} = \dfrac{}{(b + 3)(b - 1)(b + 2)}$

[5.4] *Add or subtract as indicated. Write each answer in lowest terms.*

33. $\frac{10}{x} + \frac{5}{x}$

34. $\frac{6}{3p} - \frac{12}{3p}$

35. $\frac{9}{k} - \frac{5}{k - 5}$

36. $\frac{4}{y} + \frac{7}{7 + y}$

37. $\frac{m}{3} - \frac{2 + 5m}{6}$

38. $\frac{12}{x^2} - \frac{3}{4x}$

39. $\frac{5}{a - 2b} + \frac{2}{a + 2b}$

40. $\frac{4}{k^2 - 9} - \frac{k + 3}{3k - 9}$

41. $\frac{8}{z^2 + 6z} - \frac{3}{z^2 + 4z - 12}$

42. $\frac{11}{2p - p^2} - \frac{2}{p^2 - 5p + 6}$

[5.5] *Simplify each complex fraction.*

43. $\dfrac{\frac{a^4}{b^2}}{\frac{a^3}{b}}$

44. $\dfrac{\frac{y - 3}{y}}{\frac{y + 3}{4y}}$

45. $\dfrac{\frac{3m + 2}{m}}{\frac{2m - 5}{6m}}$

46. $\dfrac{\frac{1}{p} - \frac{1}{q}}{\frac{1}{q - p}}$

47. $\dfrac{x + \frac{1}{w}}{x - \frac{1}{w}}$

48. $\dfrac{\frac{1}{r + t} - 1}{\frac{1}{r + t} + 1}$

[5.6] *Solve each equation. Check your answer.*

49. $\frac{k}{5} - \frac{2}{3} = \frac{1}{2}$

50. $\frac{4 - z}{z} + \frac{3}{2} = \frac{-4}{z}$

51. $\frac{x}{2} - \frac{x - 3}{7} = -1$

52. $\frac{3y - 1}{y - 2} = \frac{5}{y - 2} + 1$

53. $\frac{3}{m - 2} + \frac{1}{m - 1} = \frac{7}{m^2 - 3m + 2}$

Solve for the specified variable.

54. $m = \frac{Ry}{t}$ for t

55. $x = \frac{3y - 5}{4}$ for y

56. $p^2 = \frac{4}{3m - q}$ for m

[5.7] *Solve each problem.*

57. When half a number is subtracted from two-thirds of a number, the answer is 2. Find the number.

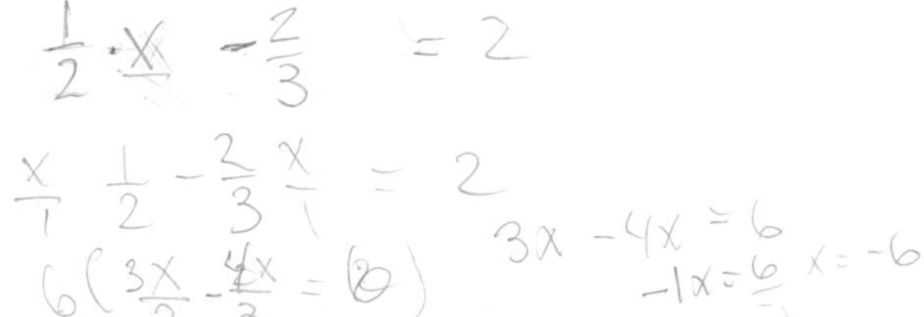

58. The commission received by a salesperson for selling a small car is $\frac{2}{3}$ that received for selling a large car. On a recent day, Linda sold one of each, earning a commission of $300. Find the commission for each type of car.

59. In 1911, at the first Indianapolis 500 (mile) race, Ray Harroun won with an average speed of 74.59 miles per hour. What was his time?

60. A man can plant his garden in 5 hours, working alone. His daughter can do the same job in 8 hours. How long would it take them if they worked together?

61. The head gardener can mow the lawns in the city park twice as fast as his assistant. Working together, they can complete the job in $1\frac{1}{3}$ hours. How long would it take the head gardener working alone?

62. The area of a circle varies directly as the length of its radius. A circle with a radius of 5 inches has an area of approximately 78.5 square inches. Find the radius of a circle with an area of 100 square inches.

63. If a parallelogram has a fixed area, the height varies inversely as the base. A parallelogram has a height of 8 centimeters and a base of 12 centimeters. Find the height if the base is changed to 24 centimeters.

Mixed Review Exercises

Perform the indicated operations.

64. $\dfrac{\dfrac{5}{x-y}+2}{3-\dfrac{2}{x+y}}$

65. $\dfrac{4}{m-1}-\dfrac{3}{m+1}$

66. $\dfrac{8p^5}{5}\div\dfrac{2p^3}{10}$

67. $\dfrac{r-3}{8}\div\dfrac{3r-9}{4}$

68. $\dfrac{\dfrac{5}{x}-1}{\dfrac{5-x}{3x}}$

69. $\dfrac{4}{z^2-2z+1}-\dfrac{3}{z^2-1}$

Solve.

70. $F = \dfrac{k}{d - D}$ for d

71. $\dfrac{2}{z} - \dfrac{z}{z + 3} = \dfrac{1}{z + 3}$

72. About $\frac{1}{10}$ as many people in the United States speak French at home as speak Spanish. A total of 19.1 million U.S. residents speak one of these two languages at home. How many speak Spanish?

73. One pipe can fill a swimming pool in 6 hours, and another pipe can do it in 9 hours. How long will it take the two pipes working together to fill the pool $\frac{3}{4}$ full?

74. Anne Kelly flew her plane 400 kilometers with the wind in the same time it took her to go 200 kilometers against the wind. The speed of the wind is 50 kilometers per hour. Find the speed of the plane in still air.

75. Phillip Morris and Ford are among the top ten U.S. companies in sales. In 1992, Phillip Morris's sales were close to $\frac{1}{2}$ Ford's sales. If total sales for the two companies were \$151 billion, what were Ford's sales that year?

NAME DATE HOUR

CHAPTER 5 TEST

1. Find any values for which $\frac{3x - 1}{x^2 - 2x - 8}$ is undefined.

Write each rational expression in lowest terms.

2. $\frac{4m^2 - 2m}{6m - 3}$

3. $\frac{6a^2 + a - 2}{2a^2 - 3a + 1}$

Multiply or divide. Write all answers in lowest terms.

4. $\frac{x^6 y}{x^3} \cdot \frac{y^2}{x^2 y^3}$

5. $\frac{5(d - 2)}{9} \div \frac{3(d - 2)}{5}$

6. $\frac{6k^2 - k - 2}{8k^2 + 10k + 3} \cdot \frac{4k^2 + 7k + 3}{3k^2 + 5k + 2}$

7. $\frac{4a^2 + 9a + 2}{3a^2 + 11a + 10} \div \frac{4a^2 + 17a + 4}{3a^2 + 2a - 5}$

Rewrite each rational expression with the given denominator.

8. $\frac{15}{4p} = \frac{\quad}{64p^3}$

9. $\frac{3}{6m - 12} = \frac{\quad}{42m - 84}$

Add or subtract as indicated. Write each answer in lowest terms.

10. $\frac{8}{c} - \frac{5}{c}$

11. $\frac{-4}{y + 2} + \frac{6}{5y + 10}$

1. ________
2. ________
3. ________
4. ________
5. ________
6. ________
7. ________
8. ________
9. ________
10. ________
11. ________

12. ______

13. ______

14. ______

15. ______

16. ______

17. ______

18. ______

19. ______

20. ______

12. $\dfrac{3}{2m^2 - 9m - 5} - \dfrac{m + 1}{2m^2 - m - 1}$

Simplify each complex fraction.

13. $\dfrac{\dfrac{2p}{k^2}}{\dfrac{3p^2}{k^3}}$

14. $\dfrac{\dfrac{1}{x + 3} - 1}{1 + \dfrac{1}{x + 3}}$

15. What values of x could not be solutions of the equation

$$\frac{2}{x + 1} - \frac{3}{x - 4} = 6?$$

Solve each equation.

16. $\dfrac{4}{3y} + \dfrac{3}{5y} = -\dfrac{11}{30}$

17. $\dfrac{2}{p^2 - 2p - 3} = \dfrac{3}{p - 3} + \dfrac{2}{p + 1}$

For each problem, write an equation, and solve it.

18. If four times a number is added to the reciprocal of twice the number, the result is 3. Find the number.

19. A boat goes 7 miles per hour in still water. It takes as long to go 20 miles upstream as 50 miles downstream. Find the speed of the current.

20. The current in a simple electrical circuit varies inversely as the resistance. If the current is 50 amperes (an *ampere* is a unit for measuring current) when the resistance is 10 ohms (an *ohm* is a unit for measuring resistance), find the current if the resistance is 5 ohms.

NAME DATE HOUR

CUMULATIVE REVIEW EXERCISES CHAPTERS R–5

Solve each of the following.

1. $\frac{5}{8}k = 4$

2. $3(2y - 5) = 2 + 5y$

3. $A = \frac{1}{2}bh$ for b

4. $\frac{2 + m}{2 - m} = \frac{3}{4}$

5. $5y \leq 6y + 8$

6. $5m - 9 > 2m + 3$

Evaluate each expression.

7. $\left(\frac{3}{4}\right)^{-2}$

8. $\frac{7^{-1}}{7}$

9. $\frac{(4^{-2})^3}{4^6 4^{-3}}$

Simplify each expression. Write with only positive exponents.

10. $\frac{(2x^3)^{-1} \cdot x}{2^3 x^5}$

11. $\frac{(m^{-2})^3 m}{m^5 m^{-4}}$

12. $\frac{2p^3q^4}{8p^5q^3}$

Perform the indicated operations.

13. $(2k^2 + 3k) - (k^2 + k - 1)$

14. $8x^2y^2(9x^4y^5)$

15. $(2a - b)^2$

16. $(y^2 + 3y + 5)(3y - 1)$

17. $\frac{12p^3 + 2p^2 - 12p + 4}{2p - 2}$

Factor completely.

18. $8t^2 + 10tv + 3v^2$

19. $8r^2 - 9rs + 12s^2$

20. $6a^2m + am - 2m$

Solve each equation.

21. $r^2 = 2r + 15$

22. $8m^2 = 64m$

23. $(r - 5)(2r + 1)(3r - 2) = 0$

Solve each problem.

24. One number is 4 more than another. The product of the numbers is 2 less than the smaller number. Find the smaller number.

25. The length of a rectangle is 2 meters less than twice the width. The area is 60 square meters. Find the width of the rectangle.

Perform each operation. Write each expression in lowest terms.

26. $\frac{x^6y^2}{y^5x^9} \cdot \frac{x^2}{y^3}$

27. $\frac{5}{q} - \frac{1}{q}$

28. $\frac{3}{7} + \frac{4}{r}$

29. $\frac{4}{5q - 20} - \frac{1}{3q - 12}$

30. $\frac{2}{k^2 + k} - \frac{3}{k^2 - k}$

31. $\frac{7z^2 + 49z + 70}{16z^2 + 72z - 40} \div \frac{3z + 6}{4z^2 - 1}$

32. Simplify the complex fraction $\dfrac{\frac{4}{a} + \frac{5}{2a}}{\frac{7}{6a} - \frac{1}{5a}}$.

Solve each equation.

33. $\frac{r + 2}{5} = \frac{r - 3}{3}$

34. $\frac{1}{x} = \frac{1}{x + 1} + \frac{1}{2}$

Solve each problem.

35. On a business trip, Arlene traveled to her destination at an average speed of 60 miles per hour. Coming home, her average speed was 50 miles per hour, and the trip took $\frac{1}{2}$ hour longer. How far did she travel each way?

36. Juanita can weed the yard in 3 hours. Benito can weed the yard in 2 hours. How long would it take them if they worked together?

Graphing Linear Equations 6

m slope

$y = mx + b$

(2, 13)

6.1 LINEAR EQUATIONS IN TWO VARIABLES

Many everyday situations involve two quantities that are related. Some examples include your grades in mathematics and the number of hours you spend studying, the interest rate on a loan and the amount of interest to be paid, and the number of items sold by a company and its revenue. To model these situations, two variables are needed. The equations we have discussed so far have had only one variable. In this chapter we discuss *linear equations,* one type of equation with two variables.

A **linear equation** in two variables is an equation that can be put in the form

$$Ax + By = C$$

where A, B, and C are real numbers and A and B are not both 0.

1 A solution of a linear equation in two variables requires two numbers, one for each variable. For example, the equation $y = 4x + 5$ is satisfied by replacing x with 2 and y with 13, because

$$13 = 4(2) + 5. \quad \text{Let } x = 2;\ y = 13.$$

The pair of numbers $x = 2$ and $y = 13$ gives one solution of the equation $y = 4x + 5$. The phrase "$x = 2$ and $y = 13$" is abbreviated

x-value → (2, 13) ← y-value
ordered pair

with the x-value, 2, and the y-value, 13, given as a pair of numbers written inside parentheses. The x-value is always given first. A pair of numbers such as (2, 13) is called an **ordered pair.** As the name indicates, the order in which the numbers are written is important. The ordered pairs (2, 13) and (13, 2) are not the same. The second pair indicates that $x = 13$ and $y = 2$.

Of course, letters other than x and y may be used in the equation, with the numbers in the ordered pair usually placed in alphabetical order. For example, one solution to the equation $3p - q = -11$ is $p = -2$ and $q = 5$. This solution is written as the ordered pair $(-2, 5)$.

WORK PROBLEM 1 AT THE SIDE.

OBJECTIVES

1. Write a solution as an ordered pair.
2. Decide whether a given ordered pair is a solution of a given equation.
3. Complete ordered pairs for a given equation.
4. Complete a table of ordered pairs.
5. Plot ordered pairs.

FOR EXTRA HELP

Tape 8

SSM pp. 219–225

MAC: A IBM: A

1. Write each solution as an ordered pair.

(a) $x = 5$ and $y = 7$

(b) $y = 6$ and $x = -1$

(c) $q = 4$ and $p = -3$

(d) $r = 3$ and $s = 12$

ANSWERS
1. (a) (5, 7) **(b)** $(-1, 6)$ **(c)** $(-3, 4)$ **(d)** (3, 12)

2 The next example shows how to decide whether an ordered pair is a solution of an equation. An ordered pair that is a solution of an equation is said to *satisfy* the equation.

■ **EXAMPLE 1** *Deciding Whether an Ordered Pair Satisfies an Equation*

Decide whether the given ordered pair is a solution of the given equation.

(a) (3, 2); $2x + 3y = 12$

To see whether (3, 2) is a solution of the equation $2x + 3y = 12$, substitute 3 for x and 2 for y in the given equation.

$$
\begin{aligned}
2x + 3y &= 12 \\
2(3) + 3(2) &= 12 && \text{Let } x = 3;\ \text{let } y = 2. \\
6 + 6 &= 12 \\
12 &= 12 && \text{True}
\end{aligned}
$$

This result is true, so (3, 2) satisfies $2x + 3y = 12$.

(b) (−2, −7); $2m + 3n = 12$

$$
\begin{aligned}
2(-2) + 3(-7) &= 12 && \text{Let } m = -2;\ \text{let } n = -7. \\
-4 + (-21) &= 12 \\
-25 &= 12 && \text{False}
\end{aligned}
$$

This result is false, so (−2, −7) does *not* satisfy $2m + 3n = 12$. ■

WORK PROBLEM 2 AT THE SIDE.

3 Choosing a number for one variable in a linear equation makes it possible to find the value of the other; the equation is used to find the value of the other variable.

■ **EXAMPLE 2** *Completing an Ordered Pair*

Complete the given ordered pairs for the equation $y = 4x + 5$.

(a) (7,)

In this ordered pair, $x = 7$. (Remember that x always comes first.) Find the corresponding value of y by replacing x with 7 in the equation $y = 4x + 5$.

$$y = 4(7) + 5 = 28 + 5 = 33$$

The ordered pair is (7, 33).

(b) (, −3)

In this ordered pair, $y = -3$. Find the value of x by replacing y with −3 in the equation; then solve for x.

$$
\begin{aligned}
y &= 4x + 5 \\
-3 &= 4x + 5 && \text{Let } y = -3. \\
-8 &= 4x && \text{Subtract 5 from each side.} \\
-2 &= x && \text{Divide each side by 4.}
\end{aligned}
$$

The ordered pair is (−2, −3). ■

WORK PROBLEM 3 AT THE SIDE.

2. Decide whether or not the ordered pairs satisfy the equation $5x + 2y = 20$.

(a) (0, 10)

(b) (2, −5)

(c) (3, 2)

(d) (−4, 20)

3. Complete the given ordered pairs for the equation $y = 2x - 9$.

(a) (5,)

(b) (2,)

(c) (, 7)

(d) (, −13)

ANSWERS

2. (a) yes (b) no (c) no (d) yes

3. (a) (5, 1) (b) (2, −5) (c) (8, 7) (d) (−2, −13)

4 Ordered pairs of an equation often are displayed in a **table of ordered pairs** as in the next example. The table may be written either vertically or horizontally. We will write these tables in both formats in this book.

EXAMPLE 3 *Completing a Table of Ordered Pairs*

Complete the given table of ordered pairs for the equation $x - 2y = 8$.

x	2	10		
y			0	-2

Complete the first two ordered pairs by letting $x = 2$ and $x = 10$, respectively.

If	$x = 2$,	If	$x = 10$,
then	$x - 2y = 8$	then	$x - 2y = 8$
becomes	$2 - 2y = 8$	becomes	$10 - 2y = 8$
	$-2y = 6$		$-2y = -2$
	$y = -3$.		$y = 1$.

Now complete the last two ordered pairs by letting $y = 0$ and $y = -2$, respectively.

If	$y = 0$,	If	$y = -2$,
then	$x - 2y = 8$	then	$x - 2y = 8$
becomes	$x - 2(0) = 8$	becomes	$x - 2(-2) = 8$
	$x - 0 = 8$		$x + 4 = 8$
	$x = 8$.		$x = 4$.

The completed table of ordered pairs is as follows.

x	2	10	8	4
y	-3	1	0	-2

WORK PROBLEM 4 AT THE SIDE.

EXAMPLE 4 *Completing a Table of Ordered Pairs*

Complete the given table of ordered pairs for the equation $x = 5$.

x			
y	-2	6	3

The given equation is $x = 5$. No matter which value of y might be chosen, the value of x is always the same, 5. Each ordered pair can be completed by placing 5 in the first position.

x	5	5	5
y	-2	6	3

Note When an equation such as $x = 5$ is discussed along with equations in two variables, think of $x = 5$ as an equation in two variables by rewriting $x = 5$ as $x + 0y = 5$. This form of the equation shows that for any value of y, the value of x is 5. Similarly, $y = -2$ is the same as $y = 0x - 2$.

Each of the equations in this section has an infinite number of ordered pairs as solutions. When we make a choice of a number for one variable, it leads to a particular real number for the other variable. Because there are an

4. Complete the table of ordered pairs for the equation $2x - 3y = 12$.

x	0		3	
y		0		-3

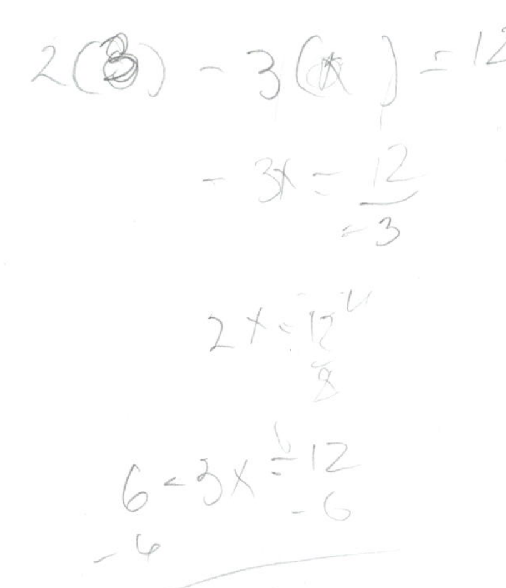

ANSWERS

4.

x	0	6	3	$\frac{3}{2}$
y	-4	0	-2	-3

infinite number of real number choices, there will be an infinite number of ordered pairs as solutions for any linear equation in two variables.

WORK PROBLEM 5 AT THE SIDE.

Earlier we used a number line to graph the solutions of an equation in one variable. Techniques for graphing the solutions of an equation in *two* variables will be shown in this section. The solutions of such an equation are *ordered pairs* of numbers of the form (x, y), so *two* number lines are needed, one for x and one for y. These two number lines are drawn as shown in Figure 1. The horizontal line is called the x-*axis* and the vertical line is called the y-axis. Together, the x-axis and y-axis form a **coordinate system.**

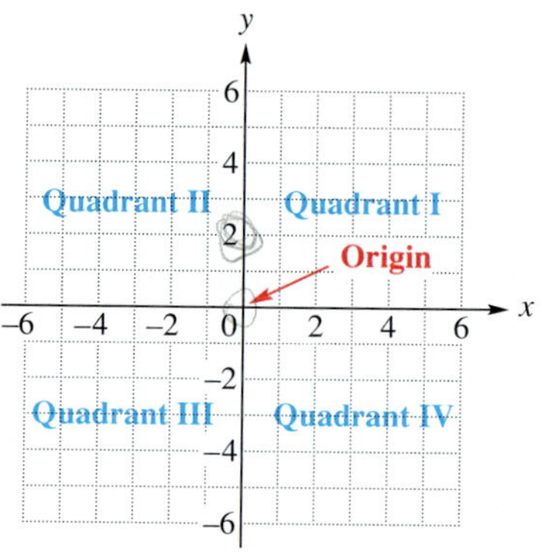

FIGURE 1

The coordinate system is divided into four regions, called **quadrants.** These quadrants are numbered counterclockwise, as shown in Figure 1. Points on the axes themselves are not in any quadrant. The point at which the x-axis and y-axis meet is called the **origin.** The origin is labeled 0 in Figure 1.

WORK PROBLEM 6 AT THE SIDE.

5 By referring to the two axes, every point in the plane can be associated with an ordered pair. The numbers in the ordered pair are called the **coordinates** of the point. For example, locate the point associated with the ordered pair (2, 3) by starting at the origin. Since the x-coordinate is 2, go 2 units to the right along the x-axis. Then, since the y-coordinate is 3, turn and go up 3 units on a line parallel to the y-axis. The point (2, 3) is **plotted** in Figure 2. From now on we will refer to the point with x-coordinate 2 and y-coordinate 3 as the point (2, 3).

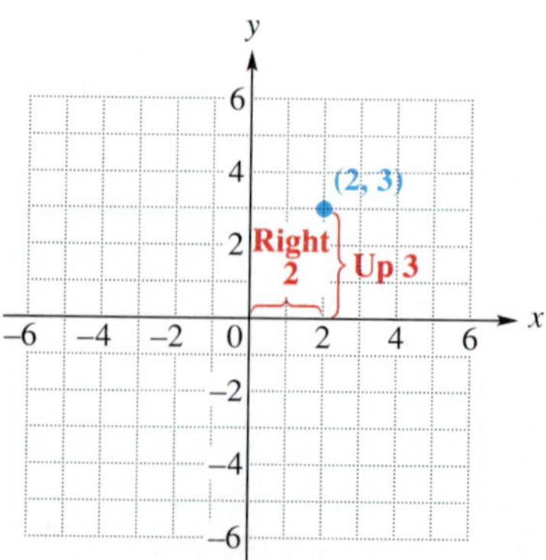

FIGURE 2

5. Complete the table of ordered pairs for each equation.

(a) $x = 3$

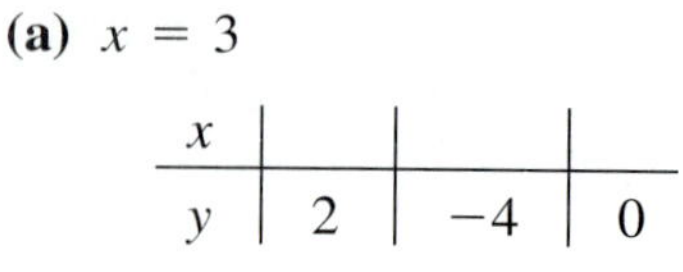

x			
y	2	-4	0

(b) $y = -4$

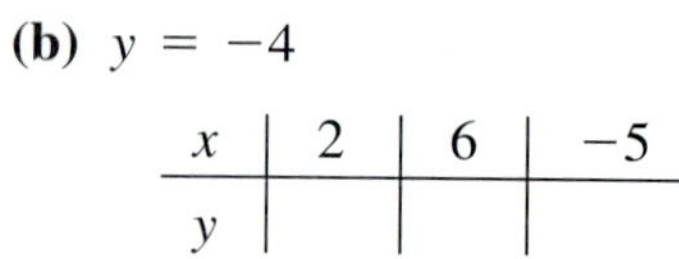

x	2	6	-5
y			

6. Name the quadrant in which each point in the figure is located.

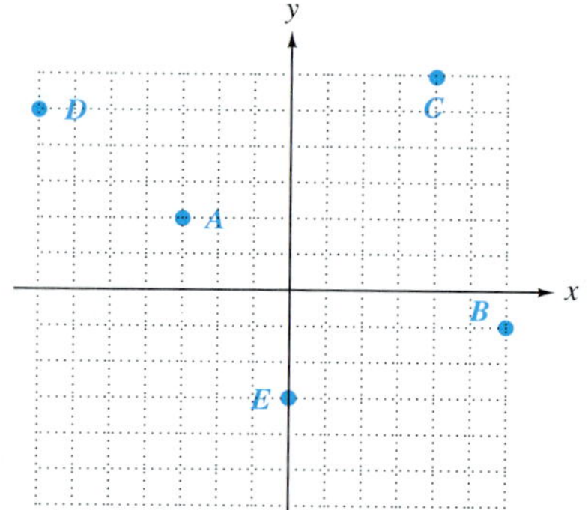

ANSWERS

5. (a)

x	3	3	3
y	2	-4	0

(b)

x	2	6	-5
y	-4	-4	-4

6. ***A*, II; *B*, IV; *C*, I; *D*, II; *E*, no quadrant**

Note When we graphed on a number line, one number corresponded to each point. On a plane, however, both numbers in the ordered pair are needed to locate a point. The ordered pair is a name for the point.

EXAMPLE 5 *Plotting Ordered Pairs*

Plot the given ordered pairs on a coordinate system.

(a) (1, 5)

(b) (−2, 3)

(c) (−1, −4)

(d) (7, −2)

(e) $\left(\frac{3}{2}, 2\right)$

(f) (5, 0)

(g) (0, −3)

Locate the point (−1, −4), for example, by first going 1 unit to the left along the x-axis. The turn and go 4 units down, paralled to the y-axis. Plot the point $(\frac{3}{2}, 2)$ by first going $\frac{3}{2}$ (or $1\frac{1}{2}$) units to the right along the x-axis. Then turn and go 2 units up, parallel to the y-axis. Figure 3 shows the graphs of the points in this example. ■

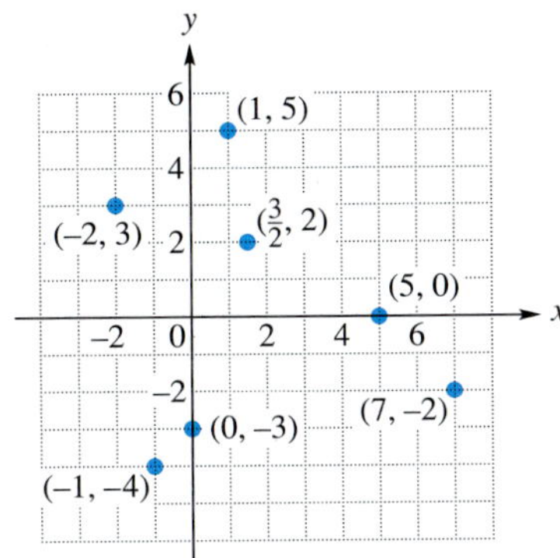

FIGURE 3

WORK PROBLEM 7 AT THE SIDE.

EXAMPLE 6 *Completing Ordered Pairs for an Application*

A company has found that the cost to produce x small calculators is

$$y = 25x + 250$$

where y represents the cost in cents. Complete the following table of values.

x	1	2	3
y			

To complete the ordered pair (1,), let $x = 1$.

$$\begin{aligned} y &= 25x + 250 \\ y &= 25(1) + 250 \quad \text{Let } x = 1. \\ y &= 25 + 250 \\ y &= 275 \end{aligned}$$

7. Plot the given ordered pairs on a coordinate system.

(a) (3, 5)

(b) (−2, 6)

(c) (−4, 0)

(d) (−5, −2)

(e) (5, −2)

(f) (0, −6)

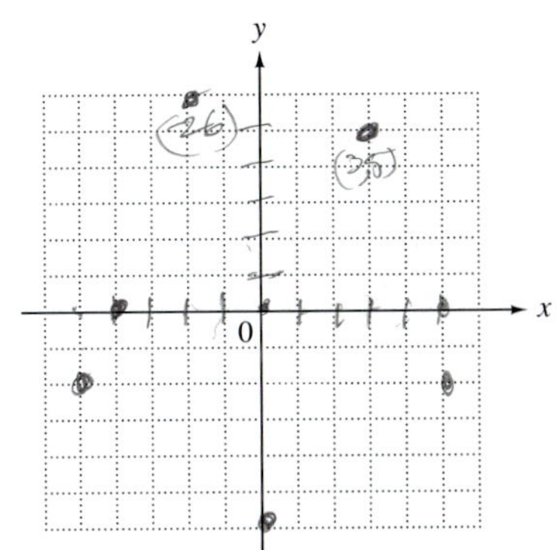

ANSWERS

7.

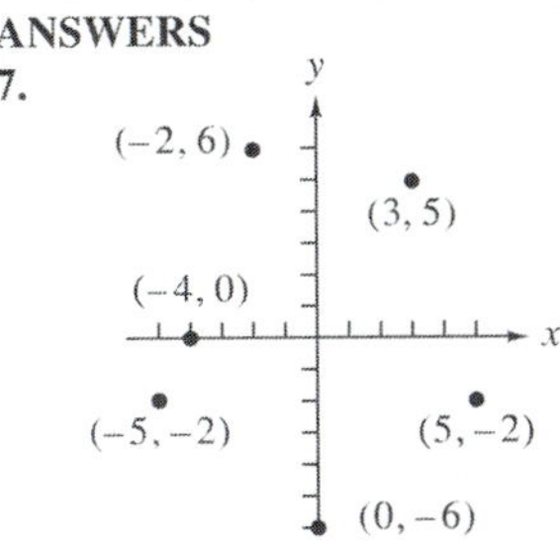

8. Complete the given ordered pairs using the equation $y = 25x + 250$ in Example 6.

(a) (4,)

(b) (5,)

(c) (6,)

(d) (7,)

(e) (8,)

(f) (9,)

This gives the ordered pair (1, 275), which says that the cost to produce 1 calculator is 275 cents or $2.75. Complete the ordered pairs (2,) and (3,) as follows.

$$\begin{aligned} y &= 25x + 250 \\ y &= 25(2) + 250 \\ y &= 50 + 250 \\ y &= 300 \end{aligned} \qquad \begin{aligned} y &= 25x + 250 \\ y &= 25(3) + 250 \\ y &= 75 + 250 \\ y &= 325 \end{aligned}$$

This gives the ordered pairs (2, 300) and (3, 325). ■

WORK PROBLEM 8 AT THE SIDE.

The ordered pairs (2, 300) and (3, 325), along with the ones given in Problem 8, are graphed in Figure 4. In this graph, different scales are used on the two axes because the y-values in the ordered pairs are much larger than the x-values. Here, each square represents 50 units in the vertical direction.

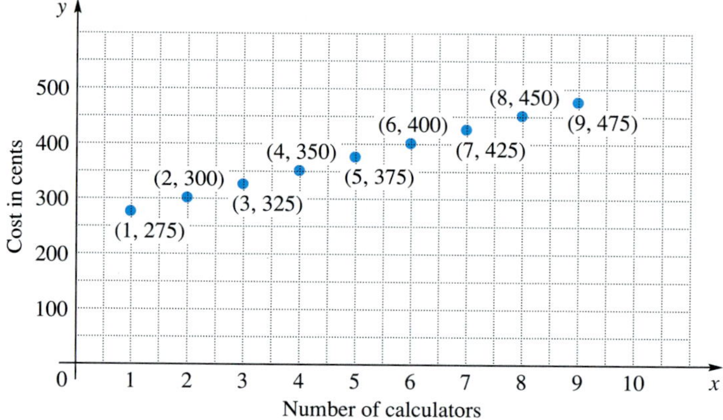

FIGURE 4

ANSWERS
8. (a) 350 (b) 375
(c) 400 (d) 425
(e) 450 (f) 475

NAME DATE HOUR

6.1 EXERCISES

Fill in the blank with the correct response.

1. The symbol (x, y) ________ (does/does not) represent an ordered pair, while the symbols $[x, y]$ and $\{x, y\}$ ________ (do/do not) represent ordered pairs.

2. The ordered pair (3, 2) is a solution of the equation $2x - 5y =$ ____.

3. The point whose graph has coordinates $(-4, 2)$ is in quadrant ____.

4. The point whose graph has coordinates (0, 5) lies along the ____-axis.

5. The ordered pair (4, ____) is a solution of the equation $y = 3$.

6. The ordered pair (____, -2) is a solution of the equation $x = 6$.

Decide whether the given ordered pair is a solution of the given equation. See Example 1.

7. $x + y = 9$; (0, 9)

8. $x + y = 8$; (0, 8)

9. $2p - q = 6$; (4, 2)

10. $2v + w = 5$; $(3, -1)$

11. $4x - 3y = 6$; (2, 1)

12. $5x - 3y = 15$; (5, 2)

13. $y = 3x$; (2, 6)

14. $x = -4y$; $(-8, 2)$

15. $x = -6$; $(5, -6)$

16. $y = 2$; (2, 4)

17. $x + 4 = 0$; $(-6, 2)$

18. $x - 6 = 0$; (4, 2)

Complete the given ordered pairs for the equation $y = 2x + 7$. See Example 2.

19. (2,)

20. (5,)

21. (0,)

22. (, 0)

23. (, -3)

24. $(-6,\ )$

Complete the given ordered pairs for the equation $y = -4x - 4$. *See Example 2.*

25. (0,) **26.** (, 0) **27.** (, 16)

28. (, 24) **29.** (10,) **30.** (5,)

31. Explain why it would be easier to find the corresponding y value for $x = \frac{1}{3}$ in the equation $y = 6x + 2$ than it would be for $x = \frac{1}{7}$.

32. For the equation $y = mx + b$, what is the y-value corresponding to $x = 0$ for *any* value of m?

Complete the given ordered pairs or table of ordered pairs for each equation. See Examples 2 and 3.

33. $y = 2x + 6$ (3,)(0,)(−1,)

34. $y = 3x + 5$ (2,)(0,)(−3,)

35. $y = -8 - 2x$ (2,)(0,)(−3,)

36. $y = -2 - 3x$ (4,)(0,)(−4,)

37. $2x + y = 5$ (0,)(, 0)(12,)

38. $-3m + n = 4$ (0,)(, 0)(11,)

39. $2p + 3q = 12$

p	0		
q		0	8

40. $4t + 3w = 24$

t	0		
w		0	4

41. $3u - 5v = -15$

u	0		
v		0	−6

42. $4x - 9y = -36$

x		0	
y	0		8

43. Plot the following ordered pairs on the grid below: (3, −2), (3, 0), (3, 4), (3, 5).

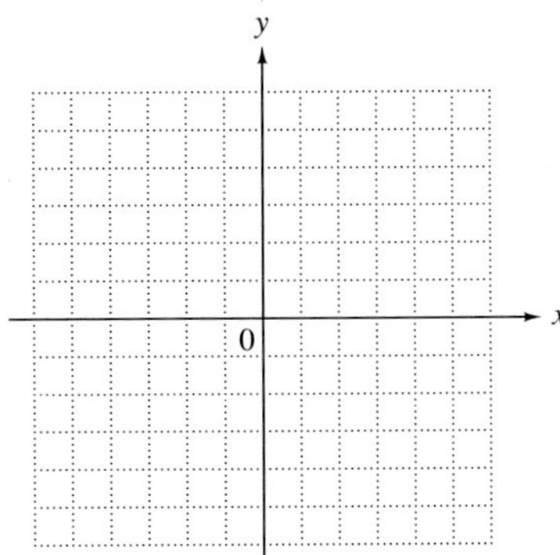

Based on this exercise, it would seem reasonable to conclude that if several ordered pairs have the same _____ value, they all lie on the same ______________ line.
(horizontal/vertical)

44. Plot the following ordered pairs on the grid below: (−4, 2), (−2, 2), (0, 2), (5, 2).

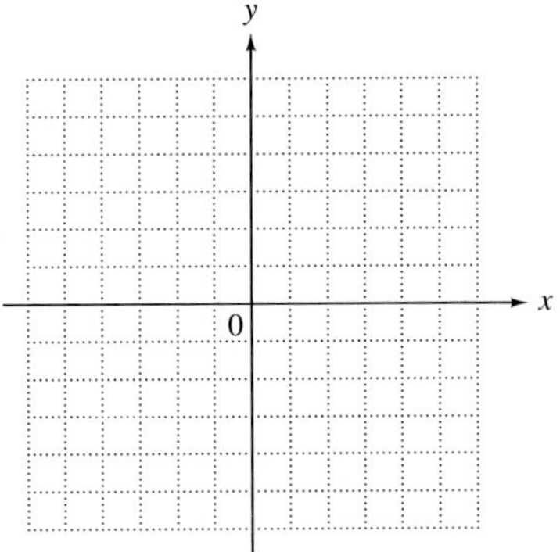

Based on this exercise, it would seem reasonable to conclude that if several ordered pairs have the same _____ value, they all lie on the same ______________ line.
(horizontal/vertical)

Complete the given table of ordered pairs for each equation. See Example 4.

45. $x = -9$

x			
y	6	2	−3

46. $x = 12$

x			
y	3	8	0

47. $y = -6$

x	8	4	−2
y			

48. $y = -10$

x	4	0	−4
y			

49. $x - 8 = 0$

x			
y	8	3	0

50. $y + 2 = 0$

x	9	2	0
y			

Give the ordered pairs for the points labeled A through F in the figure.

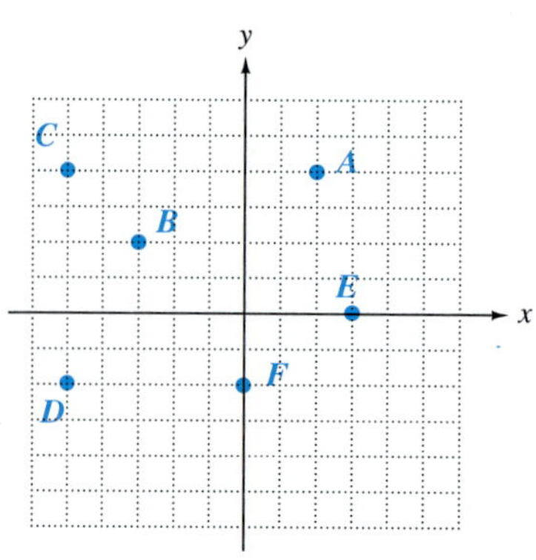

51. A

52. B

53. C

54. D

55. E

56. F

57. If the ordered pairs $(a + 1, b - 2)$ and $(4, 7)$ have the same point as their graph, what are the values of a and b?

58. Explain why $(2, 4)$ and $(4, 2)$ do not have the same point as their graphs.

Plot the ordered pairs on the coordinate system provided. See Example 5.

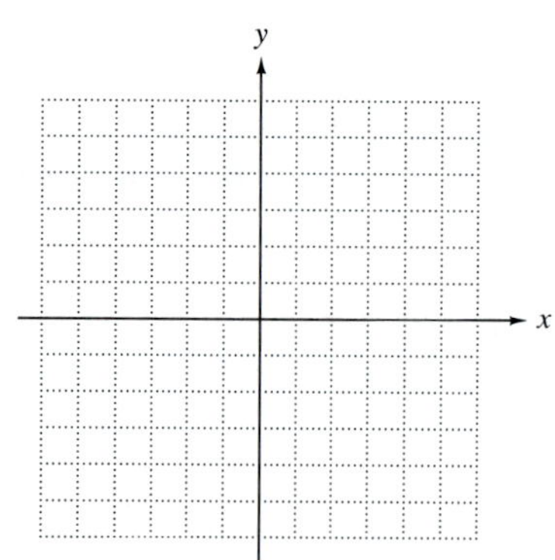

59. $(6, 2)$

60. $(5, 3)$

61. $(-4, 2)$

62. $(-3, 5)$

63. $\left(-\frac{4}{5}, -1\right)$

64. $\left(-\frac{3}{2}, -4\right)$

65. $(0, 4)$

66. $(0, -3)$

67. $(4, 0)$

68. $(-3, 0)$

Fill in the blank with the word positive *or the word* negative. *The point with coordinates* (x, y) *is in*

69. quadrant III if x is ________ and y is ________.

70. quadrant II if x is ________ and y is ________.

71. quadrant IV if x is ________ and y is ________.

72. quadrant I if x is ________ and y is ________.

Complete the table of ordered pairs for each equation, and then plot the ordered pairs. See Examples 3 and 5.

73. $x - 2y = 6$

x	y
0	
	0
2	
	−1

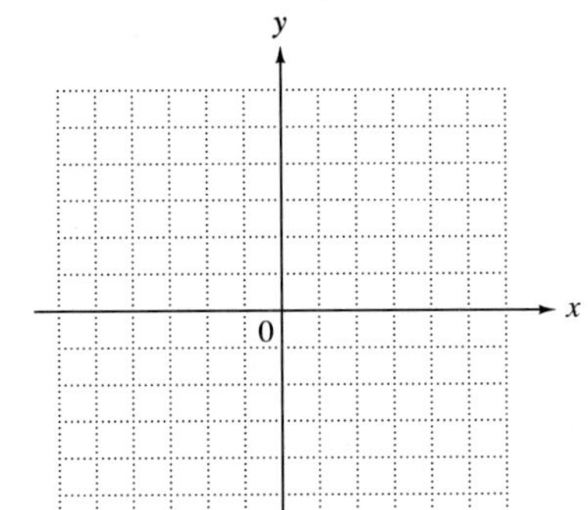

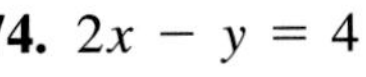

74. $2x - y = 4$

x	y
0	
	0
1	
	−6

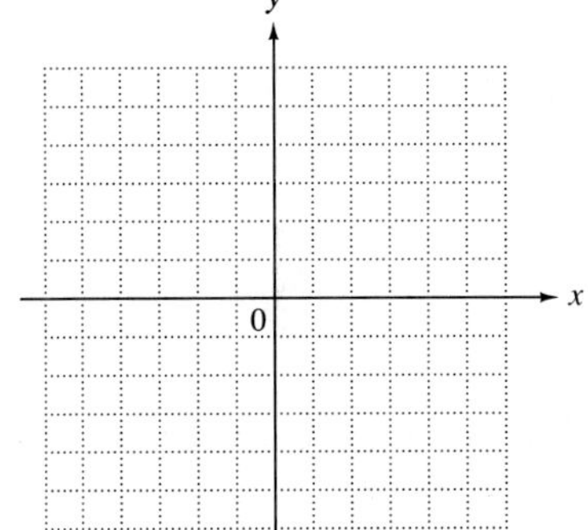

75. $3x - 4y = 12$

x	y
0	
	0
−4	
	−4

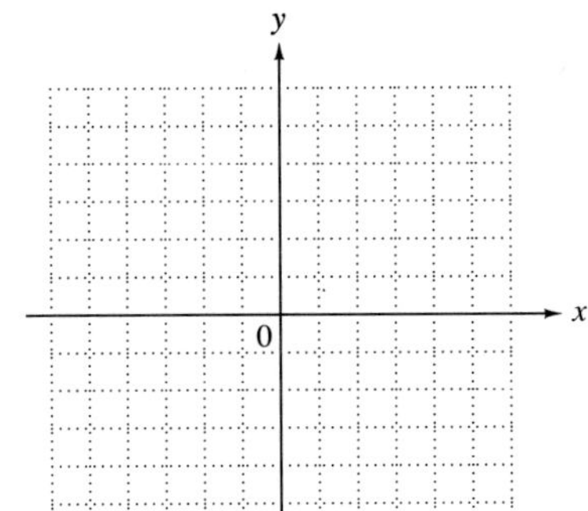

76. $2x - 5y = 10$

x	y
0	
	0
−5	
	−3

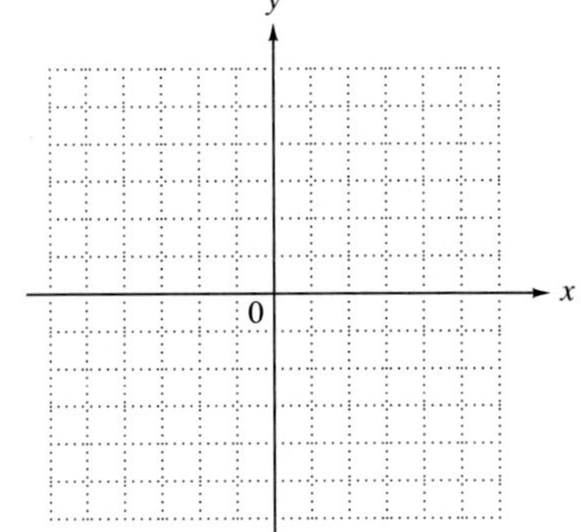

77. $y + 4 = 0$

x	y
0	
5	
−2	
−3	

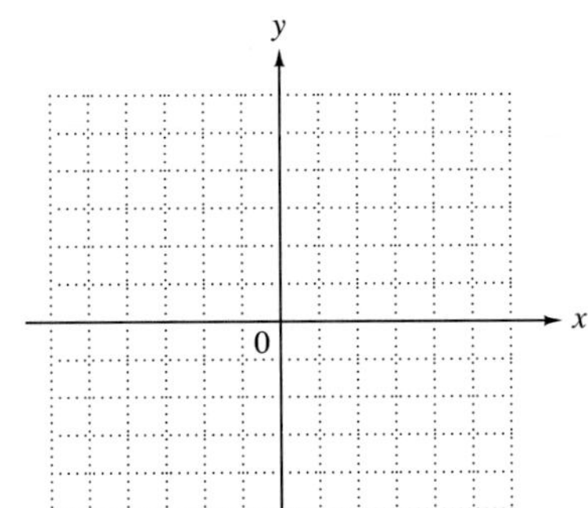

78. $x - 5 = 0$

x	y
	1
	0
	6
	−4

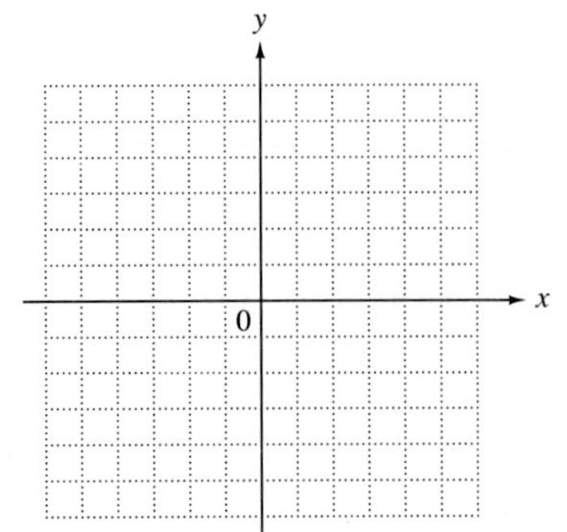

Work the following problems. See Example 6.

79. The number of U.S. jobs supported by exports to Mexico has steadily increased over the past few years. If we let $x = 0$ represent the year 1986 and let y represent the number of jobs supported (in thousands), then the following ordered pairs are formed: (0, 275), (1, 300), (2, 400), (3, 500), (4, 425), (5, 610), (6, 725). Plot these points on the axes below.

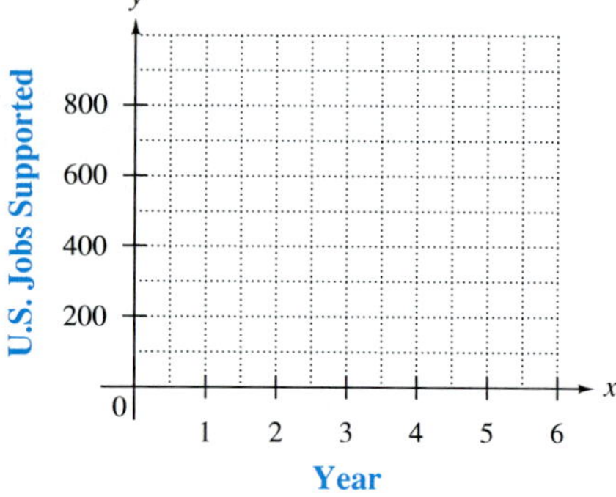

80. In statistics we often want to know whether two quantities (such as the height and weight of an individual) are related in such a way that we can predict one given the other. To find out, ordered pairs that give these quantities for a number of individuals are plotted on a graph called a *scatter diagram.* If the points lie approximately in a line, the variables have a linear relationship.

(a) Make a scatter diagram by plotting on the given axes the following pairs of heights and weights for six women: (62, 105), (65, 130), (67, 142), (63, 115), (66, 120), (60, 98). (The break on the horizontal and vertical axes shows that numbers have been skipped.)

(b) Does there seem to be a linear relationship between height and weight?

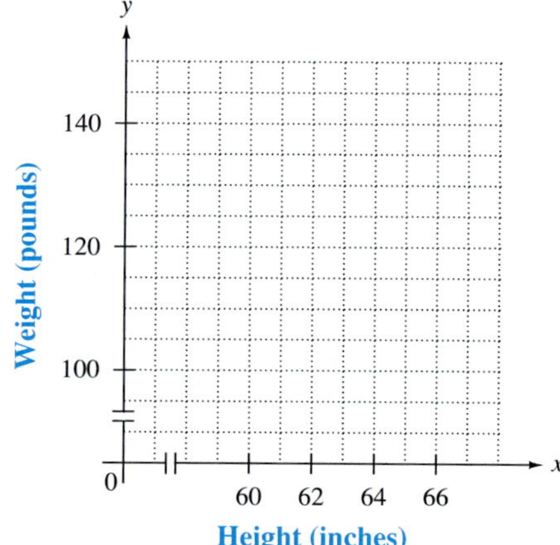

PREVIEW EXERCISES

Solve each equation. See Section 2.3.

81. $3x + 6 = 9$

82. $4 + 2y = 10$

83. $9 - y = -4$

84. $-5 + y = 3$

85. $2x - 8 = 0$

86. $-5x + 9 = 14$

6.2 GRAPHING LINEAR EQUATIONS IN TWO VARIABLES

1 There is an infinite number of ordered pairs that satisfy an equation in two variables. For example, we find ordered pairs that are solutions of the equation $x + 2y = 7$ by choosing as many values of x (or y) as we wish and then completing each ordered pair.

For example, if we choose $x = 1$, then $y = 3$, so that the ordered pair $(1, 3)$ is a solution of the equation $x + 2y = 7$.

$$1 + 2(3) = 1 + 6 = 7$$

WORK PROBLEM 1 AT THE SIDE.

Figure 5 shows a graph of all the ordered pairs found for $x + 2y = 7$ in Problem 1 at the side.

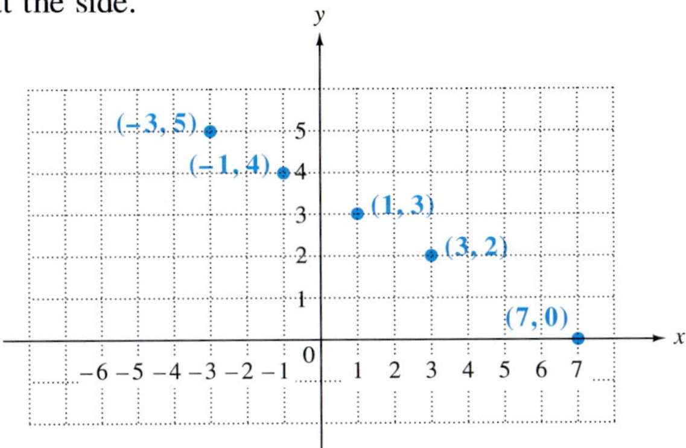

FIGURE 5

Notice that the points plotted in this figure all appear to lie on a straight line. The line that goes through these points is shown in Figure 6. In fact, all ordered pairs satisfying the equation $x + 2y = 7$ correspond to points that lie on this same straight line. This line gives a "picture" of all the solutions of the equation $x + 2y = 7$. Only a portion of the line is shown here, but it extends indefinitely in both directions, as suggested by the arrowhead on each end of the line.

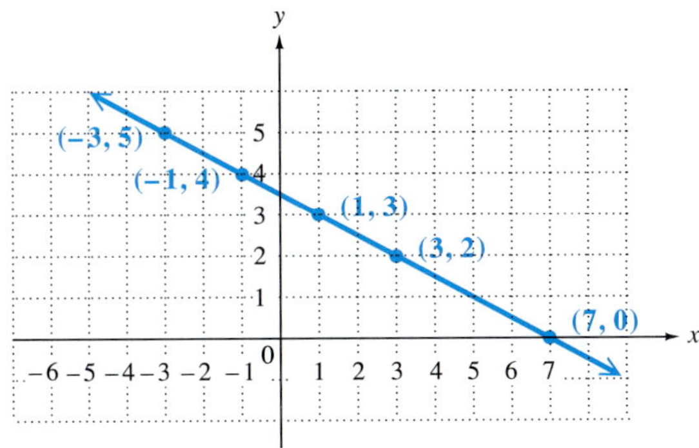

FIGURE 6

The line in Figure 6 is called the **graph** of the equation $x + 2y = 7$, and the process of plotting the ordered pairs and drawing the line through the corresponding points is called **graphing.** The preceding discussion can be generalized.

> The graph of any linear equation in two variables is a straight line.

(Notice that the word *line* appears in the name "*linear* equation.")

OBJECTIVES

1. Graph linear equations by completing ordered pairs.
2. Find intercepts.
3. Graph linear equations with just one intercept.
4. Graph linear equations of the form $y = k$ or $x = k$.

FOR EXTRA HELP

Tape 9	SSM pp. 225–229	MAC: A IBM: A

1. Complete the given ordered pairs for the equation $x + 2y = 7$.

(a) $(-3, \quad)$

(b) $(3, \quad)$

(c) $(-1, \quad)$

(d) $(7, \quad)$

ANSWERS

1. (a) $(-3, 5)$ (b) $(3, 2)$ (c) $(-1, 4)$ (d) $(7, 0)$

2. Complete the table of ordered pairs and then graph the line.

$x + y = 6$

x	y
0	
	0
2	

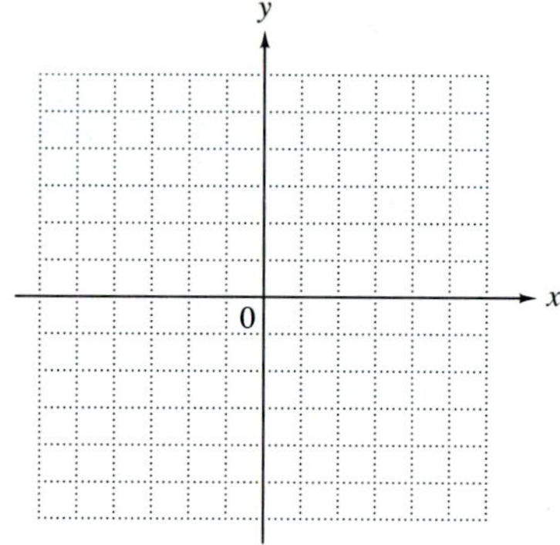

Because two distinct points determine a line, a straight line can be graphed by finding any two different points on the line. However, it is a good idea to plot a third point as a check.

■ EXAMPLE 1 *Graphing a Linear Equation*

Graph the linear equation $y = -\frac{3}{2}x + 3$.

Although this equation is not in the form $Ax + By = C$, it *could* be put in that form, and so is a linear equation. For most linear equations, two different points on the graph can be found by first letting $x = 0$ and then letting $y = 0$.

If $x = 0$,

$$y = -\frac{3}{2}x + 3$$

$$y = -\frac{3}{2}(0) + 3 \qquad \text{Let } x = 0.$$

$$y = 0 + 3$$

$$y = 3.$$

If $y = 0$,

$$y = -\frac{3}{2}x + 3$$

$$0 = -\frac{3}{2}x + 3 \qquad \text{Let } y = 0.$$

$$\frac{3}{2}x = 3$$

$$x = \frac{2}{3} \cdot 3 = 2.$$

This gives the ordered pairs (0, 3) and (2, 0). Get a third point (as a check) by letting x or y equal some other number. For example, let $x = -2$. (Any number could have been used instead.) Replace x with -2 in the given equation.

$$y = -\frac{3}{2}x + 3$$

$$y = -\frac{3}{2}(-2) + 3 \qquad \text{Let } x = -2.$$

$$y = 3 + 3 = 6$$

These three ordered pairs are shown in the table with Figure 7. Plot the corresponding points, then draw a line through them. This line, shown in Figure 7, is the graph of $y = -\frac{3}{2}x + 3$. ■

x	y
0	3
2	0
−2	6

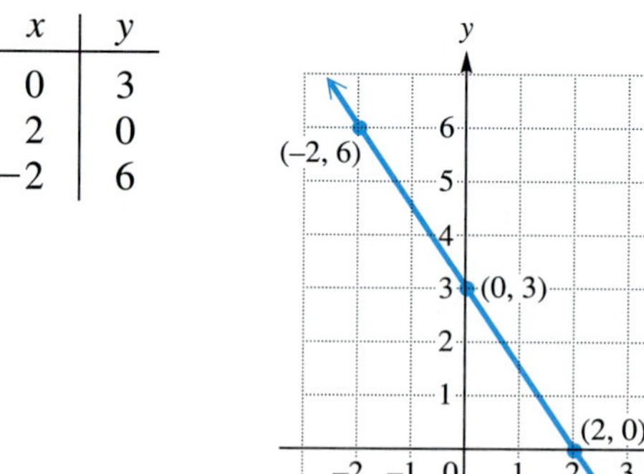

FIGURE 7

◀◀ WORK PROBLEM 2 AT THE SIDE.

ANSWERS

2. (0, 6), (6, 0), (2, 4)

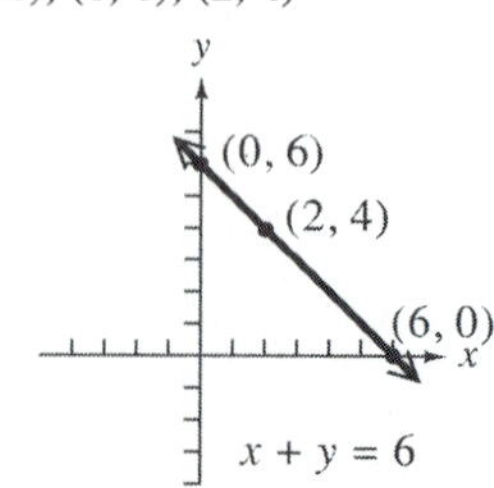

■ EXAMPLE 2 *Graphing a Linear Equation*

Graph the linear equation $4x = 5y + 20$.

As above, at least two different points are needed to draw the graph. First let $x = 0$ and then let $y = 0$ to complete two ordered pairs.

$$\begin{aligned} 4x &= 5y + 20 \\ 4(0) &= 5y + 20 \quad \text{Let } x = 0. \\ 0 &= 5y + 20 \\ -5y &= 20 \\ y &= -4 \end{aligned} \qquad \begin{aligned} 4x &= 5y + 20 \\ 4x &= 5(0) + 20 \quad \text{Let } y = 0. \\ 4x &= 20 \\ x &= 5 \end{aligned}$$

The ordered pairs are $(0, -4)$ and $(5, 0)$. Get a third ordered pair (as a check) by choosing some number other than 0 for x or y. This time, let us choose $y = 2$. Replacing y with 2 in the equation $4x = 5y + 20$ leads to the ordered pair $(\frac{15}{2}, 2)$, or $(7\frac{1}{2}, 2)$.

Plot the three ordered pairs we have found, $(0, -4)$, $(5, 0)$, and $(\frac{15}{2}, 2)$ and draw a line through them. This line, shown in Figure 8, is the graph of $4x = 5y + 20$. ■

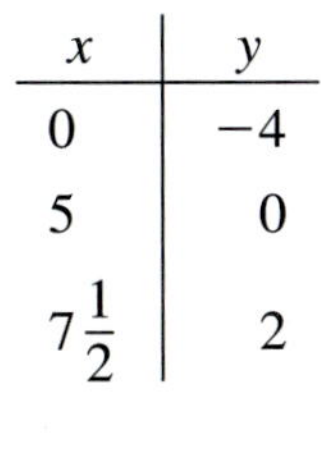

x	y
0	-4
5	0
$7\frac{1}{2}$	2

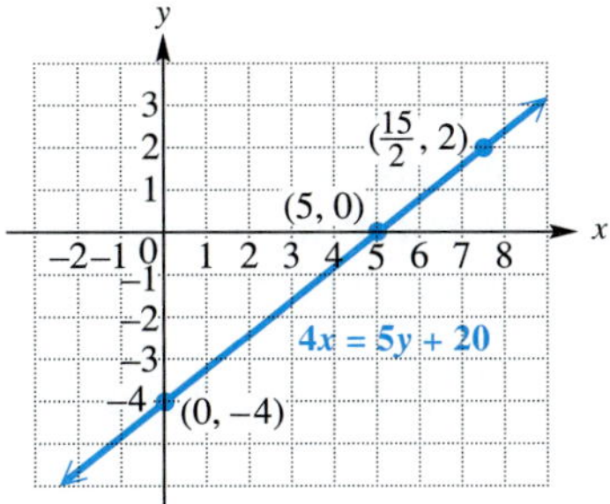

FIGURE 8

WORK PROBLEM 3 AT THE SIDE. ▶▶

2 In Figure 8 the graph crosses the y-axis at $(0, -4)$ and crosses the x-axis at $(5, 0)$. For this reason, $(0, -4)$ is called the ***y*-intercept,** and $(5, 0)$ is called the ***x*-intercept** of the graph. The intercepts are particularly useful for graphing linear equtions, as in Examples 1 and 2. The intercepts are found by replacing, in turn, each variable with 0 in the equation and solving for the value of the other variable.

FINDING INTERCEPTS

Find the x-intercept by letting $y = 0$ in the given equation and solving for x.
Find the y-intercept by letting $x = 0$ in the given equation and solving for y.

■ EXAMPLE 3 *Finding Intercepts*

Find the intercepts for the graph of $2x + y = 4$. Draw the graph.

Find the y-intercept by letting $x = 0$; find the x-intercept by letting $y = 0$.

$$\begin{aligned} 2x + y &= 4 \\ 2(0) + y &= 4 \quad \text{Let } x = 0. \\ 0 + y &= 4 \\ y &= 4 \end{aligned} \qquad \begin{aligned} 2x + y &= 4 \\ 2x + 0 &= 4 \quad \text{Let } y = 0. \\ 2x &= 4 \\ x &= 2 \end{aligned}$$

3. Complete three ordered pairs and graph the linear equation.

$2x = 3y + 6$

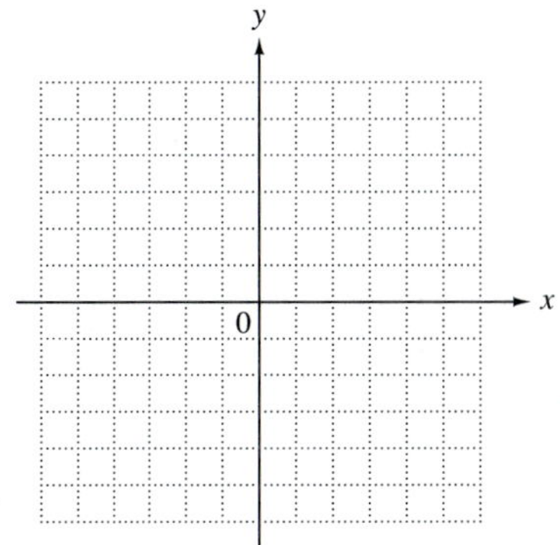

ANSWER

3.

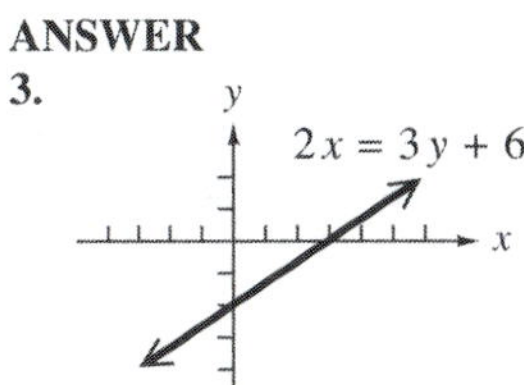

4. Find the intercepts for $5x + 2y = 10$.

The y-intercept is (0, 4). The x-intercept is (2, 0). The graph with the two intercepts shown in color is given in Figure 9. Get a third point as a check. For example, choosing $x = 1$ gives $y = 2$. Plot (0, 4), (2, 0), and (1, 2) and draw a line through them. This line, shown in Figure 9, is the graph of $2x + y = 4$. ■

x	y
0	4
2	0
1	2

FIGURE 9

◀ **WORK PROBLEM 4 AT THE SIDE.**

3 In the examples above, the x- and y-intercepts were used to help draw the graphs. This is not always possible, as the following examples show. Example 4 shows what to do when the x- and y-intercepts are same point.

■ **EXAMPLE 4** *Graphing an Equation of the Form $Ax + By = 0$*

Graph the linear equation $x - 3y = 0$.

If we let $x = 0$, then $y = 0$, giving the ordered pair (0, 0). Letting $y = 0$ also gives (0, 0). This is the same ordered pair, so choose two additional values for x or y. Choosing 2 for y gives $x - 3 \cdot 2 = 0$, giving the order pair (6, 2). For a check point, choose -6 for x getting -2 for y. This ordered pair, $(-6, -2)$, along with (0, 0) and (6, 2), was used to get the graph shown in Figure 10. ■

x	y
0	0
6	2
−6	−2

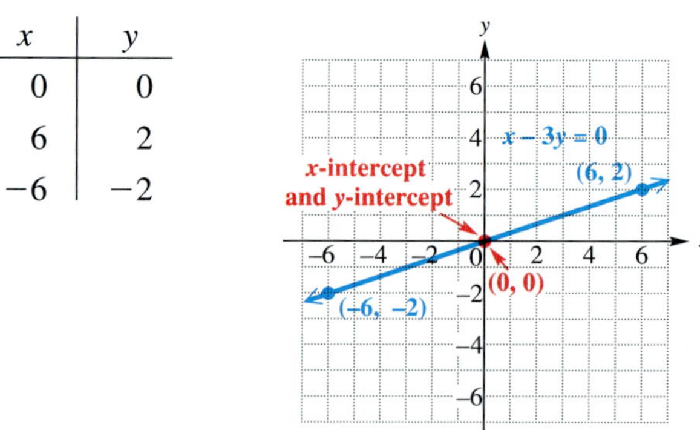

FIGURE 10

Example 4 can be generalized as follows.

If A and B are real numbers, the graph of a linear equation of the form

$$Ax + By = 0$$

goes through the origin (0, 0).

ANSWERS

4. x-intercept (2, 0); y-intercept (0, 5)

WORK PROBLEM 5 AT THE SIDE.

4 The equation $y = -4$ is a linear equation with the coefficient of x equal to 0. (To see this, write $y = -4$ as $0x + y = -4$.) Also, $x = 3$ is a linear equation with the coefficient of y equal to 0. These equations lead to horizontal or vertical straight lines, as the next examples show.

EXAMPLE 5 *Graphing an Equation of the Form $y = k$*

Graph the linear equation $y = -4$.

As the equation states, for any value of x that might be chosen, y is always equal to -4. Get ordered pairs that are solutions of this equation by choosing different numbers for x but always using -4 for y. Three ordered pairs that can be used are shown in the table of ordered pairs with Figure 11. Drawing a line through these points gives the horizontal line shown in Figure 11. ■

x	y
-2	-4
0	-4
3	-4

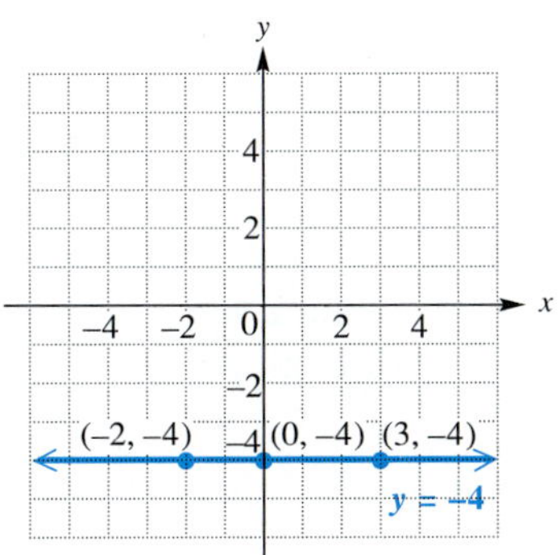

FIGURE 11

HORIZONTAL LINE

The graph of the linear equation $y = k$, where k is a real number, is the horizontal line going through the point $(0, k)$.

EXAMPLE 6 *Graphing an Equation of the Form $x = k$*

Graph the linear equation $x - 3 = 0$.

First add 3 to both sides of $x - 3 = 0$ to get $x = 3$. All the ordered pairs that satisfy this equation have an x-coordinate of 3. Any number can be used for y. Three ordered pairs that work are shown in the table of values with Figure 12. Drawing a line through these points gives the vertical line shown in Figure 12. ■

x	y
3	3
3	0
3	-2

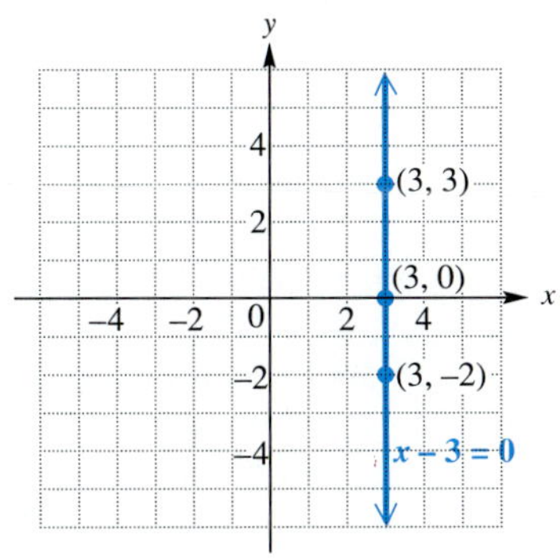

FIGURE 12

5. Graph each equation.

(a) $2x = y$

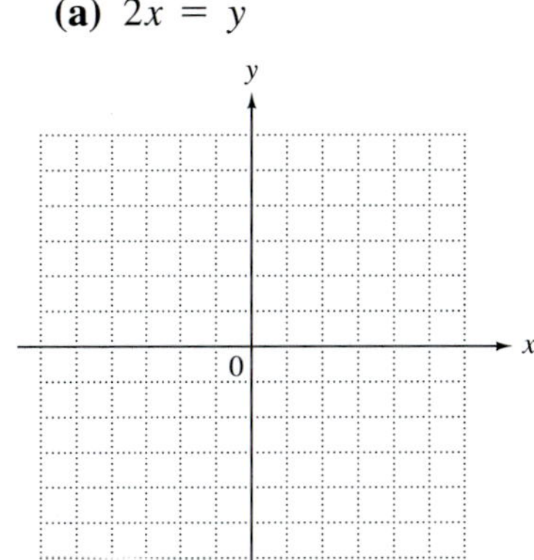

(b) $x = -4y$

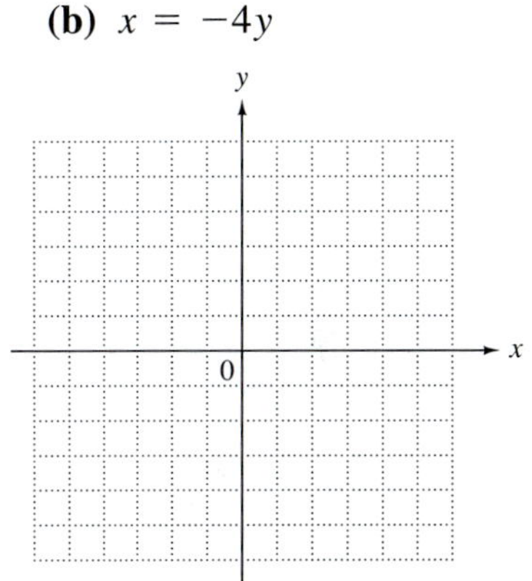

ANSWERS

5. (a)

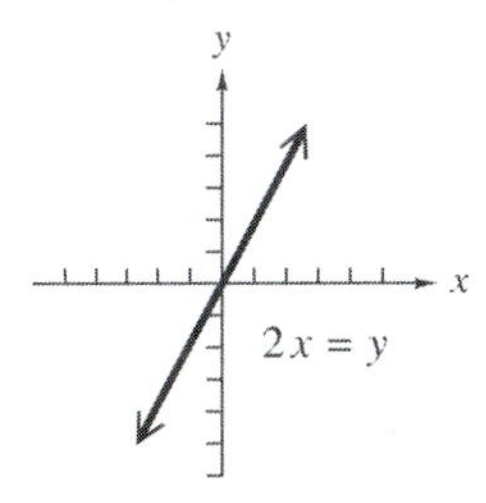

(b)

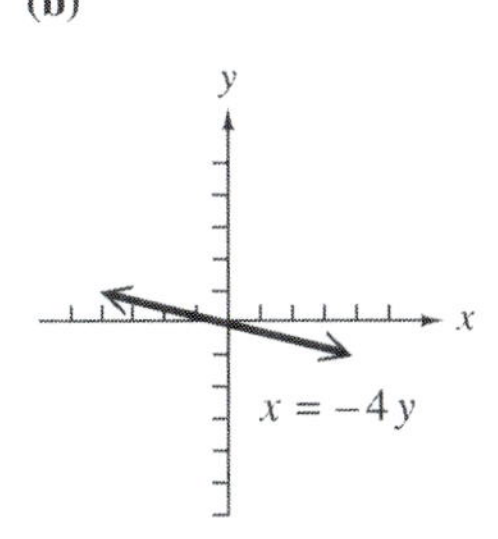

6. Graph each equation.

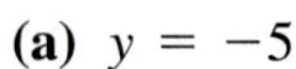

(a) $y = -5$

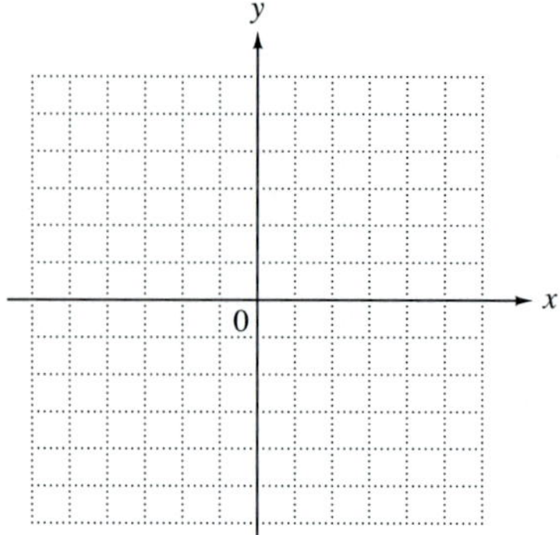

(b) $x = 2$

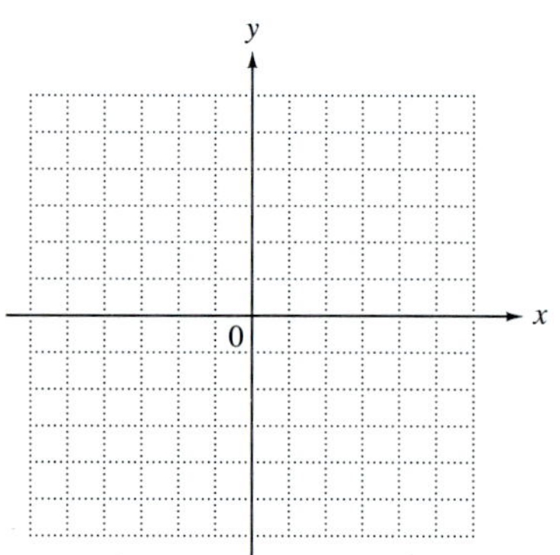

ANSWERS

6. (a) **(b)**

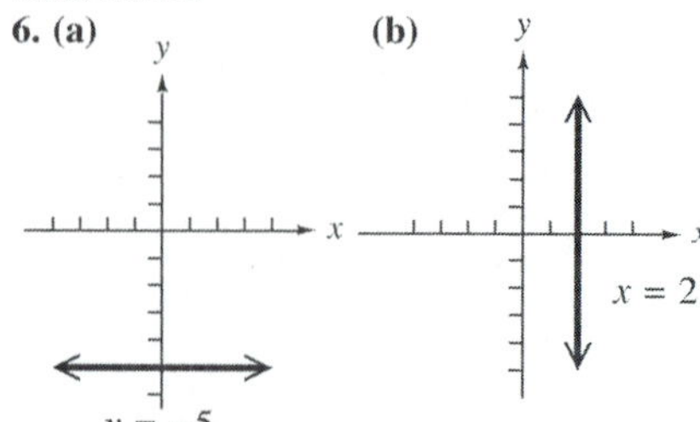

VERTICAL LINE

The graph of the linear equation $x = k$, where k is a real number, is the vertical line going through the point $(k, 0)$.

WORK PROBLEM 6 AT THE SIDE.

The different forms of straight-line equations and the methods of graphing them are summarized below.

GRAPHING LINEAR EQUATIONS

Equation	*Graphing methods*	*Example*
$y = k$	Draw a horizontal line, through $(0, k)$.	$y = -2$ 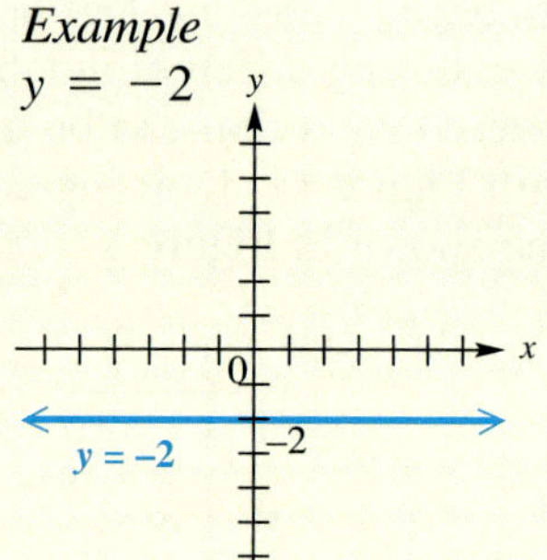
$x = k$	Draw a vertical line, through $(k, 0)$.	$x = 4$ 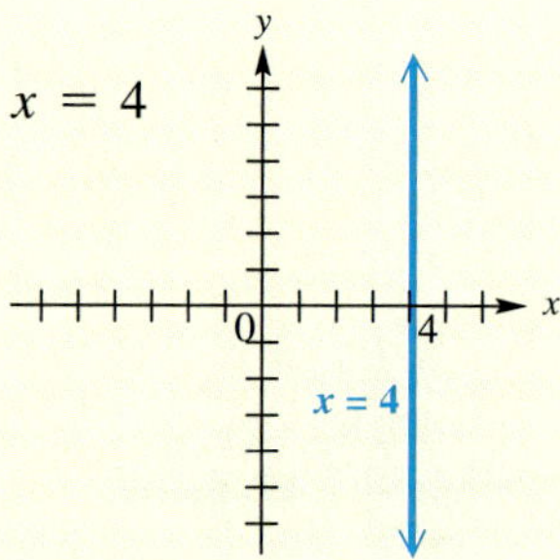
$Ax + By = 0$	Graph goes through $(0, 0)$. To get additional points that lie on the graph, choose any value of x, or y, except 0.	$3x + y = 0$ 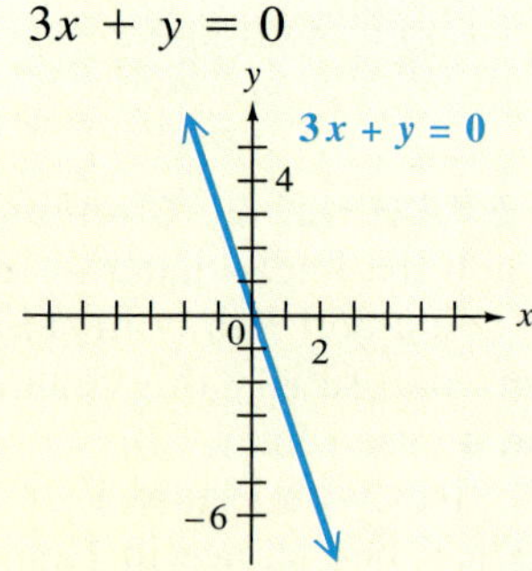
$Ax + By = C$ A, B and $C \neq 0$	Find any two points the line goes through. A good choice is to find the intercepts: let $x = 0$, and find the corresponding value of y; then let $y = 0$, and find x. As a check, get a third point by choosing a value of x or y that has not yet been used.	$3x - 2y = 6$ 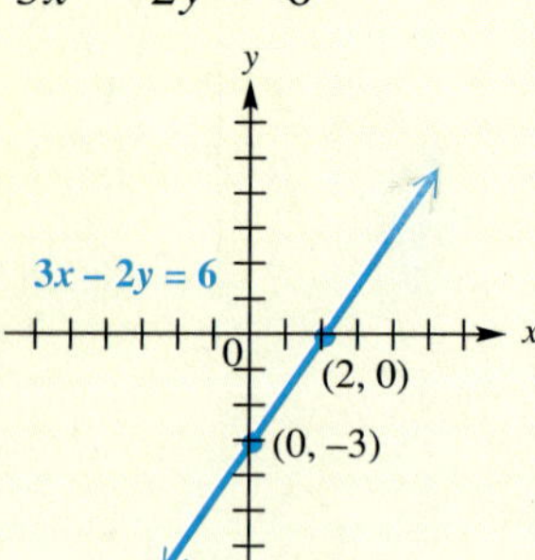

Fill in the blank with the correct response.

1. All ordered pairs that satisfy the equation $2x + 3y = 6$ lie on a straight ______.

2. To find the x-intercept for the graph of a linear equation, we let ______ $= 0$.

3. To find the y-intercept for the graph of a linear equation, we let ______ $= 0$.

4. The graph of an equation of the form $Ax + By = 0$ must go through the ______.

5. The graph of an equation of the form $y = k$ is a ______ (horizontal/vertical) line.

6. The graph of an equation of the form $x = k$ is a ______ (horizontal/vertical) line.

Complete the given ordered pairs using the given equation. Then graph the equation by plotting the points and drawing a line through them. See Examples 1 and 2.

7. $y = -x + 5$

(0,), (, 0), (2,)

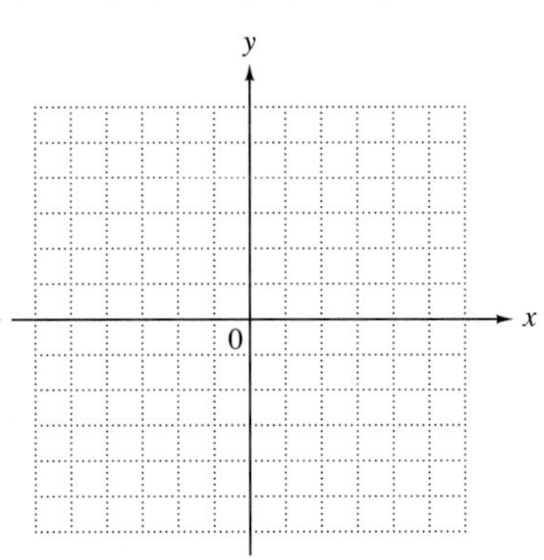

8. $y = x - 2$

(0,), (, 0), (5,)

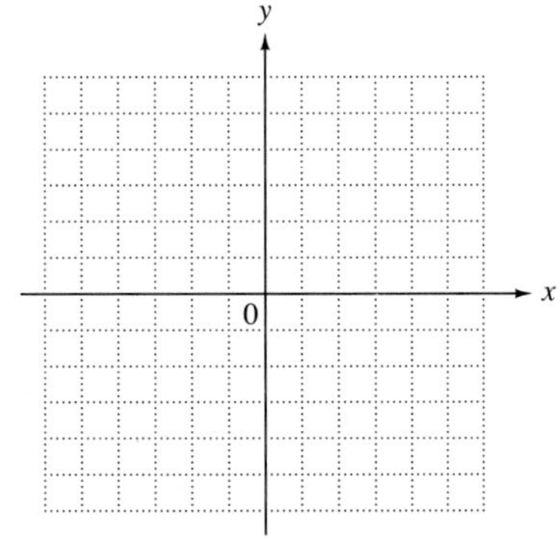

9. $y = \frac{2}{3}x + 1$

(0,), (3,), (−3,)

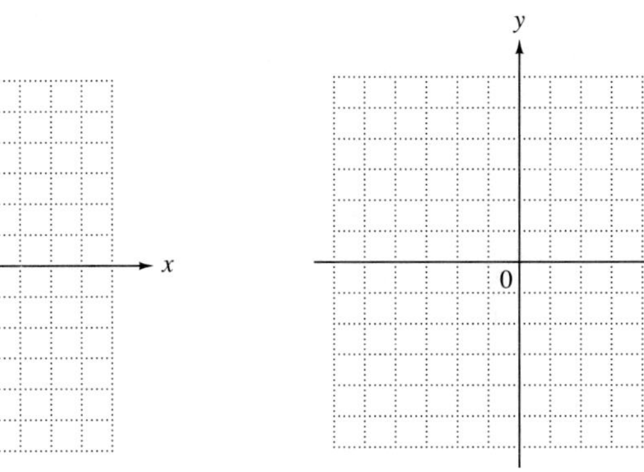

10. $y = -\frac{3}{4}x + 2$

(0,), (4,), (−4,)

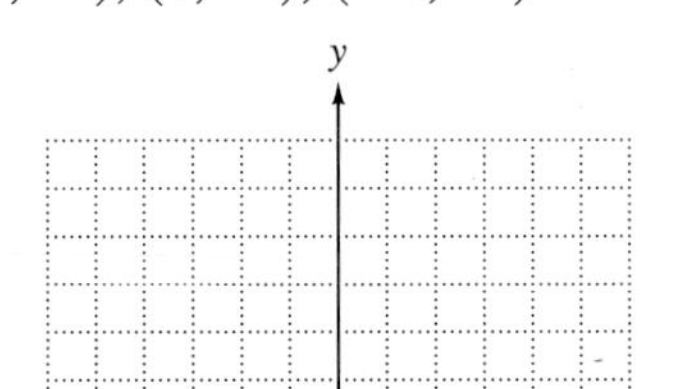

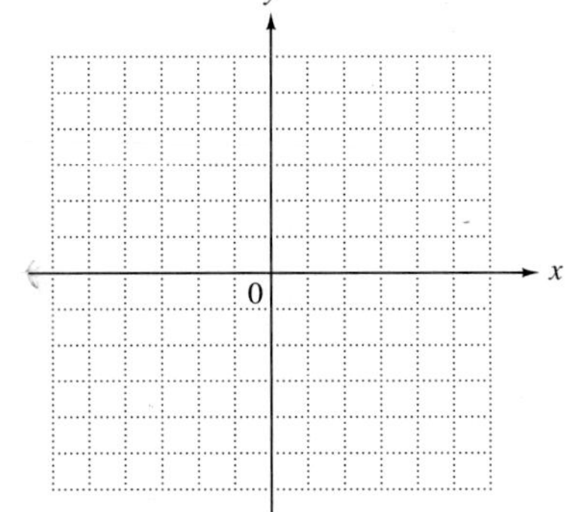

11. $3x = -y - 6$

(0,), (, 0), $\left(-\frac{1}{3}, \quad\right)$

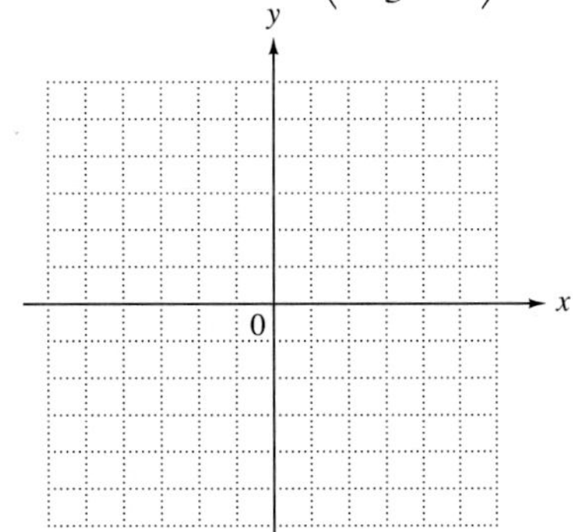

12. $x = 2y + 3$

(, 0), (0,), $\left(\quad, \frac{1}{2}\right)$

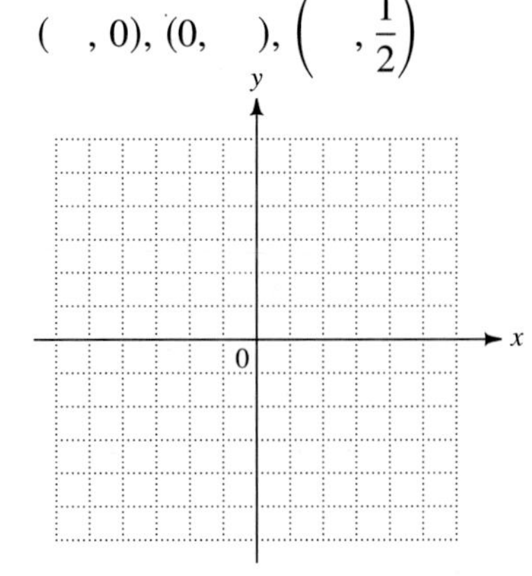

Find the intercepts for the graph of each equation. See Example 3.

13. $2x - 3y = 24$
x-intercept:
y-intercept:

14. $-3x + 8y = 48$
x-intercept:
y-intercept:

15. $x + 6y = 0$
x-intercept:
y-intercept:

16. $3x - y = 0$
x-intercept:
y-intercept:

17. A student attempted to graph $4x + 5y = 0$ by finding intercepts. She first let $x = 0$ and found y; then she let $y = 0$ and found x. In both cases, the resulting point was (0, 0). She knew that she needed at least two points to graph the line, but was unsure of what to do next because finding intercepts gave her only one point. How would you explain to her what to do next?

18. What is the equation of the x-axis? What is the equation of the y-axis?

Graph each linear equation. See Examples 1–6.

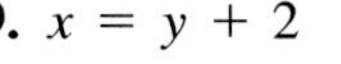
19. $x = y + 2$

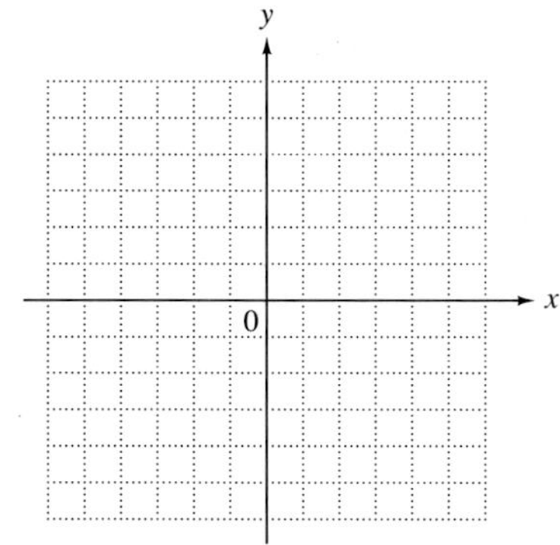

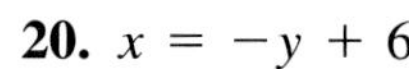
20. $x = -y + 6$

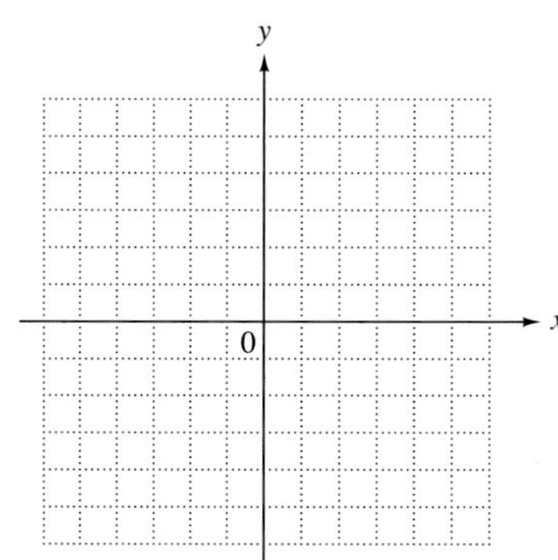

21. $x - y = 4$

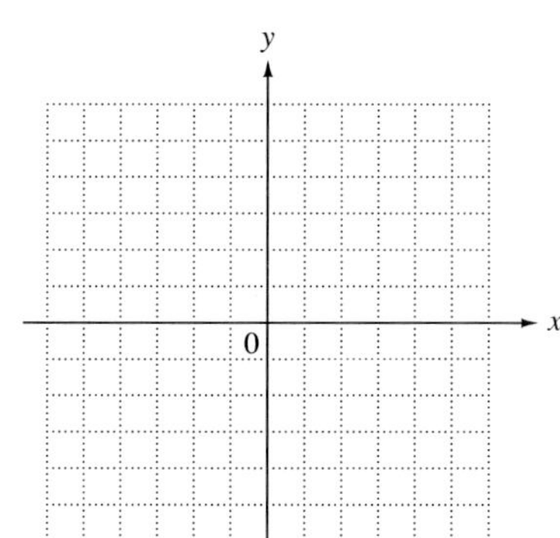

22. $x - y = 5$

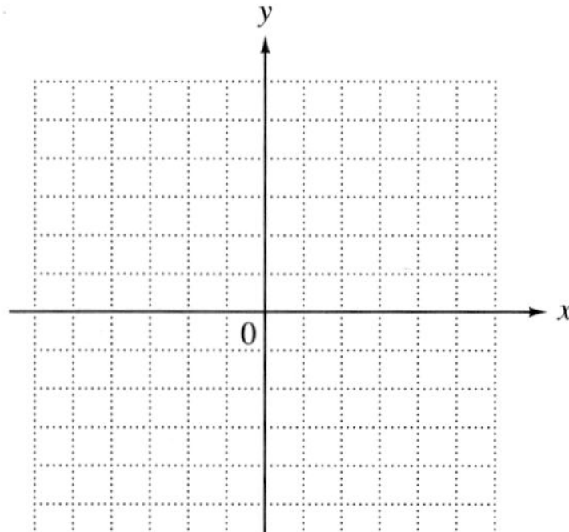

23. $2x + y = 6$

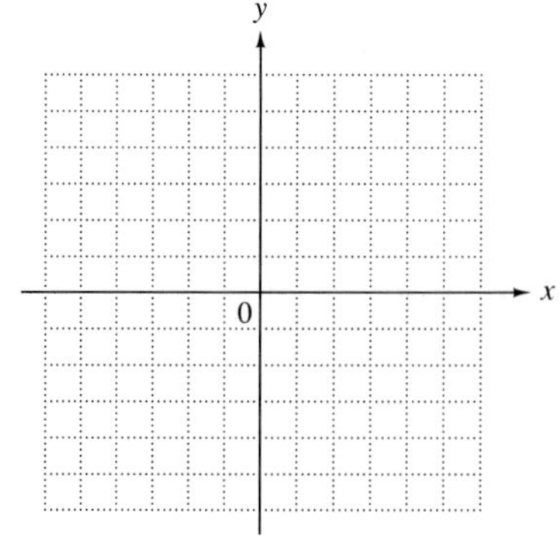

24. $-3x + y = -6$

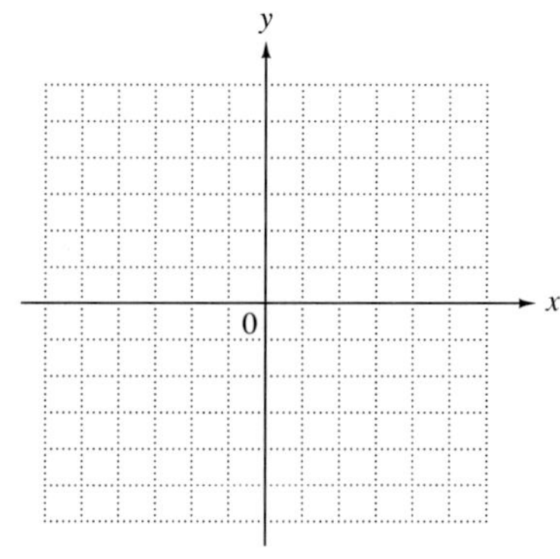

25. $3y = 4x + 12$

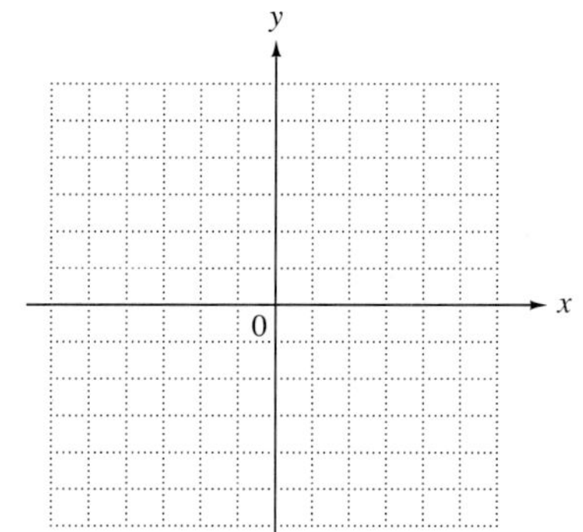

26. $2y = 5x + 10$

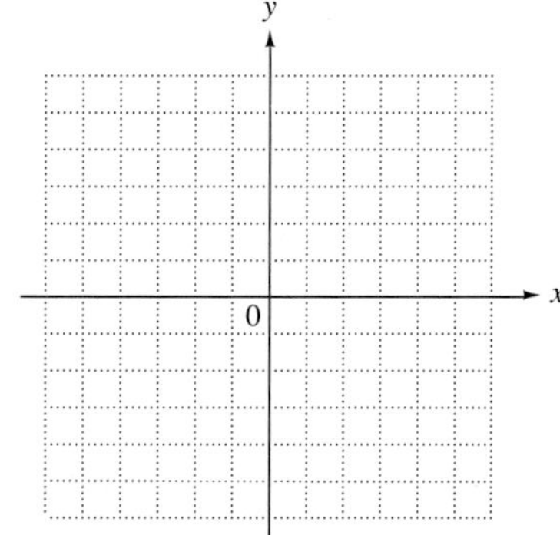

27. $3x + 7y = 14$

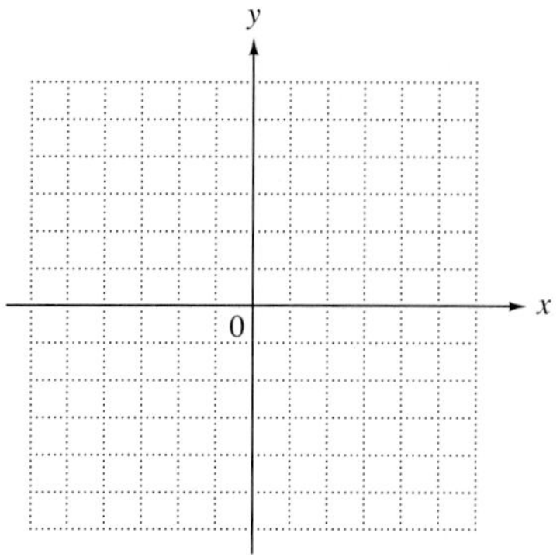

28. $6x - 5y = 18$

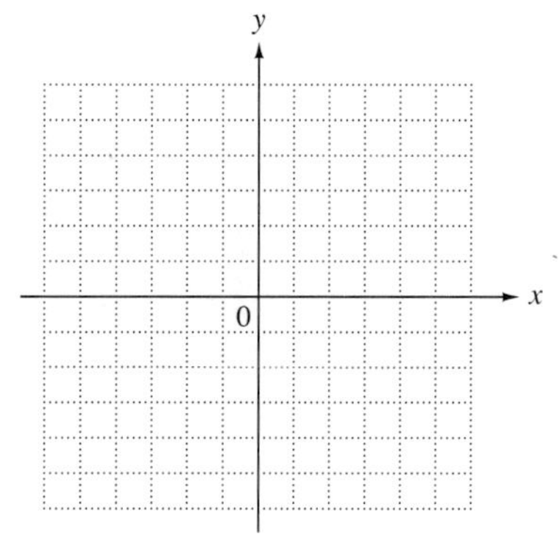

29. $y - 2x = 0$

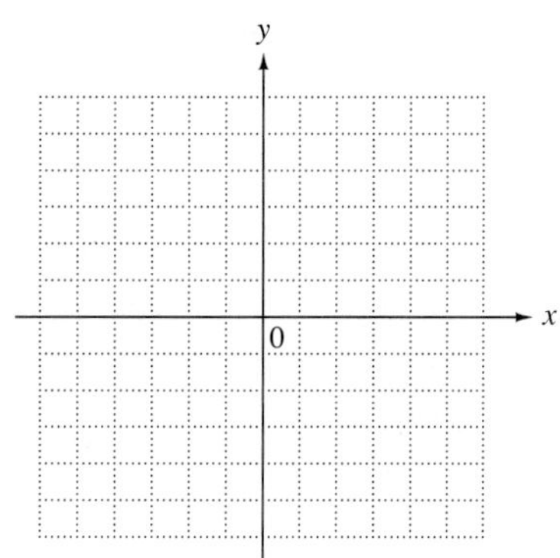

30. $y + 3x = 0$

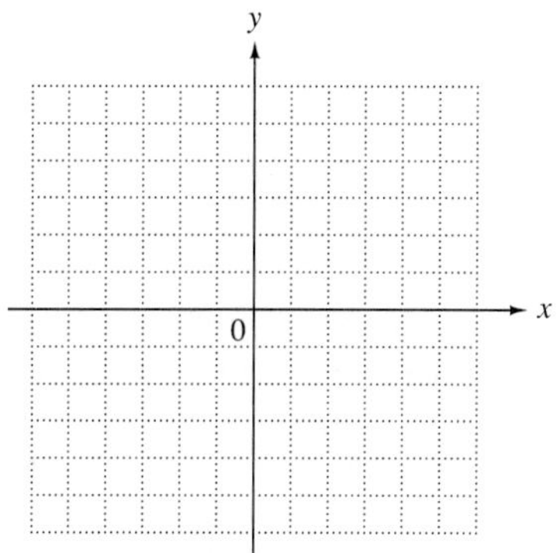

31. $y = -6x$

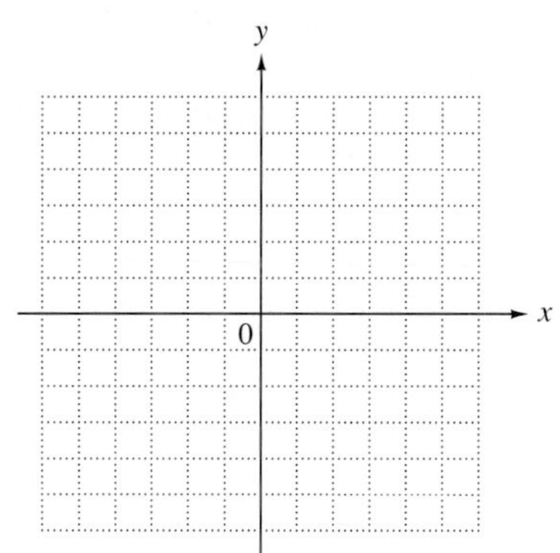

32. $y = 4x$

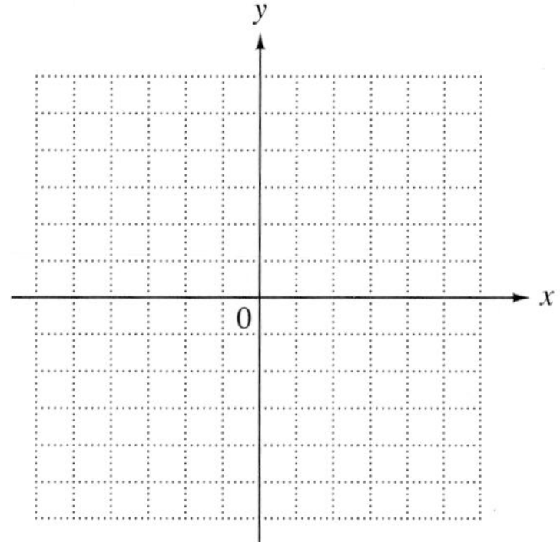

33. $y + 1 = 0$

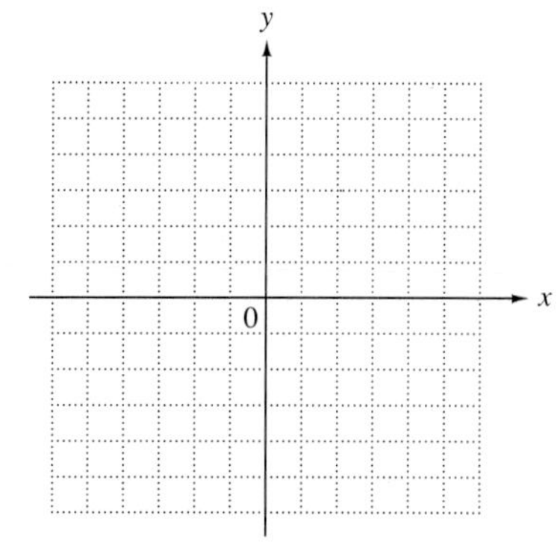

34. $y - 3 = 0$

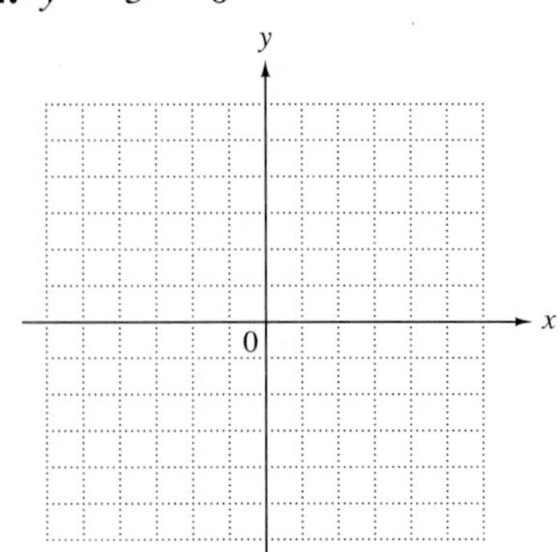

35. $x = -2$

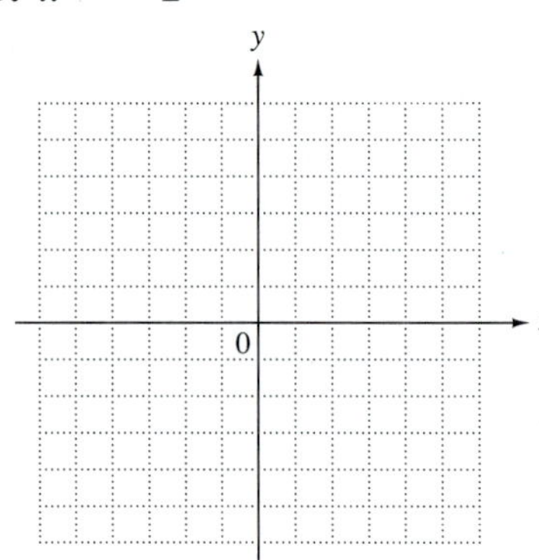

36. $x = 4$

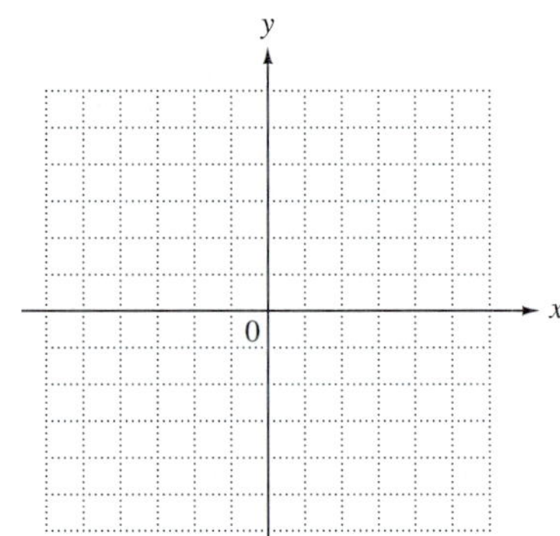

Work the following problems.

37. The graph shows the value of a certain minivan over the first five years of ownership. Use the graph to do the following.

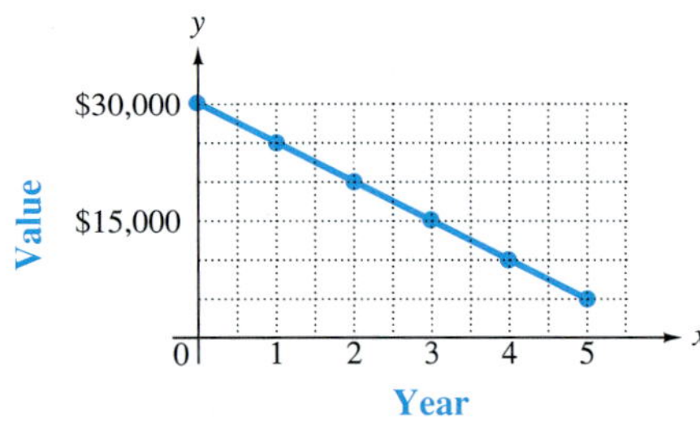

(a) Determine the initial value of the minivan.

(b) Find the depreciation (loss in value) from the original value after the first three years.

(c) What is the *yearly* depreciation in each of the first five years?

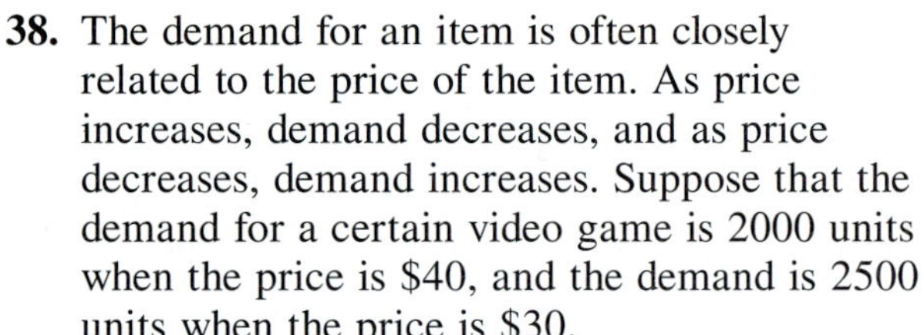

38. The demand for an item is often closely related to the price of the item. As price increases, demand decreases, and as price decreases, demand increases. Suppose that the demand for a certain video game is 2000 units when the price is \$40, and the demand is 2500 units when the price is \$30.

(a) Let x be the price and y be the demand for the game. Graph the two given pairs of prices and demands.

(b) Assume the relationship is linear. Draw a line through the two points from part (a). From your graph, estimate the demand if the price drops to \$20.

(c) Use the graph to estimate the price if the demand is 3500 units.

PREVIEW EXERCISES

Find each quotient, if possible. See Section 1.7.

39. $\dfrac{2 - 7}{-3 - 5}$

40. $\dfrac{3 - (-1)}{-2 - (-4)}$

41. $\dfrac{-2 + 2}{-9 - 4}$

42. $\dfrac{5 - (-4)}{6 - 6}$

6.3 THE SLOPE OF A LINE

We can graph a straight line if at least two different points on the line are known. We can also graph a line by using just one point on the line, along with the "steepness" of the line.

OBJECTIVES

1. Find the slope of a line given two points.
2. Find the slope from the equation of a line.
3. Use slopes to determine whether two lines are parallel, perpendicular, or neither.

FOR EXTRA HELP

Tape 9 | SSM pp. 229–233 | MAC: A IBM: A

1 One way to measure the steepness of a line is to compare the vertical change in the line (the rise) to the horizontal change (the run) while moving along the line from one fixed point to another. This measure of steepness is called the *slope* of the line.

Figure 13 shows a line with the points (x_1, y_1) and (x_2, y_2). (Read x_1 as "x-sub-one" and x_2 as "x-sub-two.") Moving along the line from the point (x_1, y_1) to the point (x_2, y_2) causes y to change by $y_2 - y_1$ units. This is the vertical change. Similarly, x changes by $x_2 - x_1$ units, the horizontal change. The ratio of the change in y to the change in x gives the slope of the line. We usually denote slope with the letter m. The slope of a line is defined as follows.

SLOPE FORMULA

The **slope** of the line through the points (x_1, y_1) and (x_2, y_2) is

$$m = \frac{\textbf{change in } y}{\textbf{change in } x} = \frac{y_2 - y_1}{x_2 - x_1} \quad \textbf{if } x_2 \neq x_1.$$

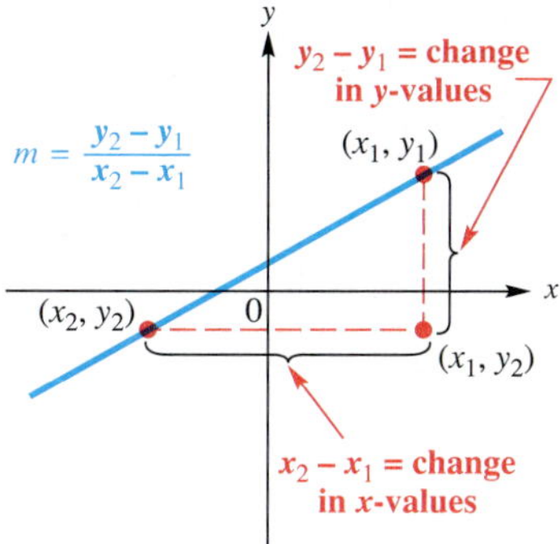

FIGURE 13

WORK PROBLEM 1 AT THE SIDE.

1. Evaluate $\frac{y_2 - y_1}{x_2 - x_1}$ for the following values.

(a) $y_2 = 4$, $y_1 = -1$
$x_2 = 3$, $x_1 = 4$

(b) $x_1 = 3$, $x_2 = -5$,
$y_1 = 7$, $y_2 = -9$

(c) $x_1 = 2$, $x_2 = 7$,
$y_1 = 4$, $y_2 = 9$

The slope of a line tells how fast y changes for each unit of change in x; that is, the slope gives the rate of change in y for each unit of change in x.

The idea of slope is useful in many everyday situations. For example, a highway with a 10% or $\frac{1}{10}$ grade (or slope) rises 1 meter for every 10 meters horizontally. Architects specify the pitch of a roof by indicating the slope; a $\frac{5}{12}$ roof means that the roof rises 5 feet for every 12 feet in the horizontal direction. The slope of a stairwell also indicates the ratio of the vertical rise to the horizontal run. See Figure 14.

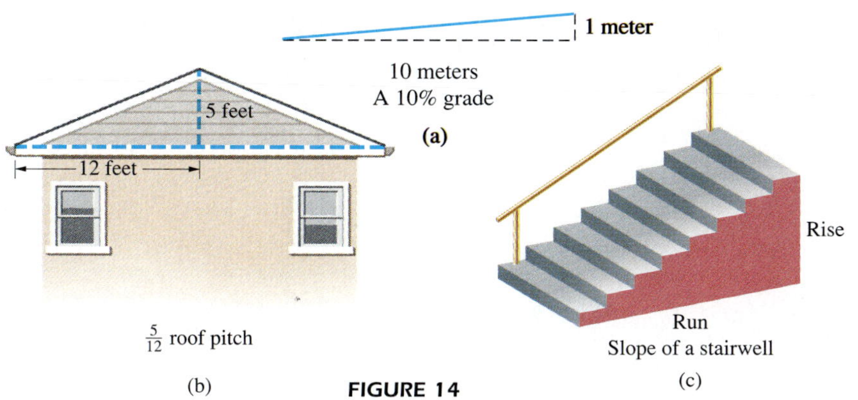

$\frac{5}{12}$ roof pitch
(b)

Slope of a stairwell
(c)

FIGURE 14

ANSWERS
1. (a) −5 (b) 2 (c) 1

2. Find the slope of each of the following lines.

(a) Through $(6, -2)$ and $(5, 4)$

(b) Through $(-3, 5)$ and $(-4, 7)$

(c) Through $(6, -8)$ and $(-2, 4)$

ANSWERS

2. (a) -6 (b) -2 (c) $-\frac{3}{2}$

■ **EXAMPLE 1** *Finding the Slope of a Line*

Find the slope of each of the following lines.

(a) The line through $(-4, 7)$ and $(1, -2)$

Use the definition given above. Let $(-4, 7) = (x_2, y_2)$ and $(1, -2) = (x_1, y_1)$. Then

$$\text{slope} = m = \frac{y_2 - y_1}{x_2 - x_1} = \frac{7 - (-2)}{-4 - 1} = \frac{9}{-5} = -\frac{9}{5}.$$

See Figure 15.

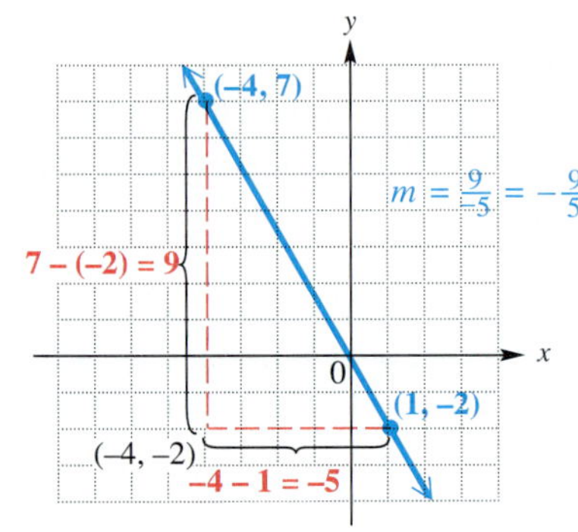

FIGURE 15

(b) The line through $(12, 5)$ and $(-9, -2)$

$$m = \frac{y_2 - y_1}{x_2 - x_1} = \frac{5 - (-2)}{12 - (-9)} = \frac{7}{21} = \frac{1}{3}$$

The same slope is obtained by subtracting in reverse order.

$$\frac{-2 - 5}{-9 - 12} = \frac{-7}{-21} = \frac{1}{3}$$ ■

Caution As shown in Example 1(b), it makes no difference which point is (x_1, y_1) or (x_2, y_2); however, it is important to be consistent; start with the x- and y-values of one point (either one) and subtract the corresponding values of the other point. Also, the slope is the same for *any* two points on the line.

◀ WORK PROBLEM 2 AT THE SIDE.

In Example 1(a) the slope was negative and the corresponding line in Figure 15 fell from left to right. The slope in Example 1(b) was positive and the corresponding line, shown in Figure 16, rises from left to right. These facts can be generalized.

A line with positive slope rises from left to right.
A line with negative slope falls from left to right.

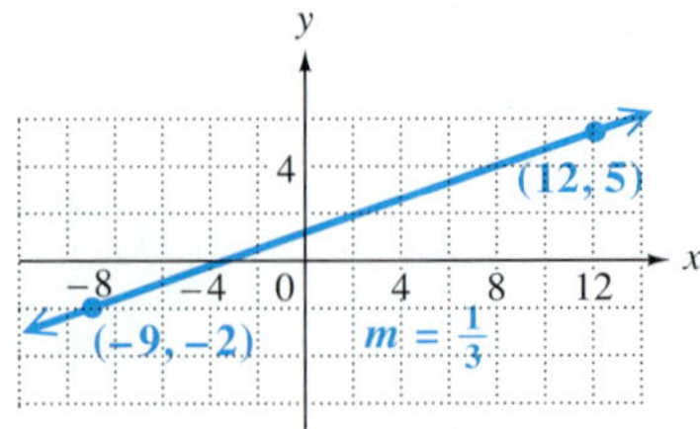

FIGURE 16

The next examples illustrate slopes of horizontal and vertical lines.

EXAMPLE 2 *Finding the Slope of a Horizontal Line*
Find the slope of the line through (−8, 4) and (2, 4).

Use the definition of slope.

$$m = \frac{y_2 - y_1}{x_2 - x_1} = \frac{4 - 4}{-8 - 2} = \frac{0}{-10} = 0$$

As shown in Figure 17 by a sketch of the graph through these two points, the line through the given points is horizontal. *All horizontal lines have a slope of 0* because the difference in y-values will always be 0.

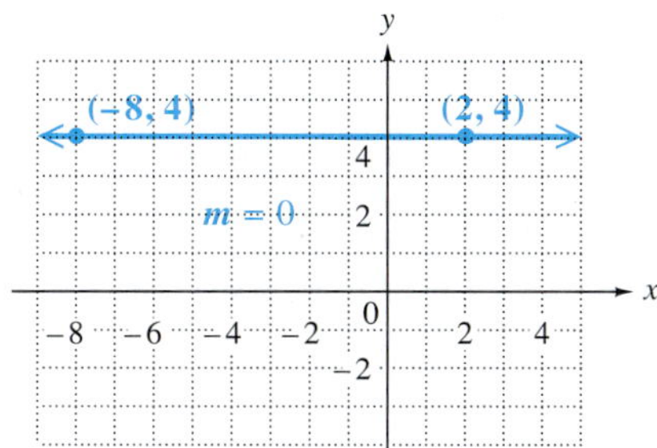

FIGURE 17

EXAMPLE 3 *Finding the Slope of a Vertical Line*
Find the slope of the line through (6, 2) and (6, −9).

$$m = \frac{y_2 - y_1}{x_2 - x_1} = \frac{2 - (-9)}{6 - 6} = \frac{11}{0} \qquad \text{Undefined}$$

Because division by zero is undefined, this line has an undefined slope. (This is why the formula for slope at the beginning of this section had the restriction $x_1 \neq x_2$.) The graph in Figure 18 shows that this line is vertical. Since all points on a vertical line have the same x-value, *vertical lines have undefined slope.*

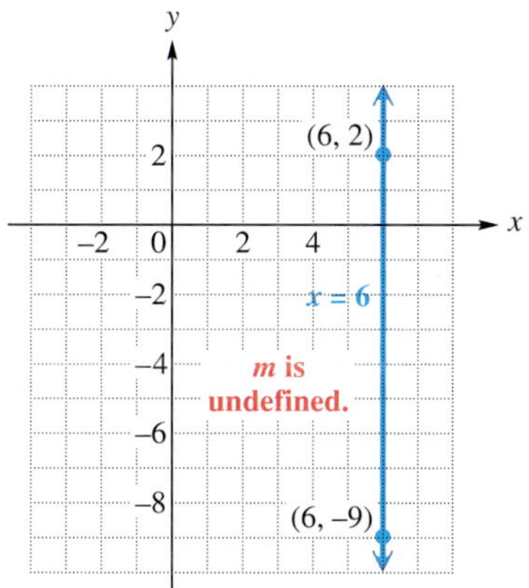

FIGURE 18

3. Find the slope of each line.

(a) Through (2, 5) and (−1, 5)

(b) Through (3, 1) and (3, −4)

(c) With equation $y = -1$

(d) With equation $x - 4 = 0$

4. Find the slope of each line.

(a) $y = -\frac{7}{2}x + 1$

(b) $3x + 2y = 9$

(c) $y + 4 = 0$

(d) $x + 3 = 7$

ANSWERS

3. (a) 0 (b) undefined slope (c) 0 (d) undefined slope

4. (a) $-\frac{7}{2}$ **(b)** $-\frac{3}{2}$ **(c) 0 (d) undefined**

SLOPES OF HORIZONTAL AND VERTICAL LINES

Horizontal lines, which have equations of the form $y = k$, have a slope of 0.
Vertical lines, which have equations of the form $x = k$, have undefined slope.

WORK PROBLEM 3 AT THE SIDE.

2 The slope of a line can be found directly from its equation. For example, let us find the slope of the line

$$y = -3x + 5.$$

The definition of slope requires two different points on the line. We can get these two points by first choosing two different values of x and then finding the corresponding values of y. Let us choose $x = -2$ and $x = 4$.

$$y = -3(-2) + 5 \quad \text{Let } x = -2. \qquad y = -3(4) + 5 \quad \text{Let } x = 4.$$
$$y = 6 + 5 \qquad\qquad y = -12 + 5$$
$$y = 11 \qquad\qquad y = -7$$

The ordered pairs are (−2, 11) and (4, −7). Now we can find the slope.

$$\text{Slope} = \frac{11 - (-7)}{-2 - 4} = \frac{18}{-6} = -3$$

The slope, −3, is the same number as the coefficient of x in the equation $y = -3x + 5$. It can be shown that this always happens, *as long as the equation is solved for y.* This fact is used to find the slope of a line from its equation.

FINDING THE SLOPE OF A LINE FROM ITS EQUATION

Step 1 Solve the equation for y.
Step 2 The slope is given by the coefficient of x.

■ EXAMPLE 4 *Finding Slope from an Equation*

Find the slope of each of the following lines.

(a) $2x - 5y = 4$
Solve the equation for y.

$$2x - 5y = 4$$
$$-5y = -2x + 4 \quad \text{Subtract } 2x \text{ from each side.}$$
$$y = \frac{2}{5}x - \frac{4}{5} \quad \text{Divide each side by } -5.$$

The slope is given by the coefficient of x, so the slope $= \frac{2}{5}$.

(b) $8x + 4y = 1$
Solve for y; you should get

$$y = -2x + \frac{1}{4}.$$

The slope of this line is given by the coefficient of x, −2. ■

WORK PROBLEM 4 AT THE SIDE.

3 Two lines in a plane that never intersect are **parallel.** We use slopes to tell whether two lines are parallel. For example, Figure 19 shows the graph of $x + 2y = 4$ and the graph of $x + 2y = -6$. These lines appear to be parallel. Solving for y we find that both $x + 2y = 4$ and $x + 2y = -6$ have a slope of $-\frac{1}{2}$. Nonvertical parallel lines always have equal slopes.

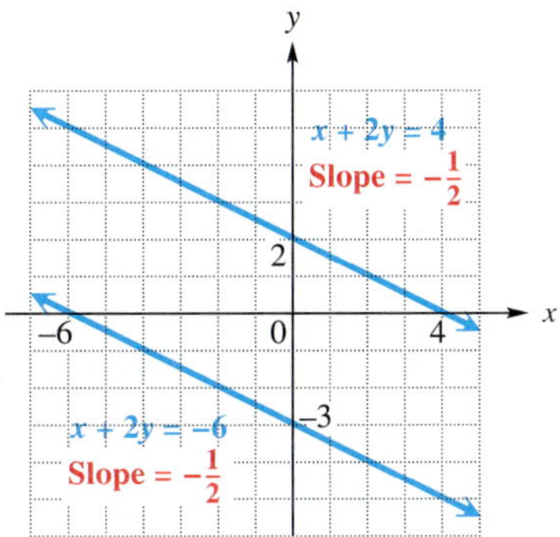

FIGURE 19

Figure 20 shows the graph of $x + 2y = 4$ and the graph of $2x - y = 6$. These lines appear to be **perpendicular** (meet at a 90° angle). Solving for y shows that the slope of $x + 2y = 4$ is $-\frac{1}{2}$, while the slope of $2x - y = 6$ is 2. The product of $-\frac{1}{2}$ and 2 is

$$\left(-\frac{1}{2}\right)(2) = -1.$$

This is true in general; the product of the slopes of two perpendicular lines is always -1.

SLOPES OF PARALLEL AND PERPENDICULAR LINES

Two lines with the same slope are parallel; two lines whose slopes have a product of -1 are perpendicular.

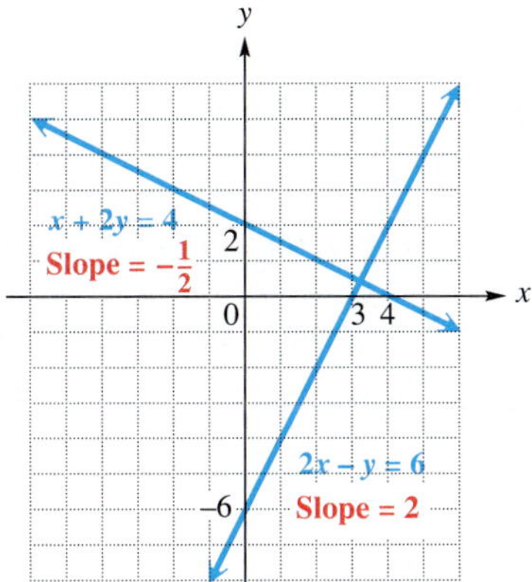

FIGURE 20

5. Write *parallel, perpendicular,* or *neither* for each pair of lines.

(a) $x + y = 6$
$x + y = 1$

(b) $3x - y = 4$
$x + 3y = 9$

(c) $2x - y = 5$
$2x + y = 3$

(d) $3x - 7y = 35$
$7x - 3y = -6$

EXAMPLE 5 *Deciding Whether Lines Are Parallel, Perpendicular, or Neither*

Decide whether each pair of lines is *parallel, perpendicular,* or *neither.*

(a) $x + 2y = 7$
$-2x + y = 3$

Find the slope of each line by first solving each equation for y.

$$x + 2y = 7 \qquad\qquad -2x + y = 3$$
$$2y = -x + 7$$
$$y = -\frac{1}{2}x + \frac{7}{2} \qquad\qquad y = 2x + 3$$

Slope is $-\frac{1}{2}$. Slope is 2.

Because the slopes are not equal, the lines are not parallel. Check the product of the slopes: $(-\frac{1}{2})(2) = -1$. The two lines are perpendicular because the product of their slopes is -1. See Figure 21.

(b) $3x - y = 4$
$6x - 2y = -12$

Find the slopes. Both lines have a slope of 3, so the lines are parallel. The graphs of these lines are shown in Figure 22.

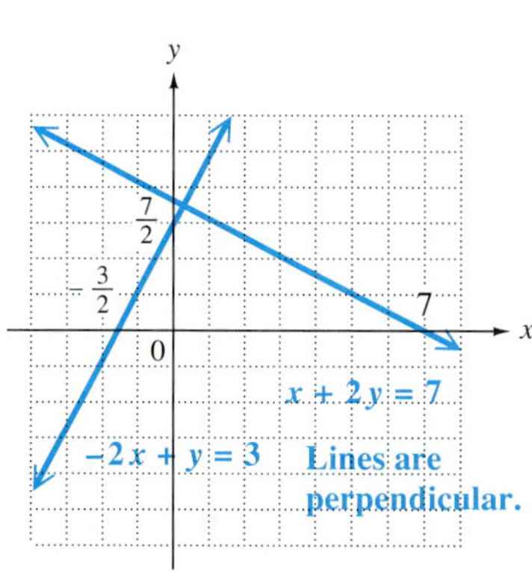

FIGURE 21

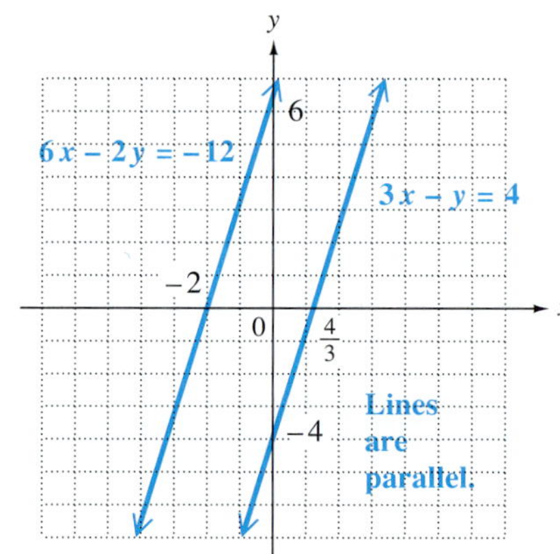

FIGURE 22

(c) $4x + 3y = 6$
$2x - y = 5$

Here the slopes are $-\frac{4}{3}$ and 2. These lines are neither parallel nor perpendicular. ■

WORK PROBLEM 5 AT THE SIDE.

ANSWERS
5. (a) parallel (b) perpendicular (c) neither (d) neither

6.3 EXERCISES

NAME DATE HOUR

On the axes provided, sketch the graph of a straight line that has the described slope.

1. negative

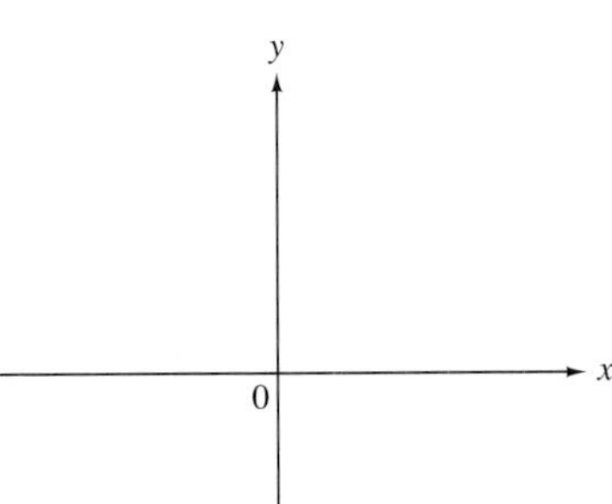

2. positive

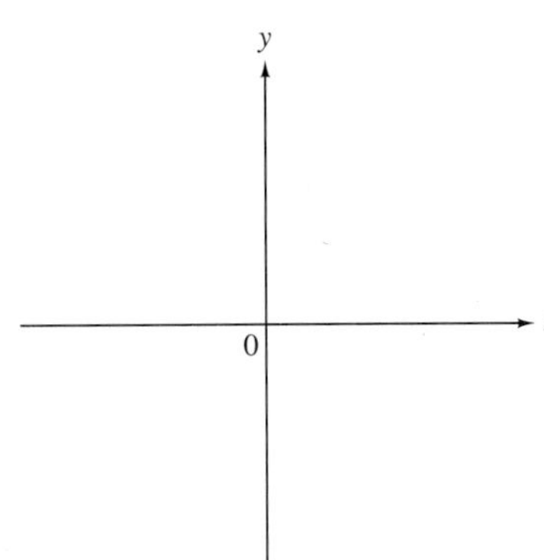

3. undefined

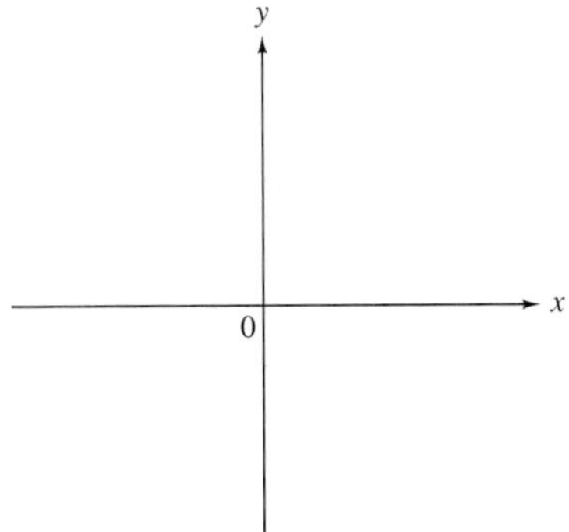

4. zero

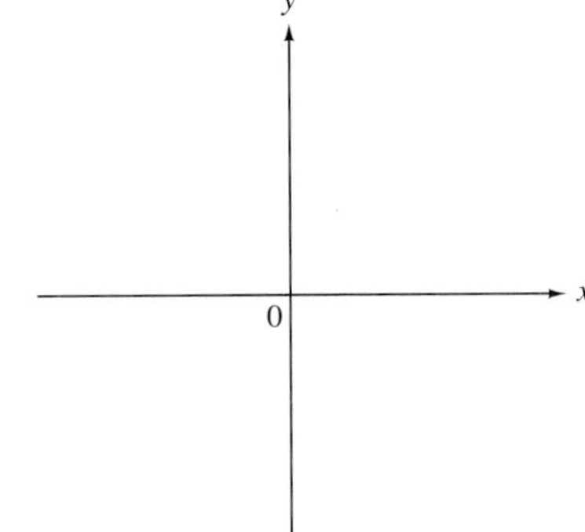

5. Explain in your own words what is meant by *slope* of a line.

6. If two nonvertical lines are parallel, what do we know about their slopes? If two lines are perpendicular, and neither is parallel to an axis, what do we know about their slopes?

7. What is the slope of a line parallel to the graph of $y = 5x - 3$?

8. What is the slope of a line perpendicular to the graph of $y = -3x + 7$?

Use the coordinates of the indicated points to find the slope of each of the following lines. See Example 1.

9.

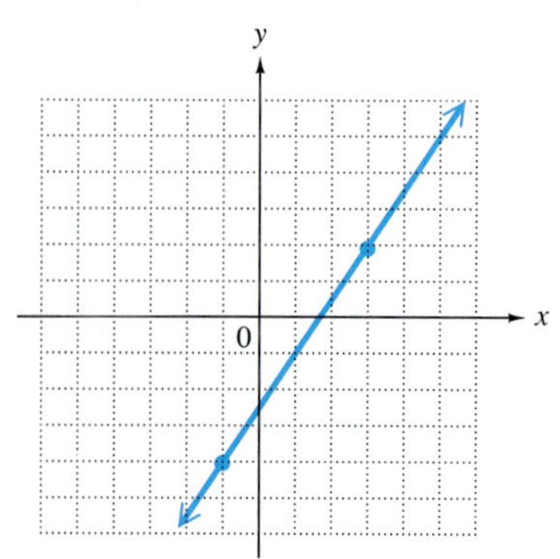

10.

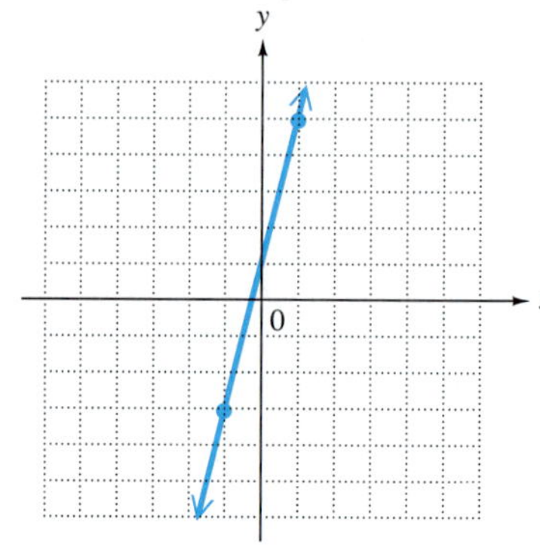

11.

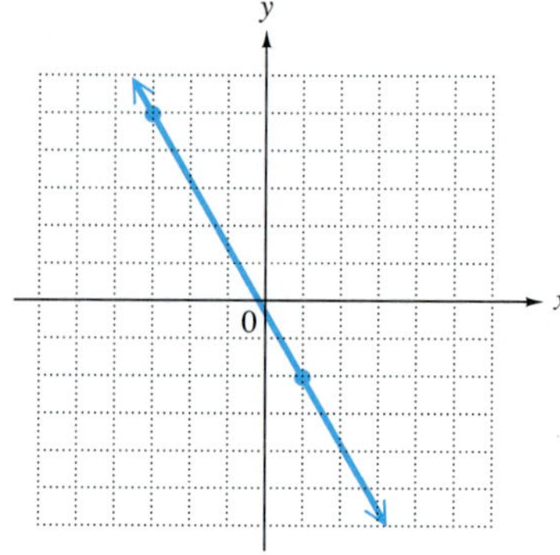

12.

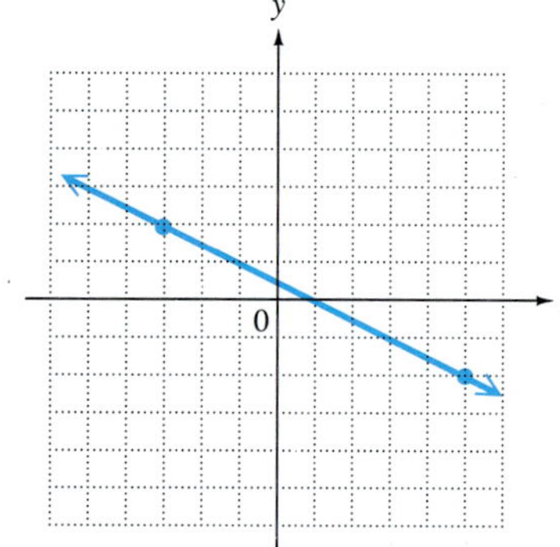

13.

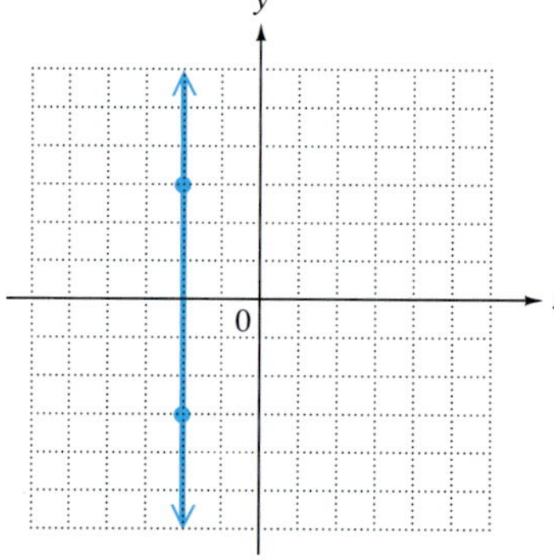

14.

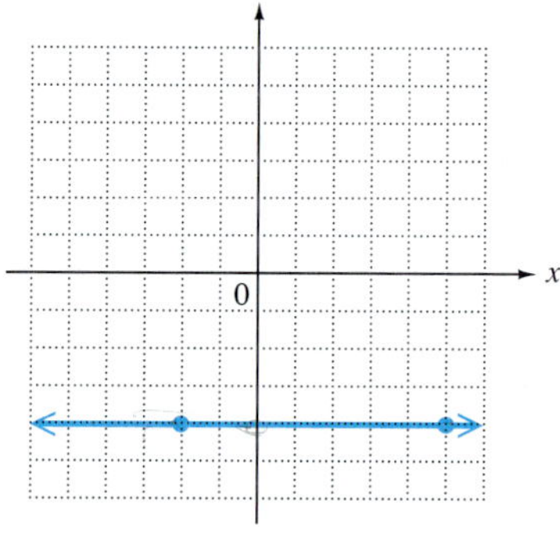

15. Look at the graph in Exercise 9, and answer the following.

(a) Start at the point $(-1, -4)$ and count vertically up to the horizontal line that goes through the other point plotted. What is this vertical change? (Remember: "up" means positive, "down" means negative.) _______

(b) From this new position, count horizontally to the other point plotted. What is this horizontal change? (Remember: "right" means positive, "left" means negative.) _______

(c) What is the quotient of the numbers found in parts (a) and (b)? _______
What do we call this number? _______

16. Refer to Exercise 15. If we were to *start* at the point $(3, 2)$ and *end* at the point $(-1, -4)$, do you think that the answer to part (c) would be the same? Explain why or why not.

Find the slope of the line going through the pair of points. See Examples 1–3.

17. $(4, -1)$ and $(-2, -8)$

18. $(1, -2)$ and $(-3, -7)$

19. $(-8, 0)$ and $(0, -5)$

20. $(0, 3)$ and $(-2, 0)$

21. $(6, -5)$ and $(-12, -5)$

22. $(4, 3)$ and $(-6, 3)$

23. $(-8, 6)$ and $(-8, -1)$

24. $(-12, 3)$ and $(-12, -7)$

25. $(3.1, 2.6)$ and $(1.6, 2.1)$

26. $\left(-\frac{7}{5}, \frac{3}{10}\right)$ and $\left(\frac{1}{5}, -\frac{1}{2}\right)$

Find the slope of each of the following lines. See Example 4.

27. $y = 2x - 3$

28. $y = 5x + 12$

29. $2y = -x + 4$

30. $4y = x + 1$

31. $y = 6 - 4x$

32. $y = 3 + 2x$

33. $-6x + 4y = 4$

34. $3x - 2y = 3$

35. $y = 4$

36. $x = 6$

The figure below shows a line that has a positive slope (because it rises from left to right) and a positive y-value for the y-intercept (because it intersects the y-axis above the origin).

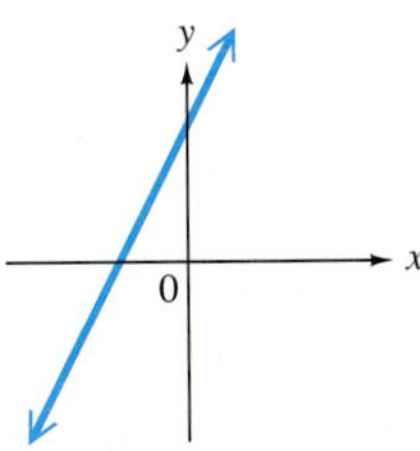

For each figure in Exercises 37–42, decide whether (a) the slope is positive, negative, or zero and whether (b) the y-value of the y-intercept is positive, negative, or zero.

37. **(a)** ______
(b) ______

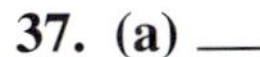

38. **(a)** ______
(b) ______

39. **(a)** ______
(b) ______

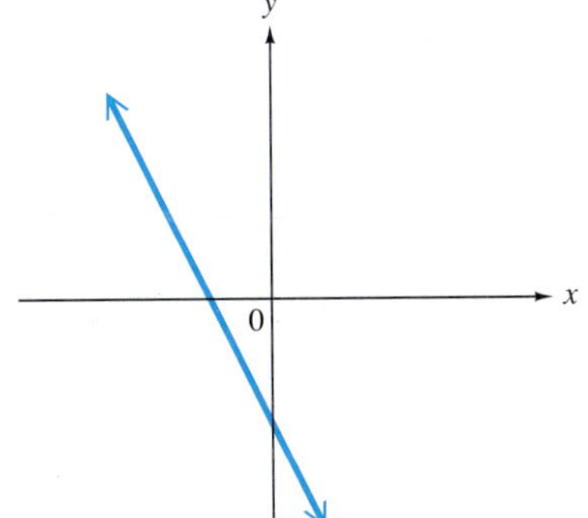

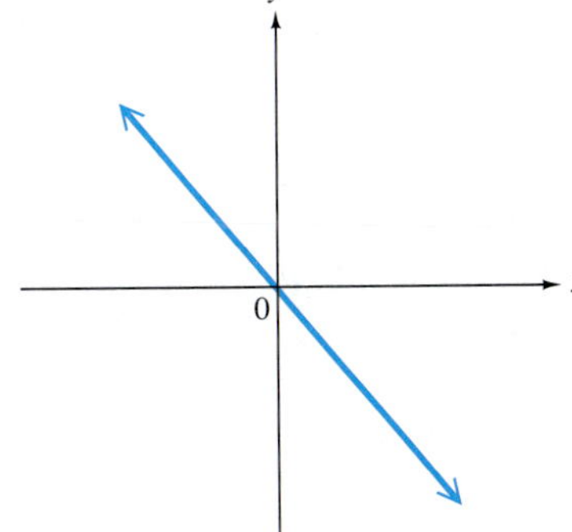

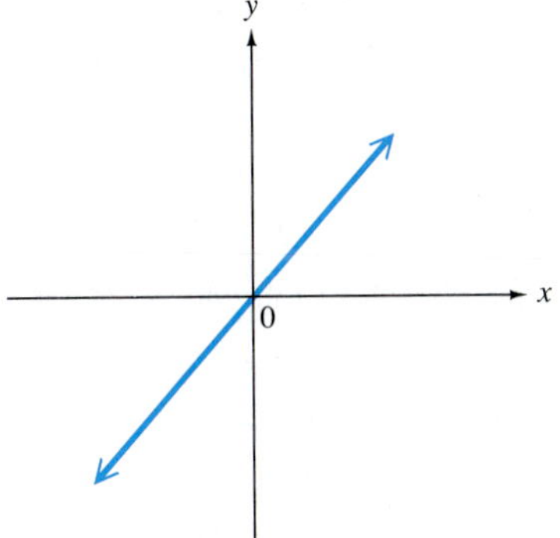

40. **(a)** ________
(b) ________

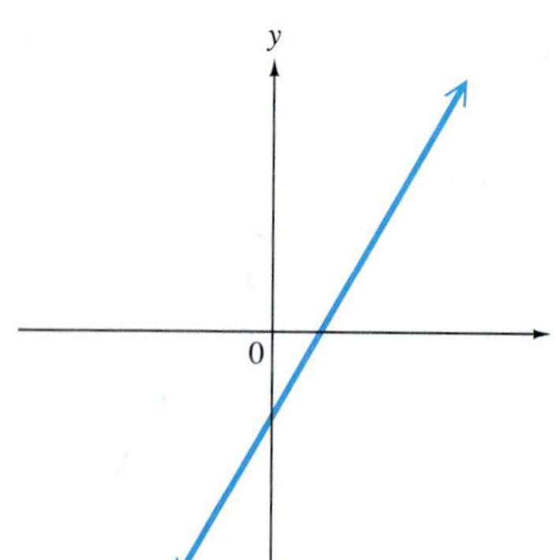

41. **(a)** ________
(b) ________

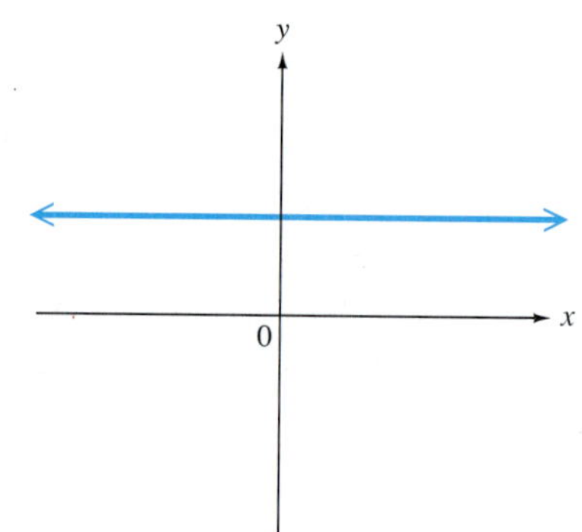

42. **(a)** ________
(b) ________

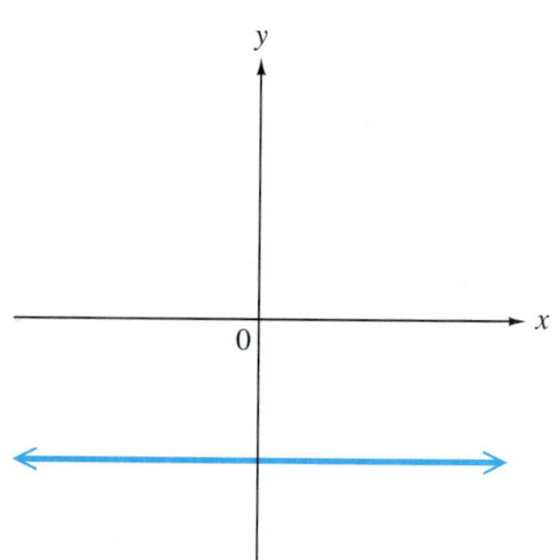

In each pair of equations, give the slope of each line, and then determine whether the two lines are parallel, perpendicular, or neither parallel nor perpendicular. See Example 5.

43. $2x + 5y = 4$
$4x + 10y = 1$

44. $-4x + 3y = 4$
$-8x + 6y = 0$

45. $8x - 9y = 6$
$8x + 6y = -5$

46. $5x - 3y = -2$
$3x - 5y = -8$

47. $3x - 2y = 6$
$2x + 3y = 3$

48. $3x - 5y = -1$
$5x + 3y = 2$

49. What is the slope (or pitch) of this roof? Measurements are given in feet.

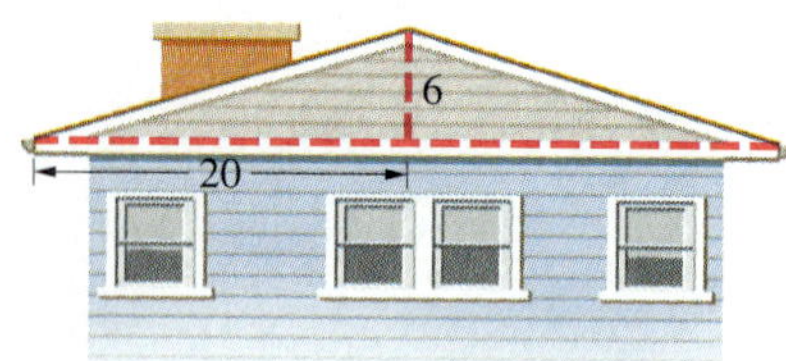

50. What is the slope (or grade) of this hill? Measurements are given in meters.

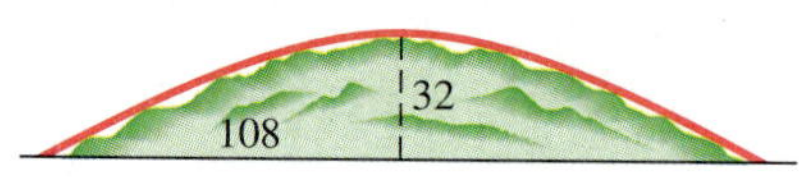

PREVIEW EXERCISES

Solve for y in each equation. See Sections 2.3 and 2.5.

51. $y - (-8) = 2(x - 4)$

52. $y - 3 = 4(x - (-6))$

6.4 EQUATIONS OF A LINE

The last section showed how to find the slope of a line from its equation. For example, the slope of the line having the equation $y = 2x + 3$ is 2, the coefficient of x. What does the number 3 represent? If we let $x = 0$, the equation becomes $y = 2(0) + 3 = 0 + 3 = 3$. Because we found $y = 3$ by letting $x = 0$, $(0, 3)$ is the y-intercept of the graph of $y = 2x + 3$. An equation solved for y is said to be in **slope-intercept form** because both the slope and the y-intercept of the line can be found from the equation.

We can generalize this discussion. Suppose a line has slope m and y-intercept $(0, b)$. If (x, y) represents any other point on the line, by the slope formula, we have

$$m = \frac{y - b}{x - 0}$$

$$mx = y - b \qquad \text{Multiply by } x.$$

$$y = mx + b. \qquad \text{Add } b.$$

SLOPE-INTERCEPT FORM

> The slope-intercept form of the equation of a line with slope m and y-intercept $(0, b)$ is
>
> $$y = mx + b.$$

1 Given the slope and y-intercept of a line, we can use the slope-intercept form to find an equation of the line.

EXAMPLE 1 *Finding an Equation of a Line*

Find an equation for each of the following lines.

(a) With slope 5 and y-intercept $(0, 3)$

Use the slope-intercept form. Let $m = 5$ and $b = 3$.

$$y = mx + b$$

$$y = 5x + 3$$

(b) With slope $\frac{2}{3}$ and y-intercept $(0, -1)$

Here $m = \frac{2}{3}$ and $b = -1$.

$$y = \frac{2}{3}x - 1$$ ■

WORK PROBLEM 1 AT THE SIDE.

2 The slope and y-intercept can be used to graph a line. Graph $y = \frac{2}{3}x - 1$ by first locating the y-intercept $(0, -1)$ on the y-axis. From the definition of slope and the fact that the slope of this line is $\frac{2}{3}$,

$$m = \frac{\text{difference in } y\text{-values}}{\text{difference in } x\text{-values}} = \frac{2}{3}.$$

We can find another point on the graph by counting from the y-intercept 2 units up and then counting 3 units to the right. We then draw the line through this point and the y-intercept, as shown in Figure 23. (Notice that

OBJECTIVES

1. Write an equation of a line, given its slope and y-intercept.
2. Graph a line, given its slope and a point on the line.
3. Write an equation of a line, given its slope and a point on the line.
4. Write an equation of a line, given two points on the line.

FOR EXTRA HELP

Tape 9 | SSM pp. 233–237 | MAC: A IBM: A

1. Find the equation of the line with the given slope and value of b.

(a) $m = \frac{1}{2}$; $b = -4$

(b) $m = -1$; $b = 8$

(c) $m = 3$; $b = 0$

(d) $m = 0$; $b = 2$

ANSWERS

1. (a) $y = \frac{1}{2}x - 4$ (b) $y = -x + 8$ (c) $y = 3x$ (d) $y = 2$

2. Graph each line.

(a) Through $(-1, 2)$ with slope $\frac{3}{2}$

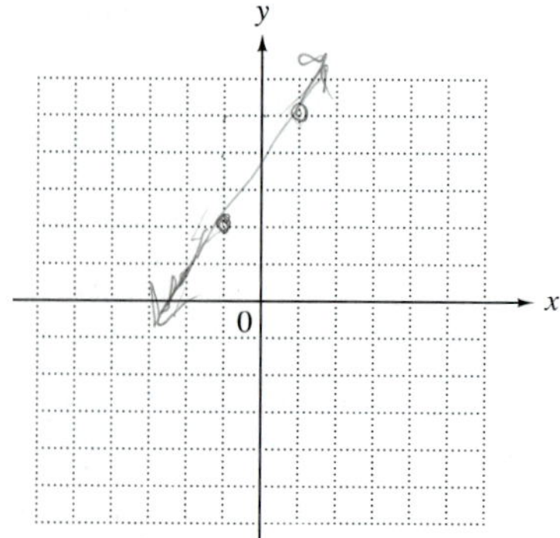

(b) Through $(2, -3)$, with slope $-\frac{1}{3}$

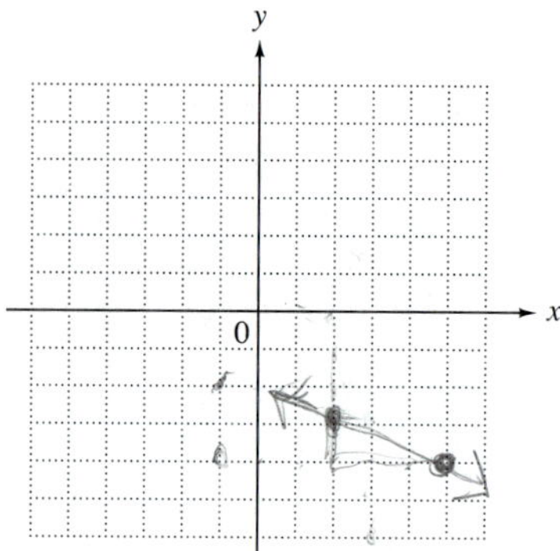

ANSWERS

2. (a)

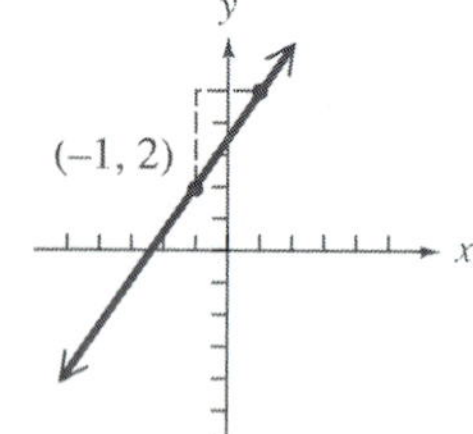

(b)

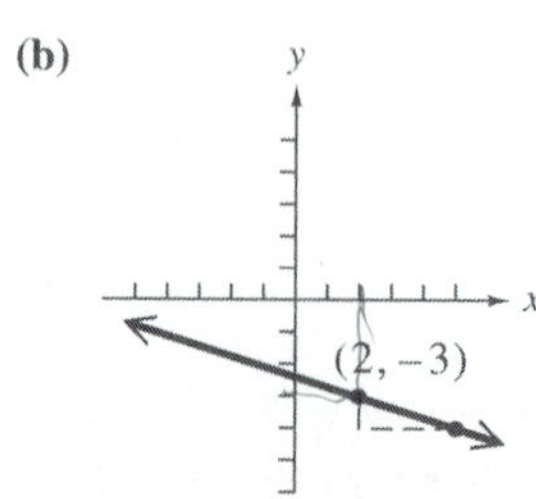

we could have counted 2 units down and 3 units to the left to obtain a different point on this same line.) This method can be extended to graph a line given its slope and a point on the line.

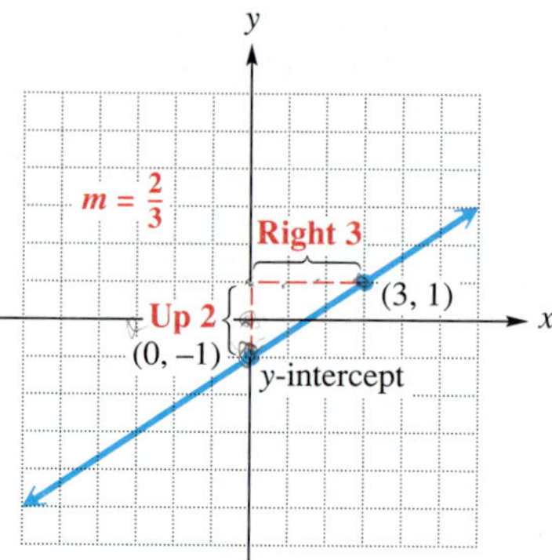

FIGURE 23

■ **EXAMPLE 2** *Graphing a Line Given Its Slope and a Point*

Graph the line through $(-2, 3)$ with slope $-\frac{5}{4}$.

First, locate the point $(-2, 3)$. Write the slope as

$$m = \frac{\text{difference in } y\text{-values}}{\text{difference in } x\text{-values}} = \frac{-5}{4}.$$

(We could have used $\frac{5}{-4}$ instead.) Locate another point on the line by counting 5 units *down* (because of the negative sign) and then 4 units to the right. Finally, draw the line through this new point and the given point $(-2, 3)$. See Figure 24. ■

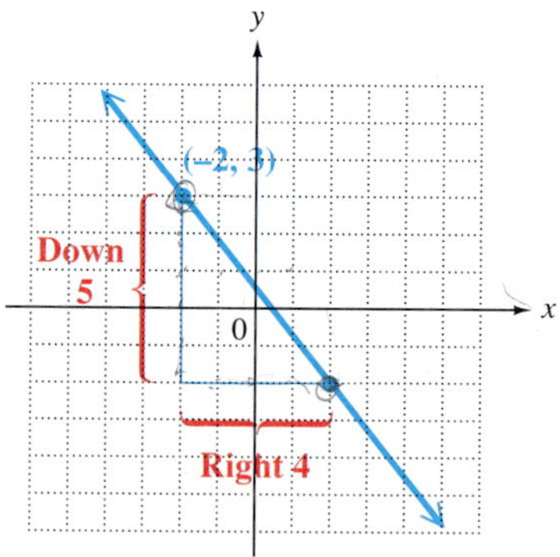

FIGURE 24

WORK PROBLEM 2 AT THE SIDE.

3 We can also find an equation for a line from any point on the line and the slope of the line. Let m represent the slope of the line, and let (x_1, y_1) represent the given point on the line. Let (x, y) represent any other point on the line. Then by the definition of slope,

$$\frac{y - y_1}{x - x_1} = m$$

$$\text{or} \quad y - y_1 = m(x - x_1).$$

This result is the **point-slope form** of the equation of a line.

POINT-SLOPE FORM

The point-slope form of the equation of a line with slope m and going through (x_1, y_1) is

$$y - y_1 = m(x - x_1).$$

This very important result should be memorized.

EXAMPLE 3 *Using the Point-Slope Form to Write an Equation*
Find an equation of each of the following lines. Write the equation in the form $Ax + By = C$.

(a) Through $(-2, 4)$, with slope -3
The given point is $(-2, 4)$ so $x_1 = -2$ and $y_1 = 4$. Also, $m = -3$. Substitute these values into the point-slope form.

$$\begin{aligned} y - y_1 &= m(x - x_1) \\ y - 4 &= -3[x - (-2)] \\ y - 4 &= -3(x + 2) \\ y - 4 &= -3x - 6 && \text{Distributive property} \\ y &= -3x - 2 && \text{Add 4.} \\ 3x + y &= -2 && \text{Add } 3x. \end{aligned}$$

(b) Through $(4, 2)$ with slope $\frac{3}{5}$
Use $x_1 = 4$, $y_1 = 2$, and $m = \frac{3}{5}$ in the point-slope form.

$$\begin{aligned} y - y_1 &= m(x - x_1) \\ y - 2 &= \frac{3}{5}(x - 4) \\ 5(y - 2) &= 5 \cdot \frac{3}{5}(x - 4) && \text{Multiply by 5 to clear fractions.} \\ 5(y - 2) &= 3(x - 4) \\ 5y - 10 &= 3x - 12 && \text{Distributive property} \\ 5y &= 3x - 2 && \text{Add 10.} \\ 3x - 5y &= 2 && Ax + By = C \text{ form} \end{aligned}$$

■

WORK PROBLEM 3 AT THE SIDE.

3. Find an equation for each line. Write answers in the form $Ax + By = C$.

(a) Through $(-1, 3)$, with slope -2

(b) Through $(5, 2)$, with slope $-\frac{1}{3}$

4 The point-slope form also can be used to find an equation of a line when two different points on the line are known.

EXAMPLE 4 *Finding the Equation of a Line from Two Points*
Find an equation for the line through the points $(-2, 5)$ and $(3, 4)$. Write the equation in the form $Ax + By = C$.

First find the slope of the line, using the definition of slope.

$$\text{Slope} = \frac{5 - 4}{-2 - 3} = \frac{1}{-5} = -\frac{1}{5}$$

ANSWERS
3. (a) $2x + y = 1$ (b) $x + 3y = 11$

4. Write an equation for the line through the following points. Write answers in the form $Ax + By = C$.

(a) $(-3, 1)$ and $(2, 4)$

(b) $(2, 5)$ and $(-1, 6)$

Now use either $(-2, 5)$ or $(3, 4)$ and the point-slope form. Using $(3, 4)$ gives

$$\begin{aligned} y - y_1 &= m(x - x_1) \\ y - 4 &= -\frac{1}{5}(x - 3) \\ 5(y - 4) &= -1(x - 3) && \text{Multiply by 5.} \\ 5y - 20 &= -x + 3 && \text{Distributive property} \\ 5y &= -x + 23 && \text{Add 20 on each side.} \\ x + 5y &= 23. && \text{Add } x \text{ on each side.} \end{aligned}$$

The same result would be found by using $(-2, 5)$ for (x_1, y_1). ■

WORK PROBLEM 4 AT THE SIDE.

Here is a summary of the types of *linear equations.*

LINEAR EQUATIONS

$Ax + By = C$	**Standard form** (neither A nor B is 0) Slope is $-\frac{A}{B}$. x-intercept is $\left(\frac{C}{A}, 0\right)$. y-intercept is $\left(0, \frac{C}{B}\right)$.
$x = k$	**Vertical line** Slope is undefined. x-intercept is $(k, 0)$.
$y = k$	**Horizontal line** Slope is 0. y-intercept is $(0, k)$.
$y = mx + b$	**Slope-intercept form** Slope is m. y-intercept is $(0, b)$.
$y - y_1 = m(x - x_1)$	**Point-slope form** Slope is m. Line goes through (x_1, y_1).

Caution The above definition of "standard form" is not the same in all texts. Also, we can write a linear equation as $Ax + By = C$ in many different (equally correct) ways. For example, $3x + 4y = 12$, $6x + 8y = 24$, and $9x + 12y = 36$ all represent the same set of ordered pairs. Let us agree that $3x + 4y = 12$ is preferable to the other forms because the greatest common factor of 3, 4 and 12 is 1.

Remember, to find an equation of a line, you always need the following.

1. a point on the line
2. the slope of the line

ANSWERS

4. (a) $3x - 5y = -14$ (b) $x + 3y = 17$

6.4 EXERCISES

NAME DATE HOUR

Use the geometric interpretation of slope (rise divided by run, from Section 6.3) to find the slope of the line. Then, by identifying the y-intercept from the graph, write the $y = mx + b$ form of the equation of the line.

1.

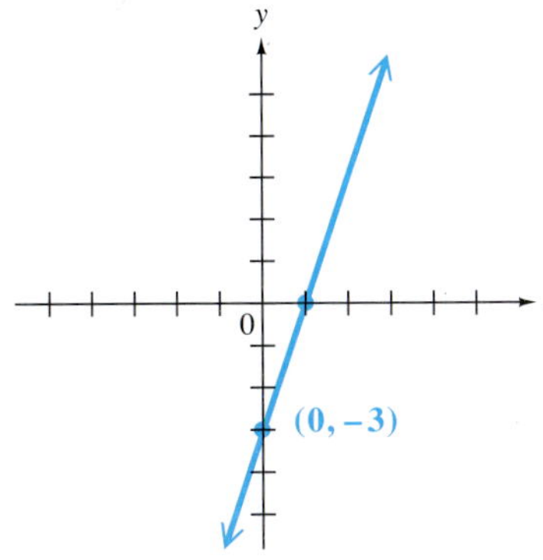

2.

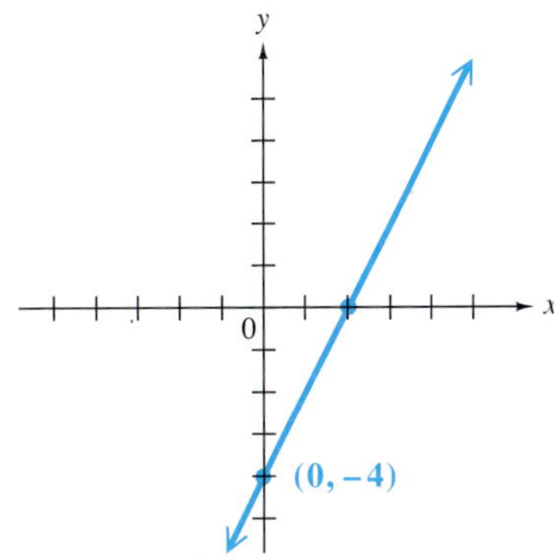

3.

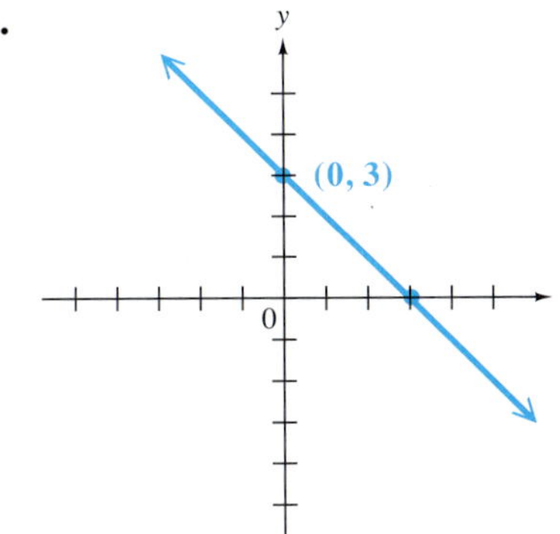

4.

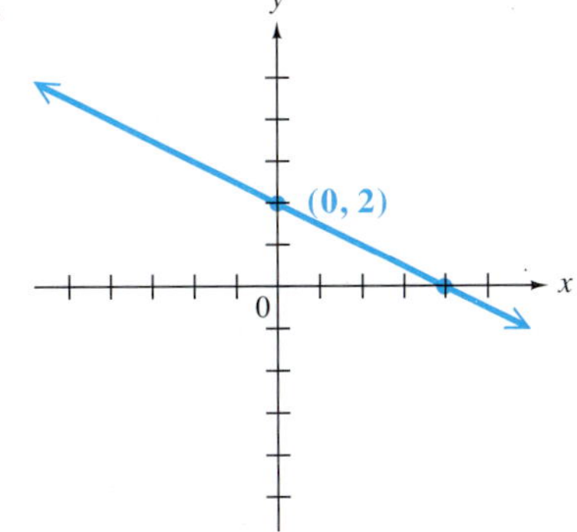

Write the equation of the line with the given slope and value of b. See Example 1.

5. $m = 4, b = -3$

6. $m = -5, b = 6$

7. $m = 0, b = 3$

8. $m = 3, b = 0$

9. Explain why the equation of a vertical line cannot be written in the form $y = mx + b$.

10. Match the equation with the graph that would most closely resemble its graph.

_______ **(a)** $y = x + 3$

_______ **(b)** $y = -x + 3$

_______ **(c)** $y = x - 3$

_______ **(d)** $y = -x - 3$

A.

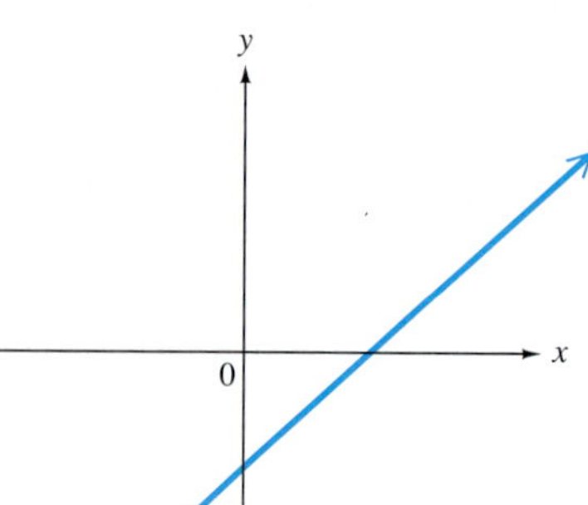

B.

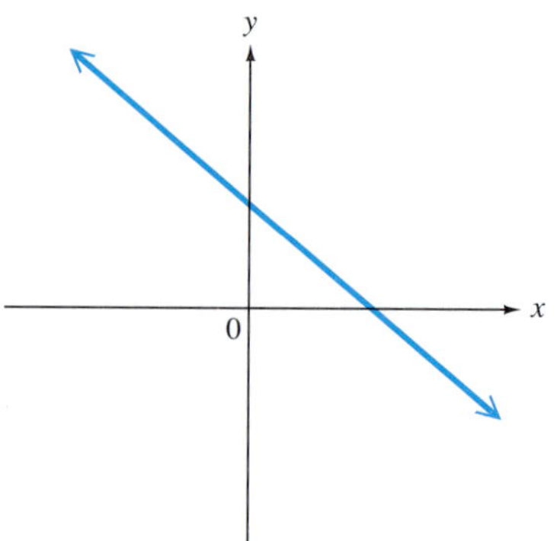

C.

D.

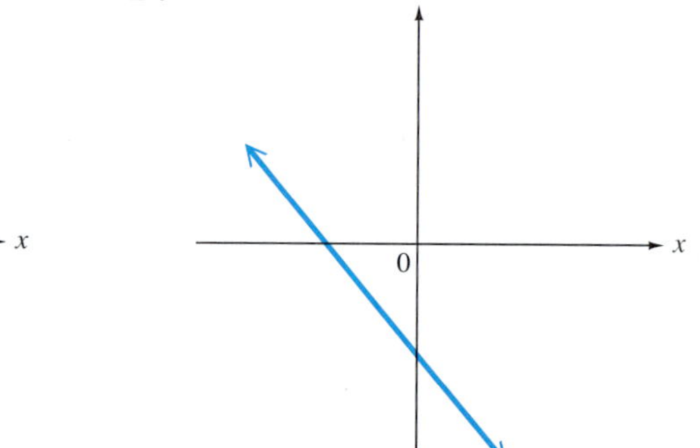

Graph the line going through the given point and having the given slope. (In Exercises 17–20, recall the types of lines having slope 0 and undefined slope.) See Example 2.

11. $(-2, 3)$, $m = \frac{1}{2}$

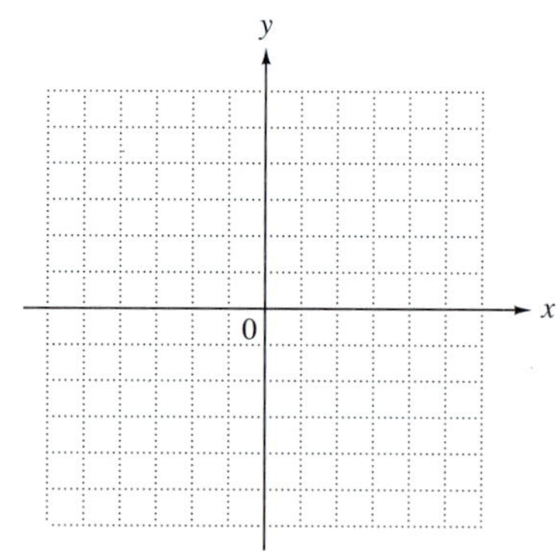

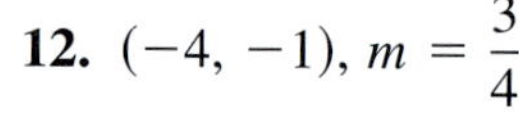

12. $(-4, -1)$, $m = \frac{3}{4}$

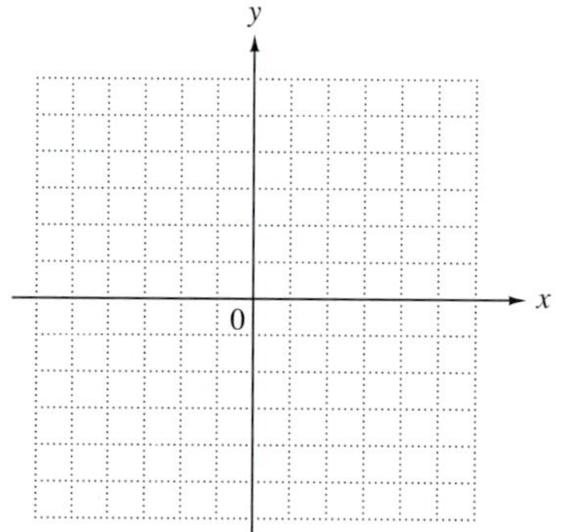

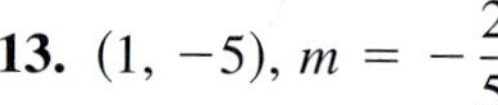

13. $(1, -5)$, $m = -\frac{2}{5}$

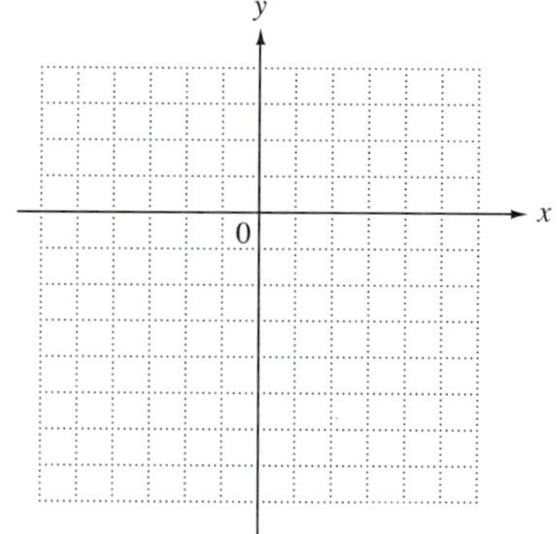

14. $(2, -1)$, $m = -\frac{1}{3}$

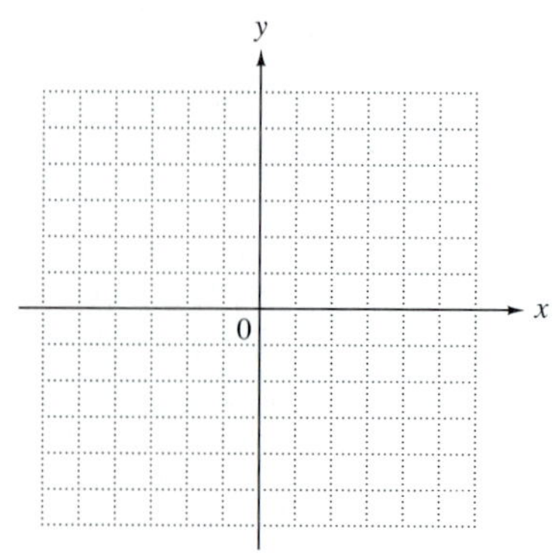

15. $(0, 2)$, $m = 3$

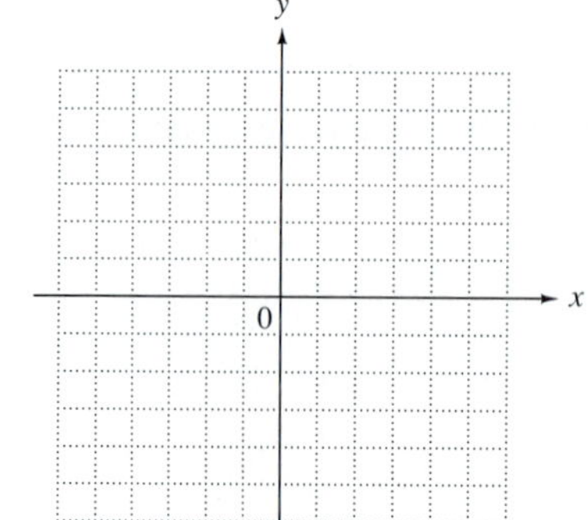

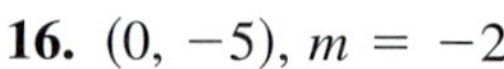

16. $(0, -5)$, $m = -2$

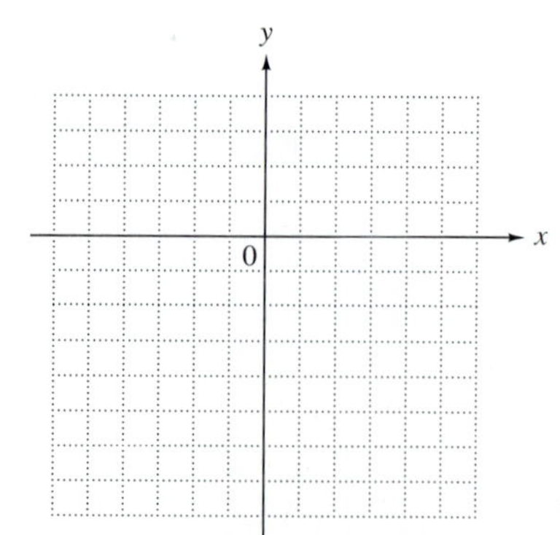

17. $(3, 2)$, $m = 0$

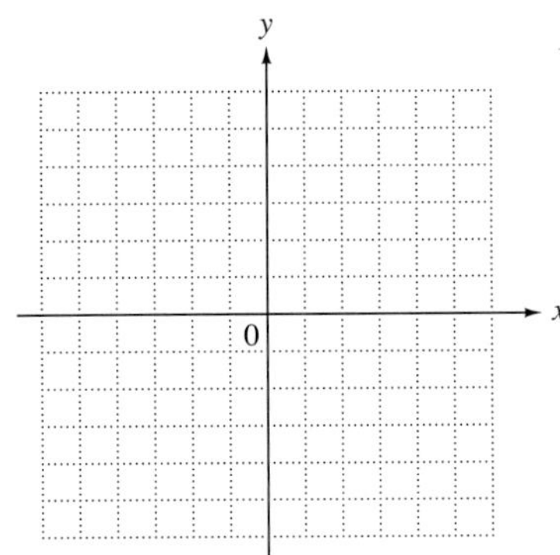

18. $(-2, 3)$, $m = 0$

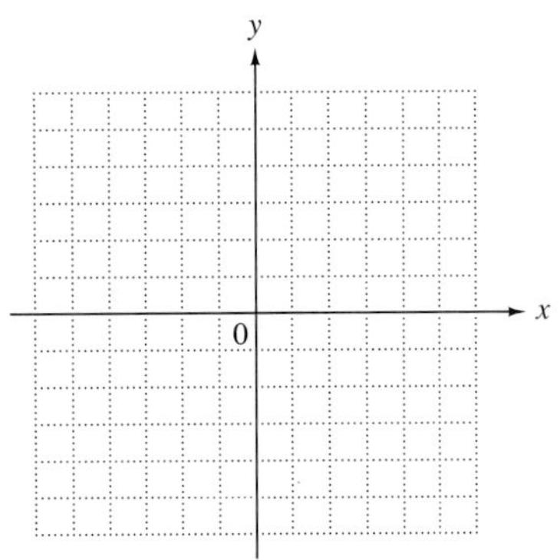

19. $(3, -2)$, undefined slope

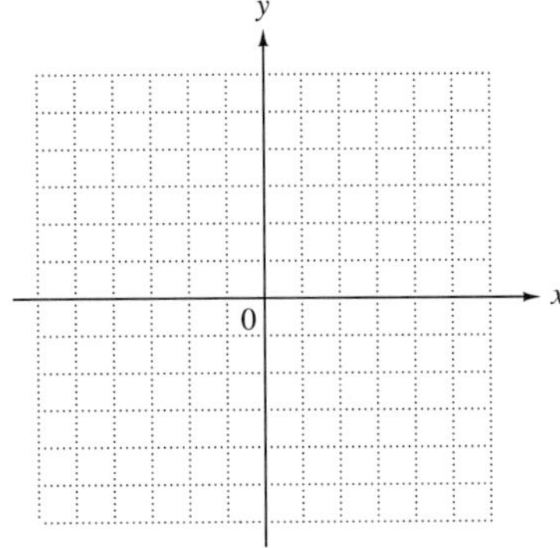

20. $(2, 4)$, undefined slope

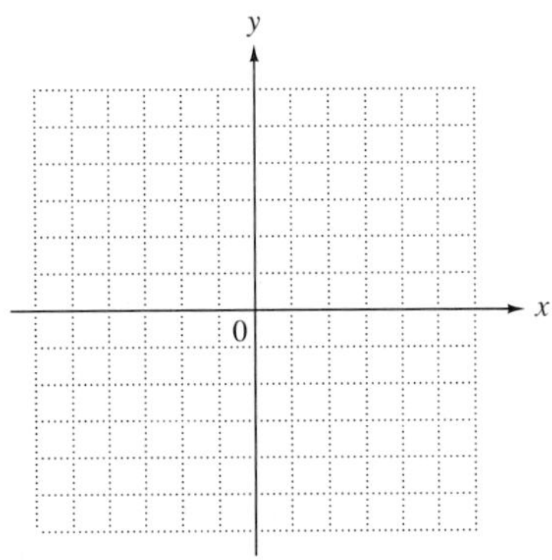

21. What is the common name given to a vertical line whose x-intercept is the origin?

22. What is the common name given to a line with slope 0 whose y-intercept is the origin?

Write an equation for the line passing through the given point and having the given slope. Write the equation in the standard form $Ax + By = C$. See Example 3.

23. $(4, 1)$, $m = 2$

24. $(2, 7)$, $m = 3$

25. $(3, -10)$, $m = -2$

26. $(2, -5)$, $m = -4$

27. $(-2, 5)$, $m = \frac{2}{3}$

28. $(-4, 1)$, $m = \frac{3}{4}$

29. $(6, -3)$, $m = -\frac{4}{5}$

30. $(7, -2)$, $m = -\frac{7}{2}$

31. If a line passes through the origin and a second point whose x- and y-coordinates are equal, what is an equation of the line?

32. What point *must* the graph of $Ax + By = 0$ pass through?

Write an equation for the line passing through each of the given pairs of points. Write the equations in the standard form $Ax + By = C$. See Example 4.

33. $(8, 5)$ and $(9, 6)$

34. $(4, 10)$ and $(6, 12)$

35. $(-1, -7)$ and $(-8, -2)$

36. $(-2, -1)$ and $(3, -4)$

37. $(0, -2)$ and $(-3, 0)$

38. $(-4, 0)$ and $(0, 2)$

39. $\left(\frac{1}{2}, \frac{3}{2}\right)$ and $\left(-\frac{1}{4}, \frac{5}{4}\right)$

40. $\left(-\frac{2}{3}, \frac{8}{3}\right)$ and $\left(\frac{1}{3}, \frac{7}{3}\right)$

The cost to produce x items is, in some cases, expressed as $y = mx + b$. The number b gives the fixed cost *(the cost that is the same no matter how many items are produced), and the number m is the* variable cost *(the cost to produce an additional item). Write the cost equation for each of the following, and answer the questions.*

41. It costs \$400 to start up a business of selling snow cones. Each snow cone costs \$.25 to produce.
 (a) What will be the cost to produce 100 snow cones, based on the cost equation?
 (b) How many snow cones will be produced if total cost is \$775?

42. It costs \$2000 to purchase a copier, and each copy costs \$.02 to make.
 (a) What will be the cost to produce 10,000 copies, based on the cost equation?
 (b) How many copies will be produced if total cost is \$2600?

The sales of a company for a given year can be written as an ordered pair in which the first number, x, gives the year (perhaps since the company started business) and the second number, y, gives the sales for that year. If the sales increase at the same rate each year, a linear equation for sales can be found. Sales for two years are given for two different companies.

Jeff's Rental Properties, Inc.

Year in operation, x	*Sales in dollars,* y
1	4800
5	24,800

Star Enterprises

Year in operation, x	*Sales in dollars,* y
1	18,000
4	93,000

43. **(a)** Write two ordered pairs in the form (year, sales) for Jeff's Rental Properties, Inc.
(b) Write the sales equation in the form $y = mx + b$.
(c) What does the slope, m, represent in this situation?

44. **(a)** Write two ordered pairs in the form (year, sales) for Star Enterprises.
(b) Write the sales equation in the form $y = mx + b$.
(c) What does the slope, m, represent in this situation?

If we think of ordered pairs of the form (C, F), then the two most common methods of measuring temperature, Celsius and Fahrenheit, can be related as follows: When $C = 0$, $F = 32$, and when $C = 100$, $F = 212$. The relationship is linear.

45. Write two ordered pairs relating these two temperature scales.

46. Write an expression for F in terms of C.

47. Use the expression in Exercise 46 to write an expression for C in terms of F.

48. For what temperature is F = C?

PREVIEW EXERCISES

Solve each inequality, and graph it on the number line. See Section 2.7.

49. $3x + 8 > -1$

50. $\frac{1}{2}x - 3 \leq 2$

51. $5 - 3x < -10$

52. $x - 4 \leq 0$

6.5 GRAPHING LINEAR INEQUALITIES IN TWO VARIABLES

Earlier, we graphed inequalities with one variable as intervals on the number line. Now we are ready to graph inequalities with two variables on the coordinate plane. Instead of intervals, we shall see that the graphs will be regions on the plane.

OBJECTIVES

1. Graph $\leq$ or $\geq$ linear inequalities.
2. Graph $>$ or $<$ linear inequalities.
3. Graph inequalities with a boundary through the origin.

FOR EXTRA HELP

Tape	SSM	MAC: A
9	pp. 238–241	IBM: A

1 The inequality $2x + 3y \leq 6$ means that

$$2x + 3y < 6 \quad \text{or} \quad 2x + 3y = 6.$$

As we found at the beginning of this chapter, the graph of $2x + 3y = 6$ is a line. This line divides the plane into two regions. The graph of the solutions of the *inequality* $2x + 3y < 6$ will include only *one* of these regions. We find the required region by solving the given inequality for y.

$$2x + 3y \leq 6$$
$$3y \leq -2x + 6 \quad \text{Subtract } 2x.$$
$$y \leq -\frac{2}{3}x + 2 \quad \text{Divide by 3.}$$

By this last statement, ordered pairs in which y is *less than or equal to* $-\frac{2}{3}x + 2$ will be solutions to the inequality. The ordered pairs in which y is equal to $-\frac{2}{3}x + 2$ are on a line, so the ordered pairs in which y is less than $-\frac{2}{3}x + 2$ will be *below* that line. We indicate the solution by shading the region below the line, as in Figure 25. The shaded region, along with the original line, is the desired graph.

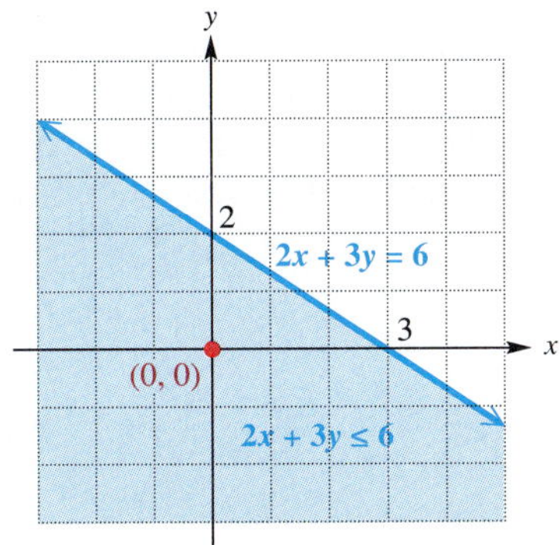

FIGURE 25

WORK PROBLEM 1 AT THE SIDE.

A test point gives a quick way to find the correct region to shade. Choose any point *not* on the line. Because (0, 0) is easy to substitute into an inequality, it is often a good choice, and we will use it here.

Substitute 0 for x and 0 for y in the given inequality to see whether the resulting statement is true or false. In the example above,

$$2x + 3y \leq 6$$
$$2(0) + 3(0) \leq 6 \quad \text{Let } x = 0 \text{ and } y = 0.$$
$$0 + 0 \leq 6$$
$$0 \leq 6. \quad \text{True}$$

Because the last statement is true, we shade the region that includes the test point (0, 0). This agrees with the result shown in Figure 25.

1. Shade the appropriate region for each linear inequality.

(a) $x + 2y \geq 6$

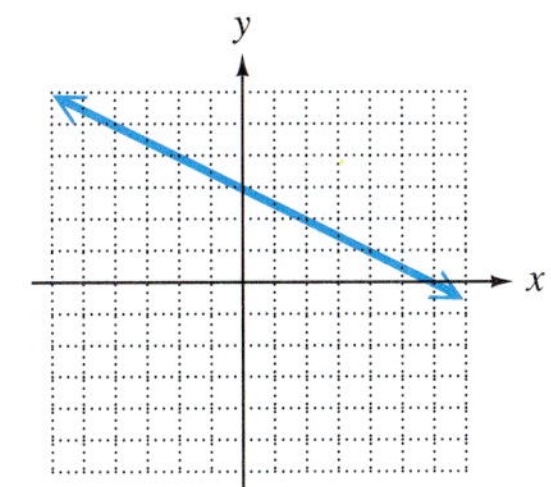

(b) $3x + 4y \leq 12$

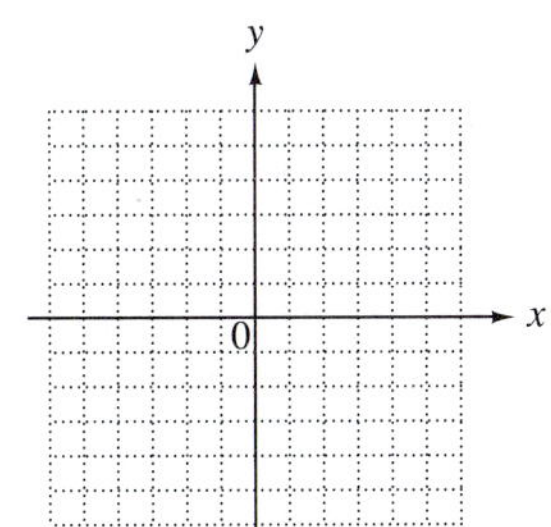

ANSWERS

1. (a)

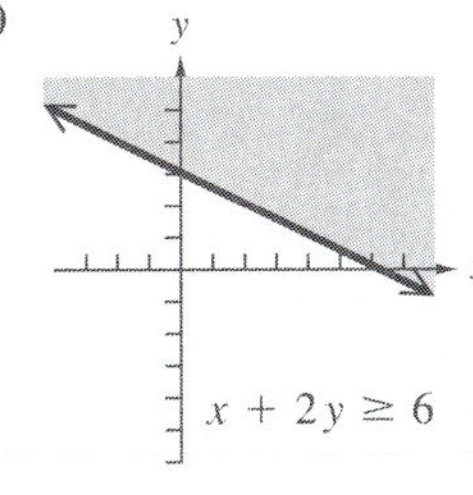

(b)

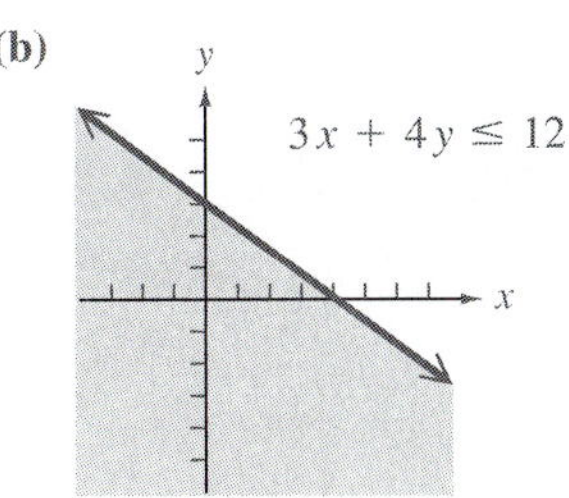

2. Use (0, 0) as a test point to shade the proper region for the inequality $4x - 5y \leq 20$.

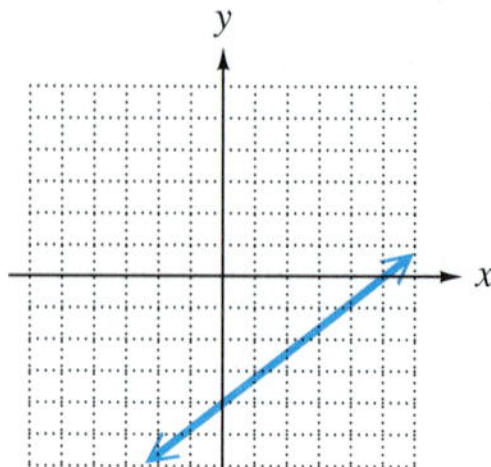

3. Use (1, 2) as a test point to shade the proper region for the inequality $3x + 5y > 15$.

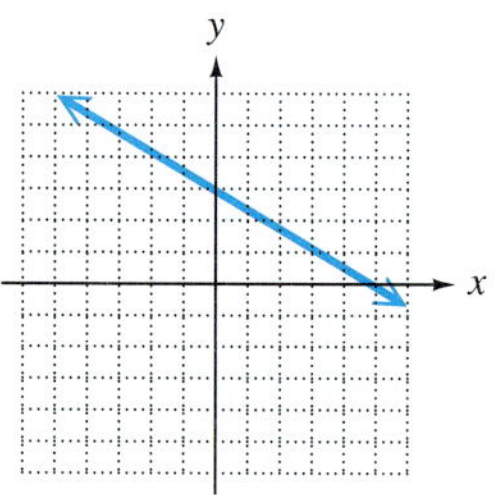

WORK PROBLEM 2 AT THE SIDE.

2 Inequalities that do not include the equals sign are graphed in a similar way.

EXAMPLE 1 *Graphing a Linear Inequality*

Graph the inequality $x - y > 5$.

This inequality does not include the equals sign. Therefore, the points on the line

$$x - y = 5$$

do not belong to the graph. However, the line still serves as a boundary for two regions, one of which satisfies the inequality. To graph the inequality, we first graph the equation $x - y = 5$. We use a dashed line to show that the points on the line are *not* solutions of the inequality $x - y > 5$. We can choose a test point not on the line to see which side of the line satisfies the inequality. Let us choose $(1, -2)$ this time.

$$x - y > 5$$
$$1 - (-2) > 5 \quad \text{Let } x = 1 \text{ and } y = -2.$$
$$3 > 5 \quad \text{False}$$

Because $3 > 5$ is false, the graph of the inequality includes the region that does *not* contain $(1, -2)$. Shade this region, as shown in Figure 26. This shaded region is the desired graph. Check that the proper region is shaded by selecting a point in the shaded region and substituting for x and y in the inequality $x - y > 5$. For example, use $(4, -3)$ from the shaded region, as follows.

$$x - y > 5$$
$$4 - (-3) > 5 \quad \text{Let } x = 4 \text{ and } y = -3.$$
$$7 > 5 \quad \text{True}$$

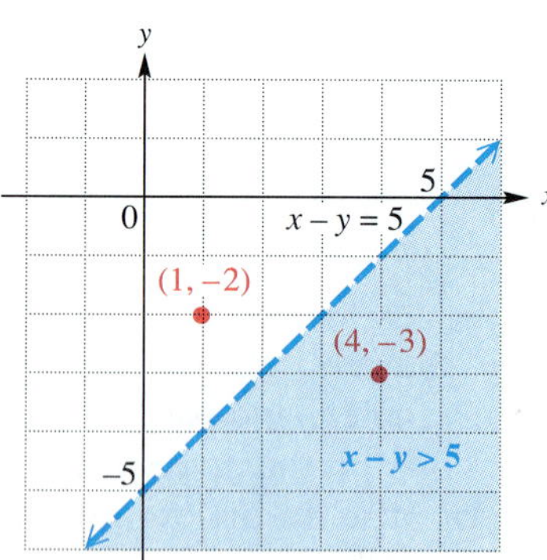

FIGURE 26

WORK PROBLEM 3 AT THE SIDE.

ANSWERS

2.

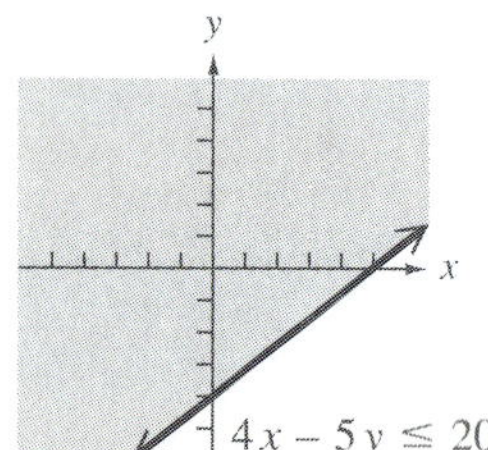

3.

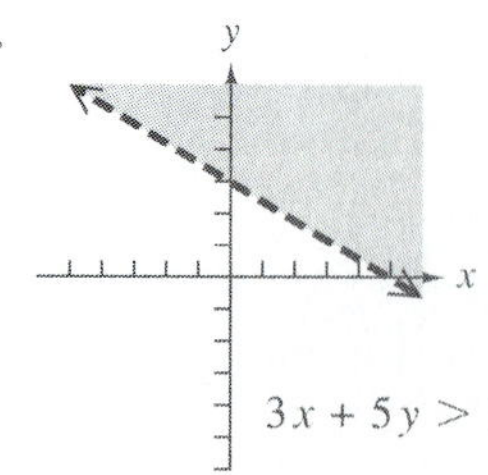

A summary of the steps used to graph a linear inequality is given below.

GRAPHING A LINEAR INEQUALITY IN TWO VARIABLES

Step 1 Graph the line that is the boundary of the region. Use the methods of Section 6.2. Make the line solid if the inequality has $\leq$ or $\geq$; make the line dashed if the inequality has $<$ or $>$.

Step 2 Use any point not on the line as a test point. Substitute for x and y in the inequality. If a true statement results, shade the side containing the test point. If a false statement results, shade the other side.

EXAMPLE 2 *Graphing a Linear Inequality*

Graph the inequality $2x - 5y \geq 10$.

Start by graphing the equation

$$2x - 5y = 10.$$

Use a solid line to show that the points on the line are solutions of the inequality $2x - 5y \geq 10$. Choose any test point not on the line. Again, we choose (0, 0).

$$2x - 5y \geq 10$$
$$2(0) - 5(0) \geq 10 \quad \text{Let } x = 0 \text{ and } y = 0.$$
$$0 - 0 \geq 10$$
$$0 \geq 10 \quad \text{False}$$

Because $0 \geq 10$ is false, shade the region *not* containing (0, 0). (See Figure 27.)

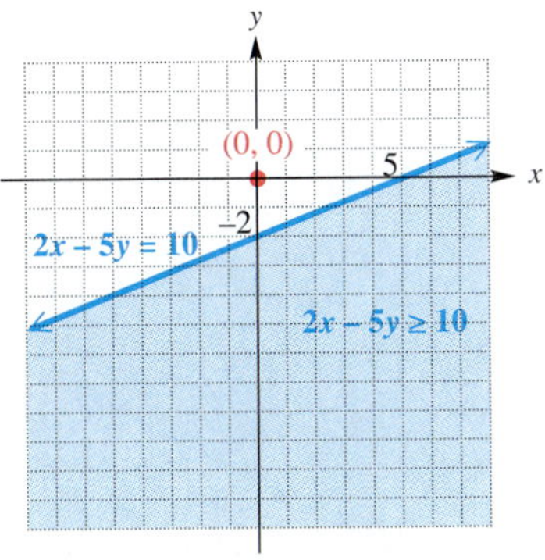

FIGURE 27

WORK PROBLEM 4 AT THE SIDE.

EXAMPLE 3 *Graphing a Linear Inequality*

Graph the inequality $x \leq 3$.

First graph $x = 3$, a vertical line going through the point (3, 0). Use a solid line. (Why?) Choose (0, 0) as a test point.

$$x \leq 3$$
$$0 \leq 3 \quad \text{Let } x = 0.$$
$$0 \leq 3 \quad \text{True}$$

Because $0 \leq 3$ is true, shade the region containing (0, 0), as in Figure 28.

4. Graph $2x - y \geq -4$.

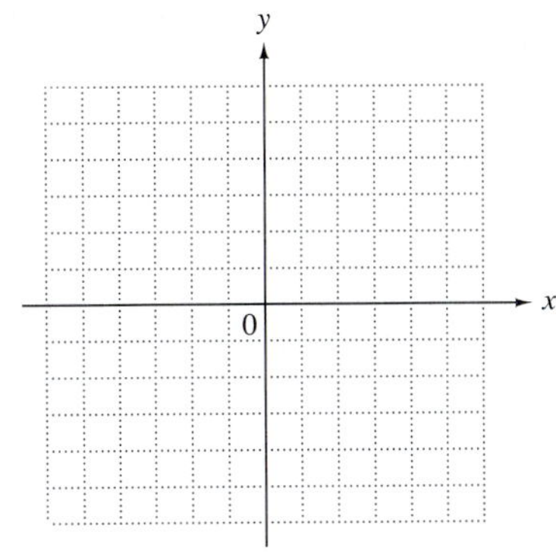

ANSWERS

4.

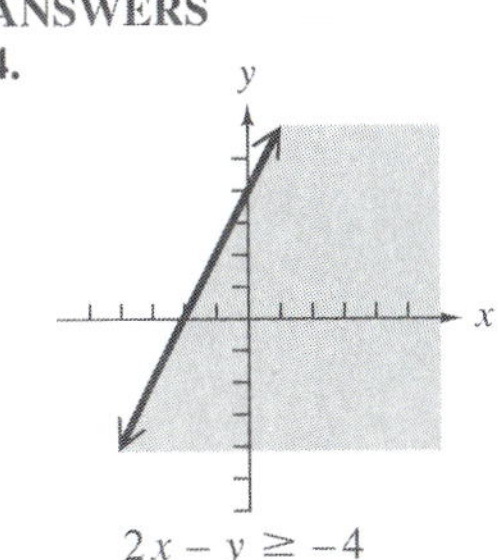

5. Graph $y < 4$.

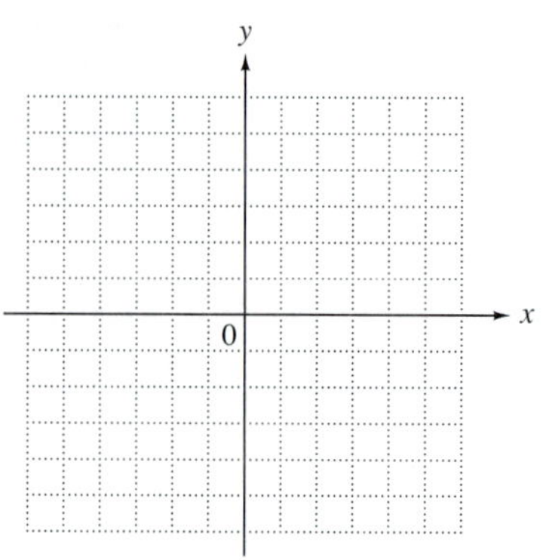

6. Graph $x \geq -3y$.

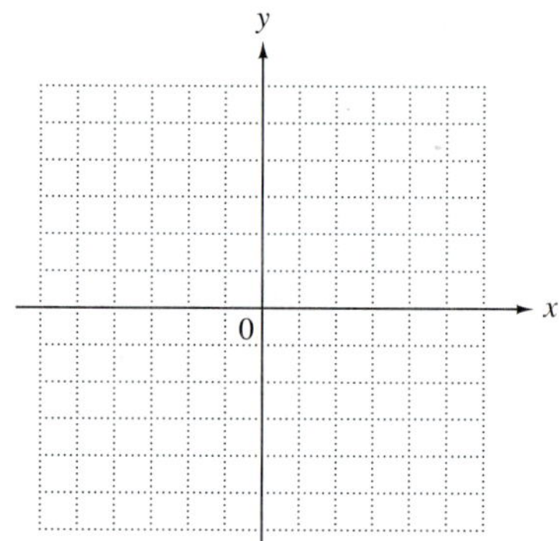

ANSWERS

5.

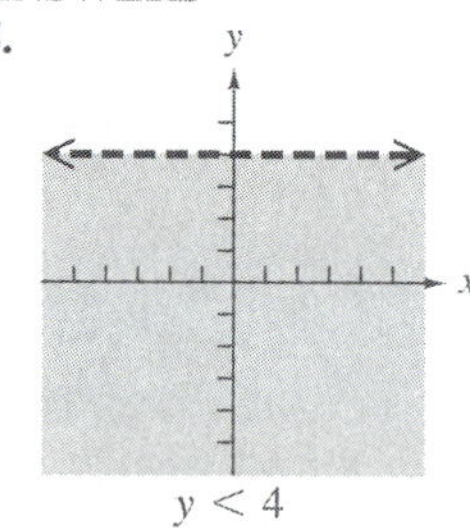

6.

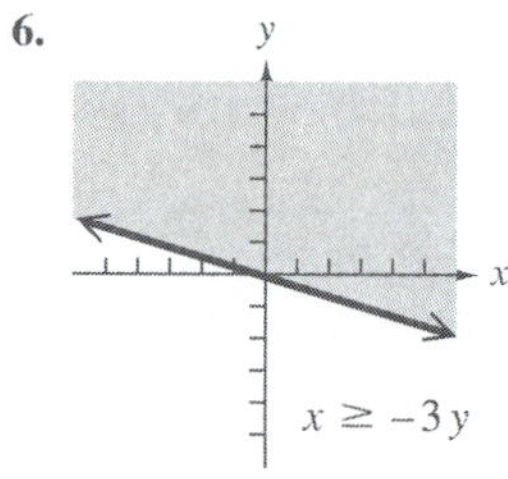

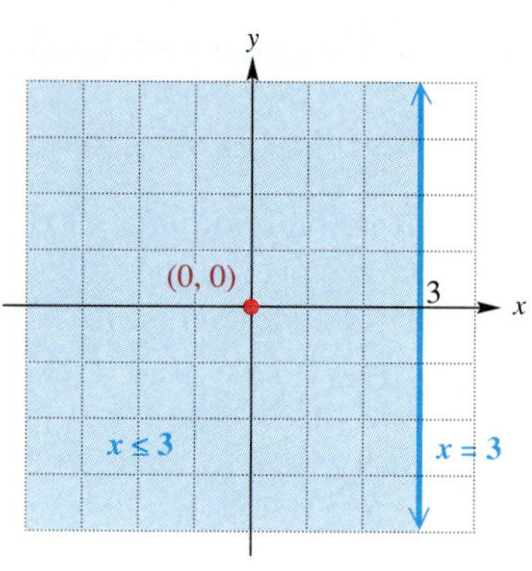

FIGURE 28

WORK PROBLEM 5 AT THE SIDE.

3 The next example shows how to graph an inequality having a boundary line going through the origin, an inequality in which (0, 0) cannot be used as a test point.

EXAMPLE 4 *Graphing a Linear Inequality*

Graph the inequality $x \leq 2y$.

Begin by graphing $x = 2y$. Some ordered pairs that can be used to graph this line are (0, 0), (6, 3), and (4, 2). Use a solid line. We cannot use (0, 0) as a test point because (0, 0) is on the line $x = 2y$. Instead, choose a test point off the line. Let us choose (1, 3), which is not on the line.

$x \leq 2y$	Original inequality
$1 \leq 2(3)$	Let $x = 1$ and $y = 3$.
$1 \leq 6$	True

Because $1 \leq 6$ is true, shade the side of the graph containing the test point (1, 3). (See Figure 29.) ■

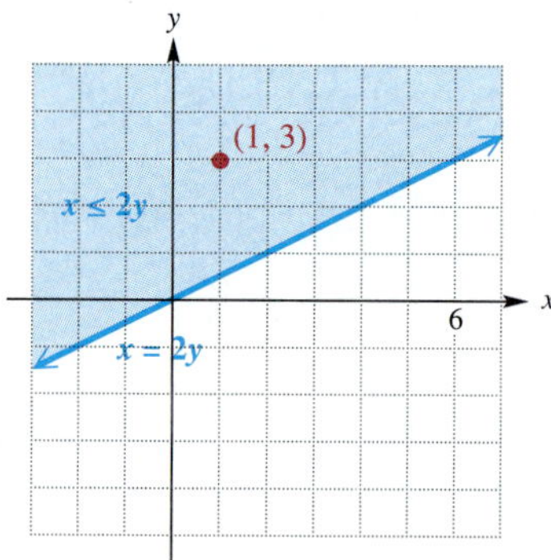

FIGURE 29

WORK PROBLEM 6 AT THE SIDE.

6.5 EXERCISES

NAME DATE HOUR

Decide whether the given ordered pairs are solutions of the inequality.

1. $3x - 4y < 12$
(a) $(2, 6)$ **(b)** $(4, 0)$

2. $2x + 5y < -10$
(a) $(4, -20)$ **(b)** $(-5, 0)$

3. $3x - 2y \geq 0$
(a) $(4, 1)$ **(b)** $(0, 0)$

4. $6x + 3y \geq 0$
(a) $(1, -3)$ **(b)** $(0, 0)$

In Exercises 5–14, the straight line boundary has been drawn. Complete each graph by shading the correct region. See Examples 1–4.

5. $x + y \geq 4$

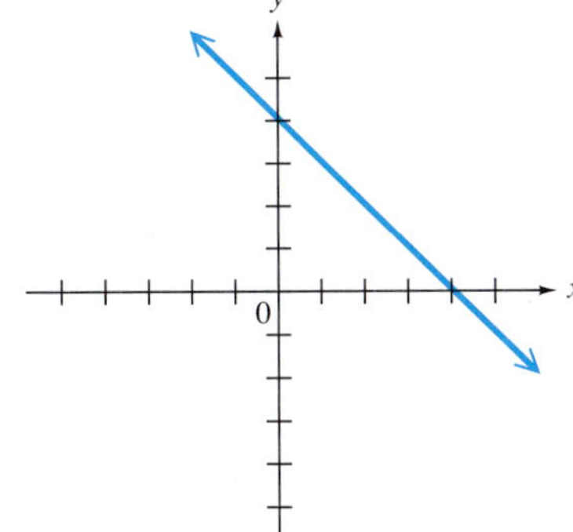

6. $x + y \leq 2$

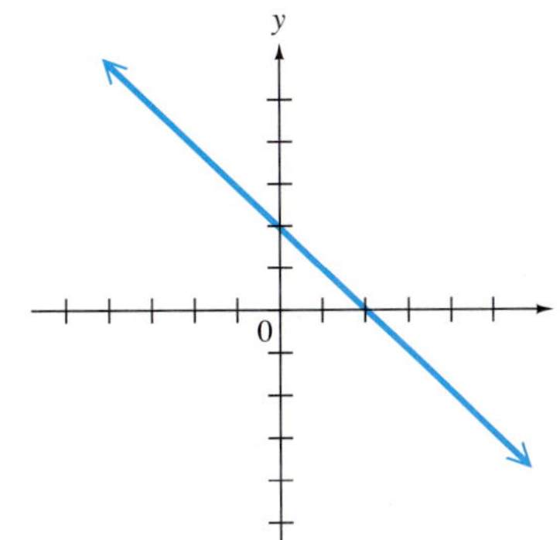

7. $x + 2y \geq 7$

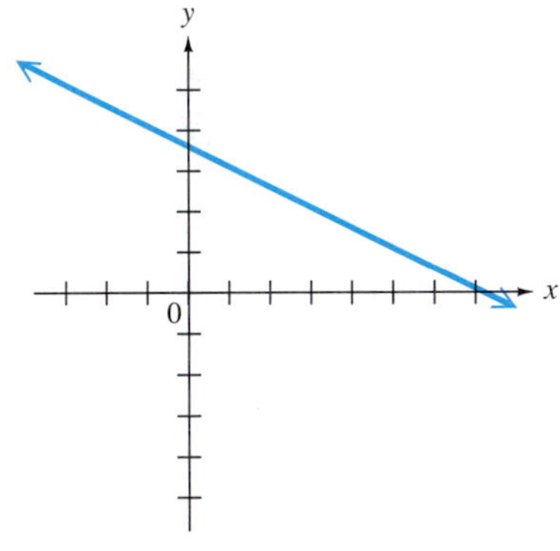

8. $2x + y \geq 5$

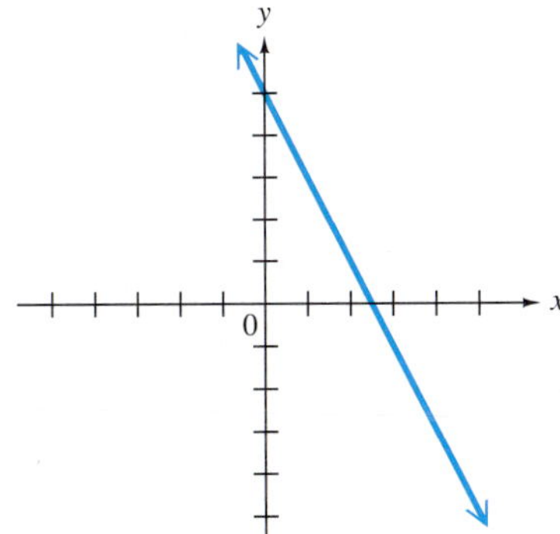

9. $-3x + 4y > 12$

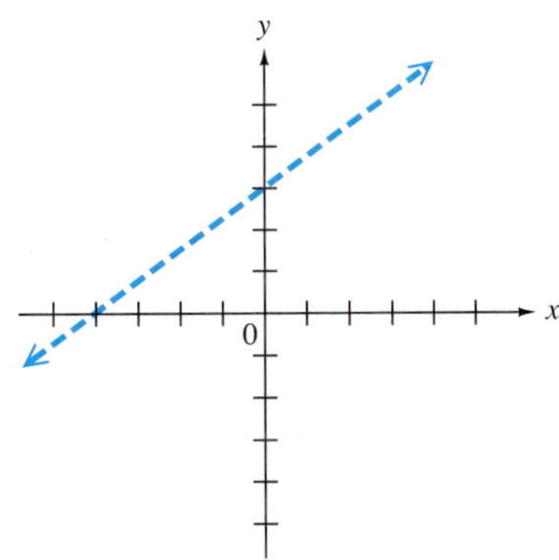

10. $4x - 5y < 20$

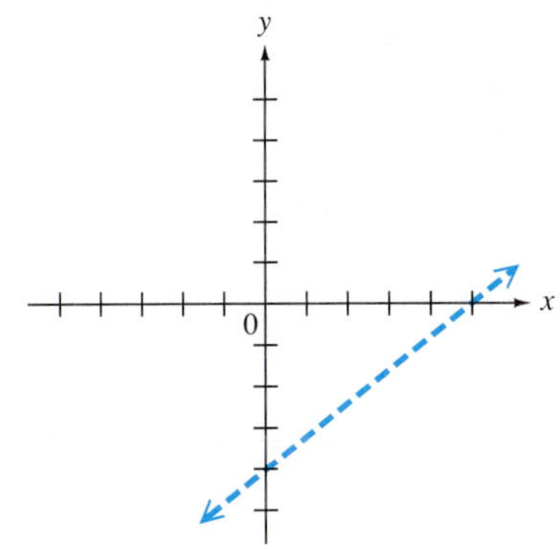

11. $x > 4$

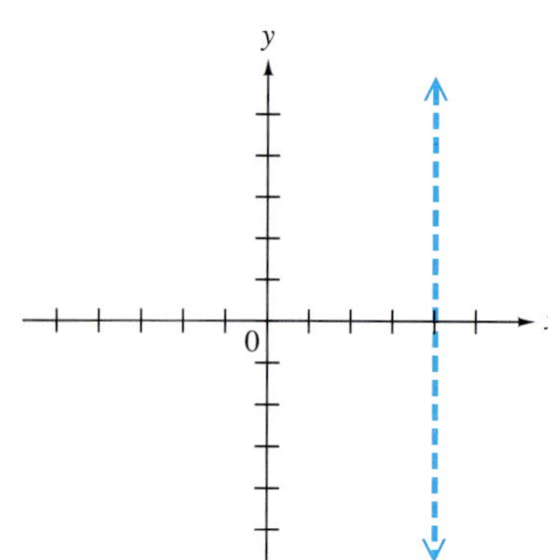

12. $y < -1$

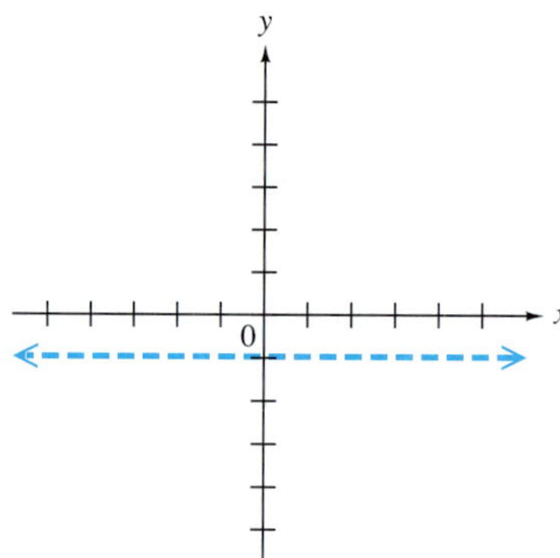

13. $x \leq 3y$

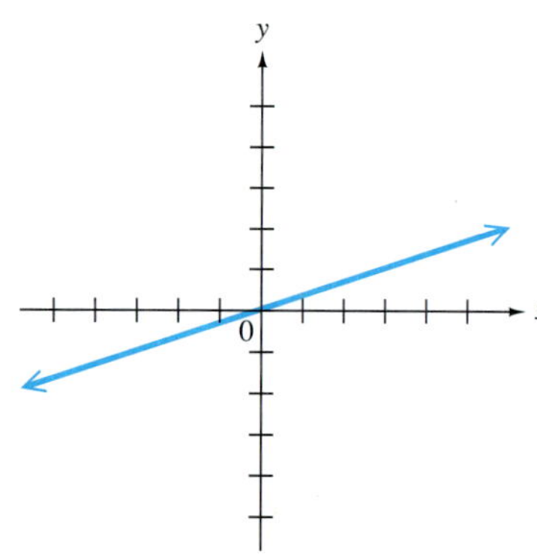

14. $x \geq -2y$

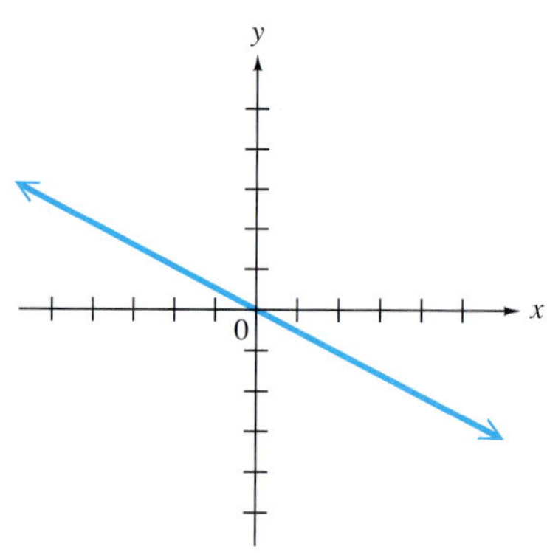

15. Explain how you will determine whether to use a dashed line or a solid line when graphing linear inequalities in two variables.

16. Explain why the point (0, 0) is not an appropriate choice for a test point when graphing an inequality whose boundary goes through the origin.

Graph each linear inequality. See Examples 1–4.

17. $x + y \leq 5$

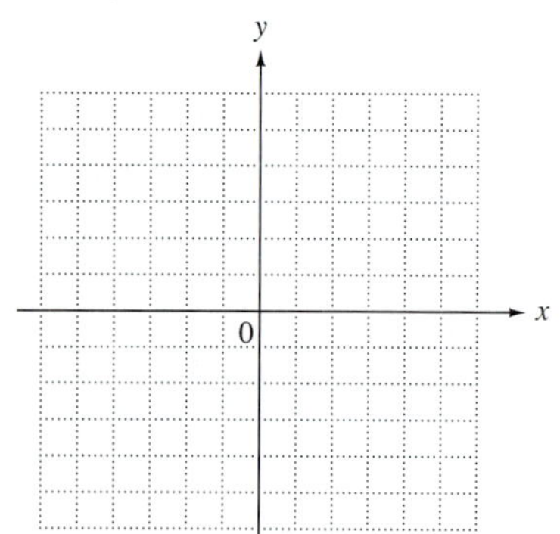

18. $x + y \geq 3$

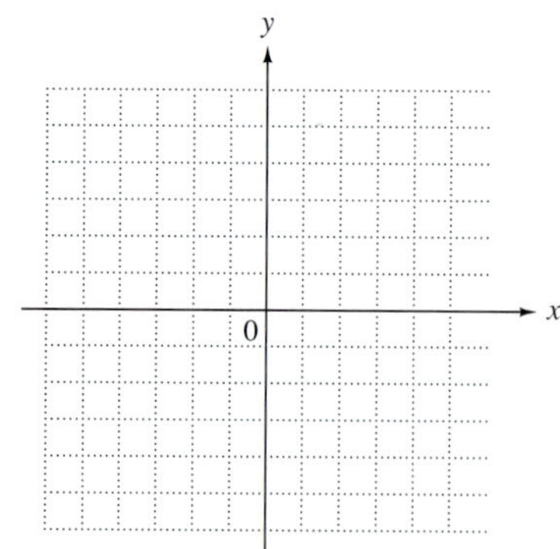

19. $x + 2y < 4$

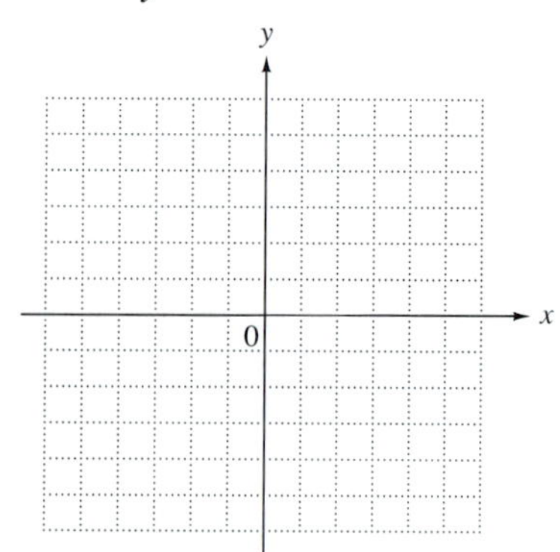

20. $x + 3y > 6$

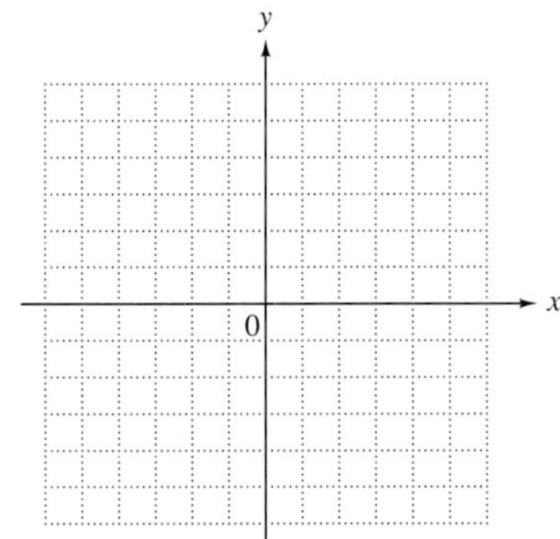

21. $2x + 3y > -6$

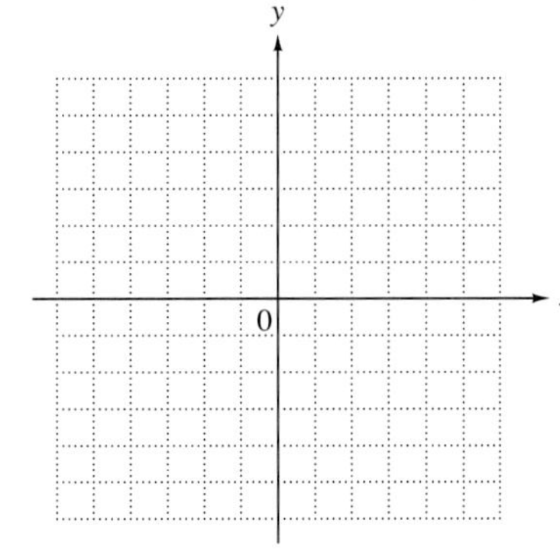

22. $3x + 4y < 12$

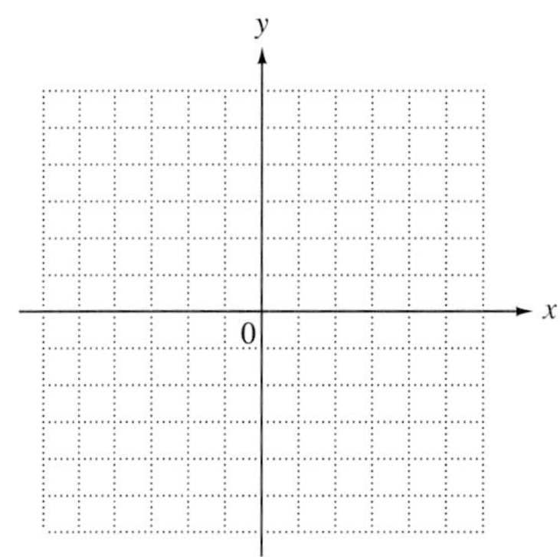

23. $y \geq 2x + 1$

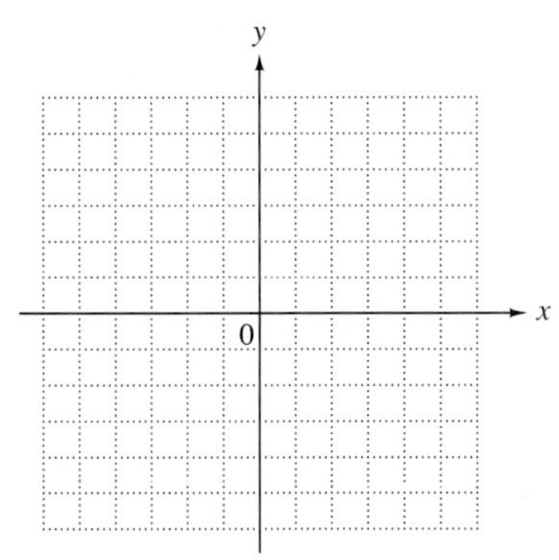

24. $y < -3x + 1$

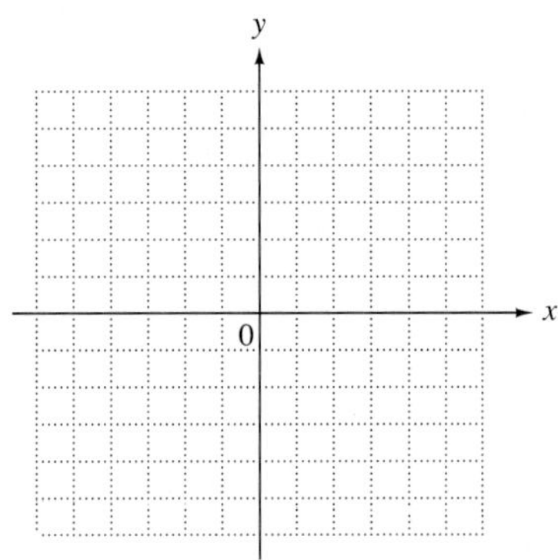

25. $x \leq -2$

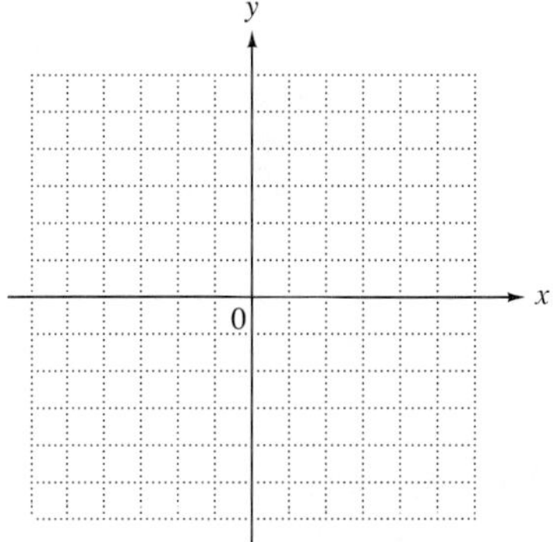

26. $x \geq 1$

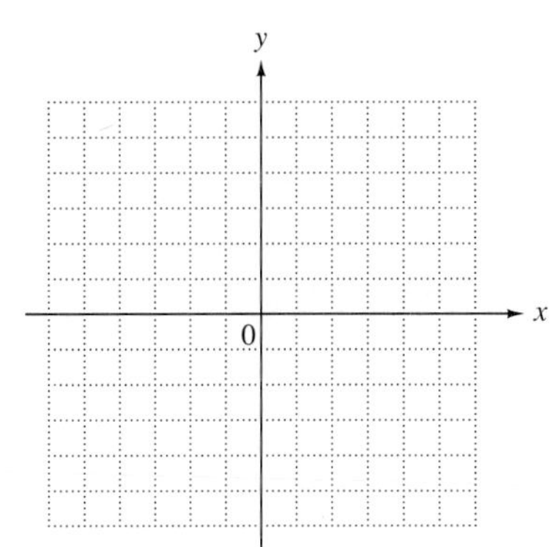

27. $y < 5$

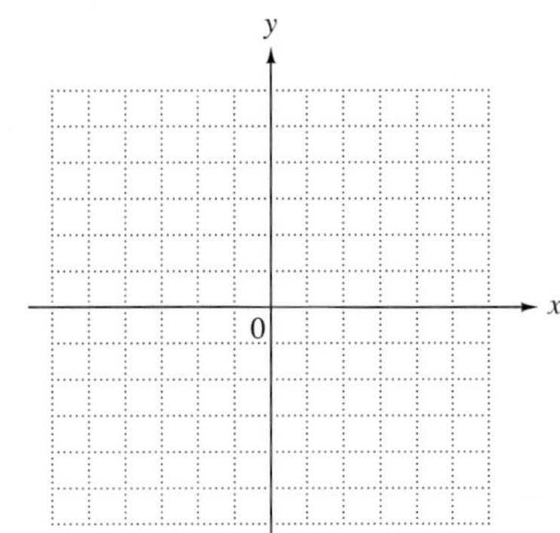

28. $y < -3$

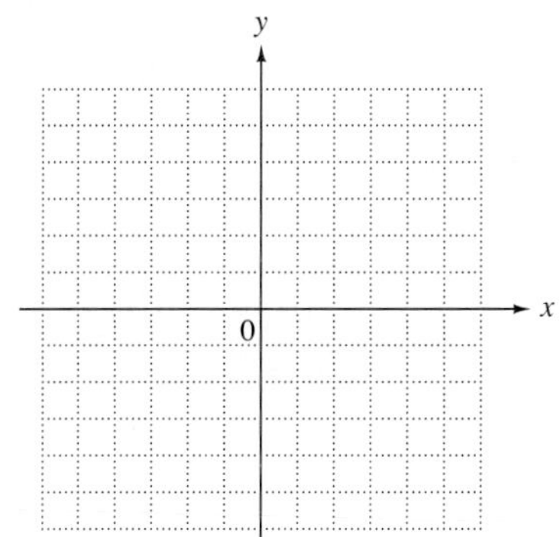

29. $y \geq 4x$

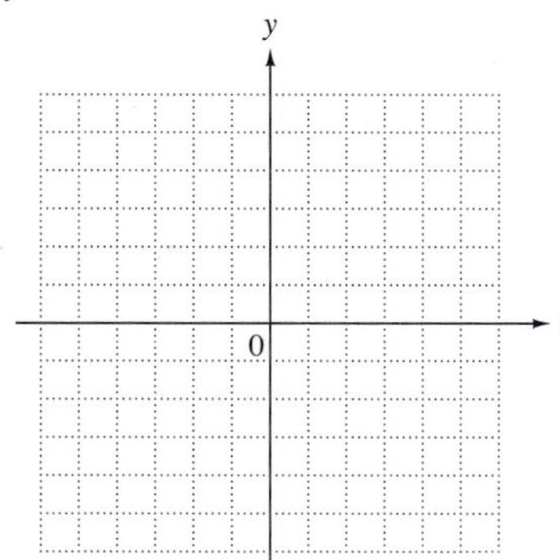

30. $y \leq 2x$

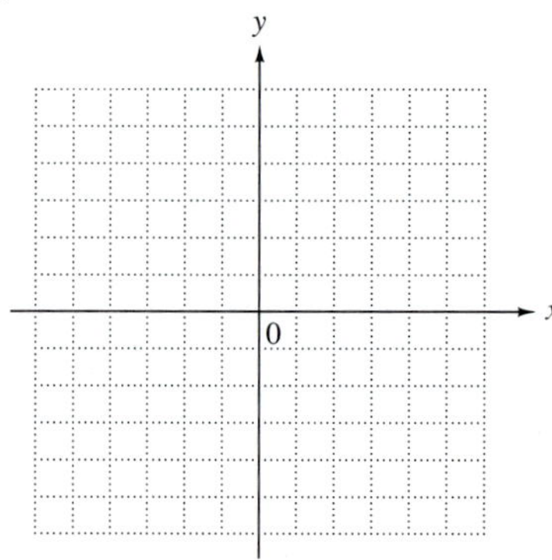

31. Explain why the graph of $y > x$ cannot lie in quadrant IV.

32. Explain why the graph of $y < x$ cannot lie in quadrant II.

33. Suppose that x represents the number of pounds of wheat and y represents the number of pounds of corn that Farmer Brown harvested this month. He harvested no more than 6000 pounds altogether. The following inequalities describe his situation:

$$x \geq 0$$
$$y \geq 0$$
$$x + y \leq 6000.$$

Graph these inequalities on the one set of axes shown, and give four ordered pairs that satisfy all three of them.

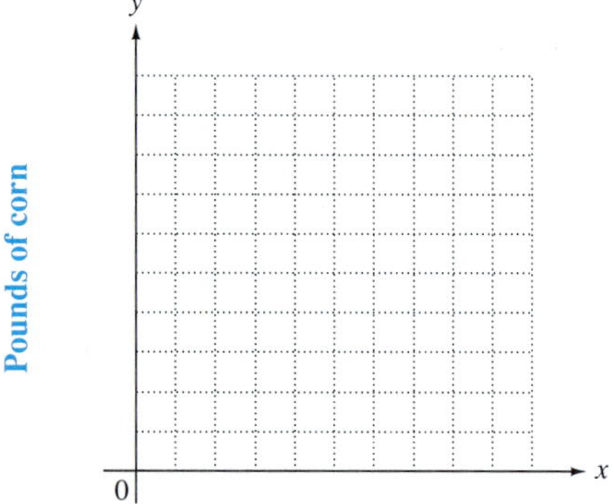

34. Gwen takes x ibuprofen pills each day at a cost of 20¢ per pill, and y vitamin pills at a cost of 30¢ per pill. She wants the total cost to be no more than \$1.50. Her situation can be described by the following inequalities:

$$x \geq 0$$
$$y \geq 0$$
$$20x + 30y \leq 150.$$

Graph these inequalities on the one set of axes shown, and give four ordered pairs that satisfy all three of them.

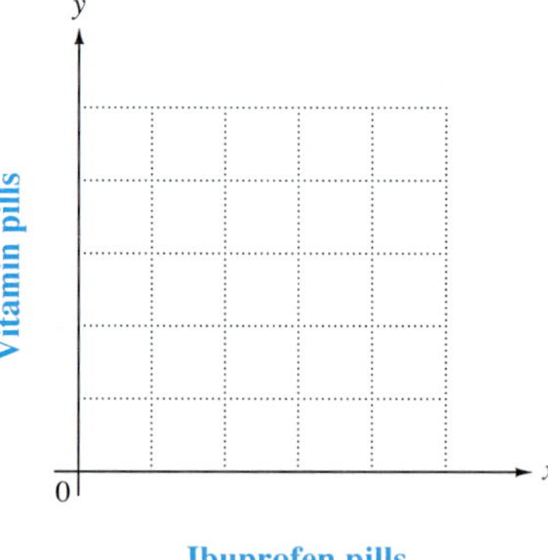

PREVIEW EXERCISES

Add the polynomials. See Section 3.4.

35. $\begin{array}{r} 2x + 3y \\ \underline{-3x - 3y} \end{array}$

36. $\begin{array}{r} 6x + 2y \\ \underline{5x - 2y} \end{array}$

37. $\begin{array}{r} -3x + 2y \\ \underline{3x - 2y} \end{array}$

38. $\begin{array}{r} x + y \\ \underline{x - y} \end{array}$

QUEST FOR NUMERACY

Do Statistics Ever Lie?

The statement that there are "lies, damned lies, and statistics" is attributed to Disraeli, Queen Victoria's prime minister. Other people have said even stronger things about statistics. This often intense distrust of statistics has come about because of a belief that "you can prove anything with numbers." It must be admitted that there is often a conscious or unconscious distortion in many published statistics. The classic book on distortion in statistics is *How to Lie with Statistics,** by Darrell Huff.

Negative attitudes toward statistics and polls are understandable when we are aware of some of the distorted methods used in surveys. Hopefully, as more people learn what statistics are and what they are not, what they can do and what they cannot do, their abuses can be curbed and their legitimate uses can be strengthened.

FOR GROUP DISCUSSION

1. The graphs shown here are found in Huff's book. He uses them to illustrate how graphs can be totally meaningless. Discuss what is wrong with these. (They come from two boxes of Grape-Nuts Flakes.)

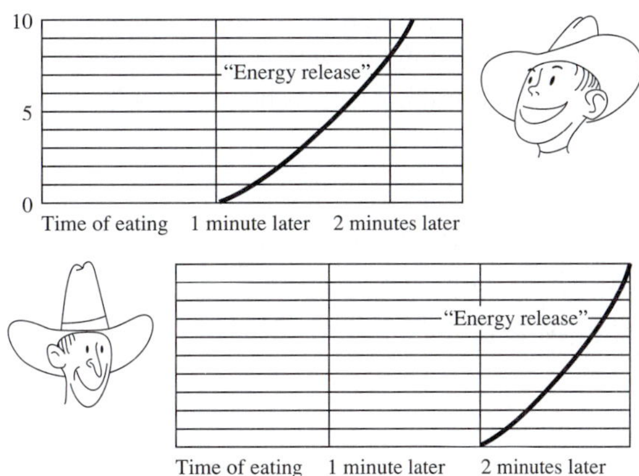

2. Relate any stories that you may have heard about statistics that were used to convey incorrect information.

*Reprinted from *How to Lie with Statistics* by Darrell Huff, pictures by Irving Geis, with permission of W.W. Norton & Company, Inc. Copyright 1954 and renewed ©1982 by Darrell Huff and Irving Geis.

CHAPTER 6 SUMMARY

KEY TERMS

6.1	**linear equation**	An equation that can be written in the form $Ax + By = C$ is a linear equation in two variables. (A and B cannot both be 0.)
	ordered pair	A pair of numbers written between parentheses in which the order is important is called an ordered pair.
	table of ordered pairs	A table showing selected ordered pairs of numbers that satisfy an equation is called a table of ordered pairs.
	coordinate system	An x-axis and y-axis at right angles form a coordinate system.
	quadrants	A coordinate system divides the plane into four regions called quadrants.
	origin	The point at which the x-axis and y-axis intersect is called the origin.
	coordinates	The numbers in an ordered pair are called the coordinates of the corresponding point.
	plot	To plot an ordered pair is to find the corresponding point on a coordinate system.
6.2	**graph**	A graph of an equation is the set of all points that correspond to the ordered pairs that satisfy the equation.
	graphing	The process of plotting the ordered pairs that satisfy a linear equation and drawing a line through them is called graphing.
	y-intercept	If a graph crosses the y-axis at k, then the y-intercept is $(0, k)$.
	x-intercept	If a graph crosses the x-axis at k, then the x-intercept is $(k, 0)$.
6.3	**slope**	The slope of a line is the ratio of the change in y compared to the change in x when moving along the line.
	parallel lines	Two lines in a plane that never intersect are parallel.
	perpendicular lines	Perpendicular lines intersect at a 90° angle.

NEW SYMBOLS

(x, y)	an ordered pair
m	slope
b	y-value of the y-intercept

QUICK REVIEW

Concepts	Examples
6.1 Linear Equations in Two Variables	
If a value of either variable in an equation is given, the other variable can be found by substitution.	Complete the ordered pair (0,) for $3x = y + 4$. $3(0) = y + 4$ $0 = y + 4$ $-4 = y$ The ordered pair is $(0, -4)$.
Plot the ordered pair $(-2, 4)$ by starting at the origin, going 2 units to the left, and from there going 4 units up.	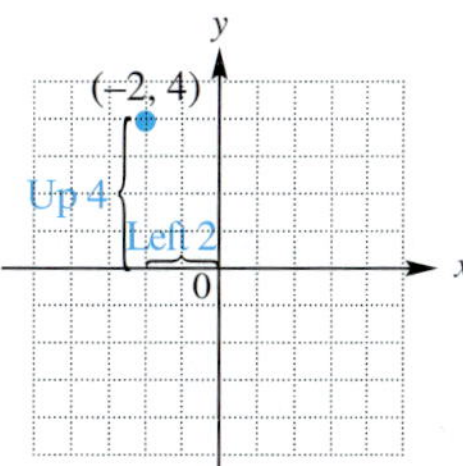
6.2 Graphing Linear Equations in Two Variables	
To graph a linear equation: **1.** Find at least two pairs that satisfy the equation. **2.** Plot the corresponding points. **3.** Draw a straight line through the points.	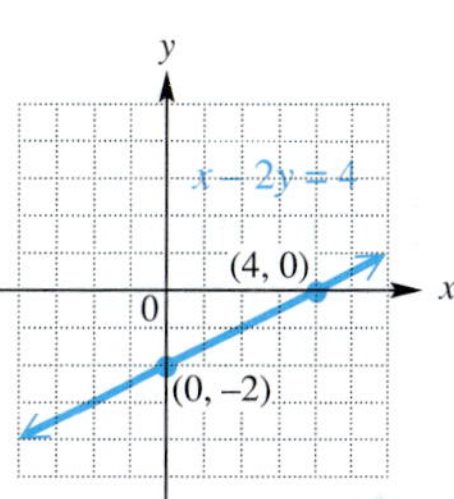
6.3 The Slope of a Line	
The slope of the line through (x_1, y_1) and (x_2, y_2) is $m = \frac{y_2 - y_1}{x_2 - x_1}, \ (x_1 \neq x_2).$ Horizontal lines have slope 0. Vertical lines have undefined slope. To find the slope of a line from its equation, solve for y. The slope is the coefficient of x.	The line through $(-2, 3)$ and $(4, -5)$ has slope $m = \frac{-5 - 3}{4 - (-2)} = \frac{-8}{6} = -\frac{4}{3}.$ The line $y = -2$ has slope 0. The line $x = 4$ has undefined slope. Given: $3x - 4y = 12$. $-4y = -3x + 12$ $y = \frac{3}{4}x - 3$ The slope is $\frac{3}{4}$.

Concepts	Examples
6.4 Equations of a Line	
Slope-Intercept Form $y = mx + b$ m is the slope. $(0, b)$ is the y-intercept.	Find an equation of the line with slope 2 and y-intercept $(0, -5)$: $y = 2x - 5$.
Point-Slope Form $y - y_1 = m(x - x_1)$ m is the slope. (x_1, y_1) is a point on the line.	Find an equation of the line with slope $-\frac{1}{2}$ through $(-4, 5)$: $y - 5 = -\frac{1}{2}(x - (-4))$ $2(y - 5) = -(x + 4)$ $2y - 10 = -x - 4$ $x + 2y = 6$
6.5 Graphing Linear Inequalities in Two Variables	
1. Graph the line that is the boundary of the region. Make it solid if the inequality is $\leq$ or $\geq$; make it dashed if the inequality is $<$ or $>$.	Graph $2x + y \leq 5$: Graph the line $2x + y = 5$. Make it solid because of $\leq$.
2. Use any point not on the line as a test point. Substitute for x and y in the inequality. If the result is true, shade the side of the line containing the test point; if the result is false, shade the other side.	Use $(1, 0)$ as a test point. $2(1) + 0 \leq 5$ $2 \leq 5$ True Shade the side of the line containing $(1, 0)$. 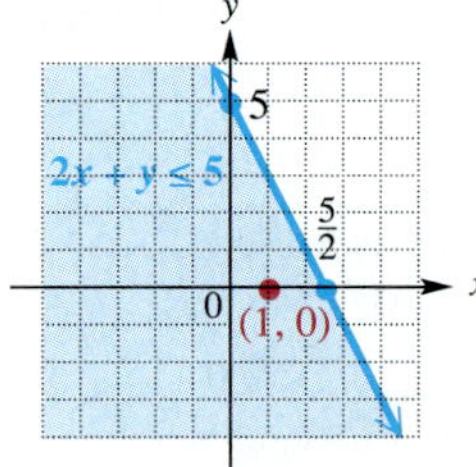

NAME DATE HOUR

CHAPTER 6 REVIEW EXERCISES

[6.1] *Complete the given ordered pairs for each equation.*

1. $y = 3x + 2$ $(-1, \quad)(0, \quad)(\quad, 5)$

2. $4x + 3y = 6$ $(0, \quad)(\quad, 0)(-2, \quad)$

3. $x = 3y$ $(0, \quad)(8, \quad)(\quad, -3)$

4. $x - 7 = 0$ $(\quad, -3)(\quad, 0)(\quad, 5)$

Decide whether the given ordered pair is a solution of the given equation.

5. $x + y = 7$; $(2, 5)$

6. $2x + y = 5$; $(-1, 3)$

7. $3x - y = 4$; $\left(\frac{1}{3}, -3\right)$

8. $5x - 3y = 16$; $\left(0, -\frac{16}{3}\right)$

Plot the ordered pairs on the given coordinate system.

9. $(2, 3)$

10. $(-4, 2)$

11. $(3, 0)$

12. $(0, -6)$

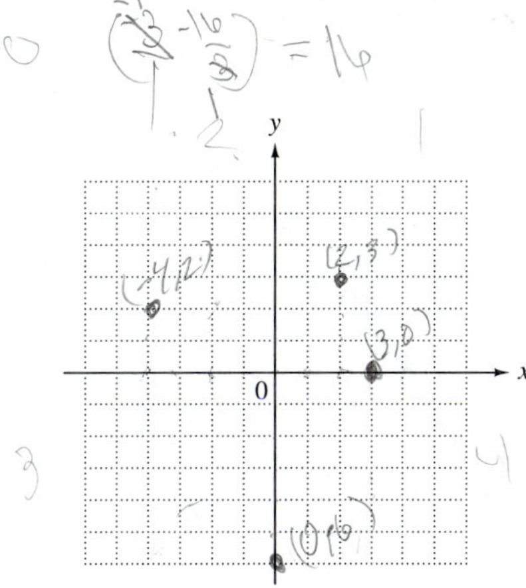

13. If $xy > 0$, in what quadrant or quadrants must (x, y) lie?

14. On what axis does the point $(k, 0)$ lie for any real value of k?

Without plotting the given point, name the quadrant in which each point lies.

15. $(-2, 3)$

16. $(-1, -4)$

[6.2] *Find the intercepts for each equation.*

17. $y = 2x + 5$
x-intercept:
y-intercept:

18. $2x + y = -7$
x-intercept:
y-intercept:

19. $3x + 2y = 8$
x-intercept:
y-intercept:

Graph each linear equation.

20. $y = -2x - 5$

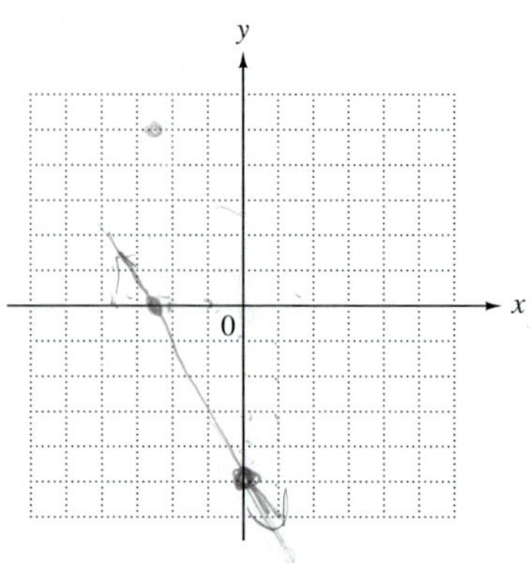

21. $x + 2y = -4$

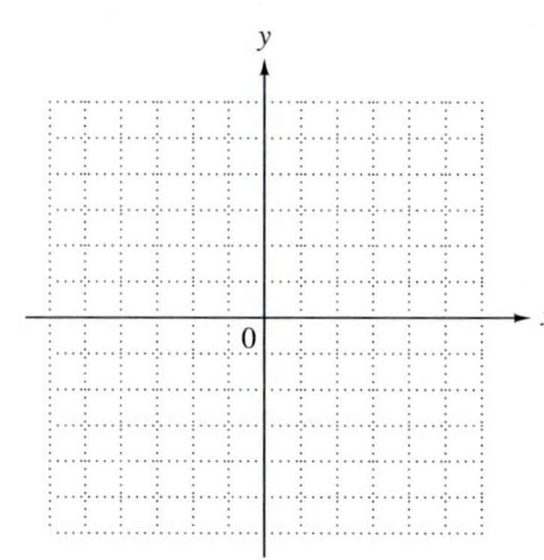

22. $x + y = 0$

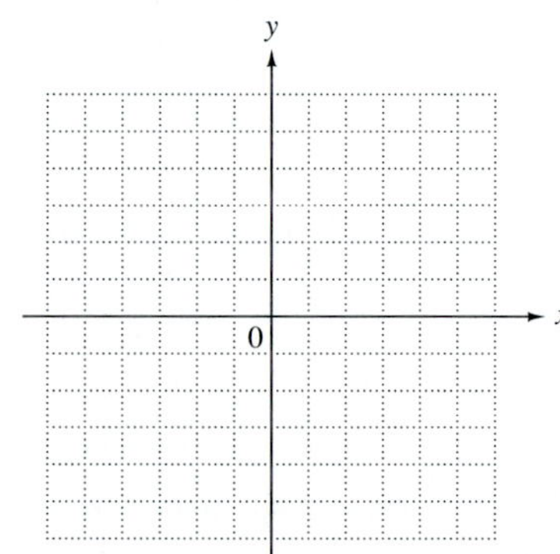

[6.3] *Find the slope of each line.*

23. Through $(2, 3)$ and $(-4, 6)$

24. Through $(0, 0)$ and $(-3, 2)$

25. Through $(0, 6)$ and $(1, 6)$

26. Through $(2, 5)$ and $(2, 8)$

27. $y = 3x - 4$

28. $y = \frac{2}{3}x + 1$

29.

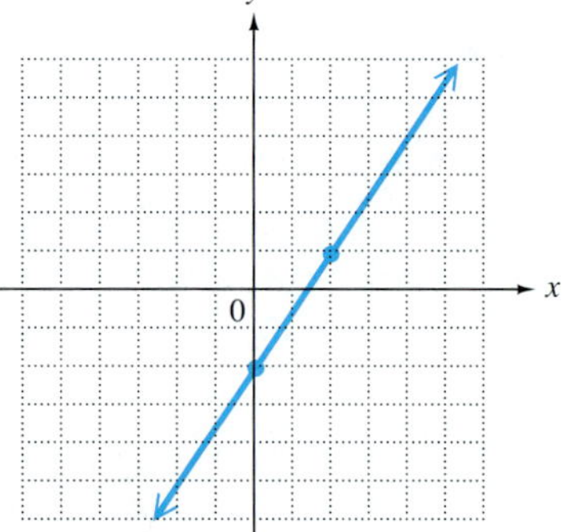

30.

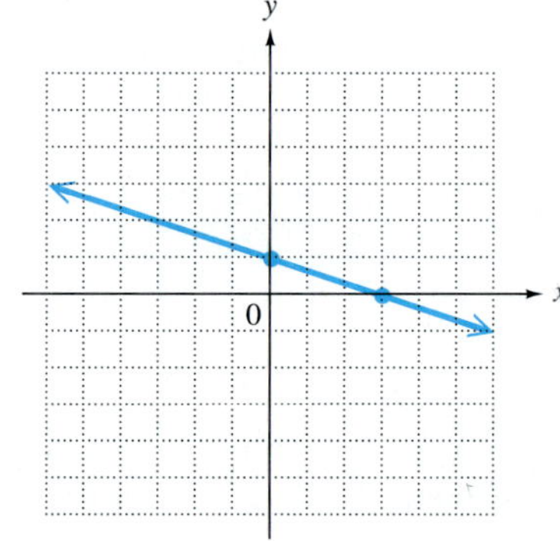

31. $y = 4$

32. $x = 0$

33. For the line $y = 4x + 6$, give the slope of a line (a) perpendicular to it and (b) parallel to it.

34. Explain why the signs of the slopes of perpendicular lines cannot be the same.

Decide whether the lines of each pair are parallel, perpendicular, or neither.

35. $3x + 2y = 6$
$6x + 4y = 8$

36. $x - 3y = 1$
$3x + y = 4$

37. $x - 2y = 8$
$x + 2y = 8$

38. What is the slope of a line perpendicular to a line with undefined slope?

[6.4] *Write an equation for each line. Express it in the standard form $Ax + By = C$.*

39. $m = -1; b = \frac{2}{3}$

40. The line in Exercise 30

41. Through $(4, -3)$; $m = 1$

42. Through $(-1, 4)$; $m = \frac{2}{3}$

43. Through $(1, -1)$; $m = -\frac{3}{4}$

44. Through $(2, 1)$ and $(-2, 2)$

45. Through $(-4, 1)$ with slope 0

46. Through $\left(\frac{1}{3}, -\frac{3}{4}\right)$ with undefined slope

[6.5] *Complete the graph of each linear inequality by shading the correct region.*

47. $x - y \geq 3$

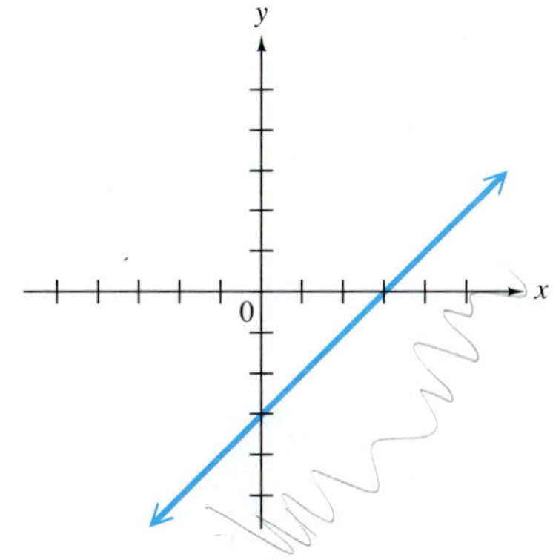

48. $3x - y \leq 5$

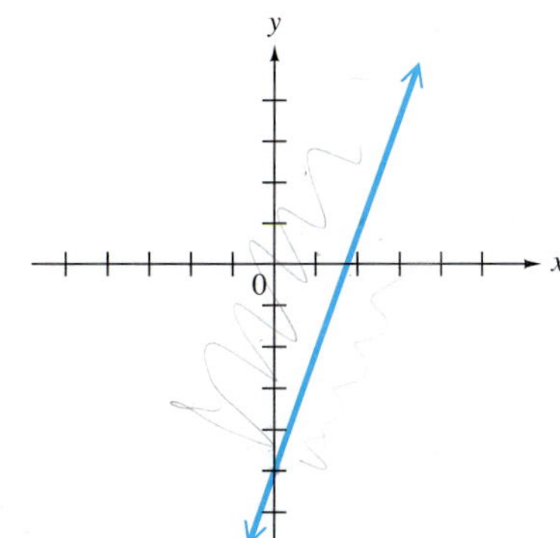

49. $x + 2y < 6$

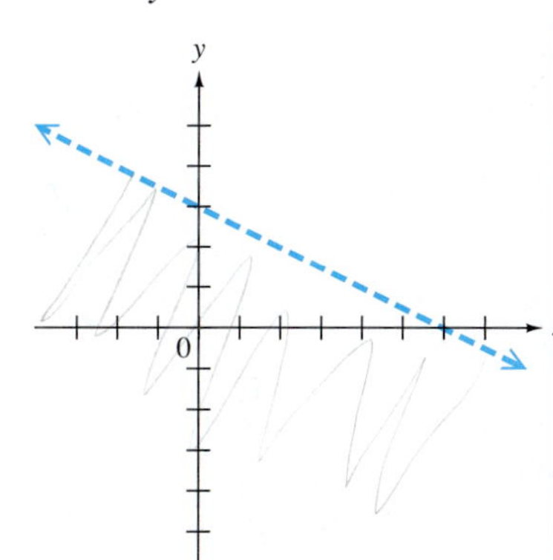

Graph each linear inequality.

50. $3x + 5y > 9$

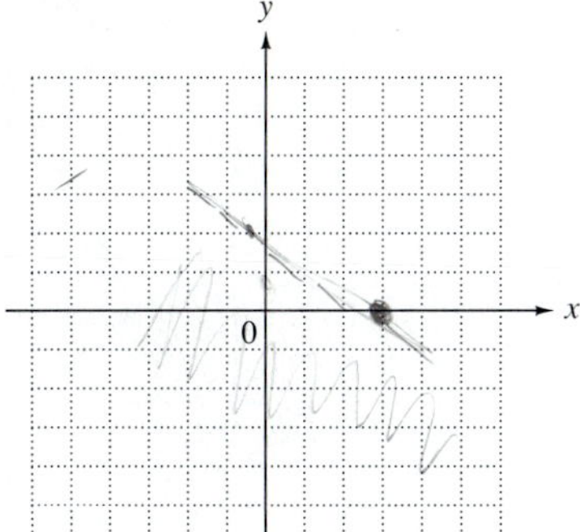

51. $2x - 3y > -6$

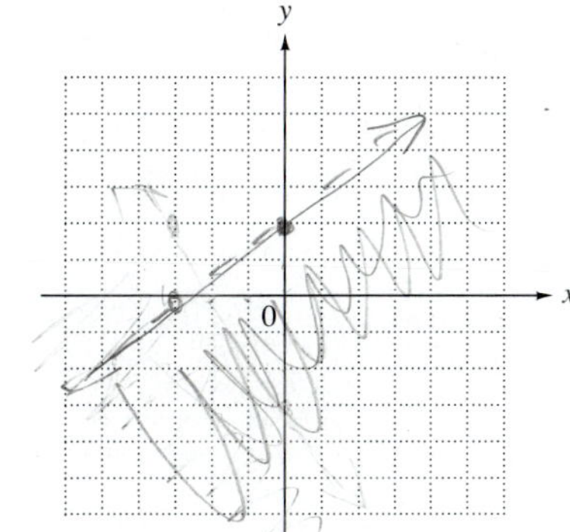

52. $x - 2y \geq 0$

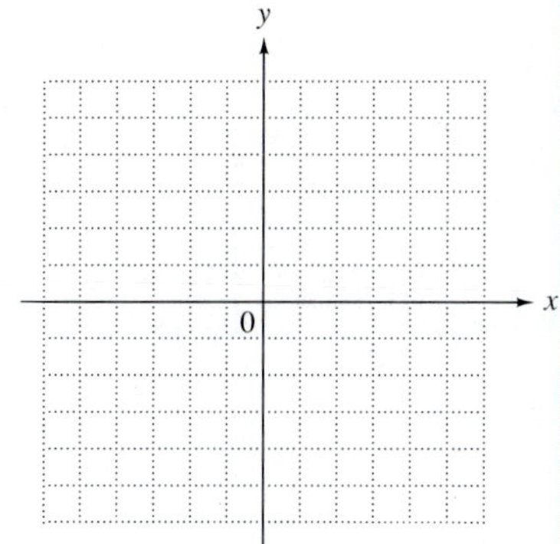

MIXED REVIEW EXERCISES

Find the intercepts and the slope of each line.

53. $11x - 3y = 4$

54. Through $(4, -1)$ and $(-2, -3)$

55. Through $(0, -1)$ and $(9, -5)$

56. $8x = 6 - 3y$

Write an equation in the standard form $Ax + By = C$ for each of the following lines.

57. Through $(5, 0)$ and $(5, -1)$

58. $m = -\frac{1}{4}$; $b = -\frac{5}{4}$

59. Through $(8, 6)$; $m = -3$

60. Through $(3, -5)$ and $(-4, -1)$

Graph each equation or inequality.

61. $x - 2y \leq 6$

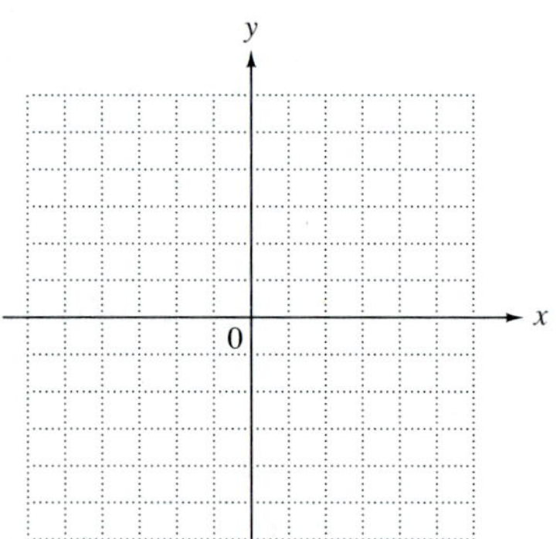

62. $x + 3y = 0$

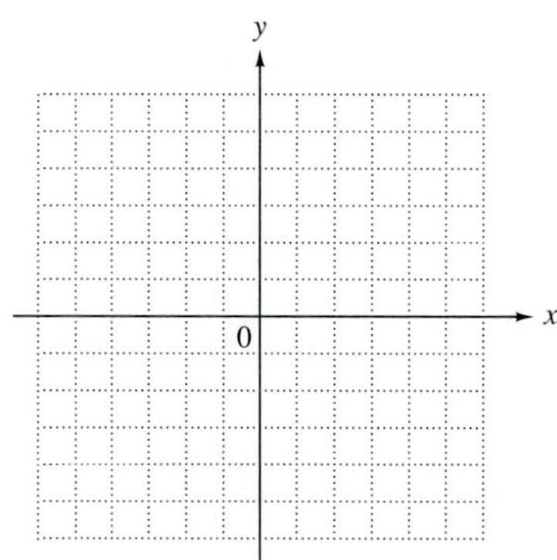

63. $y < -4x$

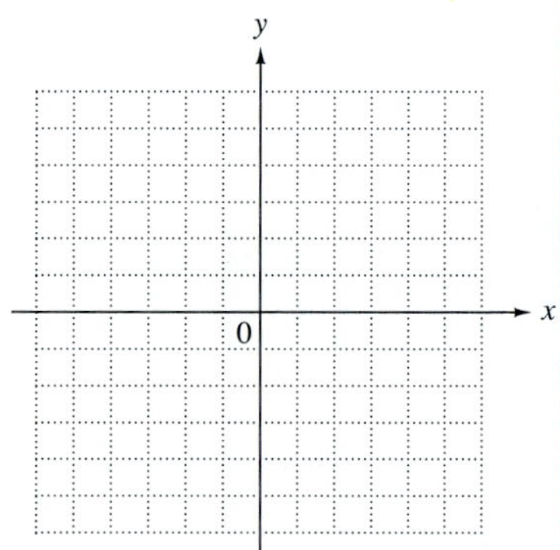

64. $y - 5 = 0$

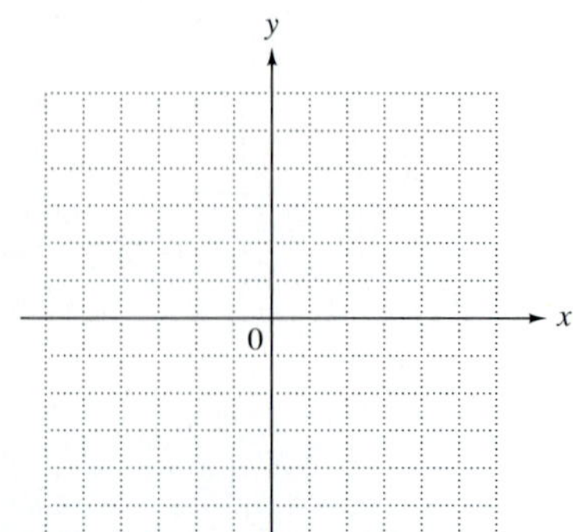

65. $2x - y = 3$

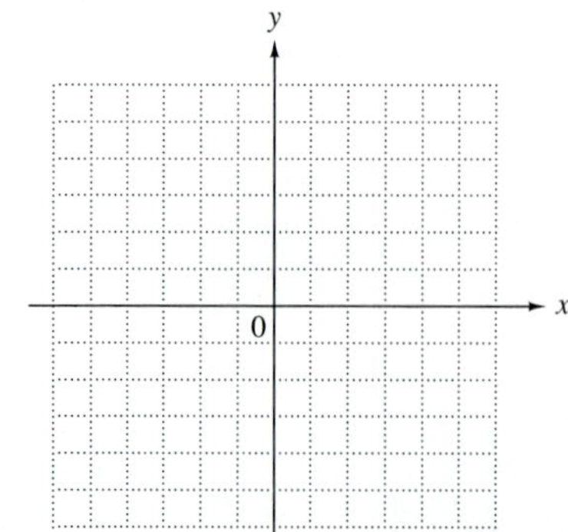

66. $x \geq -4$

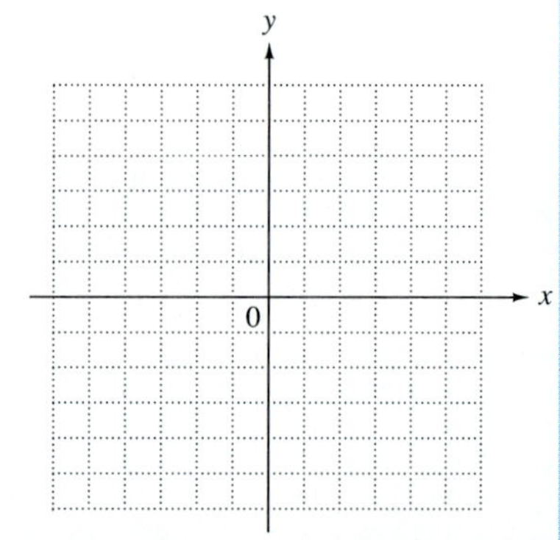

CHAPTER 6 TEST

Complete the ordered pairs for the given equations.

1. $y = 4x - 9$ $(0, \quad)$ $(1, \quad)$ $(\quad , -3)$

1. ____________________

2. $3x + 5y = -30$ $(0, \quad)$ $(\quad , 0)$ $(\quad , 3)$

2. ____________________

3. $y + 12 = 0$ $(0, \quad)$ $(-4, \quad)$ $\left(\frac{5}{2}, \quad\right)$

3. ____________________

4. How do we find the x-intercept of a linear equation in two variables? How do we find the y-intercept?

4. ____________________

Graph each linear equation. Give the x- and y-intercepts.

5. $x - y = 4$

5. x-intercept: ____________________
y-intercept: ____________________

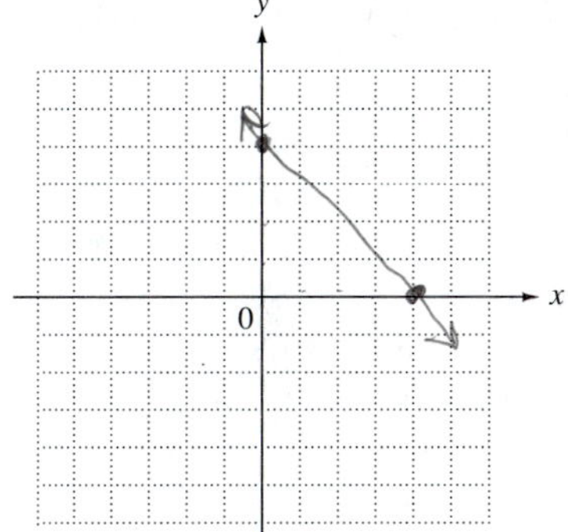

6. $3x + y = 6$

6. x-intercept: ____________________
y-intercept: ____________________

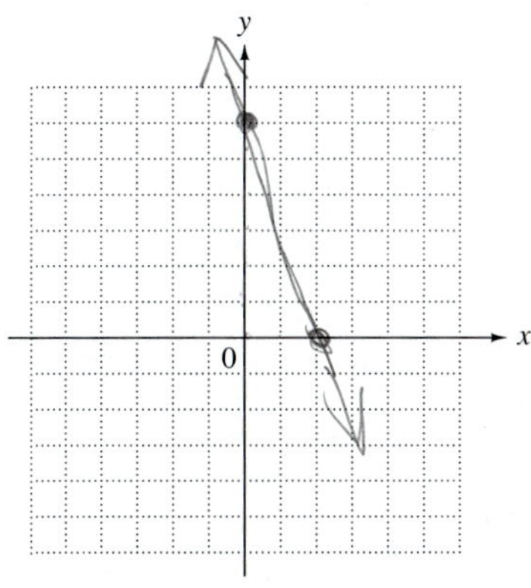

7. x-intercept: ______________
y-intercept: ______________

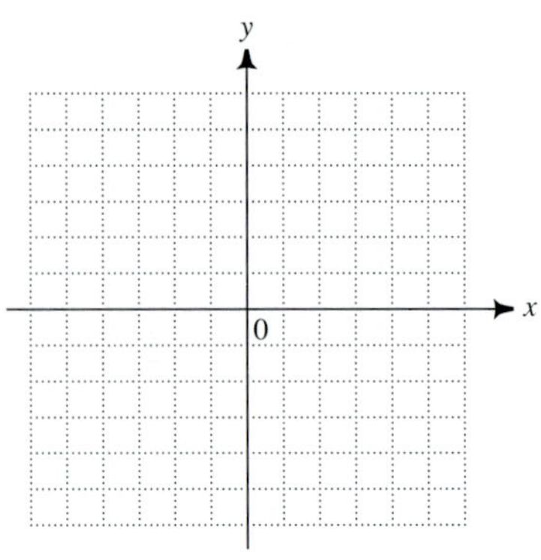

7. $y - 2x = 0$

8. x-intercept: -3
y-intercept: 0

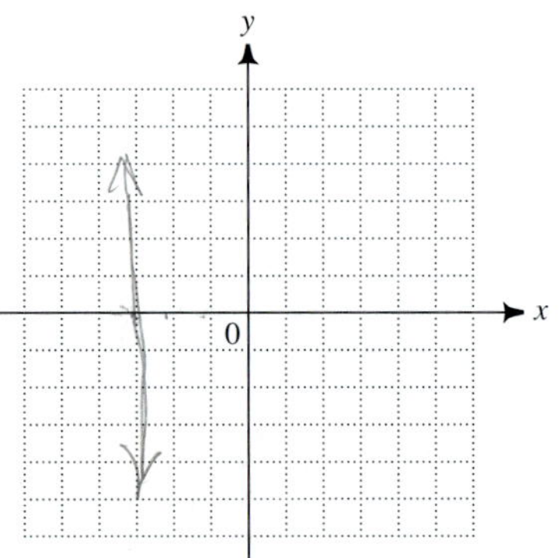

8. $x + 3 = 0$

9. x-intercept: ______________
y-intercept: ______________

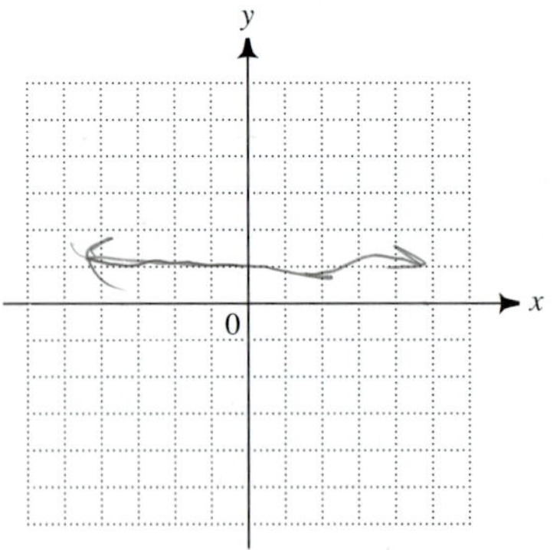

9. $y = 1$

Find the slope of each line.

10. Through $(-4, 6)$ and $(-1, -2)$ — 10. ________

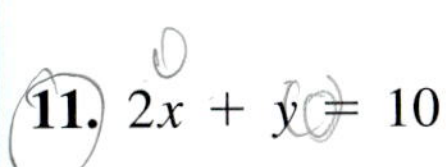

11. $2x + y = 10$ — 11. ________

12. $x + 12 = 0$ — 12. ________

13. — 13. ________

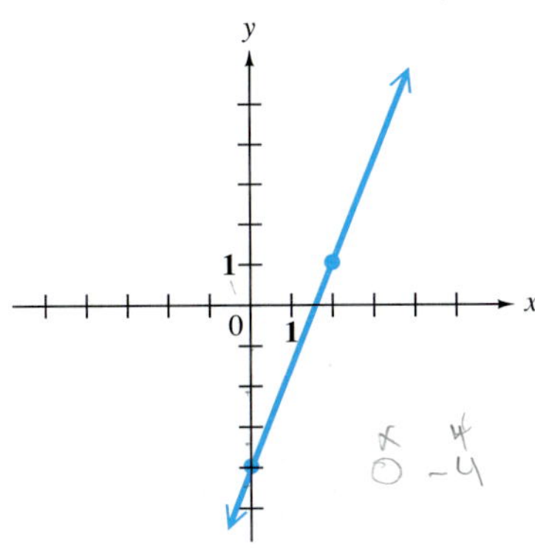

14. a line parallel to the graph of $y - 4 = 6$ — 14. ________

Write an equation for each line. Express it in the standard form $Ax + By = C$.

15. Through $(-1, 4)$; $m = 2$ — 15. ________

16. The line in Exercise 13 — 16. ________

17. Through $(2, -6)$ and $(1, 3)$ — 17. ________

18.

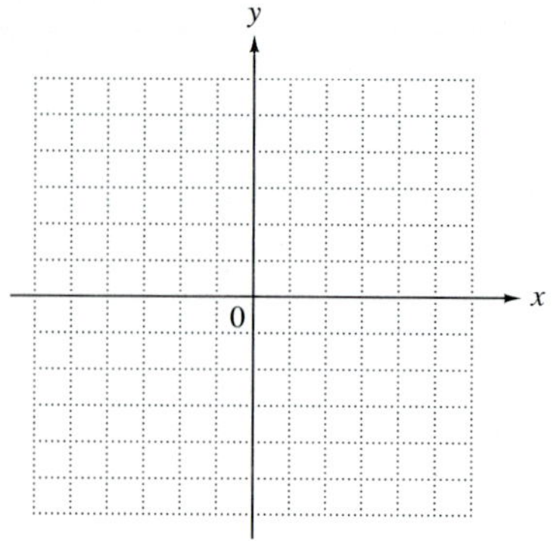

19.

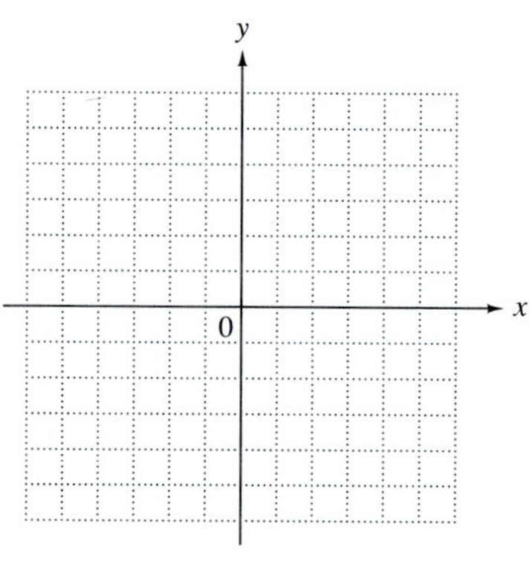

20.

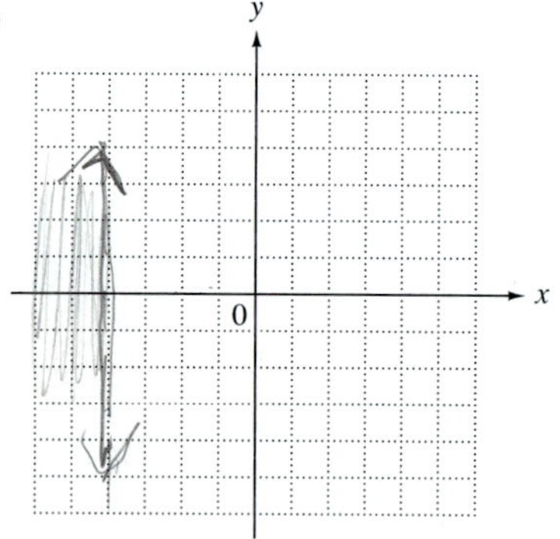

Graph each linear inequality.

18. $x + y \leq 3$

19. $3x - y > 0$

20. $x \leq -4$

NAME DATE HOUR

CUMULATIVE REVIEW EXERCISES CHAPTERS R-6

Solve each equation.

1. $-5(8 - 2z) + 4(7 - z) = 7(8 + z) - 3$

2. $A = p + prt$ for t

3. $7x^2 + 8x + 1 = 0$

4. $\dfrac{4}{x - 2} = 4$

5. $\dfrac{2}{x - 1} = \dfrac{5}{x - 1} - \dfrac{3}{4}$

Solve each inequality and graph the solutions.

6. $-2.5x < 6.5$

7. $4(x + 3) - 5x < 12$

8. $\dfrac{2}{3}y - \dfrac{1}{6}y \leq -2$

Simplify, and write with only positive exponents. Assume that all variables represent positive real numbers.

9. $(x^2y^{-3})(x^{-4}y^2)$

10. $\dfrac{x^{-6}y^3z^{-1}}{x^7y^{-4}z}$

11. $(2m^{-2}n^3)^{-3}$

Perform the indicated operations.

12. $2(3x^2 - 8x + 1) - 4(x^2 - 3x - 9)$

13. $(3x + 2y)(5x - y)$

14. $(x + 2y)(x^2 - 2xy + 4y^2)$

15. $\dfrac{m^3 - 3m^2 + 5m - 3}{m - 1}$

Factor each polynomial completely.

16. $y^2 + 4yk - 12k^2$

17. $9x^4 - 25y^2$

18. $125x^4 - 400x^3y + 195x^2y^2$

19. $f^2 + 20f + 100$

20. $100x^2 + 49$

Perform the indicated operations. Express answers in lowest terms.

21. $\dfrac{3}{2x + 6} + \dfrac{2x + 3}{2x + 6}$

22. $\dfrac{8}{x + 1} - \dfrac{2}{x + 3}$

23. $\dfrac{x^2 - 25}{3x + 6} \cdot \dfrac{4x + 8}{x^2 + 10x + 25}$

24. $\dfrac{x^2 + 2x - 3}{x^2 - 5x + 4} \cdot \dfrac{x^2 - 3x - 4}{x^2 + 3x}$

25. $\dfrac{x^2 + 5x + 6}{3x} \div \dfrac{x^2 - 4}{x^2 + x - 6}$

26. $\dfrac{6x^4y^3z^2}{8xyz^4} \div \dfrac{3x^2}{16y^2}$

Simplify the complex fraction.

27. $\dfrac{\dfrac{2}{3} - \dfrac{1}{4}}{\dfrac{1}{2} + \dfrac{1}{6}}$

28. $\dfrac{\dfrac{12}{x + 6}}{\dfrac{4}{2x + 12}}$

Find the slope of the line described.

29. Through $(-4, 5)$ and $(2, -3)$

30. Horizontal, through $(4, 5)$

Find the equation of the line described. Express it in the standard form $Ax + By = C$.

31. Through $(4, -1)$; $m = -4$

32. Through $(0, 0)$ and $(1, 4)$

Graph the equation or inequality.

33. $-3x + 4y = 12$

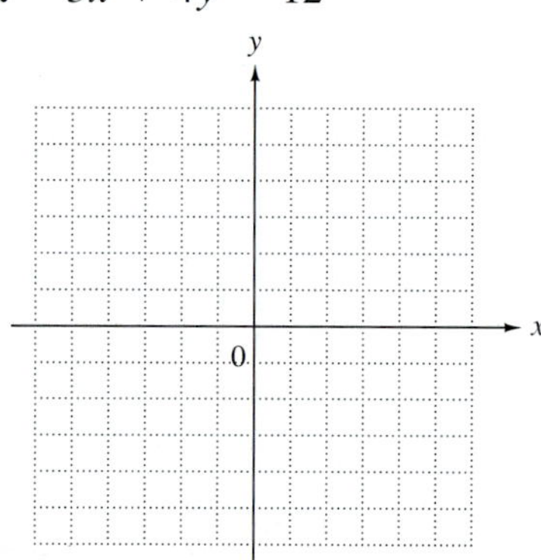

34. $y \leq 2x - 6$

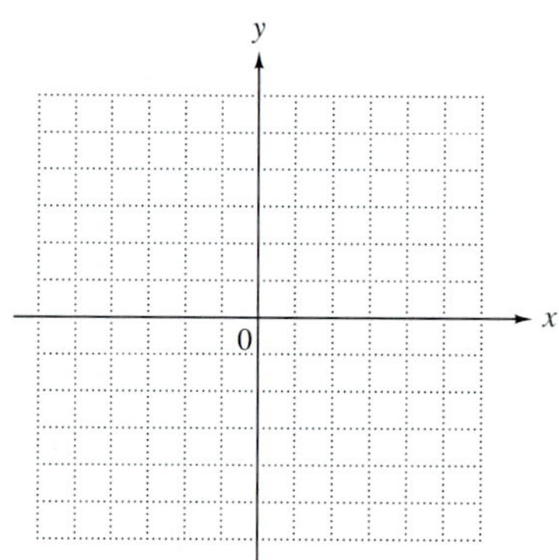

35. $3x + 2y < 0$

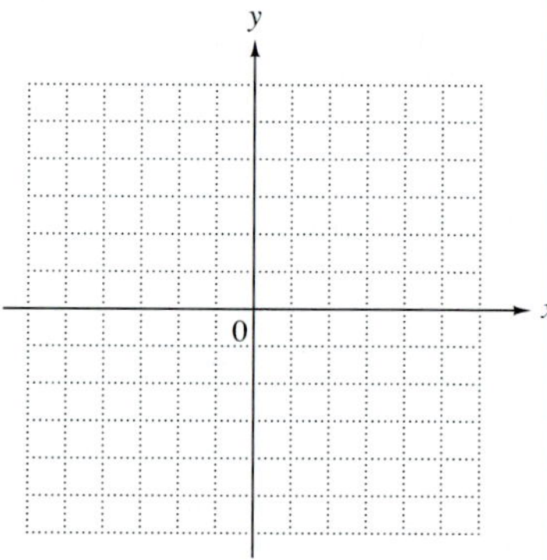

Solve each problem.

36. The audience at a concert included 36 more men than women. If there was a total of 196 men and women there, how many men and how many women attended the concert?

37. Find the measure of each angle of the triangle.

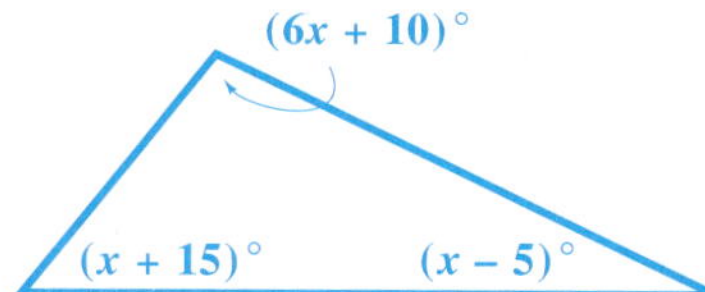

38. The length of the shorter leg of a right triangle is tripled, and 4 inches is added to the result, giving the length of the hypotenuse. The longer leg is 10 inches longer than twice the shorter leg. Find the length of the shorter leg of the triangle.

39. If x varies directly as y and $x = 4$ when $y = 12$, find x when $y = 42$.

40. If a man can mow his lawn in 3 hours, and his wife can do the same job in 1.5 hours, how long will it take them to do the job together?

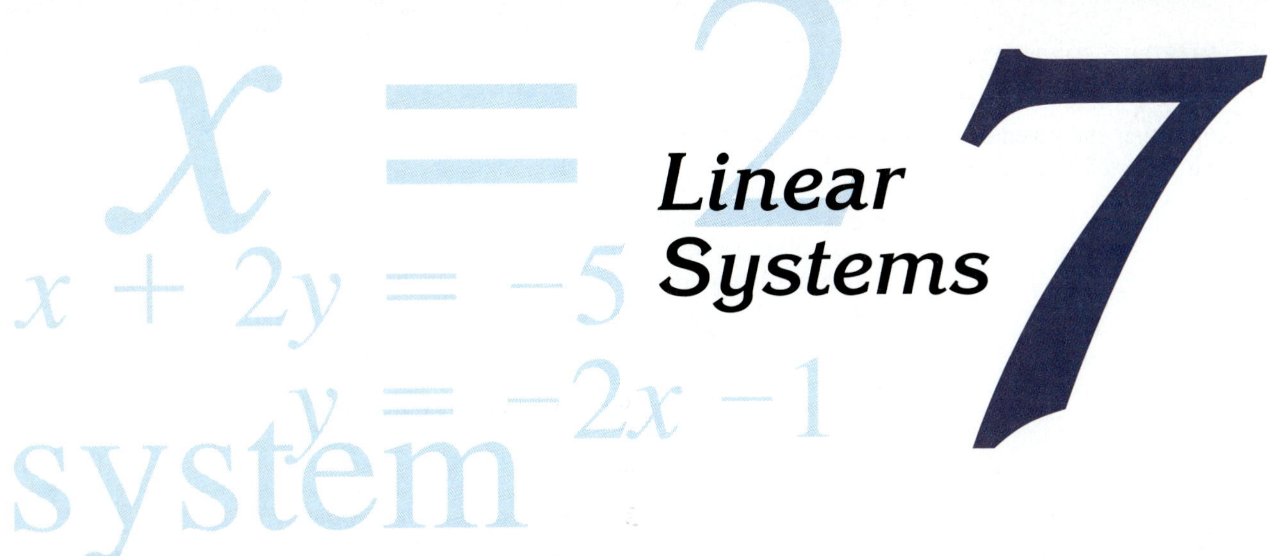

7 Linear Systems

7.1 SOLVING SYSTEMS OF LINEAR EQUATIONS BY GRAPHING

When a number of equations in several variables are considered simultaneously, we have what is known as a system of equations. A **system of linear equations** consists of two or more linear equations with the same variables. Examples of systems of two linear equations with two variables are shown below.

$$\begin{aligned} 2x + 3y &= 4 \\ 3x - y &= -5 \end{aligned} \qquad \begin{aligned} x + 3y &= 1 \\ -y &= 4 - 2x \end{aligned} \qquad \begin{aligned} x - y &= 1 \\ y &= 3 \end{aligned}$$

In the system on the right, think of $y = 3$ as an equation in two variables by writing it as $0x + y = 3$.

OBJECTIVES

1. Decide whether a given ordered pair is a solution of a system.
2. Solve linear systems by graphing.
3. Identify systems with no solutions or with an infinite number of solutions.

FOR EXTRA HELP

Tape 9	SSM pp. 262–266	MAC: B IBM: B

1 Applications often require solving a system of equations. The **solution of a system** of linear equations includes all the ordered pairs that make both equations true at the same time.

EXAMPLE 1 *Determining Whether an Ordered Pair Is a Solution*

Is $(4, -3)$ a solution of the following systems?

(a) $x + 4y = -8$
$3x + 2y = 6$

Decide whether or not $(4, -3)$ is a solution of the system by substituting 4 for x and -3 for y in each equation.

$$\begin{aligned} x + 4y &= -8 \\ 4 + 4(-3) &= -8 \quad ? \\ 4 + (-12) &= -8 \quad ? \quad \text{Multiply.} \\ -8 &= -8 \quad \text{True} \end{aligned} \qquad \begin{aligned} 3x + 2y &= 6 \\ 3(4) + 2(-3) &= 6 \quad ? \\ 12 + (-6) &= 6 \quad ? \quad \text{Multiply.} \\ 6 &= 6 \quad \text{True} \end{aligned}$$

Because $(4, -3)$ satisfies both equations, it is a solution of the system.

(b) $2x + 5y = -7$
$3x + 4y = 2$

Again, substitute 4 for x and -3 for y in both equations.

$$\begin{aligned} 2x + 5y &= -7 \\ 2(4) + 5(-3) &= -7 \quad ? \\ 8 + (-15) &= -7 \quad ? \quad \text{Multiply.} \\ -7 &= -7 \quad \text{True} \end{aligned} \qquad \begin{aligned} 3x + 4y &= 2 \\ 3(4) + 4(-3) &= 2 \quad ? \\ 12 + (-12) &= 2 \quad ? \quad \text{Multiply.} \\ 0 &= 2 \quad \text{False} \end{aligned}$$

The ordered pair $(4, -3)$ is not a solution because it does not satisfy the second equation. ■

1. Decide whether the given ordered pair is a solution of the system.

(a) (2, 5)
$3x - 2y = -4$
$5x + y = 15$

(b) (1, −2)
$x - 3y = 7$
$4x + y = 5$

ANSWERS
1. (a) yes (b) no

◀ WORK PROBLEM 1 AT THE SIDE.

2 Several methods of solving a system of two linear equations in two variables are discussed in this chapter. One way to find the solution of a system of two linear equations is to graph both equations on the same axes. The graph of each line shows points whose coordinates satisfy the equation of that line. The coordinates of any point where the lines intersect give a solution of the system. Because two different straight lines can intersect at no more than one point, there can never be more than one solution for such a system.

■ **EXAMPLE 2** *Solving Systems by Graphing*

Solve each system of equations by graphing both equations on the same axes.

(a) $2x + 3y = 4$
$3x - y = -5$

As shown in Chapter 6, we can graph these two equations by plotting points for each line.

$2x + 3y = 4$

x	y
0	$\frac{4}{3}$
2	0
−2	$\frac{8}{3}$

$3x - y = -5$

x	y
0	5
$-\frac{5}{3}$	0
−2	−1

The lines in Figure 1 suggest that the graphs intersect at the point (−1, 2). Check this by substituting −1 for x and 2 for y in both equations. Because (−1, 2) satisfies both equation, the solution of this system is (−1, 2).

(b) $2x + y = 0$
$4x - 3y = 10$

Find the solution of the system by graphing the two lines on the same axes. As suggested by Figure 2, the solution is (1, −2), the point at which the graphs of the two lines intersect. Check by substituting 1 for x and −2 for y in both equations of the system. ■

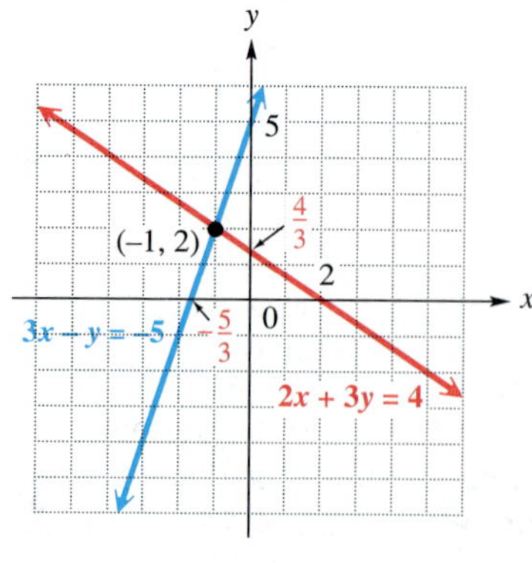

FIGURE 1

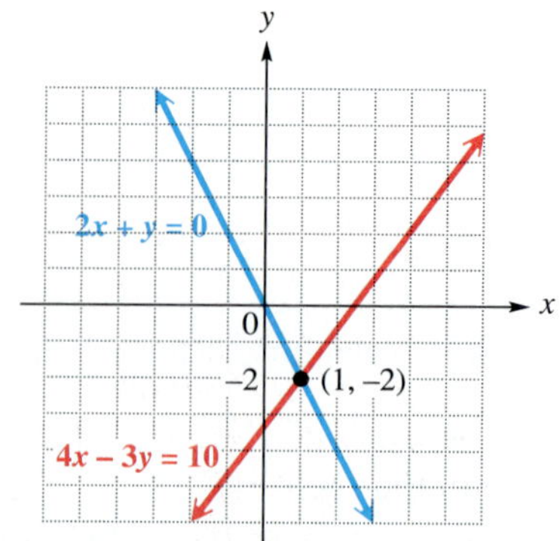

FIGURE 2

Caution A difficulty with the graphing method of solution is that it may not be possible to determine from the graph the exact coordinates of the point that represents the solution. For this reason, algebraic methods of solution are explained later in this chapter. The graphing method does, however, show geometrically how solutions are found.

WORK PROBLEM 2 AT THE SIDE.

3 Sometimes the graphs of the two equations in a system either do not intersect at all or are the same line, as in the systems in Example 3.

EXAMPLE 3 *Solving Special Systems*

Solve each system by graphing.

(a) $2x + y = 2$
$2x + y = 8$

The graphs of these lines are shown in Figure 3. The two lines are parallel with equal slopes and have no points in common. For a system whose equations lead to graphs with no points in common, we will write "no solution."

(b) $2x + 5y = 1$
$6x + 15y = 3$

The graphs of these two equations are the same line. See Figure 4. Here the second equation can be obtained by multiplying each side of the first equation by 3. In this case, every point on the line is a solution of the system, and the solution is an infinite number of ordered pairs. We will write "infinite number of solutions" or "same line" to indicate this situation. ■

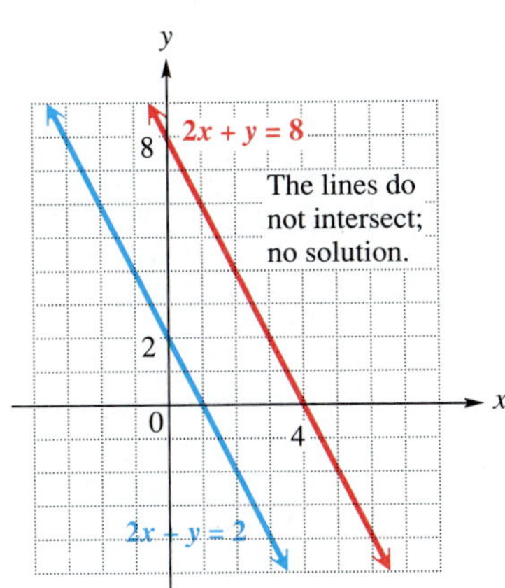

FIGURE 3

FIGURE 4

Each system in Example 2 has a solution. A system with a solution is called a **consistent system.** A system of equations with no solutions, such as the one in Example 3(a), is called an **inconsistent system.** The equations in Example 2 are independent equations. **Independent equations** have differ-

2. Solve each system of equations by graphing both equations on the same axes. Check your answers.

(a) $5x - 3y = 9$
$x + 2y = 7$
(One of the lines is already graphed.)

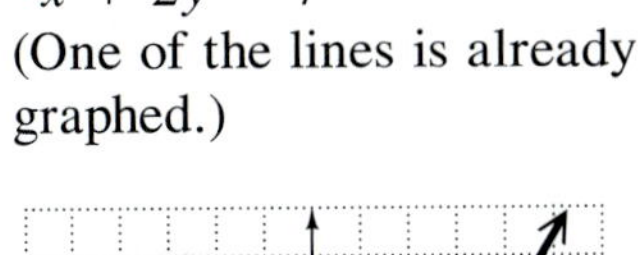

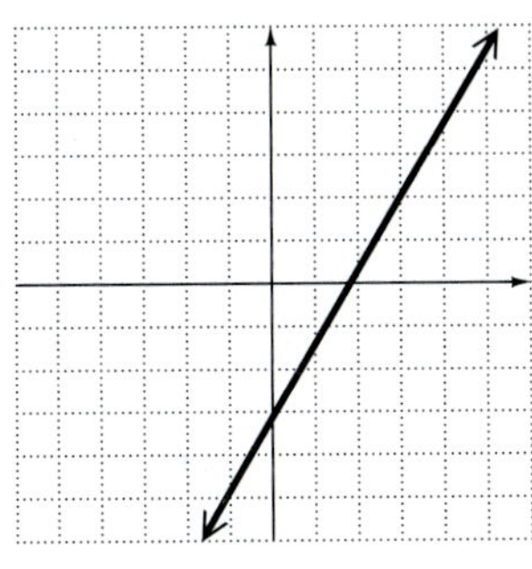

(b) $x + y = 4$
$2x - y = -1$

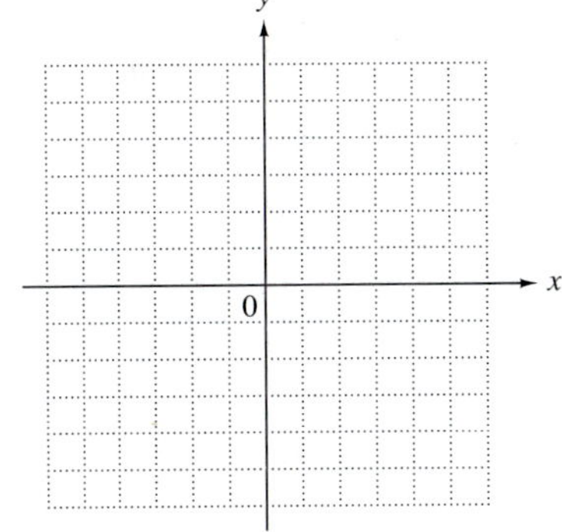

ANSWERS
2. (a) (3, 2) (b) (1, 3)

3. Solve each system of equations by graphing both equations on the same axes.

(a) $3x - y = 4$
$6x - 2y = 12$
(One of the lines is already graphed.)

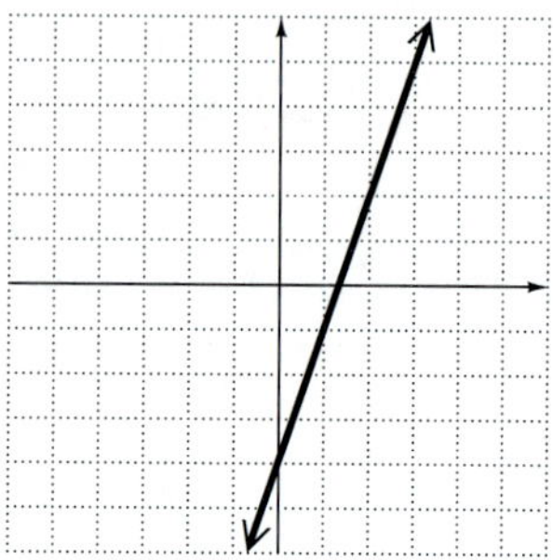

(b) $-x + 3y = 2$
$2x - 6y = -4$

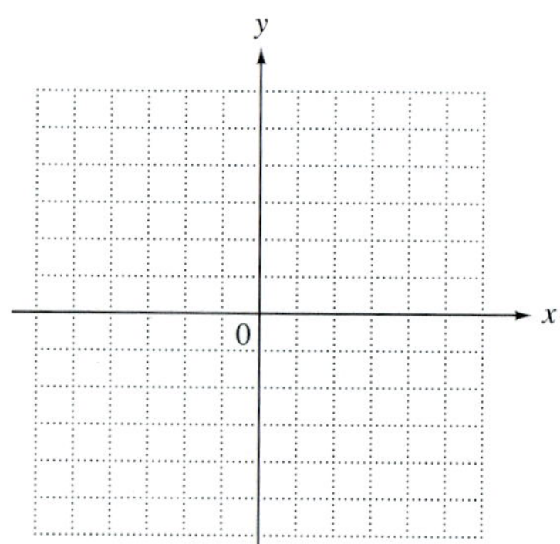

ANSWERS
3. (a) no solution
(b) infinite number of solutions

ent graphs. The equations of the system in Example 3(b) have the same graph. Because they are different forms of the same equation, these equations are called **dependent equations.**

WORK PROBLEM 3 AT THE SIDE.

Examples 2 and 3 show the three cases that may occur in a system of two equations with two unknowns.

1. The graphs intersect at exactly one point, which gives the (single) solution of the system. The system is consistent, and the equations are independent.

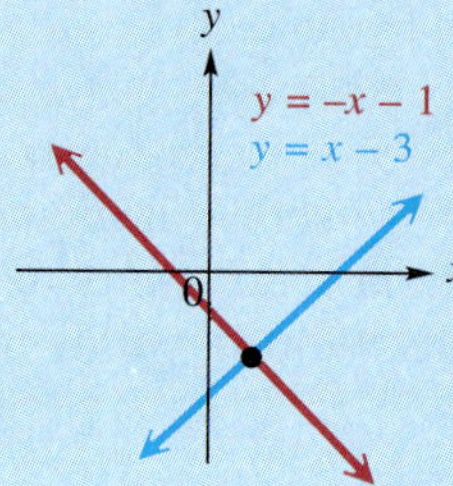

2. The graphs are parallel lines, so there is no solution. The system is inconsistent.

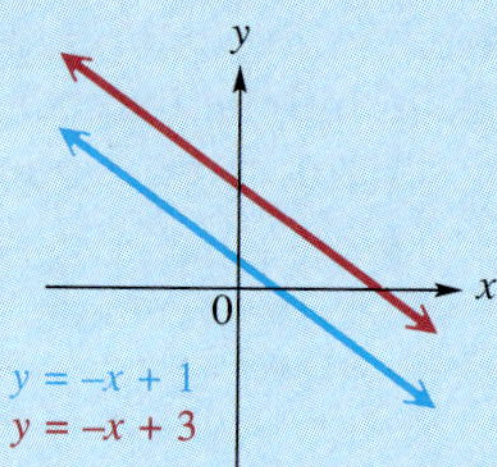

3. The graphs are the same line. The solution is an infinite set of ordered pairs. The equations are dependent.

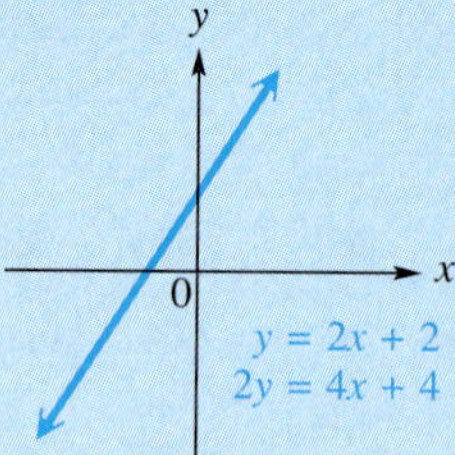

7.1 NAME DATE HOUR

EXERCISES

1. When a student was asked to determine whether the ordered pair $(1, -2)$ is a solution of the system

$$x + y = -1$$
$$2x + y = 4,$$

he answered "yes." His reasoning was that the ordered pair satisfies the equation $x + y = -1$: $1 + (-2) = -1$ is true. Why is the student's answer wrong?

2. Let a, b, c, d, e, and f represent six consecutive integers. (For example, if $a = 2$, then we have $b = 3$, $c = 4$, $d = 5$, $e = 6$, and $f = 7$.) You may start with *any* value for a. Then, substitute them into the system

$$ax + by = c$$
$$dx + ey = f.$$

(For example, starting with 2 we would get the equations $2x + 3y = 4$ and $5x + 6y = 7$.) Now show that $(-1, 2)$ is a solution of the system you just made up.

Decide whether the given ordered pair is a solution of the given system. See Example 1.

3. $(6, 2)$
$3x + y = 20$
$2x + 3y = 18$

4. $(3, 4)$
$2x + y = 10$
$3x + 2y = 17$

5. $(2, -3)$
$x + y = -1$
$2x + 5y = 19$

6. $(4, 3)$
$x + 2y = 10$
$3x + 5y = 3$

7. $(-1, -3)$
$3x + 5y = -18$
$4x + 2y = -10$

8. $(-9, -2)$
$2x - 5y = -8$
$3x + 6y = -39$

9. $(7, -2)$
$4x = 26 - y$
$3x = 29 + 4y$

10. $(9, 1)$
$2x = 23 - 5y$
$3x = 24 + 3y$

11. $(6, -8)$
$-2y = x + 10$
$3y = 2x + 30$

12. $(-5, 2)$
$5y = 3x + 20$
$3y = -2x - 4$

13. Which one of the ordered pairs below could possibly be a solution of the system graphed? Why is it the only valid choice?
(a) $(2, 2)$
(b) $(-2, 2)$
(c) $(-2, -2)$
(d) $(2, -2)$

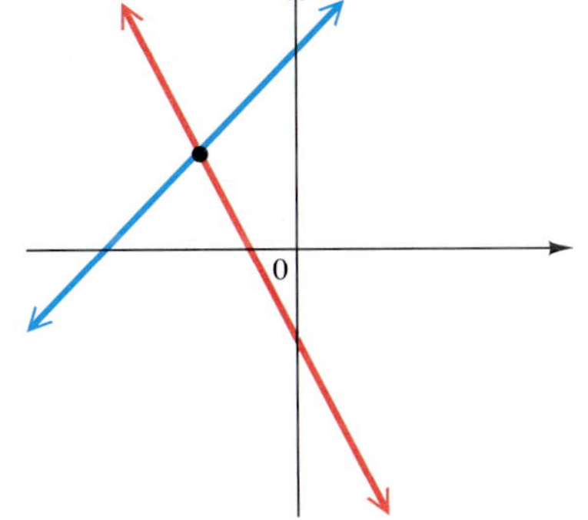

14. Which one of the ordered pairs below could possibly be a solution of the system graphed? Why is it the only valid choice?
(a) $(2, 0)$
(b) $(0, 2)$
(c) $(-2, 0)$
(d) $(0, -2)$

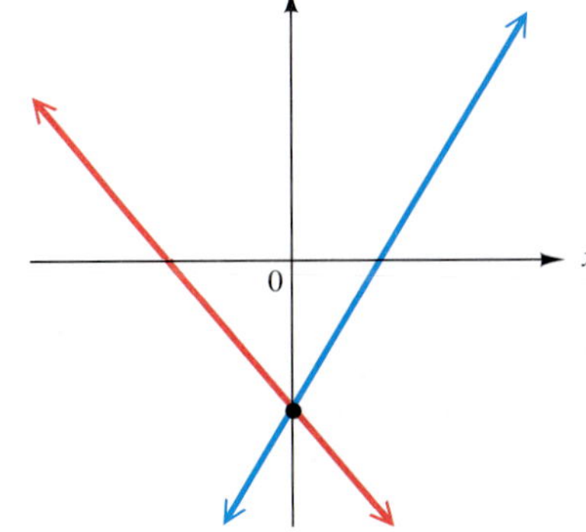

Solve each system of equations by graphing both equations on the same axes. See Example 2.

15. $x - y = 2$
$x + y = 6$

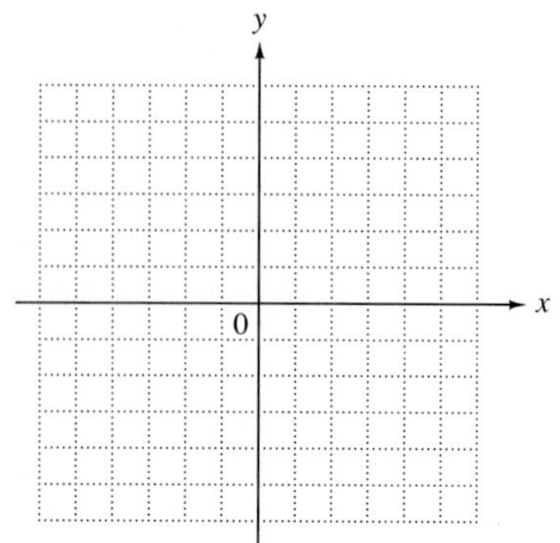

16. $x - y = 3$
$x + y = -1$

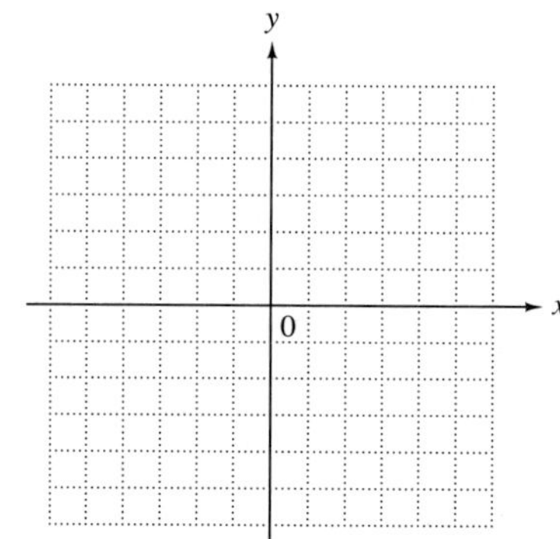

17. $x + y = 4$
$y - x = 4$

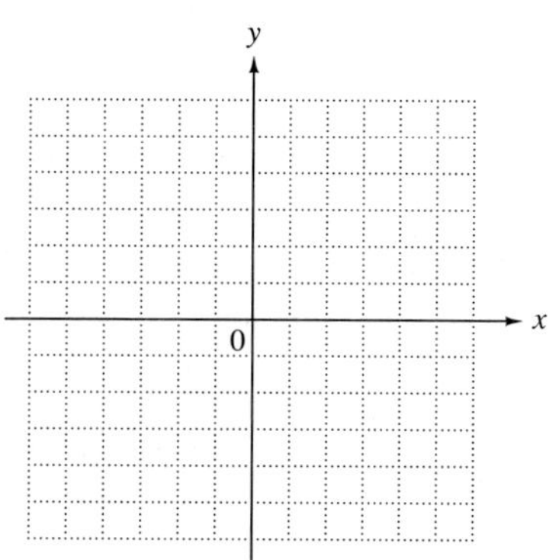

18. $x + y = -5$
$x - y = 5$

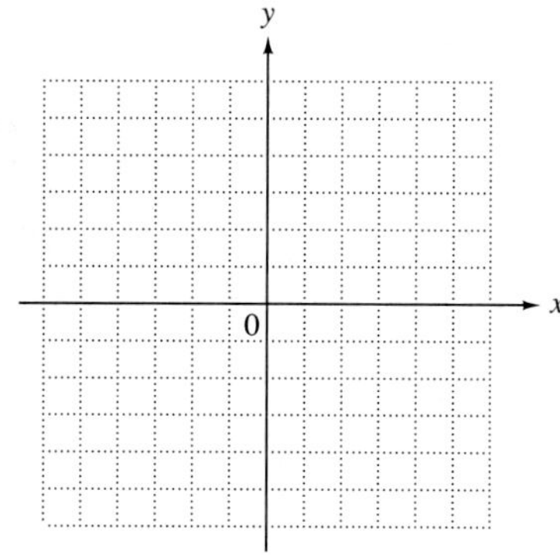

19. $x - 2y = 6$
$x + 2y = 2$

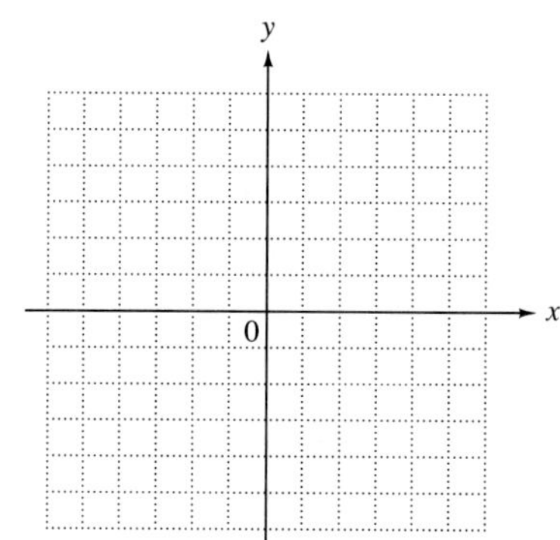

20. $2x - y = 4$
$4x + y = 2$

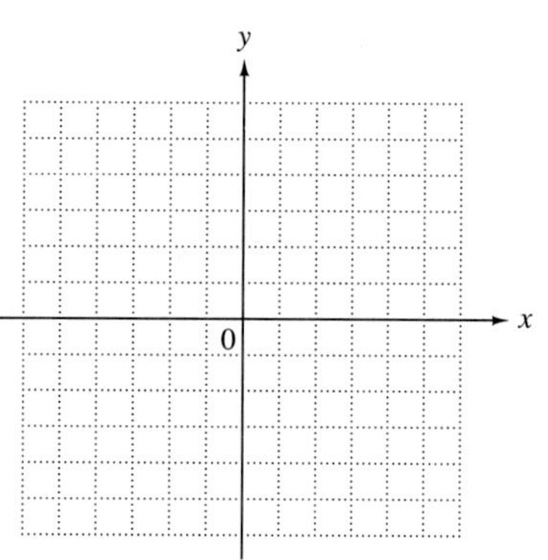

21. $3x - 2y = -3$
$-3x - y = -6$

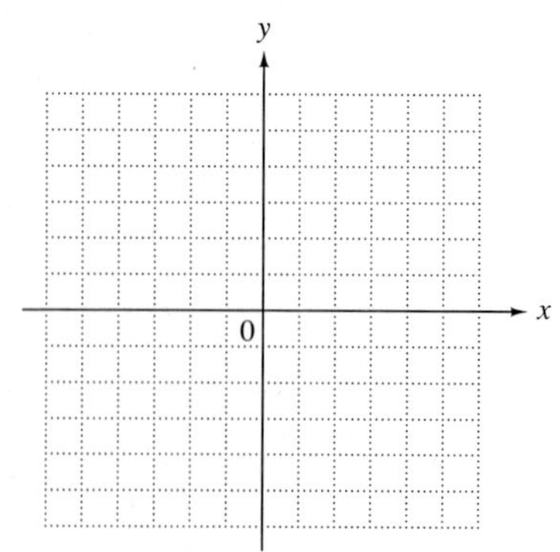

22. 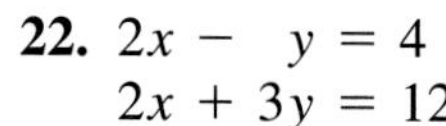
$2x - y = 4$
$2x + 3y = 12$

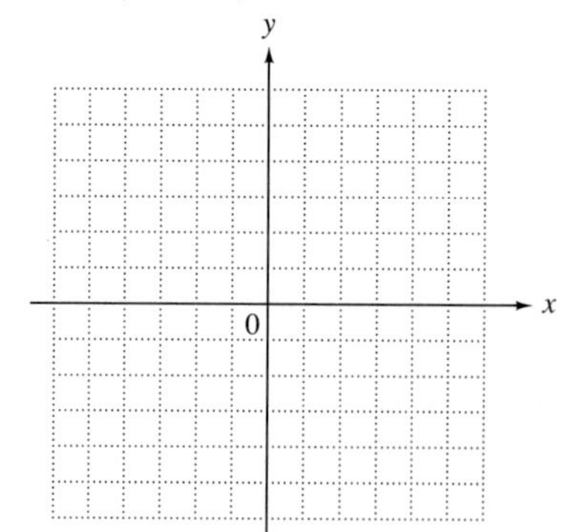

23. 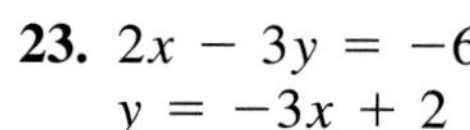
$2x - 3y = -6$
$y = -3x + 2$

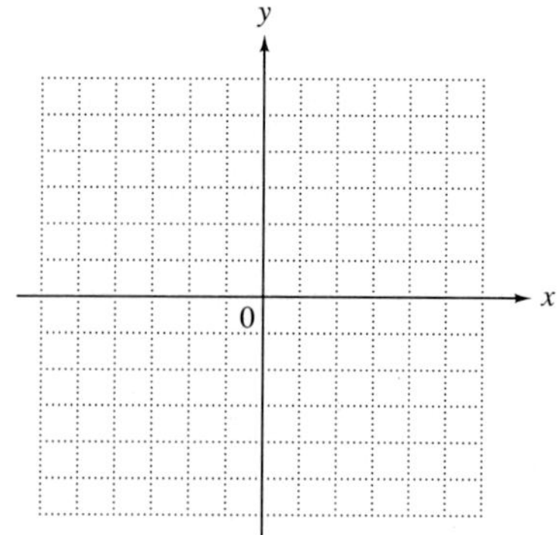

NAME ______________________________

24. $-3x + y = -3$
$y = x - 3$

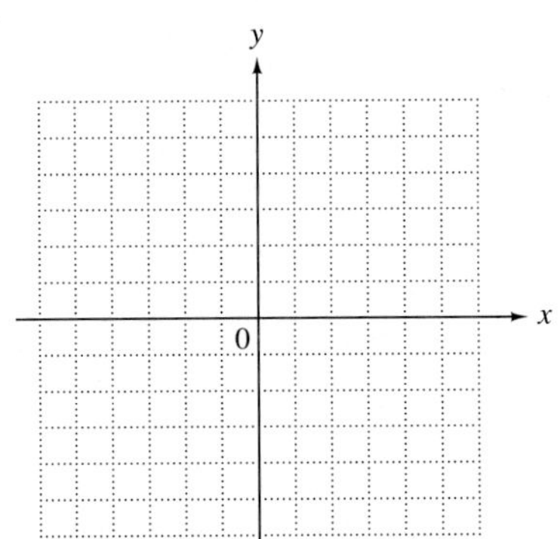

25. $3x - 4y = 24$
$y = -\frac{3}{2}x + 3$

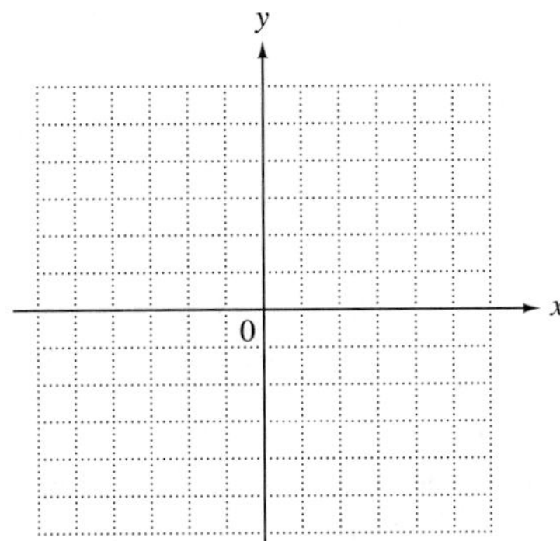

26. $3x - 2y = 12$
$y = -4x + 5$

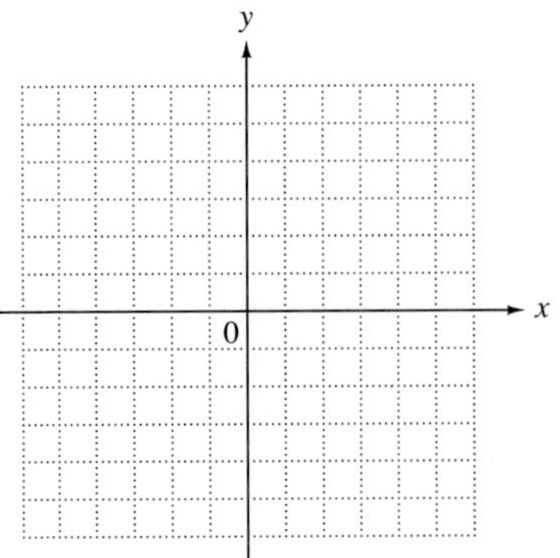

27. Explain why a system of two linear equations cannot have exactly two solutions.

28. Write each equation in the following systems in slope-intercept form. Use what you learned in Chapter 6 about slope and the y-intercept to describe the graphs of these equations.

(a) $3x + 2y = 6$
$-2y = 3x - 5$

(b) $2x - y = 4$
$x = .5y + 2$

(c) $x - 3y = 5$
$2x + y = 8$

Explain how you can use this information to determine the solution of certain equations.

29. Find a system of equations with the solution $(-2, 3)$, and show the graph.

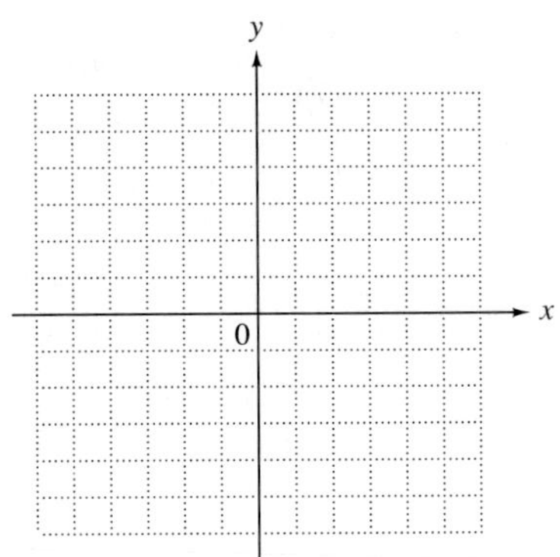

30. Graph the system

$$2x + 3y = 6$$
$$x - 3y = 5.$$

Can you check your answer? What is the problem?

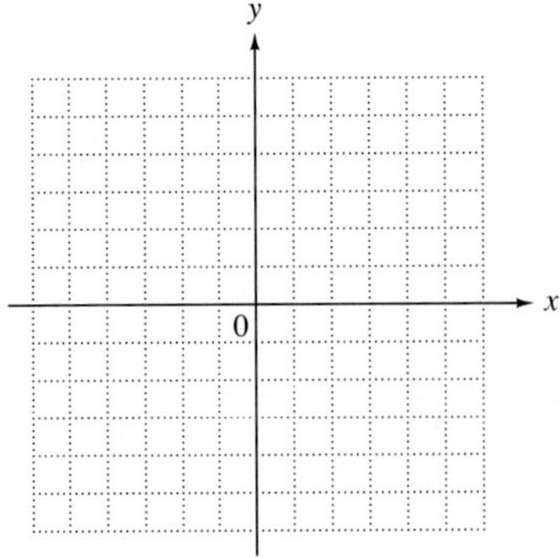

31. Explain some of the drawbacks of solving a system of equations graphically.

32. If the two lines that are the graphs of the equations in a system are parallel, how many solutions does the system have? If the two lines coincide, how many solutions does the system have?

Solve each system by graphing. If the two equations produce parallel lines, write no solution. *If the two equations produce the same line, write* infinite number of solutions. *See Example 3.*

33. $x + 2y = 6$
$2x + 4y = 8$

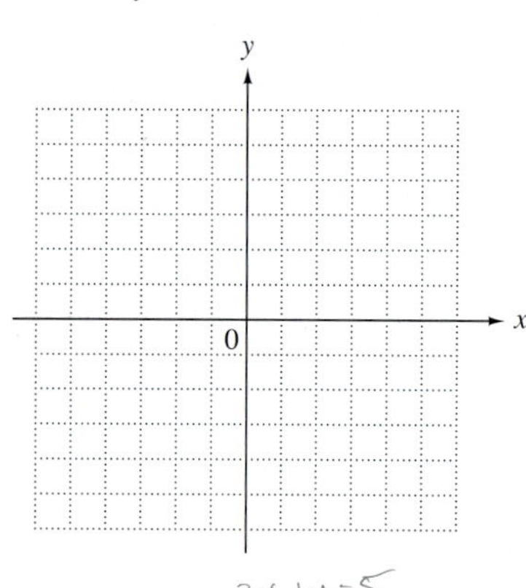

34. $2x - y = 6$
$6x - 3y = 12$

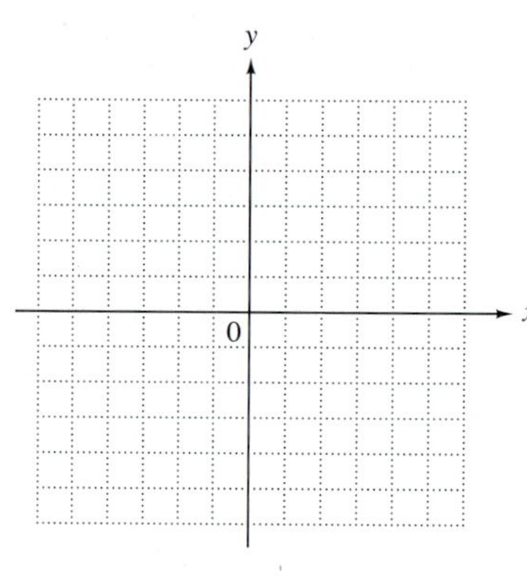

35. $-2x + y = -4$
$4x = 2y + 8$

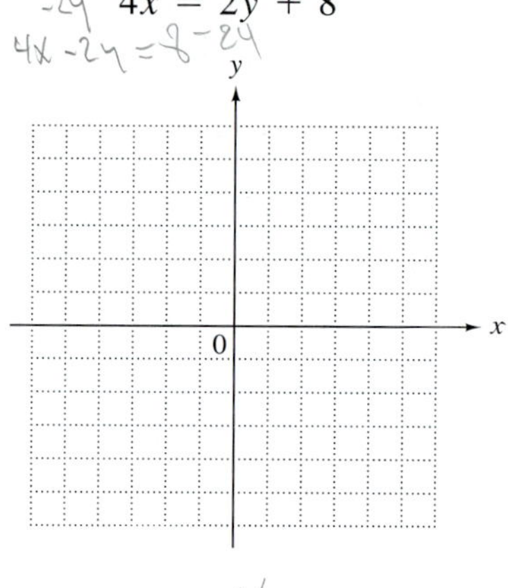

36. $3x + y = 5$
$6x = 10 - 2y$

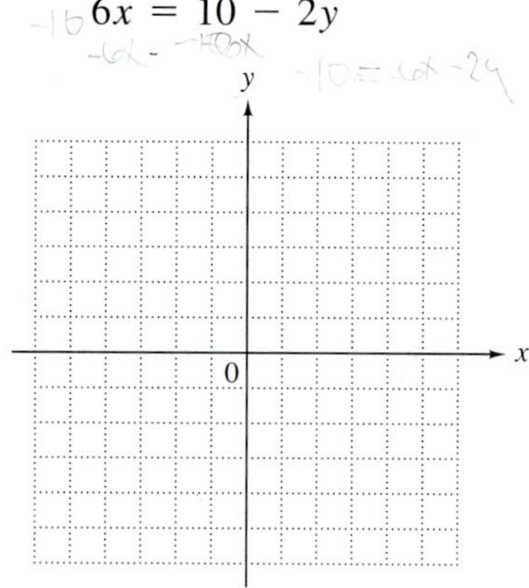

37. $3x = y + 5$
$6x - 5 = 2y$

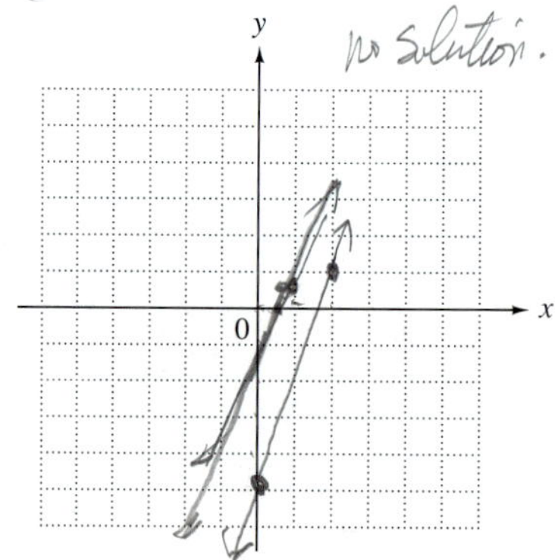

38. $2x = y - 4$
$4x - 2y = -4$

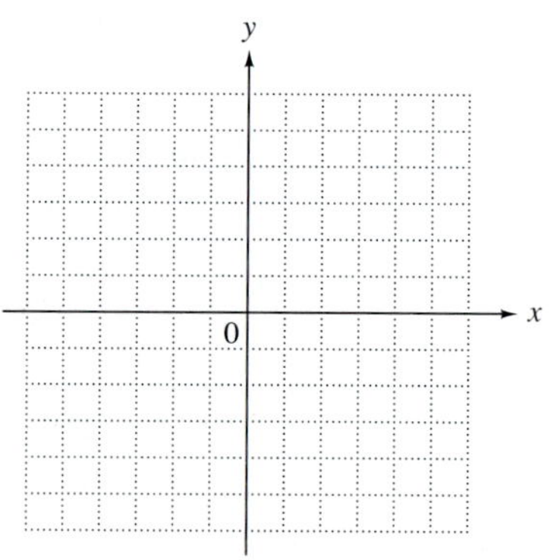

PREVIEW EXERCISES

Add. See Section 3.4.

39. $\begin{array}{r} 14x - 3y \\ \underline{2x + 3y} \end{array}$

40. $\begin{array}{r} -6x + 8y \\ \underline{6x + 2y} \end{array}$

41. $\begin{array}{r} -3x + 7y \\ \underline{3x - 7y} \end{array}$

42. What must be added to $-4x$ to get a sum of 0?

43. What must be added to $6y$ to get a sum of 0?

44. What must $4y$ be multiplied by so that when this product is added to $8y$, the sum is 0?

7.2 SOLVING SYSTEMS OF LINEAR EQUATIONS BY ADDITION

Graphing to solve a system of equations has a serious drawback: It is difficult to accurately find a solution such as $(\frac{1}{3}, -\frac{5}{6})$ from a graph.

1 An algebraic method that depends on the addition property of equality can be used to solve systems. As mentioned earlier, adding the same quantity to each side of an equation results in equal sums.

$$\text{If } A = B, \quad \text{then} \quad A + C = B + C.$$

This addition can be taken a step further. Adding *equal* quantities, rather than the *same* quantity, to both sides of an equation also results in equal sums.

$$\text{If } A = B \quad \text{and} \quad C = D, \quad \text{then} \quad A + C = B + D.$$

The use of the addition property to solve systems is called the **addition method** for solving systems of equations. For most systems, this method is more efficient than graphing.

> *Note* When using the addition method, the idea is to eliminate one of the variables. To do this, one of the variables must have coefficients that are opposites. Keep this in mind throughout the examples in this section.

OBJECTIVES

1. Solve linear systems by addition.
2. Multiply one or both equations of a system so the addition method can be used.
3. Write equations in the proper form to use the addition method.
4. Solve linear systems having parallel lines as their graphs.
5. Solve linear systems having the same line as their graphs.

FOR EXTRA HELP

Tape 10	SSM pp. 266–273	MAC: B IBM: B

EXAMPLE 1 *Using the Addition Method*

Use the addition method to solve the system

$$x + y = 5$$
$$x - y = 3.$$

Each equation in this system is a statement of equality, so, as discussed above, the sum of the right-hand sides equals the sum of the left-hand sides. Adding in this way gives

$$(x + y) + (x - y) = 5 + 3.$$

Combine terms to get

$$2x = 8$$
$$x = 4. \quad \text{Divide by 2.}$$

The result, $x = 4$, gives the x-value of the solution of the given system. Find the y-value of the solution by substituting 4 for x in either of the two equations in the system.

WORK PROBLEM 1 AT THE SIDE.

1. (a) Substitute $x = 4$ in the equation $x + y = 5$ to find the value of y.

(b) Give the solution of the system.

The solution found at the side, (4, 1), can be checked by substituting 4 for x and 1 for y into both equations in the given system.

Check:

$x + y = 5$		$x - y = 3$	
$4 + 1 = 5$	?	$4 - 1 = 3$	?
$5 = 5$	True	$3 = 3$	True

Because both results are true, the solution of the given system is (4, 1). The two equations are graphed in Figure 5 on the next page. Notice that the point of intersection is (4, 1), as indicated by the solution of the system using the addition method. ■

ANSWERS
1. (a) $y = 1$ (b) (4, 1)

2. Solve each system by the addition method. Check each solution.

(a) $x + y = 8$
$x - y = 2$

(b) $3x - y = 7$
$2x + y = 3$

3. Solve each system by the addition method. Check each solution.

(a) $2x - y = 2$
$4x + y = 10$

(b) $8x - 5y = 32$
$4x + 5y = 4$

ANSWERS
2. (a) (5, 3) (b) (2, −1)
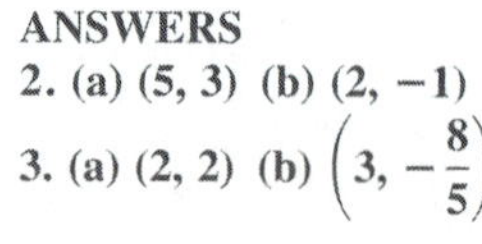
3. (a) (2, 2) (b) $\left(3, -\frac{8}{5}\right)$

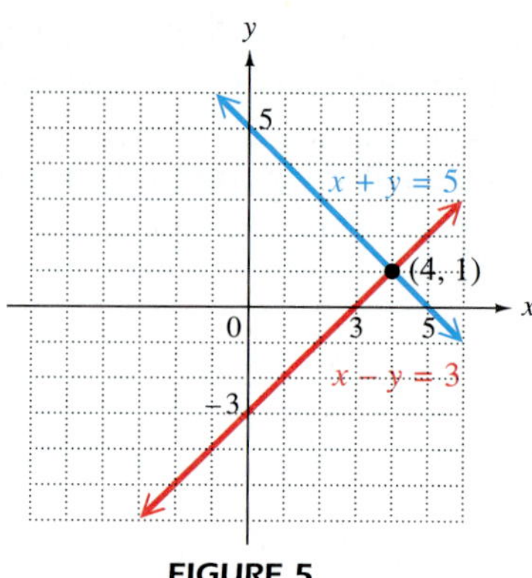

FIGURE 5

Caution A system is not completely solved until you find values for *both* x and y. Do not make the mistake of finding the value of only one variable.

WORK PROBLEM 2 AT THE SIDE.

EXAMPLE 2 *Using the Addition Method*

Solve the system

$$-2x + y = -11$$
$$5x - y = 26.$$

As above, add left-hand sides and add right-hand sides. This may be done most easily by drawing a line under the second equation and adding vertically. (Like terms must be placed in columns.)

$$\begin{aligned} -2x + y &= -11 \\ 5x - y &= 26 \\ \hline 3x \quad &= 15 && \text{Add in columns.} \\ x &= 5 && \text{Divide by 3.} \end{aligned}$$

Substitute 5 for x in either of the original equations. We choose the first.

$$\begin{aligned} -2x + y &= -11 \\ -2(5) + y &= -11 && \text{Let } x = 5. \\ -10 + y &= -11 && \text{Multiply.} \\ y &= -1 && \text{Add 10.} \end{aligned}$$

The solution is $(5, -1)$. Check this solution by substituting 5 for x and -1 for y in both of the original equations. ■

WORK PROBLEM 3 AT THE SIDE.

2 In both earlier examples, the addition step eliminated a variable. Sometimes one or both equations in a system must be multiplied by some number before the addition step will eliminate a variable.

EXAMPLE 3 *Multiplying Before Using the Addition Method*

Solve the system

$$x + 3y = 7 \qquad (1)$$
$$2x + 5y = 12. \qquad (2)$$

Adding the two equations gives $3x + 8y = 19$, which does not help to solve the system. However, if each side of equation (1) is first multiplied by -2 using the multiplication property of equality, the terms with the variable x will drop out after adding.

$$-2(x + 3y) = -2(7)$$
$$-2x - 6y = -14 \qquad \textbf{(3)}$$

Now add equations (3) and (2).

$$\begin{aligned} -2x - 6y &= -14 && \textbf{(3)} \\ 2x + 5y &= 12 && \textbf{(2)} \\ \hline -y &= -2 && \text{Add.} \\ y &= 2 && \text{Multiply by } -1. \end{aligned}$$

Substituting into equation (1) gives

$$\begin{aligned} x + 3y &= 7 \\ x + 3(2) &= 7 && \text{Let } y = 2. \\ x + 6 &= 7 && \text{Multiply.} \\ x &= 1. && \text{Subtract 6.} \end{aligned}$$

The solution of this system is (1, 2). Check that this ordered pair satisfies both of the original equations. ■

WORK PROBLEM 4 AT THE SIDE. ▶▶

■ EXAMPLE 4 *Multiplying Twice Before Using the Addition Method*

Solve the system

$$2x + 3y = -15 \qquad \textbf{(1)}$$
$$5x + 2y = 1. \qquad \textbf{(2)}$$

Here we must use the multiplication property of equality with both equations instead of just one. Multiply by numbers that will cause the coefficients of x (or of y) in the two equations to be additive inverses of each other. For example, multiply each side of equation (1) by 5, and each side of equation (2) by -2.

$$\begin{aligned} 10x + 15y &= -75 && \text{Multiply (1) by 5.} \\ -10x - 4y &= -2 && \text{Multiply (2) by } -2. \\ \hline 11y &= -77 && \text{Add.} \\ y &= -7 && \text{Divide by 11.} \end{aligned}$$

Substituting -7 for y in either equation (1) or (2) gives $x = 3$. The solution of the system is $(3, -7)$. Check this solution.

The same result would have been obtained by multiplying each side of equation (1) by 2 and each side of equation (2) by -3. This process would eliminate the y terms so that the value of x would have been found first. ■

WORK PROBLEM 5 AT THE SIDE. ▶▶

3 Before a system can be solved by the addition method, the two equations of the system must have like terms in the same positions. When this is not the case, the terms should first be rearranged, as the next example shows. This example also shows an alternative way to get the second number when finding the solution of a system.

4. Solve each system by the addition method. Check each solution.

(a) $x - 3y = -7$
$3x + 2y = 23$

(b) $8x + 2y = 2$
$3x - y = 6$

5. Solve each system of equations. Check each solution.

(a) $4x - 5y = -18$
$3x + 2y = -2$

(b) $6x + 7y = 4$
$5x + 8y = -1$

ANSWERS

4. (a) (5, 4) (b) (1, −3)
5. (a) (−2, 2) (b) (3, −2)

6. Solve each system of equations.

(a) $5x = 7 + 2y$
$5y = 5 - 3x$

(b) $3y = 8 + 4x$
$6x = 9 - 2y$

ANSWERS

6. (a) $\left(\frac{45}{31}, \frac{4}{31}\right)$ (b) $\left(\frac{11}{26}, \frac{42}{13}\right)$

EXAMPLE 5 *Rearranging Terms Before Using the Addition Method*

Solve the system

$$4x = 9 - 3y \quad (1)$$
$$5x - 2y = 8. \quad (2)$$

Rearrange the terms in equation (1) so that the like terms can be aligned in columns. Add $3y$ to each side to get the following system.

$$4x + 3y = 9 \quad (3)$$
$$5x - 2y = 8 \quad (2)$$

Let us eliminate y by multiplying each side of equation (3) by 2 and each side of equation (2) by 3, and then adding.

$$\begin{aligned} 8x + 6y &= 18 && \text{Multiply by 2.} \\ 15x - 6y &= 24 && \text{Multiply by 3.} \\ \hline 23x &= 42 && \text{Add.} \\ x &= \frac{42}{23} && \text{Divide by 23.} \end{aligned}$$

Substituting $\frac{42}{23}$ for x in one of the given equations would give y, but the arithmetic involved would be messy. Instead, solve for y by starting again with the original equations and eliminating x. Do this by multiplying each side of equation (3) by 5 and each side of equation (2) by -4, and then adding.

$$\begin{aligned} 20x + 15y &= 45 && \text{Multiply by 5.} \\ -20x + 8y &= -32 && \text{Multiply by } -4. \\ \hline 23y &= 13 && \text{Add.} \\ y &= \frac{13}{23} && \text{Divide by 23.} \end{aligned}$$

The solution is $\left(\frac{42}{23}, \frac{13}{23}\right)$. ■

When the value of the first variable is a fraction, the method used in Example 5 prevents errors that often occur when working with fractions. (Of course, this method could be used in solving any system of equations.)

WORK PROBLEM 6 AT THE SIDE.

The solution of a linear system of equations having exactly one solution can be found by the addition method. A summary of the steps is given below.

SOLVING A LINEAR SYSTEM

Step 1 Write both equations of the system in the form $Ax + By = C$.

Step 2 If necessary, multiply one or both equations by appropriate numbers so that the coefficients of x (or y) are negatives of each other.

Step 3 Add the two equations to get an equation with only one variable.

Step 4 Solve the equation from Step 3.

Step 5 Substitute the solution from Step 4 into either of the original equations.

Step 6 Solve the resulting equation from Step 5 for the remaining variable.

Step 7 Check the answer.

4 In Section 7.1 some of the systems had equations with graphs that were parallel lines (from an inconsistent system), while the equations of other systems had graphs that were the same line (dependent equations). These systems can also be solved with the addition method.

EXAMPLE 6 *Using Addition to Solve an Inconsistent System*

Solve by the addition method.

$$2x + 4y = 5$$
$$4x + 8y = -9$$

Multiply each side of $2x + 4y = 5$ by -2 and then add.

$$\begin{aligned} -4x - 8y &= -10 \\ 4x + 8y &= -9 \\ \hline 0 &= -19 \end{aligned} \qquad \text{False}$$

The false statement $0 = -19$ shows that the given system is self-contradictory. *It has no solution.* This means that the graphs of the equations of this system are parallel lines, as shown in Figure 6. Since this system has no solution, it is inconsistent. ■

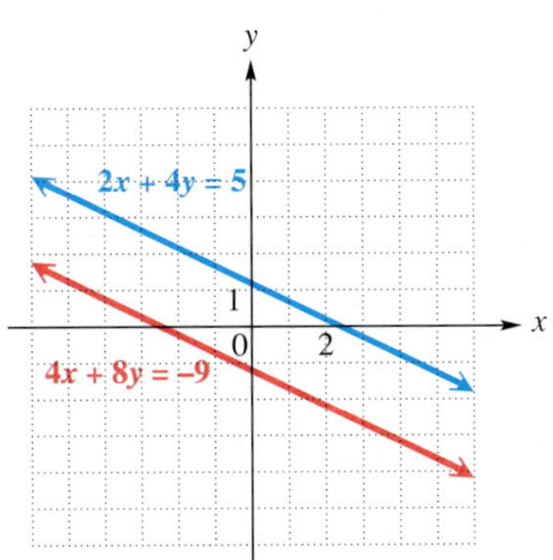

FIGURE 6

WORK PROBLEM 7 AT THE SIDE.

7. Solve each system by the addition method.

(a) $4x + 3y = 10$

$2x + \frac{3}{2}y = 12$

(b) $-2x - 4y = -1$

$5x + 10y = 15$

5 The next example shows the result of using the addition method when the equations of the system are dependent, with the graphs of the equations in the system the same line.

EXAMPLE 7 *Using Addition to Solve a System of Dependent Equations*

Solve by the addition method.

$$3x - y = 4$$
$$-9x + 3y = -12$$

Multiply each side of the first equation by 3 and then add the two equations to get

$$\begin{aligned} 9x - 3y &= 12 \\ -9x + 3y &= -12 \\ \hline 0 &= 0. \end{aligned} \qquad \text{True}$$

This result means that every solution of one equation is also a solution of the other, so the system has an infinite number of solutions: all the

ANSWERS
7. (a) no solution (b) no solution

8. Solve each system by the addition method.

(a) $6x + 3y = 9$
$-8x - 4y = -12$

(b) $4x - 6y = 10$
$-10x + 15y = -25$

ANSWERS
8. (a) infinite number of solutions
(b) infinite number of solutions

ordered pairs corresponding to points that lie on the common graph. As mentioned in Section 7.1, the equations in this system are dependent. In the answers at the back of this book, a solution of such a system of dependent equations is indicated by *infinite number of solutions.* A graph of the equations of this system is shown in Figure 7. ■

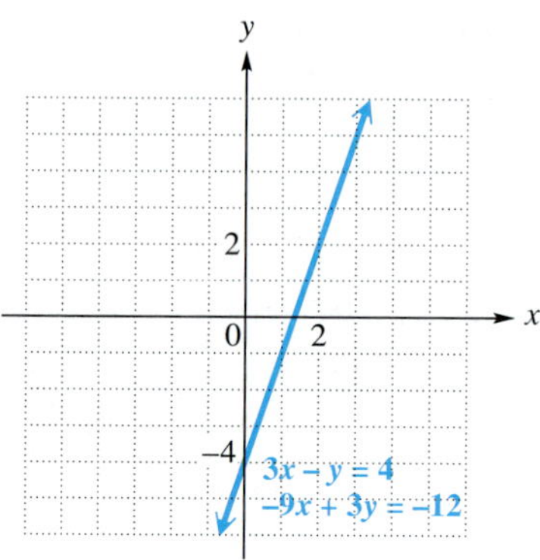

FIGURE 7

WORK PROBLEM 8 AT THE SIDE.

One of three situations may occur when the addition method is used to solve a linear system of equations.

1. The result of the addition step is a statement such as $x = 2$ or $y = -3$. The solution will be exactly one ordered pair. The graphs of the equations of the system will intersect at exactly one point.

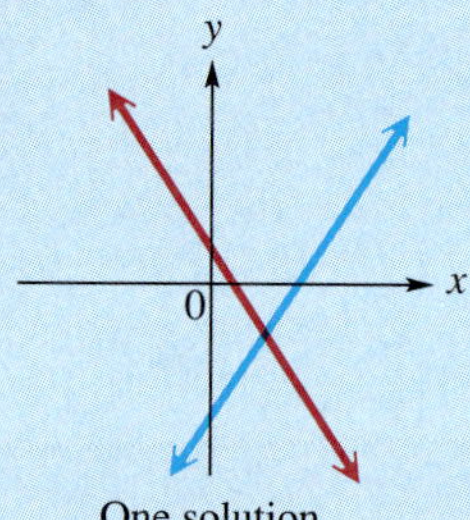

One solution

2. The result of the addition step is a false statement, such as $0 = 4$. In this case, the graphs are parallel lines, and there is no solution for the system.

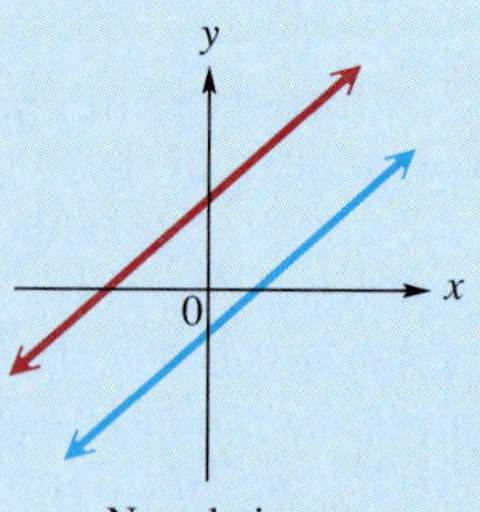

No solution

3. The result of the addition step is a true statement, such as $0 = 0$. The graphs of the equations of the system are the same line, and an infinite number of ordered pairs are solutions. These ordered pairs must satisfy the equation of the line.

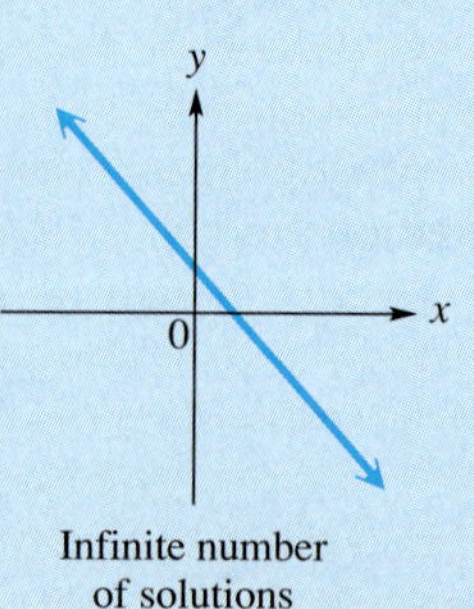

Infinite number of solutions

NAME DATE HOUR

7.2 EXERCISES

1. Only one of the following systems does not require that we multiply one or both equations by a constant in order to solve the system by the addition method. Which one is it?

(a) $-4x + 3y = 7$
$3x - 4y = 4$

(b) $5x + 8y = 13$
$12x + 24y = 36$

(c) $2x + 3y = 5$
$x - 3y = 12$

(d) $x + 2y = 9$
$3x - y = 6$

2. For the system

$$2x + 12y = 7$$
$$3x + 4y = 1,$$

if we were to multiply the first (top) equation by -3, by what number would we have to multiply the second (bottom) equation in order to

(a) eliminate the x terms when solving by the addition method?

(b) eliminate the y terms when solving by the addition method?

Solve each system by the addition method. Check your answers. See Examples 1 and 2.

3. $x - y = -2$
$x + y = 10$

4. $x + y = 10$
$x - y = -6$

5. $x + y = 2$
$2x - y = -5$

6. $3x - y = -12$
$x + y = 4$

7. $2x + y = -5$
$x - y = 2$

8. $2x + y = -15$
$-x - y = 10$

9. $3x + 2y = 0$
$-3x - y = 3$

10. $5x - y = 5$
$-5x + 2y = 0$

11. $6x - y = -1$
$-6x + 5y = 17$

12. $6x + y = 9$
$-6x + 3y = 15$

Solve each system by the addition method. Check your answers. See Example 3.

13. $2x - y = 12$
$3x + 2y = -3$

14. $x + y = 3$
$-3x + 2y = -19$

15. $x + 3y = 19$
$2x - y = 10$

16. $4x - 3y = -19$
$2x + y = 13$

17. $x + 4y = 16$
$3x + 5y = 20$

18. $2x + y = 8$
$5x - 2y = -16$

19. $5x - 3y = -20$
$-3x + 6y = 12$

20. $4x + 3y = -28$
$5x - 6y = -35$

21. $3x - 2y = 22$
$-5x + 4y = -36$

22. $-4x + 3y = -18$
$5x - 6y = 18$

23. Explain why (0, 0) *must* be a solution for the system

$$Ax + By = 0$$
$$Cx + Dy = 0$$

for any choices of A, B, C, and D.

24. Without actually solving the system, explain why

$$x + y = 1$$
$$x + y = 2$$

can have no solutions.

25. Which one of these systems would you find more difficult to solve, and why?

$$\begin{aligned} 2x + y &= 5 \\ 5x + 3y &= 11 \end{aligned} \qquad \begin{aligned} 4x - 3y &= -7 \\ 6x + 5y &= 18 \end{aligned}$$

26. Why would it be easier to solve

$$\begin{aligned} 8x + 3y &= -4 \\ 12x + 7y &= 4 \end{aligned} \quad \text{than} \quad \begin{aligned} 4y &= 2 - 3x \\ 4x &= -3y + 6 \end{aligned}$$

by the addition method?

Solve each system by the addition method. Check your answers. See Examples 4 and 5.

27. $\begin{aligned} 3x + 5y &= 7 \\ 5x + 4y &= -10 \end{aligned}$

28. $\begin{aligned} 2x + 3y &= 13 \\ 5x + 2y &= -6 \end{aligned}$

29. $\begin{aligned} 2x + 3y &= 0 \\ 7y - 29 &= 5x \end{aligned}$

30. $\begin{aligned} 2x + 9y &= 44 \\ 6y - 61 &= 5x \end{aligned}$

31. $\begin{aligned} 24x + 12y &= -7 \\ 16x - 17 &= 18y \end{aligned}$

32. $\begin{aligned} 9x + 4y &= -3 \\ 6x + 7 &= -6y \end{aligned}$

33. $\begin{aligned} 3x &= 3 + 2y \\ -\frac{4}{3}x + y &= \frac{1}{3} \end{aligned}$

34. $\begin{aligned} 3x &= 27 + 2y \\ x - \frac{7}{2}y &= -25 \end{aligned}$

Use the addition method to solve each system. See Examples 6 and 7.

35. $x + y = 7$
$x + y = -3$

36. $x - y = 4$
$x - y = -3$

37. $-x + 3y = 4$
$-2x + 6y = 8$

38. $6x - 2y = 24$
$-3x + y = -12$

39. $5x - 2y = 3$
$10x - 4y = 5$

40. $3x - 5y = 1$
$6x - 10y = 4$

41. $6x + 3y = 0$
$-18x - 9y = 0$

42. $3x - 5y = 0$
$9x - 15y = 0$

43. $2x - 8y = 0$
$4x + 5y = 0$

44. $3x - 15y = 0$
$6x + 10y = 0$

PREVIEW EXERCISES

Solve each equation, and check the solutions. See Sections 2.3 and 5.6.

45. $-2(y - 2) + 5y = -5$

46. $2m - 3(4 - m) = 13$

47. $p + 4(6 - 2p) = 24$

48. $4\left(\frac{3 - 2k}{2}\right) + 3k = -3$

49. $4x - 2\left(\frac{1 - 3x}{2}\right) = 6$

50. $a + 3\left(\frac{1 - a}{2}\right) = 5$

7.3 SOLVING SYSTEMS OF LINEAR EQUATIONS BY SUBSTITUTION

OBJECTIVES

1. Solve linear systems by substitution.
2. Solve linear systems with fractions as coefficients.

FOR EXTRA HELP

Tape 10 | SSM pp. 273–279 | MAC: B IBM: B

1 The graphical method and the addition method for solving systems of linear equations were discussed in the preceding sections. A third method, the **substitution method,** is particularly useful for solving systems where one equation is solved, or can be solved quickly, for one of the variables.

■ **EXAMPLE 1** *Using the Substitution Method*

Solve the system

$$3x + 5y = 26$$
$$y = 2x.$$

The second of these two equations is already solved for y. This equation says that $y = 2x$. Substituting $2x$ for y in the first equation gives

$$\begin{aligned} 3x + 5y &= 26 & \\ 3x + 5(2x) &= 26 & &\text{Let } y = 2x. \\ 3x + 10x &= 26 & &\text{Multiply.} \\ 13x &= 26 & &\text{Combine terms.} \\ x &= 2. & &\text{Divide by 13.} \end{aligned}$$

Because $x = 2$, we find y from the equation $y = 2x$ by substituting 2 for x.

$$y = 2(2) = 4 \qquad \text{Let } x = 2.$$

Check that the solution of the given system is (2, 4). ■

WORK PROBLEM 1 AT THE SIDE.

1. Solve by the substitution method.

(a) $3x + 5y = 69$
$y = 4x$

(b) $-x + 4y = 26$
$y = -3x$

■ **EXAMPLE 2** *Using the Substitution Method*

Use substitution to solve the system

$$2x + 5y = 7$$
$$x = -1 - y.$$

The second equation gives x in terms of y. Substitute $-1 - y$ for x in the first equation.

$$\begin{aligned} 2x + 5y &= 7 & \\ 2(-1 - y) + 5y &= 7 & &\text{Let } x = -1 - y. \\ -2 - 2y + 5y &= 7 & &\text{Distributive property} \\ -2 + 3y &= 7 & &\text{Combine terms.} \\ 3y &= 9 & &\text{Add 2.} \\ y &= 3 & &\text{Divide by 3.} \end{aligned}$$

To find x, substitute $y = 3$ in the equation $x = -1 - y$ to get $x = -1 - 3 = -4$. Check that the solution of the given system is $(-4, 3)$. ■

ANSWERS
1. (a) (3, 12) (b) (−2, 6)

2. Solve each system by substitution. Check each solution.

(a) $3x - 4y = -11$
$x = y - 2$

(b) $8x - y = 4$
$y = 8x + 4$

ANSWERS
2. (a) (3, 5) (b) no solution

■ **EXAMPLE 3** *Using the Substitution Method*
Use substitution to solve the system

$$x = 5 - 2y$$
$$2x + 4y = 6.$$

Substitute $5 - 2y$ for x in the second equation.

$$\begin{aligned} 2x + 4y &= 6 \\ 2(5 - 2y) + 4y &= 6 && \text{Let } x = 5 - 2y. \\ 10 - 4y + 4y &= 6 && \text{Distributive property} \\ 10 &= 6 && \text{False} \end{aligned}$$

As shown in the last section, this false result means that the equations in the system have graphs that are parallel lines. The system is inconsistent and has no solution. ■

◀◀ WORK PROBLEM 2 AT THE SIDE.

■ **EXAMPLE 4** *Using the Substitution Method*
Use substitution to solve the system

$$2x + 3y = 8$$
$$-4x - 2y = 0.$$

To use the substitution method, one of the equations must be solved for one of the variables. Let us choose the first equation of the system, $2x + 3y = 8$, and solve for x.

To get x alone on one side, subtract $3y$ from each side.

$$\begin{aligned} 2x + 3y &= 8 \\ 2x &= 8 - 3y && \text{Subtract } 3y. \end{aligned}$$

Now divide each side by 2.

$$x = \frac{8 - 3y}{2} \qquad \text{Divide by 2.}$$

Substitute this value for x in the second equation of the system.

$$\begin{aligned} -4x - 2y &= 0 \\ -4\left(\frac{8 - 3y}{2}\right) - 2y &= 0 && \text{Let } x = \frac{8 - 3y}{2}. \\ -2(8 - 3y) - 2y &= 0 && \text{Divide } -4 \text{ by } 2. \\ -16 + 6y - 2y &= 0 && \text{Distributive property} \\ -16 + 4y &= 0 && \text{Combine terms.} \\ 4y &= 16 && \text{Add 16.} \\ y &= 4 && \text{Divide by 4.} \end{aligned}$$

Find x by letting $y = 4$ in $x = \dfrac{8 - 3y}{2}$.

$$\begin{aligned} x &= \frac{8 - 3(4)}{2} && \text{Let } y = 4. \\ x &= \frac{8 - 12}{2} && \text{Multiply.} \\ x &= -2 && \text{Subtract and divide.} \end{aligned}$$

Check:

$$\begin{aligned} 2x + 3y &= 8 \\ 2(-2) + 3(4) &= 8 \quad ? \\ -4 + 12 &= 8 \quad ? \\ 8 &= 8 \quad \text{True} \end{aligned} \qquad \begin{aligned} -4x - 2y &= 0 \\ -4(-2) - 2(4) &= 0 \quad ? \\ 8 - 8 &= 0 \quad ? \\ 0 &= 0 \quad \text{True} \end{aligned}$$

The solution of the given system is $(-2, 4)$. ■

WORK PROBLEM 3 AT THE SIDE. ▶

■ **EXAMPLE 5** *Using the Substitution Method*

Use substitution to solve the system

$$2x = 4 - y \qquad \textbf{(1)}$$
$$6 + 3y + 4x = 16 - x. \qquad \textbf{(2)}$$

Start by simplifying the second equation by adding x and subtracting 6 on each side. This gives the simplified system

$$2x = 4 - y \qquad \textbf{(1)}$$
$$5x + 3y = 10. \qquad \textbf{(3)}$$

For the substitution method, one of the equations must be solved for either x or y. Because the coefficient of y in equation (1) is -1, avoid fractions by solving this equation for y.

$$\begin{aligned} 2x &= 4 - y && \textbf{(1)} \\ 2x - 4 &= -y && \text{Subtract 4.} \\ -2x + 4 &= y && \text{Multiply by } -1. \end{aligned}$$

Now substitute $-2x + 4$ for y in equation (3).

$$\begin{aligned} 5x + 3y &= 10 \\ 5x + 3(-2x + 4) &= 10 && \text{Let } y = -2x + 4. \\ 5x - 6x + 12 &= 10 && \text{Distributive property} \\ -x + 12 &= 10 && \text{Combine terms.} \\ -x &= -2 && \text{Subtract 12.} \\ x &= 2 && \text{Multiply by } -1. \end{aligned}$$

Since $y = -2x + 4$ and $x = 2$,

$$y = -2(2) + 4 = 0,$$

and the solution is $(2, 0)$.

Check:

$$\begin{aligned} 2x &= 4 - y \\ 2(2) &= 4 - 0 \quad ? \\ 4 &= 4 \quad \text{True} \end{aligned} \qquad \begin{aligned} 6 + 3y + 4x &= 16 - x \\ 6 + 3(0) + 4(2) &= 16 - 2 \quad ? \\ 6 + 0 + 8 &= 14 \quad \text{True} \end{aligned}$$ ■

WORK PROBLEM 4 AT THE SIDE. ▶

2 When a system includes equations with fractions as coefficients, we eliminate the fractions by multiplying each side by a common denominator. Then we solve the resulting system.

3. Solve each system by substitution. Check each solution.

(a) $x + 4y = -1$
$2x - 5y = 11$

(b) $2x + 5y = 4$
$x + y = -1$

4. Solve each system by substitution. First simplify where necessary.

(a) $x = 5 - 3y$
$2x + 3 = 5x - 4y + 14$

(b) $5x - y = -14 + 2x + y$
$7x + 9y + 4 = 3x + 8y$

ANSWERS
3. (a) $(3, -1)$ (b) $(-3, 2)$
4. (a) $(-1, 2)$ (b) $(-2, 4)$

5. Verify that the same solution is found if the addition method is used to solve the system of equations (3) and (4) in Example 6.

6. Solve the following system by any method. First clear all fractions.

$$\frac{2}{3}x + \frac{1}{2}y = 6$$
$$\frac{1}{2}x - \frac{3}{4}y = 0$$

EXAMPLE 6 *Using the Substitution Method with Fractions as Coefficients*

Solve the following system by the substitution method.

$$3x + \frac{1}{4}y = 2 \qquad \textbf{(1)}$$
$$\frac{1}{2}x + \frac{3}{4}y = -\frac{5}{2} \qquad \textbf{(2)}$$

Clear equation (1) of fractions by multiplying each side by 4.

$$4\left(3x + \frac{1}{4}y\right) = 4(2) \qquad \text{Multiply by 4.}$$
$$4(3x) + 4\left(\frac{1}{4}y\right) = 4(2) \qquad \text{Distributive property}$$
$$12x + y = 8 \qquad \textbf{(3)}$$

Now clear equation (2) of fractions by multiplying each side by the common denominator 4.

$$4\left(\frac{1}{2}x + \frac{3}{4}y\right) = 4\left(-\frac{5}{2}\right) \qquad \text{Multiply by 4.}$$
$$4\left(\frac{1}{2}x\right) + 4\left(\frac{3}{4}y\right) = 4\left(-\frac{5}{2}\right) \qquad \text{Distributive property}$$
$$2x + 3y = -10 \qquad \textbf{(4)}$$

The given system of equations has been simplified as follows.

$$12x + y = 8 \qquad \textbf{(3)}$$
$$2x + 3y = -10 \qquad \textbf{(4)}$$

Let us solve this system by the substitution method. Equation (3) can be solved for y by subtracting $12x$ from each side.

$$12x + y = 8$$
$$y = -12x + 8 \qquad \text{Subtract } 12x.$$

Now substitute the result for y in equation (4).

$$2x + 3(-12x + 8) = -10 \qquad \text{Let } y = -12x + 8.$$
$$2x - 36x + 24 = -10 \qquad \text{Distributive property}$$
$$-34x = -34 \qquad \text{Combine terms; subtract 24.}$$
$$x = 1 \qquad \text{Divide by } -34.$$

Using $x = 1$ in $y = -12x + 8$ gives $y = -12(1) + 8 = -4$. The solution is $(1, -4)$. Check by substituting 1 for x and -4 for y in both of the original equations. ■

WORK PROBLEMS 5 AND 6 AT THE SIDE.

ANSWERS
5. The solution is the same.
6. (6, 4)

7.3 EXERCISES

NAME DATE HOUR

1. If you were to solve the system

$$3x + 2y = 7$$
$$5x - y = 4$$

by substitution, which variable would probably be easier to solve for in the first step? In which equation would you solve for it? Why?

2. Which one of the following systems would be easier to solve using the substitution method? Why?

$5x - 3y = 7$ $\quad$ $7x + 2y = 4$

$2x + 8y = 3$ $\quad$ $y = -3x$

Solve each system by the substitution method. Check each solution. See Examples 1–4.

3. $x + y = 12$
$y = 3x$

4. $x + 3y = -28$
$y = -5x$

5. $3x + 2y = 27$
$x = y + 4$

6. $4x + 3y = -5$
$x = y - 3$

7. $3x + 5y = 14$
$x - 2y = -10$

8. $5x + 2y = -1$
$2x - y = -13$

9. $3x + 4 = -y$
$2x + y = 0$

10. $2x - 5 = -y$
$x + 3y = 0$

11. $7x + 4y = 13$
$x + y = 1$

12. $3x - 2y = 19$
$x + y = 8$

13. $3x - y = 5$
$y = 3x - 5$

14. $4x - y = -3$
$y = 4x + 3$

15. $6x - 8y = 6$
$-3x + 2y = -2$

16. $3x + 2y = 6$
$-6x + 4y = -8$

17. $2x + 8y = 3$
$x = 8 - 4y$

18. $2x + 10y = 3$
$x = 1 - 5y$

19. $12x - 16y = 8$
$3x = 4y + 2$

20. $6x + 9y = 6$
$2x = 2 - 3y$

In Exercises 21 and 22, (a) solve the system by the addition method, (b) then solve the system by the substitution method, (c) and finally tell which method you prefer for that particular system, and why.

21. $4x - 3y = -8$
$x + 3y = 13$

22. $2x + 5y = 0$
$x = -3y + 1$

Solve each system either by the addition method or the substitution method. First simplify equations where necessary. Check each solution. See Example 5.

23. $4 + 4x - 3y = 34 + x$
$4x = -y - 2 + 3x$

24. $5x - 4y = 42 - 8y - 2$
$2x + y = x + 1$

25. $2x - 8y + 3y + 2 = 5y + 16$
$8x - 2y = 4x + 28$

26. $7x - 9 + 2y - 8 = -3y + 4x + 13$
$4y - 8x = -8 + 9x + 32$

27. $-2x + 3y = 12 + 2y$
$2x - 5y + 4 = -8 - 4y$

28. $2x + 5y = 7 + 4y - x$
$5x + 3y + 8 = 22 - x + y$

29. $5x + y = 12 - x - 7y$
$3x + 2y = 10 - 6x - 10y$

30. $-2x + 3y = 7 - 5x - y$
$-4x + 2y = 1 - 10x - 6y$

Exercises 31 and 32 refer to the system

$$\frac{1}{3}x - \frac{1}{2}y = 7$$
$$\frac{1}{6}x + \frac{1}{3}y = 0.$$

31. One student solved the system by multiplying both equations by 6 to clear fractions, and another student multiplied by 12. Assuming they do all other work correctly, should they both get the same answer?

32. One student solved the system and wrote as his answer "$x = 12$," while another solved it and wrote as her answer "$y = -6$." Who, if either, was correct? Why?

Solve each system either by the addition method or the substitution method. First clear all fractions. Check each solution. See Example 6.

33. $x + \frac{1}{3}y = y - 2$

$\frac{1}{4}x + y = x + y$

34. $\frac{5}{3}x + 2y = \frac{1}{3} + y$

$3x - 3 + \frac{y}{3} = -2 + 2x$

35. $\frac{x}{6} + \frac{y}{6} = 2$

$-\frac{1}{2}x - \frac{1}{3}y = -8$

36. $\frac{x}{2} - \frac{y}{3} = 9$

$\frac{x}{5} - \frac{y}{4} = 5$

37. $\frac{x}{3} - \frac{3y}{4} = -\frac{1}{2}$

$\frac{x}{6} + \frac{y}{8} = \frac{3}{4}$

38. $\frac{x}{5} + 2y = \frac{16}{5}$

$\frac{3x}{5} + \frac{y}{2} = -\frac{7}{5}$

PREVIEW EXERCISES

Solve each applied problem. See Section 2.4.

39. In the 1991–1992 National Hockey League season, Mario Lemieux of Pittsburgh had a total of 131 points (goals plus assists). He had 43 more assists than goals. How many goals and how many assists did he have?

40. In the 1992 Olympic Games in Barcelona, the Unified Team earned 4 more medals than the United States. Together these two teams earned 220 medals. How many medals did each team earn?

41. (See the answer to Exercise 40.) The United States had the same number of gold and bronze medals, and three fewer silver medals. How many of each kind of medal did it earn?

42. (See the answer to Exercise 40.) The Unified Team had 7 more gold than silver medals, and 9 more silver than bronze medals. How many of each medal did it earn?

7.4 APPLICATIONS OF LINEAR SYSTEMS

Many practical problems are more easily translated into equations if two variables are used. With two variables, a system of two equations is needed to find the desired solution. The examples in this section illustrate the method of solving applied problems using two equations and two variables.

Recall from Chapter 2 the steps used in solving applied problems. The steps presented there can be modified as follows to allow for two variables and two equations.

OBJECTIVES

1. Use linear systems to solve applied problems about two numbers.
2. Use linear systems to solve applied problems about money.
3. Use linear systems to solve applied problems about mixtures.
4. Use linear systems to solve applied problems about rate or speed using the distance formula.

FOR EXTRA HELP

Tape 10

SSM pp. 279–286

MAC: B IBM: B

SOLVING AN APPLIED PROBLEM WITH TWO VARIABLES

Step 1 Choose a variable to represent each of the unknown values that must be found. Write down what each variable is to represent.
Step 2 Translate the problem into two equations using both variables.
Step 3 Solve the system of two equations.
Step 4 Answer the question or questions asked in the problem.
Step 5 Check the solution in the words of the original problem.

1 The first example shows how to use two variables to solve a problem about two unknown numbers.

EXAMPLE 1 *Solving a Problem About Two Numbers*
The sum of two numbers is 63. Their difference is 19. Find the two numbers.

Step 1 Let x = one number; y = the other number.
Step 2 Set up a system of equations from the information in the problem.

$$x + y = 63 \quad \text{The sum is 63.}$$
$$x - y = 19 \quad \text{The difference is 19.}$$

Step 3 Solve the system from Step 2.

WORK PROBLEM 1 AT THE SIDE.

Step 4 The numbers required in the problem are 41 and 22.
Step 5 Check: The sum of 41 and 22 is 63, and their difference is 19. The solution satisfies the conditions of the problem. ■

WORK PROBLEM 2 AT THE SIDE.

2 The next example illustrates how to set up and solve a common type of applied problem that involves two quantities and their costs.

EXAMPLE 2 *Solving a Problem About Quantities and Costs*
Admission prices at a football game were \$6 for adults and \$2 for children. The total value of the tickets sold was \$2528, and 454 tickets were sold. How many adults and how many children attended the game?

Step 1 Let a = the number of adults' tickets sold; c = the number of children's tickets sold.

---

1. Solve the system of equations

$$x + y = 63$$
$$x - y = 19.$$

2. Set up a system of equations to solve each problem. Do not solve.

(a) The sum of two numbers is 97. Their difference is 41. What are the numbers?

(b) The sum of two numbers is 38. If twice the first is added to three times the second, the result is 99. Find the numbers.

ANSWERS
1. $x = 41, y = 22$
2. (a) $x + y = 97$
$x - y = 41$
(b) $x + y = 38$
$2x + 3y = 99$

3. The value of the tickets sold for a concert was \$1850. The price for a regular ticket was \$5.00, and student tickets were \$3.50. A total of 400 tickets were sold.

(a) Complete this table.

Kind of ticket	*Number sold*	*Cost of each*	*Total value*
Regular	r		
Student	s		
Totals			

(b) Write a system of equations.

(c) Solve the system, and check your solution in the words of the original problem.

Step 2 The information given in the problem is summarized in the table. The entries in the "total value" column were found by multiplying the number of tickets sold by the price per ticket.

Kind of ticket	*Number sold*	*Cost of each (in dollars)*	*Total value (in dollars)*
Adult	a	6	$6a$
Child	c	2	$2c$
Total	454	—	2528

The total number of tickets sold was 454, so

$$a + c = 454.$$

Because the total value was \$2528, the right-hand column leads to

$$6a + 2c = 2528.$$

Step 3 These two equations give the following system.

$$a + c = 454 \qquad (1)$$
$$6a + 2c = 2528 \qquad (2)$$

Solve the system of equations with the addition method.

$$\begin{aligned} -2a - 2c &= -908 && \text{Multiply (1) by } -2. \\ 6a + 2c &= 2528 && \text{Equation (2)} \\ 4a \quad &= 1620 && \text{Add.} \\ a &= 405 && \text{Divide by 4.} \end{aligned}$$

Now find the value of c.

$$\begin{aligned} a + c &= 454 && \text{Equation (1)} \\ 405 + c &= 454 && \text{Let } a = 405. \\ c &= 49. && \text{Subtract 405.} \end{aligned}$$

Step 4 There were 405 adults and 49 children at the game.

Step 5 Because 405 adults paid \$6 each and 49 children paid \$2 each, the value of tickets sold should be $405(6) + 49(2) = 2528$, or \$2528. This result agrees with the given information. ■

◀◀ WORK PROBLEM 3 AT THE SIDE.

In the rest of the examples try to identify each step in the solution of the problems.

3 Mixture problems occur in many fields of application of mathematics. One important type of mixture problem occurs in the study of chemistry. We solve this kind of problem as shown in the next example.

■ **EXAMPLE 3** *Solving a Problem About Mixture (Involving Percent)*

A pharmacist needs 100 liters of 50% alcohol solution. She has on hand 30% alcohol solution and 80% alcohol solution, which she can mix. How many liters of each will be required to make the 100 liters of 50% alcohol solution?

ANSWERS

3. (a)

Kind of ticket	*Number sold*	*Cost of each*	*Total value*
Regular	r	5	$5r$
Student	s	3.50	$3.50s$
Total	400	—	1850

(b) $r + s = 400$
$5r + 3.50s = 1850$

(c) 300 regular, 100 student

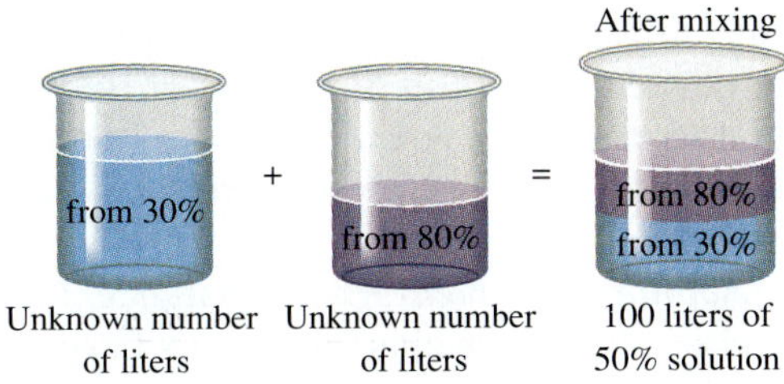

FIGURE 8

A 30% solution means that 30% of the solution is alcohol, and the rest is water. The information given above was used for the sketch in Figure 8. Let x represent the number of liters of 30% alcohol needed. Let y represent the number of liters of 80% alcohol needed.

The information given in the problem is summarized in the table below. (Each percent has been changed to its decimal form in the third column.)

Percent	*Liters of solution*	*Liters of pure alcohol*
30	x	$.30x$
80	y	$.80y$
50	100	$.50(100)$

The pharmacist will have $.30x$ liters of alcohol from the x liters of 30% solution and $.80y$ liters of alcohol from the y liters of 80% solution, for a total of $.30x + .80y$ liters of pure alcohol. In the mixture, she wants 100 liters of 50% solution. This 100 liters would contain $.50(100) = 50$ liters of pure alcohol. Since the amounts of pure alcohol must be equal,

$$.30x + .80y = 50.$$

The total number of liters is 100, or

$$x + y = 100.$$

These two equations give the system

$$\begin{aligned} .30x + .80y &= 50 \\ x + \quad y &= 100. \end{aligned}$$

The first of these equations involves decimal fractions. Recall from Example 6 in Section 2.3 that we may clear an equation of decimals by multiplying each side by a power of 10. In the equation $.30x + .80y = 50$, start by multiplying by 100. The system then becomes

$$\begin{aligned} 30x + 80y &= 5000 \\ x + \quad y &= 100. \end{aligned}$$

WORK PROBLEM 4 AT THE SIDE.

From Problem 4 at the side, $x = 60$ and $y = 40$. The pharmacist should use 60 liters of the 30% solution and 40 liters of the 80% solution. Since $.30(60) + (.80)40 = 50$, this mix will give 100 liters of 50% solution, as required in the original problem. ■

WORK PROBLEMS 5 AND 6 AT THE SIDE.

4. Solve the system

$$\begin{aligned} 30x + 80y &= 5000 \\ x + \quad y &= 100. \end{aligned}$$

5. How many liters of 25% alcohol solution must be mixed with 12% solution to get 13 liters of 15% solution?

(a) Complete the table.

Percent	*Liters*	*Liters of pure alcohol*
25	x	$.25x$
12	y	
15	13	

(b) Write a system of equations, and solve it.

6. Joe needs 100 cc (cubic centimeters) of 20% acid solution for a chemistry experiment. The lab has on hand only 10% and 25% solutions. How much of each should he mix to get the desired amount of 20% solution?

ANSWERS

4. $x = 60, y = 40$

5. (a)

Percent	*Liters*	*Liters of pure alcohol*
25	x	$.25x$
12	y	$.12y$
15	13	$.15(13)$

(b) $x + y = 13$
$.25x + .12y = .15(13)$
3 liters of 25%, 10 liters of 12%

6. $33\frac{1}{3}$ cc of 10%, $66\frac{2}{3}$ cc of 25%

7. Use substitution to solve the system

$$4x + 4y = 400$$
$$x = 20 + y.$$

8. (a) Two cars that were 450 miles apart traveled toward each other. They met after 5 hours. If one car traveled twice as fast as the other, what were their speeds? Complete this table.

	r	t	d
Faster car	x	5	
Slower car	y	5	

Write a system, and solve it.

(b) In one hour Ann can row 2 miles against the current or 10 miles with the current. Find the speed of the current and Ann's speed in still water.

ANSWERS

7. $x = 60$, $y = 40$

8. (a)

	r	t	d
Faster car	x	5	$5x$
Slower car	y	5	$5y$

$$5x + 5y = 450$$
$$x = 2y$$

$x = 60$ miles per hour,
$y = 30$ miles per hour

(b) 4 miles per hour, 6 miles per hour

4 Problems that use the distance formula, relating distance, rate, and time, were first introduced in Section 5.7. Many of these problems can be solved with a system of two linear equations. Keep in mind that setting up a chart and drawing a sketch will help you solve such problems.

EXAMPLE 4 *Solving a Problem About Distance, Rate, and Time*

Two executives in cities 400 miles apart drive to a business meeting at a location between their cities. They meet after 4 hours. (See Figure 9.) Find the average speed of each car if one travels 20 miles per hour faster than the other.

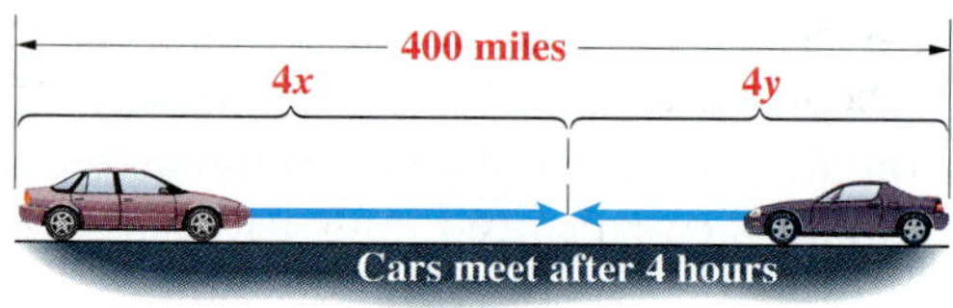

FIGURE 9

Use the formula that relates distance, rate, and time, $d = rt$. Let x be the speed of the faster car and y the speed of the slower car. Because each car travels for 4 hours, the time, t, for each car is 4. This information is shown in the chart. The distance is found by using the formula $d = rt$ and the expressions already entered in the chart.

	r	t	d
Faster car	x	4	$4x$
Slower car	y	4	$4y$

Find d from $d = rt$.

Since the total distance traveled by both cars is 400 miles, $4x + 4y = 400$. The faster car goes 20 miles per hour faster than the slower, giving $x = 20 + y$. We can solve this system of equations,

$$4x + 4y = 400$$
$$x = 20 + y,$$

by substitution.

WORK PROBLEM 7 AT THE SIDE.

Problem 7 at the side gives the solution of the system: $x = 60$, $y = 40$. Thus, the speeds of the two cars were 40 miles per hour and 60 miles per hour. If each car travels for 4 hours, the total distance is

$$4(40) + 4(60) = 160 + 240 = 400$$

miles, as required. ■

WORK PROBLEM 8 AT THE SIDE.

The problems in this section also could be solved using only one variable, but for most of them the solution is simpler with two variables.

Caution Be careful: Don't forget that two variables require two equations.

EXERCISES

1. Using the list of steps for solving an applied problem with two variables, write a short paragraph describing the general procedure you will use to solve the problems that follow in this exercise set.

2. Write an expression that illustrates each of the following quantities.

_______ **(a)** The monetary value of x five-dollar bills

_______ **(b)** The cost of y pounds of candy that sells for \$1.30 per pound

_______ **(c)** The amount of pure acid in a solution of x liters of 40% acid

_______ **(d)** The actual speed of a plane that flies 300 miles per hour against a wind of 40 miles per hour

Write a system of equations for each problem, and then solve the system. See Example 1.

3. Find two numbers whose sum is 113 and whose difference is 71.

4. The sum of two numbers is 60, and their difference is 8. Find the numbers.

5. According to the 1990 United States census, a total of 1096 people lived in the New Mexico counties of Harding and Los Alamos. Harding County had 878 more people than Los Alamos. What was the population of each of these counties?

6. The Terminal Tower in Cleveland, Ohio, is 240 feet shorter than the Society Center, also in Cleveland. The total of the heights of the two buildings is 1656 feet. Find the heights of the buildings.

7. The perimeter of a rectangle is 36 centimeters. The length is 8 centimeters longer than the width. Find the dimensions of the rectangle.

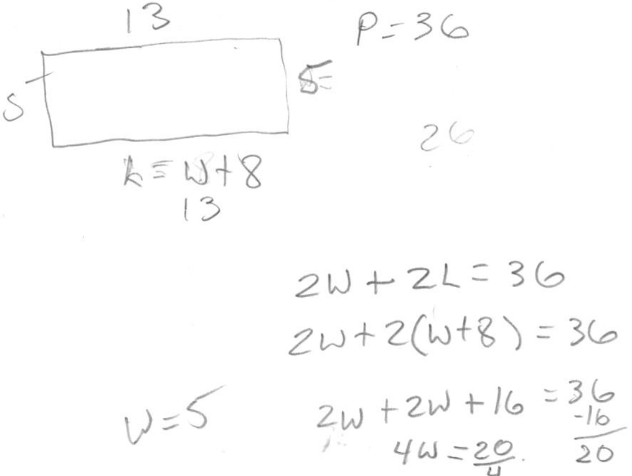

8. The perimeter of a triangle is 53 inches. If two sides are of equal length, and the third side measures 4 inches less than each of the equal sides, what are the lengths of the three sides?

Write a system of equations for each problem and then solve the system. See Example 2.

9. A motel clerk counts his \$1 and \$10 bills at the end of a day. He finds that he has a total of 74 bills having a combined monetary value of \$326. Find the number of bills of each denomination that he has.

Denomination of bill	*Number of bills*	*Total value*
\$1	x	
\$10	y	
Totals	74	\$326

10. Deanne Matlock is a bank teller. At the end of a day, she has a total of 69 \$5 and \$10 bills. The total value of the money is \$590. How many of each denomination does she have?

Denomination of bill	*Number of bills*	*Total value*
\$5	x	$\$5x$
\$10	y	
Totals		

11. At a recent soccer game, 386 tickets were sold, and a total of \$523 was collected. Student tickets cost \$.50 each, and nonstudent tickets cost \$2.00 each. How many of each type of ticket were sold?

12. For the school band fundraiser, Juanita sold license plates at \$4.00 each and fruitcakes at \$6.00 each. She sold a total of 62 items and collected \$304.00. How many of each item did she sell?

13. Maria Lopez has twice as much money invested at 5% simple annual interest as she does at 4%. If her yearly income from these two investments is \$350, how much does she have invested at each rate?

14. Charles invested his textbook royalty income in two accounts, one paying 3% annual simple interest and the other paying 2% interest. He earned a total of \$11 interest. If he invested three times as much in the 3% account as he did in the 2% account, how much did he invest at each rate?

15. A book collector bought some paperbacks at \$.25 each and some hardbacks at \$1.50 each. He bought a total of 20 books and paid \$21.25. How many of each type of book did he buy?

16. A nursing home bought 38 bottles of disinfectant. Small bottles cost \$5.00 each, and large ones cost \$7.00 each. A total of \$250.00 was spent. How many of each size bottle were bought?

Write a system of equations for each problem, and then solve the system. See Example 3.

17. A 40% dye solution is to be mixed with a 70% dye solution to get 120 liters of a 50% solution. How many liters of the 40% and 70% solutions will be needed?

Liters of solution	*Percent (as a decimal)*	*Liters of pure dye*
x	.40	$.40x$
y	.70	
120	.50	

18. A 90% antifreeze solution is to be mixed with a 75% solution to make 120 liters of a 78% solution. How many liters of the 90% and 75% solutions will be used?

Liters of solution	*Percent (as a decimal)*	*Liters of pure antifreeze*
x	.90	
y	.75	
120	.78	

19. A merchant wishes to mix coffee worth \$6 per pound with coffee worth \$3 per pound to get 90 pounds of a mixture worth \$4 per pound. How many pounds of the \$6 and the \$3 coffee will be needed?

Pounds	*Dollars per pound*	*Cost*
x	6	
y	3	
90		

20. A grocer wishes to blend candy selling for \$1.20 a pound with candy selling for \$1.80 a pound to get a mixture that will be sold for \$1.40 a pound. How many pounds of the \$1.20 and the \$1.80 candy should be used to get 45 pounds of the mixture?

Pounds	*Dollars per pound*	*Cost*
x	1.20	
y	1.80	
45		

21. How many barrels of pickles worth \$40 per barrel and pickles worth \$60 per barrel must be mixed to obtain 50 barrels of a mixture worth \$48 per barrel?

22. The owner of a nursery wants to mix some fertilizer worth \$70 per bag with some worth \$90 per bag to obtain 40 bags of mixture worth \$77.50 per bag. How many bags of each type should she use?

Write a system of equations for each problem, and then solve the system. See Example 4.

23. A boat takes 3 hours to go 24 miles upstream. It can go 36 miles downstream in the same time. Find the speed of the current and the speed of the boat in still water if x = the speed of the boat in still water and y = the speed of the current.

	d	*r*	*t*
Downstream	36	$x + y$	3
Upstream	24	$x - y$	3

24. It takes a boat $1\frac{1}{2}$ hours to go 12 miles downstream, and 6 hours to return. Find the speed of the boat in still water and the speed of the current. Let x = the speed of the boat in still water and y = the speed of the current.

	d	*r*	*t*
Downstream	12	$x + y$	$\frac{3}{2}$
Upstream	12	$x - y$	6

25. If a plane can travel 440 miles per hour into the wind and 500 miles per hour with the wind, find the speed of the wind and the speed of the plane in still air.

26. A small plane travels 200 miles per hour with the wind and 120 miles per hour against it. Find the speed of the wind and the speed of the plane in still air.

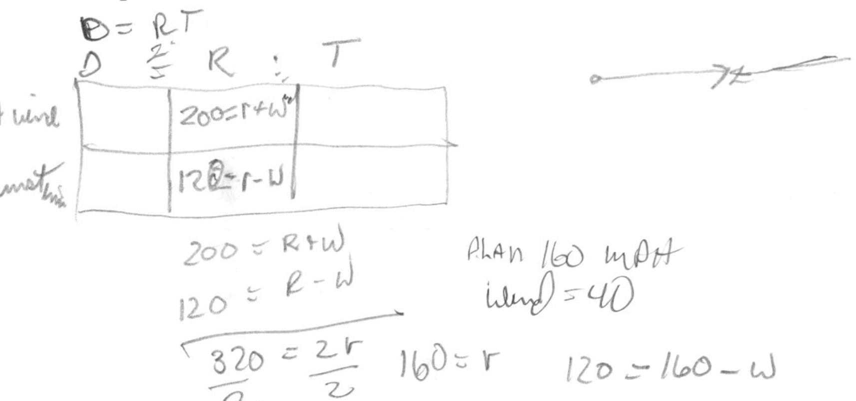

27. Two trains start from stations 1000 miles apart and travel toward each other. They meet after 5 hours. Find the rate of each train if one travels 20 miles per hour faster than the other.

28. Two cars start from towns 300 miles apart and travel toward each other. They meet after 3 hours. Find the rate of each car if one car travels 30 miles per hour slower than the other.

29. At the beginning of a walk for charity, Roberto and Juana are 30 miles apart. If they leave at the same time and walk in the same direction, Roberto overtakes Juana in 60 hours. If they walk toward each other, they meet in 5 hours. What are their speeds?

30. Mr. Abbot left Farmersville in a plane at noon to travel to Exeter. Mr. Baker left Exeter in his automobile at 2 P.M. to travel to Farmersville. It is 400 miles from Exeter to Farmersville. If the sum of their speeds was 120 miles per hour, and if they crossed paths at 4 P.M., find the speed of each.

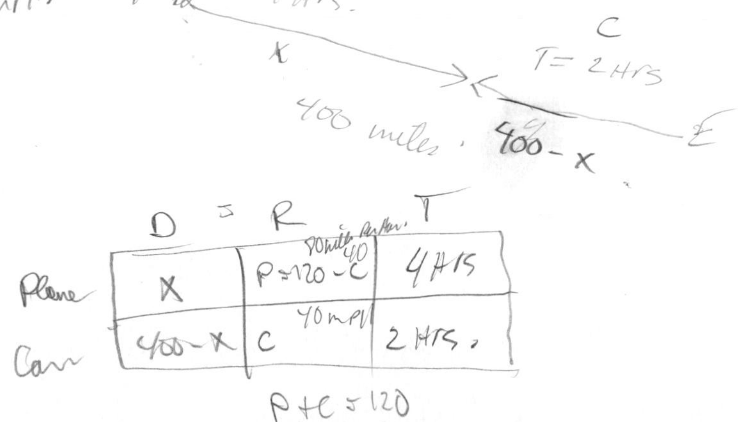

PREVIEW EXERCISES

Graph each linear inequality. See Section 6.5.

31. $x + y \leq 4$

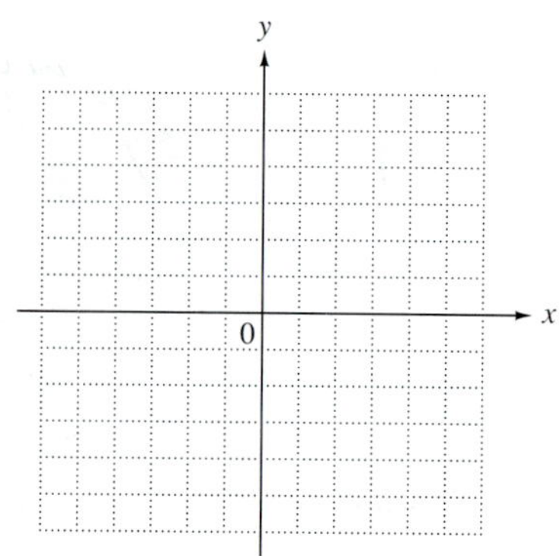

32. $2x - y > 4$

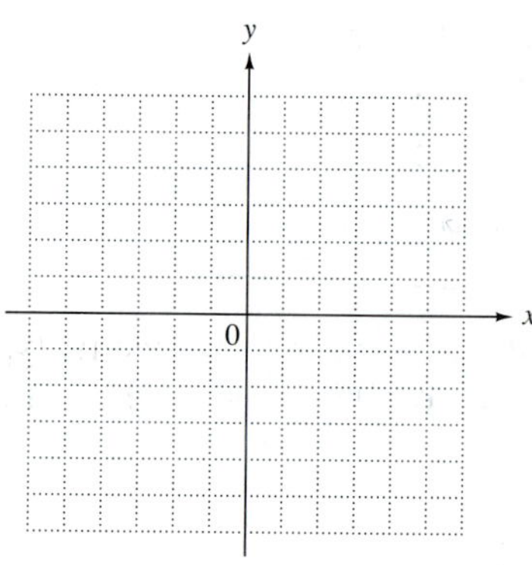

33. $y \geq -3x + 2$

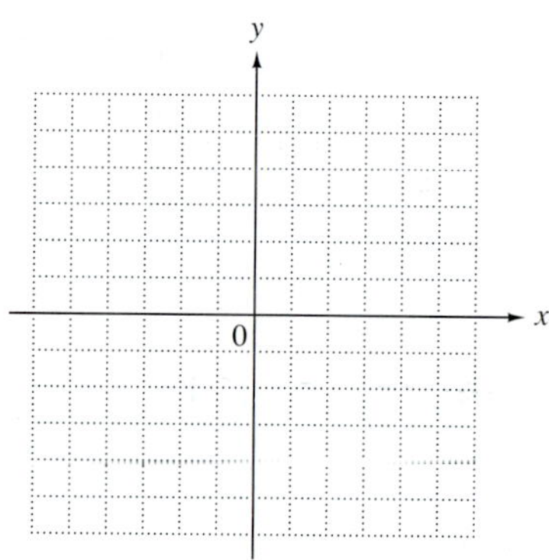

34. $x - 3y < 6$

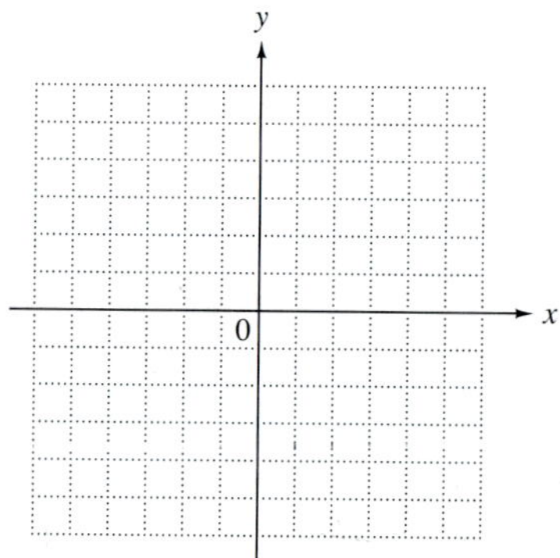

35. $2x + 4y > 8$

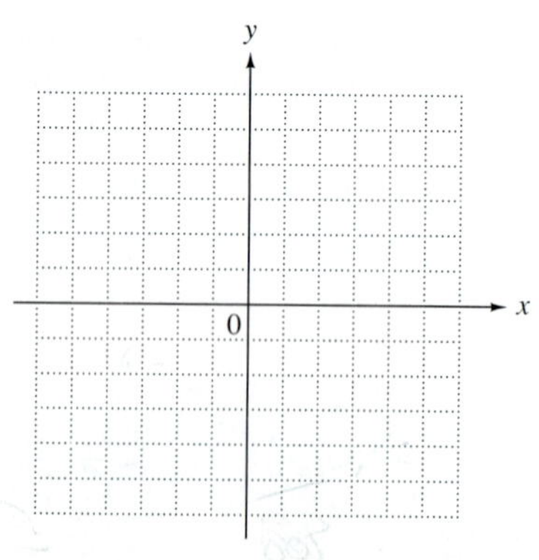

36. $3x + 2y \leq 0$

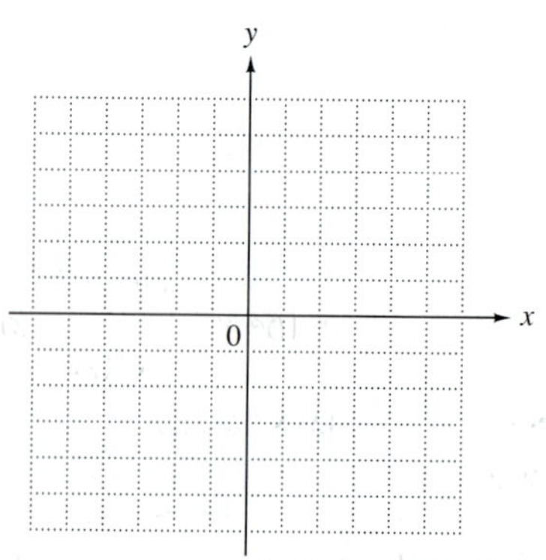

7.5 SOLVING SYSTEMS OF LINEAR INEQUALITIES

Graphing the solution of a linear inequality was discussed in Section 6.5. Let us review the method. To graph the solution of $x + 3y > 12$, we first graph the line $x + 3y = 12$ by finding a few ordered pairs that satisfy the equation. Because the points on the line do not satisfy the inequality, we use a dashed line. To decide which side of the line should be shaded, we choose any test point not on the line, such as (0, 0). Then we substitute 0 for x and 0 for y in the given inequality.

$$x + 3y > 12$$
$$0 + 3(0) > 12 \quad \text{Let } x = 0 \text{ and let } y = 0.$$
$$0 > 12 \quad \text{False}$$

Since the test point does not satisfy the inequality, we shade the region on the side of the line that does not include (0, 0), as shown in Figure 10.

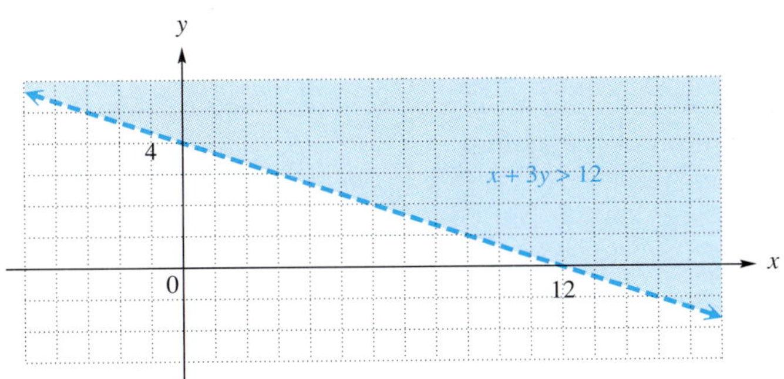

FIGURE 10

WORK PROBLEM 1 AT THE SIDE.

OBJECTIVE

1 Solve systems of linear inequalities by graphing.

FOR EXTRA HELP

Tape 10

SSM pp. 286–288

MAC: B IBM: B

1. Graph each inequality.

(a) $3x + 2y \geq 6$
The line $3x + 2y = 6$ is graphed already.

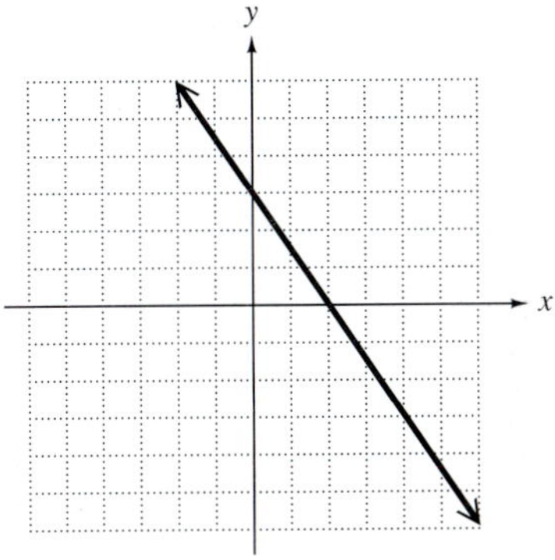

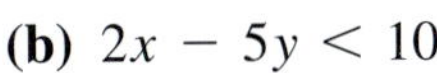
(b) $2x - 5y < 10$

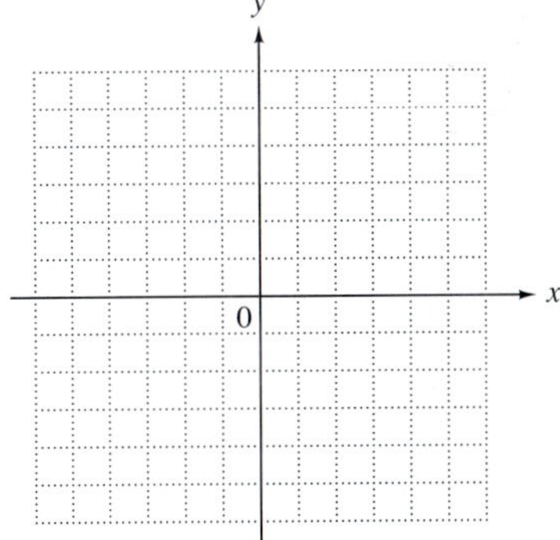

1 We use the same method to find the solution of a system of two linear inequalities, as shown in the following examples. A **system of linear inequalities** contains two or more linear inequalities (and no other kinds of inequalities). The **solution of a system of linear inequalities** includes all points that make all inequalities of the system true *at the same time.*

EXAMPLE 1 *Solving a System of Linear Inequalities*

Graph the solution of the system

$$3x + 2y \leq 6$$
$$2x - 5y \geq 10.$$

First graph $3x + 2y \leq 6$. Graph $3x + 2y = 6$ as a solid line. Because (0, 0) makes the inequality true, shade the region containing (0, 0), as shown in Figure 11 on the next page. Now, on the same axes, graph $2x - 5y \geq 10$. The boundary is the line $2x - 5y = 10$. Here, (0, 0) makes the inequality false, so shade the region that does not contain (0, 0), as shown in Figure 11. The solution of the system in shown by the intersection (overlap), the region in Figure 11 with the darkest shading, and includes the portions of the two boundary lines that bound the region. ■

ANSWERS

1. (a) (b)

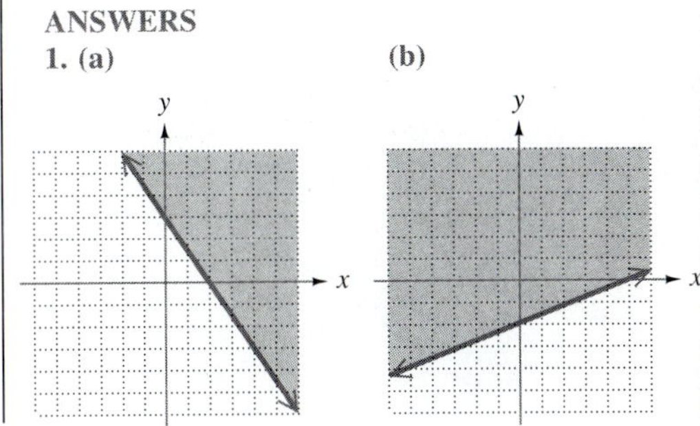

2. Graph the solution of the system

$$x - 2y \leq 8$$
$$3x + y \geq 6.$$

To get you started, the graphs of $x - 2y = 8$ and $3x + y = 6$ are included.

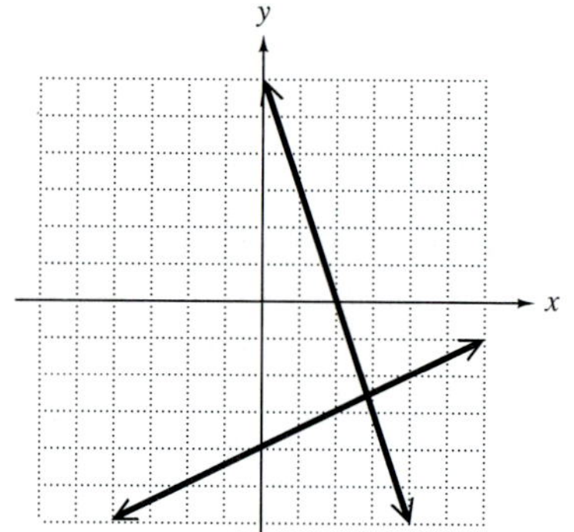

3. Graph the solution of the system

$$x + 2y < 0$$
$$3x - 4y < 12.$$

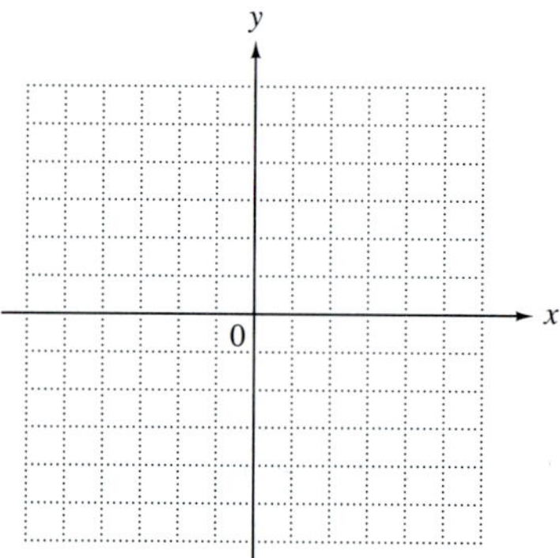

ANSWERS

2.

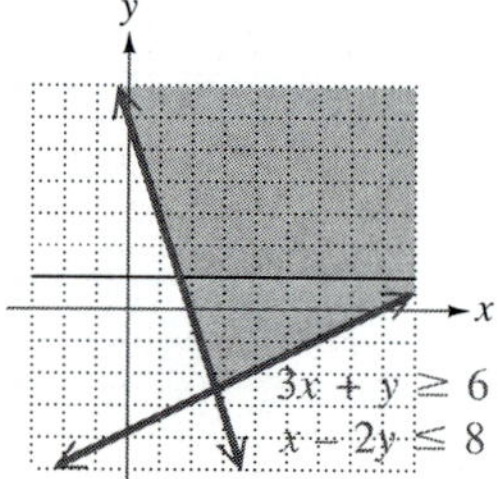

3.

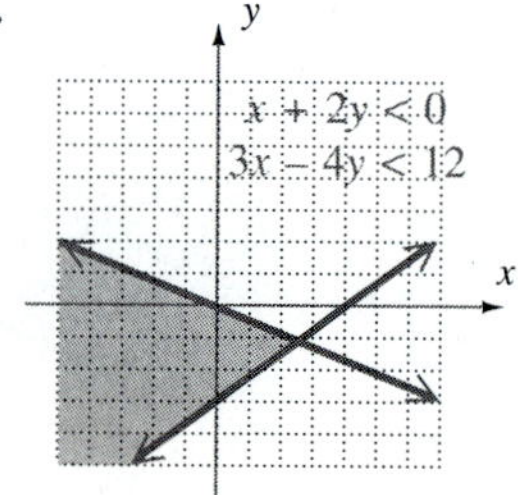

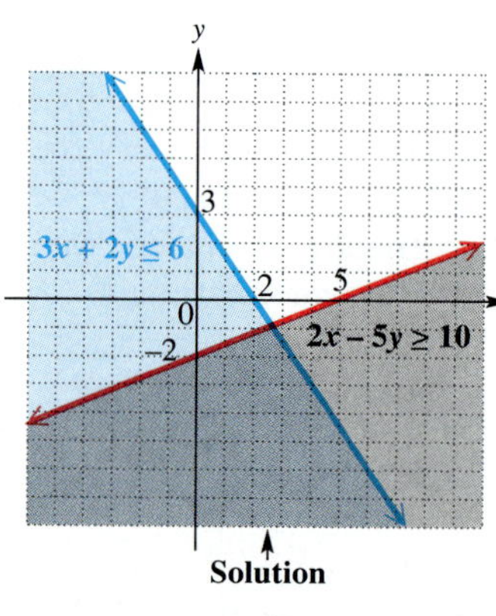

FIGURE 11

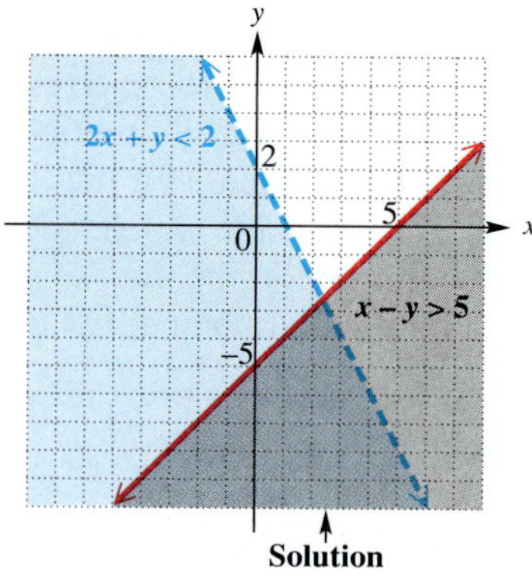

FIGURE 12

WORK PROBLEM 2 AT THE SIDE.

EXAMPLE 2 *Solving a System of Linear Inequalities*

Graph the solution of the system

$$x - y > 5$$
$$2x + y < 2.$$

Figure 12 shows the graphs of both $x - y > 5$ and $2x + y < 2$. Dashed lines show that the graphs of the inequalities do not include their boundary lines. The solution of the system is the region with the darkest shading. The solution does not include either boundary line. ■

EXAMPLE 3 *Solving A System of Linear Inequalities*

Graph the solution of the system

$$4x - 3y \leq 8$$
$$x \geq 2.$$

Recall that $x = 2$ is a vertical line through the point (2, 0). The graph of the solution of the system is the region in Figure 13 with the darkest shading. ■

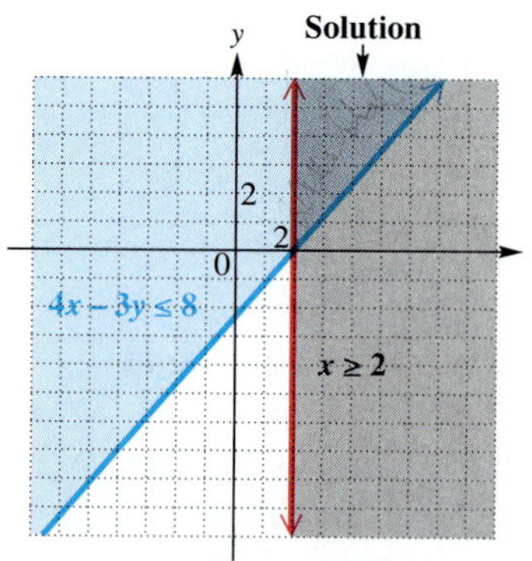

FIGURE 13

WORK PROBLEM 3 AT THE SIDE.

7.5 EXERCISES

NAME DATE HOUR

1. Every system of equations illustrated in the examples of this section has infinitely many solutions. Explain why this is so. Does this mean that *any* ordered pair is a solution?

2. Explain how you will determine in the exercises that follow in this section whether to draw a solid boundary line or a dashed boundary line.

Graph the solution of each system of linear inequalities. See Examples 1–3.

3. $x + y \leq 6$
   $x - y \geq 1$

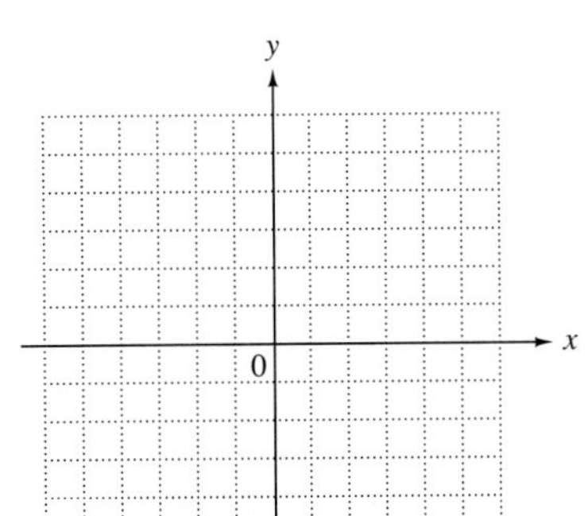

4. $x + y \leq 2$
   $x - y \geq 3$

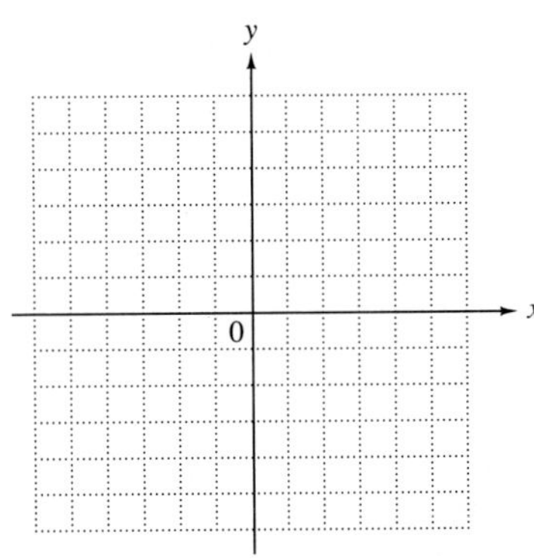

5. $4x + 5y \geq 20$
   $x - 2y \leq 5$

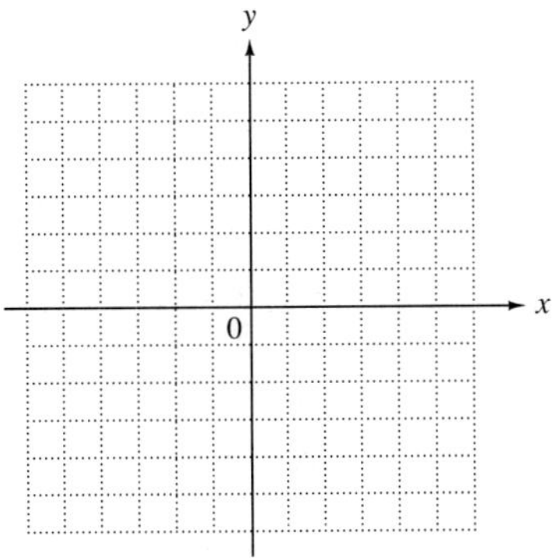

6. $x + 4y \leq 8$
   $2x - y \geq 4$

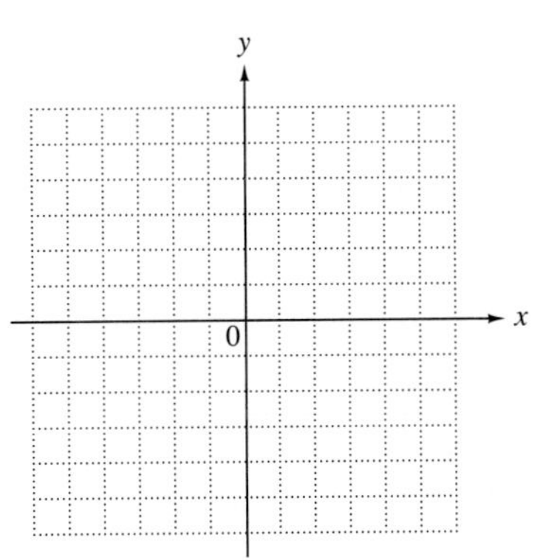

7. $2x + 3y < 6$
   $x - y < 5$

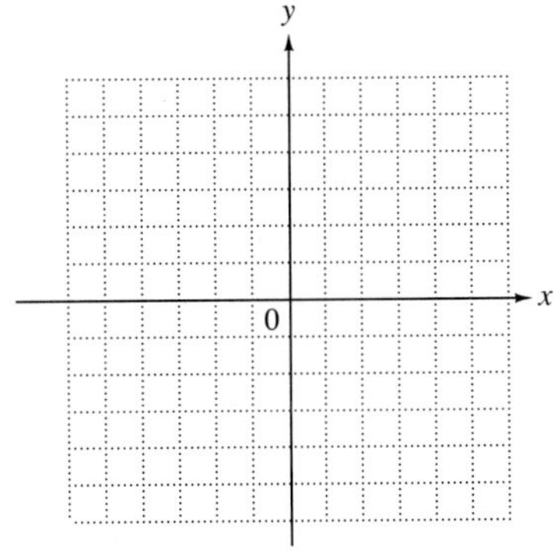

8. $x + 2y < 4$
   $x - y < -1$

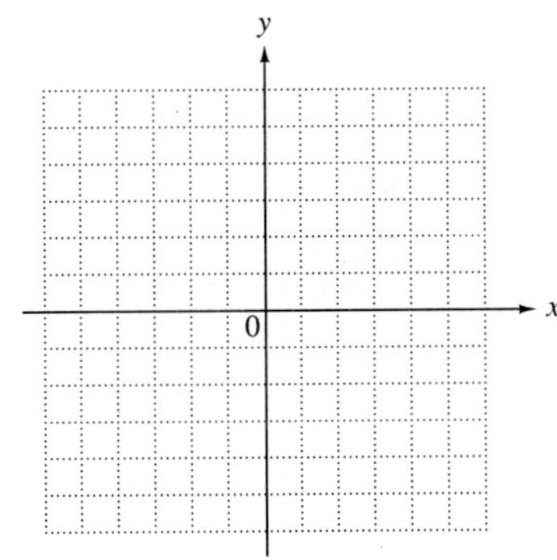

9. $y \le 2x - 5$
$x < 3y + 2$

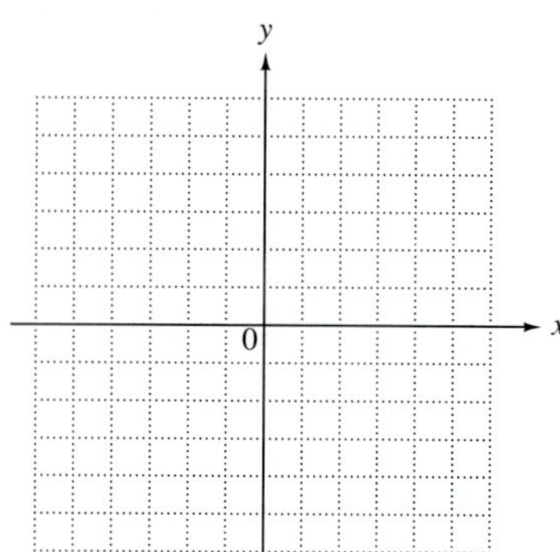

10. $x \ge 2y + 6$
$y > -2x + 4$

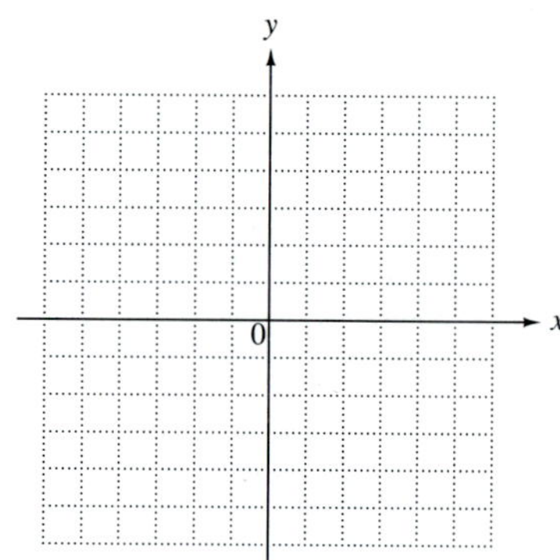

11. $4x + 3y < 6$
$x - 2y > 4$

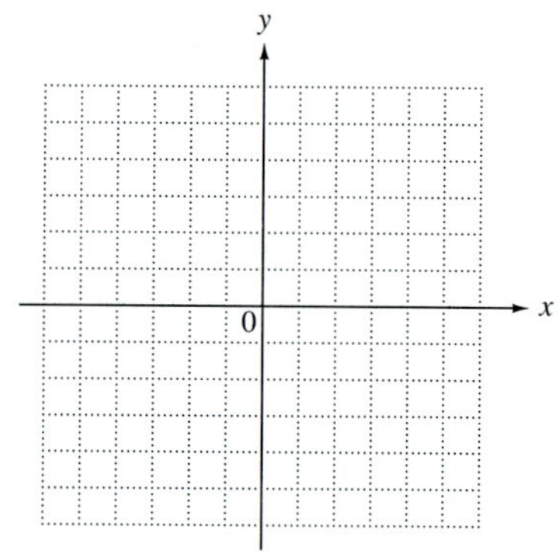

12. $3x + y > 4$
$x + 2y < 2$

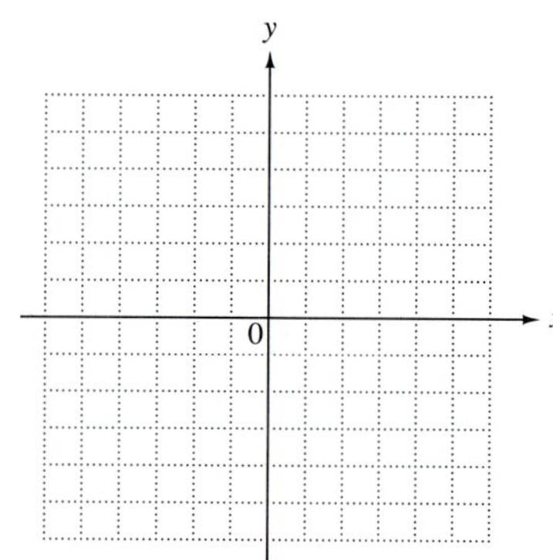

13. $x \le 2y + 3$
$x + y < 0$

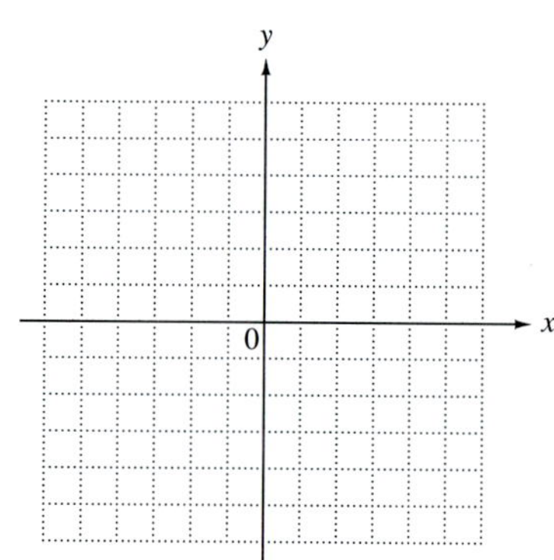

14. $x \le 4y + 3$
$x + y > 0$

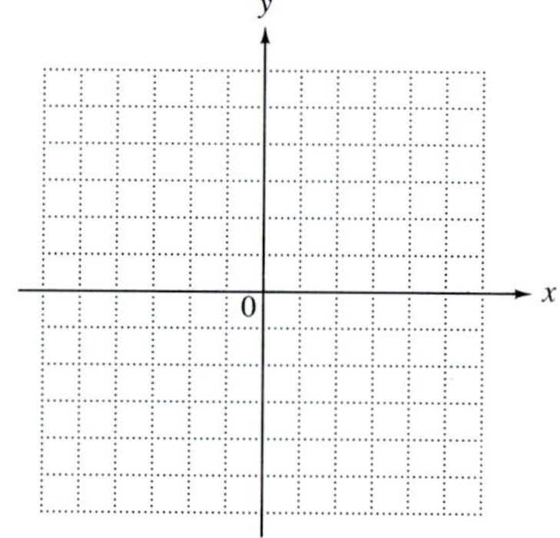

15. $-3x + y \ge 4$
$6x - 2y \ge -10$

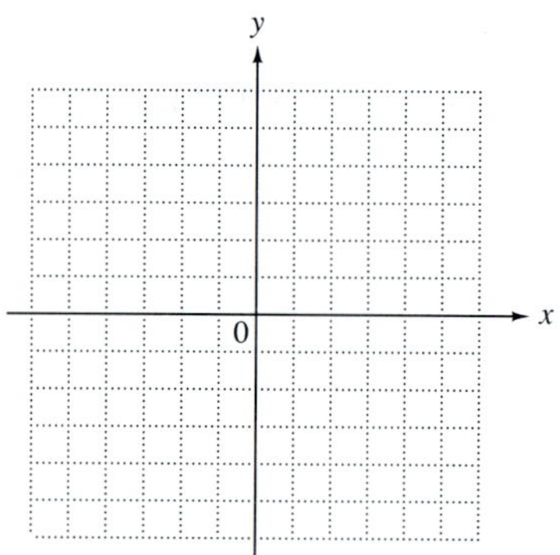

16. $2x + 3y < 6$
$4x + 6y > 18$

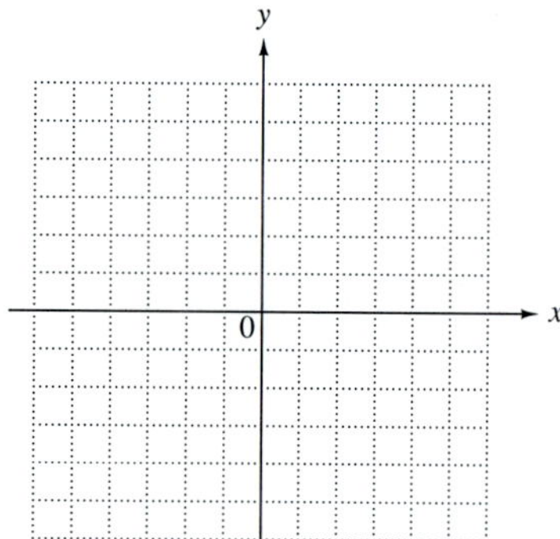

17. $x - 3y \le 6$
$x \ge -4$

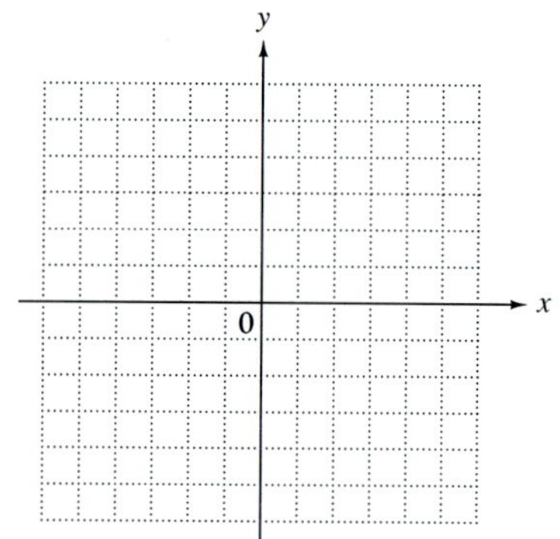

PREVIEW EXERCISES

Find each power. See Section 1.1.

18. 6^2 **19.** 8^2 **20.** 12^2 **21.** 16^2

Find the value of $a^2 + b^2$ for the given values of a and b. See Section 1.2.

22. $a = 5, b = 12$ **23.** $a = 8, b = 15$

QUEST FOR NUMERACY

Happy Birthday to You, and You, and You . . .

It is highly probable that in your extended (or even immediate) family, two people share the same birthday. One of the strange but true facts of mathematics is this: In a randomly chosen group of thirty people, the probability of at least one duplication of birthdays is over 70 percent! The chart below gives more information about this phenomenon.

Number of People	*Probability of at Least One Duplication*
2	.003
4	.016
6	.040
8	.074
10	.117
12	.167
14	.223
16	.284
18	.347
20	.411
22	.476
24	.538
26	.598
28	.654
30	.706
32	.753
34	.795
36	.832
38	.864
40	.891
42	.914
44	.933
46	.948
48	.961
50	.970

FOR GROUP DISCUSSION

1. Discuss birthdates with your class members. Stop when there is a duplication of dates (if there is one). Based on the table above, would the odds have been in your favor if you had bet that there was at least one duplication?
2. What is the largest number of people that would have to be considered for the probability of at least one duplication to be 100%?
3. Discuss with your class any interesting, common birthday stories you know.

KEY TERMS

7.1	**system of linear equations**	A system of linear equations consists of two or more linear equations with the same variables.
	solution of a system	The solution of a system of linear equations includes all the ordered pairs that make all the equations of the system true at the same time.
	consistent system	A system of equations with a solution is a consistent system.
	inconsistent system	An inconsistent system of equations is a system with no solutions.
	independent equations	Equations of a system that have different graphs are called independent equations.
	dependent equations	Equations of a system that have the same graph (because they are different forms of the same equation) are called dependent equations.
7.5	**system of linear inequalities**	A system of linear inequalities contains two or more linear inequalities (and no other kinds of inequalities).
	solution of a system of linear inequalities	The solution of a system of linear inequalities includes all points that make all inequalities of the system true at the same time.

QUICK REVIEW

Concepts	Examples
7.1 Solving Systems of Linear Equations by Graphing	
An ordered pair is a solution of a system if it makes all equations of the system true at the same time.	Is $(4, -1)$ a solution of the following system? $$x + y = 3$$ $$2x - y = 9$$ Yes because $4 + (-1) = 3$, and $2(4) - (-1) = 9$ are both true.
If the graphs of the equations of a system are both sketched on the same axes, the points of intersection, if any, are solutions of the system.	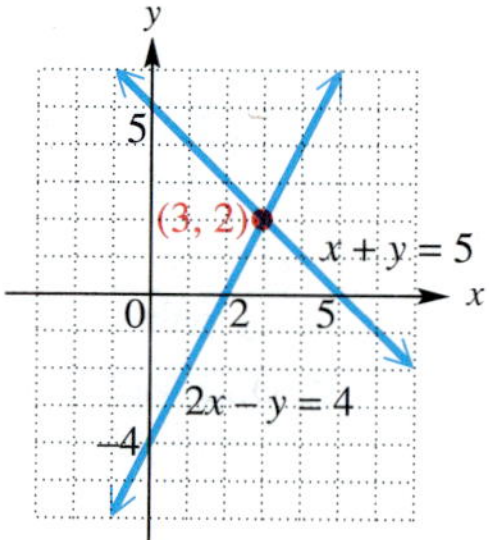 (3, 2) is the solution of the system $$x + y = 5$$ $$2x - y = 4.$$

Concepts	Examples
7.2 Solving Systems of Linear Equations by Addition	
Step 1 Write both equations in the form $Ax + By = C$.	Solve by addition. $x + 3y = 7$, $3x - y = 1$
Step 2 If necessary, multiply one or both equations by appropriate numbers so that the coefficients of x (or y) are negatives of each other.	Multiply the top equation by -3 to eliminate the x terms. $-3x - 9y = -21$, $3x - y = 1$
Step 3 Add the equations to get an equation with only one variable.	$-10y = -20$ Add.
Step 4 Solve the equation from Step 3.	$y = 2$ Divide by -10.
Step 5 Substitute the solution from Step 4 into either of the original equations.	Substitute to get the value of x. $x + 3y = 7$
Step 6 Solve the resulting equation from Step 5 for the remaining variable.	$x + 3(2) = 7$ Let $y = 2$. $x + 6 = 7$ Multiply. $x = 1$ Subtract 6.
Step 7 Check the answer.	The solution, (1, 2), checks.
If the result of the addition step is a false statement, such as $0 = 4$, the graphs are parallel lines, and there is *no solution* for the system.	$x - 2y = 6$, $-x + 2y = -2$; $\mathbf{0 = 4}$ No solution
If the result is a true statement, such as $0 = 0$, the graphs are the *same line,* and an infinite number of ordered pairs are solutions.	$x - 2y = 6$, $-x + 2y = -6$; $\mathbf{0 = 0}$ Infinite number of solutions
7.3 Solving Systems of Linear Equations by Substitution	
Solve one equation for one variable, and substitute the expression into the other equation to get an equation in one variable. Solve the equation, and then substitute the solution into either of the original equations to obtain the other variable. Check the answer.	Solve by substitution. $x + 2y = -5$ **(1)**, $y = -2x - 1$ **(2)**. Substitute $-2x - 1$ for y in equation (1). $x + 2(-2x - 1) = -5$. Solve to get $x = 1$. To find y, let $x = 1$ in equation (2): $y = -2(1) - 1 = -3$. The solution is $(1, -3)$.

Concepts	Examples
7.4 Applications of Linear Systems	
Step 1 Choose a variable to represent each unknown value.	The sum of two numbers is 30. Their difference is 6. Find the numbers. Let x represent one number. Let y represent the other number.
Step 2 Translate the problem into two equations using both variables. *Step 3* Solve the system.	$x + y = 30$ $x - y = 6$ $2x = 36$ Add. $x = 18$ Divide by 2.
Step 4 Answer the question or questions asked.	Let $x = 18$ in the top equation: $18 + y = 30$. Solve to get $y = 12$. The numbers are 18 and 12.
Step 5 Check the solution in the words of the problem.	$18 + 12 = 30$ and $18 - 12 = 6$, so the solution checks.
7.5 Solving Systems of Linear Inequalities	
To solve a system of linear inequalities, graph both inequalities on the same axes. (This was explained in Section 6.5.) The solution of the system is the overlap of the regions of the two graphs. The portions of the boundary lines that bound the region of solutions are included for a $\leq$ or $\geq$ inequality and are excluded for a $<$ or $>$ inequality.	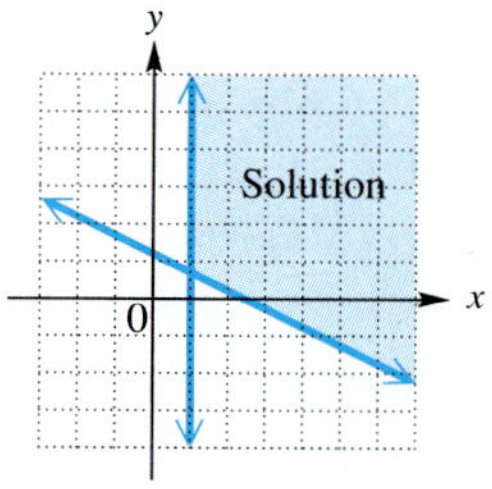 The shaded region is the solution of the system $2x + 4y \geq 5$, $x \geq 1$.

NAME DATE HOUR

CHAPTER 7 REVIEW EXERCISES

[7.1] *Decide whether the given ordered pair is a solution of the given system.*

1. $(3, 4)$
$4x - 2y = 4$
$5x + y = 19$

2. $(1, -3)$
$5x + 3y = -4$
$2x - 3y = 11$

3. $(-5, 2)$
$x - 4y = -13$
$2x + 3y = 4$

4. $(0, 1)$
$3x + 8y = 8$
$4x - 3y = 3$

Solve each system by graphing.

5. $x + y = 4$
$2x - y = 5$

6. $x - 2y = 4$
$2x + y = -2$

7. $x - 2 = 2y$
$2x - 4y = 4$

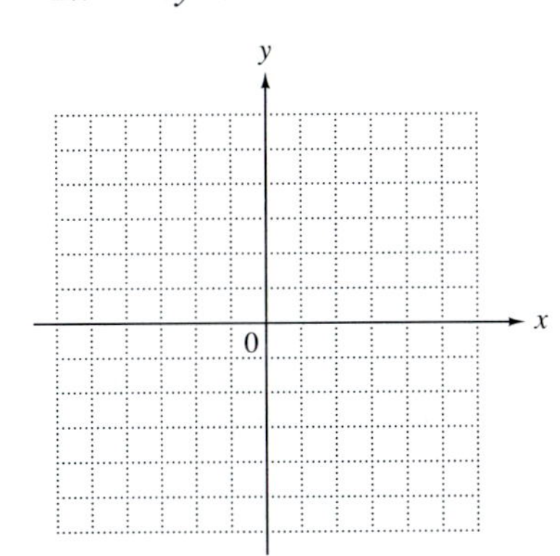

8. $2x + 4 = 2y$
$y - x = -3$

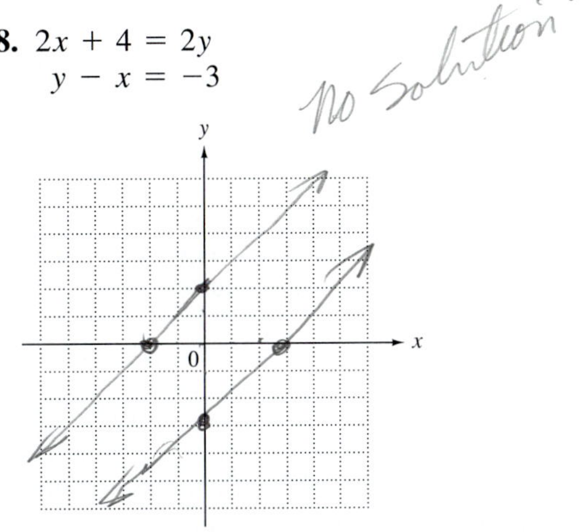

9. Why would a system of linear equations having the solution $(-\frac{1}{2}, \frac{2}{3})$ be difficult to solve using the graphing method?

10. Explain the three methods of solving systems of linear equations described in this chapter.

[7.2] *Solve each system by the addition method.*

11. $2x - y = 13$
$x + y = 8$

12. $3x - y = -13$
$x - 2y = -1$

13. $5x + 4y = -7$
$3x - 4y = -17$

14. $-4x + 3y = 25$
$6x - 5y = -39$

15. $3x - 4y = 9$
$6x - 8y = 18$

16. $2x + y = 3$
$-4x - 2y = 6$

17. After solving a system of linear equations by the addition method, a student obtained the result "0 = 0." He gave the solution of the system as (0, 0). Comment on this answer.

18. What are the three methods of solving systems discussed in this chapter? Choose one, and discuss its drawbacks and advantages.

[7.3] *Solve each system by the substitution method.*

19. $3x + y = 7$
$x = 2y$

20. $2x - 5y = -19$
$y = x + 2$

21. $4x + 5y = 44$
$x + 2 = 2y$

22. $5x + 15y = 3$
$x + 3y = 2$

Solve each system by any method. First simplify equations, and clear them of fractions where necessary.

23. $2x + 3y = -5$
$3x + 4y = -8$

24. $6x - 9y = 0$
$2x - 3y = 0$

25. $2x + y - x = 3y + 5$
$y + 2 = x - 5$

26. $5x - 3 + y = 4y + 8$
$2y + 1 = x - 3$

27. $\frac{x}{2} + \frac{y}{3} = 7$
$\frac{x}{4} + \frac{2y}{3} = 8$

28. $\frac{3x}{4} - \frac{y}{3} = \frac{7}{6}$
$\frac{x}{2} + \frac{2y}{3} = \frac{5}{3}$

[7.4] *Solve each applied problem by any method. Use two variables.*

29. The sum of two numbers is 42, and their difference is 6. Find the numbers.

30. One number is 2 more than twice as large as another. Their sum is 11. Find the numbers.

31. The perimeter of a rectangle is 90 meters. Its length is $1\frac{1}{2}$ times its width. Find the length and width of the rectangle.

32. A cashier has 20 bills, all of which are \$10 or \$20 bills. The total value of the money is \$330. How many of each type does the cashier have?

33. Candy that sells for \$1.30 a pound is to be mixed with candy selling for \$.90 a pound to get 100 pounds of a mix that will sell for \$1 per pound. How much of each type should be used?

34. A 40% antifreeze solution is to be mixed with a 70% solution to get 90 liters of a 50% solution. How many liters of the 40% and 70% solutions will be needed?

35. Mrs. Dawkins can buy 6 apples and 5 bananas for \$5.55. She can also buy 12 apples and 2 bananas for \$7.50. Find the cost of one apple and the cost of one banana.

36. A certain plane flying with the wind travels 540 miles in 2 hours. Later, flying against the same wind, the plane travels 690 miles in 3 hours. Find the speed of the plane in still air and the speed of the wind.

[7.5] *Graph the solution for each system of linear inequalities.*

37. $x + y \geq 2$
$x - y \leq 4$

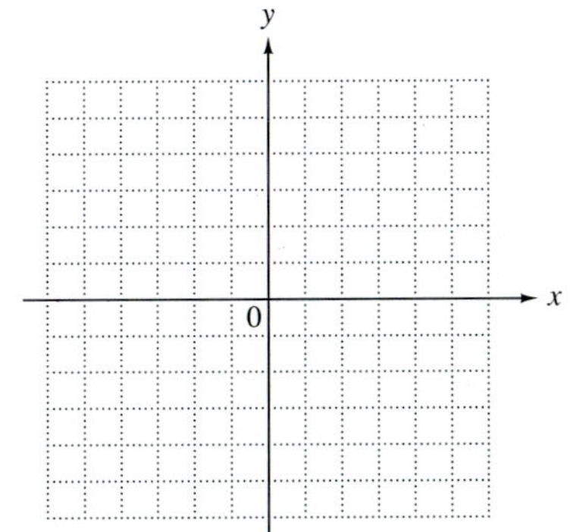

38. $y \geq 2x$
$2x + 3y \leq 6$

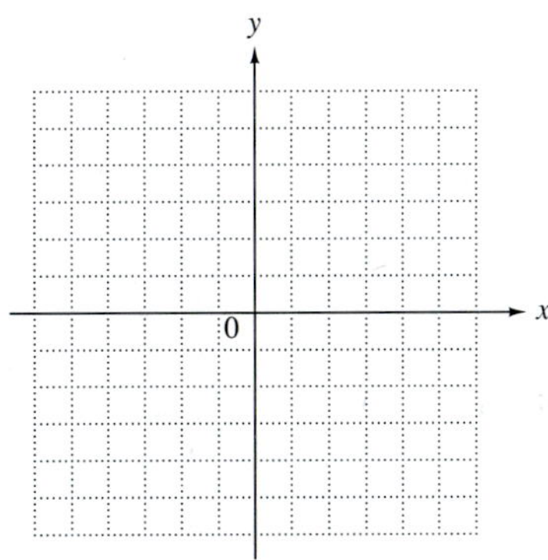

39. $x + y < 3$
$2x > y$

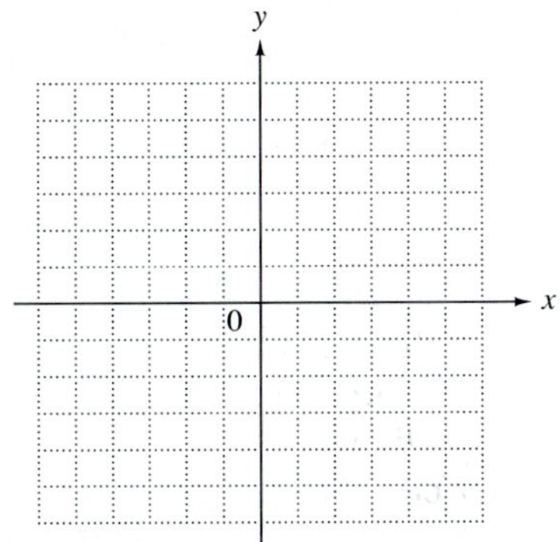

40. $y + 2 \leq 2x$
$x \geq 4$

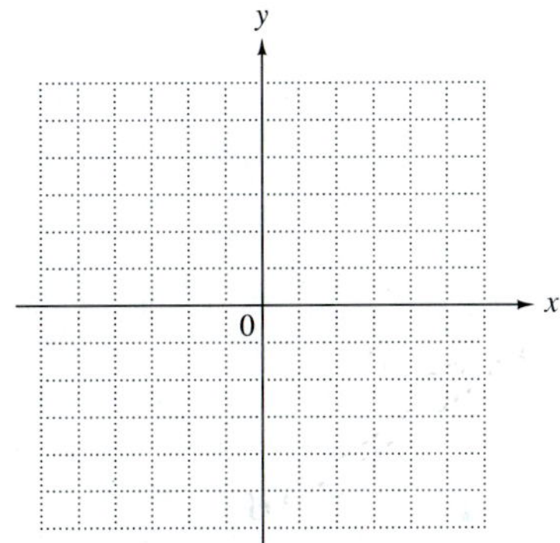

MIXED REVIEW EXERCISES

Solve each of the following exercises using the methods of this chapter.

41. $\frac{2x}{3} + \frac{y}{4} = \frac{14}{3}$
$\frac{x}{2} + \frac{y}{12} = \frac{8}{3}$

42. $x + y = 2y + 6$
$y + 8 = x + 2$

43. $3x + 4y = 6$
$4x - 5y = 8$

44. $\frac{3x}{2} + \frac{y}{5} = -3$
$4x + \frac{y}{3} = -11$

45. $x + y < 5$
$x - y \geq 2$

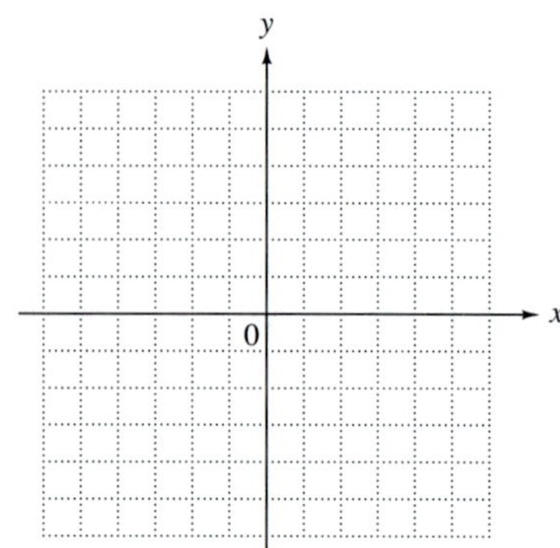

46. $y \leq 2x$
$x + 2y > 4$

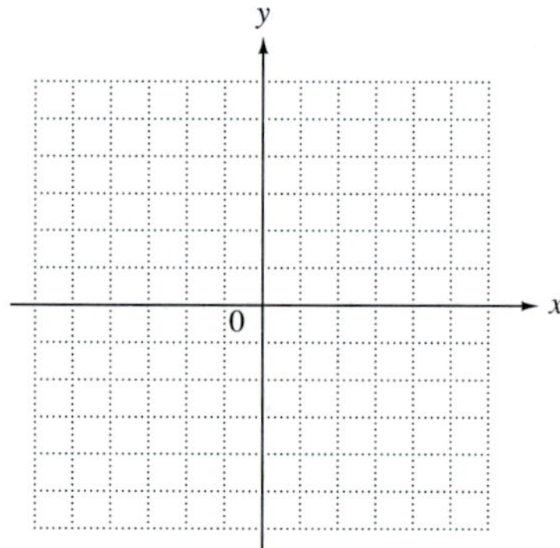

47. $x + 6y = 3$
$2x + 12y = 2$

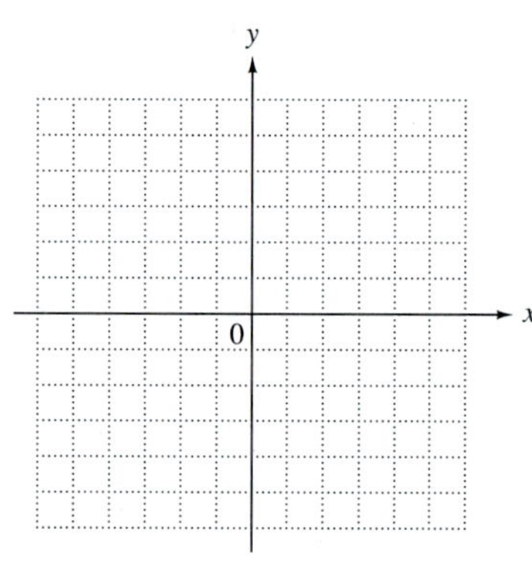

48. Dr. Rolfsen has \$18,000 to invest. He wants the total annual income from the money to be \$650. He can invest part of it at 3% interest and the rest at 4%. How much should he invest at each rate?

49. The perimeter of an isosceles triangle is 29 inches. One side of the triangle is 5 inches longer than each of the two equal sides. Find the lengths of the sides of the triangle.

50. The sum of two numbers is -12, and their difference is 24. Find the numbers.

51. Can the problem in Exercise 49 be worked using a single variable? If so, explain how it can be done, and then solve it.

52. Without actually graphing, determine which one of the following systems of inequalities has no solution.

(a) $x \geq 4$, $y \leq 3$ **(b)** $x + y > 4$, $x + y < 3$ **(c)** $x > 2$, $y < 1$ **(d)** $x + y > 4$, $x - y < 3$

NAME DATE HOUR

CHAPTER 7 TEST

Solve each system by graphing.

1. $2x + y = 1$
   $3x - y = 9$

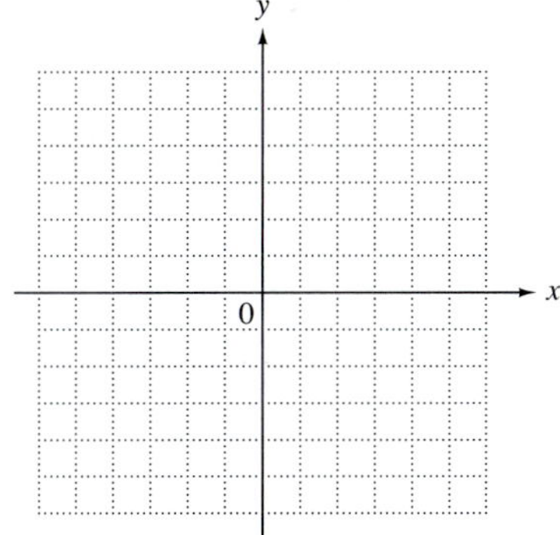

1. ____________________

2. $x + 2y = 6$
   $-2x + y = -7$

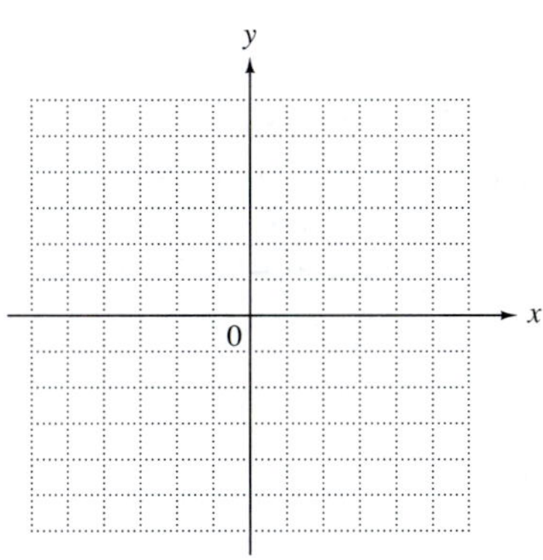

2. ____________________

3. Make up a system of linear equations having $(-3, 4)$ as its only solution. (Hint: start with the solution and work backwards.)

3. ____________________

Solve each system by the addition method.

4. $2x - y = 4$
   $3x + y = 21$

4. ____________________

5. $4x + 2y = 2$
   $5x + 4y = 7$

5. ____________________

6. ______________________

6. $$\begin{aligned} 6x + 5y &= 13 \\ 3x + 2y &= 4 \end{aligned}$$

7. ______________________

7. $$\begin{aligned} 4x + 5y &= 2 \\ -8x - 10y &= 6 \end{aligned}$$

8. ______________________

8. $$\begin{aligned} 6x - 5y &= 0 \\ -2x + 3y &= 0 \end{aligned}$$

9. ______________________

9. $$\begin{aligned} \frac{6}{5}x - \frac{1}{3}y &= -20 \\ -\frac{2}{3}x + \frac{1}{6}y &= 11 \end{aligned}$$

Solve each system by substitution.

10. ______________________

10. $$\begin{aligned} 2x + y &= -4 \\ x &= y + 7 \end{aligned}$$

11. ______________________

11. $$\begin{aligned} 4x + 3y &= -35 \\ x + y &= 0 \end{aligned}$$

Solve each system by any method.

12. ______________________

12. $$\begin{aligned} 8 + 3x - 4y &= 14 - 3y \\ 3x + y + 12 &= 9x - y \end{aligned}$$

13. $\frac{x}{2} - \frac{y}{4} = 7$

$\frac{2x}{3} + \frac{5y}{4} = 3$

13. ______________________

Use a system of equations to solve each applied problem.

14. The sum of two numbers is 18. If the smaller number is doubled, the result is 72 less than the larger number. Find the numbers.

14. ______________________

15. Wally bought a total of 15 compact discs. Some of them cost \$12 each, and the rest cost \$16 each. He paid a total of \$228. How many of each did he buy?

15. ______________________

16. A 40% solution of acid is to be mixed with a 65% solution to get 200 liters of a 45% solution. How many liters of each solution should be used?

16. ______________________

17. ______________________

17. Two cars leave from the same place and travel in opposite directions. One car travels 30 miles per hour faster than the other. After $2\frac{1}{2}$ hours they are 265 miles apart. What are the rates of the cars?

Graph the solution of each system of inequalities.

18.

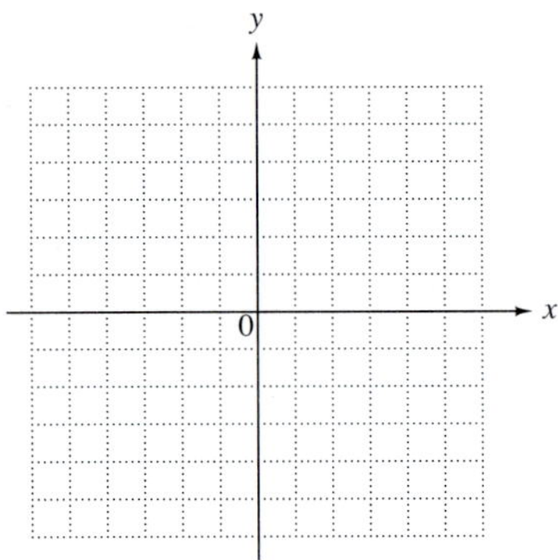

18. $2x + 7y \leq 14$
$x - y \geq 1$

19.

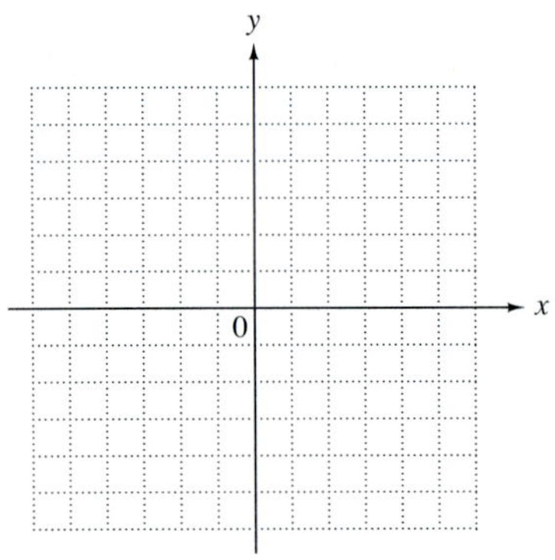

19. $2x - y > 6$
$4y + 12 \geq -3x$

20. ______________________

20. Give an example of a system of linear inequalities that has no solution.

NAME DATE HOUR

CUMULATIVE REVIEW EXERCISES CHAPTERS R-7

1. List all integer factors of 40.

2. Find the value of the expression if $x = 1$ and $y = 5$.

$$\frac{3x^2 + 2y^2}{10y + 3}$$

Name the property that justifies each of the following statements.

3. $5 + (-4) = (-4) + 5$

4. $r(s - k) = rs - rk$

5. $-\frac{2}{3} + \frac{2}{3} = 0$

6. Evaluate $-2 + 6[3 - (4 - 9)]$.

Solve each of the following linear equations.

7. $2 - 3(6x + 2) = 4(x + 1) + 18$

8. $\frac{3}{2}\left(\frac{1}{3}x + 4\right) = 6\left(\frac{1}{4} + x\right)$

Solve each of the following linear inequalities.

9. $-\frac{5}{6}x < 15$

10. $-8 < 2x + 3$

11. If 15 yards of cloth are needed for 18 nurses' smocks, how much cloth would be needed for 45 smocks?

Perform each of the following operations.

12. $-3(-5x^2 + 3x - 10) - (x^2 - 4x + 7)$

13. $(3x - 7)(2y + 4)$

14. $\dfrac{3k^3 + 17k^2 - 27k + 7}{k + 7}$

15. Write in scientific notation: 36,500,000,000.

16. Simplify, and write the answer using only positive exponents: $\left(\dfrac{x^{-4}y^3}{x^2y^4}\right)^{-1}$.

Factor completely.

17. $10m^2 + 7mp - 12p^2$

18. $64t^2 - 48t + 9$

Solve each quadratic equation.

19. $6x^2 - 7x - 3 = 0$

20. $r^2 - 121 = 0$

Perform each operation, and express answers in lowest terms.

21. $\dfrac{-3x + 6}{2x + 4} - \dfrac{-3x - 8}{2x + 4}$

22. $\dfrac{16k^2 - 9}{8k + 6} \div \dfrac{16k^2 - 24k + 9}{6}$

23. Solve the equation $\dfrac{4}{x+1} + \dfrac{3}{x-2} = 4$.

24. Solve the formula $P = \dfrac{kT}{V}$ for T.

Graph each linear equation or inequality.

25. $x - y = 4$

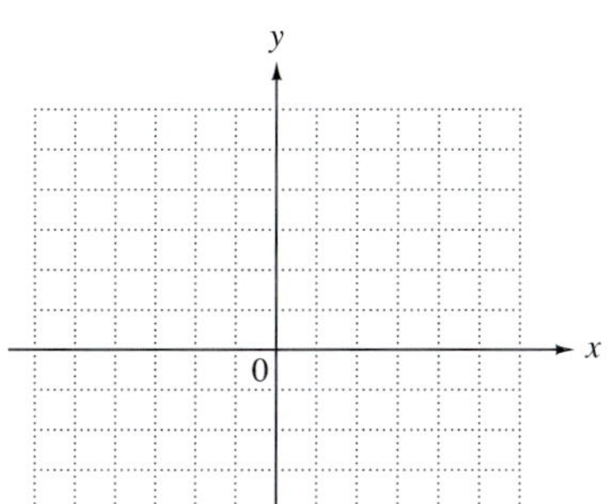

26. $3x + y = 6$

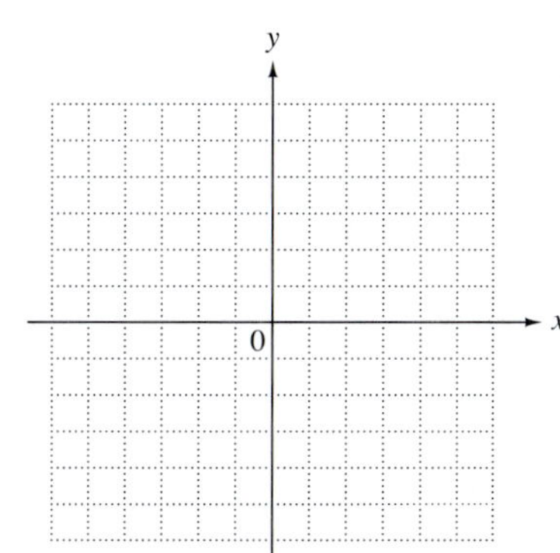

27. $x - 3y > 6$

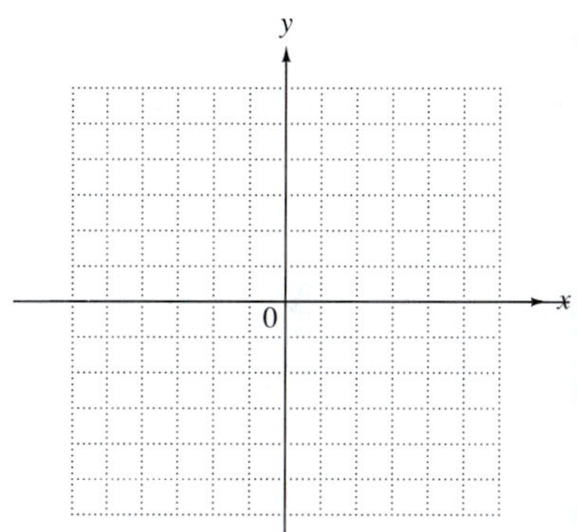

Find the slope of each of the lines described.

28. Through $(-5, 6)$ and $(1, -2)$

29. Perpendicular to the line $y = 4x - 3$

Write an equation for each line described. Express in the form $Ax + By = C$.

30. Through $(2, -5)$ with slope 3

31. Through the points $(0, 4)$ and $(2, 4)$

32. **(a)** Write an equation of the vertical line through $(9, -2)$.

(b) Write an equation of the horizontal line through $(4, -1)$.

33. Is $(-6, 2)$ a solution of the system

$$\begin{aligned} 3x + 2y &= 14 \\ 4x + y &= -26 \end{aligned} \quad ?$$

Solve each system by any method.

34.
$$\begin{aligned} 2x - y &= -8 \\ x + 2y &= 11 \end{aligned}$$

35.
$$\begin{aligned} 4x + 5y &= -8 \\ 3x + 4y &= -7 \end{aligned}$$

36.
$$\begin{aligned} 3x + 5y &= 1 \\ x &= y + 3 \end{aligned}$$

37.
$$\begin{aligned} 3x + 4y &= 2 \\ 6x + 8y &= 1 \end{aligned}$$

Use a system of equations to solve each problem.

38. Two numbers have a sum of 24 and a difference of 6. Find the numbers.

39. A rectangle has a perimeter of 40 meters. The length is 2 meters more than the width. Find the width of the rectangle.

40. The cashier at a small motel has 17 bills, all of which are fives and tens. The value of the money is \$125. How many of each denomination of bill are there?

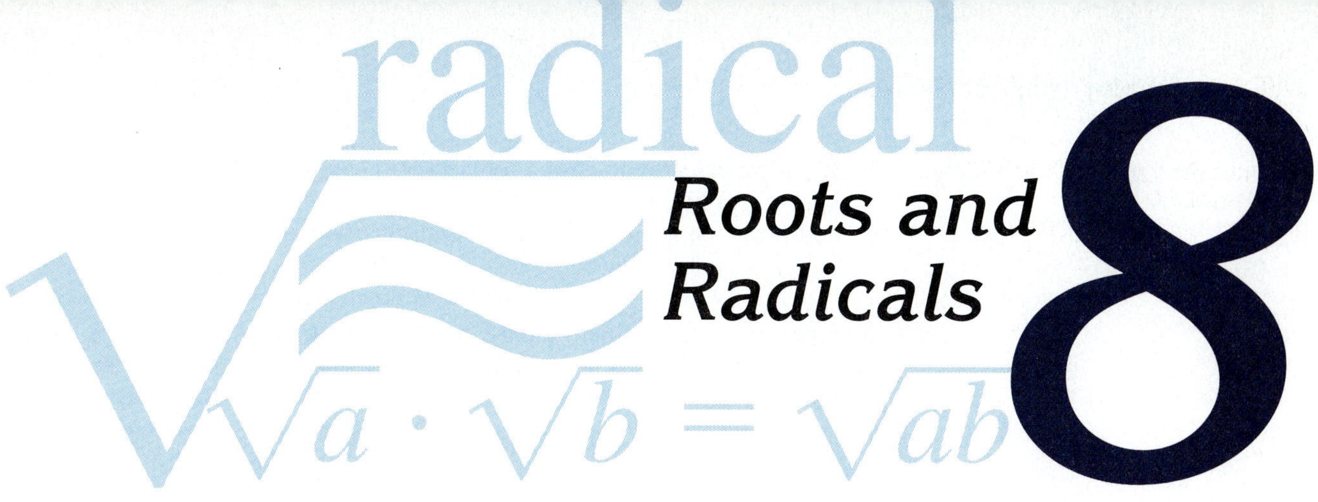

Roots and Radicals 8

8.1 FINDING ROOTS

In Section 1.1, we discussed the idea of the *square* of a number. Recall that squaring a number means multiplying the number by itself.

$$\text{If } a = 8, \quad \text{then} \quad a^2 = 8 \cdot 8 = 64.$$
$$\text{If } a = -4, \quad \text{then} \quad a^2 = (-4)(-4) = 16.$$
$$\text{If } a = -\frac{1}{2}, \quad \text{then} \quad a^2 = \left(-\frac{1}{2}\right)\left(-\frac{1}{2}\right) = \frac{1}{4}.$$

In this chapter, the opposite problem is considered.

$$\text{If } a^2 = 49, \text{ then } \quad a = ?$$
$$\text{If } a^2 = 100, \text{ then } \quad a = ?$$
$$\text{If } a^2 = 25, \text{ then } \quad a = ?$$

OBJECTIVES

1. Find square roots.
2. Decide whether a given root is rational, irrational, or not a real number.
3. Find decimal approximations for irrational square roots.
4. Use the Pythagorean formula.
5. Find higher roots.

FOR EXTRA HELP

Tape 10	SSM pp. 316–319	MAC: B IBM: B

1 Finding a in the three statements above requires finding a number that can be multiplied by itself to result in the given number. The number a is called a **square root** of the number a^2.

EXAMPLE 1 *Finding the Square Roots of a Number*

Find all square roots of 49.

Find a square root of 49 by thinking of a number that multiplied by itself gives 49. One square root is 7 because $7 \cdot 7 = 49$. Another square root of 49 is -7 because $(-7)(-7) = 49$. The number 49 has two square roots 7 and -7. One is positive, and one is negative. ■

WORK PROBLEM 1 AT THE SIDE. ▶▶

1. Find all square roots.

(a) 100

(b) 25

(c) 36

(d) 64

All numbers that have rational number square roots are called **perfect squares.** The positive square root of a number is written with the symbol $\sqrt{}$. For example, the positive square root of 121 is 11, written

$$\sqrt{121} = 11.$$

The symbol $-\sqrt{}$ is used for the negative square root of a number. For example, the negative square root of 121 is -11, written

$$-\sqrt{121} = -11.$$

ANSWERS

1. (a) 10, −10 (b) 5, −5 (c) 6, −6 (d) 8, −8

2. Find the *square* of each radical expression.

(a) $\sqrt{41}$

(b) $-\sqrt{39}$

(c) $\sqrt{2x^2 + 3}$

ANSWERS
2. (a) 41 (b) 39 (c) $2x^2 + 3$

Most calculators have a square root key, usually labeled $\sqrt{x}$, that will allow us to find the square root of a number. For example, if we enter 121 and press the square root key, the display will show 11.

The symbol $\sqrt{}$ is called a **radical sign** and, used alone, always represents the positive square root (except that $\sqrt{0} = 0$). The number inside the radical sign is called the **radicand,** and the entire expression, radical sign and radicand, is called a **radical.** An algebraic expression containing a radical is called a **radical expression.**

If a is a nonnegative real number,

$\sqrt{a}$ is the positive square root of a,

$-\sqrt{a}$ is the negative square root of a.

Also, for nonnegative a,

$$\sqrt{a} \cdot \sqrt{a} = (\sqrt{a})^2 = a \quad \text{and} \quad -\sqrt{a} \cdot -\sqrt{a} = (-\sqrt{a})^2 = a.$$

When the square root of a positive real number is squared, the result is that positive real number. (Also, $(\sqrt{0})^2 = 0$.) This is illustrated in the next example.

EXAMPLE 2 *Squaring Radical Expressions*

Find the *square* of each radical expression.

(a) $\sqrt{13}$

$(\sqrt{13})^2 = 13$, by the definition of square root.

(b) $-\sqrt{29}$

$(-\sqrt{29})^2 = 29$ The square of a *negative* number is positive.

(c) $\sqrt{p^2 + 1}$

$(\sqrt{p^2 + 1})^2 = p^2 + 1$ ■

WORK PROBLEM 2 AT THE SIDE.

EXAMPLE 3 *Finding Square Roots*

Find each square root.

(a) $\sqrt{36}$

We find square roots by factoring to prime factors and looking for pairs of factors.

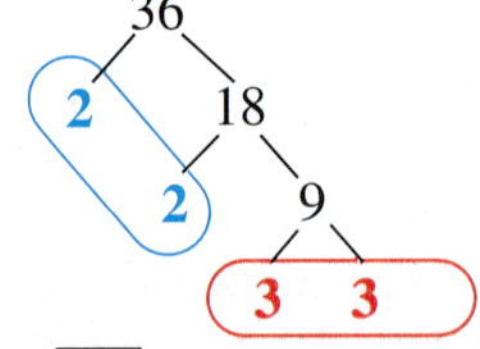

Because the root is 2, circle like numbers in *pairs.* Each pair gives a factor of the square root: $\sqrt{36} = 2 \cdot 3 = 6$.

(b) $-\sqrt{144}$

This symbol represents the negative square root of 144. By finding pairs of prime factors of 144, you should get

$$144 = \underbrace{2 \cdot 2}\cdot\underbrace{2 \cdot 2}\cdot\underbrace{3 \cdot 3}$$

$$-\sqrt{144} = -(2 \cdot 2 \cdot 3) = -12.$$

(c) $-\sqrt{\frac{16}{49}}$

Because $\frac{16}{49} = \frac{4}{7} \cdot \frac{4}{7}$, $-\sqrt{\frac{16}{49}} = -\frac{4}{7}$. ■

WORK PROBLEM 3 AT THE SIDE.

2 A number that is not a perfect square has a square root that is not a rational number. For example, $\sqrt{5}$ is not a rational number because it cannot be written as the ratio of two integers. However, $\sqrt{5}$ is a real number and corresponds to a point on the number line. As mentioned in Chapter 1, a real number that is not rational is called an *irrational number*. The number $\sqrt{5}$ is irrational. Many square roots of integers are irrational.

Not every number has a *real number* square root. For example, there is no real number that can be squared to get -36. (The square of a real number can never be negative.) Because of this, $\sqrt{-36}$ is not a real number. (A calculator will show an error message in a case like this.)

> If a is a negative real number, $\sqrt{a}$ is not a real number.

EXAMPLE 4 *Identifying Types of Square Roots*

Tell whether each square root is rational, irrational, or not a real number.

(a) $\sqrt{17}$

Because 17 is not a perfect square, $\sqrt{17}$ is irrational.

(b) $\sqrt{64}$

The number 64 is a perfect square, 8^2, so $\sqrt{64} = 8$, a rational number.

(c) $\sqrt{-25}$

There is no real number whose square is -25. Therefore, $\sqrt{-25}$ is not a real number. ■

WORK PROBLEM 4 AT THE SIDE.

Note Not all irrational numbers are square roots of integers. For example, π (approximately 3.14159) is an irrational number that is not a square root of any integer.

3 Even if a number is irrational, a decimal that approximates the number can be found by using a calculator.

For example, if we use a calculator to find $\sqrt{10}$, the display will show 3.16227766, which is only a rational approximation of $\sqrt{10}$.

3. Find each square root.

(a) $\sqrt{16}$

(b) $-\sqrt{169}$

(c) $-\sqrt{225}$

(d) $\sqrt{729}$

(e) $\sqrt{\frac{36}{25}}$

4. Tell whether each square root is *rational, irrational,* or *not a real number.*

(a) $\sqrt{9}$

(b) $\sqrt{7}$

(c) $\sqrt{\frac{4}{9}}$

(d) $\sqrt{72}$

(e) $\sqrt{-43}$

ANSWERS

3. (a) 4 (b) -13 (c) -15 (d) 27 (e) $\frac{6}{5}$

4. (a) rational (b) irrational (c) rational (d) irrational (e) not a real number

5. Find a decimal approximation for each square root.

(a) $\sqrt{28}$

(b) $\sqrt{63}$

(c) $\sqrt{190}$

(d) $\sqrt{1000}$

ANSWERS
5. (a) 5.292 (b) 7.937 (c) 13.784 (d) 31.623

■ **EXAMPLE 5** *Approximating Irrational Square Roots*

Use a calculator to find a decimal approximation for each square root. Round answers to the nearest thousandth.

(a) $\sqrt{11}$

Using the square root key of a calculator gives $3.31662479 \approx 3.317$, where $\approx$ means "is approximately equal to."

(b) $\sqrt{39} \approx 6.245$ Use a calculator.

(c) $\sqrt{740} \approx 27.203$ ■

WORK PROBLEM 5 AT THE SIDE.

4 One application of square roots comes from the Pythagorean formula. Recall from Section 4.6 that by this formula if c is the length of the hypotenuse of a right triangle, and a and b are the lengths of the two legs, then

$$a^2 + b^2 = c^2.$$

(See Figure 1.)

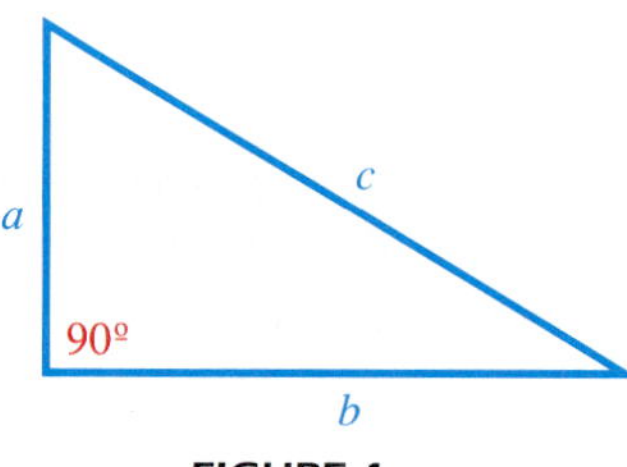

FIGURE 1

■ **EXAMPLE 6** *Using the Pythagorean Formula*

Find the third side of each right triangle with sides a, b, and c, where c is the hypotenuse.

(a) $a = 3, b = 4$

Use the formula to find c^2 first.

$$c^2 = a^2 + b^2$$
$$c^2 = 3^2 + 4^2 \qquad \text{Let } a = 3 \text{ and } b = 4.$$
$$c^2 = 9 + 16 = 25 \qquad \text{Square and add.}$$

Now find the positive square root of 25 to get c.

$$c = \sqrt{25} = 5$$

(Although -5 is also a square root of 25, the length of a side of a triangle must be a positive number.)

(b) $c = 9, b = 5$

Substitute the given values in the formula, $c^2 = a^2 + b^2$. Then solve for a^2.

$$9^2 = a^2 + 5^2 \qquad \text{Let } c = 9 \text{ and } b = 5.$$
$$81 = a^2 + 25 \qquad \text{Square.}$$
$$56 = a^2$$

Use a calculator to find $a = \sqrt{56} \approx 7.483$. ■

Caution Be careful not to make the common mistake of thinking that $\sqrt{a^2 + b^2}$ equals $a + b$. As Example 6(a) shows,

$$\sqrt{9 + 16} = \sqrt{25} \neq \sqrt{9} + \sqrt{16} = 3 + 4,$$

so that, in general,

$$\sqrt{a^2 + b^2} \neq a + b.$$

WORK PROBLEM 6 AT THE SIDE.

The Pythagorean formula can be used to solve applied problems that involve right triangles.

EXAMPLE 7 *Using the Pythagorean Formula*

A ladder 10 feet long leans against a wall. The foot of the ladder is 6 feet from the base of the wall. How high up the wall does the top of the ladder rest?

As shown in Figure 2, a right triangle is formed with the ladder as the hypotenuse.

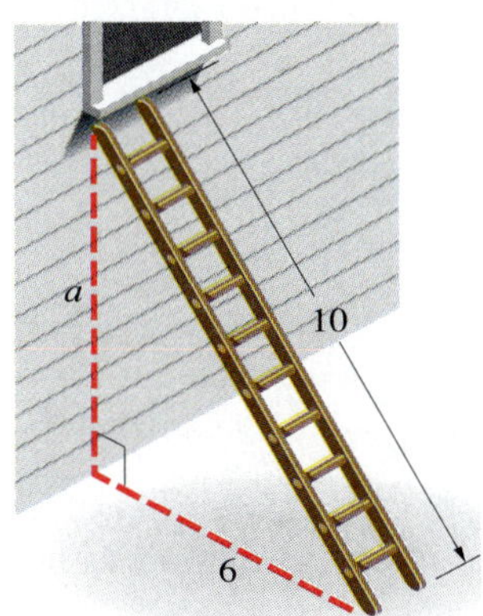

FIGURE 2

Let a represent the height of the top of the ladder. By the Pythagorean formula,

$$\begin{aligned} c^2 &= a^2 + b^2 \\ 10^2 &= a^2 + 6^2 && \text{Let } c = 10 \text{ and } b = 6. \\ 100 &= a^2 + 36 && \text{Square.} \\ 64 &= a^2 && \text{Subtract 36.} \\ \sqrt{64} &= a \\ a &= 8. && \sqrt{64} = 8 \end{aligned}$$

Choose the positive square root of 64 because a represents a length. The top of the ladder rests 8 feet up the wall. ■

WORK PROBLEM 7 AT THE SIDE.

5 Finding the square root of a number is the inverse of squaring a number. In a similar way, there are inverses to finding the cube of a number, or finding the fourth or higher power of a number. These inverses are called finding the **cube root,** written $\sqrt[3]{a}$, the **fourth root,** written $\sqrt[4]{a}$, and so on. In $\sqrt[n]{a}$, the number n is the **index** or **order** of the radical. It would be possible to write $\sqrt[2]{a}$, instead of $\sqrt{a}$, but the simpler symbol $\sqrt{a}$ is customary because the square root is the most commonly used root. A calculator

6. Find the unknown side in each right triangle.

(a) $a = 7$, $b = 24$

(b) $c = 15$, $b = 13$

(c) $c = 11$, $a = 8$

7. A rectangle has dimensions 5 feet by 12 feet. Find the length of its diagonal.

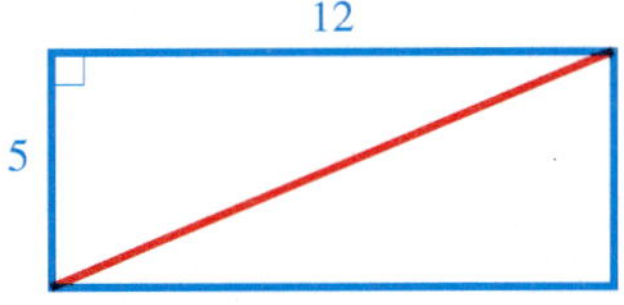

ANSWERS

6. (a) 25 **(b)** $\sqrt{56} \approx 7.483$ **(c)** $\sqrt{57} \approx 7.550$

7. 13 feet

8. Find each cube root.

(a) $\sqrt[3]{27}$

(b) $\sqrt[3]{64}$

(c) $\sqrt[3]{-125}$

9. Find each root.

(a) $\sqrt[4]{81}$

(b) $\sqrt[4]{-81}$

(c) $-\sqrt[4]{625}$

(d) $\sqrt[5]{243}$

(e) $\sqrt[5]{-32}$

ANSWERS
8. (a) 3 (b) 4 (c) −5
9. (a) 3
(b) not a real number
(c) −5 (d) 3 (e) −2

that has a key marked $\sqrt[x]{\ }$ or x^y can be used to find these roots. When working with cube roots or fourth roots, it is helpful to memorize the first few *perfect cubes* ($2^3 = 8$, $3^3 = 27$, and so on), and the first few perfect fourth powers.

■ EXAMPLE 8 *Finding Cube Roots*

Find each cube root.

(a) $\sqrt[3]{8}$

Look for a number that can be cubed to give 8. Because $2^3 = 8$, $\sqrt[3]{8} = 2$.

(b) $\sqrt[3]{-8}$

$\sqrt[3]{-8} = -2$ because $(-2)^3 = -8$.

(c) $\sqrt[3]{216}$

Factor prime factors as we did with square roots.

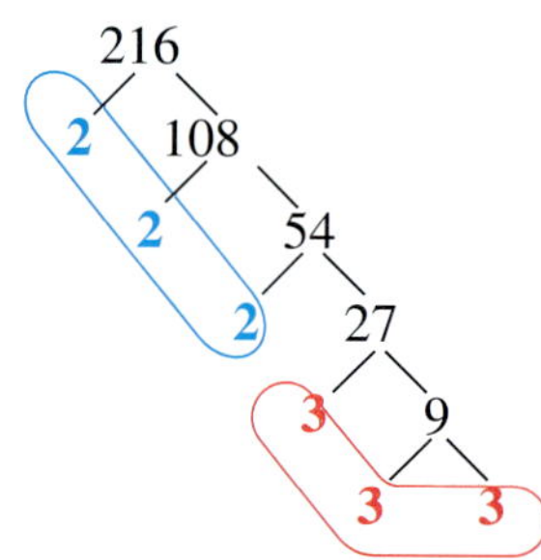

Because the root is 3, circle groups of 3 like factors. Each group gives a factor of the cube root.

Therefore, $\sqrt[3]{216} = 2 \cdot 3 = 6$. ■

As these examples suggest, the cube root of a positive number is positive, and the cube root of a negative number is negative. *There is only one real number cube root for each real number.*

WORK PROBLEM 8 AT THE SIDE.

When the index of the radical is even (square root, fourth root, and so on), the radicand must be nonnegative to get a real number root. Also, for even indexes the symbols $\sqrt{\ }$, $\sqrt[4]{\ }$, $\sqrt[6]{\ }$, and so on are used for the *nonnegative* roots, which are called **principal roots.**

■ EXAMPLE 9 *Finding Higher Roots*

Find each root.

(a) $\sqrt[4]{16}$

$\sqrt[4]{16} = 2$ because 2 is positive and $2^4 = 16$.

(b) $-\sqrt[4]{16} = -2$

(c) $\sqrt[4]{-16}$

There is no real number that equals $\sqrt[4]{-16}$ because a fourth power of a real number must be nonnegative.

(d) $-\sqrt[5]{32}$

First find $\sqrt[5]{32}$. Because 2 is the number whose fifth power is 32, $\sqrt[5]{32} = 2$. If $\sqrt[5]{32} = 2$, then $-\sqrt[5]{32} = -2$. ■

WORK PROBLEM 9 AT THE SIDE.

8.1 EXERCISES

NAME DATE HOUR

1. How many square roots does any positive number have?

2. How many square roots does 0 have?

3. How many real number square roots does any negative number have?

4. How many real number cube roots does any positive number have?

5. How many real number cube roots does any negative number have?

6. How many cube roots does 0 have?

Find all square roots of each number. See Example 1.

7. 16

8. 9

9. 144

10. 225

11. $\frac{25}{196}$

12. $\frac{81}{400}$

13. 900

14. 1600

Find the square of each radical expression. See Example 2.

15. $\sqrt{100}$

16. $\sqrt{36}$

17. $-\sqrt{19}$

18. $-\sqrt{99}$

19. $\sqrt{3x^2 + 4}$

20. $\sqrt{9y^2 + 3}$

What must be true about a *for each statement in Exercises 21–24 to be true?*

21. $\sqrt{a}$ represents a positive number.

22. $-\sqrt{a}$ represents a negative number.

23. $\sqrt{a}$ is not a real number.

24. $-\sqrt{a}$ is not a real number.

Find each square root that is a real number. See Examples 3 and 4(c).

25. $\sqrt{49}$

26. $\sqrt{81}$

27. $-\sqrt{121}$

28. $\sqrt{196}$

29. $-\sqrt{\frac{144}{121}}$

30. $-\sqrt{\frac{49}{36}}$

31. $\sqrt{-121}$

32. $\sqrt{-25}$

Write rational, irrational, *or* not a real number *for each number. If a number is rational, give its exact value. If a number is irrational, give a decimal approximation to the nearest thousandth. Use a calculator as necessary. See Examples 4 and 5.*

33. $\sqrt{25}$

34. $\sqrt{169}$

35. $\sqrt{29}$

36. $\sqrt{33}$

37. $-\sqrt{64}$

38. $-\sqrt{900}$

39. $-\sqrt{300}$

40. $-\sqrt{500}$

41. $\sqrt{-29}$

42. $\sqrt{-47}$

43. Explain why the answers to Exercises 27 and 31 are different.

44. Explain why $\sqrt[3]{-8}$ and $-\sqrt[3]{8}$ represent the same number.

Find the length of the unknown side of each right triangle with sides a, b, and c, where c is the hypotenuse. See Figure 1 and Example 6.

45. $a = 8, b = 15$

46. $a = 24, b = 10$

47. $a = 6, c = 10$

48. $b = 12, c = 13$

49. $a = 11, b = 4$

50. $a = 13, b = 9$

Use the Pythagorean formula to solve each problem. In Exercises 57 and 58, round the answer to the nearest thousandth. See Example 7.

51. The diagonal of a rectangle measures 25 centimeters. The width of the rectangle is 7 centimeters. Find the length of the rectangle.

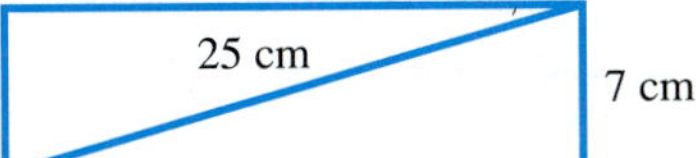

52. The length of a rectangle is 40 meters, and the width is 9 meters. Find the measure of the diagonal of the rectangle.

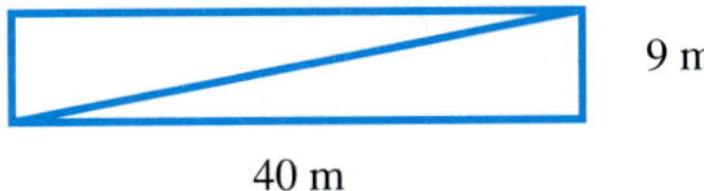

53. Margaret is flying a kite on 100 feet of string. How high is it above her hand (vertically) if the horizontal distance between Margaret and the kite is 60 feet?

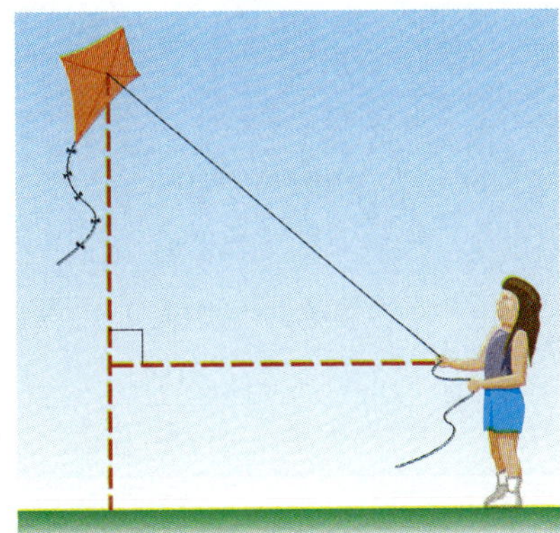

54. A guy wire is attached to the mast of a short-wave transmitting antenna. It is attached 96 feet above ground level. If the wire is staked to the ground 72 feet from the base of the mast, how long is the wire?

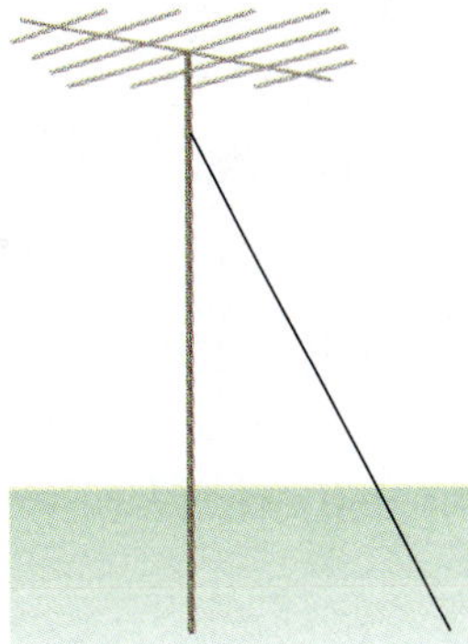

55. Two cars leave Tomball, Texas, at the same time. One travels north at 25 miles per hour, and the other travels west at 60 miles per hour. How far apart are they after 3 hours?

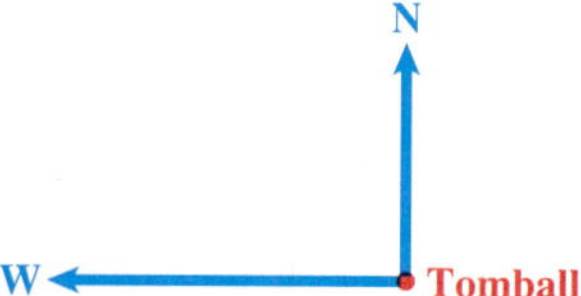

56. A boat is being pulled toward a dock with a rope attached at water level. When the boat is 24 feet from the dock, 30 feet of rope is extended. What is the height of the dock above the water?

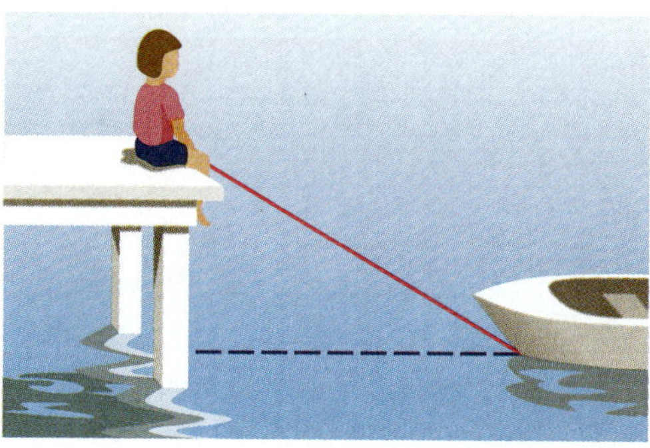

57. What is the value of x in the figure?

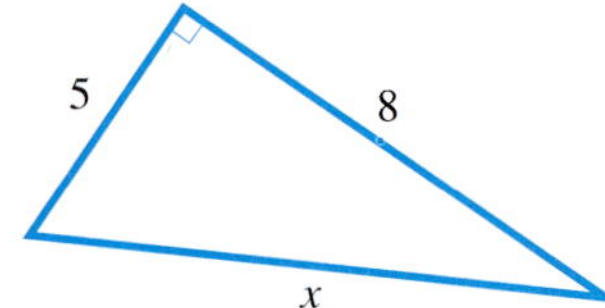

58. What is the value of y in the figure?

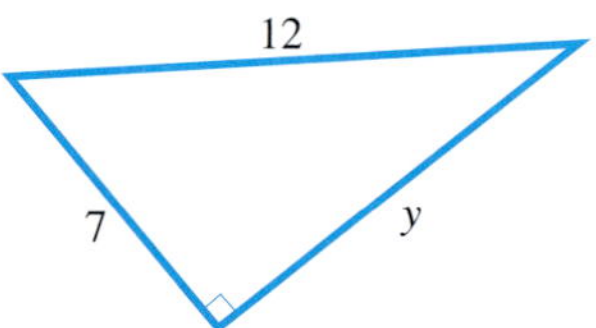

59. Use specific values for a and b different from those given in the "Caution" following Example 6 to show that $\sqrt{a^2 + b^2} \neq a + b$.

60. Why would the values $a = 0$ and $b = 1$ *not* be satisfactory in Exercise 59?

Find each of the following roots that are real numbers. See Examples 8 and 9.

61. $\sqrt[3]{1000}$ **62.** $\sqrt[3]{8}$ **63.** $\sqrt[3]{125}$ **64.** $\sqrt[3]{216}$

65. $\sqrt[3]{-27}$ **66.** $\sqrt[3]{-64}$ **67.** $\sqrt[4]{625}$ **68.** $\sqrt[4]{10{,}000}$

69. $\sqrt[4]{-1}$ **70.** $\sqrt[4]{-625}$ **71.** $-\sqrt[5]{243}$ **72.** $-\sqrt[5]{100{,}000}$

Preview Exercises

Write each number in prime factored form. See Section 4.1.

73. 72 **74.** 100 **75.** 40 **76.** 242

8.2 MULTIPLICATION AND DIVISION OF RADICALS

OBJECTIVES

1. Multiply radicals.
2. Simplify radicals using the product rule.
3. Simplify radical quotients.
4. Simplify higher roots.

FOR EXTRA HELP

Tape 11

SSM pp. 319–322

MAC: B IBM: B

1 We develop several useful rules for finding products and quotients of radicals in this section. To illustrate the rule for products, notice that

$$\sqrt{4} \cdot \sqrt{9} = 2 \cdot 3 = 6 \quad \text{and} \quad \sqrt{4 \cdot 9} = \sqrt{36} = 6,$$

showing that

$$\sqrt{4} \cdot \sqrt{9} = \sqrt{4 \cdot 9}.$$

This result is a particular case of the more general *product rule for radicals.*

PRODUCT RULE FOR RADICALS

For nonnegative real numbers a and b,

$$\sqrt{a} \cdot \sqrt{b} = \sqrt{a \cdot b} \quad \text{and} \quad \sqrt{a \cdot b} = \sqrt{a} \cdot \sqrt{b}.$$

The product of two radicals is the radical of the product, and the radical of a product is the product of the radicals.

EXAMPLE 1 *Using the Product Rule to Multiply Radicals*

Use the product rule for radicals to find each product.

(a) $\sqrt{2} \cdot \sqrt{3} = \sqrt{2 \cdot 3} = \sqrt{6}$

(b) $\sqrt{7} \cdot \sqrt{5} = \sqrt{35}$

(c) $\sqrt{11} \cdot \sqrt{a} = \sqrt{11a}$ Assume $a > 0$. ■

WORK PROBLEM 1 AT THE SIDE. ▶▶

1. Use the product rule for radicals to find each product.

(a) $\sqrt{6} \cdot \sqrt{11}$

(b) $\sqrt{2} \cdot \sqrt{5}$

(c) $\sqrt{10} \cdot \sqrt{r},\ r > 0$

2 A very important use of the product rule is in simplifying radical expressions. As a first step, a radical expression is *simplified* when no perfect square factor remains under the radical sign. This is accomplished by using the product rule in the form $\sqrt{a \cdot b} = \sqrt{a} \cdot \sqrt{b}$. Example 2 shows how a radical may be simplified using the product rule.

EXAMPLE 2 *Using the Product Rule to Simplify Radicals*

Simplify each radical.

(a) $\sqrt{20}$

Because 20 has a perfect square factor of 4, we can write

$$\begin{aligned} \sqrt{20} &= \sqrt{\mathbf{4} \cdot 5} && \text{4 is a perfect square.} \\ &= \sqrt{\mathbf{4}} \cdot \sqrt{5} && \text{Product rule} \\ &= \mathbf{2}\sqrt{5}. && \sqrt{4} = 2 \end{aligned}$$

Thus, $\sqrt{20} = 2\sqrt{5}$. Because 5 has no perfect square factor (other than 1), $2\sqrt{5}$ is called the *simplified form* of $\sqrt{20}$.

We could also factor 20 to prime factors and look for pairs of like factors.

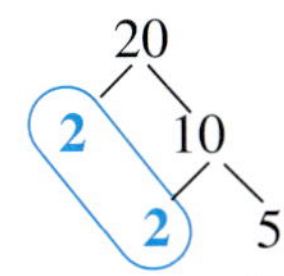

Each pair of like factors produces a rational factor of the simplified form.

Therefore, $\sqrt{20} = \sqrt{2 \cdot 2 \cdot 5} = 2\sqrt{5}$.

ANSWERS
1. (a) $\sqrt{66}$ (b) $\sqrt{10}$ (c) $\sqrt{10r}$

2. Simplify each radical.

(a) $\sqrt{8}$

(b) $\sqrt{27}$

(c) $\sqrt{50}$

(d) $\sqrt{60}$

3. Find each product and simplify.

(a) $\sqrt{3} \cdot \sqrt{15}$

(b) $\sqrt{10} \cdot \sqrt{50}$

(c) $\sqrt{12} \cdot \sqrt{2}$

(d) $\sqrt{7} \cdot \sqrt{14}$

ANSWERS
2. (a) $2\sqrt{2}$ (b) $3\sqrt{3}$ (c) $5\sqrt{2}$ (d) $2\sqrt{15}$
3. (a) $3\sqrt{5}$ (b) $10\sqrt{5}$ (c) $2\sqrt{6}$ (d) $7\sqrt{2}$

(b) $\sqrt{72}$

Notice that 9 is a perfect square factor of 72. We could begin by factoring 72 as $9 \cdot 8$, to get

$$\sqrt{72} = \sqrt{9 \cdot 8} = 3\sqrt{8},$$

but then we would have to factor 8 as $4 \cdot 2$ in order to complete the simplification.

$$\sqrt{72} = 3\sqrt{8} = 3\sqrt{4 \cdot 2} = 3\sqrt{4} \cdot \sqrt{2} = 3 \cdot 2\sqrt{2} = 6\sqrt{2}$$

We could also factor 72 by finding pairs of like factors (not necessarily prime).

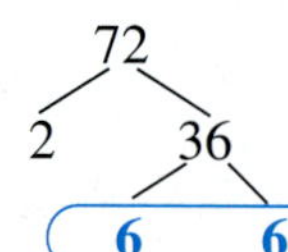

Because $36 = 6 \cdot 6$, we can shortcut the prime factor process.

Thus, $\sqrt{72} = \sqrt{6 \cdot 6 \cdot 2} = 6\sqrt{2}$. In either case, we obtain $6\sqrt{2}$ as the simplified form of $\sqrt{72}$; however, our work is simpler if we begin with the largest perfect square factor.

(c)
$$\begin{aligned} \sqrt{300} &= \sqrt{100 \cdot 3} && \text{100 is a perfect square.} \\ &= \sqrt{100} \cdot \sqrt{3} && \text{Product rule} \\ &= 10\sqrt{3} && \sqrt{100} = 10 \end{aligned}$$

(d) $\sqrt{15}$

The number 15 has no perfect square factors (except 1), so $\sqrt{15}$ cannot be simplified further. ■

WORK PROBLEM 2 AT THE SIDE

Sometimes the product rule can be used to simplify an answer, as Example 3 shows.

■ **EXAMPLE 3** *Multiplying and Simplifying Radicals*

Find each product and simplify.

(a)
$$\begin{aligned} \sqrt{9} \cdot \sqrt{75} &= 3\sqrt{75} && \sqrt{9} = 3 \\ &= 3\sqrt{25 \cdot 3} && \text{25 is a perfect square.} \\ &= 3\sqrt{25} \cdot \sqrt{3} && \text{Product rule} \\ &= 3 \cdot 5\sqrt{3} && \sqrt{25} = 5 \\ &= 15\sqrt{3} && \text{Multiply.} \end{aligned}$$

Notice that we could have used the product rule to get $\sqrt{9} \cdot \sqrt{75} = \sqrt{675}$, and then simplified. However, the product rule as used above allows us to obtain the final answer without using a number as large as 675.

(b)
$$\begin{aligned} \sqrt{8} \cdot \sqrt{12} &= \sqrt{8 \cdot 12} && \text{Product rule} \\ &= \sqrt{8 \cdot 2 \cdot 6} && \text{Factor 12.} \\ &= \sqrt{16 \cdot 6} && \text{16 is a perfect square.} \\ &= \sqrt{16} \cdot \sqrt{6} && \text{Product rule} \\ &= 4\sqrt{6} && \sqrt{16} = 4 \end{aligned}$$ ■

WORK PROBLEM 3 AT THE SIDE.

3 The quotient rule for radicals is very similar to the product rule. It, too, can be used either way.

QUOTIENT RULE FOR RADICALS

If a and b are nonnegative real numbers and b is not 0,

$$\sqrt{\frac{a}{b}} = \frac{\sqrt{a}}{\sqrt{b}} \quad \text{and} \quad \frac{\sqrt{a}}{\sqrt{b}} = \sqrt{\frac{a}{b}}.$$

The radical of a quotient is the quotient of the radicals, and the quotient of two radicals is the radical of the quotient.

EXAMPLE 4 *Using the Quotient Rule to Simplify Radicals*

Simplify each radical.

(a) $\sqrt{\frac{25}{9}} = \frac{\sqrt{25}}{\sqrt{9}} = \frac{5}{3}$ Quotient rule

(b) $\frac{\sqrt{288}}{\sqrt{2}} = \sqrt{\frac{288}{2}} = \sqrt{144} = 12$ Quotient rule

(c) $\sqrt{\frac{3}{4}} = \frac{\sqrt{3}}{\sqrt{4}} = \frac{\sqrt{3}}{2}$ Quotient rule ■

EXAMPLE 5 *Using the Quotient Rule*

Simplify $\frac{27\sqrt{15}}{9\sqrt{3}}$.

Use multiplication of fractions and the quotient rule as follows.

$$\frac{27\sqrt{15}}{9\sqrt{3}} = \frac{27}{9} \cdot \frac{\sqrt{15}}{\sqrt{3}} = \frac{27}{9} \cdot \sqrt{\frac{15}{3}} = 3\sqrt{5}$$ ■

WORK PROBLEM 4 AT THE SIDE.

Some problems require both the product and the quotient rules, as Example 6 shows.

EXAMPLE 6 *Using Both the Product and Quotient Rules*

Simplify $\sqrt{\frac{3}{5}} \cdot \sqrt{\frac{1}{5}}$.

Use the product and quotient rules.

$$\sqrt{\frac{3}{5}} \cdot \sqrt{\frac{1}{5}} = \sqrt{\frac{3}{5} \cdot \frac{1}{5}}$$ Product rule

$$= \sqrt{\frac{3}{25}}$$ Multiply fractions.

$$= \frac{\sqrt{3}}{\sqrt{25}}$$ Quotient rule

$$= \frac{\sqrt{3}}{5}$$ $\sqrt{25} = 5$ ■

WORK PROBLEM 5 AT THE SIDE.

4. Use the quotient rule to simplify each radical.

(a) $\sqrt{\frac{81}{16}}$

(b) $\frac{\sqrt{192}}{\sqrt{3}}$

(c) $\sqrt{\frac{10}{49}}$

(d) $\frac{8\sqrt{50}}{4\sqrt{5}}$

5. Multiply and then simplify each product.

(a) $\sqrt{\frac{5}{6}} \cdot \sqrt{120}$

(b) $\sqrt{\frac{3}{8}} \cdot \sqrt{\frac{7}{2}}$

ANSWERS

4. (a) $\frac{9}{4}$ (b) 8 (c) $\frac{\sqrt{10}}{7}$ (d) $2\sqrt{10}$

5. (a) 10 (b) $\frac{\sqrt{21}}{4}$

Finally, the properties of this section are also valid when variables appear under the radical sign, as long as all the variables represent only nonnegative real numbers. For example, $\sqrt{5^2} = 5$, but $\sqrt{(-5)^2} = \sqrt{25} \neq -5$. This means that the square root of a squared number is always nonnegative. We can use absolute value to express this.

For any real number a,

$$\sqrt{a^2} = |a|$$

In examples and exercises where variables are assumed to be nonnegative, absolute value bars are not necessary, because for $x \geq 0$, $|x| = x$.

EXAMPLE 7 *Simplifying Radicals Involving Variables*

Simplify each radical. Assume all variables represent nonnegative real numbers.

(a) $\sqrt{25m^4} = \sqrt{25} \cdot \sqrt{m^4}$ Product rule

$= 5m^2$

(b) $\sqrt{64p^{10}} = 8p^5$ Product rule

(c) $\sqrt{r^9} = \sqrt{r^8 \cdot r}$

$= \sqrt{r^8} \cdot \sqrt{r} = r^4\sqrt{r}$ Product rule ■

WORK PROBLEM 6 AT THE SIDE.

4 The product rule and the quotient rule for radicals also work for other roots. To simplify cube roots, look for factors that are *perfect cubes.* A **perfect cube** is a number with a rational cube root. For example, $\sqrt[3]{64} = 4$, and because 4 is a rational number, 64 is a perfect cube. Higher roots are handled in a similar manner.

PROPERTIES OF RADICALS

For all real numbers where the indicated roots exist,

$$\sqrt[n]{x} \cdot \sqrt[n]{y} = \sqrt[n]{xy} \quad \text{and} \quad \frac{\sqrt[n]{x}}{\sqrt[n]{y}} = \sqrt[n]{\frac{x}{y}} \quad (y \neq 0).$$

EXAMPLE 8 *Simplifying Higher Roots*

Simplify each radical.

(a) $\sqrt[3]{32} = \sqrt[3]{8 \cdot 4}$ 8 is a perfect cube.

$= \sqrt[3]{8} \cdot \sqrt[3]{4} = 2\sqrt[3]{4}$

(b) $\sqrt[4]{32} = \sqrt[4]{16} \cdot \sqrt[4]{2} = 2\sqrt[4]{2}$ 16 is a perfect fourth power.

(c) $\sqrt[3]{\frac{8}{125}} = \frac{\sqrt[3]{8}}{\sqrt[3]{125}} = \frac{2}{5}$ ■

WORK PROBLEM 7 AT THE SIDE.

6. Simplify each radical. Assume all variables represent nonnegative real numbers.

(a) $\sqrt{36y^6}$

(b) $\sqrt{100p^8}$

(c) $\sqrt{a^5}$

7. Simplify each root.

(a) $\sqrt[3]{108}$

(b) $\sqrt[4]{160}$

(c) $\sqrt[4]{\frac{16}{625}}$

ANSWERS

6. (a) $6y^3$ (b) $10p^4$ (c) $a^2\sqrt{a}$

7. (a) $3\sqrt[3]{4}$ (b) $2\sqrt[4]{10}$ (c) $\frac{2}{5}$

8.2 EXERCISES

NAME DATE HOUR

Decide whether each statement is true or false.

1. $\sqrt{9} \cdot \sqrt{16} = \sqrt{9 \cdot 16}$

2. $\sqrt{9 + 16} = \sqrt{9} + \sqrt{16}$

3. $\sqrt{.5} = \sqrt{\frac{1}{2}}$

4. For nonnegative real numbers x and y, $\sqrt{xy} = \sqrt{x} \cdot \sqrt{y}$.

5. $\sqrt{(-6)^2} = -6$

6. $\sqrt[3]{(-6)^3} = -6$

Use the product rule for radicals to find each product. See Example 1.

7. $\sqrt{3} \cdot \sqrt{27}$

8. $\sqrt{2} \cdot \sqrt{8}$

9. $\sqrt{6} \cdot \sqrt{15}$

10. $\sqrt{10} \cdot \sqrt{15}$

11. $\sqrt{13} \cdot \sqrt{13}$

12. $\sqrt{17} \cdot \sqrt{17}$

13. $\sqrt{13} \cdot \sqrt{r}$, $r \geq 0$

14. $\sqrt{19} \cdot \sqrt{k}$, $k \geq 0$

15. Which one of the following radicals is simplified according to the guidelines of Objective 2?
(a) $\sqrt{47}$ **(b)** $\sqrt{45}$ **(c)** $\sqrt{48}$ **(d)** $\sqrt{44}$

16. If p is a prime number, is $\sqrt{p}$ in simplified form? Explain your answer.

Simplify each radical. See Example 2.

17. $\sqrt{45}$ **18.** $\sqrt{27}$ **19.** $\sqrt{90}$ **20.** $\sqrt{56}$

21. $\sqrt{75}$ **22.** $\sqrt{18}$ **23.** $\sqrt{125}$ **24.** $\sqrt{80}$

25. $-\sqrt{700}$ **26.** $-\sqrt{600}$ **27.** $3\sqrt{27}$ **28.** $9\sqrt{8}$

Find each product and simplify. See Example 3.

29. $\sqrt{3} \cdot \sqrt{18}$ **30.** $\sqrt{3} \cdot \sqrt{21}$ **31.** $\sqrt{12} \cdot \sqrt{48}$

32. $\sqrt{50} \cdot \sqrt{72}$ **33.** $\sqrt{12} \cdot \sqrt{30}$ **34.** $\sqrt{30} \cdot \sqrt{24}$

35. Simplify the product $\sqrt{8} \cdot \sqrt{32}$ in two different ways. First, multiply 8 by 32 and simplify the square root of this product. Second, simplify $\sqrt{8}$ and simplify $\sqrt{32}$ and then multiply. Do you get the same answer? Make a conjecture (an educated guess) about whether the correct answer can always be obtained using either method in a simplification such as this.

36. Simplify the radical $\sqrt{288}$ in two ways. First, factor 288 as $144 \cdot 2$ and then simplify. Second, factor 288 as $48 \cdot 6$ and then simplify, performing any additional steps as needed. Do you get the same answer? Make a conjecture concerning the quickest way to simplify such a radical.

Use the quotient rule and the product rule, as necessary, to simplify each of the following. See Examples 4–6.

37. $\sqrt{\dfrac{16}{225}}$

38. $\sqrt{\dfrac{9}{100}}$

39. $\sqrt{\dfrac{7}{16}}$

40. $\sqrt{\dfrac{13}{25}}$

41. $\sqrt{\dfrac{5}{7}} \cdot \sqrt{35}$

42. $\sqrt{\dfrac{10}{13}} \cdot \sqrt{130}$

43. $\sqrt{\dfrac{5}{2}} \cdot \sqrt{\dfrac{125}{8}}$

44. $\sqrt{\dfrac{8}{3}} \cdot \sqrt{\dfrac{512}{27}}$

45. $\dfrac{30\sqrt{10}}{5\sqrt{2}}$

46. $\dfrac{50\sqrt{20}}{2\sqrt{10}}$

Simplify each radical. Assume that all variables represent nonnegative real numbers. See Example 7.

47. $\sqrt{m^2}$

48. $\sqrt{k^2}$

49. $\sqrt{y^4}$

50. $\sqrt{s^4}$

51. $\sqrt{36z^2}$

52. $\sqrt{49n^2}$

53. $\sqrt{400x^6}$

54. $\sqrt{900y^8}$

55. $\sqrt{z^5}$

56. $\sqrt{a^{13}}$

57. $\sqrt{x^6y^{12}}$

58. $\sqrt{a^8b^{10}}$

Simplify each radical. See Example 8.

59. $\sqrt[3]{40}$

60. $\sqrt[3]{48}$

61. $\sqrt[3]{54}$

62. $\sqrt[3]{135}$

63. $\sqrt[3]{128}$

64. $\sqrt[3]{192}$

65. $\sqrt[4]{80}$

66. $\sqrt[4]{243}$

67. $\sqrt[3]{\frac{8}{27}}$

68. $\sqrt[3]{\frac{64}{125}}$

69. $\sqrt[3]{-\frac{216}{125}}$

70. $\sqrt[3]{-\frac{1}{64}}$

71. In Example 2(a) we showed *algebraically* that $\sqrt{20}$ is equal to $2\sqrt{5}$. To give *numerical support* to this result, use a calculator to do the following:

(a) Find a decimal approximation for $\sqrt{20}$ using your calculator. Record as many digits as the calculator shows.

(b) Find a decimal approximation for $\sqrt{5}$ using your calculator, and then multiply the result by 2. Record as many digits as the calculator shows.

(c) Your results in parts (a) and (b) should be the same. A mathematician would not accept this numerical exercise as *proof* that $\sqrt{20}$ is equal to $2\sqrt{5}$. Can you explain why?

72. On your calculator, multiply the approximations for $\sqrt{3}$ and $\sqrt{5}$. Now, predict what your calculator will show when you find an approximation for $\sqrt{15}$. What rule stated in this section justifies your answer?

PREVIEW EXERCISES

Combine like terms. See Section 1.9.

73. $4x + 7 - 9x + 12$

74. $9x^2 + 3x^2 - 2x + 4x - 8 + 1$

75. $2xy + 3x^2y - 9xy + 8x^2y$

76. $x + 3y + 12z$

8.3 ADDITION AND SUBTRACTION OF RADICALS

1 We add or subtract radical expressions by using the distributive property. For example,

$$8\sqrt{3} + 6\sqrt{3} = (8 + 6)\sqrt{3} = 14\sqrt{3}.$$

Also,

$$2\sqrt{11} - 7\sqrt{11} = -5\sqrt{11}.$$

Only **like radicals,** those that are multiples of the same root of the same number, can be combined in this way.

EXAMPLE 1 *Adding and Subtracting Like Radicals*
Add or subtract, as indicated.

(a) $3\sqrt{6} + 5\sqrt{6} = (3 + 5)\sqrt{6} = 8\sqrt{6}$ Distributive property

(b) $5\sqrt{10} - 7\sqrt{10} = (5 - 7)\sqrt{10} = -2\sqrt{10}$

(c) $\sqrt{7} + 2\sqrt{7} = 1\sqrt{7} + 2\sqrt{7} = 3\sqrt{7}$

(d) $\sqrt{5} + \sqrt{5} = 1\sqrt{5} + 1\sqrt{5} = (1 + 1)\sqrt{5} = 2\sqrt{5}$

(e) $\sqrt{3} + \sqrt{7}$ Cannot be further simplified ■

WORK PROBLEM 1 AT THE SIDE.

2 Sometimes each radical expression in a sum or difference must be simplified first. Doing this might cause like radicals to appear, which then can be added or subtracted.

EXAMPLE 2 *Adding and Subtracting Radicals that Require Simplification*
Add or subtract as indicated.

(a)
$$\begin{aligned} 3\sqrt{2} + \sqrt{8} &= 3\sqrt{2} + \sqrt{4 \cdot 2} && \text{Simplify } \sqrt{8}. \\ &= 3\sqrt{2} + \sqrt{4} \cdot \sqrt{2} && \text{Product rule} \\ &= 3\sqrt{2} + 2\sqrt{2} && \sqrt{4} = 2 \\ &= 5\sqrt{2} && \text{Add like radicals.} \end{aligned}$$

(b)
$$\begin{aligned} \sqrt{18} - \sqrt{27} &= \sqrt{9 \cdot 2} - \sqrt{9 \cdot 3} && \text{Simplify } \sqrt{18} \text{ and } \sqrt{27}. \\ &= \sqrt{9} \cdot \sqrt{2} - \sqrt{9} \cdot \sqrt{3} && \text{Product rule} \\ &= 3\sqrt{2} - 3\sqrt{3} && \sqrt{9} = 3 \end{aligned}$$

Because $\sqrt{2}$ and $\sqrt{3}$ are unlike radicals, this difference cannot be simplified further.

(c)
$$\begin{aligned} 2\sqrt{12} + 3\sqrt{75} &= 2(\sqrt{4} \cdot \sqrt{3}) + 3(\sqrt{25} \cdot \sqrt{3}) && \text{Product rule} \\ &= 2(2\sqrt{3}) + 3(5\sqrt{3}) && \sqrt{4} = 2 \text{ and } \sqrt{25} = 5 \\ &= 4\sqrt{3} + 15\sqrt{3} && \text{Multiply.} \\ &= 19\sqrt{3} && \text{Add like radicals.} \end{aligned}$$ ■

WORK PROBLEM 2 AT THE SIDE.

3 Some radical expressions require both multiplication and addition (or subtraction). The order of operations presented earlier still applies.

OBJECTIVES

1. Add and subtract radical expressions.
2. Simplify radical expressions before adding or subtracting.
3. Simplify radical expressions involving multiplication.

FOR EXTRA HELP

Tape	SSM	MAC: B
11	pp. 323–324	IBM: B

1. Add or subtract, as indicated.

(a) $8\sqrt{5} + 2\sqrt{5}$

(b) $-4\sqrt{3} + 9\sqrt{3}$

(c) $12\sqrt{11} - 3\sqrt{11}$

(d) $\sqrt{15} + \sqrt{15}$

(e) $2\sqrt{7} + 2\sqrt{10}$

2. Add or subtract as indicated.

(a) $\sqrt{8} + 4\sqrt{2}$

(b) $\sqrt{27} + \sqrt{12}$

(c) $5\sqrt{200} - 6\sqrt{18}$

ANSWERS
1. (a) $10\sqrt{5}$ (b) $5\sqrt{3}$ (c) $9\sqrt{11}$ (d) $2\sqrt{15}$ (e) cannot be simplified further
2. (a) $6\sqrt{2}$ (b) $5\sqrt{3}$ (c) $32\sqrt{2}$

3. Multiply and combine terms. Assume all variables represent nonnegative real numbers.

(a) $\sqrt{7} \cdot \sqrt{21} + 2\sqrt{27}$

(b) $\sqrt{3r} \cdot \sqrt{6} + \sqrt{8r}$

(c) $\sqrt[3]{81x^4} + 5\sqrt[3]{24x^4}$

ANSWERS
3. **(a)** $13\sqrt{3}$ **(b)** $5\sqrt{2r}$ **(c)** $13x\sqrt[3]{3x}$

■ **EXAMPLE 3** *Multiplying and Combining Terms in Radical Expressions*

Multiply and combine terms. Assume all variables represent nonnegative real numbers.

(a)

$$\begin{aligned}
\sqrt{5} \cdot \sqrt{15} + 4\sqrt{3} &= \sqrt{5 \cdot 15} + 4\sqrt{3} && \text{Product rule} \\
&= \sqrt{75} + 4\sqrt{3} && \text{Multiply.} \\
&= \sqrt{25 \cdot 3} + 4\sqrt{3} && \text{25 is a perfect square.} \\
&= \sqrt{25} \cdot \sqrt{3} + 4\sqrt{3} && \text{Product rule} \\
&= 5\sqrt{3} + 4\sqrt{3} && \sqrt{25} = 5 \\
&= 9\sqrt{3} && \text{Add like radicals.}
\end{aligned}$$

(b)

$$\begin{aligned}
\sqrt{2} \cdot \sqrt{6k} + \sqrt{27k} &= \sqrt{12k} + \sqrt{27k} && \text{Product rule} \\
&= \sqrt{4 \cdot 3k} + \sqrt{9 \cdot 3k} && \text{Factor.} \\
&= \sqrt{4} \cdot \sqrt{3k} + \sqrt{9} \cdot \sqrt{3k} && \text{Product rule} \\
&= 2\sqrt{3k} + 3\sqrt{3k} && \sqrt{4} = 2 \text{ and } \sqrt{9} = 3 \\
&= 5\sqrt{3k} && \text{Add like radicals.}
\end{aligned}$$

(c)

$$\begin{aligned}
2\sqrt[3]{32m^3} - \sqrt[3]{108m^3} &= 2\sqrt[3]{(8m^3)4} - \sqrt[3]{(27m^3)4} && \text{Product rule} \\
&= 2(2m)\sqrt[3]{4} - 3m\sqrt[3]{4} && \sqrt[3]{8m^3} = 2m;\ \sqrt[3]{27m^3} = 3m \\
&= 4m\sqrt[3]{4} - 3m\sqrt[3]{4} && \text{Multiply.} \\
&= m\sqrt[3]{4} && \text{Subtract like radicals.}
\end{aligned}$$

■

Caution A sum or difference of radicals can be simplified only if the radicals are **like radicals.** For example, $\sqrt{5} + 3\sqrt{5} = 4\sqrt{5}$, but $\sqrt{5} + 5\sqrt{3}$ cannot be simplified further. Also, $2\sqrt{3} + 5\sqrt[3]{3}$ cannot be simplified further.

◀ **WORK PROBLEM 3 AT THE SIDE.**

8.3 EXERCISES

NAME DATE HOUR

1. The distributive property, which says $a(b + c) = ab + ac$ and $ba + ca = (b + c)a$, provides the justification for adding and subtracting like radicals. While we usually skip the step that indicates this property, we could not make the statement $2\sqrt{3} + 4\sqrt{3} = 6\sqrt{3}$ without it. Write an equation showing how the distributive property is actually used in this statement.

2. In Example 1(e), we state that $\sqrt{3} + \sqrt{7}$ cannot be further simplified. Why is this so? Show, by using calculator approximations, that $\sqrt{3} + \sqrt{7}$ is *not* equal to $\sqrt{10}$.

Simplify and add or subtract wherever possible. See Examples 1 and 2.

3. $2\sqrt{3} + 6\sqrt{3}$

4. $6\sqrt{5} + 18\sqrt{5}$

5. $14\sqrt{7} - 19\sqrt{7}$

6. $16\sqrt{2} - 18\sqrt{2}$

7. $\sqrt{17} + 4\sqrt{17}$

8. $5\sqrt{19} + \sqrt{19}$

9. $6\sqrt{7} - \sqrt{7}$

10. $11\sqrt{14} - \sqrt{14}$

11. $\sqrt{45} + 4\sqrt{20}$

12. $\sqrt{24} + 6\sqrt{54}$

13. $5\sqrt{72} - 3\sqrt{50}$

14. $6\sqrt{18} - 5\sqrt{32}$

15. $-5\sqrt{32} + 2\sqrt{98}$

16. $-4\sqrt{75} + 3\sqrt{12}$

17. $5\sqrt{7} - 3\sqrt{28} + 6\sqrt{63}$

18. $3\sqrt{11} + 5\sqrt{44} - 8\sqrt{99}$

19. $2\sqrt{8} - 5\sqrt{32} - 2\sqrt{48}$

20. $5\sqrt{72} - 3\sqrt{48} + 4\sqrt{128}$

21. $4\sqrt{50} + 3\sqrt{12} - 5\sqrt{45}$

22. $6\sqrt{18} + 2\sqrt{48} + 6\sqrt{28}$

23. $\frac{1}{4}\sqrt{288} + \frac{1}{6}\sqrt{72}$

24. $\frac{2}{3}\sqrt{27} + \frac{3}{4}\sqrt{48}$

Perform the indicated operations. Assume that all variables represent nonnegative real numbers. See Example 3.

25. $\sqrt{3} \cdot \sqrt{7} + 4\sqrt{21}$

26. $\sqrt{13} \cdot \sqrt{2} + 7\sqrt{26}$

27. $\sqrt{6} \cdot \sqrt{2} + 9\sqrt{3}$

28. $4\sqrt{15} \cdot \sqrt{3} + 4\sqrt{5}$

29. $\sqrt{9x} + \sqrt{49x} - \sqrt{25x}$

30. $\sqrt{4a} - \sqrt{16a} + \sqrt{100a}$

31. $\sqrt{6x^2} + x\sqrt{24}$

32. $\sqrt{75x^2} + x\sqrt{108}$

33. $3\sqrt{8x^2} - 4x\sqrt{2} - x\sqrt{8}$

34. $\sqrt{2b^2} + 3b\sqrt{18} - b\sqrt{200}$

35. $-8\sqrt{32k} + 6\sqrt{8k}$

36. $4\sqrt{12x} + 2\sqrt{27x}$

37. $2\sqrt{125x^2z} + 8x\sqrt{80z}$

38. $\sqrt{48x^2y} + 5x\sqrt{27y}$

39. $4\sqrt[3]{16} - 3\sqrt[3]{54}$

40. $5\sqrt[3]{128} + 3\sqrt[3]{250}$

41. $6\sqrt[3]{8p^2} - 2\sqrt[3]{27p^2}$

42. $8k\sqrt[3]{54k} + 6\sqrt[3]{16k^4}$

43. $5\sqrt[4]{m^3} + 8\sqrt[4]{16m^3}$

44. $5\sqrt[4]{m^5} + 3\sqrt[4]{81m^5}$

45. Despite the fact that $\sqrt{25}$ and $\sqrt[3]{8}$ are radicals that have different root indexes, they can be added to obtain a single term: $\sqrt{25} + \sqrt[3]{8} = 5 + 2 = 7$. Make up a similar sum of radicals that leads to an answer of 10.

46. In the directions for Exercises 25–44, we made the assumption that all variables represent nonnegative real numbers. However, in Exercises 41 and 42, variables actually *may* represent negative numbers. Explain why this is so.

PREVIEW EXERCISES

Perform each operation. See Section 8.1.

47. $(\sqrt{6})^2$

48. $(\sqrt{25})^2$

49. $(-\sqrt{37})^2$

Simplify each radical. See Section 8.2.

50. $\sqrt{288}$

51. $\sqrt{7500}$

52. $\sqrt{x^2y^6}$, $x \geq 0$, $y \geq 0$

8.4 RATIONALIZING THE DENOMINATOR

OBJECTIVES

1. Rationalize denominators with square roots.
2. Write radicals in simplified form.
3. Rationalize denominators with cube roots.

FOR EXTRA HELP

Tape	SSM	MAC: B
11	pp. 324–327	IBM: B

1 We found decimal approximations for radicals in the first section of this chapter. For more complicated radical expressions, it is easier to find these decimals if the denominators do not contain any radicals. For example, the radical in the denominator of

$$\frac{\sqrt{3}}{\sqrt{2}}$$

can be eliminated by multiplying the numerator and the denominator by $\sqrt{2}$.

$$\frac{\sqrt{3}}{\sqrt{2}} = \frac{\sqrt{3} \cdot \sqrt{2}}{\sqrt{2} \cdot \sqrt{2}} = \frac{\sqrt{6}}{2} \qquad \sqrt{2} \cdot \sqrt{2} = \sqrt{4} = 2$$

This process of changing the denominator from a radical (irrational number) to a rational number is called **rationalizing the denominator.** The value of the radical expression is not changed; only the form is changed, because the expression has been multiplied by 1 in the form of $\sqrt{2}/\sqrt{2}$.

EXAMPLE 1 *Rationalizing Denominators*

Rationalize each denominator.

(a) $\dfrac{9}{\sqrt{6}}$

$$\frac{9}{\sqrt{6}} = \frac{9 \cdot \sqrt{6}}{\sqrt{6} \cdot \sqrt{6}} \qquad \text{Multiply numerator and denominator by } \sqrt{6}.$$

$$= \frac{9\sqrt{6}}{6} \qquad \sqrt{6} \cdot \sqrt{6} = 6$$

$$= \frac{3\sqrt{6}}{2} \qquad \text{Lowest terms}$$

(b) $\dfrac{12}{\sqrt{8}}$

The denominator could be rationalized here by multiplying by $\sqrt{8}$. However, the result can be found more directly by multiplying numerator and denominator by $\sqrt{2}$. This is because $\sqrt{8} \cdot \sqrt{2} = \sqrt{16} = 4$, a rational number.

$$\frac{12}{\sqrt{8}} = \frac{12 \cdot \sqrt{2}}{\sqrt{8} \cdot \sqrt{2}} \qquad \text{Multiply by } \sqrt{2} \text{ in numerator and denominator.}$$

$$= \frac{12\sqrt{2}}{\sqrt{16}} \qquad \text{Product rule}$$

$$= \frac{12\sqrt{2}}{4} \qquad \sqrt{16} = 4$$

$$= 3\sqrt{2} \qquad \text{Lowest terms}$$

1. Rationalize each denominator.

(a) $\frac{3}{\sqrt{5}}$

(b) $\frac{-6}{\sqrt{11}}$

(c) $-\frac{\sqrt{7}}{\sqrt{2}}$

(d) $\frac{20}{\sqrt{18}}$

2. Simplify by rationalizing each denominator.

(a) $\sqrt{\frac{16}{11}}$

(b) $\sqrt{\frac{5}{18}}$

(c) $\sqrt{\frac{8}{32}}$

ANSWERS

1. (a) $\frac{3\sqrt{5}}{5}$ (b) $\frac{-6\sqrt{11}}{11}$ (c) $-\frac{\sqrt{14}}{2}$ (d) $\frac{10\sqrt{2}}{3}$

2. (a) $\frac{4\sqrt{11}}{11}$ (b) $\frac{\sqrt{10}}{6}$ (c) $\frac{1}{2}$

WORK PROBLEM 1 AT THE SIDE.

2 A radical is considered to be in simplified form if the following three conditions are met.

SIMPLIFIED FORM OF A RADICAL

1. The radicand contains no factor (except 1) that is a perfect square.
2. The radicand has no fractions.
3. No denominator contains a radical.

In the following examples, radicals are simplified according to these conditions.

EXAMPLE 2 *Simplifying a Radical*

Simplify $\sqrt{\frac{27}{5}}$.

This violates condition 2. To begin, use the quotient rule for radicals.

$$\sqrt{\frac{27}{5}} = \frac{\sqrt{27}}{\sqrt{5}} \qquad \text{Quotient rule}$$

$$\frac{\sqrt{27}}{\sqrt{5}} = \frac{\sqrt{27}\cdot\sqrt{5}}{\sqrt{5}\cdot\sqrt{5}} \qquad \text{Multiply both numerator and denominator by } \sqrt{5}.$$

$$= \frac{\sqrt{9\cdot 3\cdot\sqrt{5}}}{5} \qquad \text{Product rule; } \sqrt{5}\cdot\sqrt{5} = 5$$

$$= \frac{\sqrt{9}\cdot\sqrt{3}\cdot\sqrt{5}}{5} \qquad \text{Product rule}$$

$$= \frac{3\cdot\sqrt{3}\cdot\sqrt{5}}{5} \qquad \sqrt{9} = 3$$

$$= \frac{3\sqrt{15}}{5} \qquad \text{Product rule}$$

WORK PROBLEM 2 AT THE SIDE.

EXAMPLE 3 *Simplifying a Product of Radicals*

Simplify $\sqrt{\frac{5}{8}}\cdot\sqrt{\frac{1}{6}}$.

Use both the quotient rule and the product rule.

$$\sqrt{\frac{5}{8}}\cdot\sqrt{\frac{1}{6}} = \sqrt{\frac{5}{8}\cdot\frac{1}{6}} \qquad \text{Product rule}$$

$$= \sqrt{\frac{5}{48}} \qquad \text{Multiply fractions.}$$

$$= \frac{\sqrt{5}}{\sqrt{48}} \qquad \text{Quotient rule}$$

Now rationalize the denominator by multiplying both the numerator and the denominator by $\sqrt{3}$ (because $\sqrt{48}\cdot\sqrt{3} = \sqrt{48\cdot 3} = \sqrt{144} = 12$). One way to find the smallest radical to multiply by is to first factor the denominator: $\sqrt{48} = \sqrt{3}\cdot\sqrt{16} = \sqrt{3}\cdot 4$. Since $\sqrt{3}\cdot\sqrt{3} = 3$, multiplying by $\sqrt{3}$ will produce a rational denominator, as required.

$$\frac{\sqrt{5}}{\sqrt{48}} = \frac{\sqrt{5} \cdot \sqrt{3}}{\sqrt{48} \cdot \sqrt{3}}$$

$$= \frac{\sqrt{15}}{\sqrt{144}} \quad \text{Product rule}$$

$$= \frac{\sqrt{15}}{12} \quad \sqrt{144} = 12 \quad ■$$

WORK PROBLEM 3 AT THE SIDE.

■ **EXAMPLE 4** *Simplifying a Quotient of Radicals*

Simplify $\frac{\sqrt{4x}}{\sqrt{y}}$. Assume that x and y are positive real numbers.

Multiply numerator and denominator by $\sqrt{y}$.

$$\frac{\sqrt{4x}}{\sqrt{y}} = \frac{\sqrt{4x} \cdot \sqrt{y}}{\sqrt{y} \cdot \sqrt{y}} = \frac{\sqrt{4xy}}{y} = \frac{2\sqrt{xy}}{y} \quad ■$$

WORK PROBLEM 4 AT THE SIDE.

■ **EXAMPLE 5** *Simplifying a Radical Quotient*

Simplify the expression $\sqrt{\frac{2x^2y}{3}}$. Assume that x and y are nonnegative real numbers.

First use the quotient rule.

$$\sqrt{\frac{2x^2y}{3}} = \frac{\sqrt{2x^2y}}{\sqrt{3}} \quad \text{Quotient rule}$$

Next, multiply both the numerator and denominator by $\sqrt{3}$ to rationalize the denominator.

$$\frac{\sqrt{2x^2y}}{\sqrt{3}} = \frac{\sqrt{2x^2y} \cdot \sqrt{3}}{\sqrt{3} \cdot \sqrt{3}}$$

$$= \frac{\sqrt{6x^2y}}{3} \quad \text{Product rule}$$

$$= \frac{\sqrt{x^2}\sqrt{6y}}{3} \quad \text{Product rule}$$

$$= \frac{x\sqrt{6y}}{3} \quad \sqrt{x^2} = x, \text{ since } x \geq 0 \quad ■$$

WORK PROBLEM 5 AT THE SIDE.

3 A denominator with a cube root is rationalized by changing the radicand in the denominator to a perfect cube, as shown in the next example.

---

3. Simplify.

(a) $\sqrt{\frac{1}{2}} \cdot \sqrt{\frac{5}{6}}$

(b) $\sqrt{\frac{1}{10}} \cdot \sqrt{20}$

(c) $\sqrt{\frac{5}{8}} \cdot \sqrt{\frac{24}{10}}$

4. Simplify $\frac{\sqrt{5p}}{\sqrt{q}}$. Assume that p and q are positive real numbers.

5. Simplify $\sqrt{\frac{5r^2t^2}{7}}$. Assume that r and t represent nonnegative real numbers.

ANSWERS

3. (a) $\frac{\sqrt{15}}{6}$ (b) $\sqrt{2}$ (c) $\frac{\sqrt{6}}{2}$

4. $\frac{\sqrt{5pq}}{q}$

5. $\frac{rt\sqrt{35}}{7}$

6. Simplify. Rationalize each denominator.

(a) $\sqrt[3]{\frac{5}{7}}$

(b) $\frac{\sqrt[3]{5}}{\sqrt[3]{9}}$

ANSWERS

6. (a) $\frac{\sqrt[3]{245}}{7}$

(b) $\frac{\sqrt[3]{15}}{3}$

■ **EXAMPLE 6** *Rationalizing a Cube Root Denominator*

Simplify. Rationalize each denominator.

(a) $\sqrt[3]{\frac{3}{2}}$

Multiply the numerator and the denominator by enough factors of 2 to make the denominator a perfect cube. This will eliminate the radical in the denominator. Here, multiply by $\sqrt[3]{2^2}$.

$$\sqrt[3]{\frac{3}{2}} = \frac{\sqrt[3]{3}}{\sqrt[3]{2}} = \frac{\sqrt[3]{3} \cdot \sqrt[3]{2^2}}{\sqrt[3]{2} \cdot \sqrt[3]{2^2}} = \frac{\sqrt[3]{3 \cdot 2^2}}{\sqrt[3]{2^3}} = \frac{\sqrt[3]{12}}{2}$$

Since $\sqrt[3]{2^3} = \sqrt[3]{8} = 2$

(b) $\frac{\sqrt[3]{3}}{\sqrt[3]{4}}$

Since $4 \cdot 2 = 2^2 \cdot 2 = 2^3$, multiply numerator and denominator by $\sqrt[3]{2}$.

$$\frac{\sqrt[3]{3}}{\sqrt[3]{4}} = \frac{\sqrt[3]{3} \cdot \sqrt[3]{2}}{\sqrt[3]{4} \cdot \sqrt[3]{2}} = \frac{\sqrt[3]{6}}{\sqrt[3]{8}} = \frac{\sqrt[3]{6}}{2}$$ ■

Caution A common error in Example 6(a) is to multiply by $\sqrt[3]{2}$ instead of $\sqrt[3]{2^2}$ or $\sqrt[3]{4}$. Notice that this would give a denominator of $\sqrt[3]{2} \cdot \sqrt[3]{2} = \sqrt[3]{4}$. Because 4 is not a perfect cube, the denominator is still not rationalized.

◀◀ WORK PROBLEM 6 AT THE SIDE.

EXERCISES

1. When we rationalize the denominator of an expression such as $\frac{4}{\sqrt{3}}$, we multiply both the numerator and the denominator by $\sqrt{3}$. By what number are we actually multiplying the given expression, and what property of real numbers justifies the fact that our result is equal to the given expression?

2. In Example 1(a), we show algebraically that $\frac{9}{\sqrt{6}}$ is equal to $\frac{3\sqrt{6}}{2}$. Give numerical support to this result by finding the decimal approximation of $\frac{9}{\sqrt{6}}$ on your calculator, and then finding the decimal approximation of $\frac{3\sqrt{6}}{2}$. Are they the same?

Rationalize each denominator. See Examples 1–3.

3. $\frac{7}{\sqrt{5}}$ **4.** $\frac{4}{\sqrt{3}}$ **5.** $\frac{8}{\sqrt{2}}$ **6.** $\frac{12}{\sqrt{3}}$

7. $\frac{-\sqrt{11}}{\sqrt{3}}$ **8.** $\frac{-\sqrt{13}}{\sqrt{5}}$ **9.** $\frac{7\sqrt{3}}{\sqrt{5}}$ **10.** $\frac{4\sqrt{6}}{\sqrt{5}}$

11. $\frac{24\sqrt{10}}{16\sqrt{3}}$ **12.** $\frac{18\sqrt{15}}{12\sqrt{2}}$ **13.** $\frac{16}{\sqrt{27}}$ **14.** $\frac{24}{\sqrt{18}}$

15. $\frac{-3}{\sqrt{50}}$ **16.** $\frac{-5}{\sqrt{75}}$ **17.** $\frac{63}{\sqrt{45}}$ **18.** $\frac{27}{\sqrt{32}}$

19. $\frac{\sqrt{24}}{\sqrt{8}}$ **20.** $\frac{\sqrt{36}}{\sqrt{18}}$ **21.** $\frac{-\sqrt{80}}{\sqrt{6}}$ **22.** $\frac{-\sqrt{24}}{\sqrt{15}}$

23. $\sqrt{\frac{1}{2}}$ **24.** $\sqrt{\frac{1}{3}}$ **25.** $\sqrt{\frac{13}{5}}$ **26.** $\sqrt{\frac{17}{11}}$

27. $\sqrt{\frac{7}{13}} \cdot \sqrt{\frac{13}{3}}$ **28.** $\sqrt{\frac{19}{20}} \cdot \sqrt{\frac{20}{3}}$ **29.** $\sqrt{\frac{21}{7}} \cdot \sqrt{\frac{21}{8}}$ **30.** $\sqrt{\frac{5}{8}} \cdot \sqrt{\frac{5}{6}}$

31. $\sqrt{\frac{1}{12}} \cdot \sqrt{\frac{1}{3}}$ **32.** $\sqrt{\frac{1}{8}} \cdot \sqrt{\frac{1}{2}}$ **33.** $\sqrt{\frac{2}{9}} \cdot \sqrt{\frac{9}{2}}$ **34.** $\sqrt{\frac{4}{3}} \cdot \sqrt{\frac{3}{4}}$

Simplify each radical. Assume that all variables represent positive real numbers. See Examples 4 and 5.

35. $\sqrt{\frac{7}{x}}$ **36.** $\sqrt{\frac{19}{y}}$ **37.** $\sqrt{\frac{4x^3}{y}}$ **38.** $\sqrt{\frac{9t^3}{s}}$

39. $\sqrt{\frac{18x^3}{6y}}$ **40.** $\sqrt{\frac{24t^3}{8p}}$ **41.** $\sqrt{\frac{9a^2r^5}{7t}}$ **42.** $\sqrt{\frac{16x^3y^2}{13z}}$

43. Which one of the following would be an appropriate choice for multiplying the numerator and the denominator of $\frac{\sqrt[3]{2}}{\sqrt[3]{5}}$ in order to rationalize the denominator?
(a) $\sqrt[3]{5}$ **(b)** $\sqrt[3]{25}$ **(c)** $\sqrt[3]{2}$
(d) $\sqrt[3]{3}$

44. In Example 6(b), we multiply numerator and denominator of $\frac{\sqrt[3]{3}}{\sqrt[3]{4}}$ by $\sqrt[3]{2}$ to rationalize the denominator. Suppose we had chosen to multiply by $\sqrt[3]{16}$ instead. Would we have obtained the correct answer after all simplifications were done?

Simplify. Rationalize each denominator. Assume that variables in the denominator are nonzero. See Example 6.

45. $\sqrt[3]{\frac{3}{2}}$ **46.** $\sqrt[3]{\frac{2}{5}}$ **47.** $\frac{\sqrt[3]{4}}{\sqrt[3]{7}}$ **48.** $\frac{\sqrt[3]{5}}{\sqrt[3]{10}}$

49. $\sqrt[3]{\frac{3}{4y^2}}$ **50.** $\sqrt[3]{\frac{3}{25x^2}}$ **51.** $\frac{\sqrt[3]{7m}}{\sqrt[3]{36n}}$ **52.** $\frac{\sqrt[3]{11p}}{\sqrt[3]{49q}}$

PREVIEW EXERCISES

Find each product. See Sections 3.5 and 3.6.

53. $(4x + 7)(8x - 3)$ **54.** $ab(3a^2b - 2ab^2 + 7)$ **55.** $(6x - 1)(6x + 1)$

56. $(r + 7)(r - 7)$ **57.** $(p + q)(a - m)$ **58.** $(3w - 8)^2$

8.5 SIMPLIFYING RADICAL EXPRESSIONS

It can be difficult to decide on the "simplest" form of a radical. The conditions for which a square root radical is in simplest form were listed in the previous section. Below is a set of guidelines you should follow when you are simplifying radical expressions.

SIMPLIFYING RADICAL EXPRESSIONS

1. If a radical represents a rational number, then that rational number should be used in place of the radical.

   For example, $\sqrt{49}$ is simplified by writing 7; $\sqrt{64}$ by writing 8; $\sqrt{\frac{169}{9}}$ by writing $\frac{13}{3}$.

2. If a radical expression contains products of radicals, the product rule for radicals, $\sqrt{x} \cdot \sqrt{y} = \sqrt{xy}$, should be used to get a single radical.

   For example, $\sqrt{3} \cdot \sqrt{2}$ is simplified to $\sqrt{6}$; $\sqrt{5} \cdot \sqrt{x}$ to $\sqrt{5x}$.

3. If a radicand has a factor that is a perfect square, the radical should be expressed as the product of the positive square root of the perfect square and the remaining radical factor. A similar statement applies to higher roots.

   For example, $\sqrt{20}$ is simplified to $\sqrt{20} = 2\sqrt{5}$; $\sqrt[3]{16} = 2\sqrt[3]{2}$.

4. If a radical expression contains sums or differences of radicals, the distributive property should be used to combine like radicals.

   For example, $3\sqrt{2} + 4\sqrt{2}$ is combined as $7\sqrt{2}$, but $3\sqrt{2} + 4\sqrt{3}$ cannot be further combined.

5. Any denominator containing a radical should be rationalized.

   For example, $\frac{5}{\sqrt{3}}$ is rationalized as $\frac{5}{\sqrt{3}} = \frac{5\sqrt{3}}{\sqrt{3} \cdot \sqrt{3}} = \frac{5\sqrt{3}}{3}$.

1 The first example involves sums of radical expressions.

EXAMPLE 1 *Adding Radical Expressions*

Perform the indicated operations.

(a) $\sqrt{16} + \sqrt{9}$

Here $\sqrt{16} + \sqrt{9} = 4 + 3 = 7$.

(b) $5\sqrt{2} + 2\sqrt{18}$

First simplify $\sqrt{18}$.

$$\begin{aligned} 5\sqrt{2} + 2\sqrt{18} &= 5\sqrt{2} + 2(\sqrt{9} \cdot \sqrt{2}) && \text{9 is a perfect square.} \\ &= 5\sqrt{2} + 2(3\sqrt{2}) && \sqrt{9} = 3 \\ &= 5\sqrt{2} + 6\sqrt{2} && \text{Multiply.} \\ &= 11\sqrt{2} && \text{Add like radicals.} \end{aligned}$$

WORK PROBLEM 1 AT THE SIDE.

2 The next examples show how to simplify radical expressions with products.

OBJECTIVES

1. Simplify radical expressions with sums.
2. Simplify radical expressions with products.
3. Simplify radical expressions with quotients.
4. Write radical expressions with quotients in lowest terms.

FOR EXTRA HELP

Tape 11 | SSM pp. 327–329 | MAC: B IBM: B

1. Perform the indicated operations.

(a) $\sqrt{36} + \sqrt{25}$

(b) $3\sqrt{3} + 2\sqrt{27}$

(c) $4\sqrt{8} - 2\sqrt{32}$

(d) $2\sqrt{12} - 5\sqrt{48}$

ANSWERS

1. (a) 11 (b) $9\sqrt{3}$ (c) 0 (d) $-16\sqrt{3}$

2. Find each product. Simplify the answers.

(a) $\sqrt{7}(\sqrt{2} + \sqrt{5})$

(b) $\sqrt{2}(\sqrt{8} + \sqrt{20})$

(c) $(\sqrt{2} + 5\sqrt{3})(\sqrt{3} - 2\sqrt{2})$

(d) $(\sqrt{2} - \sqrt{5})(\sqrt{10} + \sqrt{2})$

ANSWERS
2. (a) $\sqrt{14} + \sqrt{35}$
(b) $4 + 2\sqrt{10}$
(c) $11 - 9\sqrt{6}$
(d) $2\sqrt{5} + 2 - 5\sqrt{2} - \sqrt{10}$

■ EXAMPLE 2 *Multiplying Radical Expressions*

Multiply $\sqrt{5}(\sqrt{8} - \sqrt{32})$ and simplify the product.

Start by simplifying $\sqrt{8}$ and $\sqrt{32}$.

$$\sqrt{8} = 2\sqrt{2} \quad \text{and} \quad \sqrt{32} = 4\sqrt{2}$$

Now simplify inside the parentheses.

$$\begin{aligned}\sqrt{5}(\sqrt{8} - \sqrt{32}) &= \sqrt{5}(2\sqrt{2} - 4\sqrt{2}) \\ &= \sqrt{5}\,(-2\sqrt{2}) && \text{Subtract like radicals.} \\ &= -2\sqrt{5 \cdot 2} && \text{Product rule} \\ &= -2\sqrt{10} && \text{Multiply.}\end{aligned}$$

■

■ EXAMPLE 3 *Multiplying Radical Expressions*

Find each product and simplify the answers.

(a) $(\sqrt{3} + 2\sqrt{5})(\sqrt{3} - 4\sqrt{5})$

We can find the products of these sums of radicals in the same way that we found the product of binomials in Chapter 3. The pattern of multiplication is the same, using the FOIL method.

$$\begin{aligned}&(\sqrt{3} + 2\sqrt{5})(\sqrt{3} - 4\sqrt{5}) \\ &= \underbrace{\sqrt{3} \cdot \sqrt{3}}_{\text{First}} + \underbrace{\sqrt{3}(-4\sqrt{5})}_{\text{Outside}} + \underbrace{2\sqrt{5}(\sqrt{3})}_{\text{Inside}} + \underbrace{2\sqrt{5}(-4\sqrt{5})}_{\text{Last}} \\ &= 3 - 4\sqrt{15} + 2\sqrt{15} - 8 \cdot 5 && \text{Product rule} \\ &= 3 - 2\sqrt{15} - 40 && \text{Add like radicals.} \\ &= -37 - 2\sqrt{15} && \text{Combine terms.}\end{aligned}$$

(b) $(\sqrt{3} + \sqrt{21})(\sqrt{3} - \sqrt{7})$

$$\begin{aligned}&= \sqrt{3}(\sqrt{3}) + \sqrt{3}(-\sqrt{7}) + \sqrt{21}(\sqrt{3}) \\ &\quad + \sqrt{21}(-\sqrt{7}) && \text{FOIL} \\ &= 3 - \sqrt{21} + \sqrt{63} - \sqrt{147} && \text{Product rule} \\ &= 3 - \sqrt{21} + \sqrt{9} \cdot \sqrt{7} - \sqrt{49} \cdot \sqrt{3} && \text{9 and 49 are perfect squares.} \\ &= 3 - \sqrt{21} + 3\sqrt{7} - 7\sqrt{3} && \sqrt{9} = 3 \text{ and } \sqrt{49} = 7\end{aligned}$$

Since there are no like radicals, no terms may be combined. ■

◀ WORK PROBLEM 2 AT THE SIDE.

Since radicals represent real numbers, the special products of binomials discussed in Chapter 3 can be used to find products of radicals. Example 4 uses the rule for the product of the sum and difference of two terms,

$$(a + b)(a - b) = a^2 - b^2.$$

■ EXAMPLE 4 *Using a Special Product with Radicals*

Find each product.

(a) $(4 + \sqrt{3})(4 - \sqrt{3})$

Follow the pattern given above. Let $a = 4$ and $b = \sqrt{3}$.

$$\begin{aligned}(4 + \sqrt{3})(4 - \sqrt{3}) &= (4)^2 - (\sqrt{3})^2 \\ &= 16 - 3 && 4^2 = 16 \text{ and } (\sqrt{3})^2 = 3 \\ &= 13\end{aligned}$$

(b) $(\sqrt{12} - \sqrt{6})(\sqrt{12} + \sqrt{6}) = (\sqrt{12})^2 - (\sqrt{6})^2$

$= 12 - 6$ $\quad (\sqrt{12})^2 = 12$ and $(\sqrt{6})^2 = 6$

$= 6$ ■

WORK PROBLEM 3 AT THE SIDE.

Both products in Example 4 resulted in rational numbers. The pairs of expressions in those products, such as $4 + \sqrt{3}$ and $4 - \sqrt{3}$, and $\sqrt{12} - \sqrt{6}$ and $\sqrt{12} + \sqrt{6}$, are called **conjugates** of each other.

3 Products of conjugates similar to those in Example 4 can be used to rationalize the denominators in more complicated quotients, such as

$$\frac{2}{4 - \sqrt{3}}.$$

By Example 4(a), if this denominator, $4 - \sqrt{3}$, is multiplied by $4 + \sqrt{3}$, then the product $(4 - \sqrt{3})(4 + \sqrt{3})$ is the rational number 13. Multiplying numerator and denominator of the quotient by $4 + \sqrt{3}$ gives

$$\frac{2}{4 - \sqrt{3}} = \frac{2(4 + \sqrt{3})}{(4 - \sqrt{3})(4 + \sqrt{3})} = \frac{2(4 + \sqrt{3})}{13}.$$

The denominator now has been rationalized; it contains no radical signs.

USING CONJUGATES TO SIMPLIFY A RADICAL EXPRESSION

To simplify a radical expression with two terms in the denominator, where at least one of those terms is a radical, multiply both the numerator and the denominator by the conjugate of the denominator.

■ EXAMPLE 5 *Using Conjugates to Rationalize a Denominator*

Rationalize the denominator in the quotient

$$\frac{5}{3 + \sqrt{5}}.$$

The radical in the denominator can be eliminated by multiplying both numerator and denominator by $3 - \sqrt{5}$.

$$\frac{5}{3 + \sqrt{5}} = \frac{5(3 - \sqrt{5})}{(3 + \sqrt{5})(3 - \sqrt{5})}$$

$$= \frac{5(3 - \sqrt{5})}{3^2 - (\sqrt{5})^2} \qquad (a + b)(a - b) = a^2 - b^2$$

$$= \frac{5(3 - \sqrt{5})}{9 - 5} \qquad 3^2 = 9 \text{ and } (\sqrt{5})^2 = 5$$

$$= \frac{5(3 - \sqrt{5})}{4} \quad ■$$

3. Find each product. Simplify the answers.

(a) $(3 + \sqrt{5})(3 - \sqrt{5})$

(b) $(\sqrt{3} - 2)(\sqrt{3} + 2)$

(c) $(\sqrt{5} + \sqrt{3})(\sqrt{5} - \sqrt{3})$

ANSWERS

3. (a) 4 (b) −1 (c) 2

4. Rationalize each denominator.

(a) $\dfrac{5}{4 + \sqrt{2}}$

(b) $\dfrac{\sqrt{5} + 3}{2 - \sqrt{5}}$

(c) $\dfrac{1}{\sqrt{6} + \sqrt{3}}$

5. Write each quotient in lowest terms.

(a) $\dfrac{5\sqrt{3} - 15}{10}$

(b) $\dfrac{8\sqrt{5} + 12}{16}$

ANSWERS

4. (a) $\dfrac{5(4 - \sqrt{2})}{14}$ (b) $-11 - 5\sqrt{5}$

(c) $\dfrac{\sqrt{6} - \sqrt{3}}{3}$

5. (a) $\dfrac{\sqrt{3} - 3}{2}$ (b) $\dfrac{2\sqrt{5} + 3}{4}$

EXAMPLE 6 *Using Conjugates to Rationalize a Denominator*

Simplify $\dfrac{6 + \sqrt{2}}{\sqrt{2} - 5}$.

Multiply numerator and denominator by $\sqrt{2} + 5$.

$$\frac{6 + \sqrt{2}}{\sqrt{2} - 5} = \frac{(6 + \sqrt{2})(\sqrt{2} + 5)}{(\sqrt{2} - 5)(\sqrt{2} + 5)}$$

$$= \frac{6\sqrt{2} + 30 + 2 + 5\sqrt{2}}{2 - 25} \qquad \text{FOIL}$$

$$= \frac{11\sqrt{2} + 32}{-23} \qquad \text{Combine terms.}$$

$$= \frac{-11\sqrt{2} - 32}{23} \qquad \frac{a}{-b} = \frac{-a}{b}$$ ■

WORK PROBLEM 4 AT THE SIDE.

4 The final example shows how to write certain quotients in lowest terms.

EXAMPLE 7 *Writing a Radical Quotient in Lowest Terms*

Write $\dfrac{3\sqrt{3} + 9}{12}$ in lowest terms.

Factor the numerator and denominator, and then use the fundamental property from Section 5.1 to replace common factors with 1.

$$\frac{3\sqrt{3} + 9}{12} = \frac{3(\sqrt{3} + 3)}{3(4)} = 1 \cdot \frac{\sqrt{3} + 3}{4} = \frac{\sqrt{3} + 3}{4}$$ ■

Caution A common error is to try to reduce an expression like the one in Example 7 to lowest terms before factoring. For example,

$$\frac{4 + 8\sqrt{5}}{4} \neq 1 + 8\sqrt{5}.$$

The correct simplification is $1 + 2\sqrt{5}$. Do you see why?

WORK PROBLEM 5 AT THE SIDE.

8.5 EXERCISES

NAME DATE HOUR

Based on the work so far in this section, many simple operations involving radicals should now be performed mentally. In Exercises 1–8, perform the operations mentally, and write the answers without doing intermediate steps.

1. $\sqrt{49} + \sqrt{36}$

2. $\sqrt{100} - \sqrt{81}$

3. $\sqrt{2} \cdot \sqrt{8}$

4. $\sqrt{8} \cdot \sqrt{8}$

5. $\sqrt{2}(\sqrt{32} - \sqrt{8})$

6. $\sqrt{3}(\sqrt{27} - \sqrt{3})$

7. $\sqrt[3]{8} + \sqrt[3]{27}$

8. $\sqrt{4} - \sqrt[3]{64} + \sqrt[4]{16}$

Simplify each expression. Use the five guidelines given in the text. See Examples 1–4.

9. $3\sqrt{5} + 2\sqrt{45}$

10. $2\sqrt{2} + 4\sqrt{18}$

11. $8\sqrt{50} - 4\sqrt{72}$

12. $4\sqrt{80} - 5\sqrt{45}$

13. $\sqrt{5}(\sqrt{3} - \sqrt{7})$

14. $\sqrt{7}(\sqrt{10} + \sqrt{3})$

15. $2\sqrt{5}(\sqrt{2} + 3\sqrt{5})$

16. $3\sqrt{7}(2\sqrt{7} + 4\sqrt{5})$

17. $3\sqrt{14} \cdot \sqrt{2} - \sqrt{28}$

18. $7\sqrt{6} \cdot \sqrt{3} - 2\sqrt{18}$

19. $(2\sqrt{6} + 3)(3\sqrt{6} + 7)$

20. $(4\sqrt{5} - 2)(2\sqrt{5} - 4)$

21. $(5\sqrt{7} - 2\sqrt{3})(3\sqrt{7} + 4\sqrt{3})$

22. $(2\sqrt{10} + 5\sqrt{2})(3\sqrt{10} - 3\sqrt{2})$

23. $(2\sqrt{7} + 3)^2$

24. $(4\sqrt{5} + 5)^2$

25. $(5 - \sqrt{2})(5 + \sqrt{2})$

26. $(3 - \sqrt{5})(3 + \sqrt{5})$

27. $(\sqrt{8} - \sqrt{7})(\sqrt{8} + \sqrt{7})$

28. $(\sqrt{12} - \sqrt{11})(\sqrt{12} + \sqrt{11})$

29. $(\sqrt{2} + \sqrt{3})(\sqrt{6} - \sqrt{2})$

30. $(\sqrt{3} + \sqrt{5})(\sqrt{15} - \sqrt{5})$

31. $(\sqrt{10} - \sqrt{5})(\sqrt{5} + \sqrt{20})$

32. $(\sqrt{6} - \sqrt{3})(\sqrt{3} + \sqrt{18})$

33. $(\sqrt{5} + \sqrt{30})(\sqrt{6} + \sqrt{3})$

34. $(\sqrt{10} - \sqrt{20})(\sqrt{2} - \sqrt{5})$

35. In Example 3(a), the original expression simplifies to $-37 - 2\sqrt{15}$. Students often try to simplify expressions like this by combining the -37 and the -2 to get $-39\sqrt{15}$, which is incorrect. Explain why this is incorrect.

36. If you were to attempt to rationalize the denominator of $\frac{2}{4 + \sqrt{3}}$ by multiplying the numerator and the denominator by $4 + \sqrt{3}$, what problem would arise? What should you multiply by?

Rationalize the denominators. See Examples 5 and 6.

37. $\frac{1}{3 + \sqrt{2}}$

38. $\frac{1}{4 - \sqrt{3}}$

39. $\frac{14}{2 - \sqrt{11}}$

40. $\frac{19}{5 - \sqrt{6}}$

41. $\frac{\sqrt{2}}{2 - \sqrt{2}}$

42. $\frac{\sqrt{7}}{7 - \sqrt{7}}$

43. $\frac{\sqrt{5}}{\sqrt{2} + \sqrt{3}}$

44. $\frac{\sqrt{3}}{\sqrt{2} + \sqrt{3}}$

45. $\frac{\sqrt{12}}{\sqrt{3} + 1}$

46. $\frac{\sqrt{18}}{\sqrt{2} - 1}$

47. $\frac{\sqrt{5} + 2}{2 - \sqrt{3}}$

48. $\frac{\sqrt{7} + 3}{4 - \sqrt{5}}$

Write each quotient in lowest terms. See Example 7.

49. $\frac{6\sqrt{11} - 12}{6}$

50. $\frac{12\sqrt{5} - 24}{12}$

51. $\frac{2\sqrt{3} + 10}{16}$

52. $\frac{4\sqrt{6} + 24}{20}$

53. $\frac{12 - \sqrt{40}}{4}$

54. $\frac{9 - \sqrt{72}}{12}$

PREVIEW EXERCISES

Solve each equation. See Section 4.5.

55. $y^2 + 4y + 3 = 0$

56. $x^2 - 6x + 9 = 0$

57. $(k - 1) = (k - 1)^2$

58. $x(x + 4) = 21$

8.6 EQUATIONS WITH RADICALS

The addition and multiplication properties of equality are not enough to solve an equation with radicals such as

$$\sqrt{x + 1} = 3.$$

1 Solving equations that have radicals requires a new property, the *squaring property*.

SQUARING PROPERTY OF EQUALITY

If each side of a given equation is squared, all solutions of the original equation are *among* the solutions of the squared equation.

Caution Be very careful with the squaring property: Using this property can give a new equation with *more* solutions than the original equation. For example, starting with the equation $y = 4$ and squaring each side gives

$$y^2 = 4^2, \quad \text{or} \quad y^2 = 16.$$

This last equation, $y^2 = 16$, has *two* solutions, 4 or -4, while the original equation, $y = 4$, has only *one* solution, 4. Because of this possibility, checking is more than just a guard against algebraic errors when solving an equation with radicals. It is an essential part of the solution process. *All potential solutions from the squared equation must be checked in the original equation.*

EXAMPLE 1 *Using the Squaring Property of Equality*

Solve the equation $\sqrt{p + 1} = 3$.

Use the squaring property of equality to square each side of the equation.

$$(\sqrt{p + 1})^2 = 3^2$$

$$p + 1 = 9 \qquad (\sqrt{p + 1})^2 = p + 1$$

$$p = 8 \qquad \text{Subtract 1.}$$

Now check this answer in the original equation.

Check:

$$\sqrt{p + 1} = 3$$

$$\sqrt{8 + 1} = 3 \quad ? \qquad \text{Let } p = 8.$$

$$\sqrt{9} = 3 \quad ?$$

$$3 = 3 \qquad \text{True}$$

Because this statement is true, 8 is the solution of $\sqrt{p + 1} = 3$. In this case the squared equation had just one solution, which also satisfied the original equation. ■

WORK PROBLEM 1 AT THE SIDE.

EXAMPLE 2 *Using the Squaring Property of Equality*

Solve $3\sqrt{x} = \sqrt{x + 8}$.

Squaring each side gives

$$(3\sqrt{x})^2 = (\sqrt{x + 8})^2$$

$$3^2(\sqrt{x})^2 = (\sqrt{x + 8})^2 \qquad (ab)^2 = a^2b^2$$

$$9x = x + 8 \qquad (\sqrt{x})^2 = x;\ (\sqrt{x + 8})^2 = x + 8$$

$$8x = 8 \qquad \text{Subtract } x.$$

$$x = 1. \qquad \text{Divide by 8.}$$

OBJECTIVES

1. Solve equations with radicals.
2. Identify equations with no solutions.
3. Solve equations by squaring a binomial.

FOR EXTRA HELP

Tape 11

SSM pp. 330–335

MAC: B IBM: B

1. Solve each equation.

(a) $\sqrt{k} = 3$

(b) $\sqrt{m - 2} = 4$

(c) $\sqrt{9 - y} = 4$

ANSWERS

1. (a) 9 (b) 18 (c) -7

2. Solve each equation.

(a) $\sqrt{3x+9} = 2\sqrt{x}$

(b) $5\sqrt{a} = \sqrt{20a+5}$

3. Solve each equation that has a solution. (*Hint:* In (**a**) subtract 4 from each side.)

(a) $\sqrt{y} + 4 = 0$

(b) $m = \sqrt{m^2 - 4m - 16}$

ANSWERS
2. (a) 9 (b) 1
3. (a) no solution (b) no solution

Check:

$$3\sqrt{x} = \sqrt{x+8}$$
$$3\sqrt{1} \stackrel{?}{=} \sqrt{1+8} \quad \text{Let } x = 1.$$
$$3(1) \stackrel{?}{=} \sqrt{9} \quad \sqrt{1} = 1$$
$$3 = 3 \quad \text{True}$$

The solution of $3\sqrt{x} = \sqrt{x+8}$ is 1. ■

Caution Avoid the common error of writing the solution as the final result obtained in the check. In Example 2, the solution is 1, *not* 3.

WORK PROBLEM 2 AT THE SIDE.

2 Not all equations with radicals have a solution, as shown by the equations in Examples 3 and 4.

■ EXAMPLE 3 *Using the Squaring Property of Equality*

Solve the equation $\sqrt{y} = -3$.

Square each side.

$$(\sqrt{y})^2 = (-3)^2$$
$$y = 9$$

Check this proposed answer in the original equation.

Check:

$$\sqrt{y} = -3$$
$$\sqrt{9} \stackrel{?}{=} -3 \quad \text{Let } y = 9.$$
$$3 = -3 \quad \text{False}$$

Because the statement $3 = -3$ is false, the number 9 is not a solution of the given equation and is said to be *extraneous*. In fact, $\sqrt{y} = -3$ has no solution. Because $\sqrt{y}$ represents the *nonnegative* square root of y, we might have seen immediately that there is no solution. ■

■ EXAMPLE 4 *Using the Squaring Property of Equality*

Solve $a = \sqrt{a^2 + 5a + 10}$.

Square each side.

$$(a)^2 = (\sqrt{a^2 + 5a + 10})^2$$
$$a^2 = a^2 + 5a + 10 \quad (\sqrt{a^2 + 5a + 10})^2 = a^2 + 5a + 10$$
$$0 = 5a + 10 \quad \text{Subtract } a^2.$$
$$-10 = 5a \quad \text{Subtract 10.}$$
$$a = -2 \quad \text{Divide by 5.}$$

Check this potential solution in the original equation.

Check:

$$a = \sqrt{a^2 + 5a + 10}$$
$$-2 \stackrel{?}{=} \sqrt{(-2)^2 + 5(-2) + 10} \quad \text{Let } a = -2.$$
$$-2 \stackrel{?}{=} \sqrt{4 - 10 + 10} \quad \text{Multiply.}$$
$$-2 = 2 \quad \text{False}$$

Because $a = -2$ leads to a false result, the equation has no solution. ■

WORK PROBLEM 3 AT THE SIDE.

The steps to use in solving an equation with radicals are summarized below.

SOLVING AN EQUATION WITH RADICALS

Step 1 Arrange the terms so that a radical is alone on one side of the equation.
Step 2 Square each side.
Step 3 Combine like terms. Repeat Steps 1–3 if a radical is still present.
Step 4 Solve the equation for potential solutions.
Step 5 Check all solutions from Step 4 in the *original* equation.

3 The next examples use the following facts from Section 3.6.

$$(a + b)^2 = a^2 + 2ab + b^2$$

and

$$(a - b)^2 = a^2 - 2ab + b^2.$$

By these patterns, for example,

$$\begin{aligned}(y - 3)^2 &= y^2 - 2(y)(3) + (3)^2\\ &= y^2 - 6y + 9.\end{aligned}$$

WORK PROBLEM 4 AT THE SIDE.

4. Square each expression.

(a) $m - 5$

(b) $2k - 5$

(c) $3m - 2p$

EXAMPLE 5 *Using the Squaring Property of Equality*

Solve the equation $\sqrt{2y - 3} = y - 3$.

Square each side, using the result above on the right side of the equation.

$$\begin{aligned}(\sqrt{2y - 3})^2 &= (y - 3)^2\\ 2y - 3 &= y^2 - 6y + 9\end{aligned}$$

This equation is quadratic because of the y^2 term. As shown in Section 4.5, solving this equation requires that one side be equal to 0. Subtract $2y$ and add 3, getting

$$0 = y^2 - 8y + 12.$$

Solve this equation by factoring.

$$0 = (y - 6)(y - 2)$$

$$y - 6 = 0 \quad \text{or} \quad y - 2 = 0 \qquad \text{Set each factor equal to 0.}$$

$$y = 6 \quad \text{or} \quad y = 2 \qquad \text{Solve.}$$

Check both of these potential solutions in the original equation.

Check:

If $y = 6$,

$$\begin{aligned}\sqrt{2y - 3} &= y - 3\\ \sqrt{2(6) - 3} &= 6 - 3 \quad ? \quad \text{Let } y = 6.\\ \sqrt{12 - 3} &= 3 \quad ? \quad \text{Multiply.}\\ \sqrt{9} &= 3 \quad ?\\ 3 &= 3. \quad \text{True}\end{aligned}$$

If $y = 2$,

$$\begin{aligned}\sqrt{2y - 3} &= y - 3\\ \sqrt{2(2) - 3} &= 2 - 3 \quad ? \quad \text{Let } y = 2.\\ \sqrt{4 - 3} &= -1 \quad ? \quad \text{Multiply.}\\ \sqrt{1} &= -1 \quad ?\\ 1 &= -1. \quad \text{False}\end{aligned}$$

Only 6 is a valid solution of the equation. ■

ANSWERS

4. (a) $m^2 - 10m + 25$
(b) $4k^2 - 20k + 25$
(c) $9m^2 - 12mp + 4p^2$

5. Solve each equation.

(a) $\sqrt{6w + 6} = w + 1$

(b) $2u - 1 = \sqrt{10u + 9}$

6. Solve each equation.

(a) $\sqrt{x} - 3 = x - 15$

(b) $\sqrt{z + 5} + 2 = z + 5$

ANSWERS
5. (a) 5, −1 (b) 4
6. (a) 16 (b) −1

WORK PROBLEM 5 AT THE SIDE.

Sometimes it is necessary to write an equation in a different form before squaring each side. The next example shows why.

EXAMPLE 6 *Using the Squaring Property of Equality*

Solve the equation $\sqrt{x} + 1 = 2x$.

Squaring each side gives

$$(\sqrt{x} + 1)^2 = (2x)^2$$
$$x + 2\sqrt{x} + 1 = 4x^2,$$

an equation that is more complicated, and still contains a radical. It would be better instead to rewrite the original equation so that the radical is alone on one side of the equals sign. Get the radical alone by subtracting 1 from each side.

$$\sqrt{x} = 2x - 1$$
$$(\sqrt{x})^2 = (2x - 1)^2 \qquad \text{Square each side.}$$
$$x = 4x^2 - 4x + 1$$
$$0 = 4x^2 - 5x + 1 \qquad \text{Subtract } x.$$
$$0 = (4x - 1)(x - 1) \qquad \text{Factor.}$$
$$4x - 1 = 0 \quad \text{or} \quad x - 1 = 0 \qquad \text{Set each factor equal to 0.}$$
$$x = \frac{1}{4} \quad \text{or} \quad x = 1 \qquad \text{Solve.}$$

Check:

If $x = \frac{1}{4}$,

$$\sqrt{x} + 1 = 2x$$
$$\sqrt{\frac{1}{4}} + 1 = 2\left(\frac{1}{4}\right) \quad ? \qquad \text{Let } x = \frac{1}{4}.$$
$$\frac{1}{2} + 1 = \frac{1}{2}. \qquad \text{False}$$

If $x = 1$,

$$\sqrt{x} + 1 = 2x$$
$$\sqrt{1} + 1 = 2(1) \quad ? \qquad \text{Let } x = 1.$$
$$2 = 2. \qquad \text{True}$$

The only solution to the original equation is 1. ■

Caution Errors often occur when each side of an equation is squared. For example, in Example 6 after the equation is rewritten as

$$\sqrt{x} = 2x - 1,$$

it would be *incorrect* to write the next step as

$$x = 4x^2 + 1.$$

Don't forget that the binomial $2x - 1$ must be squared to get $4x^2 - 4x + 1$.

WORK PROBLEM 6 AT THE SIDE.

8.6 EXERCISES

NAME DATE HOUR

1. How can you tell that the equation $\sqrt{x} = -8$ has no real number solution without performing any algebraic steps?

2. Explain why the equation $x^2 = 36$ has two real number solutions, while the equation $\sqrt{x} = 6$ has only one real number solution.

Find all solutions for each of the following equations. See Examples 1–4.

3. $\sqrt{x} = 7$

4. $\sqrt{k} = 10$

5. $\sqrt{y + 2} = 3$

6. $\sqrt{x + 7} = 5$

7. $\sqrt{r - 4} = 9$

8. $\sqrt{k - 12} = 3$

9. $\sqrt{4 - t} = 7$

10. $\sqrt{9 - s} = 5$

11. $\sqrt{2t + 3} = 0$

12. $\sqrt{5x - 4} = 0$

13. $\sqrt{3x - 8} = -2$

14. $\sqrt{6y + 4} = -3$

15. $\sqrt{m} - 4 = 7$

16. $\sqrt{t} + 3 = 10$

17. $\sqrt{10x - 8} = 3\sqrt{x}$

18. $\sqrt{17t - 4} = 4\sqrt{t}$

19. $5\sqrt{x} = \sqrt{10x + 15}$

20. $4\sqrt{y} = \sqrt{20y - 16}$

21. $\sqrt{3x - 5} = \sqrt{2x + 1}$

22. $\sqrt{5y + 2} = \sqrt{3y + 8}$

23. $k = \sqrt{k^2 - 5k - 15}$

24. $s = \sqrt{s^2 - 2s - 6}$

25. $7x = \sqrt{49x^2 + 2x - 10}$

26. $6m = \sqrt{36m^2 + 5m - 5}$

27. The first step in solving the equation $\sqrt{2x + 1} = x - 7$ is to square both sides of the equation. Errors often occur in solving equations such as this one when the right side of the equation is squared incorrectly. Why is the square of the right side *not* equal to $x^2 + 49$? What is the correct answer for the square of the right side?

28. Explain why the equation $x = 3\sqrt{x + 13}$ cannot have a negative solution.

Find all solutions for each equation. See Examples 5 and 6.

29. $\sqrt{2x + 1} = x - 7$

30. $\sqrt{3x + 3} = x - 5$

31. $\sqrt{3k + 10} + 5 = 2k$

32. $\sqrt{4t + 13} + 1 = 2t$

33. $\sqrt{5x + 1} - 1 = x$

34. $\sqrt{x + 1} - x = 1$

35. $\sqrt{6t + 7} + 3 = t + 5$

36. $\sqrt{10x + 24} = x + 4$

37. $x - 4 - \sqrt{2x} = 0$

38. $x - 3 - \sqrt{4x} = 0$

39. $\sqrt{x} + 6 = 2x$

40. $\sqrt{k} + 12 = k$

Solve each problem.

41. Police sometimes use the following procedure to estimate the speed at which a car was traveling at the time of an accident. A police officer drives the car involved in the accident under conditions similar to those during which the accident took place and then skids to a stop. If the car is driven at 30 miles per hour, then the speed at the time of the accident is given by

$$s = 30\sqrt{\frac{a}{p}}$$

where a is the length of the skid marks left at the time of the accident and p is the length of the skid marks in the police test. Find s for the following values of a and p. Round to the nearest tenth.

(a) $a = 862$ feet; $p = 156$ feet
(b) $a = 382$ feet; $p = 96$ feet
(c) $a = 84$ feet; $p = 26$ feet

42. A formula for calculating the distance, d, one can see from an airplane to the horizon on a clear day is

$$d = 1.22\sqrt{x}$$

where x is the altitude of the plane in feet and d is given in miles. How far can one see to the horizon in a plane flying at the following altitudes? Round to the nearest tenth.

(a) 15,000 feet
(b) 18,000 feet
(c) 24,000 feet

PREVIEW EXERCISES

Find all square roots of each number. Simplify where possible. See Section 8.1.

43. 121 **44.** 625 **45.** 23 **46.** 160

CHAPTER 8 SUMMARY

KEY TERMS

8.1	**square root**	The square roots of a^2 are a and $-a$ (a is nonnegative).
	perfect square	A number with a rational square root is called a perfect square.
	radicand	The number or expression under a radical sign is called the radicand.
	radical	A radical sign with a radicand is called a radical.
	radical expression	An algebraic expression containing a radical is called a radical expression.
8.3	**like radicals**	Like radicals are multiples of the same radical.
8.4	**rationalizing the denominator**	The process of changing the denominator of a fraction from a radical (irrational number) to a rational number is called rationalizing the denominator.
	simplified form	A radical is in simplified form if the radicand contains no factor (except 1) that is a perfect square, the radicand contains no fractions, and no denominator contains a radical.
8.5	**conjugates**	The conjugate of $a + b$ is $a - b$.

NEW SYMBOLS

$\sqrt{}$ radical sign $\qquad$ $\approx$ is approximately equal to

QUICK REVIEW

Concepts	Examples
8.1 Finding Roots	
If a is a positive real number, $\sqrt{a}$ is the positive square root of a; $-\sqrt{a}$ is the negative square root of a; $\sqrt{0} = 0$.	$\sqrt{49} = 7$ $-\sqrt{81} = -9$
If a is a negative real number, $\sqrt{a}$ is not a real number.	$\sqrt{-25}$ is not a real number.
If a is a positive rational number, $\sqrt{a}$ is rational if a is a perfect square. $\sqrt{a}$ is irrational if a is not a perfect square.	$\sqrt{\frac{4}{9}}$, $\sqrt{16}$ are rational. $\quad \sqrt{\frac{2}{3}}$, $\sqrt{21}$ are irrational.
Each real number has exactly one real cube root.	$\sqrt[3]{27} = 3$; $\sqrt[3]{-8} = -2$
8.2 Multiplication and Division of Radicals	
Product Rule for Radicals For nonnegative real numbers a and b, $\sqrt{a} \cdot \sqrt{b} = \sqrt{ab}$ and $\sqrt{ab} = \sqrt{a} \cdot \sqrt{b}$.	$\sqrt{5} \cdot \sqrt{7} = \sqrt{5 \cdot 7} = \sqrt{35}$ $\sqrt{8} \cdot \sqrt{2} = \sqrt{16} = 4$ $\sqrt{48} = \sqrt{16} \cdot \sqrt{3} = 4\sqrt{3}$
Quotient Rule for Radicals If a and b are nonnegative real numbers and b is not 0, $\frac{\sqrt{a}}{\sqrt{b}} = \sqrt{\frac{a}{b}}$ and $\sqrt{\frac{a}{b}} = \frac{\sqrt{a}}{\sqrt{b}}$.	$\sqrt{\frac{25}{64}} = \frac{\sqrt{25}}{\sqrt{64}} = \frac{5}{8}$ $\frac{\sqrt{8}}{\sqrt{2}} = \sqrt{\frac{8}{2}} = \sqrt{4} = 2$

Concepts	Examples
8.3 Addition and Subtraction of Radicals Add and subtract like radicals by using the distributive property. Only like radicals can be combined in this way.	$2\sqrt{5} + 4\sqrt{5} = (2 + 4)\sqrt{5} = 6\sqrt{5}$ $\sqrt{8} + \sqrt{32} = 2\sqrt{2} + 4\sqrt{2} = 6\sqrt{2}$
8.4 Rationalizing the Denominator The denominator of a radical is rationalized by multiplying both the numerator and denominator by the same number.	$\dfrac{2}{\sqrt{3}} = \dfrac{2 \cdot \sqrt{3}}{\sqrt{3} \cdot \sqrt{3}} = \dfrac{2\sqrt{3}}{3}$ $\sqrt[3]{\dfrac{5}{121}} = \dfrac{\sqrt[3]{5} \cdot \sqrt[3]{11}}{\sqrt[3]{11^2} \cdot \sqrt[3]{11}} = \dfrac{\sqrt[3]{55}}{11}$
8.5 Simplifying Radical Expressions When appropriate, use the rules for adding and multiplying polynomials to simplify radical expressions. If a radical expression contains two terms in the denominator and at least one of those terms is a radical, multiply both the numerator and the denominator by the conjugate of the denominator.	$\sqrt{6}(\sqrt{5} - \sqrt{7}) = \sqrt{30} - \sqrt{42}$ $(\sqrt{5} - \sqrt{3})(\sqrt{5} + \sqrt{3}) = 5 - 3 = 2$ $\dfrac{6}{\sqrt{7} - \sqrt{2}} = \dfrac{6}{\sqrt{7} - \sqrt{2}} \cdot \dfrac{\sqrt{7} + \sqrt{2}}{\sqrt{7} + \sqrt{2}}$ $= \dfrac{6(\sqrt{7} + \sqrt{2})}{7 - 2}$ Multiply fractions. $= \dfrac{6(\sqrt{7} + \sqrt{2})}{5}$ Simplify.
8.6 Equations with Radicals **Solving an Equation with Radicals** *Step 1* Arrange the terms so that a radical is alone on one side of the equation. *Step 2* Square each side. (By the squaring property of equality, all solutions of the original equation are *among* the solutions of the squared equation.) *Step 3* Combine like terms. Repeat Steps 1–3 if necessary. *Step 4* Solve the equation for potential solutions. *Step 5* Check all potential solutions from Step 4 in the original equation.	Solve: $\sqrt{2x - 3} + x = 3$. $\sqrt{2x - 3} = 3 - x$ Isolate the radical. $(\sqrt{2x - 3})^2 = (3 - x)^2$ Square. $2x - 3 = 9 - 6x + x^2$ $0 = x^2 - 8x + 12$ Put in standard form. $0 = (x - 2)(x - 6)$ Factor. $x - 2 = 0$ or $x - 6 = 0$ Set each factor = 0. $x = 2$ $x = 6$ Solve. A check is essential here. Verify that 2 is the only solution (6 is extraneous).

NAME DATE HOUR

CHAPTER 8 REVIEW EXERCISES

[8.1] *Find all square roots of each number.*

1. 49 **2.** 81 **3.** 196 **4.** 121

5. 225 **6.** 729

Find each root that is a real number.

7. $\sqrt{16}$ **8.** $-\sqrt{36}$ **9.** $\sqrt[3]{1000}$ **10.** $\sqrt[4]{81}$

11. $\sqrt{-8100}$ **12.** $-\sqrt{4225}$ **13.** $\sqrt{\frac{49}{36}}$ **14.** $\sqrt{\frac{100}{81}}$

15. If $\sqrt{a}$ is not a real number, then what kind of number must a be?

16. Find the value of x:

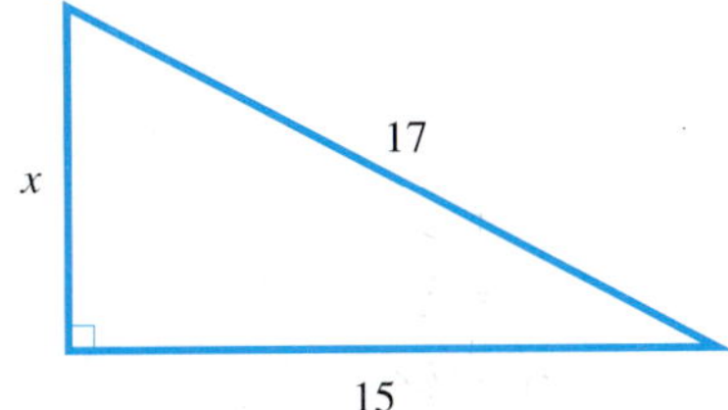

Write rational, irrational, *or* not a real number *for each number. If a number is rational, give its exact value. If a number is irrational, give a decimal approximation for the number. Round approximations to the nearest thousandth.*

17. $\sqrt{23}$ **18.** $\sqrt{169}$ **19.** $-\sqrt{25}$ **20.** $\sqrt{-4}$

[8.2] *Use the product rule to simplify each expression.*

21. $\sqrt{2} \cdot \sqrt{7}$

22. $\sqrt{12} \cdot \sqrt{3}$

23. $\sqrt{5} \cdot \sqrt{15}$

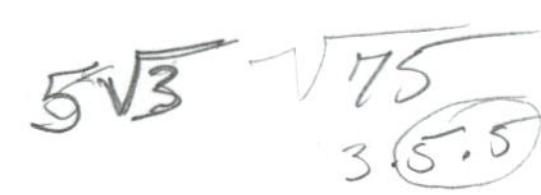

24. $\sqrt{12} \cdot \sqrt{12}$

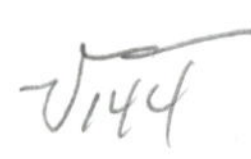

25. $-\sqrt{27}$

26. $\sqrt{48}$

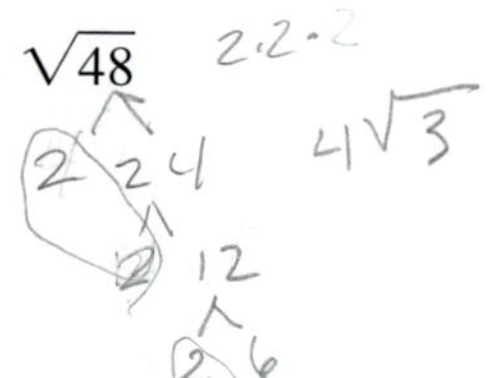

27. $\sqrt{160}$

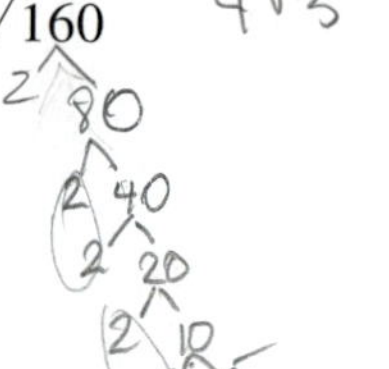

28. $\sqrt[3]{-125}$

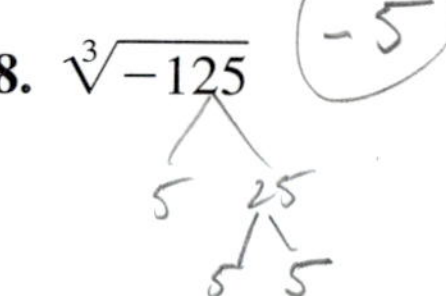

29. $\sqrt[3]{1728}$

30. $\sqrt{12} \cdot \sqrt{27}$

31. $\sqrt{32} \cdot \sqrt{48}$

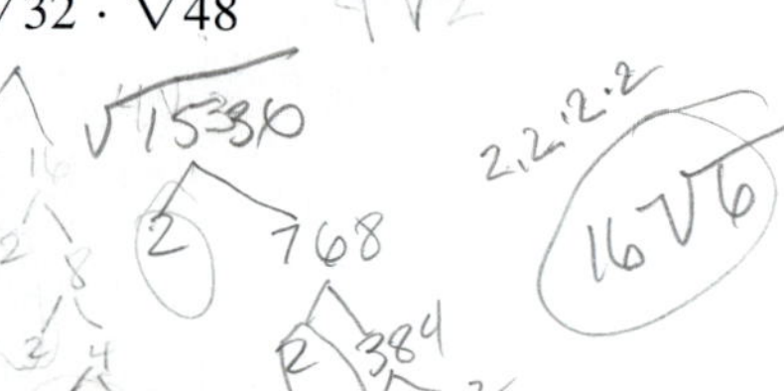

32. $\sqrt{50} \cdot \sqrt{125}$

Use the product rule and the quotient rule, as necessary, to simplify each expression.

33. $\sqrt{\frac{9}{4}}$

34. $-\sqrt{\frac{121}{400}}$

35. $\sqrt{\frac{3}{49}}$

36. $\sqrt{\frac{7}{169}}$

37. $\sqrt{\frac{1}{6}} \cdot \sqrt{\frac{5}{6}}$

38. $\sqrt{\frac{2}{5}} \cdot \sqrt{\frac{2}{45}}$

39. $\frac{3\sqrt{10}}{\sqrt{5}}$

40. $\frac{24\sqrt{12}}{6\sqrt{3}}$

41. $\frac{8\sqrt{150}}{4\sqrt{75}}$

Simplify each expression. Assume that all variables represent nonnegative real numbers.

42. $\sqrt{p} \cdot \sqrt{p}$

43. $\sqrt{k} \cdot \sqrt{m}$

44. $\sqrt{r^{18}}$

45. $\sqrt{x^{10}y^{16}}$

46. $\sqrt{x^{9}}$

47. $\sqrt{\dfrac{36}{p^2}}, p \neq 0$

48. $\sqrt{a^{15}b^{21}}$

49. $\sqrt{121x^6y^{10}}$

50. Use a calculator to find approximations for $\sqrt{.5}$ and $\dfrac{\sqrt{2}}{2}$. Based on your results, do you think that these two expressions represent the same number? If so, verify it *algebraically*.

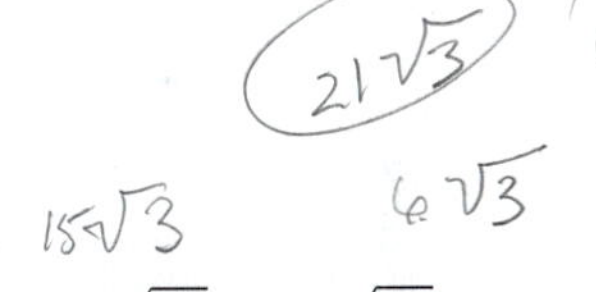

[8.3] *Simplify and combine terms where possible.*

51. $\sqrt{11} + \sqrt{11}$

52. $3\sqrt{2} + 6\sqrt{2}$

53. $3\sqrt{75} + 2\sqrt{27}$

54. $4\sqrt{12} + \sqrt{48}$

55. $4\sqrt{24} - 3\sqrt{54} + \sqrt{6}$

56. $2\sqrt{7} - 4\sqrt{28} + 3\sqrt{63}$

57. $\dfrac{2}{5}\sqrt{75} + \dfrac{3}{4}\sqrt{160}$

58. $\dfrac{1}{3}\sqrt{18} + \dfrac{1}{4}\sqrt{32}$

59. $\sqrt{15} \cdot \sqrt{2} + 5\sqrt{30}$

Simplify each expression. Assume that all variables represent nonnegative real numbers.

60. $\sqrt{4x} + \sqrt{36x} - \sqrt{9x}$

61. $\sqrt{16p} + 3\sqrt{p} - \sqrt{49p}$

62. $\sqrt{20m^2} - m\sqrt{45}$

63. $3k\sqrt{8k^2n} + 5k^2\sqrt{2n}$

[8.4] *Perform the indicated operations, and write all answers in simplest form. Rationalize all denominators. Assume that all variables represent nonnegative real numbers.*

64. $\dfrac{10}{\sqrt{3}}$

65. $\dfrac{15}{\sqrt{2}}$

66. $\dfrac{8\sqrt{2}}{\sqrt{5}}$

67. $\dfrac{5}{\sqrt{5}}$

68. $\dfrac{12}{\sqrt{24}}$

69. $\dfrac{\sqrt{2}}{\sqrt{15}}$

70. $\sqrt{\dfrac{2}{5}}$

71. $\sqrt{\dfrac{5}{14}} \cdot \sqrt{28}$

72. $\sqrt{\dfrac{2}{7}} \cdot \sqrt{\dfrac{1}{3}}$

73. $\sqrt{\dfrac{r^2}{16x}}$, $x \neq 0$

74. $\sqrt[3]{\dfrac{1}{3}}$

75. $\sqrt[3]{\dfrac{2}{7}}$

76. Explain how you would show, without using a calculator, that $\dfrac{\sqrt{6}}{4}$ and $\sqrt{\dfrac{48}{128}}$ represent the exact same number. Then actually perform the necessary steps.

[8.5] *Simplify each expression.*

77. $-\sqrt{3}(\sqrt{5} + \sqrt{27})$

78. $3\sqrt{2}(\sqrt{3} + 2\sqrt{2})$

79. $(2\sqrt{3} - 4)(5\sqrt{3} + 2)$

80. $(5\sqrt{7} + 2)^2$

81. $(\sqrt{5} - \sqrt{7})(\sqrt{5} + \sqrt{7})$

82. $(2\sqrt{3} + 5)(2\sqrt{3} - 5)$

83. $(\sqrt{7} + 2\sqrt{6})(\sqrt{12} - \sqrt{2})$

Rationalize the denominators.

84. $\dfrac{1}{2 + \sqrt{5}}$

85. $\dfrac{2}{\sqrt{2} - 3}$

86. $\dfrac{\sqrt{8}}{\sqrt{2} + 6}$

87. $\dfrac{\sqrt{3}}{1 + \sqrt{3}}$

88. $\dfrac{\sqrt{5} - 1}{\sqrt{2} + 3}$

89. $\dfrac{2 + \sqrt{6}}{\sqrt{3} - 1}$

Write each quotient in lowest terms.

90. $\dfrac{15 + 10\sqrt{6}}{15}$

91. $\dfrac{3 + 9\sqrt{7}}{12}$

92. $\dfrac{6 + \sqrt{192}}{2}$

[8.6] *Find all solutions for each equation.*

93. $\sqrt{m} + 5 = 0$

94. $\sqrt{p} + 4 = 0$

95. $\sqrt{k + 1} = 7$

96. $\sqrt{5m + 4} = 3\sqrt{m}$

97. $\sqrt{2p + 3} = \sqrt{5p - 3}$

98. $\sqrt{4y + 1} = y - 1$

99. $\sqrt{-2k - 4} = k + 2$

100. $\sqrt{2 - x} + 3 = x + 7$

101. $\sqrt{x} - x + 2 = 0$

102. $\sqrt{2 - x} + x = 0$

103. $\sqrt{4y - 2} = \sqrt{3y + 1}$

104. $\sqrt{2x + 3} = x + 2$

MIXED REVIEW EXERCISES

Simplify each expression if possible. Assume all variables represent nonnegative real numbers.

105. $\sqrt{3} \cdot \sqrt{27}$

106. $2\sqrt{27} + 3\sqrt{75} - \sqrt{300}$

107. $\sqrt{\frac{121}{t^2}}$, $t \neq 0$

108. $\frac{1}{5 + \sqrt{2}}$

109. $\sqrt{\frac{1}{3}} \cdot \sqrt{\frac{24}{5}}$

110. $\sqrt{50y^2}$

111. $\sqrt[3]{-125}$

112. $-\sqrt{5}(\sqrt{2} + \sqrt{75})$

113. $\sqrt{\frac{16r^3}{3s}}$, $s \neq 0$

114. $\frac{12 + 6\sqrt{13}}{12}$

115. $-\sqrt{162} + \sqrt{8}$

116. $(\sqrt{5} - \sqrt{2})^2$

117. $(6\sqrt{7} + 2)(4\sqrt{7} - 1)$

118. $-\sqrt{121}$

119. $\sqrt{98}$

Solve.

120. $\sqrt{x + 2} = x - 4$

121. $\sqrt{k} + 3 = 0$

122. $\sqrt{1 + 3t} - t = -3$

NAME DATE HOUR

CHAPTER 8 TEST

On this test, assume that all variables represent nonnegative real numbers.

1. Find all square roots of 196.

2. Consider $\sqrt{142}$.
   (a) Determine whether it is rational or irrational.
   (b) Find a decimal approximation to the nearest thousandth.

3. Simplify $\sqrt[3]{216}$.

Simplify where possible.

4. $-\sqrt{27}$

5. $\sqrt{\dfrac{128}{25}}$

6. $\sqrt[3]{32}$

7. $\dfrac{20\sqrt{18}}{5\sqrt{3}}$

8. $3\sqrt{28} + \sqrt{63}$

9. $3\sqrt{27x} - 4\sqrt{48x} + 2\sqrt{3x}$

10. $\sqrt[3]{32x^2y^3}$

11. $(6 - \sqrt{5})(6 + \sqrt{5})$

12. $(2 - \sqrt{7})(3\sqrt{2} + 1)$

13. $(\sqrt{5} + \sqrt{6})^2$

1. ______
2. (a) ______
   (b) ______
3. ______
4. ______
5. ______
6. ______
7. ______
8. ______
9. ______
10. ______
11. ______
12. ______
13. ______

14. (a) ______________________

(b) ______________________

15. ______________________

16. ______________________

17. ______________________

18. ______________________

19. ______________________

20. ______________________

Solve the following problem.

14. The hypotenuse of a right triangle measures 9 inches, and one leg measures 3 inches. Find the measure of the other leg.

(a) Give its length in simplified radical form.

(b) Round the answer to the nearest thousandth.

Rationalize each denominator.

15. $\frac{5\sqrt{2}}{\sqrt{7}}$

16. $\sqrt{\frac{2}{3x}}$ $(x > 0)$

17. $\frac{-2}{\sqrt[3]{4}}$

18. $\frac{-3}{4 - \sqrt{3}}$

Solve each equation.

19. $\sqrt{x + 1} = 5 - x$

20. $3\sqrt{x} - 1 = 2x$

NAME DATE HOUR

CUMULATIVE REVIEW EXERCISES CHAPTERS R–8

Simplify each of the following.

1. $3(6 + 7) + 6 \cdot 4 - 3^2$

2. $\dfrac{3(6 + 7) + 3}{2(4) - 1}$

3. $|-6| - |-3|$

4. $-9 + 14 + 11 + (-3 + 5)$

5. $13 - [-4 - (-2)]$

6. $-2.523 + 8.674 - 1.928$

Solve each equation or inequality.

7. $5(k - 4) - k = k - 11$

8. $-\dfrac{3}{4}y \leq 12$

9. $5z + 3 - 4 > 2z + 9 + z$

Solve the following problem.

10. The perimeter of a rectangle is 56 meters. The length of the rectangle is 7 meters more than the width. Find the length and the width of the rectangle.

Simplify and write each expression without negative exponents. Assume that variables represent positive numbers.

11. $(3x^6)(2x^2y)^2$

12. $\left(\dfrac{3^2y^{-2}}{2^{-1}y^3}\right)^{-3}$

13. Subtract $7x^3 - 8x^2 + 4$ from $10x^3 + 3x^2 - 9$.

14. Divide: $\dfrac{8t^3 - 4t^2 - 14t + 15}{2t + 3}$.

Factor each polynomial completely.

15. $m^2 + 12m + 32$

16. $25t^4 - 36$

17. $12a^2 + 4ab - 5b^2$

18. $81z^2 + 72z + 16$

Solve each quadratic equation.

19. $x^2 - 7x = -12$

20. $(x + 4)(x - 1) = -6$

21. For what real number(s) is the expression $\dfrac{3}{x^2 + 5x - 14}$ undefined?

Multiply or divide as indicated. Express answers in lowest terms.

22. $\dfrac{x^2 - 3x - 4}{x^2 + 3x} \cdot \dfrac{x^2 + 2x - 3}{x^2 - 5x + 4}$

23. $\dfrac{t^2 + 4t - 5}{t + 5} \div \dfrac{t - 1}{t^2 + 8t + 15}$

24. Simplify the complex fraction $\dfrac{\frac{2}{3} + \frac{1}{2}}{\frac{1}{9} - \frac{1}{6}}$.

Add or subtract as indicated. Express answers in lowest terms.

25. $\dfrac{y}{y^2 - 1} + \dfrac{y}{y + 1}$

26. $\dfrac{2}{x + 3} - \dfrac{4}{x - 1}$

Graph each of the following.

27. $-4x + 5y = -20$

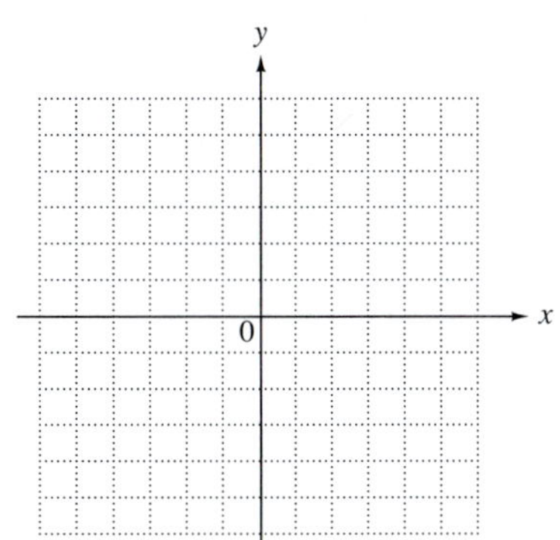

28. $x = 2$

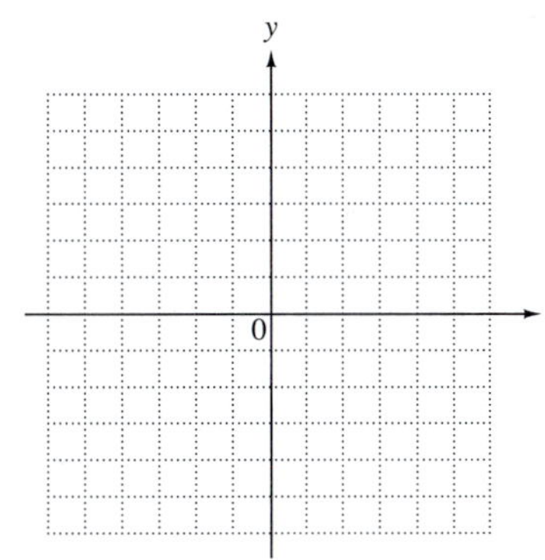

29. $2x - 5y > 10$

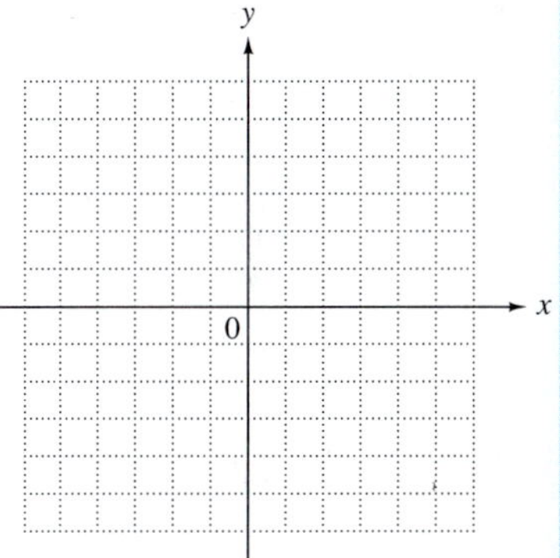

30. Find the slope of the line through the points $(9, -2)$ and $(-3, 8)$.

Solve each of the following systems of equations.

31. $4x - y = 19$
$3x + 2y = -5$

32. $2x - y = 6$
$3y = 6x - 18$

Solve the following problem.

33. A cashier has 20 bills, all of which are ten-dollar or twenty-dollar bills. The total value of the money is \$250. How many of each denomination does the cashier have?

Simplify each expression if possible. Assume all variables represent nonnegative real numbers.

34. $\frac{\sqrt{56}}{\sqrt{7}}$

35. $\sqrt{27} - 2\sqrt{12} + 6\sqrt{75}$

36. $\frac{2}{\sqrt{3} + \sqrt{5}}$

37. $\sqrt{200x^2y^5}$

38. $\frac{5 + \sqrt{75}}{10}$

39. $(3\sqrt{2} + 1)(4\sqrt{2} - 3)$

40. Solve the equation $\sqrt{x} + 2 = x - 10$.

parabola

$2x^2 + 4x - 1 = 0$ $\pm$

9 Quadratic Equations

9.1 SOLVING QUADRATIC EQUATIONS BY THE SQUARE ROOT PROPERTY

OBJECTIVES

1. Solve equations of the form x^2 = a number.
2. Solve equations of the form $(ax + b)^2$ = a number.

FOR EXTRA HELP

Tape 12

SSM pp. 351–353

MAC: B IBM: B

Recall that a *quadratic equation* is an equation that can be written in the standard form $ax^2 + bx + c = 0$ for real numbers a, b, and c, with $a \neq 0$. In Chapter 4 these equations were solved by factoring. As mentioned there, not all quadratic equations can easily be solved by factoring (for instance, $x^2 - x + 1 = 0$ cannot). We show other ways to solve quadratic equations in this chapter. A quadratic equation such as $(x - 3)^2 = 16$, with the square of a binomial equal to a number, can be solved by taking square roots.

1 The *square root property of equations* justifies taking square roots of both sides of an equation.

SQUARE ROOT PROPERTY OF EQUATIONS

If b is a positive number and if $a^2 = b$, then

$$a = \sqrt{b} \quad \text{or} \quad a = -\sqrt{b}.$$

Note When we solve an equation, we want to find *all* values of the variable that satisfy the equation. Therefore, we want both the positive and the negative square roots of b.

EXAMPLE 1 *Solving a Quadratic Equation by the Square Root Property*

Solve each equation. Write radicals in simplified form.

(a) $x^2 = 16$

By the square root property, if $x^2 = 16$, then

$$x = \sqrt{16} = 4 \quad \text{or} \quad x = -\sqrt{16} = -4.$$

An abbreviation for $x = 4$ or $x = -4$ is written $x = \pm 4$ (read "plus or minus 4"). Check each solution by substituting back in the original equation.

(b) $z^2 = 5$

The solutions are $z = \sqrt{5}$ or $z = -\sqrt{5}$, which may be written $\pm\sqrt{5}$.

1. Solve each equation. Write radicals in simplified form.

(a) $k^2 = 49$

(b) $b^2 = 11$

(c) $c^2 = 12$

(d) $x^2 = -9$

2. Solve each equation.

(a) $(m + 2)^2 = 36$

(b) $(p - 4)^2 = 3$

3. Solve each equation.

(a) $(2x - 5)^2 = 18$

(b) $(7z - 1)^2 = -1$

ANSWERS
1. (a) $-7, 7$ (b) $-\sqrt{11}, \sqrt{11}$
(c) $-2\sqrt{3}, 2\sqrt{3}$
(d) no real number solution
2. (a) $-8, 4$ (b) $4 + \sqrt{3}, 4 - \sqrt{3}$
3. (a) $\frac{5 + 3\sqrt{2}}{2}, \frac{5 - 3\sqrt{2}}{2}$
(b) no real number solution

(c) $m^2 = 8$

Use the square root property to get $m = \sqrt{8}$ or $m = -\sqrt{8}$. Simplify $\sqrt{8}$ as $\sqrt{8} = 2\sqrt{2}$, so $m = 2\sqrt{2}$ or $m = -2\sqrt{2}$.

(d) $y^2 = -4$

Because -4 is a negative number and because the square of a real number cannot be negative, there is no real number solution for this equation. (The square root property cannot be used because of the requirement that b must be positive.) ■

WORK PROBLEM 1 AT THE SIDE.

2 The equation $(x - 3)^2 = 16$ also can be solved with the square root property of equations. If $(x - 3)^2 = 16$, then

$$x - 3 = 4 \quad \text{or} \quad x - 3 = -4 \qquad \text{Square root property}$$
$$x = 7 \quad \text{or} \quad x = -1. \qquad \text{Add 3.}$$

Check both answers in the original equation.

$$(7 - 3)^2 = 4^2 = 16 \quad \text{and} \quad (-1 - 3)^2 = (-4)^2 = 16$$

Both 7 and -1 are solutions.

■ EXAMPLE 2 *Solving a Quadratic Equation by the Square Root Property*

Solve $(x - 1)^2 = 6$.

$$x - 1 = \sqrt{6} \quad \text{or} \quad x - 1 = -\sqrt{6} \qquad \text{Square root property}$$
$$x = 1 + \sqrt{6} \quad \text{or} \quad x = 1 - \sqrt{6} \qquad \text{Add 1.}$$

Check: $(1 + \sqrt{6} - 1)^2 = (\sqrt{6})^2 = 6;$
$(1 - \sqrt{6} - 1)^2 = (-\sqrt{6})^2 = 6.$

The solutions are $1 + \sqrt{6}$ and $1 - \sqrt{6}$. ■

WORK PROBLEM 2 AT THE SIDE.

■ EXAMPLE 3 *Solving a Quadratic Equation by the Square Root Property*

Solve the equation $(3r - 2)^2 = 27$.

$$3r - 2 = \sqrt{27} \quad \text{or} \quad 3r - 2 = -\sqrt{27} \qquad \text{Square root property}$$

Now simplify the radical: $\sqrt{27} = \sqrt{9 \cdot 3} = \sqrt{9} \cdot \sqrt{3} = 3\sqrt{3}$.

$$3r - 2 = 3\sqrt{3} \quad \text{or} \quad 3r - 2 = -3\sqrt{3}$$
$$3r = 2 + 3\sqrt{3} \quad \text{or} \quad 3r = 2 - 3\sqrt{3} \qquad \text{Add 2.}$$
$$r = \frac{2 + 3\sqrt{3}}{3} \quad \text{or} \quad r = \frac{2 - 3\sqrt{3}}{3} \qquad \text{Divide by 3.}$$

The solutions are $\frac{2 + 3\sqrt{3}}{3}$ and $\frac{2 - 3\sqrt{3}}{3}$. ■

■ EXAMPLE 4 *Recognizing a Quadratic Equation with No Solution*

Solve $(x + 3)^2 = -9$.

The square root of -9 is not a real number. There is no real number solution for this equation. ■

WORK PROBLEM 3 AT THE SIDE.

EXERCISES

Answer true or false to each of the following. If false, tell why.

1. If $k > 0$, then $x^2 = k$ has exactly two real solutions.

2. If $k < 0$, then $x^2 = k$ has no real solutions.

3. If $k = 0$, then $x^2 = k$ has no real solutions.

4. If k is a positive perfect square, then $x^2 = k$ has two rational solutions.

5. If k is a prime number, then $x^2 = k$ has two irrational solutions.

6. If k is a positive integer, then $x^2 = k$ must have two rational solutions.

7. When a student was asked to solve $x^2 = 81$, she wrote as her answer "9." Her teacher did not give her full credit, and the student argued, saying that because $9^2 = 81$, her answer had to be correct. Why was her answer not completely correct?

8. Explain why $x^2 = -6$ has no real solutions.

Solve each equation by using the square root property. Express all radicals in simplest form. See Example 1.

9. $x^2 = 81$

10. $y^2 = 121$

11. $k^2 = 14$

12. $m^2 = 22$

13. $t^2 = 48$

14. $x^2 = 54$

15. $y^2 = \frac{25}{4}$

16. $m^2 = \frac{36}{121}$

17. $z^2 = 2.25$

18. $w^2 = 56.25$

19. $r^2 - 3 = 0$

20. $x^2 - 13 = 0$

Solve each equation by using the square root property. Express all radicals in simplest form. See Examples 2–4.

21. $(x - 3)^2 = 25$

22. $(y - 7)^2 = 16$

23. $(z + 5)^2 = -13$

24. $(m + 2)^2 = -17$

25. $(x - 8)^2 = 27$

26. $(y - 5)^2 = 40$

27. $(3k + 2)^2 = 49$

28. $(5t + 3)^2 = 36$

29. $(4x - 3)^2 = 9$

30. $(7y - 5)^2 = 25$

31. $(5 - 2x)^2 = 30$

32. $(3 - 2a)^2 = 70$

33. $(3k + 1)^2 = 18$

34. $(5z + 6)^2 = 75$

35. $\left(\frac{1}{2}x + 5\right)^2 = 12$

36. $\left(\frac{1}{3}y + 4\right)^2 = 27$

37. $(4k - 1)^2 - 48 = 0$

38. $(2s - 5)^2 - 180 = 0$

39. Johnny solved the equation in Exercise 31 and wrote his answer as $\frac{5 + \sqrt{30}}{2}, \frac{5 - \sqrt{30}}{2}$. Linda solved the same equation and wrote her answer as $\frac{-5 + \sqrt{30}}{-2}, \frac{-5 - \sqrt{30}}{-2}$. The teacher gave them both full credit. Explain why both students were correct, although their answers seem to differ.

40. In the solutions found in Example 3 of this section, why is it not valid to reduce the answers by dividing out the threes in the numerator and denominator?

If one side of a quadratic equation is a perfect square trinomial, then we can factor the trinomial and use the method of Examples 2–4 to solve it. Solve each of the following equations in this way.

41. $x^2 + 4x + 4 = 25$

42. $x^2 + 6x + 9 = 100$

43. The area A of a circle with radius r is given by the formula

$$A = \pi r^2.$$

If a circle has area 36π square inches, what is its radius?

44. The surface area S of a sphere with radius r is given by the formula

$$S = 4\pi r^2.$$

If a sphere has surface area 64π square feet, what is its radius?

PREVIEW EXERCISES

Simplify all radicals, and combine like terms. Express fractions in lowest terms. See Sections 8.3 and 8.4.

45. $\frac{4}{5} + \sqrt{\frac{48}{25}}$

46. $12 + \sqrt{\frac{2}{3}}$

47. $\frac{6 + \sqrt{24}}{8}$

Factor these perfect square trinomials. See Section 4.4.

48. $y^2 - 10y + 25$

49. $x^2 - 7x + \frac{49}{4}$

50. $z^2 + z + \frac{1}{4}$

9.2 SOLVING QUADRATIC EQUATIONS BY COMPLETING THE SQUARE

1 The properties studied so far are not enough to solve the equation

$$x^2 + 6x + 7 = 0.$$

If we could write the equation in a form like $(x + 3)^2 = 2$, we could solve it with the square root property discussed in the previous section. To do that, we need to have a perfect square trinomial on one side. The next example shows how to rewrite the equation $x^2 + 6x + 7$ so it can be solved by that method.

WORK PROBLEM 1 AT THE SIDE.

OBJECTIVES

1. Solve quadratic equations by completing the square when the coefficient of the squared term is 1.
2. Solve quadratic equations by completing the square when the coefficient of the squared term is not 1.
3. Simplify an equation before solving.
4. Solve applied problems that require quadratic equations.

FOR EXTRA HELP

Tape 12 | SSM pp. 353–357 | MAC: B IBM: B

1. As a review, factor each of these perfect square trinomials.

(a) $x^2 + 6x + 9$

(b) $q^2 - 20q + 100$

EXAMPLE 1 *Rewriting an Equation to Use the Square Root Property*

Solve $x^2 + 6x + 7 = 0$.

Start by subtracting 7 from each side of the equation.

$$x^2 + 6x = -7$$

The quantity on the left-hand side of $x^2 + 6x = -7$ must be made into a perfect square trinomial. The expression $x^2 + 6x + 9$ is a perfect square, since

$$x^2 + 6x + 9 = (x + 3)^2.$$

Therefore, if 9 is added to each side, the equation will have a perfect square trinomial on the left-hand side, as needed.

$$x^2 + 6x + 9 = -7 + 9 \quad \text{Add 9.}$$
$$(x + 3)^2 = 2 \quad \text{Factor.}$$

Now use the square root property to complete the solution.

$$x + 3 = \sqrt{2} \quad \text{or} \quad x + 3 = -\sqrt{2}$$
$$x = -3 + \sqrt{2} \quad \text{or} \quad x = -3 - \sqrt{2}$$

The solutions of the original equation are $-3 + \sqrt{2}$ and $-3 - \sqrt{2}$. Check by substituting $-3 + \sqrt{2}$ and $-3 - \sqrt{2}$ for x in the original equation. ■

The process of changing the form of the equation in Example 1 from

$$x^2 + 6x + 7 = 0 \quad \text{to} \quad (x + 3)^2 = 2$$

is called **completing the square.** When completing the square, only the *form* of the equation is changed. To see this, simplify $(x + 3)^2 = 2$; the result will be $x^2 + 6x + 7 = 0$.

EXAMPLE 2 *Solving a Quadratic Equation by Completing the Square*

Solve $m^2 - 8m = 5$.

A suitable number must be added to each side to make the left side a perfect square. This number can be found as follows: Recall from Chapter 3 that

$$(m + a)^2 = m^2 + 2am + a^2.$$

ANSWERS

1. (a) $(x + 3)^2$ (b) $(q - 10)^2$

2. Solve $a^2 + 4a = 1$ by completing the square.

We want to find the value of a^2, the number to be added to each side. First we must find a. Here, the middle term of the trinomial $m^2 - 8m + a^2$ is $-8m$, so

$$2am = -8m$$
$$2a = -8 \quad \text{Divide each side by } m.$$
$$a = -4. \quad \text{Divide each side by 2.}$$

Then $a^2 = (-4)^2 = 16$, and 16 should be added to each side of the given equation.

$$m^2 - 8m = 5$$
$$m^2 - 8m + 16 = 5 + 16 \quad \text{Add 16.} \qquad \textbf{(1)}$$

The trinomial $m^2 - 8m + 16$ is a perfect square trinomial. Factor this trinomial to get

$$m^2 - 8m + 16 = (m - 4)^2.$$

Equation (1) becomes

$$(m - 4)^2 = 21.$$

Now use the square root property.

$$m - 4 = \sqrt{21} \quad \text{or} \quad m - 4 = -\sqrt{21}$$
$$m = 4 + \sqrt{21} \quad \text{or} \quad m = 4 - \sqrt{21}$$

The solutions are

$$4 + \sqrt{21} \quad \text{and} \quad 4 - \sqrt{21}. \; \blacksquare$$

Let us summarize what we did to find $a = -4$ above. The coefficient of m in the middle term was -8.

1. We multiplied -8 by $\frac{1}{2}$ (took half of -8) to get -4.
2. We squared -4 to get 16.
3. We added 16 to each side of the given equation.

Thus, to find the number to add to both sides, we take half the coefficient of the first degree term, square it, and add it to both sides.

WORK PROBLEM 2 AT THE SIDE.

2 The process of completing the square discussed above requires the coefficient of the squared term to be 1. With an equation of the form $ax^2 + bx + c = 0$, to get 1 as a coefficient of x^2, first divide each side of the equation by a. The next examples illustrate this approach.

■ EXAMPLE 3 *Solving a Quadratic Equation by Completing the Square*

Solve $4y^2 + 16y = 9$.

Before completing the square, the coefficient of y^2 must be 1. Here the coefficient of y^2 is 4. Make the coefficient 1 by dividing each side of the equation by 4.

$$y^2 + 4y = \frac{9}{4} \quad \text{Divide by 4.}$$

ANSWERS
2. $-2 + \sqrt{5}, -2 - \sqrt{5}$

Next, complete the square by taking half the coefficient of y, or $(\frac{1}{2})(4) = 2$, and squaring the result: $2^2 = 4$. Add 4 to each side of the equation, and perform the addition on the right-hand side.

$$y^2 + 4y + 4 = \frac{9}{4} + 4 \quad \text{Add 4.}$$

$$y^2 + 4y + 4 = \frac{25}{4} \quad \text{Combine terms.}$$

$$(y + 2)^2 = \frac{25}{4} \quad \text{Factor.}$$

Use the square root property of equations and solve for y.

$$y + 2 = \frac{5}{2} \quad \text{or} \quad y + 2 = -\frac{5}{2} \quad \text{Square root property}$$

$$y = -2 + \frac{5}{2} \quad \text{or} \quad y = -2 - \frac{5}{2} \quad \text{Subtract 2.}$$

$$y = \frac{1}{2} \quad \text{or} \quad y = -\frac{9}{2} \quad \text{Combine terms.}$$

Check:

$$4y^2 + 16y = 9 \qquad\qquad 4y^2 + 16y = 9$$

$$4\left(\frac{1}{2}\right)^2 + 16\left(\frac{1}{2}\right) = 9 \;? \qquad 4\left(-\frac{9}{2}\right)^2 + 16\left(-\frac{9}{2}\right) = 9 \;?$$

$$4\left(\frac{1}{4}\right) + 8 = 9 \;? \qquad 4\left(\frac{81}{4}\right) - 72 = 9 \;?$$

$$1 + 8 = 9 \quad \text{True} \qquad 81 - 72 = 9 \quad \text{True}$$

The two solutions are $\frac{1}{2}$ and $-\frac{9}{2}$. ■

The steps in solving a quadratic equation by completing the square are summarized below.

COMPLETING THE SQUARE

Use *completing the square* to solve the quadratic equation $ax^2 + bx + c = 0$ as follows.

Step 1 If the coefficient of the squared term is 1, proceed to Step 2. If the coefficient of the squared term is not 1 but some other nonzero number, divide each side of the equation by this coefficient. This gives an equation that has 1 as coefficient of x^2.

Step 2 Make sure that all terms with variables are on one side of the equals sign and that all constants are on the other side.

Step 3 Take half the coefficient of x and square the result. Add the square to each side of the equation. By factoring, the side containing the variables can now be written as a perfect square.

Step 4 Apply the square root property of equations.

WORK PROBLEM 3 AT THE SIDE.

3. Solve by completing the square.

(a) $9m^2 + 18m + 5 = 0$

(b) $4k^2 - 24k + 11 = 0$

ANSWERS

3. (a) $-\frac{1}{3}, -\frac{5}{3}$ **(b)** $\frac{11}{2}, \frac{1}{2}$

4. Solve $3x^2 + 5x - 2 = 0$ by completing the square.

EXAMPLE 4 *Solving a Quadratic Equation by Completing the Square*

Solve $2x^2 - 7x = 9$.

Divide each side of the equation by 2 to get a coefficient of 1 for the x^2 term.

$$x^2 - \frac{7}{2}x = \frac{9}{2} \quad \text{Divide by 2.}$$

Now take half the coefficient of x and square it. Half of $-\frac{7}{2}$ is $-\frac{7}{4}$, and $(-\frac{7}{4})^2 = \frac{49}{16}$. Add $\frac{49}{16}$ to each side of the equation, and write the left side as a perfect square.

$$x^2 - \frac{7}{2}x + \frac{49}{16} = \frac{9}{2} + \frac{49}{16} \quad \text{Add } \tfrac{49}{16}.$$

$$\left(x - \frac{7}{4}\right)^2 = \frac{121}{16} \quad \text{Factor.}$$

Use the square root property.

$$x - \frac{7}{4} = \sqrt{\frac{121}{16}} \quad \text{or} \quad x - \frac{7}{4} = -\sqrt{\frac{121}{16}}$$

Because $\sqrt{\frac{121}{16}} = \frac{11}{4}$,

$$x - \frac{7}{4} = \frac{11}{4} \quad \text{or} \quad x - \frac{7}{4} = -\frac{11}{4}$$

$$x = \frac{18}{4} \quad \text{or} \quad x = -\frac{4}{4} \quad \text{Add } \tfrac{7}{4}.$$

$$x = \frac{9}{2} \quad \text{or} \quad x = -1.$$

Check that the solutions are $\frac{9}{2}$ and -1. ■

WORK PROBLEM 4 AT THE SIDE.

EXAMPLE 5 *Solving a Quadratic Equation by Completing the Square*

Solve $4p^2 + 8p + 5 = 0$.

$$p^2 + 2p + \frac{5}{4} = 0 \quad \text{Divide each side by 4.}$$

$$p^2 + 2p = -\frac{5}{4} \quad \text{Subtract } \tfrac{5}{4} \text{ from each side.}$$

The coefficient of p is 2. Take half of 2, square the result, and add this square to each side. The left-hand side can then be written as a perfect square.

$$p^2 + 2p + 1 = -\frac{5}{4} + 1 \quad \text{Add 1 on each side.}$$

$$(p + 1)^2 = -\frac{1}{4} \quad \text{Factor.}$$

ANSWERS

4. $-2, \frac{1}{3}$

The square root of $-\frac{1}{4}$ is not a real number, so the square root property does not apply. This equation has no real number solutions.* ■

WORK PROBLEM 5 AT THE SIDE.

5. Solve $5v^2 + 3v + 1 = 0$ by completing the square.

3 The next example shows how to simplify an equation before solving it.

■ EXAMPLE 6 *Solving a Quadratic Equation by Completing the Square*

Solve $(m - 2)(m + 1) = 5$.

Before we can use the method of completing the square, the equation must be in the form $ax^2 + bx + c = 0$. Start by multiplying on the left.

$$(m - 2)(m + 1) = 5$$
$$m^2 - m - 2 = 5 \quad \text{Use FOIL.}$$
$$m^2 - m = 7 \quad \text{Add 2.}$$

Now complete the square. Half of -1 is $-\frac{1}{2}$, and $(-\frac{1}{2})^2 = \frac{1}{4}$. Add $\frac{1}{4}$ to each side.

$$m^2 - m + \frac{1}{4} = 7 + \frac{1}{4} \quad \text{Add } \tfrac{1}{4}.$$
$$\left(m - \frac{1}{2}\right)^2 = \frac{29}{4} \quad \text{Factor; combine terms.}$$
$$m - \frac{1}{2} = \sqrt{\frac{29}{4}} \quad \text{or} \quad m - \frac{1}{2} = -\sqrt{\frac{29}{4}} \quad \text{Square root property}$$
$$m - \frac{1}{2} = \frac{\sqrt{29}}{2} \quad \text{or} \quad m - \frac{1}{2} = -\frac{\sqrt{29}}{2}$$
$$m = \frac{1}{2} + \frac{\sqrt{29}}{2} \quad \text{or} \quad m = \frac{1}{2} - \frac{\sqrt{29}}{2} \quad \text{Add } \tfrac{1}{2}.$$
$$m = \frac{1 + \sqrt{29}}{2} \quad \text{or} \quad m = \frac{1 - \sqrt{29}}{2}$$

We check each of these solutions by substituting it in the original equation. For example, let $m = \frac{1 + \sqrt{29}}{2}$.

Check:
$$(m - 2)(m + 1) = 5$$
$$\left(\frac{1 + \sqrt{29}}{2} - 2\right)\left(\frac{1 + \sqrt{29}}{2} + 1\right) = 5 \quad ?$$
$$\left(\frac{1 + \sqrt{29}}{2} - \frac{4}{2}\right)\left(\frac{1 + \sqrt{29}}{2} + \frac{2}{2}\right) = 5 \quad ?$$
$$\left(\frac{-3 + \sqrt{29}}{2}\right)\left(\frac{3 + \sqrt{29}}{2}\right) = 5 \quad ?$$
$$\frac{(-3 + \sqrt{29})(3 + \sqrt{29})}{2 \cdot 2} = 5 \quad ?$$
$$\frac{-9 - 3\sqrt{29} + 3\sqrt{29} + 29}{4} = 5 \quad ?$$
$$\frac{20}{4} = 5 \quad \text{True} \ ■$$

*The equation in Example 5 has no *real number* solution. In the context of another number system, called the *complex numbers,* however, this equation does have a solution. The complex numbers include numbers whose squares are negative. These numbers are discussed in Intermediate and College Algebra courses.

ANSWER
5. no real number solution

6. Solve each equation.

(a) $r^2 + 1 = 3r$

(b) $(x + 2)(x + 1) = 5$

7. Suppose a ball is propelled upward with an initial velocity of 128 feet per second. Its height at time t is $s = -16t^2 + 128t$. At what times will it be 48 feet above the ground? Give answers to the nearest tenth.

ANSWERS

6. (a) $\frac{3 + \sqrt{5}}{2}, \frac{3 - \sqrt{5}}{2}$

(b) $\frac{-3 + \sqrt{21}}{2}, \frac{-3 - \sqrt{21}}{2}$

7. .4 and 7.6 seconds

WORK PROBLEM 6 AT THE SIDE.

4 There are many practical applications of quadratic equations. The next example illustrates an application from physics.

EXAMPLE 7 *Solving a Velocity Problem*

If a ball is thrown into the air from ground level with an initial velocity of 64 feet per second, its height s (in feet) in t seconds is given by the formula $s = -16t^2 + 64t$. How long will it take the ball to reach a height of 48 feet?

Since s represents the height, let $s = 48$ in the formula to get

$$48 = -16t^2 + 64t.$$

To solve this equation for the time, t, by completing the square, we should divide both sides by -16. Let us also reverse the sides of the equation.

$-3 = t^2 - 4t$	Divide by -16.
$t^2 - 4t = -3$	Reverse the sides.
$t^2 - 4t + 4 = -3 + 4$	Add $[(\frac{1}{2})4]^2 = 4$.
$(t - 2)^2 = 1$	Factor.
$t - 2 = 1$ or $t - 2 = -1$	Square root property
$t = 3$ or $t = 1$	Add 2.

You may wonder how we can get two correct answers for the time required for the ball to reach a height of 48 feet. The ball reaches that height twice, once on the way up and again on the way down. So it takes 1 second to reach 48 feet on the way up, and then after 3 seconds, the ball reaches 48 feet again on the way down. ■

WORK PROBLEM 7 AT THE SIDE.

NAME DATE HOUR

9.2 EXERCISES

1. What is the first step that you would perform in order to solve the equation $4x^2 + 8x = 3$ by completing the square?

2. Why is it not possible to solve the equation $2x^3 - x - 1 = 0$ by completing the square?

Solve each equation by completing the square. See Examples 1 and 2.

3. $x^2 - 4x = -3$

4. $y^2 - 2y = 8$

5. $x^2 + 2x - 5 = 0$

6. $r^2 + 4r + 1 = 0$

7. $z^2 + 6z + 9 = 0$

8. $k^2 - 8k + 16 = 0$

Find the number that should be added to each expression to make it a perfect square. See Example 2.

9. $y^2 + 14y$

10. $z^2 + 18z$

11. $k^2 - 5k$

12. $m^2 - 9m$

13. $r^2 + \frac{1}{2}r$

14. $s^2 - \frac{1}{3}s$

15. Which one of the following steps is an appropriate way to begin solving the quadratic equation

$$2x^2 - 4x = 9$$

by completing the square?

(a) Add 4 to both sides of the equation.
(b) Factor the left side as $2x(x - 2)$.
(c) Factor the left side as $x(2x - 4)$.
(d) Divide both sides by 2.

16. In Example 3 of Section 4.5, we solved the quadratic equation

$$4p^2 - 26p + 40 = 0$$

by the factoring method. If we were to solve this by the method of completing the square, would we get the same solutions, $\frac{5}{2}$ and 4?

Solve each equation by completing the square. See Examples 3–6.

17. $2x^2 - 4x - 5 = 0$

18. $2x^2 - 6x - 3 = 0$

19. $4y^2 + 4y = 3$

20. $9x^2 + 3x = 2$

21. $2p^2 - 2p + 3 = 0$

22. $3q^2 - 3q + 4 = 0$

23. $3k^2 + 7k = 4$

24. $2k^2 + 5k = 1$

25. $(x + 3)(x - 1) = 5$

26. $(y - 8)(y + 2) = 24$

27. $-x^2 + 2x = -5$

28. $-r^2 + 3r = -2$

Solve each problem. See Example 7.

29. If an object is thrown upward from ground level with an initial velocity of 96 feet per second, its height, s, (in feet) in t seconds is given by the formula $s = -16t^2 + 96t$. How long will take for the object to be at a height of 80 feet?

30. How long will it take the object described in Exercise 29 to be at a height of 100 feet? Round your answers to the nearest tenth.

PREVIEW EXERCISES

Write each quotient in lowest terms. Simplify the radicals. See Section 8.5.

31. $\frac{8 - 6\sqrt{3}}{6}$

32. $\frac{4 + \sqrt{28}}{2}$

33. $\frac{6 - \sqrt{45}}{6}$

34. $\frac{8 + \sqrt{32}}{4}$

9.3 SOLVING QUADRATIC EQUATIONS BY THE QUADRATIC FORMULA

Completing the square can be used to solve any quadratic equation, but the method is tedious. In this section we complete the square on the quadratic equation $ax^2 + bx + c = 0$ to get the *quadratic formula,* a formula that gives the solution for any quadratic equation. (Note that $a \neq 0$, or we would have a linear, not a quadratic, equation.)

OBJECTIVES

1. Identify the values of *a, b,* and *c* in a quadratic equation.
2. Use the quadratic formula to solve quadratic equations.
3. Solve quadratic equations with only one solution.
4. Solve quadratic equations with fractions.

FOR EXTRA HELP

Tape 12

SSM pp. 357–362

MAC: B
IBM: B

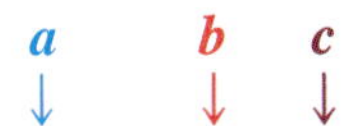
1 The first step in solving a quadratic equation by this new method is to identify the values of *a, b,* and *c* in the standard form of the quadratic equation.

EXAMPLE 1 *Determining Values of a, b, and c in a Quadratic Equation*

Match the coefficients of each of the following quadratic equations with the letters *a, b,* and *c* of the standard quadratic equation

$$ax^2 + bx + c = 0.$$

(a) $\overset{a}{2}x^2 + \overset{b}{3}x \overset{c}{-5} = 0$

In this example $a = 2$, $b = 3$, and $c = -5$.

(b) $-x^2 + 2 = 6x$

First rewrite the equation with 0 on one side to match the standard form $ax^2 + bx + c = 0$.

$$-x^2 + 2 = 6x$$

$$-x^2 - 6x + 2 = 0 \qquad \text{Subtract } 6x.$$

Now identify $a = -1$, $b = -6$, and $c = 2$.

(c) $(2x - 7)(x + 4) = -23$

Put the equation in standard form.

$$(2x - 7)(x + 4) = -23$$

$$2x^2 + x - 28 = -23 \qquad \text{Use FOIL on the left.}$$

$$2x^2 + x - 5 = 0 \qquad \text{Add 23 on each side.}$$

Now, identify the values: $a = 2$, $b = 1$, $c = -5$. ■

WORK PROBLEM 1 AT THE SIDE.

1. Match the coefficients of each of the following quadratic equations with the letters *a, b,* and *c* of the standard quadratic equation $ax^2 + bx + c = 0$.

(a) $5x^2 + 2x - 1 = 0$

(b) $3m^2 = m - 2$

(c) $p(p + 5) = 4$

To develop the quadratic formula, we follow the steps given in the previous section for completing the square on $ax^2 + bx + c = 0$.

$$x^2 + \frac{b}{a}x + \frac{c}{a} = 0 \qquad \text{Divide by } a.$$

$$x^2 + \frac{b}{a}x = -\frac{c}{a} \qquad \text{Subtract } \frac{c}{a}.$$

$$x^2 + \frac{b}{a}x + \frac{b^2}{4a^2} = -\frac{c}{a} + \frac{b^2}{4a^2} \qquad \text{Add } \left(\frac{1}{2} \cdot \frac{b}{a}\right)^2 = \frac{b^2}{4a^2}.$$

$$\left(x + \frac{b}{2a}\right)^2 = -\frac{c}{a} + \frac{b^2}{4a^2} \qquad \text{Factor.}$$

ANSWERS

1. (a) $a = 5, b = 2, c = -1$
(b) $a = 3, b = -1, c = 2$
(c) $a = 1, b = 5, c = -4$

2. Complete these steps to simplify the right side of the equation.

(a) $-\frac{c}{a} + \frac{b^2}{4a^2} = \frac{b^2}{4a^2} - ?$

(b) $\frac{b^2}{4a^2} - \frac{c}{a} = \frac{b^2}{4a^2} - \frac{?}{4a^2}$

(c) $\frac{b^2}{4a^2} - \frac{4ac}{4a^2} = \frac{?}{4a^2}$

ANSWERS

2. (a) $\frac{c}{a}$ (b) $4ac$ (c) $b^2 - 4ac$

Now simplify the right-hand side of the equation.

WORK PROBLEM 2 AT THE SIDE.

$$\left(x + \frac{b}{2a}\right)^2 = \frac{b^2 - 4ac}{4a^2} \qquad \text{Problem 2 at the side}$$

$$x + \frac{b}{2a} = \pm\sqrt{\frac{b^2 - 4ac}{4a^2}} \qquad \text{Square root property}$$

Simplify the radical.

$$\sqrt{\frac{b^2 - 4ac}{4a^2}} = \frac{\sqrt{b^2 - 4ac}}{\sqrt{4a^2}} = \frac{\sqrt{b^2 - 4ac}}{2a}$$

Write the solutions as follows.

$$x + \frac{b}{2a} = \pm\frac{\sqrt{b^2 - 4ac}}{2a}$$

$$x = -\frac{b}{2a} \pm \frac{\sqrt{b^2 - 4ac}}{2a}$$

$$x = \frac{-b \pm \sqrt{b^2 - 4ac}}{2a}$$

The result is called the **quadratic formula,** a key formula that should be memorized.

THE QUADRATIC FORMULA

The solutions of the quadratic equation $ax^2 + bx + c = 0$ are

$$\frac{-b + \sqrt{b^2 - 4ac}}{2a} \quad \text{and} \quad \frac{-b - \sqrt{b^2 - 4ac}}{2a}$$

or

$$x = \frac{-b \pm \sqrt{b^2 - 4ac}}{2a}, \quad a \neq 0.$$

Caution Notice that the fraction bar is under $-b$ as well as the radical. In using this formula, be sure to find the values of $-b \pm \sqrt{b^2 - 4ac}$ first, then divide those results by the value of $2a$.

2 The following examples show how to use the quadratic formula.

EXAMPLE 2 *Solving a Quadratic Equation by the Quadratic Formula*

Use the quadratic formula to solve $2x^2 - 7x - 9 = 0$.

Match the coefficients of the variables with those of the standard quadratic equation

$$ax^2 + bx + c = 0.$$

Here, $a = 2$, $b = -7$, and $c = -9$. Substitute these numbers into the quadratic formula, and simplify the result.

$$x = \frac{-b \pm \sqrt{b^2 - 4ac}}{2a}$$

$$x = \frac{-(-7) \pm \sqrt{(-7)^2 - 4(2)(-9)}}{2(2)} \qquad \text{Let } a = 2,\ b = -7,\ c = -9.$$

$$x = \frac{7 \pm \sqrt{49 + 72}}{4} = \frac{7 \pm \sqrt{121}}{4}$$

$$x = \frac{7 \pm 11}{4} \qquad \sqrt{121} = 11$$

Write the two separate solutions by first using the plus sign, and then using the minus sign:

$$x = \frac{7 + 11}{4} = \frac{18}{4} = \frac{9}{2} \quad \text{or} \quad x = \frac{7 - 11}{4} = \frac{-4}{4} = -1.$$

The solutions of $2x^2 - 7x - 9 = 0$ are $\frac{9}{2}$ and -1. Check by substituting back in the original equation. ■

WORK PROBLEM 3 AT THE SIDE.

■ **EXAMPLE 3** *Solving a Quadratic Equation by the Quadratic Formula*

Solve $x^2 = 2x + 1$.

Find a, b, and c by rewriting the equation in standard form (with 0 on one side). Add $-2x - 1$ to each side of the equation to get

$$x^2 - 2x - 1 = 0.$$

Then $a = 1$, $b = -2$, and $c = -1$. The solution is found by substituting these values into the quadratic formula.

$$x = \frac{-b \pm \sqrt{b^2 - 4ac}}{2a}$$

$$x = \frac{-(-2) \pm \sqrt{(-2)^2 - 4(1)(-1)}}{2(1)} \qquad \text{Let } a = 1,\ b = -2,\ c = -1.$$

$$x = \frac{2 \pm \sqrt{4 + 4}}{2} = \frac{2 \pm \sqrt{8}}{2}$$

$$x = \frac{2 \pm 2\sqrt{2}}{2} \qquad \sqrt{8} = \sqrt{4 \cdot 2} = \sqrt{4} \cdot \sqrt{2} = 2\sqrt{2}$$

Write these solutions in lowest terms by factoring $2 \pm 2\sqrt{2}$ as $2(1 \pm \sqrt{2})$ to get

$$x = \frac{2(1 \pm \sqrt{2})}{2} = 1 \pm \sqrt{2}.$$

The two solutions of the original equation are

$$1 + \sqrt{2} \quad \text{and} \quad 1 - \sqrt{2}.$$ ■

WORK PROBLEM 4 AT THE SIDE.

3. Solve by using the quadratic formula.

(a) $2x^2 + 3x - 5 = 0$

(b) $6p^2 + p = 1$

4. Solve $-y^2 = 8y + 1$ by using the quadratic formula.

ANSWERS

3. (a) $1, -\frac{5}{2}$ **(b)** $-\frac{1}{2}, \frac{1}{3}$

4. $-4 + \sqrt{15}, -4 - \sqrt{15}$

5. Solve $9y^2 - 12y + 4 = 0$.

6. Solve $x^2 - \frac{4}{3}x + \frac{2}{3} = 0$.

ANSWERS

5. $\frac{2}{3}$

6. no real number solution

3 When the quantity under the radical, $b^2 - 4ac$, equals 0, the equation has just one rational number solution. In this case, the trinomial $ax^2 + bx + c$ is a perfect square.

EXAMPLE 4 *Solving a Quadratic Equation by the Quadratic Formula*

Solve $4x^2 + 25 = 20x$.

Write the equation as

$$4x^2 - 20x + 25 = 0. \qquad \text{Subtract } 20x.$$

Here, $a = 4$, $b = -20$, and $c = 25$. By the quadratic formula,

$$x = \frac{-(-20) \pm \sqrt{(-20)^2 - 400}}{8} = \frac{20 \pm 0}{8} = \frac{5}{2}.$$

Since there is just one solution, the trinomial $4x^2 - 20x + 25$ is a perfect square. ■

WORK PROBLEM 5 AT THE SIDE.

4 The final example shows how to solve quadratic equations with fractions.

EXAMPLE 5 *Solving a Quadratic Equation with Fractions by the Quadratic Formula*

Solve the equation

$$\frac{1}{10}t^2 = \frac{2}{5}t - \frac{1}{2}.$$

Eliminate the denominators by multiplying each side of the equation by the common denominator, 10.

$$10\left(\frac{1}{10}t^2\right) = 10\left(\frac{2}{5}t - \frac{1}{2}\right)$$

$$t^2 = 4t - 5$$

$$t^2 - 4t + 5 = 0. \qquad \text{Add } -4t \text{ and } 5.$$

From this form, identify $a = 1$, $b = -4$, and $c = 5$. Use the quadratic formula to complete the solution.

$$t = \frac{4 \pm \sqrt{(-4)^2 - 4(1)(5)}}{2(1)} = \frac{4 \pm \sqrt{16 - 20}}{2} = \frac{4 \pm \sqrt{-4}}{2}$$

The radical $\sqrt{-4}$ is not a real number, so the equation has no real number solution. ■

WORK PROBLEM 6 AT THE SIDE.

9.3 EXERCISES

NAME DATE HOUR

For each equation, write in the form $ax^2 + bx + c = 0$, if necessary. Then identify the values of a, b, and c. Do not actually solve the equation. See Example 1.

1. $4x^2 + 5x - 9 = 0$

$a =$ ______

$b =$ ______

$c =$ ______

2. $8x^2 + 3x - 4 = 0$

$a =$ ______

$b =$ ______

$c =$ ______

3. $3x^2 = 4x + 2$

$a =$ ______

$b =$ ______

$c =$ ______

4. $5x^2 = 3x - 6$

$a =$ ______

$b =$ ______

$c =$ ______

5. $3x^2 = -7x$

$a =$ ______

$b =$ ______

$c =$ ______

6. $9x^2 = 8x$

$a =$ ______

$b =$ ______

$c =$ ______

Use the quadratic formula to solve each equation. Write all radicals in simplified form, and write all answers in lowest terms. See Examples 2–4.

7. $3x^2 + 5x + 1 = 0$

8. $6y^2 - 6y + 1 = 0$

9. $k^2 = -12k + 13$

10. $r^2 = 8r + 9$

11. $p^2 - 4p + 4 = 0$

12. $9x^2 + 6x + 1 = 0$

13. $2x^2 + 12x = -5$

14. $5m^2 + m = 1$

15. $2y^2 = 5 + 3y$

16. $2z^2 = 30 + 7z$

17. $6x^2 + 6x = 0$

18. $4n^2 - 12n = 0$

19. $7x^2 = 12x$

20. $9r^2 = 11r$

21. $x^2 - 24 = 0$

22. $z^2 - 96 = 0$

23. $25x^2 - 4 = 0$

24. $16x^2 - 9 = 0$

25. $3x^2 - 2x + 5 = 10x + 1$

26. $4x^2 - x + 4 = x + 7$

27. $-2x^2 = -3x + 2$

28. $-x^2 = -5x + 20$

29. $2x^2 + x + 5 = 0$

30. $3x^2 + 2x + 8 = 0$

31. If we apply the quadratic formula and find that the value of $b^2 - 4ac$ is negative, what can we conclude about the solutions?

32. If we were to solve the quadratic equation $-2x^2 - 4x + 3 = 0$, we might choose to use $a = -2$, $b = -4$, and $c = 3$. On the other hand, we might decide to multiply both sides by -1 to begin, obtaining the equation $2x^2 + 4x - 3 = 0$, and then use $a = 2$, $b = 4$, and $c = -3$. Show that in either case, we obtain the same solutions.

Use the quadratic formula to solve each equation. See Example 5.

33. $\frac{3}{2}k^2 - k - \frac{4}{3} = 0$

34. $\frac{2}{5}x^2 - \frac{3}{5}x - 1 = 0$

35. $\frac{1}{2}x^2 + \frac{1}{6}x = 1$

36. $\frac{2}{3}y^2 - \frac{4}{9}y = \frac{1}{3}$

37. $.5x^2 = x + .5$

38. $.25x^2 = -1.5x - 1$

39. $\frac{3}{8}x^2 - x + \frac{17}{24} = 0$

40. $\frac{1}{3}y^2 + \frac{8}{9}y + \frac{7}{9} = 0$

41. Solve the formula $S = 2\pi rh + \pi r^2$ for r by first writing it in the form $ar^2 + br + c = 0$, and then using the quadratic formula.

42. Solve the formula $V = \pi r^2 h + \pi R^2 h$ for r, using the method described in Exercise 41.

Use a calculator and the Pythagorean formula to find the lengths of the sides of each right triangle. Round to the nearest thousandth.

43.

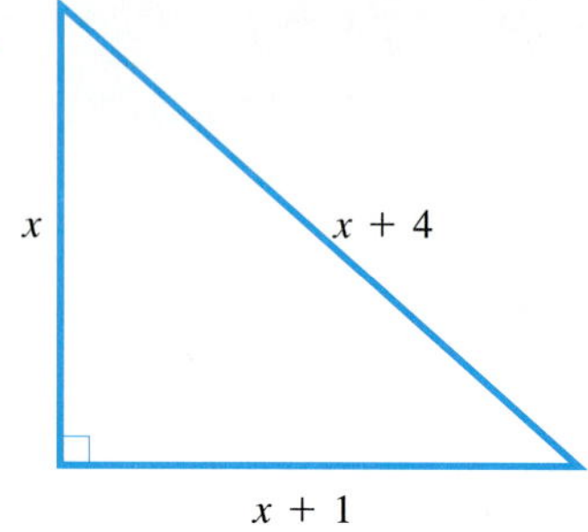

44.

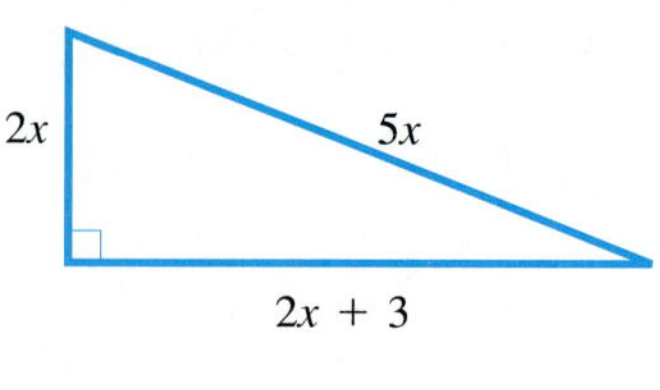

PREVIEW EXERCISES

Graph each linear equation. See Section 6.2.

45. $2x - 3y = 6$

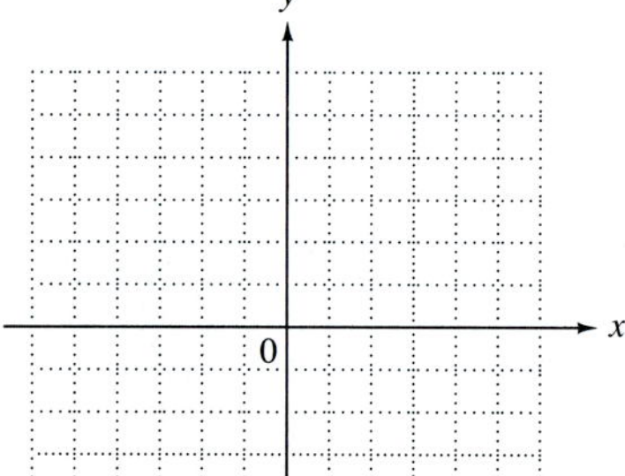

46. $y = 4x - 3$

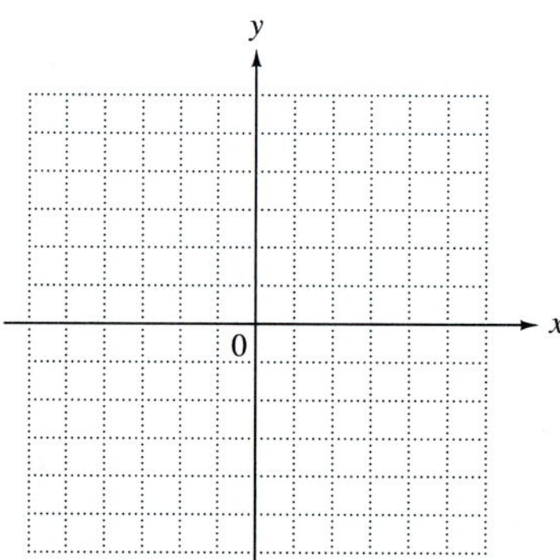

47. $3x + 5y = 15$

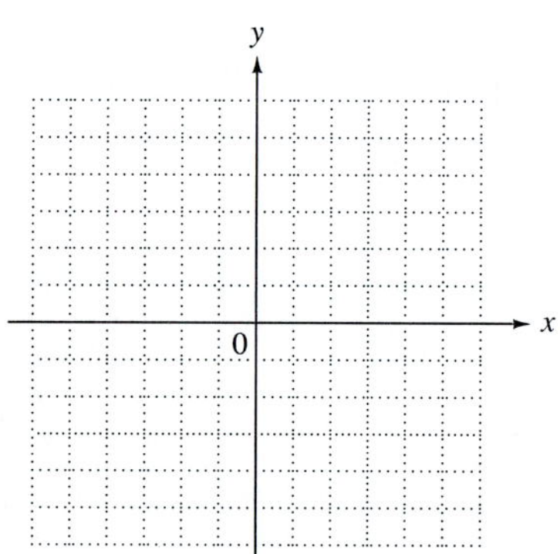

Evaluate each expression if $x = 3$. See Section 1.9.

48. $x^2 - 8$

49. $2x^2 - x + 1$

50. $(x - 1)^2$

NAME DATE HOUR

SUMMARY EXERCISES ON QUADRATIC EQUATIONS

Four methods have now been introduced for solving quadratic equations written in the form $ax^2 + bx + c = 0$. The chart below shows some advantages and some disadvantages of each method.

Method	*Advantages*	*Disadvantages*
1. Factoring	Usually the fastest method.	Not all equations can be solved by factoring. Some factorable polynomials are hard to factor.
2. Completing the square	Can always be used (also, the procedure is useful in other areas of mathematics).	Requires more steps than other methods.
3. Quadratic formula	Can always be used.	More difficult than factoring because of the $\sqrt{b^2 - 4ac}$ expression.
4. Square root method	Simplest method for solving equations of the form $(ax + b)^2 =$ a number.	Few equations are given in this form.

Solve each quadratic equation by the method of your choice.

1. $s^2 = 36$

2. $x^2 + 3x = -1$

3. $y^2 - \frac{100}{81} = 0$

4. $81t^2 = 49$

5. $z^2 - 4z + 3 = 0$

6. $w^2 + 3w + 2 = 0$

7. $z(z - 9) = -20$

8. $x^2 + 3x - 2 = 0$

9. $(3k - 2)^2 = 9$

10. $(2s - 1)^2 = 10$

11. $(x + 6)^2 = 121$

12. $(5k + 1)^2 = 36$

13. $(3r - 7)^2 = 24$

14. $(7p - 1)^2 = 32$

15. $(5x - 8)^2 = -6$

16. $2t^2 + 1 = t$

17. $-2x^2 = -3x - 2$

18. $-2x^2 + x = -1$

19. $8z^2 = 15 + 2z$

20. $3k^2 = 3 - 8k$

21. $0 = -x^2 + 2x + 1$

22. $3x^2 + 5x = -1$

23. $5y^2 - 22y = -8$

24. $y(y + 6) + 4 = 0$

25. $(x + 2)(x + 1) = 10$

26. $16x^2 + 40x + 25 = 0$

27. $4x^2 = -1 + 5x$

28. $2p^2 = 2p + 1$

29. $3m(3m + 4) = 7$

30. $5x - 1 + 4x^2 = 0$

31. $\frac{r^2}{2} + \frac{7r}{4} + \frac{11}{8} = 0$

32. $t(15t + 58) = -48$

33. $9k^2 = 16(3k + 4)$

34. $\frac{1}{5}x^2 + x + 1 = 0$

35. $y^2 - y + 3 = 0$

36. $4m^2 - 11m + 8 = -2$

37. $-3x^2 + 4x = -4$

38. $z^2 - \frac{5}{12}z = \frac{1}{6}$

39. $5k^2 + 19k = 2k + 12$

40. $\frac{1}{2}n^2 - n = \frac{15}{2}$

41. $k^2 - \frac{4}{15} = -\frac{4}{15}k$

42. If $D > 0$ and $\frac{5 + \sqrt{D}}{3}$ is a solution of $ax^2 + bx + c = 0$, what must be another solution of the equation?

9.4 GRAPHING QUADRATIC EQUATIONS IN TWO VARIABLES

OBJECTIVES

1. Graph quadratic equations.
2. Find the vertex of a parabola.

FOR EXTRA HELP

Tape 12 | SSM pp. 364–367 | MAC: B IBM: B

1 In Chapter 6 we saw that the graph of a linear equation in two variables is a straight line that represents all the solutions of the equation. Quadratic equations in two variables, of the form $y = ax^2 + bx + c$, are graphed in this section. Perhaps the simplest quadratic equation is $y = x^2$ (or $y = 1x^2 + 0x + 0$). The graph of this equation cannot be a straight line because only linear equations of the form $Ax + By = C$ have graphs that are straight lines. However, $y = x^2$ can be graphed in much the same way that straight lines were graphed, by finding ordered pairs that satisfy the equation $y = x^2$.

EXAMPLE 1 *Graphing a Quadratic Equation*

Graph $y = x^2$.

Select several values for x; then find the corresponding y-values. For example, selecting $x = 2$ gives

$$y = 2^2 = 4,$$

and so the point (2, 4) is on the graph of $y = x^2$. (Recall that in an ordered pair such as (2, 4), the x-value comes first and the y-value second.)

WORK PROBLEM 1 AT THE SIDE.

1. Complete the following table of values for $y = x^2$.

x	y
3	
2	4
1	
0	
−1	
−2	
−3	

If the points from Problem 1 at the side are plotted on a coordinate system and a smooth curve drawn through them, the graph is as shown in Figure 1. The table of values completed in Problem 1 is shown with the graph. ■

x	y
3	9
2	4
1	1
0	0
−1	1
−2	4
−3	9

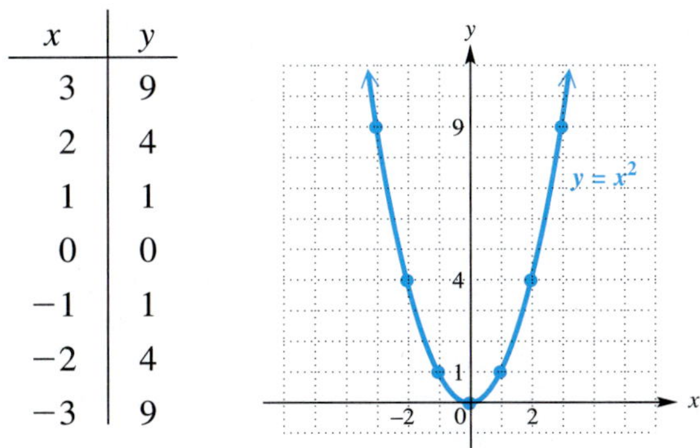

FIGURE 1

The curve in Figure 1 is called a **parabola.** The point (0, 0), the lowest point on this graph, is called the **vertex** of the parabola. The vertical line through the vertex (the y-axis here) is called the **axis** of the parabola. The axis of a parabola is a **line of symmetry** for the graph. If the graph is folded on this line, the two halves will match.

Every equation of the form

$$y = ax^2 + bx + c,$$

with $a \neq 0$, has a graph that is a parabola. Because of its many useful properties, the parabola occurs frequently in real-life applications. For example, if an object is thrown into the air, the path that the object follows is a parabola (ignoring wind resistance). The cross sections of radar, spotlight, and telescope reflectors also form parabolas.

ANSWERS

1.

x	3	1	0	−1	−2	−3
y	9	1	0	1	4	9

EXAMPLE 2 *Graphing a Parabola*

Graph $y = -x^2 + 3$.

Find several ordered pairs. Begin with the intercepts. If $x = 0$,

$$y = -x^2 + 3 = -0^2 + 3 = 3,$$

giving the ordered pair $(0, 3)$. If $y = 0$,

$$\begin{aligned} y &= -x^2 + 3 \\ 0 &= -x^2 + 3 \\ x^2 &= 3 \\ x &= \sqrt{3} \quad \text{or} \quad -\sqrt{3}, \end{aligned}$$

giving the two ordered pairs $(-\sqrt{3}, 0)$ and $(\sqrt{3}, 0)$. Now choose additional x-values near the x-values of these three points.

WORK PROBLEM 2 AT THE SIDE.

2. Complete each ordered pair for $y = -x^2 + 3$.

$(-2, \quad)$

$(-1, \quad)$

$(1, \quad)$

$(2, \quad)$

The ordered pairs from above and from Problem 2 at the side are listed in the table of values shown with Figure 2. Plot all these points and connect them with a smooth curve as shown in Figure 2. The vertex of this parabola is $(0, 3)$. The vertex is the *highest* point of this graph. The graph opens downward because x^2 has a negative coefficient. ■

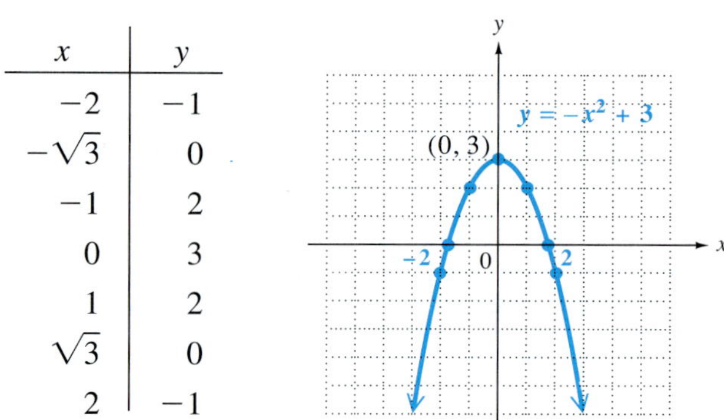

x	y
-2	-1
$-\sqrt{3}$	0
-1	2
0	3
1	2
$\sqrt{3}$	0
2	-1

FIGURE 2

WORK PROBLEM 3 AT THE SIDE.

3. Graph $y = -x^2 - 1$. Identify the vertex.

2 As the graphs we found above suggest, the vertex is the most important point to locate when you are graphing a quadratic equation. The next example shows how to find the vertex in a more general case.

EXAMPLE 3 *Finding the Vertex to Graph a Parabola*

Graph $y = x^2 - 2x - 3$.

We want to find the vertex of the graph. Note in Figure 2 that the vertex is exactly halfway between the x-intercepts. If a parabola has two x-intercepts this is always the case. Therefore, let us begin by finding the x-intercepts. Let $y = 0$ in the equation, and solve for x.

$$0 = x^2 - 2x - 3$$

$$0 = (x + 1)(x - 3) \qquad \text{Factor.}$$

$$x + 1 = 0 \quad \text{or} \quad x - 3 = 0 \qquad \text{Set each factor equal to 0.}$$

$$x = -1 \quad \text{or} \quad x = 3$$

There are two x-intercepts, $(-1, 0)$ and $(3, 0)$. Now find any y-intercepts.

$$y = 0^2 - 2(0) - 3 = -3 \qquad \text{Let } x = 0.$$

There is one y-intercept, $(0, -3)$.

ANSWERS

2. $(-2, -1)$, $(-1, 2)$, $(1, 2)$, $(2, -1)$

3.

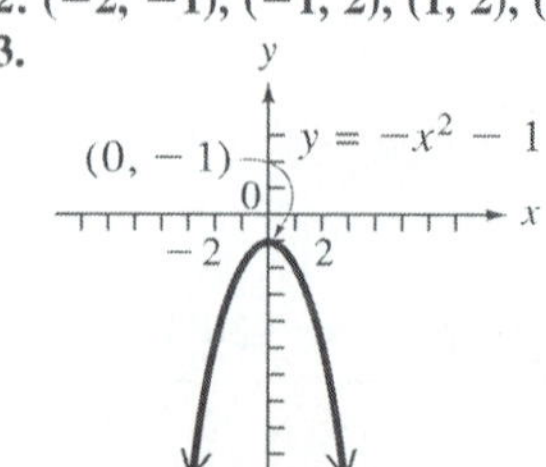

As mentioned above, the x-value of the vertex is halfway between the x-values of the two x-intercepts. Thus, it is $\frac{1}{2}$ their sum.

$$x = \frac{1}{2}(-1 + 3) = 1$$

Find the corresponding y-value by substituting 1 for x in the equation.

$$y = 1^2 - 2(1) - 3 = -4$$

The vertex is $(1, -4)$. Here, the axis is the line $x = 1$. Plot the three intercepts and the vertex. Find additional ordered pairs as needed. For example, if $x = 2$,

$$y = 2^2 - 2(2) - 3 = -3,$$

leading to the ordered pair $(2, -3)$. A table of ordered pairs with the ordered pairs we have found is shown with the graph in Figure 3. ■

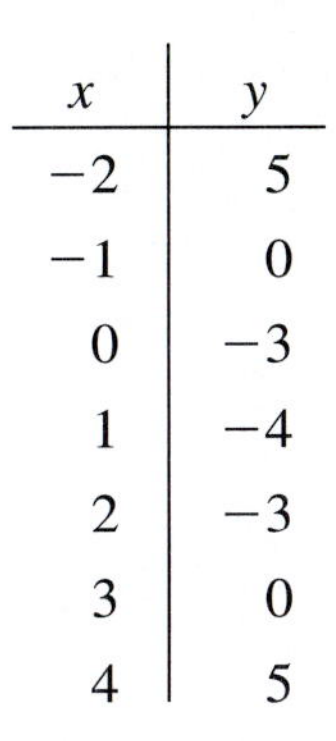

x	y
-2	5
-1	0
0	-3
1	-4
2	-3
3	0
4	5

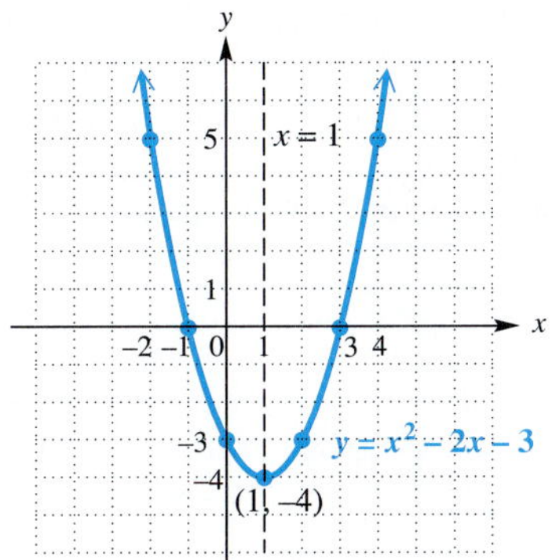

FIGURE 3

WORK PROBLEM 4 AT THE SIDE.

We can generalize from Example 3. The x-intercepts for the equation $0 = ax^2 + bx + c$, by the quadratic formula, are

$$x = \frac{-b + \sqrt{b^2 - 4ac}}{2a} \quad \text{and} \quad x = \frac{-b - \sqrt{b^2 - 4ac}}{2a}.$$

Thus, the x-value of the vertex is

$$x = \frac{1}{2}\left(\frac{-b + \sqrt{b^2 - 4ac}}{2a} + \frac{-b - \sqrt{b^2 - 4ac}}{2a}\right)$$

$$x = \frac{1}{2}\left(\frac{-b + \sqrt{b^2 - 4ac} - b - \sqrt{b^2 - 4ac}}{2a}\right)$$

$$x = \frac{1}{2}\left(\frac{-2b}{2a}\right) = -\frac{b}{2a}.$$

For the equation in Example 3, $y = x^2 - 2x - 3$, $a = 1$, and $b = -2$. Thus, the x-value of the vertex is

$$x = -\frac{b}{2a} = -\frac{-2}{2(1)} = 1,$$

which is the same x-value for the vertex we found in Example 3. (The x-value of the vertex is $x = -\frac{b}{2a}$, even if the graph has no x-intercepts.) A procedure for graphing quadratic equations is given on the next page.

4. Graph $y = x^2 + 2x - 8$.

ANSWER

4.

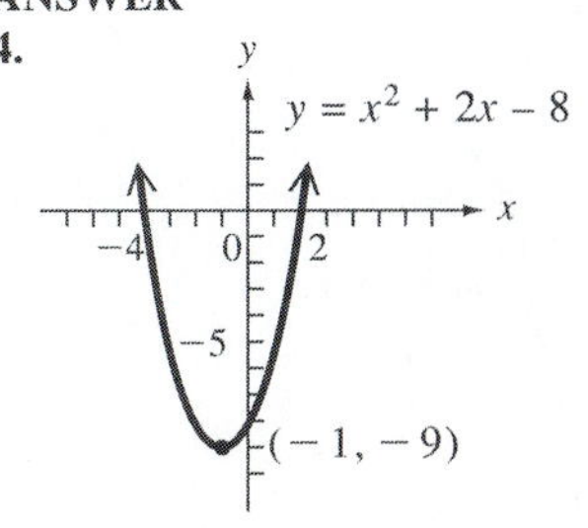

5. Complete the following ordered pairs for $y = x^2 - 4x + 1$.

(5,)(1,)(4,)

(3,)(−1,)

GRAPHING THE PARABOLA $y = ax^2 + bx + c$

Step 1 Find the y-intercept.
Step 2 Find the x-intercepts.
Step 3 Find the vertex. Let $x = -\frac{b}{2a}$, and find the corresponding y-value by substituting for x in the equation.
Step 4 Plot the intercepts and the vertex.
Step 5 Find and plot additional ordered pairs near the vertex and intercepts as needed, using the symmetry about the axis of the parabola.

EXAMPLE 4 *Graphing a Parabola*

Graph $y = x^2 - 4x + 1$.

Find the intercepts. Let $x = 0$ in $y = x^2 - 4x + 1$ to get the y-intercept (0, 1). Let $y = 0$ to get the x-intercepts. If $y = 0$, the equation is $0 = x^2 - 4x + 1$, which cannot be factored. Use the quadratic formula to solve for x.

$$x = \frac{4 \pm \sqrt{16 - 4}}{2} \qquad \text{Let } a = 1, b = -4, c = 1.$$

$$x = \frac{4 \pm \sqrt{12}}{2}$$

$$x = \frac{4 \pm 2\sqrt{3}}{2} \qquad \sqrt{12} = 2\sqrt{3}$$

$$x = \frac{2(2 \pm \sqrt{3})}{2} = 2 \pm \sqrt{3}$$

Use a calculator to find that the x-intercepts are (3.7, 0) and (.3, 0) to the nearest tenth. The x-value of the vertex is

$$x = -\frac{b}{2a} = -\frac{-4}{2(1)} = 2.$$

The y-value of the vertex is

$$y = 2^2 - 4(2) + 1 = -3,$$

so the vertex is $(2, -3)$. The axis is the line $x = 2$.

WORK PROBLEM 5 AT THE SIDE.

6. Graph the parabola

$$y = -x^2 + 2x + 4.$$

Identify the vertex.

Plot the intercepts, vertex, and the points found in Problem 5. Connect these points with a smooth curve. The graph is shown in Figure 4. ■

x	y
−1	6
0	1
1	−2
2	−3
3	−2
4	1
5	6

FIGURE 4

WORK PROBLEM 6 AT THE SIDE.

ANSWERS

5. (5, 6), (1, −2), (4, 1), (3, −2), (−1, 6)

6.

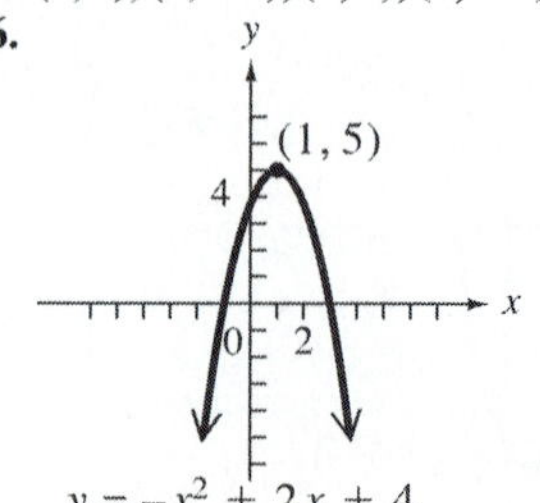

EXERCISES

1. In your own words, explain what is meant by the vertex of a parabola.

2. In your own words, explain what is meant by the line of symmetry of a parabola.

Sketch the graph of each equation. Give the coordinates of the vertex in each case. See Examples 1–4.

3. $y = 2x^2$

4. $y = 3x^2$

5. $y = x^2 + 2x + 3$

6. $y = x^2 - 4x + 3$

7. $y = x^2 - 4$

8. $y = x^2 - 6$

9. $y = 2 - x^2$

10. $y = 4 - x^2$

11. $y = (x + 3)^2$

12. $y = (x - 4)^2$

13. $y = -x^2 + 6x - 5$

14. $y = -x^2 - 4x - 3$

15. $y = -x^2 + 4x - 4$

16. $y = -x^2 - 2x - 1$

17. Based on your work in these exercises, what seems to be the direction in which the parabola $y = ax^2 + bx + c$ opens if $a > 0$? If $a < 0$?

CHAPTER 9 SUMMARY

KEY TERMS

9.4	**parabola**	The graph of a quadratic equation is called a parabola.
	vertex	The vertex of a parabola is the highest or lowest point on the graph.
	axis	The axis of a parabola is a vertical line through the vertex.
	line of symmetry	If a graph is folded on its line of symmetry, the two sides match.

NEW SYMBOLS

$\pm$ plus or minus

QUICK REVIEW

Concepts	Examples
9.1 Solving Quadratic Equations by the Square Root Property	
Square Root Property of Equations If b is positive, and if $a^2 = b$, then $a = \sqrt{b}$ or $a = -\sqrt{b}$.	Solve $(2x + 1)^2 = 5$. $2x + 1 = \pm\sqrt{5}$ $2x = -1 \pm \sqrt{5}$ $x = \dfrac{-1 \pm \sqrt{5}}{2}$
9.2 Solving Quadratic Equations by Completing the Square	
Completing the Square	Solve $2x^2 + 4x - 1 = 0$.
1. If the coefficient of the squared term is 1, go to Step 2. If it is not 1, divide each side of the equation by this coefficient.	**1.** $x^2 + 2x - \dfrac{1}{2} = 0$
2. Make sure that all variable terms are on one side of the equation, and all constant terms are on the other.	**2.** $x^2 + 2x = \dfrac{1}{2}$
3. Take half the coefficient of x, square it, and add the square to each side of the equation.	**3.** $x^2 + 2x + 1 = \dfrac{1}{2} + 1$
4. Factor the variable side.	**4.** $(x + 1)^2 = \dfrac{3}{2}$
5. Use the square root property to solve the equation.	**5.** $x + 1 = \pm\sqrt{\dfrac{3}{2}} = \pm\dfrac{\sqrt{6}}{2}$ $x = -1 \pm \dfrac{\sqrt{6}}{2}$ $x = \dfrac{-2 \pm \sqrt{6}}{2}$

Concepts	Examples
9.3 Solving Quadratic Equations by the Quadratic Formula	
Quadratic Formula The solutions of $ax^2 + bx + c = 0$, $(a \neq 0)$, are $x = \dfrac{-b \pm \sqrt{b^2 - 4ac}}{2a}$.	Solve $3x^2 - 4x - 2 = 0$. $x = \dfrac{-(-4) \pm \sqrt{(-4)^2 - 4(3)(-2)}}{2(3)}$ $x = \dfrac{4 \pm \sqrt{16 + 24}}{6}$ $x = \dfrac{4 \pm \sqrt{40}}{6} = \dfrac{4 \pm 2\sqrt{10}}{6}$ $x = \dfrac{2(2 \pm \sqrt{10})}{6} = \dfrac{2 \pm \sqrt{10}}{3}$
9.4 Graphing Quadratic Equations in Two Variables	
Graphing $y = ax^2 + bx + c$	Graph $y = 2x^2 - 5x - 3$.
1. Find the y-intercept.	**1.** $y = 2(0)^2 - 5(0) - 3 = -3$ The y-intercept is $(0, -3)$.
2. Find the x-intercepts.	**2.** $0 = 2x^2 - 5x - 3$ $0 = (2x + 1)(x - 3)$ $2x + 1 = 0$ or $x - 3 = 0$ $2x = -1$ or $x = 3$ $x = -\frac{1}{2}$ or $x = 3$ The x-intercepts are $(-\frac{1}{2}, 0)$ and $(3, 0)$.
3. Find the vertex: $x = -\frac{b}{2a}$; find y by substituting this value for x in the equation.	**3.** $x = -\dfrac{b}{2a} = -\dfrac{-5}{2(2)} = \dfrac{5}{4}$ $y = 2\left(\dfrac{5}{4}\right)^2 - 5\left(\dfrac{5}{4}\right) - 3$ $y = 2\left(\dfrac{25}{16}\right) - \dfrac{25}{4} - 3$ $y = \dfrac{25}{8} - \dfrac{50}{8} - \dfrac{24}{8} = -\dfrac{49}{8}$ The vertex is $(\frac{5}{4}, -\frac{49}{8})$.
4. Plot the intercepts and the vertex. **5.** Find and plot additional ordered pairs near the vertex and intercepts as needed.	**4.** and **5.** 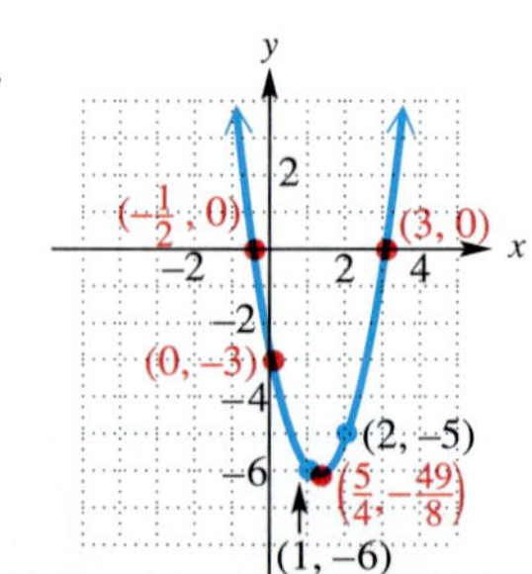

NAME DATE HOUR

CHAPTER 9 REVIEW EXERCISES

[9.1] *In Exercises 1–7, solve each equation by using the square root property. Express all radicals in simplest form.*

1. $y^2 = 144$

2. $x^2 = 37$

3. $m^2 = 128$

4. $(k + 2)^2 = 25$

5. $(r - 3)^2 = 10$

6. $(2p + 1)^2 = 14$

7. $(3k + 2)^2 = -3$

8. The square root property can be applied only to equations of the form __________.

[9.2] *Solve each equation by completing the square.*

9. $m^2 + 6m + 5 = 0$

10. $p^2 + 4p = 7$

11. $-x^2 + 5 = 2x$

12. $2y^2 - 3 = -8y$

13. $5k^2 - 3k - 2 = 0$

14. $(4a + 1)(a - 1) = -7$

Solve each problem.

15. If an object is thrown upward from a height of 50 feet, with an initial velocity of 32 feet per second, then its height after t seconds is given by $h = -16t^2 + 32t + 50$, where h is in feet. After how many seconds will it reach a height of 30 feet?

16. A certain projectile is located $d = 2t^2 - 5t + 2$ feet from the ground after t seconds have elapsed. How many seconds will it take the projectile to be 14 feet from the ground?

[9.3] *Solve each equation by using the quadratic formula.*

17. $3k^2 + 2k + 3 = 0$

18. $x(5x - 1) = 1$

19. $2p^2 + 8 = 4p + 11$

20. $-4a^2 + 7 = 2a$

21. $\frac{1}{4}x^2 = 2 - \frac{3}{4}x$

22. Why is this not the statement of the quadratic formula? $x = -b \pm \frac{\sqrt{b^2 - 4ac}}{2a}$

[9.4] *Sketch the graph of each equation. Identify each vertex.*

23. $y = -3x^2$

24. $y = x^2 - 2x + 1$

25. $y = -x^2 + 5$

26. $y = -x^2 + 2x + 3$

27. $y = x^2 + 4x + 2$

28. $y = (x + 4)^2$

Mixed Review Exercises

Solve by any method.

29. $(2t - 1)(t + 1) = 54$

30. $(2p + 1)^2 = 100$

31. $(k + 2)(k - 1) = 3$

32. $6t^2 + 7t - 3 = 0$

33. $2x^2 + 3x + 2 = x^2 - 2x$

34. $x^2 + 2x + 5 = 7$

35. $m^2 - 4m + 10 = 0$

36. $k^2 - 9k + 10 = 0$

37. $(3x + 5)^2 = 0$

38. $\frac{1}{2}r^2 = \frac{7}{2} - r$

39. $x^2 + 4x = 1$

40. $7x^2 - 8 = 5x^2 + 8$

NAME DATE HOUR

CHAPTER 9 TEST

Solve by using the square root property.

1. $x^2 = 39$

2. $(y + 3)^2 = 64$

3. $(4x + 3)^2 = 24$

Solve by completing the square.

4. $x^2 - 4x = 6$

5. $2x^2 + 12x - 3 = 0$

6. For a quadratic equation to have two real solutions, what must be true about the quantity under the radical in the quadratic formula?

Solve by the quadratic formula.

7. $2x^2 + 5x - 3 = 0$

8. $3w^2 + 2 = 6w$

9. $4x^2 + 8x + 11 = 0$

10. $t^2 - \frac{5}{3}t + \frac{1}{3} = 0$

Solve by the method of your choice.

11. $p^2 - 2p - 1 = 0$

12. $(2x + 1)^2 = 18$

13. $(x - 5)(2x - 1) = 1$

1. ______
2. ______
3. ______
4. ______
5. ______
6. ______
7. ______
8. ______
9. ______
10. ______
11. ______
12. ______
13. ______

14. ______________________

14. $t^2 + 25 = 10t$

Solve the following problem.

15. ______________________

15. If a ball is thrown into the air from ground level with an initial velocity of 64 feet per second, its height s (in feet) after t seconds is given by the formula $s = -16t^2 + 64t$. After how many seconds will the ball reach a height of 64 feet?

16. ______________________

16. Which of these equations has exactly one real number solution?

(a) $x^2 = 4$ **(b)** $y^2 = -4$

(c) $(x - 4)^2 = 1$ **(d)** $t^2 = 0$

17. ______________________

17. Find the value of x:

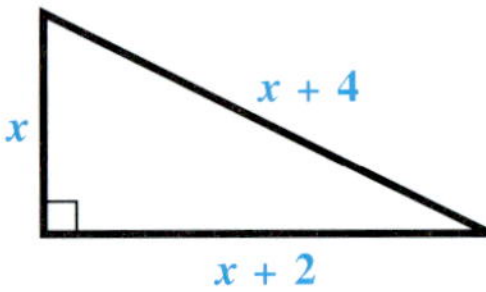

Sketch the graph of each equation. Identify each vertex.

18.

18. $y = (x - 3)^2$

19.

19. $y = -x^2 - 2x - 4$

20. ______________________

20. How can you tell, without doing any work, that $(4x - 7)^2 = -4$ has no real solutions?

NAME DATE HOUR

CUMULATIVE REVIEW EXERCISES CHAPTERS R–9

Note: This cumulative review exercise set may be considered as a final examination for the course.

Perform each operation, wherever possible.

1. $\dfrac{-4 \cdot 3^2 + 2 \cdot 3}{2 - 4 \cdot 1}$

2. $-9 - (-8)(2) + 6 - (6 + 2)$

Perform the indicated operations.

3. $|-3| - |1 - 6|$

4. $-4r + 14 + 3r - 7$

5. $13k - 4k + k - 14k + 2k$

6. $5(4m - 2) - (m + 7)$

Solve each equation.

7. $6x - 5 = 13$

8. $3k - 9k - 8k + 6 = -64$

9. $2(m - 1) - 6(3 - m) = -4$

10. Sprague and Hartmann were opposing candidates in a mayoral election. Sprague received 80 more votes than Hartmann, with 346 total votes cast. How many votes did Hartmann receive?

11. Solve the formula $I = prt$ for p.

12. Solve the formula $P = 2L + 2W$ for L.

Solve each inequality. Graph the solution.

13. $-8m < 16$

14. $-9p + 2(8 - p) - 6 \geq 4p - 50$

1. ____________

2. ____________

3. ____________

4. ____________

5. ____________

6. ____________

7. ____________

8. ____________

9. ____________

10. ____________

11. ____________

12. ____________

13. ←————————

14. ————————→

15. ______
16. ______
17. ______
18. ______
19. ______
20. ______
21. ______
22. ______
23. ______
24. ______
25. ______
26. ______
27. ______
28. ______
29. ______
30. ______
31. ______
32. ______
33. ______

Simplify each of the following. Write answers with positive exponents.

15. $(3^2 \cdot x^{-4})^{-1}$ **16.** $\left(\frac{b^{-3}c^4}{b^5c^3}\right)^{-2}$ **17.** $\left(\frac{5}{3}\right)^{-3}$

Perform the indicated operations.

18. $(5x^5 - 9x^4 + 8x^2) - (9x^2 + 8x^4 - 3x^5)$

19. $(2x - 5)(x^3 + 3x^2 - 2x - 4)$

20. $(5t + 9)^2$ **21.** $\frac{3x^3 + 10x^2 - 7x + 4}{x + 4}$

Factor each of the following as completely as possible.

22. $16x^3 - 48x^2y$ **23.** $16x^4 - 1$

24. $2a^2 - 5a - 3$ **25.** $25m^2 - 20m + 4$

Solve each of the following equations by factoring.

26. $x^2 + 3x - 54 = 0$ **27.** $3x^2 = x + 4$

28. The length of a rectangle is 2.5 times its width. The area is 1000 square meters. Find the length.

Perform the following operations. Write all answers in lowest terms.

29. $\frac{2}{a - 3} \div \frac{5}{2a - 6}$ **30.** $\frac{1}{k} - \frac{2}{k - 1}$

31. $\frac{2}{a^2 - 4} + \frac{3}{a^2 - 4a + 4}$ **32.** $\frac{6 + \frac{1}{x}}{3 - \frac{1}{x}}$

33. Solve $\frac{1}{x + 3} + \frac{1}{x} = \frac{7}{10}$.

Graph each of the following.

34. $2x + 3y = 6$

34.

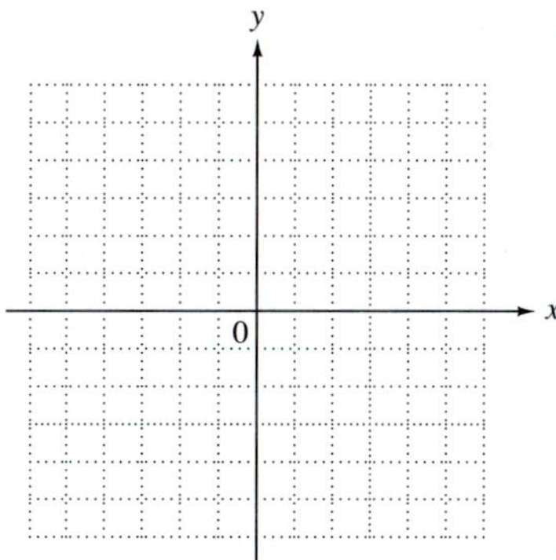

35. $y = 3$

35.

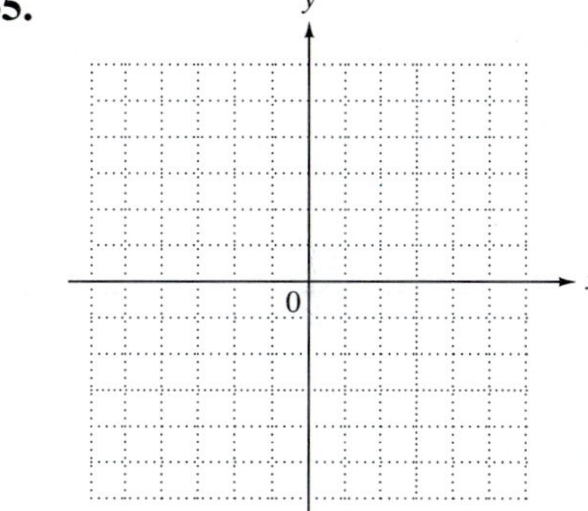

36. $2x - 5y < 10$

36.

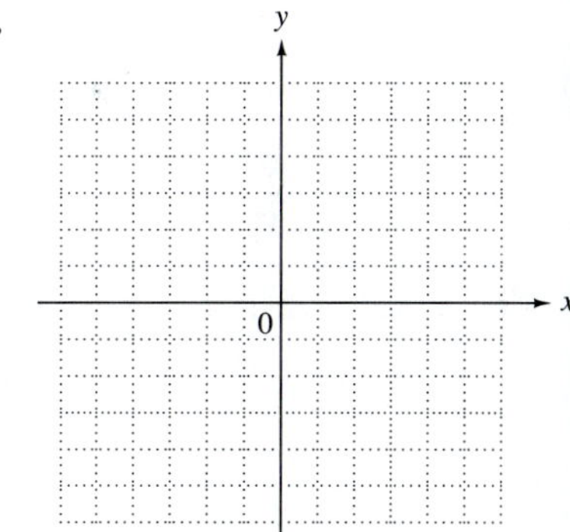

37. Find the slope of the line through $(-1, 4)$ and $(5, 2)$.

37. ______________________

38. Write an equation of a line with slope 2 and y-intercept $(0, 3)$. Give it in the form $Ax + By = C$.

38. ______________________

Solve each system of equations.

39. $2x + y = -4$
$-3x + 2y = 13$

40. $3x - 5y = 8$
$-6x + 10y = 16$

39. ______________________

40. ______________________

Graph of the solution of the system of inequalities.

41. 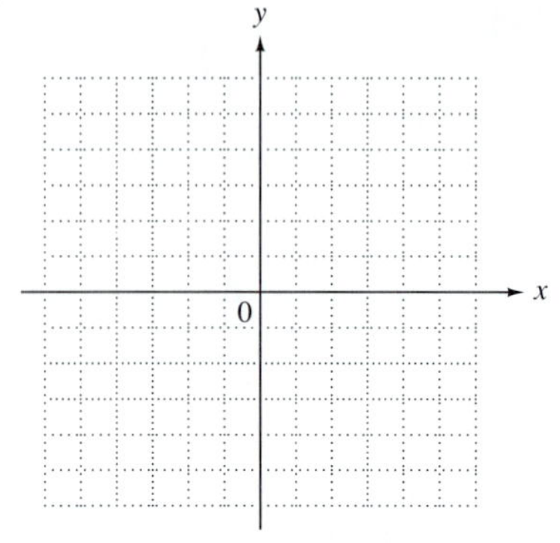

41. $2x + y \leq 4$
$x - y > 2$

42. ______________________

42. Give all square roots of 289.

Simplify each of the following as much as possible.

43. ______________________

44. ______________________

45. ______________________

46. ______________________

47. ______________________

43. $\sqrt{100}$

44. $\dfrac{6\sqrt{6}}{\sqrt{5}}$

45. $\sqrt[3]{\dfrac{7}{16}}$

46. $3\sqrt{5} - 2\sqrt{20} + \sqrt{125}$

47. $\sqrt[3]{16a^3b^4} - \sqrt[3]{54a^3b^4}$

Solve the radical equation.

48. ______________________

48. $\sqrt{x + 2} = x - 4$

Solve the quadratic equation.

49. ______________________

49. $2a^2 - 2a = 1$

50.

50. Graph the parabola $y = x^2 - 4$. Identify the vertex.

Answers to Selected Exercises

The solutions to selected odd-numbered exercises are given in the section beginning on page A-25.

In this section we provide the answers that we think most students will obtain when they work the exercises using the methods explained in the text. If your answer does not look exactly like the one given here, it is not necessarily wrong. In many cases there are equivalent forms of the answer that are correct. For example, if the answer section shows $\frac{3}{4}$ and your answer is .75, you have obtained the right answer but written it in a different (yet equivalent) form. Unless the directions specify otherwise, .75 is just as valid an answer as $\frac{3}{4}$.

In general, if your answer does not agree with the one given in the text, see whether it can be transformed into the other form. If it can, then it is the correct answer. If you still have doubts, talk with your instructor.

CHAPTER R

SECTION R.1 (page 9)

1. 3; 8 **3.** $\frac{19}{5}$ **5.** product; quotient **7.** prime **9.** composite **11.** composite **13.** neither **15.** $2 \cdot 3 \cdot 5$ **17.** $2 \cdot 2 \cdot 5 \cdot 5 \cdot 5$ **19.** $2 \cdot 2 \cdot 31$ **21.** 29 **23.** $\frac{1}{2}$ **25.** $\frac{5}{6}$ **27.** $\frac{1}{3}$ **29.** $\frac{6}{5}$ **31.** (c) **33.** $\frac{24}{35}$ **35.** $\frac{6}{25}$ **37.** $\frac{6}{5}$ **39.** $\frac{232}{15}$ or $15\frac{7}{15}$ **41.** $\frac{10}{3}$ or $3\frac{1}{3}$ **43.** 12 **45.** $\frac{1}{16}$ **47.** $\frac{84}{47}$ or $1\frac{37}{47}$ **51.** $\frac{2}{3}$ **53.** $\frac{8}{9}$ **55.** $\frac{27}{8}$ or $3\frac{3}{8}$ **57.** $\frac{17}{36}$ **59.** $\frac{11}{12}$ **61.** $\frac{4}{3}$ **63.** $2\left(\frac{10}{100}\right) + 3\left(\frac{10}{100}\right) = \frac{50}{100}$ **65.** $618\frac{3}{4}$ feet **67.** $\frac{9}{16}$ inch **69.** $\frac{1}{3}$ cup

SECTION R.2 (page 19)

1. **(a)** 6 **(b)** 9 **(c)** 1 **(d)** 7 **(e)** 4 **3.** (c) **5.** (b) **7.** $\frac{4}{10}$ **9.** $\frac{64}{100}$ **11.** $\frac{138}{1000}$ **13.** $\frac{3805}{1000}$ **15.** 151.806 **17.** 25.61 **19.** 15.33 **21.** 81.716 **23.** 4.771 **25.** 248.08 **27.** 19.0652 **29.** 7.15 **31.** 2.55 **35.** **(a)** 46.25 **(b)** 46.2 **(c)** 46 **(d)** 50 **37.** .125 **39.** .25 **41.** .1875 **43.** $.\overline{428571}$; .429 **45.** $.\overline{5}$; .556 **47.** $.1\overline{6}$; .167 **51.** .54 **53.** 1.17 **55.** .024 **57.** .008 **59.** 75% **61.** .4% **63.** 128% **65.** 30% **67.** 75% **69.** 150% **71.** $83.\overline{3}$% **73.** 31.25%

ANSWERS

CHAPTER 1

SECTION 1.1 (page 27)

1. false; 3^5 means $3 \cdot 3 \cdot 3 \cdot 3 \cdot 3$ **3.** true **5.** 49 **7.** 144 **9.** 64 **11.** 1000 **13.** 81 **15.** 1024 **17.** $\frac{16}{81}$ **19.** .000064 **23.** 32 **25.** $\frac{49}{30}$ **27.** 12 **29.** 23.01 **31.** 95 **33.** 90 **35.** 14 **37.** 9 **41.** true **43.** false **45.** true **47.** true **49.** false **51.** false **53.** true **55.** $15 = 5 + 10$ **57.** $9 > 5 - 4$ **59.** $16 \neq 19$ **61.** $2 \leq 3$ **63.** Seven is less than nineteen. True **65.** Three is not equal to six. True **67.** Eight is greater than or equal to eleven. False **69.** Answers will vary. One example is $5 + 3 \geq 2 \cdot 2$. **71.** $30 > 5$ **73.** $3 \leq 12$ **75.** .2809 **77.** 5.13

SECTION 1.2 (page 33)

1. expression **3.** equation **5.** equation **7.** $2x^3 = 2 \cdot x \cdot x \cdot x$ **9.** **(a)** 13 **(b)** 15 **11.** **(a)** 20 **(b)** 30 **13.** **(a)** 64 **(b)** 144 **15.** **(a)** $\frac{5}{3}$ **(b)** $\frac{7}{3}$ **17.** **(a)** $\frac{7}{8}$ **(b)** $\frac{13}{12}$ **19.** **(a)** 52 **(b)** 114 **21.** **(a)** 25.836 **(b)** 38.754 **23.** **(a)** 24 **(b)** 28 **25.** **(a)** 12 **(b)** 33 **27.** **(a)** 6 **(b)** $\frac{9}{5}$ **29.** **(a)** $\frac{4}{3}$ **(b)** $\frac{13}{6}$ **31.** **(a)** $\frac{2}{7}$ **(b)** $\frac{16}{27}$ **33.** **(a)** 12 **(b)** 55 **35.** **(a)** 1 **(b)** $\frac{28}{17}$ **37.** **(a)** 3.684 **(b)** 8.841 **39.** $12x$ **41.** $x + 7$ **43.** $x - 2$ **45.** $7 - x$

47. $x - 6$ **49.** $\frac{12}{x}$ **51.** $6(x - 4)$
55. Two possible pairs are $x = 0$, $y = 6$ and $x = 1$, $y = 4$.
57. no **59.** yes **61.** no **63.** yes **65.** yes
67. yes **69.** $x + 8 = 18$; 10 **71.** $16 - \frac{3}{4}x = 13$; 4
73. $2x + 5 = 5$; 0 **75.** $3x = 2x + 8$; 8

SECTION 1.3 (page 43)

1. true **3.** true **5.** false **7.** true **9.** false
11. −1760 **13.** −8 **15.** 8300
17. −66,000,000 (dollars)
19. 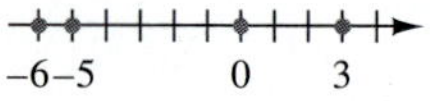

21. 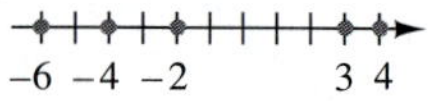

23. 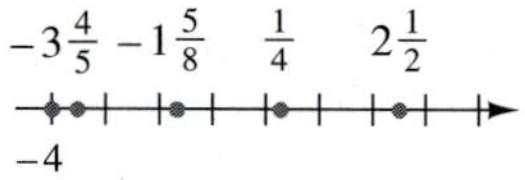

25. −11 **27.** −21 **29.** −100
31. $-\frac{2}{3}$ **33.** false **35.** true **37.** (a) 2 (b) 2
39. (a) −6 (b) 6 **41.** (a) $\frac{3}{4}$ (b) $\frac{3}{4}$ **43.** 7
45. 4 **47.** −12 **49.** −14 **51.** 9
55. $\frac{33}{20}$ **57.** $\frac{5}{12}$

SECTION 1.4 (page 49)

1. negative **3.** −3, 5 **5.** 2 **7.** −3 **9.** −10
11. −13 **13.** −15.9 **15.** −1 **17.** 13
19. 5 **21.** 0 **23.** −8 **25.** $\frac{1}{2}$ **27.** $-\frac{19}{24}$
29. $-\frac{3}{4}$ **31.** −1.6 **33.** −8.7 **35.** −25
37. −24 **39.** true **41.** true **43.** false
45. true **47.** false **49.** true **51.** false
53. $-5 + 12 + 6$; 13 **55.** $[-19 + (-4)] + 14$; −9
57. $[-4 + (-10)] + 12$; −2
59. $[8 + (-18)] + 4$; −6 **61.** $47
63. −184 meters **65.** 112°F **67.** 37 yards
69. $107 **71.** $286.60 **73.** −2 **75.** −3
77. −1 **79.** 0 **81.** 1 **83.** 56 **85.** 5

SECTION 1.5 (page 57)

1. −8; −6 **3.** 7 **5.** −4 **7.** −3 **9.** −4
11. −10 **13.** −16 **15.** 11 **17.** 19 **19.** −4
21. 5 **23.** 0 **25.** $\frac{3}{4}$ **27.** $-\frac{11}{8}$ **29.** $\frac{15}{8}$
31. 13.6 **33.** −11.9 **35.** −2.8 **37.** −6.3
41. −14 **43.** −24 **45.** −16 **47.** $-\frac{17}{8}$
49. −48.98 **51.** For example, $-8 - (-2) = -6$
53. $4 - (-8)$; 12 **55.** $-2 - 8$; −10
57. $[9 + (-4)] - 7$; −2 **59.** $[8 - (-5)] - 12$; 1
61. −15°F **63.** 14,776 feet **65.** −$80
67. $105,000 **69.** −2 **71.** −3 **73.** −3
75. positive **77.** positive **79.** 16 **81.** 56

SECTION 1.6 (page 65)

1. true **3.** false **5.** true **7.** 20 **9.** −30
11. −28 **13.** 80 **15.** 0 **17.** $\frac{5}{6}$ **19.** −2.38
21. $\frac{3}{2}$ **23.** −11 **25.** −2 **27.** −60 **29.** 35
31. 6 **33.** −18 **35.** 67 **37.** −8 **39.** 64
43. 47 **45.** 72 **47.** $-\frac{78}{25}$ **49.** 0
51. −23 **53.** −9 **55.** $9 + (-9)(2)$; −9
57. $-4 - 2(-1)(6)$; 8 **59.** $7(-12) - 9$; −93
61. $12[9 - (-8)]$; 204 **63.** $\frac{4}{5}[-8 + (-2)]$; −8
65. 3 **67.** 0 **69.** −2 **71.** −2 **73.** −1
75. −3 **77.** 28 **79.** 6 **81.** $\frac{212}{103}$ or $2\frac{6}{103}$

SECTION 1.7 (page 75)

1. positive **3.** less than **5.** positive
7. $\frac{1}{11}$ **9.** $-\frac{1}{5}$ **11.** $\frac{6}{5}$ **13.** no reciprocal
15. $-\frac{7}{8}$ **17.** 2.5 **19.** (c) **21.** −3
23. −2 **25.** 16 **27.** 0 **29.** 25.63
31. $\frac{3}{2}$ **33.** 3.4 **35.** −3 **37.** −5
39. 4 **41.** 3 **43.** 7 **45.** 4 **47.** −3
49. $\frac{1}{2}$ **51.** 10 **53.** $\frac{-36}{-9}$; 4
55. $\frac{-12}{-5 + (-1)}$; 2 **57.** $\frac{15 + (-3)}{4(-3)}$; −1
59. $\frac{-34(7)}{-14}$; 17 **61.** −8 **63.** 4 **65.** −4
67. $6x = -42$; −7 **69.** $\frac{x}{3} = -3$; −9
71. $x - 6 = 4$; 10 **73.** $x + 5 = -5$; −10
75. 0 **77.** 1

SECTION 1.8 (page 87)

1. true **3.** false **5.** true **7.** false
9. commutative property **11.** associative property
13. commutative property **15.** associative property
17. inverse property **19.** inverse property
21. identity property **23.** commutative property
25. distributive property **27.** identity property

29. distributive property **31.** identity property
35. $7 + r$ **37.** s
39. $-6x + (-6) \cdot 7$; $-6x - 42$
41. $w + [5 + (-3)]$; $w + 2$ **43.** $3x + 16$
45. 11 **47.** 7 **49.** 0 **51.** $-.38$ **53.** 1
55. $(5 + 1)x$; $6x$ **57.** $4t + 12$ **59.** $-8r - 24$
61. $-5y + 20$ **63.** $-16y - 20z$ **65.** $8(z + w)$
67. $7(2v + 5r)$ **69.** $24r + 32s - 40y$
71. $(1 + 1 + 1)q$; $3q$ **73.** $(-5 + 1)x$; $-4x$
75. $-4t - 5m$ **77.** $5c + 4d$ **79.** $3q - 5r + 8s$
81. Answers will vary. For example, "putting on your socks" and "putting on your shoes."
83. 31 **85.** 10

SECTION 1.9 (page 95)

1. (c) **3.** (a) **5.** $4r + 11$ **7.** $32q - 24t$
9. $5 + 2x - 6y$ **11.** $-7 + 3p$ **13.** 14
15. -12 **17.** 5 **19.** 1 **21.** -1 **23.** 74
25. Answers will vary. For example, $-3x$ and $4x$.
27. like **29.** unlike **31.** like **33.** unlike
37. $9k - 5$ **39.** $-\frac{1}{3}t - \frac{28}{3}$
41. $-4.1r + 5.6$ **43.** $-2y^2 + 3y^3$
45. $-19p + 16$ **47.** $-4y + 22$ **49.** $-16y + 63$
51. $4k - 7$ **53.** $-23.7y - 12.6$
55. $(x + 3) + 5x$; $6x + 3$
57. $(13 + 6x) - (-7x)$; $13 + 13x$
59. $2(3x + 4) - (-4 + 6x)$; 12
61. Wording may vary. One example is "the difference between 9 times a number and the sum of the number and 2."
63. -5 **65.** 15 **67.** 2 **69.** -14

CHAPTER 1 REVIEW EXERCISES (page 101)

1. 625 **2.** $\frac{27}{125}$ **3.** .0000000032 **4.** .000000001
5. 27 **6.** 399 **7.** 39 **8.** 5 **9.** true
10. true **11.** false **12.** true **13.** $13 < 17$
14. $5 + 2 \neq 10$ **15. (a)** 12 **(b)** 60 **16. (a)** $\frac{5}{4}$
(b) $\frac{17}{16}$ **17. (a)** 18 **(b)** 270 **18. (a)** 29.8 **(b)** 119.2
19. 30 **20.** 60 **21.** 14 **22.** 13 **23.** $x + 6$
24. $8 - x$ **25.** $6x - 9$ **26.** $12 + \frac{3}{5}x$ **27.** yes
28. no **29.** $2x - 6 = 10$; 8 **30.** $4x = 8$; 2
31.
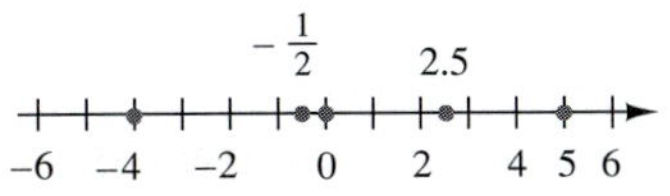

32.
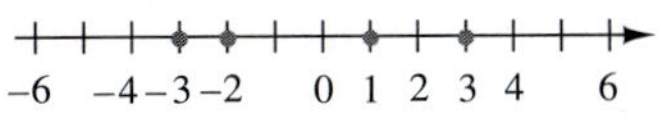

33.
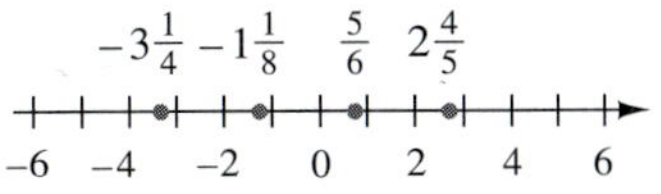

34.
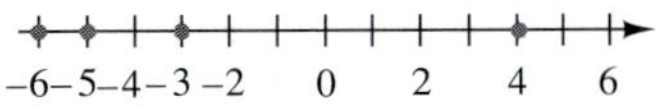

35. -10 **36.** -9 **37.** $-\frac{3}{4}$ **38.** $-|23|$
39. true **40.** true **41.** true **42.** false
43. (a) 9 **(b)** 9 **44. (a)** 0 **(b)** 0 **45. (a)** -6
(b) 6 **46. (a)** $\frac{5}{7}$ **(b)** $\frac{5}{7}$ **47.** 12 **48.** -3
49. -19 **50.** -7 **51.** -6 **52.** -4 **53.** -17
54. $-\frac{29}{36}$ **55.** -21.8 **56.** -14 **57.** -10
58. -19 **59.** $(-31 + 12) + 19$; 0
60. $[-4 + (-8)] + 13$; 1 **61.** $-\$8$ **62.** 87°
63. -2 **64.** -1 **65.** -11 **66.** -1 **67.** 7
68. $-\frac{43}{35}$ **69.** 10.31 **70.** -12 **71.** 2
72. -3 **73.** $-4 - (-6)$; 2 **74.** $[4 + (-8)] - 5$; -9 **75.** $-\$29$ **76.** $-10°$ **79.** 36 **80.** -105
81. $\frac{1}{2}$ **82.** 10.08 **83.** -20 **84.** -10
85. -24 **86.** -35 **87.** -18 **88.** -18
89. 125 **90.** -423 **91.** $-4(5) - 9$; -29
92. $\frac{5}{6}[12 + (-6)]$; 5 **93.** 4 **94.** -20 **95.** $-\frac{3}{4}$
96. 11.3 **97.** -1 **98.** 2 **99.** 1 **100.** .5
101. $\frac{12}{8 + (-4)}$; 3 **102.** $\frac{-20(12)}{15 - (-15)}$; -8
103. $8x = -24$; -3 **104.** $\frac{x}{3} = -2$; -6
105. $x - 3 = -7$; -4 **106.** $x + 5 = -6$; -11
107. identity property **108.** identity property
109. inverse property **110.** inverse property
111. associative property **112.** associative property
113. distributive property **114.** commutative property **115.** $(7 + 1)y$; $8y$ **116.** $-12 \cdot 4 - (-12)t$; $-48 + 12t$ **117.** $3(2s + 4y)$; $6s + 12y$
118. $-1(-4r + 5s) = -1(-4r) + (-1)(5s)$; $4r - 5s$
121. $17p^2$ **122.** $16r^2 + 7r$ **123.** $-19k + 54$
124. $5s - 6$ **125.** $-45t - 23$ **126.** $-45t^2 - 23.4t$
127. $-2(3x) - 7x$; $-13x$ **128.** $\frac{x + 9}{x - 6}$
130. Answers may vary. For example, "3 times the difference between 4 times a number and 6"
131. 16 **132.** $\frac{25}{36}$ **133.** -26 **134.** $\frac{8}{3}$
135. $-\frac{1}{24}$ **136.** $\frac{7}{2}$ **137.** 2 **138.** 77.6
139. $-1\frac{1}{2}$ **140.** 11 **141.** $-\frac{28}{15}$ **142.** 24
143. -11 **144.** -6 **145.** $\$13,600 - \$1400$; $\$12,200$ **146.** $\frac{x}{3x - 14}$

ANSWERS

CHAPTER 1 TEST (page 107)

[1.1] **1.** true **2.** false
[1.3] **3.** 

4. $-|-8|$ (or -8) **5.** -1.277
[1.1] **6.** $\frac{-6}{2+(-8)}$; 1 **[1.4]–[1.7]** **7.** negative
[1.4]–[1.5] **8.** 4 **9.** $-2\frac{5}{6}$ **10.** 2 **11.** 6
[1.6] **12.** 108 **[1.7]** **13.** 3 **14.** $\frac{30}{7}$
[1.2] **15.** 4 **16.** 4 **17.** -70
18. 3 **[1.5]** **19.** 178°F **[1.8]** **20.** B
21. D **22.** E **23.** A **24.** C
[1.9] **25.** $-9x^2 - 6x - 8$

CHAPTER 2

SECTION 2.1 (page 113)

1. 12 **3.** -10 **5.** 6.6 **7.** 4.2 **9.** 3
11. -2 **13.** 4 **15.** 0 **17.** no solution
19. all real numbers **23.** 13 **25.** 4
27. no solution **29.** all real numbers
31. $\frac{7}{15}$ **33.** 7 **35.** -4 **37.** 13
39. all real numbers **41.** 18 **43.** 12
47. m **49.** z **51.** $\frac{1}{7}$ **53.** $-\frac{7}{9}$

SECTION 2.2 (page 119)

1. $\frac{3}{2}$ **3.** 10 **5.** $-\frac{2}{9}$ **7.** -1 **9.** 6
11. -4 **13.** .12 **15.** -1 **17.** 6 **19.** $\frac{15}{2}$
21. -5 **23.** $-\frac{18}{5}$ **25.** 12 **27.** 0 **29.** -48
31. -12 **33.** $\frac{4}{7}$ **35.** 40 **37.** 3 **39.** 7
41. -5 **43.** -35 **45.** 9 **47.** $\frac{35}{2}$ **49.** $-\frac{27}{35}$
51. -12.2 **53.** Answers will vary. For example, $\frac{3}{2}x = -6$.
55. $24q + 32$ **57.** $-28p + 29$ **59.** $-8 + 56p$

SECTION 2.3 (page 125)

1. 2 **3.** -1 **5.** 5 **7.** $-\frac{5}{3}$ **9.** -1
11. no solution **13.** all real numbers **15.** 1
19. 5 **21.** 0 **23.** $-\frac{7}{5}$ **25.** 120 **27.** 6
29. 15,000 **31.** 8 **33.** $-\frac{13}{8}$ **35.** 0 **37.** 4
39. 20 **41.** all real numbers **43.** no solution
47. $-6 + x$ **49.** $x + 2x$ **51.** $\frac{-6}{x}$
53. $12(x - 9)$

SECTION 2.4 (page 133)

1. (c) **5.** 3 **7.** -4
9. 57 Democrats, 43 Republicans **11.** 1037
13. shorter piece: 15 inches; longer piece: 24 inches
15. Federal Express: 9; Airborne Express: 3; United Parcel Service: 1 **17.** 36 million miles
19. Gant: 141; Justice: 124; Cabrera: 3
21. A and B: 40 degrees; C: 100 degrees
23. antilock brakes: $800; power door locks: $240
25. active: 7 centigrams; inert: 665 centigrams
27. 80° **29.** 26° **31.** 55° **33.** 121 and 122
35. 20 and 22 **37.** 24 **39.** 320

SECTION 2.5 (page 143)

3. area **5.** perimeter
7. area **9.** area **11.** $P = 20$ **13.** $P = 24$
15. $A = 70$ **17.** $c = 5$ **19.** $r = 40$
21. $I = 875$ **23.** $A = 91$ **25.** $r = 1.3$
27. $A = 452.16$ **29.** $V = 384$ **31.** $V = 48$
33. $V = 904.32$ **37.** 1029.92 feet
39. perimeter: 172 inches; area: 1785 square inches
41. $\frac{729}{32}$ or $22\frac{25}{32}$ cubic inches
43. 23,800.10 square feet **45.** 107°, 73°
47. 75°, 75° **49.** 139°, 139° **51.** $L = \frac{A}{W}$
53. $r = \frac{d}{t}$ **55.** $p = \frac{I}{rt}$ **57.** $a = P - b - c$
59. $b = \frac{2A}{h}$ **61.** $r = \frac{A - p}{pt}$
63. $h = \frac{V}{\pi r^2}$ **65.** $m = \frac{y - b}{x}$ or $m = \frac{y}{x} - \frac{b}{x}$
67. 5% **69.** .5% **71.** 28

SECTION 2.6 (page 153)

1. 4.56 **3.** 124.8 **5.** 120% **7.** 600 **9.** 1.4%
11. $533 **13.** 15 **15.** $16\frac{2}{3}$% **17.** $384

19. 180% **21.** .560 **23.** .452
27. $\frac{4}{3}$ **29.** $\frac{4}{3}$ **31.** $\frac{15}{2}$
33. $\frac{1}{5}$ **35.** $\frac{24}{5}$ **39.** 35
41. 27 **43.** -1 **45.** 10 **47.** 510 calories
49. $14.58 **51.** $25\frac{2}{3}$ inches **53.** 12,500 fish
55. 20-count size **57.** 32-ounce size
59. 32-ounce size **61.** $<$ **63.** $<$ **65.** $>$

SECTION 2.7 (page 163)

1. 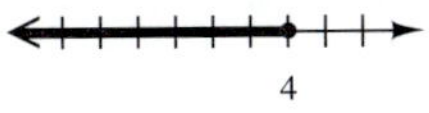 4 **3.** -3 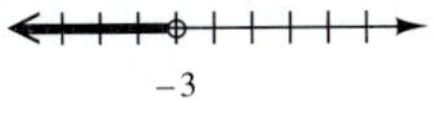
5. 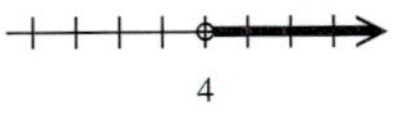 4 **7.** 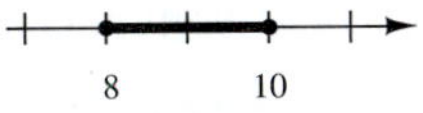 8 10
9. 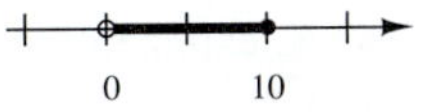 0 10
11. Use an open circle if the symbol is $>$ or $<$. Use a closed circle if the symbol is $\geq$ or $\leq$.
13. It would imply that $3 < -2$, which is false.
15. $z \geq 1$ 1 **17.** $k \geq 5$ 5
19. $n < -11$ -11
21. It must be reversed when multiplying or dividing by a negative number.
25. $x < 6$ 6 **27.** $y \geq -10$ -10
29. $t < -3$ -3 **31.** $x \leq 0$ 0
33. $r > 20$ 20 **35.** $x \geq -3$ -3 
37. $r \geq -5$ -5 **39.** $x < 1$ 1
41. $x \leq 0$ 0 **43.** $x \geq 4$ 4
45. $p < 32$ 32 **47.** $x \geq \frac{5}{12}$ $\frac{5}{12}$ 0

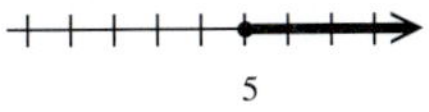

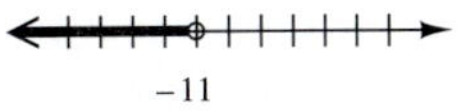

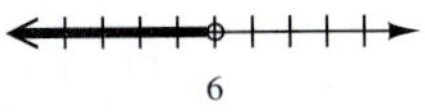

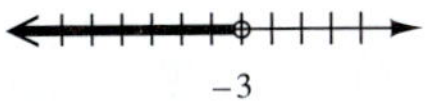

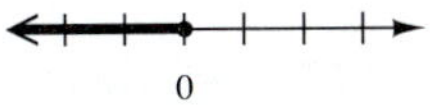

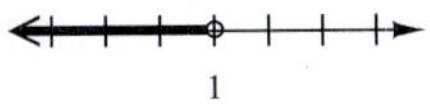

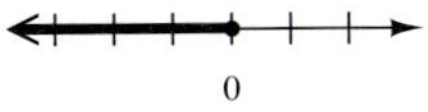

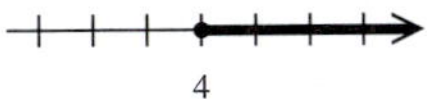

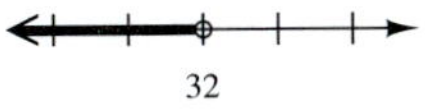

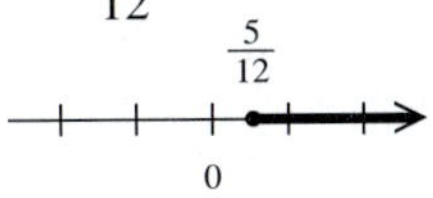

49. $k > -21$ -21 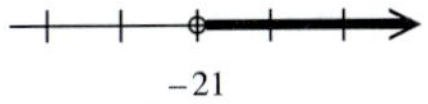
51. all numbers less than 3 **53.** 83 or more
55. It is never more than 86 degrees Fahrenheit.
57. $x \geq 500$ **59.** at least $275 **61.** 64 **63.** 64
65. 3 **67.** -5 **69.** 39

CHAPTER 2 REVIEW EXERCISES (page 171)

1. 9 **2.** 4 **3.** -6 **4.** $\frac{3}{2}$ **5.** 20 **6.** $-\frac{61}{2}$
7. 15 **8.** 0 **9.** no solution **10.** 20
11. $-\frac{7}{2}$ **12.** 20 **13.** Hawaii: 6425 square miles; Rhode Island: 1212 square miles **14.** Seven Falls: 300 feet; Twin Falls: 120 feet **15.** 80°
16. Saberhagen: 257 innings; Morris: 266 innings; Langston: 272 innings **17.** $h = 11$ **18.** $A = 28$
19. $r = 4.75$ **20.** $V = 904.32$ **21.** $L = \frac{A}{W}$
22. $h = \frac{2A}{b + B}$ **23.** 135°, 45° **24.** 100°, 100°
25. perimeter: 326.5 feet; area: 6538.875 square feet
26. diameter: approximately 19.9 feet; radius: approximately 9.95 feet; area: approximately 311 square feet
27. 17.48 **28.** 175% **29.** $33\frac{1}{3}\%$ **30.** 2500
31. $1118 **32.** 437 miles per tank **33.** $\frac{3}{2}$
34. $\frac{5}{14}$ **35.** $\frac{3}{4}$ **36.** $\frac{1}{12}$ **37.** $\frac{7}{2}$ **38.** $-\frac{8}{3}$
39. $\frac{25}{19}$ **41.** $6\frac{2}{3}$ pounds **42.** 36 ounces
43. $3.06 **44.** 375 kilometers
45. 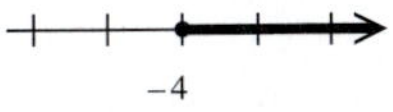 -4 **46.** 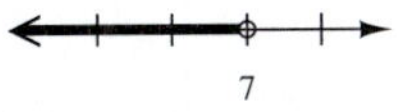 7
47.  -5 6 **48.** 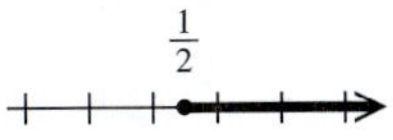 $\frac{1}{2}$
49. $y \geq -3$ -3 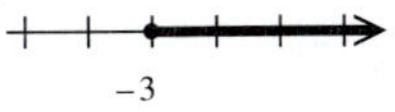 **50.** $t < 2$ 2
51. $x \geq 3$ 3 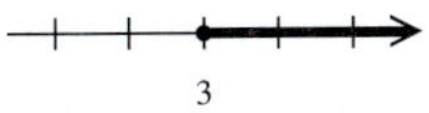 **52.** $k \geq 46$ 46
53. $x < -5$ -5 **54.** $w < -37$ -37
55. 88 or more **56.** all numbers less than or equal to $-\frac{1}{3}$ **57.** 7 **58.** $r = \frac{I}{pt}$

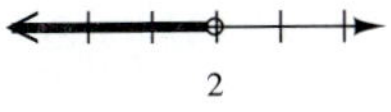

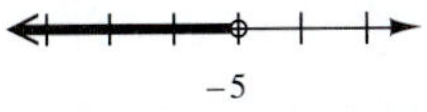

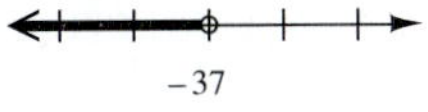

59. $x < 2$ **60.** -9 **61.** 70 **62.** $\frac{13}{4}$
63. no solution **64.** all real numbers **65.** 6
66. 12 **67.** Buddy: 1200 votes; Bob: 600 votes **68.** John: 84 miles; Gwen: 28 miles
69. 12 pounds **70.** 125 miles **71.** 35 geometry tests **72.** 37 small cars **73.** 70 feet
74. 44 meters **75.** 26 inches **76.** $20\frac{1}{2}$ inches
77. 51°, 51° **78.** 92 or more

CHAPTER 2 TEST (page 175)

[2.2] **1.** 6 **2.** -6
[2.3] **3.** $\frac{13}{4}$ **[2.1]** **4.** -10.8
5. no solution **[2.2]** **6.** 21
[2.3] **7.** 30 **[2.1]** **8.** all real numbers
[2.4] **9.** 7 **10.** Golden Gate Bridge: 4200 feet; Brooklyn Bridge: 1595 feet
[2.4] **11.** 4 centimeters, 9 centimeters, 27 centimeters
[2.5] **12.** 18 **13.** $W = \frac{P - 2L}{2}$ or $W = \frac{P}{2} - L$
14. 100°, 80° **15.** 75°, 75° **16.** 50°
[2.6] **17.** 6 **18.** -29 **19.** 8 slices for \$2.19
20. 2300 miles **21.** 14
[2.7] **22.** $x < 11$ 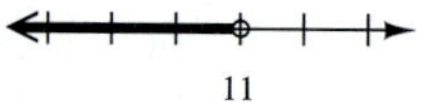

23. $x \le 4$ 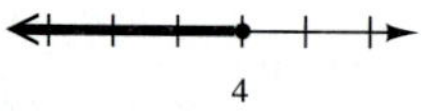

24. 86 or more **25.** When an inequality is multiplied or divided by a negative number, the direction of the symbol must be reversed.

CUMULATIVE REVIEW R-2 (page 177)

1. $\frac{3}{8}$ **2.** $\frac{3}{4}$ **3.** $\frac{31}{20}$ **4.** $\frac{551}{40}$ or $13\frac{31}{40}$ **5.** 6
6. $\frac{6}{5}$ **7.** 34.03 **8.** 27.31 **9.** 30.51
10. 56.3 **11.** 35 yards **12.** $4\frac{1}{6}$ cups
13. $20\frac{23}{24}$ pounds **14.** \$1187.65 **15.** true
16. true **17.** 7 **18.** 1 **19.** 13 **20.** -40
21. -12 **22.** undefined **23.** -6 **24.** 28 **25.** 1
26. 0 **27.** $\frac{73}{18}$ **28.** -64 **29.** -134 **30.** $-\frac{29}{6}$
31. distributive property **32.** commutative property
33. inverse property **34.** identity property
35. $7p - 14$ **36.** $2k - 11$ **37.** 7 **38.** -4
39. -1 **40.** $-\frac{3}{5}$ **41.** 2 **42.** -13 **43.** 26

44. -12 **45.** $c = P - a - b$ **46.** $s = \frac{P}{4}$
47. $z \le 2$ 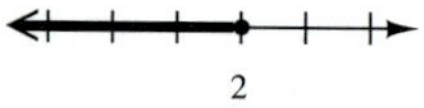

48. $r \le 1$ 

49. \$1261.88 **50.** \$3750 **51.** \$49.50 **52.** \$98.45
53. 30 centimeters **54.** 16 inches
55. 450 miles or more **56.** \$100

CHAPTER 3

SECTION 3.1 (page 187)

1. 3^7 **3.** $(-6)^4$ **5.** w^6 **7.** $\frac{1}{4^4}$ **9.** $(-7x)^4$
11. $\left(\frac{1}{2}\right)^6$ **15.** base: 3; exponent: 5; 243
17. base: -3; exponent: 5; -243
19. base: $-6x$; exponent: 4 **21.** base: x; exponent: 4
25. 5^8 **27.** 4^{12} **29.** $(-7)^9$ **31.** t^{24}
33. $-56r^7$ **35.** $42p^{10}$ **39.** $14x^4$; $45x^8$
41. $5a^2$; $-140a^6$ **43.** 4^6 **45.** t^{20} **47.** 7^3r^3
49. $5^5x^5y^5$ **51.** 5^{12} **53.** -8^{15} **55.** $8q^3r^3$
57. $\frac{1}{2^3}$ **59.** $\frac{a^3}{b^3}$ **61.** $\frac{9^8}{5^8}$ **63.** $\frac{5^5}{2^5}$ **65.** $\frac{9^5}{8^3}$
67. $2^{12}x^{12}$ **69.** $(-6)^5p^5$ **71.** $6^5x^{10}y^{15}$ **73.** x^{21}
75. $2^2w^4x^{26}y^7$ **77.** $-r^{18}s^{17}$ **79.** $\frac{5^3a^6b^{15}}{c^{18}}$
83. 10 **85.** -3

SECTION 3.2 (page 195)

1. 1 **3.** 1 **5.** -1 **7.** 0 **9.** 0 **11.** 1
13. 2 **15.** $\frac{1}{64}$ **17.** 16 **19.** $\frac{49}{36}$ **21.** $\frac{1}{81}$
23. $\frac{8}{15}$ **25.** negative **27.** negative **29.** positive
31. zero **33.** 5^3 **35.** $\frac{1}{9}$ **37.** $\frac{1}{6^5}$ **39.** x^{15}
41. 6^3 **43.** $2r^4$ **45.** $\frac{5^2}{4^3}$ **47.** $\frac{p^5}{q^8}$ **49.** r^9
51. $\frac{x^5}{6}$ **53.** 7^3 **55.** $\frac{1}{x^2}$ **57.** $\frac{4^3x}{3^2}$ **59.** $\frac{x^2z^4}{y^2}$
61. $6x$ **63.** $\frac{1}{m^{10}n^5}$ **65.** 5687 **67.** .0012

SECTION 3.3 (page 199)

1. in scientific notation
3. not in scientific notation; 5.6×10^6
5. not in scientific notation; 8×10^1
7. not in scientific notation; 4×10^{-3}

11. 5.876×10^9 **13.** 8.235×10^4 **15.** 7×10^{-6} **17.** -2.03×10^{-3} **19.** 750,000 **21.** 5,677,000,000,000 **23.** -6.21 **25.** .00078 **27.** .000000005134 **29.** 600,000,000,000 **31.** 15,000,000 **33.** 60,000 **35.** 3×10^{-4} **37.** 4×10^1 **39.** 1.3×10^{-5} **41.** 635,000,000,000 **43.** 1.15×10^6 **45.** .00023; .006 **47.** $12x$ **49.** $12x - 13$

SECTION 3.4 (page 207)

1. 1; 6 **3.** 1; 1 **5.** 2; -19, -1 **7.** 2; 1, 8 **9.** $2m^5$ **11.** $-r^5$ **13.** cannot be simplified **15.** $-5x^5$ **17.** $5p^9 + 4p^7$ **19.** $-2y^2$ **21.** sometimes **23.** always **25.** never **27.** already simplified; 4; binomial **29.** $6m^5 + 5m^4 - 7m^3 - 3m^2$; 5; none of these **31.** $x^4 + \frac{1}{3}x^2 - 4$; 4; trinomial **33.** 7; 0; monomial **35.** $5m^2 + 3m$ **37.** $4x^4 - 4x^2$ **39.** $\frac{7}{6}x^2 - \frac{2}{15}x + \frac{5}{6}$ **41.** $15m^3 - 13m^2 + 8m + 11$ **45.** $8r^2 + 5r - 12$ **47.** $5m^2 - 14m + 6$ **49.** $4x^3 + 2x^2 + 5x$ **51.** $-18y^5 + 7y^4 + 5y^3 + 3y^2 + y$ **53.** $-2m^3 + 7m^2 + 8m - 10$ **55.** $8x^2 + 8x + 6$ **57.** $8t^2 + 8t + 13$ **59.** $-11x^2 - 3x - 3$ **63.** $13a^2b - 7a^2 - b$; degree 3 **65.** $c^4d - 5c^2d^2 + d^2$; degree 5 **67.** $12m^3n - 11m^2n^2 - 4mn^2$; degree 4 **69.** $-24n^7$ **71.** $32y^4$

SECTION 3.5 (page 215)

1. associative; commutative; associative **3.** $40a^{14}$ **5.** $-6m^2 - 4m$ **7.** $24p - 18p^2 + 36p^4$ **9.** $6y^5 + 4y^6 + 10y^9$ **11.** $n^2 + n - 6$ **13.** $8r^2 - 10r - 3$ **15.** $9x^2 - 4$ **17.** $9q^2 + 6q + 1$ **19.** $6t^2 + 23st + 20s^2$ **21.** $-3t^2 - 14t + 24$ **23.** $36x^2 + 24x + 4$ **27.** $12x^3 + 26x^2 + 10x + 1$ **29.** $20m^4 - m^3 - 8m^2 - 17m - 15$ **31.** $5x^4 - 13x^3 + 20x^2 + 7x + 5$ **33.** $x^2 + 24x + 144$ **35.** $9t^2 + 6t + 1$ **37.** $25x^2 - 20xy + 4y^2$ **39.** $h^3 - 15h^2 + 75h - 125$ **41.** $r^4 + 4r^3 + 6r^2 + 4r + 1$ **43.** $9x^4 + 6x^3 - 23x^2 - 8x + 16$ **45.** 2, 4 **47.** -8, 7

SECTION 3.6 (page 221)

1. (a) $2x^2$ **(b)** $-10x$ **(c)** $3x$ **(d)** -15 **(e)** $-7x$ **(f)** $2x^2 - 7x - 15$ **3.** $r^2 + 4r + 3$ **5.** $w^2 - 10w + 24$ **7.** $s^2 - 8s - 48$ **9.** $10x^2 + x - 3$ **11.** $27x^2 + 69x + 14$ **13.** $9m^2 + 36m + 35$ **15.** $15 - 19x + 6x^2$ **17.** $-15 + 23z - 6z^2$ **19.** $16 + 2k - 3k^2$ **21.** $27w^2 + 15wz - 2z^2$ **23.** $-16p^2 - 2ps + 3s^2$ **25.** $x^2 + .5x - .24$ **27.** $x^2 - \frac{5}{12}x - \frac{1}{6}$ **29. (a)** $4x^2$ **(b)** $12x$ **(c)** 9 **(d)** $4x^2 + 12x + 9$ **31.** $p^2 + 4p + 4$ **33.** $a^2 - 2ac + c^2$ **35.** $16x^2 - 24x + 9$ **37.** $64t^2 + 112ts + 49s^2$ **39.** $25x^2 + 4xy + \frac{4}{25}y^2$ **41. (a)** $49x^2$ **(b)** 0 **(c)** $-9y^2$ **(d)** $49x^2 - 9y^2$; Because 0 is the identity element for addition, it is not necessary to write "+0." **43.** $q^2 - 4$ **45.** $4w^2 - 25$ **47.** $100x^2 - 9y^2$ **49.** $4x^4 - 25$ **51.** $49x^2 - \frac{9}{49}$ **53.** $3x^4$ **55.** $8m^7$

SECTION 3.7 (page 225)

1. $-6x^4$ **3.** $5x^7y$ **7.** $30m^3 - 10m + 5$ **9.** $5m^4 - 8m^3 + 4m^2$ **11.** $4m^4 - 2m^2 + 2m$ **13.** $\frac{m^4}{2} - 2m + \frac{4}{m}$ **15.** $4x^3 - 3x^2 + 2x$ **17.** $1 + 5x - 9x^2$ **19.** $\frac{12}{x} + 8 + 2x$ **21.** $\frac{4x^2}{3} + x + \frac{2}{3x}$ **23.** $9r^3 - 12r^2 - 2r + \frac{26}{3} - \frac{2}{3r}$ **25.** $-m^2 + 3m - \frac{4}{m}$ **27.** $4 - 3a + \frac{5}{a}$ **29.** $\frac{12}{x} - \frac{6}{x^2} + \frac{14}{x^3} - \frac{10}{x^4}$ **33.** $15x^5 - 35x^4 + 35x^3$ **35.** $-24k^3 + 36k^2 - 6k$ **37.** $-16k^3 - 10k^2 + 3k + 3$ **39.** $10x^2 - 4x + 11$

SECTION 3.8 (page 231)

1. $x + 2$ **3.** $2y - 5$ **5.** $p - 4 + \frac{44}{p + 6}$ **7.** $r - 5$ **9.** $6m - 1$ **11.** $2a - 14 + \frac{74}{2a + 3}$ **13.** $4x^2 - 7x + 3$ **15.** The divisor is $2x + 5$; the quotient is $2x^3 - 4x^2 + 3x + 2$. **17.** $3y^2 - 2y + 2$ **19.** $3k - 4 + \frac{2}{k^2 - 2}$ **21.** $x^2 + 1$ **23.** $2p^2 - 5p + 4 + \frac{6}{3p^2 + 1}$ **25.** $x^3 + 3x^2 - x + 5$ **27.** $x^2 + 1$ **29.** 1, 2, 3, 6, 9, 18 **31.** 1, 2, 3, 4, 6, 8, 12, 16, 24, 48

CHAPTER 3 REVIEW EXERCISES (page 235)

1. 4^{11} **2.** $(-5)^{11}$ **3.** $-72x^7$ **4.** $10x^{14}$
5. 19^5x^5 **6.** $(-4)^7y^7$ **7.** $5p^4t^4$ **8.** $\frac{7^6}{5^6}$
9. $3^3x^6y^9$ **10.** t^{42} **11.** $6^2x^{16}y^4z^{16}$ **13.** 2
14. $\frac{1}{32}$ **15.** $\frac{5^2}{6^2}$ or $\frac{25}{36}$
16. $-\frac{3}{16}$ **17.** 6^2 **18.** x^2
19. $\frac{1}{p^{12}}$ **20.** r^4 **21.** 2^8 **22.** $\frac{1}{9^6}$ **23.** 5^8
24. $\frac{1}{8^{12}}$ **25.** $\frac{1}{m^2}$ **26.** y^7 **27.** r^{13} **28.** $(-5)^2m^6$
29. $\frac{y^{12}}{2^3}$ **30.** $\frac{1}{a^3b^5}$ **31.** $2 \cdot 6^2 \cdot r^5$ **32.** $\frac{2^3n^{10}}{3m^{13}}$
33. 4.8×10^7 **34.** 2.8988×10^{10} **35.** 6.5×10^{-5}
36. 8.24×10^{-8} **37.** 24,000 **38.** 78,300,000
39. .000000897 **40.** .00000000000995 **41.** 800
42. 4,000,000 **43.** .025 **44.** .01
45. $22m^2$; 2; monomial
46. $p^3 - p^2 + 4p + 2$; 3; none of these
47. $8r^5 + 4r^4$; 5; binomial **48.** $15a^5 + 10a^4 + 10a^3$; 5; trinomial **49.** $-5a^3 + 4a^2$ **50.** $2r^3 - 3r^2 + 9r$
51. $11y^2 - 10y + 9$ **52.** $-13k^4 - 15k^2 - 4k - 6$
53. $10m^3 - 6m^2 - 3$ **54.** $-y^2 - 4y + 26$
55. $10p^2 - 3p - 11$ **56.** $7r^4 - 4r^3 - 1$
57. $10x^2 + 70x$ **58.** $-6p^5 + 15p^4$
59. $m^2 - 7m - 18$ **60.** $6k^2 - 9k - 6$
61. $6r^3 + 8r^2 - 17r + 6$ **62.** $8y^3 + 27$
63. $r^3 + 6r^2 + 12r + 8$ **65.** $6k^2 + 11k + 3$
66. $2a^2 + 5ab - 3b^2$ **67.** $12k^2 - 48kq + 21q^2$
68. $a^2 + 8a + 16$ **69.** $9p^2 - 12p + 4$
70. $4r^2 + 20rs + 25s^2$ **71.** $36m^2 - 25$
72. $4z^2 - 49$ **73.** $25a^2 - 36b^2$ **74.** $4x^4 - 25$
75. $\frac{5y^2}{3}$ **76.** $-2x^2y$ **77.** $-y^3 + 2y - 3$
78. $p - 3 + \frac{5}{2p}$ **79.** $-x^9 + 2x^8 - 4x^3 + 7x$
80. $-2m^2n + mn^2 + \frac{6n^3}{5}$ **81.** $2r + 7$
82. $4m + 3 + \frac{5}{3m - 5}$ **83.** $2a + 1 + \frac{-8a + 12}{5a^2 - 3}$
84. $k^2 + 2k + 4 + \frac{-2k - 12}{2k^2 + 1}$ **85.** 0 **86.** $\frac{3^5}{p^3}$
87. $\frac{1}{7^2}$ **88.** $49 - 28k + 4k^2$ **89.** $y^2 + 5y + 1$
90. $\frac{6^4r^8s^4}{5^4}$ **91.** $-8m^7 - 10m^6 - 6m^5$
92. 2^5 **93.** $5xy^3 - \frac{8y^2}{5} + 3x^2y$ **94.** $\frac{r^2}{6}$
95. $8x^3 + 12x^2y + 6xy^2 + y^3$ **96.** $\frac{3}{4}$
97. $a^3 - 2a^2 - 7a + 2$ **98.** $8y^3 - 9y^2 + 5$
99. $10r^2 + 21r - 10$ **100.** $144a^2 - 1$

CHAPTER 3 TEST (page 239)

[3.2] 1. $\frac{1}{625}$ **[3.1–3.2] 2.** 2 **3.** $\frac{7}{12}$
4. 8^5 **5.** x^2y^6 **[3.3] 6.** **(a)** 3.44×10^{11} **(b)** 5.57×10^{-6} **7.** **(a)** 29,600,000 **(b)** .0000000607
[3.4] 8. $-7x^2 + 8x$; 2; binomial
9. $4n^4 + 13n^3 - 10n^2$; 4; trinomial
10. $4t^4 + t^3 - 6t^2 - t$ **11.** $-2y^2 - 9y + 17$
12. $-12t^2 + 5t + 8$
[3.5] 13. $-27x^5 + 18x^4 - 6x^3 + 3x^2$
[3.6] 14. $t^2 - 5t - 24$ **15.** $8x^2 + 2xy - 3y^2$
16. $25x^2 - 20xy + 4y^2$ **17.** $100v^2 - 9w^2$
[3.5] 18. $2r^3 + r^2 - 16r + 15$
19. $x^3 + 3x^2 + 3x + 1$
[3.7] 20. $4y^2 - 3y + 2 + \frac{5}{y}$
21. $-3xy^2 + 2x^3y^2 + 4y^2$
[3.8] 22. $2x + 9$ **23.** $3x^2 + 6x + 11 + \frac{26}{x - 2}$
[3.5] 24. $9x^2 + 54x + 81$
[3.4] 25. Answers will vary. One example is $(-4x^4 + 3x^3 + 2x + 1) + (4x^4 - 8x^3 + 2x + 7) = -5x^3 + 4x + 8$.

CUMULATIVE REVIEW R–3 (page 241)

1. $\frac{7}{4}$ **2.** 5 **3.** $\frac{19}{24}$ **4.** $-\frac{1}{20}$ **5.** 3.72
6. 62.006 **7.** $31\frac{1}{4}$ cubic yards **8.** \$1836
9. -8 **10.** 24 **11.** $\frac{1}{2}$ **12.** -4
13. associative property **14.** inverse property
15. distributive property **16.** 10 **17.** $\frac{13}{4}$
18. no solution **19.** $r = \frac{d}{t}$ **20.** -5 **21.** -12
22. 20 **23.** all real numbers **24.** Janet: \$48; Louis: \$8 **25.** 4 **26.** 109 hens; 81 roosters
27. 11 feet and 22 feet **28.** $x \geq 10$ **29.** $x < -\frac{14}{5}$
30. $-4 \leq x < 2$ **31.** $\frac{5}{4}$ or $1\frac{1}{4}$ **32.** 2 **33.** 1
34. $\frac{2b}{a^{10}}$ **35.** 3.45×10^4 **36.** $11x^3 - 14x^2 - x + 14$
37. $18x^7 - 54x^6 + 60x^5$ **38.** $63x^2 + 57x + 12$
39. $25x^2 + 80x + 64$ **40.** $y^2 - 2y + 6$

CHAPTER 4

SECTION 4.1 (page 249)

1. 4 **3.** 4 **5.** 6 **7.** 1
11. 8 **13.** $10x^3$ **15.** $6m^3n^2$ **17.** xy^2 **19.** 6
21. yes; x^3y^2 **23.** 2 **25.** x **27.** $3m^2$
29. $2z^4$ **31.** $2mn^4$ **33.** $-7x^3y^2$ **35.** $12(y - 2)$

37. $10a(a-2)$ **39.** $5y^6(13y^4+7)$
41. no common factor (except 1) **43.** $8m^2n^2(n+3)$
45. $13y^2(y^6+2y^2-3)$ **47.** $9qp^3(5q^3p^2+4p^3+9q)$
49. $a^3(a^2+2b^2-3a^2b^2+4ab^3)$
51. $(p+4)(p+3)$ **53.** $(a-2)(a+5)$
55. $(z+2)(7z-a)$ **57.** $(3r+2y)(6r-x)$
59. $(a^2+b^2)(3a+2b)$ **61.** $(1-a)(1-b)$
63. $(4m-p^2)(4m^2-p)$
65. (a) yes **67.** $x^2-3x-54$
69. $x^2+9x+14$

SECTION 4.2 (page 255)

1. 1 and 12, −1 and −12, 2 and 6, −2 and −6, 3 and 4, −3 and −4; the pair with a sum of 7 is 3 and 4.
3. 1 and −24, −1 and 24, 2 and −12, −2 and 12, 3 and −8, −3 and 8, 4 and −6, −4 and 6; the pair with a sum of −5 is 3 and −8.
5. 1 and 27, −1 and −27, 3 and 9, −3 and −9; the pair with a sum of 28 is 1 and 27.
7. 1 and −48, −1 and 48, 2 and −24, −2 and 24, 3 and −16, −3 and 16, 4 and −12, −4 and 12, 6 and −8, −6 and 8; the pair with a sum of 2 is −6 and 8.
9. (c) **11.** $p+6$ **13.** $x+11$ **15.** $x-8$
17. $y-5$ **19.** $x+11$ **21.** $y-9$
23. $(y+8)(y+1)$ **25.** $(b+3)(b+5)$
27. $(m+5)(m-4)$ **29.** $(y-5)(y-3)$
31. $(r-6)(r+5)$ **33.** $(t-4)^2$ or $(t-4)(t-4)$
35. $(r+2a)(r+a)$ **37.** $(t+2z)(t-3z)$
39. $(x+y)(x+3y)$ **41.** $(v-5w)(v-6w)$
43. $4(x+5)(x-2)$ **45.** $2t(t+1)(t+3)$
47. $2x^4(x-3)(x+7)$ **49.** $mn(m-6n)(m-4n)$
53. $a^2+13a+36$ **55.** $2y^2+y-28$
57. $15z^2-4z-4$ **59.** $8p^2-10p-3$

SECTION 4.3 (page 261)

1. (b) **3.** (a) **5.** (a) **7.** $2a+5b$
9. x^2+3x-4; $x+4$, $x-1$, or $x-1$, $x+4$
11. $2z^2-5z-3$; $2z+1$, $z-3$, or $z-3$, $2z+1$
The order of the factors is irrelevant in Exercises 15–61.
15. $(2x+1)(x+3)$
17. $(3a+7)(a+1)$ **19.** $(4r-3)(r+1)$
21. $(3m-1)(5m+2)$ **23.** $(4m+1)(2m-3)$
25. $(4x+3)(5x-1)$ **27.** $(3m+1)(7m+2)$
29. $(4y-1)(5y+11)$ **31.** $(2b+1)(3b+2)$
33. $3(4x-1)(2x-3)$ **35.** $q(5m+2)(8m-3)$
37. $2m(m-4)(m+5)$ **39.** $3n^2(5n-3)(n-2)$
41. $3x^3(2x+5)(3x-5)$ **43.** $y^2(5x-4)(3x+1)$
45. $(3p+4q)(4p-3q)$ **47.** $(5a+2b)(5a+3b)$
49. $(3a-5b)(2a+b)$ **51.** $m^4n(3m+2n)(2m+n)$
53. $(5-x)(1-x)$ **55.** $(4+3x)(4+x)$
57. $-1(x+7)(x-3)$ **59.** $-1(3x+4)(x-1)$
61. $-1(a+2b)(2a+b)$
65. $49p^2-9$ **67.** $r^4-\frac{1}{4}$ **69.** $9t^2+24t+16$

SECTION 4.4 (page 269)

1. 1; 4; 9; 16; 25; 36; 49; 64; 81; 100; 121; 144; 169; 196; 225; 256; 289; 324; 361; 400
3. 2 **5.** $(y+5)(y-5)$ **7.** $(3r+2)(3r-2)$
9. $\left(6m+\frac{4}{5}\right)\left(6m-\frac{4}{5}\right)$ **11.** $4(3x+2)(3x-2)$
13. $(14p+15)(14p-15)$ **15.** $(4r+5a)(4r-5a)$
17. prime **19.** $(p^2+7)(p^2-7)$
21. $(x^2+1)(x+1)(x-1)$
23. $(p^2+16)(p+4)(p-4)$
27. $(w+1)^2$ **29.** $(x-4)^2$ **31.** $\left(t+\frac{1}{2}\right)^2$
33. $(x-.5)^2$ **35.** $2(x+6)^2$
37. $(4x-5)^2$ **39.** $(7x-2y)^2$ **41.** $(8x+3y)^2$
43. $-2(5h-2y)^2$ **47.** 9 **49.** 4
51. $\frac{9}{4}$ **53.** $\frac{2}{3}$

SUMMARY EXERCISES ON FACTORING (page 271)

1. $8m^3(4m^6+2m^2+3)$ **3.** $7k(2k+5)(k-2)$
5. $(6z+1)(z+5)$ **7.** $(7z+4y)(7z-4y)$
9. $4(4x+5)$ **11.** $(5y-6z)(2y+z)$
13. $(m-3)(m+5)$ **15.** $8z(4z-1)(z+2)$
17. $(z-6)^2$ **19.** $(y-6k)(y+2k)$
21. $6(y-2)(y+1)$ **23.** $(p-6)(p-11)$
25. prime **27.** $(z+2a)(z-5a)$ **29.** $(2k-3)^2$
31. $(4r+3m)^2$ **33.** $3k(k+1)(k-5)$
35. $4(2k-3)^2$ **37.** $6y^4(3y+4)(2y-5)$
39. $5z(z-2)(z-7)$ **41.** $6(3m+2z)(3m-2z)$
43. $2(3a-1)(a+2)$ **45.** $(m+9)(m-9)$
47. $5m^2(5m-13n)(5m-3n)$ **49.** $(m-2)^2$
51. $6k^2p(4k^2+10kp+25p^2)$
53. $(3p-2q)(4p+3q)$ **55.** $4(4p+5m)(4p-5m)$
57. $(10a+9y)(10a-9y)$ **59.** $(a+4)^2$

SECTION 4.5 (page 279)

1. 4, −5 **3.** $-\frac{8}{3}$, −7 **5.** 0, −4 **7.** 0, $\frac{4}{3}$
13. −2, −1 **15.** 1, 2 **17.** −8, 3 **19.** −1, 3
21. −2, −1 **23.** −4 **25.** −2, $\frac{1}{3}$ **27.** $-\frac{4}{3}$, $\frac{1}{2}$
29. $-\frac{2}{3}$ **31.** −3, 3 **33.** $-\frac{7}{4}$, $\frac{7}{4}$ **35.** −11, 11
37. −11 is another solution. **39.** 0, 7 **41.** 0, $\frac{1}{2}$
43. 2, 5 **45.** −4, $\frac{1}{2}$ **47.** −12, $\frac{11}{2}$ **49.** −1, 3
51. $-\frac{5}{2}$, $\frac{1}{3}$, 5 **53.** $-\frac{7}{2}$, −3, 1 **55.** $-\frac{7}{3}$, 0, $\frac{7}{3}$
57. Florida: 67 counties; California: 58 counties

ANSWERS

SECTION 4.6 (page 285)

1. **(a)** $80 = (x + 8)(x - 8)$ **(b)** 12 **(c)** length: 20 units; width: 4 units
3. **(a)** $45 = (2x + 1)(x + 1)$ **(b)** 4 **(c)** base: 9 units; height: 5 units
5. **(a)** $36\pi = \pi(x + 2)^2$ **(b)** 4 **(c)** radius: 6 units
7. length: 7 inches; width: 4 inches
9. length: 11 inches; width: 8 inches
11. poster: 7 feet; sign: 9 feet
13. base: 12 inches; height: 5 inches
15. height: 13 inches; width: 10 inches
17. $-3, -2$ or 4, 5
19. 7, 9, 11 **21.** $-2, 0, 2$ or 6, 8, 10
23. 12 miles **25.** 8 feet
27. height: 6 meters; area of the base: 16 square meters
29. **(a)** 144 feet **(b)** 10 seconds
33. $y^2 - 4xy + 4x^2$ **35.** $\frac{1}{6}$ **37.** -3

CHAPTER 4 REVIEW EXERCISES (page 293)

1. $7(t + 2)$ **2.** $30z(2z^2 + 1)$ **3.** $35x^2(x + 2)$
4. $(x - 4)(2y + 3)$ **5.** $50m^2n^2(2n - mn^2 + 3)$
6. $(3y + 2)(2y + 3)$ **7.** $(x + 3)(x + 2)$
8. $(y - 5)(y - 8)$ **9.** $(q + 9)(q - 3)$
10. $(r - 8)(r + 7)$ **11.** $(r + 8s)(r - 12s)$
12. $(p + 12q)(p - 10q)$ **13.** $8p(p + 2)(p - 5)$
14. $3x^2(x + 2)(x + 8)$ **15.** $(m + 3n)(m - 6n)$
16. $(y - 3z)(y - 5z)$ **17.** $p^5(p - 2q)(p + q)$
18. $3r^3(r + 3s)(r - 5s)$ **19.** prime
20. $3(x^2 + 2x + 2)$
21. r and $6r$, $2r$ and $3r$ **22.** Factor out z.
23. $(2k - 1)(k - 2)$ **24.** $(3r - 1)(r + 4)$
25. $(3r + 2)(2r - 3)$ **26.** $(5z + 1)(2z - 1)$
27. $(v + 3)(8v - 7)$ **28.** $4x^3(3x - 1)(2x - 1)$
29. $-3(x + 2)(2x - 5)$ **30.** $rs(5r + 6s)(2r + s)$
31. (b) **32.** (d) **33.** $(n + 7)(n - 7)$
34. $(5b + 11)(5b - 11)$ **35.** $(7y + 5w)(7y - 5w)$
36. $36(2p + q)(2p - q)$ **37.** prime **38.** $(z + 5)^2$
39. $(r - 6)^2$ **40.** $(3t - 7)^2$ **41.** $(4m + 5n)^2$
42. $6x(3x - 2)^2$ **43.** $-\frac{3}{4}, 1$ **44.** $-7, 4, -3$
45. $0, \frac{5}{2}$ **46.** $-3, -1$ **47.** 1, 4 **48.** 3, 5
49. $-\frac{4}{3}, 5$ **50.** $-\frac{8}{9}, \frac{8}{9}$ **51.** 0, 8 **52.** $-1, 6$
53. 7 **54.** 6 **55.** $-\frac{2}{5}, -2, -1$ **56.** $-3, 3$
57. length: 10 meters; width: 4 meters
58. length: 6 meters; width: 2 meters
59. length: 10 centimeters; width: 8 centimeters
60. length: 6 meters; height: 5 meters
61. length: 7 feet; width: 3 feet
62. length: 9 inches; width: 4 inches
63. 6, 7 or $-5, -4$ **64.** 6 inches, 8 inches, 10 inches
65. 112 feet **66.** 192 feet
67. 256 feet **68.** after 8 seconds **69.** 2 inches
70. 2 centimeters **71.** $(z - x)(z - 10x)$
72. $(3k + 5)(k + 2)$ **73.** $(3m + 4p)(5m - 4p)$
74. $(y^2 + 25)(y + 5)(y - 5)$ **75.** $3m(2m + 3)(m - 5)$
76. $8abc(3b^2c - 7ac^2 + 9ab)$ **77.** prime
78. $6xyz(2xz^2 + 2y - 5x^2yz^3)$ **79.** $2a^3(a + 2)(a - 6)$
80. $(2r + 3q)(6r - 5q)$ **81.** $(10a + 3)(10a - 3)$
82. $(7t + 4)^2$ **83.** 0, 7 **84.** $-5, 2$ **85.** $-\frac{2}{5}$
86. 15 meters, 36 meters, 39 meters
87. length: 6 meters; width: 4 meters
88. $-5, -4, -3$ or 5, 6, 7 **89.** $-6, -4$ or 6, 8
90. width: 10 meters; length: 17 meters
91. 6 meters

CHAPTER 4 TEST (page 299)

[4.3] **1.** (d) **[4.1]** **2.** $6x(2x - 5)$
3. $m^2n(2mn + 3m - 5n)$
[4.2] **4.** $(x + 3)(x - 8)$ **5.** $(x - 7)(x - 2)$
[4.3] **6.** $(2x + 3)(x - 1)$ **7.** $(3x + 1)(2x - 7)$
8. $3(x + 1)(x - 5)$ **9.** $(5z - 1)(2z - 3)$
10. prime **[4.4]** **11.** prime **12.** $(y + 7)(y - 7)$
13. $(3y + 8)(3y - 8)$ **14.** $(x + 8)^2$
15. $(2x - 7y)^2$ **16.** $-2(x + 1)^2$
17. $3t^2(2t + 9)(t - 4)$ **18.** prime
19. $4t(t + 4)^2$ **20.** $(x^2 + 9)(x + 3)(x - 3)$
21. $(p + 3)(p + 3) = p^2 + 6p + 9 \neq p^2 + 9$
[4.5] **22.** $-3, 9$ **23.** $\frac{1}{2}, 6$ **24.** $-\frac{2}{5}, \frac{2}{5}$ **25.** 10
26. 0, 3 **27.** $-\frac{2}{3}$ is also a solution.
[4.6] **28.** 6 feet by 9 feet **29.** 17 feet
30. -2 and -1

CUMULATIVE REVIEW R-4 (page 301)

1. 0 **2.** .05 **3.** 6 **4.** $P = \frac{A}{1 + rt}$ **5.** $-\frac{9}{5}$
6. antibiotics: 36; tranquilizers: 54 **7.** \$450 per week
8. 35 pounds **9.** 4 **10.** $\frac{16}{9}$ **11.** 1 **12.** 256
13. $\frac{1}{p^2}$ **14.** $\frac{1}{m^6}$ **15.** $-4k^2 - 4k + 8$
16. $6m^8 - 15m^6 + 3m^4$ **17.** $3y^3 + 8y^2 + 12y - 5$
18. $9p^2 + 12p + 4$ **19.** $4p^2 - 9q^2$
20. $45x^2 + 3x - 18$ **21.** $4x^3 + 6x^2 - 3x + 10$
22. $6p^2 + 7p + 1 + \frac{6}{2p - 2}$ **23.** $(2a - 1)(a + 4)$
24. $(2m + 3)(5m + 2)$ **25.** $(5x + 3y)(3x - 2y)$
26. $(4t + 3v)(2t + v)$ **27.** $(3x + 1)^2$
28. $(2p - 3)^2$ **29.** $-2(4t + 7z)^2$
30. $(5r + 9t)(5r - 9t)$ **31.** $25(4x^2 + 1)$
32. $m(2a - 1)(3a + 2)$ **33.** $2pq(3p + 1)(p + 1)$

34. $(2x + y)(a - b)$ **35.** $\frac{3}{2}, -2, 6$ **36.** $-\frac{2}{3}, \frac{1}{2}$
37. 0, 8 **38.** 3 centimeters **39.** 2, 4
40. 5 meters, 12 meters, 13 meters

CHAPTER 5

SECTION 5.1 (page 309)

1. 0 **3.** $\frac{5}{3}$ **5.** $-3, 2$ **7.** never undefined
9. (a) 1 **(b)** $\frac{17}{12}$ **11. (a)** 0 **(b)** $-\frac{10}{3}$
13. (a) $\frac{9}{5}$ **(b)** undefined **15. (a)** $\frac{2}{7}$ **(b)** $\frac{13}{3}$
19. $3r^2$ **21.** $\frac{2}{5}$ **23.** $\frac{x - 1}{x + 1}$ **25.** $\frac{7}{5}$ **27.** $m - n$
29. $\frac{3(2m + 1)}{4}$ **31.** $\frac{3m}{5}$ **33.** $\frac{3r - 2s}{3}$ **35.** $\frac{z - 3}{z + 5}$
37. $\frac{x + 1}{x - 1}$ **39.** -1 **41.** $-(m + 1)$ **43.** -1
45. (d) **47.** $\frac{5}{9}$ **49.** $\frac{10}{3}$ **51.** 4

SECTION 5.2 (page 315)

1. $\frac{3a}{2}$ **3.** $-\frac{4x^4}{3}$ **5.** $\frac{2}{c + d}$ **7.** $\frac{16q}{3p^3}$
9. $\frac{7}{r^2 + rp}$ **11.** $\frac{z^2 - 9}{z^2 + 7z + 12}$ **13.** 5 **15.** $-\frac{3}{2t^4}$
17. $\frac{1}{4}$ **21.** $\frac{10}{9}$ **23.** $-\frac{3}{4}$ **25.** $-\frac{9}{2}$
27. $\frac{-9(m - 2)}{(m + 4)}$ **29.** $\frac{p + 4}{p + 2}$ **31.** $\frac{(k - 1)^2}{(k + 1)(2k - 1)}$
33. $\frac{4k - 1}{3k - 2}$ **35.** $\frac{m + 4p}{m + p}$ **37.** $\frac{10}{x + 10}$ **39.** $2 \cdot 3^2$
41. $2^2 \cdot 3^3$ **43.** 6 **45.** $6q^3$

SECTION 5.3 (page 321)

1. 840 **3.** 30 **5.** x^7 **7.** $72q$ **9.** $84r^5$
13. $28m^2(3m - 5)$ **15.** $30(b - 2)$
17. $c - d$ or $d - c$ **19.** $k(k + 5)(k - 2)$
21. $(p + 3)(p + 5)(p - 6)$ **23.** $\frac{20}{55}$ **25.** $\frac{-45}{9k}$
27. $\frac{26y^2}{80y^3}$ **29.** $\frac{35t^2r^3}{42r^4}$ **31.** $\frac{20}{8(m + 3)}$ **33.** $\frac{-8t}{6t - 12}$
35. $\frac{14(z - 2)}{z(z - 3)(z - 2)}$ **37.** $\frac{2(b - 1)(b + 2)}{b^3 + 3b^2 + 2b}$
39. $\frac{5}{2}$ **41.** $\frac{11}{8}$ **43.** $\frac{13}{20}$ **45.** $\frac{13}{12}$

SECTION 5.4 (page 329)

3. $\frac{11}{m}$ **5.** b **7.** x **9.** $y - 6$
13. $\frac{3z + 5}{15}$ **15.** $\frac{10 - 7r}{14}$
17. $\frac{-3x - 2}{4x}$ **19.** $\frac{11(1 + k)}{8}$ **21.** $\frac{4(b + 4)}{3b}$
23. $\frac{7 - 6p}{3p^2}$ **25.** $\frac{-2(p - 3)}{p(p - 2)}$ **27.** $\frac{-2}{x - 5}$ or $\frac{2}{5 - x}$
29. $\frac{-2}{y - 1}$ or $\frac{2}{1 - y}$ **31.** $\frac{-2(c - 14)}{(c - 2)(c + 2)}$
33. $\frac{-(m + n)}{2(m - n)}$ **35.** $\frac{7x - 22}{4(x + 2)(x - 2)}$
37. $\frac{-a^2 + a + 5}{(a + 1)(a - 1)(a + 4)}$ **39.** $\frac{43m + 1}{5m(m - 2)}$
41. $\frac{11y^2 - y - 11}{(2y - 1)(y + 3)(3y + 2)}$ **43.** 6 **45.** $\frac{3}{25}$

SECTION 5.5 (page 335)

1. division **3.** -6 **5.** $\frac{1}{pq}$ **7.** $\frac{1}{xy}$
9. $\frac{2a^2b}{3}$ **11.** $\frac{m(m + 2)}{3(m - 4)}$ **13.** $\frac{2}{x}$ **15.** $\frac{8}{x}$
17. $\frac{a^2 - 5}{a^2 + 1}$ **19.** $\frac{3(p + 2)}{2(2p + 3)}$ **21.** $\frac{t(t - 2)}{4}$
23. $\frac{-k}{2 + k}$ **25.** $\frac{3m(m - 3)}{(m - 1)(m - 8)}$ **27.** $\frac{6}{5}$ **29.** -2
31. $-\frac{3}{2}$ **33.** $\frac{7}{3}$

SECTION 5.6 (page 341)

1. (d) **3.** $\frac{1}{4}$ **5.** $-\frac{3}{4}$ **7.** 24 **9.** -15
11. 7 **13.** -15 **15.** -5 **17.** -6
19. no solution **21.** 5 **23.** 12 **25.** 2
27. 0 and 4 **29.** 3 **31.** 3 **33.** $-2, 12$
35. no solution **37.** $-6, \frac{1}{2}$ **39.** $-\frac{1}{5}, 3$
41. $-\frac{3}{5}, 3$ **43.** $F = \frac{ma}{k}$ **45.** $a = \frac{kF}{m}$
47. $R = \frac{E - Ir}{I}$ **49.** $A = \frac{h(B + b)}{2}$
51. $a = \frac{2S - ndL}{nd}$ **53.** $\frac{288}{t}$ **55.** $\frac{1}{x}$

ANSWERS

SUMMARY EXERCISES ON RATIONAL EXPRESSIONS (page 345)

1. $\frac{10}{p}$ **3.** $\frac{1}{2x^2(x + 2)}$ **5.** $\frac{y + 2}{y - 1}$ **7.** 39
9. $\frac{13}{3(p + 2)}$ **11.** $\frac{1}{7}$, 2 **13.** $\frac{7}{12z}$
15. $\frac{3m + 5}{(m + 2)(m + 3)(m + 1)}$

SECTION 5.7 (page 355)

1. 12 **3.** $\frac{9}{5}$ **5.** \$99.7 million
7. Germany: \$.18 billion; United States: \$1.82 billion
9. 24.04 kilometers per hour
11. $\frac{D}{R} = \frac{d}{r}$
13. Hussein: 12.02 miles per hour; McDermott: 8.77 miles per hour
15. 900 miles **17.** 8 miles per hour
19. $\frac{1}{10}$ job per hour **21.** $\frac{72}{17}$ or 4.24 hours **23.** $\frac{60}{11}$ or 5.45 hours **25.** 3 hours **27.** increases **29.** \$40.32
31. 20 miles per hour **33.** 100 pounds per square inch
35. 20 pounds per square foot **37. (a)** -7 **(b)** 23
39. (a) -14 **(b)** 22 **41. (a)** $-\frac{14}{5}$ **(b)** $-\frac{2}{5}$

CHAPTER 5 REVIEW EXERCISES (page 365)

1. 3 **2.** 0 **3.** -1, 3 **4.** -5, $-\frac{2}{3}$
5. (a) $-\frac{4}{7}$ **(b)** -16 **6. (a)** $\frac{11}{8}$ **(b)** $\frac{13}{22}$
7. (a) undefined **(b)** 1 **8. (a)** undefined **(b)** $\frac{1}{2}$
9. $\frac{6}{3a}$ **10.** -1 **11.** $\frac{-(2x + 3)}{2}$ **12.** $\frac{2p + 5q}{5p + q}$
13. $\frac{72}{p}$ **14.** 2 **15.** $\frac{2}{3m^6}$ **16.** $\frac{5}{8}$ **17.** $\frac{3}{2}$
18. $\frac{r + 4}{3}$ **19.** $\frac{3a - 1}{a + 5}$ **20.** $\frac{y - 2}{y - 3}$ **21.** $\frac{p + 5}{p + 1}$
22. $\frac{3z + 1}{z + 3}$ **23.** 96 **24.** $108y^4$
25. $m(m + 2)(m + 5)$ **26.** $(x + 3)(x + 1)(x + 4)$
27. $\frac{35}{56}$ **28.** $\frac{40}{4k}$ **29.** $\frac{15a}{10a^4}$ **30.** $\frac{-54}{18 - 6x}$
31. $\frac{15y}{50 - 10y}$ **32.** $\frac{4b(b + 2)}{(b + 3)(b - 1)(b + 2)}$ **33.** $\frac{15}{x}$
34. $-\frac{2}{p}$ **35.** $\frac{4k - 45}{k(k - 5)}$ **36.** $\frac{28 + 11y}{y(7 + y)}$ **37.** $\frac{-2 - 3m}{6}$
38. $\frac{3(16 - x)}{4x^2}$ **39.** $\frac{7a + 6b}{(a - 2b)(a + 2b)}$
40. $\frac{-k^2 - 6k + 3}{3(k + 3)(k - 3)}$ **41.** $\frac{5z - 16}{z(z + 6)(z - 2)}$
42. $\frac{-13p + 33}{p(p - 2)(p - 3)}$ **43.** $\frac{a}{b}$ **44.** $\frac{4(y - 3)}{y + 3}$
45. $\frac{6(3m + 2)}{2m - 5}$ **46.** $\frac{(q - p)^2}{pq}$ **47.** $\frac{xw + 1}{xw - 1}$
48. $\frac{1 - r - t}{1 + r + t}$ **49.** $\frac{35}{6}$ **50.** -16 **51.** -4
52. no solution **53.** 3 **54.** $t = \frac{R_y}{m}$
55. $y = \frac{4x + 5}{3}$ **56.** $m = \frac{4 + p^2q}{3p^2}$ **57.** 12
58. small car: \$120; large car: \$180
59. about 6.7 hours **60.** $\frac{40}{13}$ or about 3.1 hours
61. 2 hours **62.** about 6.37 inches
63. 4 centimeters **64.** $\frac{(5 + 2x - 2y)(x + y)}{(3x + 3y - 2)(x - y)}$
65. $\frac{m + 7}{(m - 1)(m + 1)}$ **66.** $8p^2$ **67.** $\frac{1}{6}$
68. 3 **69.** $\frac{z + 7}{(z + 1)(z - 1)^2}$ **70.** $d = \frac{k + FD}{F}$ or $d = \frac{k}{F} + D$ **71.** -2, 3 **72.** about 17.4 million
73. $\frac{27}{10}$ or $2\frac{7}{10}$ hours **74.** 150 kilometers per hour
75. about \$101 billion

CHAPTER 5 TEST (page 371)

[5.1] 1. -2, 4 **2.** $\frac{2m}{3}$ **3.** $\frac{3a + 2}{a - 1}$
[5.2] 4. x **5.** $\frac{25}{27}$ **6.** $\frac{3k - 2}{3k + 2}$ **7.** $\frac{a - 1}{a + 4}$
[5.3] 8. $\frac{240p^2}{64p^3}$ **9.** $\frac{21}{42m - 84}$ **[5.4] 10.** $\frac{3}{c}$
11. $\frac{-14}{5(y + 2)}$ **12.** $\frac{-m^2 + 7m + 2}{(2m + 1)(m - 5)(m - 1)}$
[5.5] 13. $\frac{2k}{3p}$ **14.** $\frac{-2 - x}{4 + x}$
[5.6] 15. -1, 4 **16.** $-\frac{58}{11}$ **17.** 1
[5.7] 18. $\frac{1}{4}$ or $\frac{1}{2}$ **19.** 3 miles per hour
20. 100 amperes

CUMULATIVE REVIEW R–5 (page 373)

1. $\frac{32}{5}$ **2.** 17 **3.** $b = \frac{2A}{h}$ **4.** $-\frac{2}{7}$ **5.** $y \geq -8$
6. $m > 4$ **7.** $\frac{16}{9}$ **8.** $\frac{1}{49}$ **9.** $\frac{1}{4^9}$
10. $\frac{1}{2^4x^7}$ **11.** $\frac{1}{m^6}$ **12.** $\frac{q}{4p^2}$ **13.** $k^2 + 2k + 1$

14. $72x^6y^7$ **15.** $4a^2 - 4ab + b^2$

16. $3y^3 + 8y^2 + 12y - 5$ **17.** $6p^2 + 7p + 1 + \dfrac{3}{p - 1}$

18. $(4t + 3v)(2t + v)$ **19.** prime

20. $m(2a - 1)(3a + 2)$ **21.** $-3, 5$ **22.** $0, 8$

23. $-\frac{1}{2}, \frac{2}{3}, 5$ **24.** -2 or -1 **25.** 6 meters

26. $\dfrac{1}{xy^6}$ **27.** $\dfrac{4}{q}$ **28.** $\dfrac{3r + 28}{7r}$ **29.** $\dfrac{7}{15(q - 4)}$

30. $\dfrac{-k - 5}{k(k + 1)(k - 1)}$ **31.** $\dfrac{7(2z + 1)}{24}$ **32.** $\dfrac{195}{29}$

33. $\dfrac{21}{2}$ **34.** $-2, 1$ **35.** 150 miles

36. $\frac{6}{5}$ or $1\frac{1}{5}$ hours

CHAPTER 6

SECTION 6.1 (page 381)

1. does; do not **3.** II **5.** 3 **7.** yes
9. yes **11.** no **13.** yes **15.** no
17. no **19.** 11 **21.** 7 **23.** -5
25. -4 **27.** -5 **29.** -44 **33.** 12; 6; 4
35. -12; -8; -2 **37.** 5; $\frac{5}{2}$; -19 **39.** 4; 6; -6
41. 3; -5; -15 **43.** x; vertical **45.** -9; -9; -9
47. -6; -6; -6 **49.** 8; 8; 8 **51.** (2, 4)
53. $(-5, 4)$ **55.** (3, 0) **57.** $a = 3, b = 9$
59.–67.

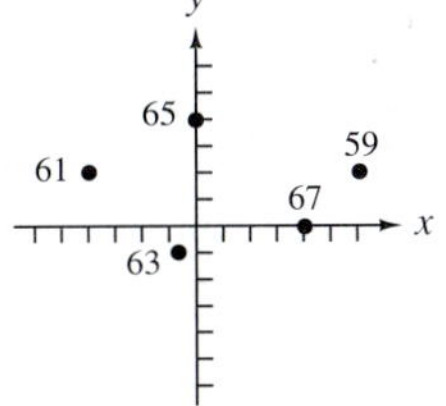

69. negative; negative **71.** positive; negative

73. -3; 6; -2; 4

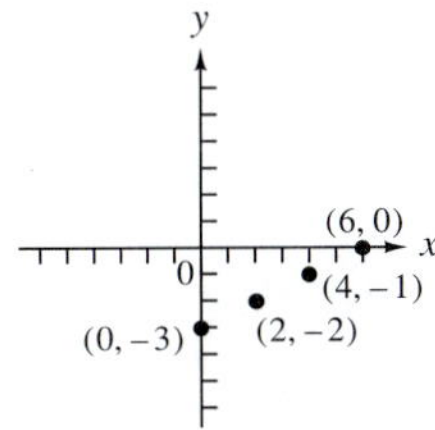

75. -3; 4; -6; $-\frac{4}{3}$

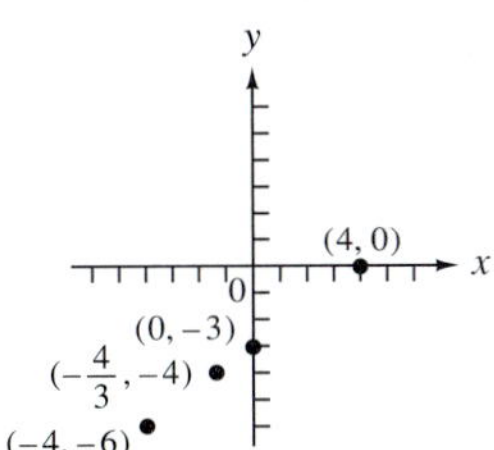

77. -4; -4; -4; -4

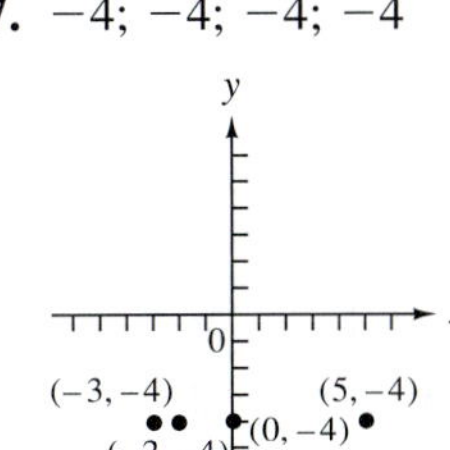

79.

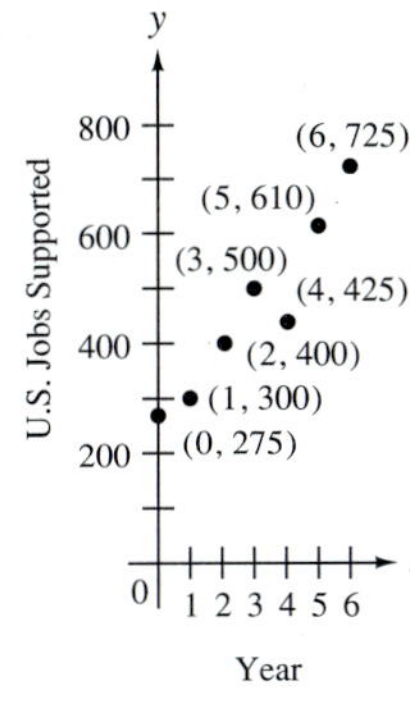

81. 1 **83.** 13 **85.** 4

SECTION 6.2 (page 393)

1. line **3.** x **5.** horizontal
7. 5; 5; 3 **9.** 1; 3; -1

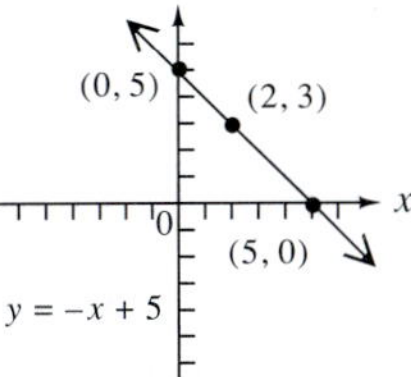

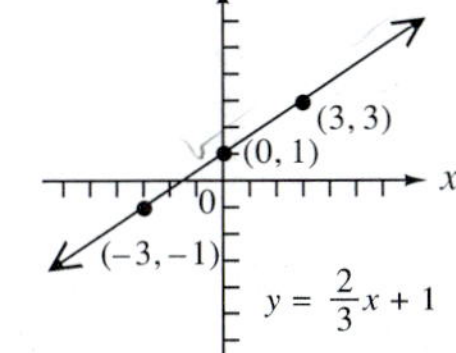

11. -6; -2; -5

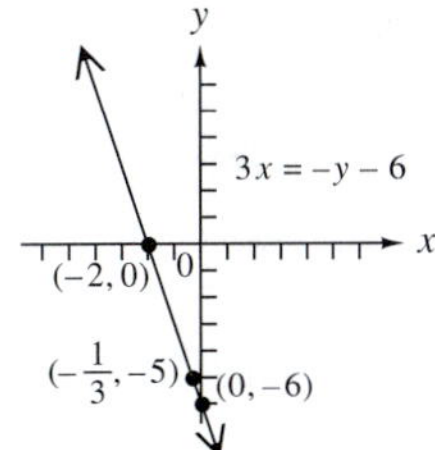

13. (12, 0); (0, -8) **15.** (0, 0); (0, 0)

19.

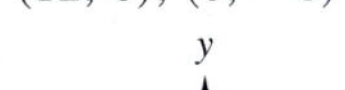

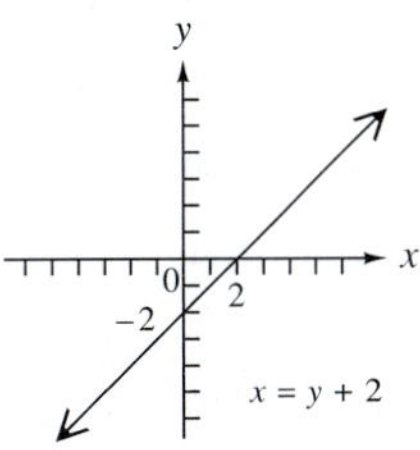

21.

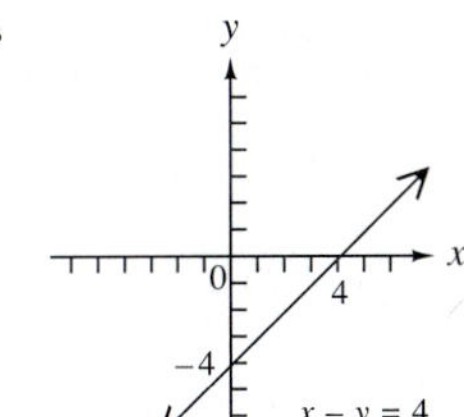

23. **25.**

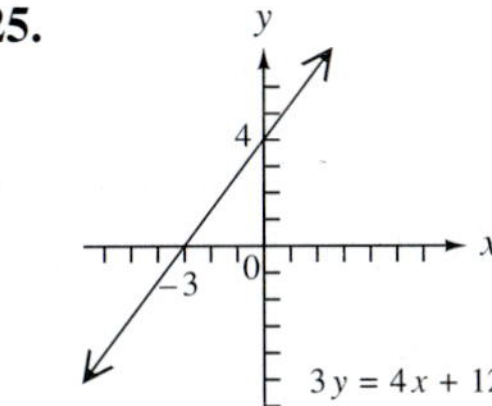

ANSWERS

27.

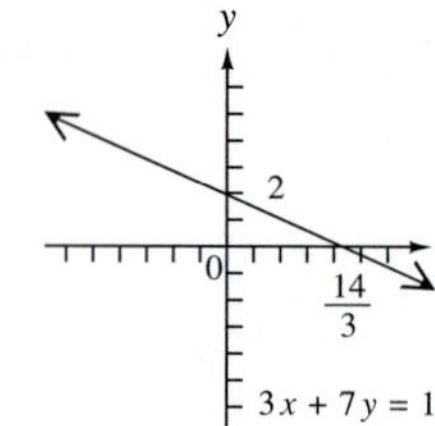

29.

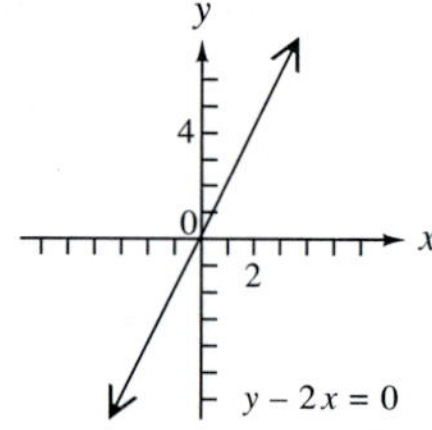

31.

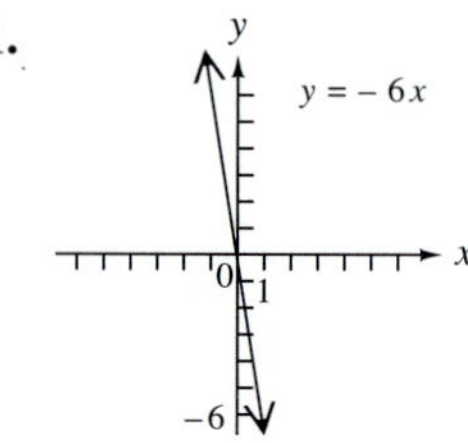

33.

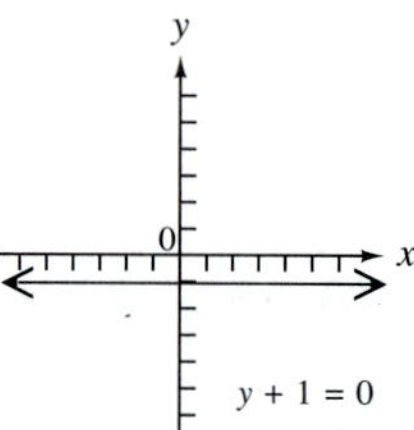

35.

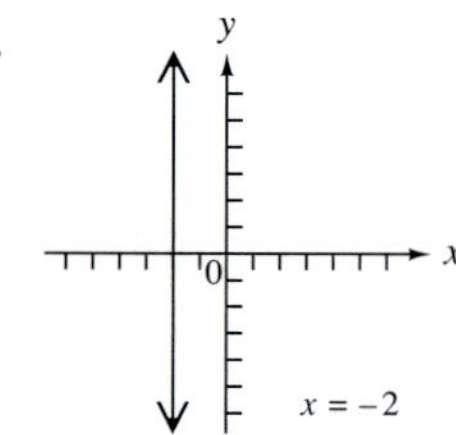

37. **(a)** \$30,000 **(b)** \$15,000 **(c)** \$5000

39. $\frac{5}{8}$ **41.** 0

SECTION 6.3 (page 403)

1. Sketches will vary. The line must fall from left to right.
3. Sketches will vary. The line must be vertical.

7. 5 **9.** $\frac{3}{2}$ **11.** $-\frac{7}{4}$ **13.** undefined

15. **(a)** 6 **(b)** 4 **(c)** $\frac{6}{4}$ or $\frac{3}{2}$; slope of the line

17. $\frac{7}{6}$ **19.** $-\frac{5}{8}$ **21.** 0 **23.** undefined

25. $\frac{1}{3}$ **27.** 2 **29.** $-\frac{1}{2}$ **31.** −4

33. $\frac{3}{2}$ **35.** 0 **37.** **(a)** negative **(b)** negative

39. **(a)** positive **(b)** zero **41.** **(a)** zero **(b)** positive

43. $-\frac{2}{5}$; $-\frac{2}{5}$; parallel **45.** $\frac{8}{9}$; $-\frac{4}{3}$; neither

47. $\frac{3}{2}$; $-\frac{2}{3}$; perpendicular **49.** $\frac{3}{10}$ **51.** $y = 2x - 16$

SECTION 6.4 (page 411)

1. $y = 3x - 3$ **3.** $y = -x + 3$
5. $y = 4x - 3$ **7.** $y = 3$

11.

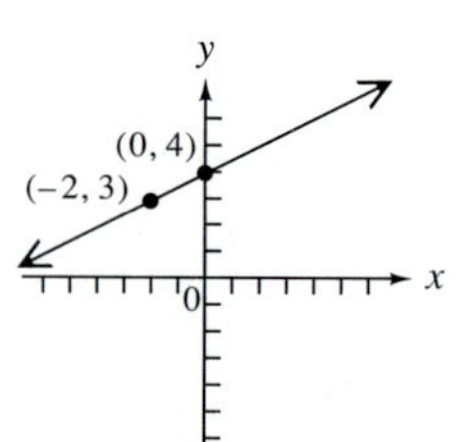

13.

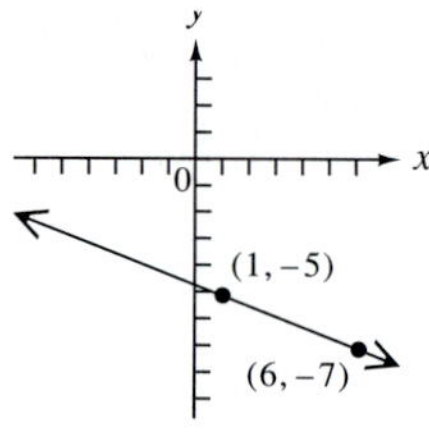

15.

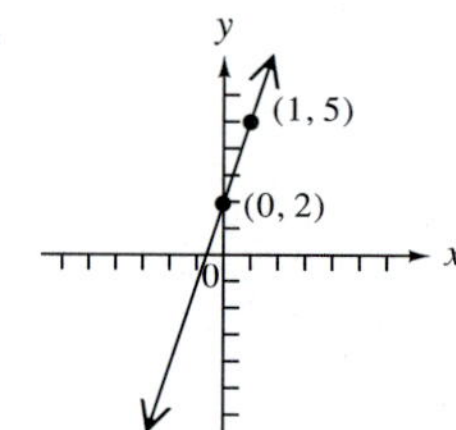

17.

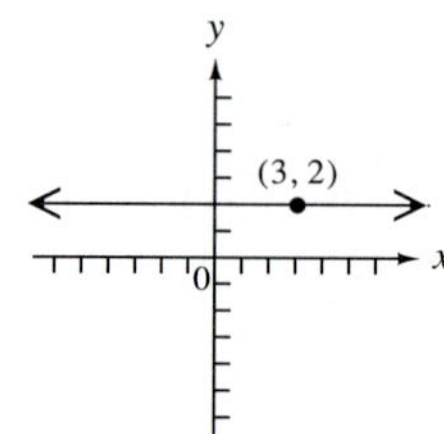

19.

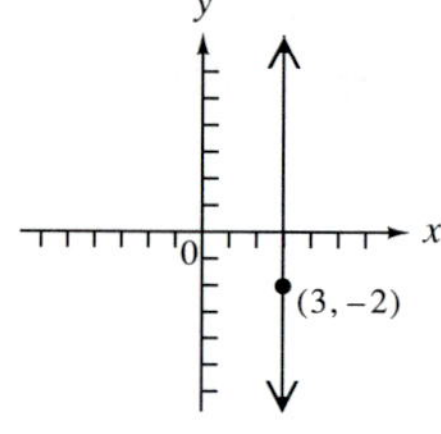

21. the y-axis **23.** $2x - y = 7$
25. $2x + y = -4$ **27.** $2x - 3y = -19$
29. $4x + 5y = 9$
31. $y = x$ (There are other forms as well.)
33. $x - y = 3$ **35.** $5x + 7y = -54$
37. $2x + 3y = -6$ **39.** $x - 3y = -4$
41. cost equation: $y = .25x + 400$
(a) \$425 **(b)** 1500
43. **(a)** (1, 4800), (5, 24,800) **(b)** $y = 5000x - 200$
(c) change in sales per year **45.** (0, 32), (100, 212)

47. $C = \frac{5}{9}(F - 32)$ or $C = \frac{5}{9}F - \frac{160}{9}$

49. $x > -3$ **51.** $x > 5$

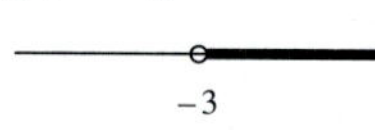

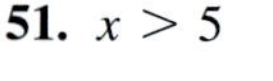

SECTION 6.5 (page 421)

1. **(a)** yes **(b)** no **3.** **(a)** yes **(b)** yes

5.

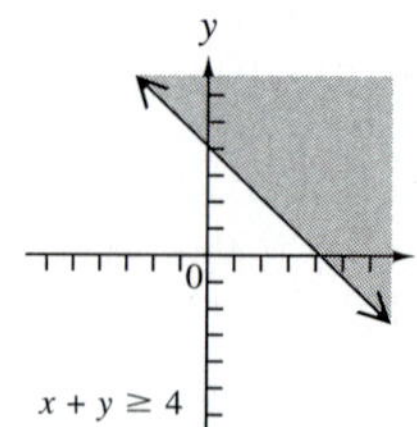

7. y x 0 $x + 2y \geq 7$

9. 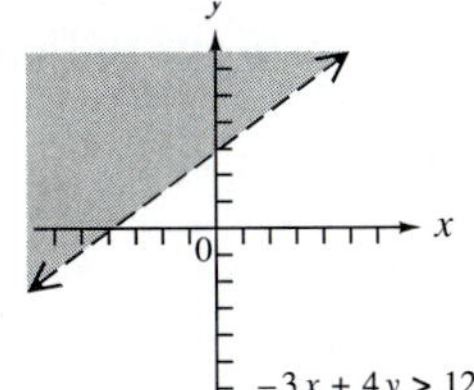

11. 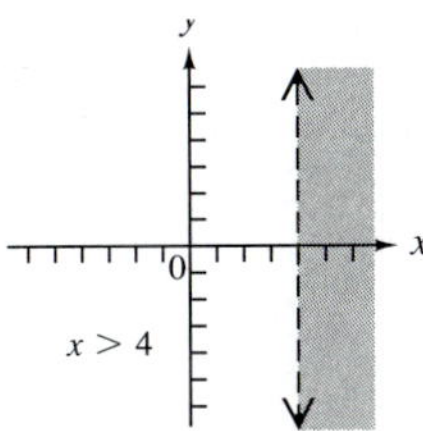

13. 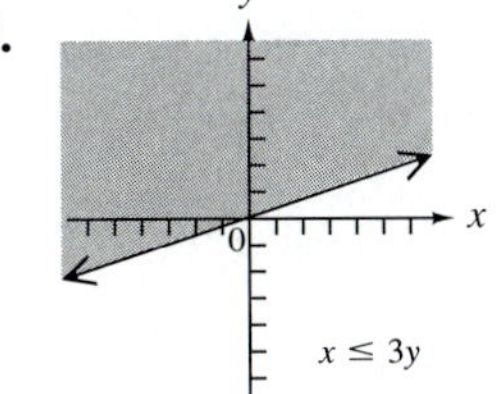

17. 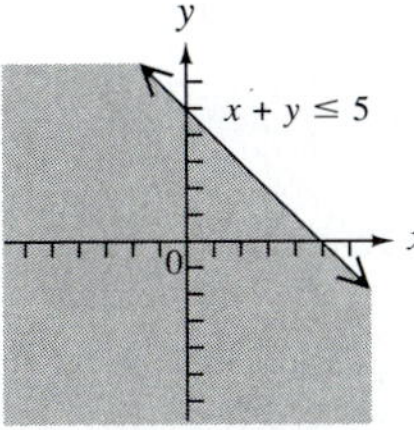

19. 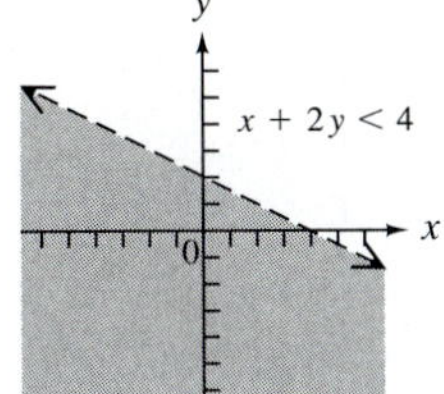

21. 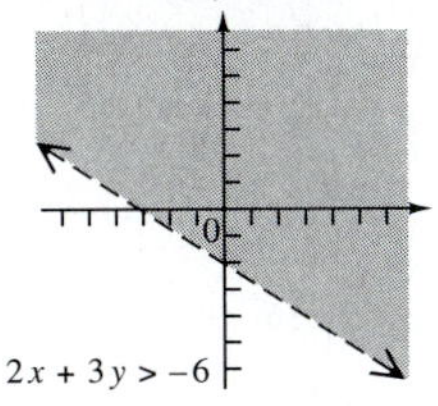

23. 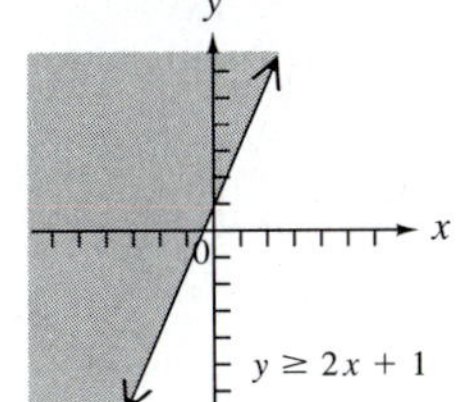

25. 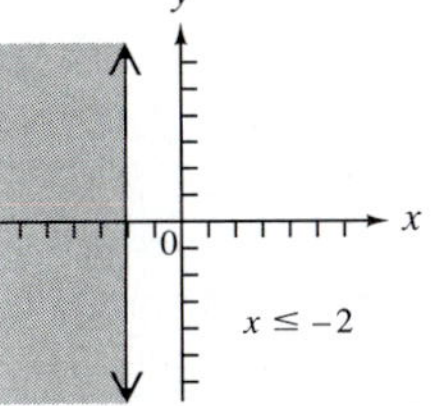

27. 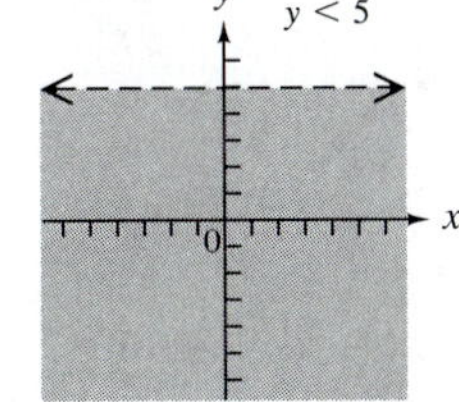

29. 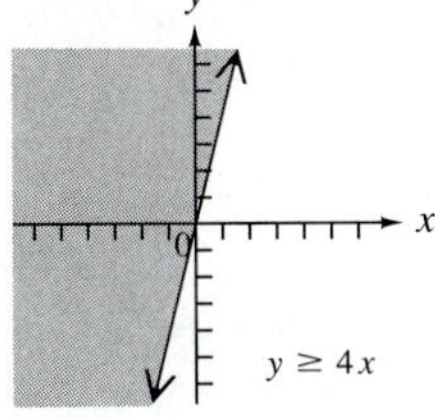

33. Answers will vary. For example, (2000, 3000), (1000, 3000), (1500, 1500), (6000, 0).

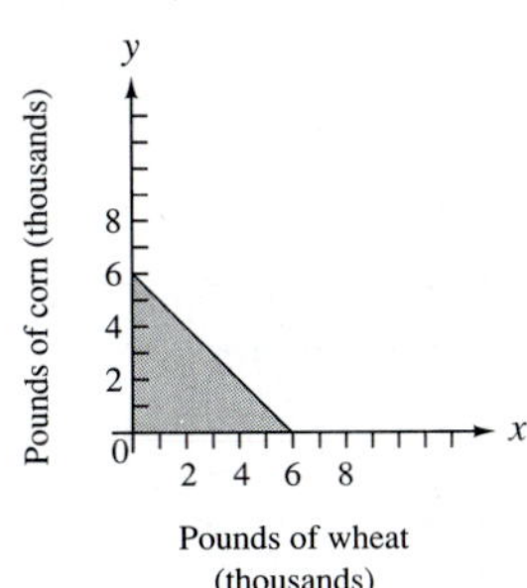

35. $-x$ **37.** 0

CHAPTER 6 REVIEW EXERCISES (page 429)

1. -1; 2; 1 **2.** 2; $\frac{3}{2}$; $\frac{14}{3}$ **3.** 0; $\frac{8}{3}$; -9
4. 7; 7; 7 **5.** yes **6.** no **7.** yes **8.** yes
9.–12.

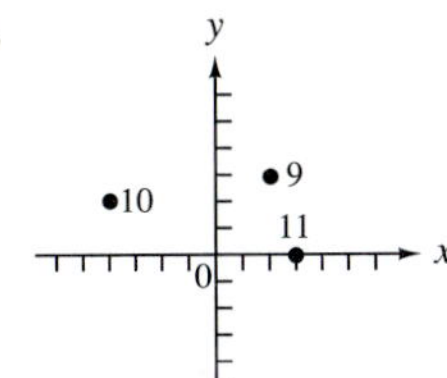

13. I or III **14.** x-axis **15.** II **16.** III
17. $\left(-\frac{5}{2}, 0\right)$; (0, 5) **18.** $\left(-\frac{7}{2}, 0\right)$; (0, -7)
19. $\left(\frac{8}{3}, 0\right)$; (0, 4)

20. 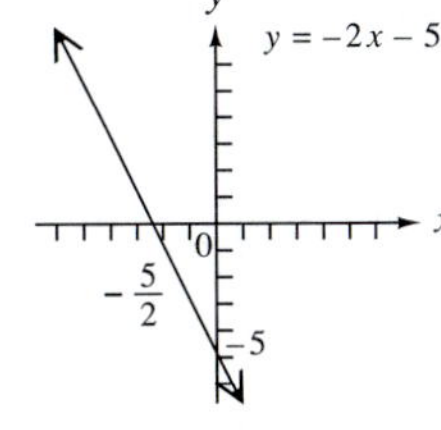

21. 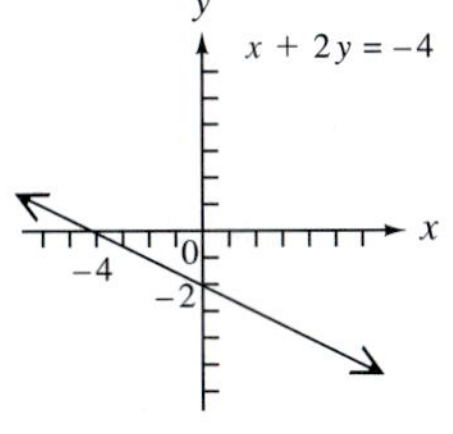

22. 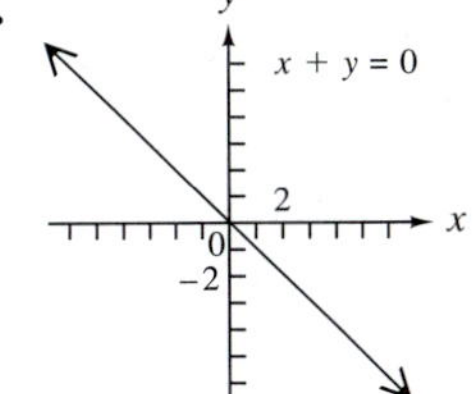

23. $-\frac{1}{2}$ **24.** $-\frac{2}{3}$ **25.** 0 **26.** undefined
27. 3 **28.** $\frac{2}{3}$ **29.** $\frac{3}{2}$ **30.** $-\frac{1}{3}$ **31.** 0
32. undefined **33. (a)** $-\frac{1}{4}$ **(b)** 4
35. parallel **36.** perpendicular
37. neither **38.** 0 **39.** $3x + 3y = 2$
40. $x + 3y = 3$ **41.** $x - y = 7$
42. $2x - 3y = -14$ **43.** $3x + 4y = -1$
44. $x + 4y = 6$ **45.** $y = 1$ **46.** $x = \frac{1}{3}$

47. 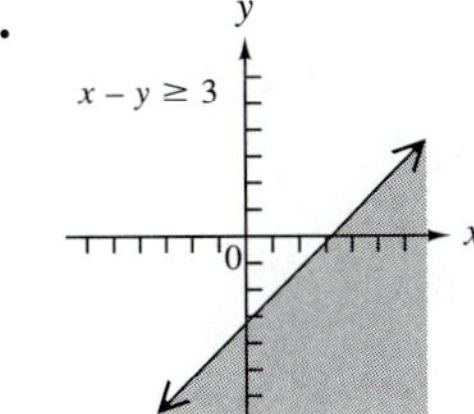

48. 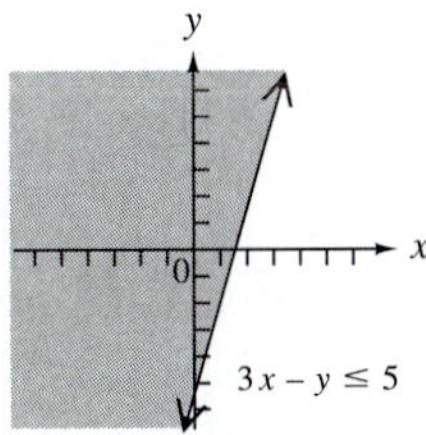

ANSWERS

49.

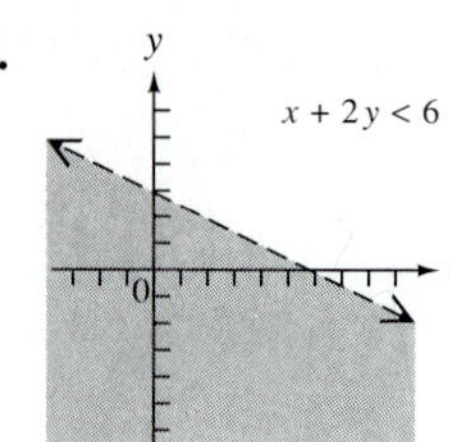

50.

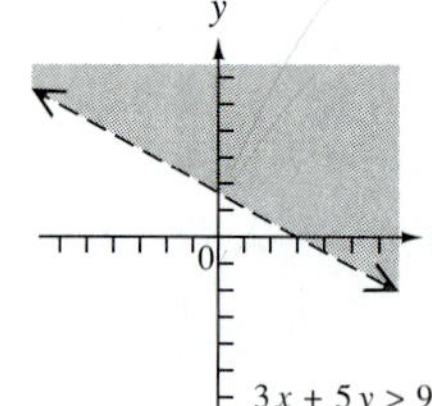

51.

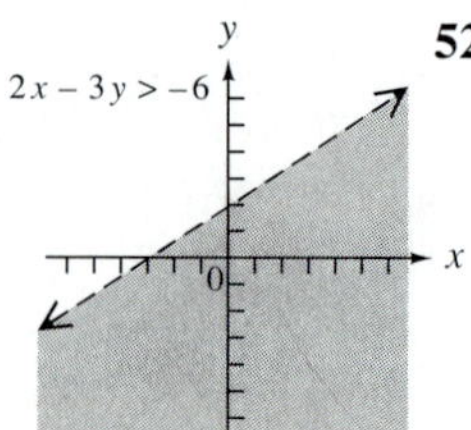

52.

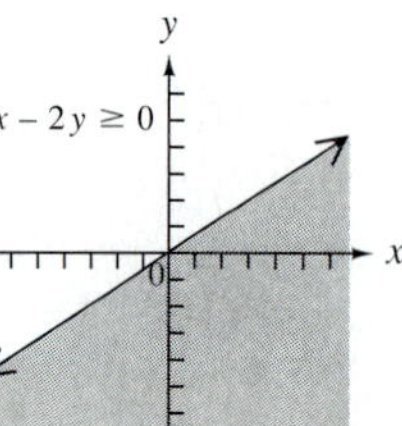

53. x-intercept: $\left(\frac{4}{11}, 0\right)$; y-intercept: $\left(0, -\frac{4}{3}\right)$; $m = \frac{11}{3}$

54. x-intercept: $(7, 0)$; y-intercept: $\left(0, -\frac{7}{3}\right)$; $m = \frac{1}{3}$

55. x-intercept: $\left(-\frac{9}{4}, 0\right)$; y-intercept: $(0, -1)$; $m = -\frac{4}{9}$

56. x-intercept: $\left(\frac{3}{4}, 0\right)$; y-intercept: $(0, 2)$; $m = -\frac{8}{3}$

57. $x = 5$ **58.** $x + 4y = -5$ **59.** $3x + y = 30$

60. $4x + 7y = -23$

61.

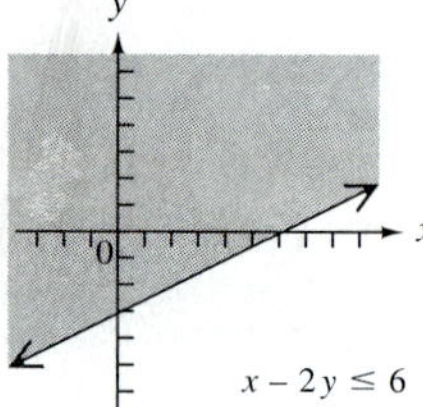

62.

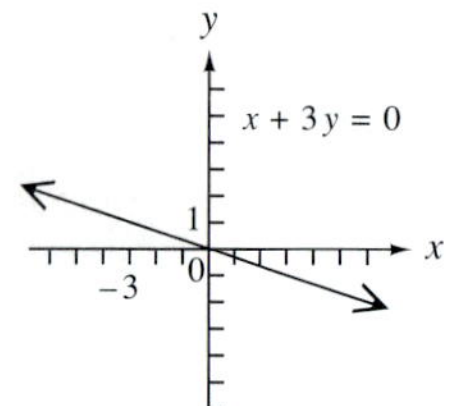

63.

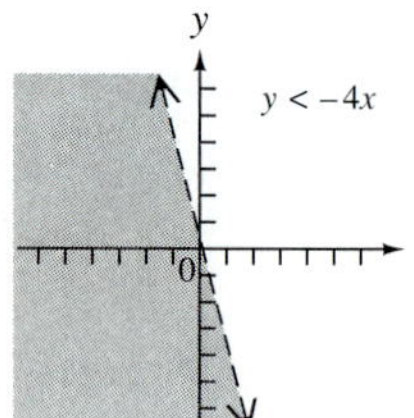

64.

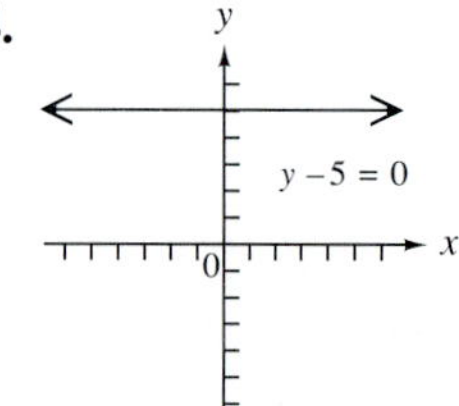

65.

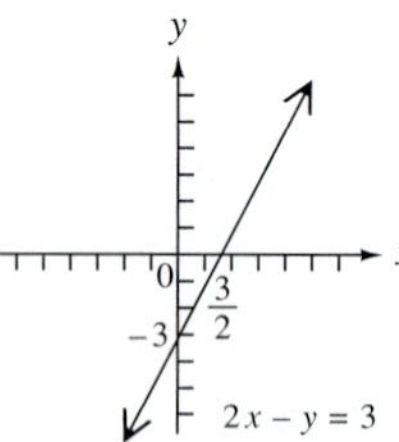

66.

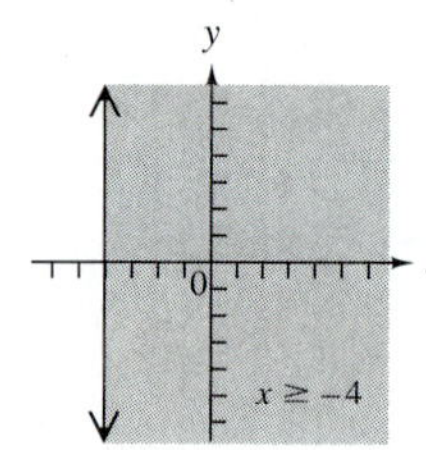

CHAPTER 6 TEST (page 433)

[6.1] **1.** $-9, -5, \frac{3}{2}$

2. $-6, -10, -15$

3. $-12, -12, -12$

[6.2] **4.** To find the x-intercept, let $y = 0$, and to find the y-intercept, let $x = 0$.

5. $(4, 0)$; $(0, -4)$ **6.** $(2, 0)$; $(0, 6)$

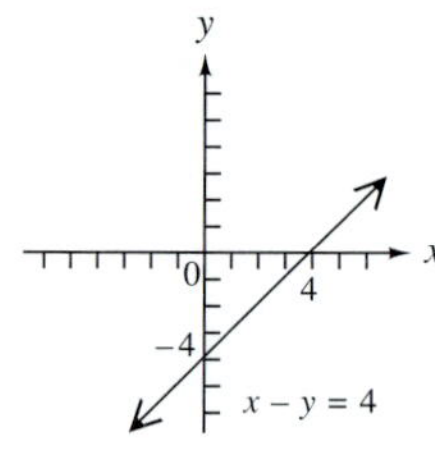

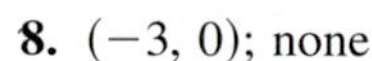

7. $(0, 0)$; $(0, 0)$ **8.** $(-3, 0)$; none

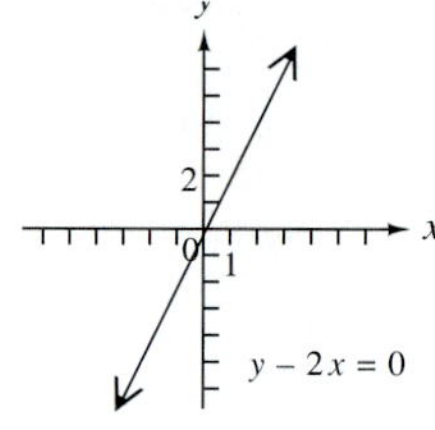

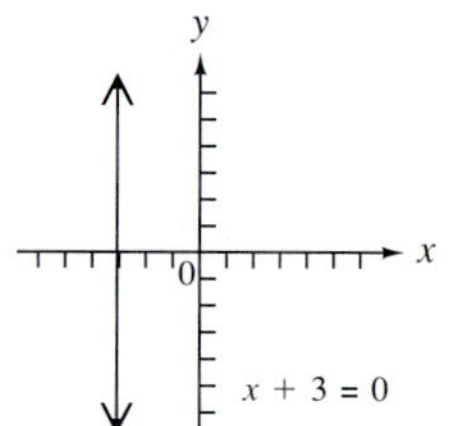

9. none; $(0, 1)$

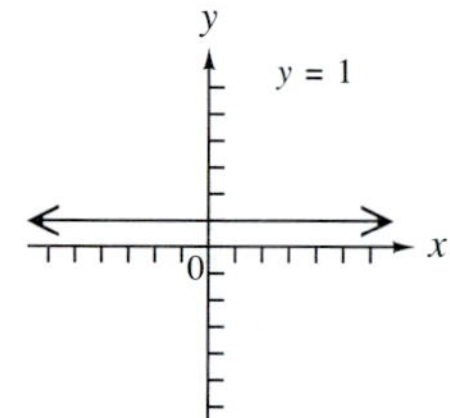

[6.3] **10.** $-\frac{8}{3}$ **11.** -2 **12.** undefined

13. $\frac{5}{2}$ **14.** 0 **[6.4]** **15.** $2x - y = -6$

16. $5x - 2y = 8$ **17.** $9x + y = 12$

[6.5] **18.**

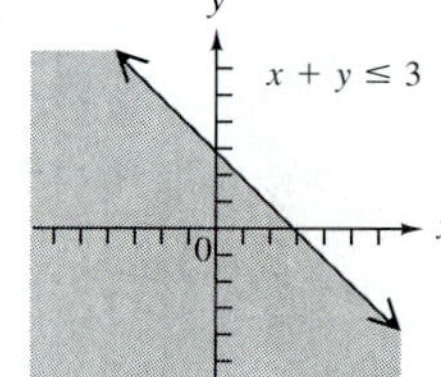

19.

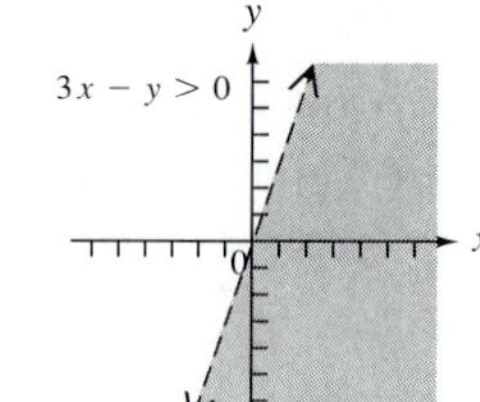

20.

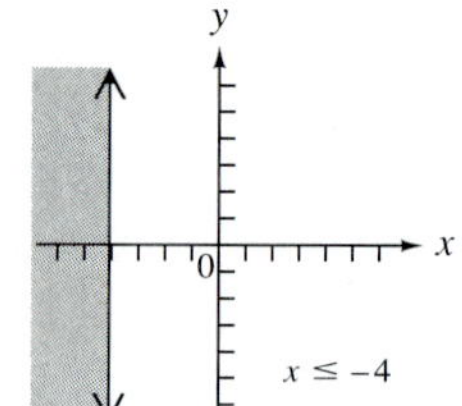

CUMULATIVE REVIEW R-6 (page 437)

1. -65 **2.** $t = \frac{A - p}{pr}$

3. $-1, -\frac{1}{7}$ **4.** 3 **5.** 5

6. $x > -2.6$

−2.6

7. $x > 0$

0

8. $y \leq -4$

−4

9. $\frac{1}{x^2y}$ **10.** $\frac{y^7}{x^{13}z^2}$ **11.** $\frac{m^6}{2^3n^9}$ **12.** $2x^2 - 4x + 38$
13. $15x^2 + 7xy - 2y^2$ **14.** $x^3 + 8y^3$
15. $m^2 - 2m + 3$ **16.** $(y + 6k)(y - 2k)$
17. $(3x^2 - 5y)(3x^2 + 5y)$ **18.** $5x^2(5x - 13y)(5x - 3y)$
19. $(f + 10)^2$ **20.** prime **21.** 1
22. $\frac{6x + 22}{(x + 1)(x + 3)}$ or $\frac{2(3x + 11)}{(x + 1)(x + 3)}$ **23.** $\frac{4(x - 5)}{3(x + 5)}$
24. $\frac{x + 1}{x}$ **25.** $\frac{(x + 3)^2}{3x}$ **26.** $\frac{4xy^4}{z^2}$ **27.** $\frac{5}{8}$
28. 6 **29.** $-\frac{4}{3}$ **30.** 0 **31.** $4x + y = 15$
32. $4x - y = 0$
33.

$-3x + 4y = 12$

34.

$y \leq 2x - 6$

35.

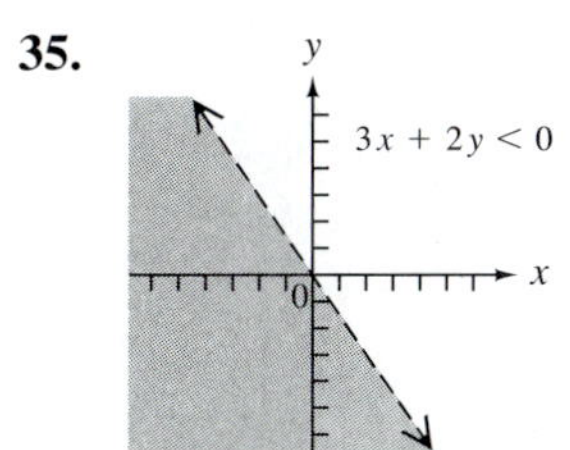

36. 116 men and 80 women **37.** 15°, 35°, 130°
38. 7 inches **39.** 14 **40.** 1 hour

CHAPTER 7

SECTION 7.1 (page 443)

1. It is not a solution of the system because it is not a solution of the second equation, $2x + y = 4$.
3. yes **5.** no **7.** yes **9.** yes **11.** no
13. (b), because the ordered pair must be in quadrant II
We show the graphs here only for Exercises 15–19.
15. (4, 2)

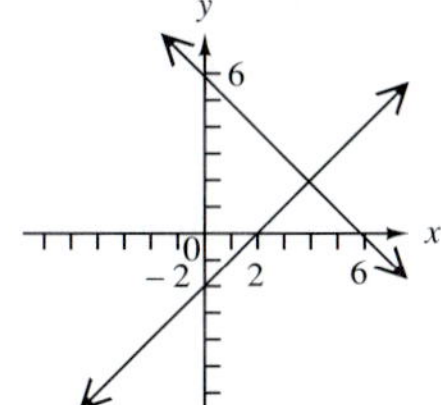

17. (0, 4)

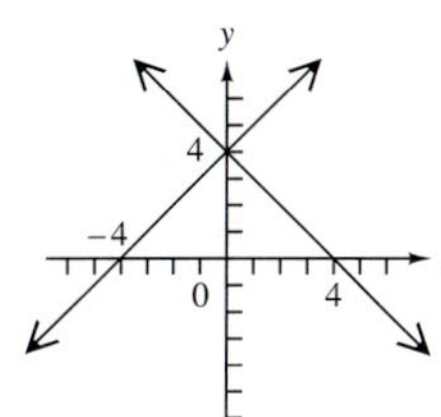

19. (4, −1)

21. (1, 3) **23.** (0, 2) **25.** (4, −3)
29. Answers will vary, but the lines must intersect at (−2, 3).
33. no solution **35.** infinite number of solutions
37. no solution **39.** $16x$ **41.** 0 **43.** $-6y$

SECTION 7.2 (page 453)

1. (c) **3.** (4, 6) **5.** (−1, 3) **7.** (−1, −3)
9. (−2, 3) **11.** $\left(\frac{1}{2}, 4\right)$ **13.** (3, −6) **15.** (7, 4)
17. (0, 4) **19.** (−4, 0) **21.** (8, 1)
27. (−6, 5) **29.** (−3, 2) **31.** $\left(\frac{1}{8}, -\frac{5}{6}\right)$
33. (11, 15) **35.** no solution
37. infinite number of solutions **39.** no solution
41. infinite number of solutions **43.** (0, 0)
45. −3 **47.** 0 **49.** 1

SECTION 7.3 (page 461)

3. (3, 9) **5.** (7, 3) **7.** (−2, 4) **9.** (−4, 8)
11. (3, −2) **13.** infinite number of solutions
15. $\left(\frac{1}{3}, -\frac{1}{2}\right)$ **17.** no solution
19. infinite number of solutions
21. **(a)** (1, 4) **(b)** (1, 4) **23.** (4, −6) **25.** (7, 0)
27. infinite number of solutions **29.** no solution
31. yes **33.** (0, 3) **35.** (24, −12) **37.** (3, 2)
39. 44 goals; 87 assists
41. gold: 37; silver: 34; bronze: 37

SECTION 7.4 (page 469)

3. 92 and 21 **5.** Harding: 987; Los Alamos: 109
7. length: 13 centimeters; width: 5 centimeters
9. 46 ones; 28 tens
11. 166 student tickets; 220 nonstudent tickets
13. $2500 at 4%; $5000 at 5%
15. 7 paperbacks; 13 hardbacks
17. 80 liters of 40% solution; 40 liters of 70% solution
19. 30 pounds at $6 per pound; 60 pounds at $3 per pound
21. 30 barrels at $40 per barrel; 20 barrels at $60 per barrel
23. boat: 10 miles per hour; current: 2 miles per hour
25. plane: 470 miles per hour; wind: 30 miles per hour
27. faster train: 110 miles per hour; slower train: 90 miles per hour

29. Roberto: $3\frac{1}{4}$ miles per hour; Juana: $2\frac{3}{4}$ miles per hour

31.

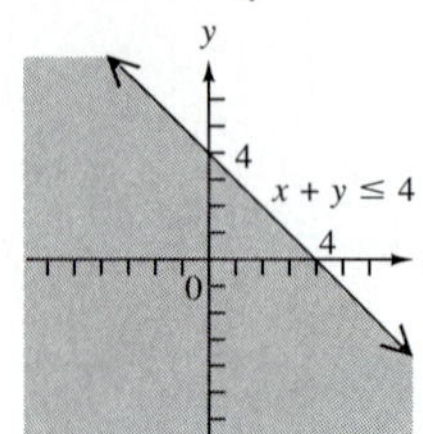

33.

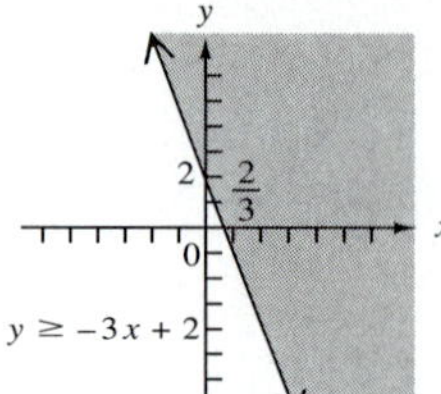

35.

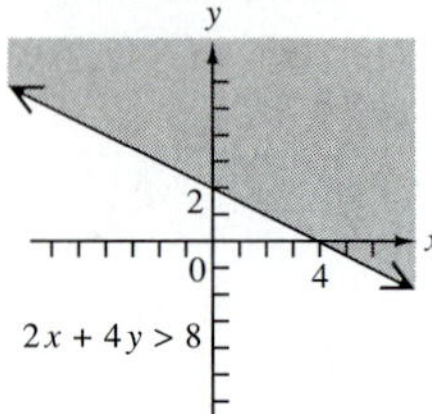

SECTION 7.5 (page 477)

3.

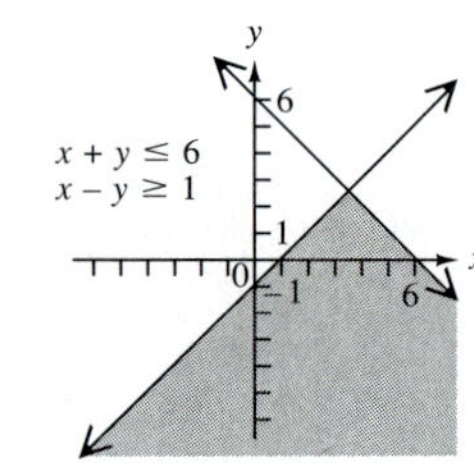

5.

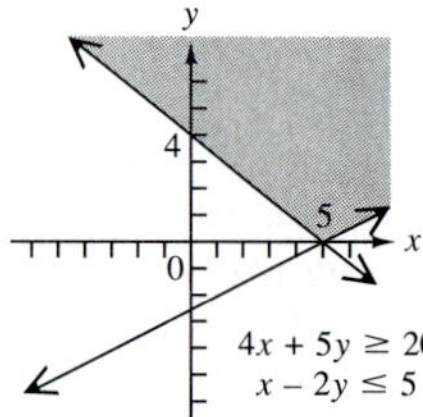

7.

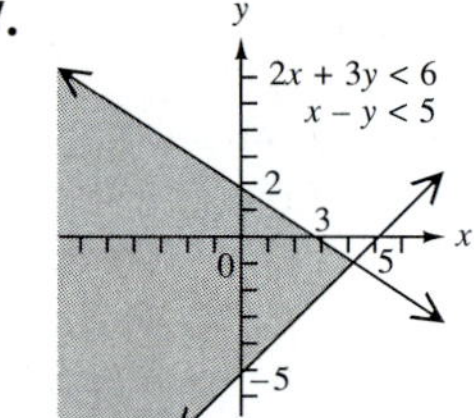

9.

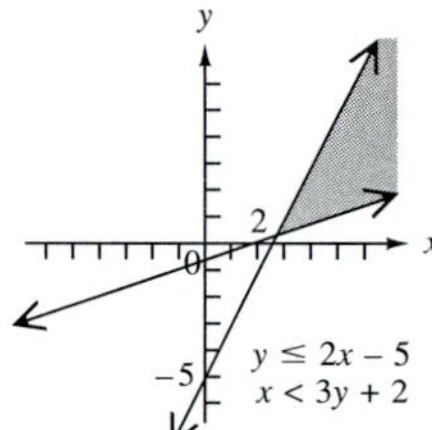

11.

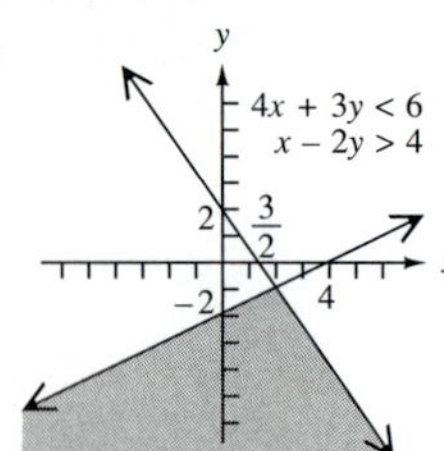

13.

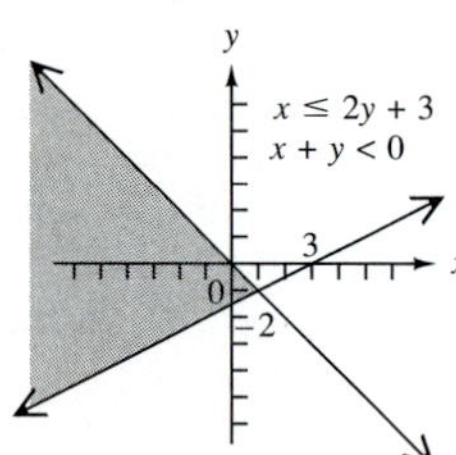

15.

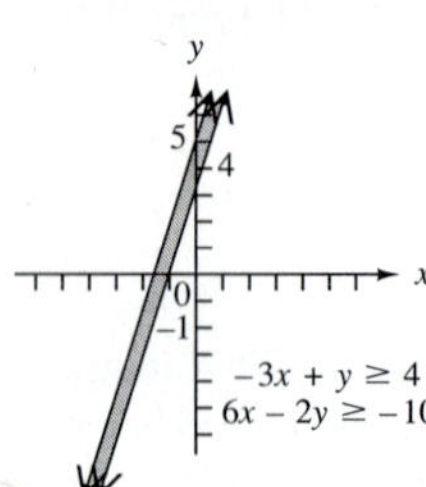

17.

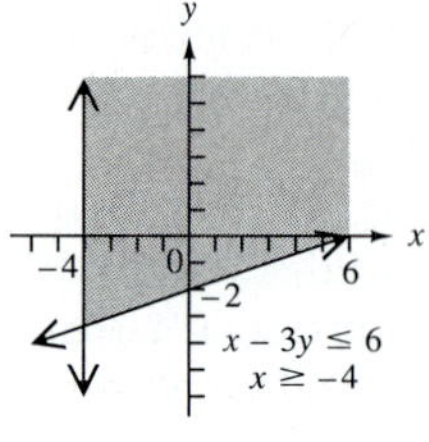

19. 64 **21.** 256 **23.** 289

CHAPTER 7 REVIEW EXERCISES (page 483)

1. yes **2.** yes **3.** no **4.** no **5.** (3, 1) **6.** (0, −2) **7.** infinite number of solutions **8.** no solution **9.** It would be difficult to read the exact coordinates from the graph. **11.** (7, 1) **12.** (−5, −2) **13.** (−3, 2) **14.** (−4, 3) **15.** infinite number of solutions **16.** no solution **19.** (2, 1) **20.** (3, 5) **21.** (6, 4) **22.** no solution **23.** (−4, 1) **24.** infinite number of solutions **25.** (9, 2) **26.** $\left(\frac{10}{7}, -\frac{9}{7}\right)$ **27.** (8, 9) **28.** (2, 1) **29.** 24 and 18 **30.** 3 and 8 **31.** length: 27 meters; width: 18 meters **32.** 13 twenties; 7 tens **33.** 25 pounds of $1.30 candy; 75 pounds of $.90 candy **34.** 60 liters of 40% solution; 30 liters of 70% solution **35.** apple: $.55; banana: $.45 **36.** plane: 250 miles per hour; wind: 20 miles per hour

37.

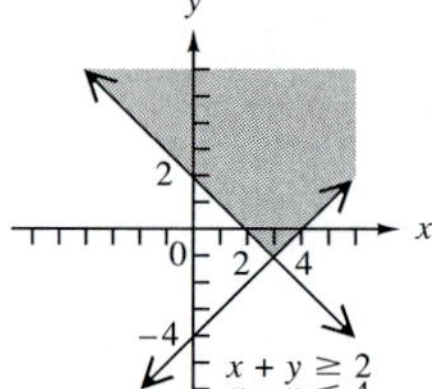

38.

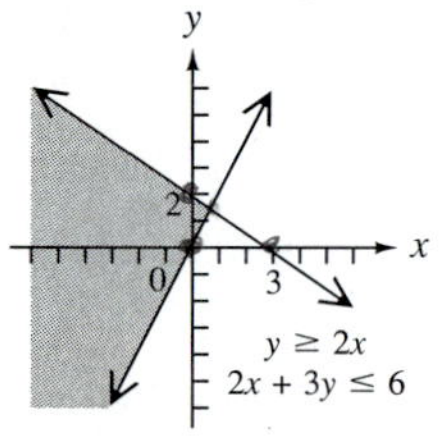

39.

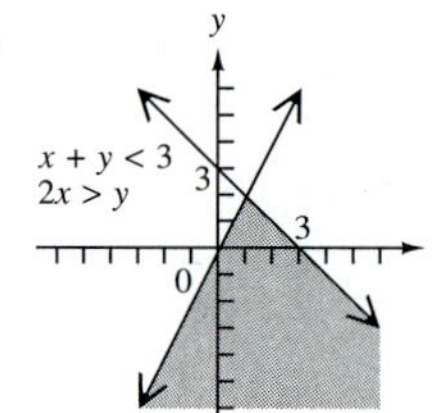

40.

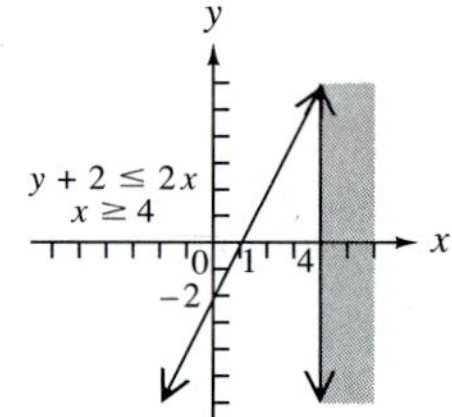

41. (4, 8) **42.** infinite number of solutions **43.** (2, 0) **44.** (−4, 15)

45.

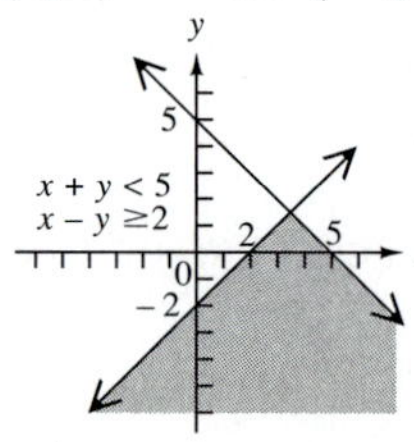

46.

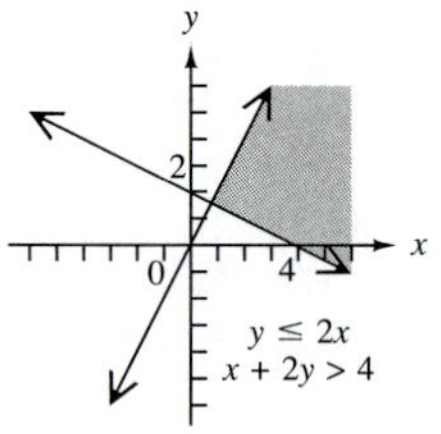

47. no solution **48.** $7000 at 3%; $11,000 at 4% **49.** 8 inches, 8 inches, and 13 inches **50.** 6 and −18 **52.** (b)

CHAPTER 7 TEST (page 489)

[7.1] **1.** (2, −3) **2.** (4, 1) **[7.1–7.3]** **3.** Answers will vary. (−3, 4) should satisfy the system. One example is $x + y = 1$, $x - y = -7$. **[7.2]** **4.** (5, 6) **5.** (−1, 3) **6.** (−2, 5) **7.** no solution **8.** (0, 0) **9.** (−15, 6) **[7.3]** **10.** (1, −6)

11. $(-35, 35)$ **[7.1–7.3]** **12.** infinite number of solutions **13.** $(12, -4)$ **[7.4]** **14.** -18 and 36 **15.** 3 at \$12 each; 12 at \$16 each **16.** 160 liters of 40% solution; 40 liters of 65% solution **17.** slower car: 38 miles per hour; faster car: 68 miles per hour

[7.5] **18.** **19.**

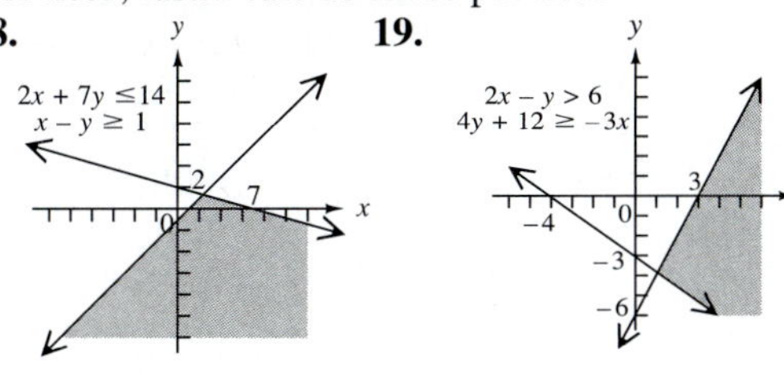

20. Answers will vary. One example is $x + y > 1$, $x + y < -1$.

CUMULATIVE REVIEW R-7 (page 493)

1. $-1, 1, -2, 2, -4, 4, -5, 5, -8, 8, -10, 10, -20, 20, -40, 40$ **2.** 1 **3.** commutative property **4.** distributive property **5.** inverse property **6.** 46 **7.** $-\frac{13}{11}$ **8.** $\frac{9}{11}$ **9.** $x > -18$ **10.** $x > -\frac{11}{2}$ **11.** $37\frac{1}{2}$ yards **12.** $14x^2 - 5x + 23$ **13.** $6xy + 12x - 14y - 28$ **14.** $3k^2 - 4k + 1$ **15.** 3.65×10^{10} **16.** x^6y **17.** $(5m - 4p)(2m + 3p)$ **18.** $(8t - 3)^2$ **19.** $-\frac{1}{3}, \frac{3}{2}$ **20.** $-11, 11$ **21.** $\frac{7}{x + 2}$ **22.** $\frac{3}{4k - 3}$ **23.** $-\frac{1}{4}, 3$ **24.** $T = \frac{PV}{k}$

25. **26.**

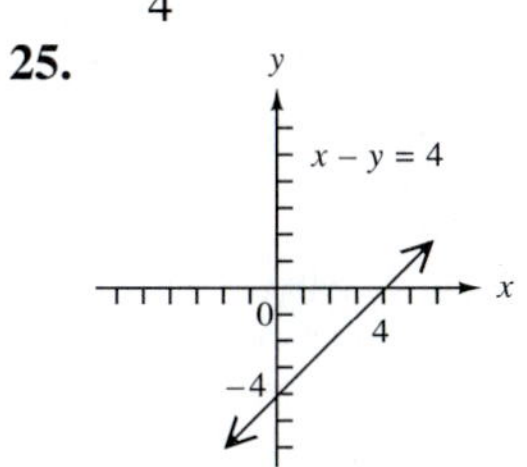

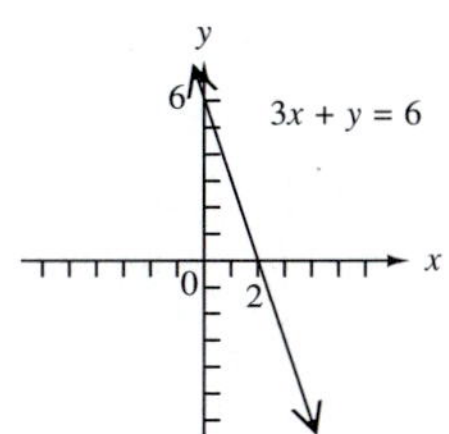

27.

y
x − 3y > 6
6
0
−2
x

28. $-\frac{4}{3}$ **29.** $-\frac{1}{4}$ **30.** $3x - y = 11$ **31.** $y = 4$ **32.** **(a)** $x = 9$ **(b)** $y = -1$ **33.** no **34.** $(-1, 6)$ **35.** $(3, -4)$ **36.** $(2, -1)$ **37.** no solution **38.** 15 and 9 **39.** 9 meters **40.** 9 fives; 8 tens

CHAPTER 8

SECTION 8.1 (page 503)

1. two **3.** none **5.** one **7.** $-4, 4$ **9.** $-12, 12$ **11.** $-\frac{5}{14}, \frac{5}{14}$ **13.** $-30, 30$ **15.** 100 **17.** 19 **19.** $3x^2 + 4$ **21.** a must be positive. **23.** a must be negative. **25.** 7 **27.** -11 **29.** $-\frac{12}{11}$ **31.** not a real number **33.** rational; 5 **35.** irrational; 5.385 **37.** rational; -8 **39.** irrational; -17.321 **41.** not a real number **45.** $c = 17$ **47.** $b = 8$ **49.** $c = 11.705$ **51.** 24 centimeters **53.** 80 feet **55.** 195 miles **57.** 9.434 **59.** Answers will vary. For example, if $a = 2$ and $b = 7$, $\sqrt{a^2 + b^2} = \sqrt{2^2 + 7^2} = \sqrt{53}$, while $\sqrt{a^2 + b^2} \neq a + b$, because $2 + 7 = 9$ and $\sqrt{53} \neq 9$. **61.** 10 **63.** 5 **65.** -3 **67.** 5 **69.** not a real number **71.** -3 **73.** $2^3 \cdot 3^2$ **75.** $2^3 \cdot 5$

SECTION 8.2 (page 511)

1. true **3.** true **5.** false **7.** 9 **9.** $3\sqrt{10}$ **11.** 13 **13.** $\sqrt{13r}$ **15.** (a) **17.** $3\sqrt{5}$ **19.** $3\sqrt{10}$ **21.** $5\sqrt{3}$ **23.** $5\sqrt{5}$ **25.** $-10\sqrt{7}$ **27.** $9\sqrt{3}$ **29.** $3\sqrt{6}$ **31.** 24 **33.** $6\sqrt{10}$ **37.** $\frac{4}{15}$ **39.** $\frac{\sqrt{7}}{4}$ **41.** 5 **43.** $\frac{25}{4}$ **45.** $6\sqrt{5}$ **47.** m **49.** y^2 **51.** $6z$ **53.** $20x^3$ **55.** $z^2\sqrt{z}$ **57.** x^3y^6 **59.** $2\sqrt[3]{5}$ **61.** $3\sqrt[3]{2}$ **63.** $4\sqrt[3]{2}$ **65.** $2\sqrt[4]{5}$ **67.** $\frac{2}{3}$ **69.** $-\frac{6}{5}$ In Exercise 71, the number of displayed digits will vary among calculator models. Also, less sophisticated models may exhibit roundoff error in the final decimal place. **71.** **(a)** 4.472135955 **(b)** 4.472135955 **73.** $-5x + 19$ **75.** $-7xy + 11x^2y$

SECTION 8.3 (page 517)

1. $2\sqrt{3} + 4\sqrt{3} = (2 + 4)\sqrt{3} = 6\sqrt{3}$ **3.** $8\sqrt{3}$ **5.** $-5\sqrt{7}$ **7.** $5\sqrt{17}$ **9.** $5\sqrt{7}$ **11.** $11\sqrt{5}$ **13.** $15\sqrt{2}$ **15.** $-6\sqrt{2}$ **17.** $17\sqrt{7}$ **19.** $-16\sqrt{2} - 8\sqrt{3}$ **21.** $20\sqrt{2} + 6\sqrt{3} - 15\sqrt{5}$ **23.** $4\sqrt{2}$ **25.** $5\sqrt{21}$ **27.** $11\sqrt{3}$ **29.** $5\sqrt{x}$ **31.** $3x\sqrt{6}$ **33.** 0 **35.** $-20\sqrt{2k}$ **37.** $42x\sqrt{5z}$ **39.** $-\sqrt[3]{2}$ **41.** $6\sqrt[3]{p^2}$ **43.** $21\sqrt[4]{m^3}$ **45.** Answers will vary. One example is $\sqrt{36} + \sqrt[3]{64} = 6 + 4 = 10$. **47.** 6 **49.** 37 **51.** $50\sqrt{3}$

SECTION 8.4 (page 523)

1. We are actually multiplying by 1. The identity property of multiplication justifies our result.
3. $\frac{7\sqrt{5}}{5}$ **5.** $4\sqrt{2}$ **7.** $\frac{-\sqrt{33}}{3}$ **9.** $\frac{7\sqrt{15}}{5}$
11. $\frac{\sqrt{30}}{2}$ **13.** $\frac{16\sqrt{3}}{9}$ **15.** $\frac{-3\sqrt{2}}{10}$ **17.** $\frac{21\sqrt{5}}{5}$
19. $\sqrt{3}$ **21.** $\frac{-2\sqrt{30}}{3}$ **23.** $\frac{\sqrt{2}}{2}$ **25.** $\frac{\sqrt{65}}{5}$
27. $\frac{\sqrt{21}}{3}$ **29.** $\frac{3\sqrt{14}}{4}$ **31.** $\frac{1}{6}$ **33.** 1
35. $\frac{\sqrt{7x}}{x}$ **37.** $\frac{2x\sqrt{xy}}{y}$ **39.** $\frac{x\sqrt{3xy}}{y}$
41. $\frac{3ar^2\sqrt{7rt}}{7t}$ **43.** (b) **45.** $\frac{\sqrt[3]{12}}{2}$ **47.** $\frac{\sqrt[3]{196}}{7}$
49. $\frac{\sqrt[3]{6y}}{2y}$ **51.** $\frac{\sqrt[3]{42mn^2}}{6n}$ **53.** $32x^2 + 44x - 21$
55. $36x^2 - 1$ **57.** $pa - pm + qa - qm$

SECTION 8.5 (page 529)

1. 13 **3.** 4 **5.** 4 **7.** 5 **9.** $9\sqrt{5}$
11. $16\sqrt{2}$ **13.** $\sqrt{15} - \sqrt{35}$ **15.** $2\sqrt{10} + 30$
17. $4\sqrt{7}$ **19.** $57 + 23\sqrt{6}$ **21.** $81 + 14\sqrt{21}$
23. $37 + 12\sqrt{7}$ **25.** 23 **27.** 1
29. $2\sqrt{3} - 2 + 3\sqrt{2} - \sqrt{6}$ **31.** $15\sqrt{2} - 15$
33. $\sqrt{30} + \sqrt{15} + 6\sqrt{5} + 3\sqrt{10}$ **37.** $\frac{3 - \sqrt{2}}{7}$
39. $-4 - 2\sqrt{11}$ **41.** $1 + \sqrt{2}$ **43.** $-\sqrt{10} + \sqrt{15}$
45. $3 - \sqrt{3}$ **47.** $2\sqrt{5} + \sqrt{15} + 4 + 2\sqrt{3}$
49. $\sqrt{11} - 2$ **51.** $\frac{\sqrt{3} + 5}{8}$ **53.** $\frac{6 - \sqrt{10}}{2}$
55. $-3, -1$ **57.** 1, 2

SECTION 8.6 (page 537)

1. Because $\sqrt{x}$ must be greater than or equal to 0 for any real number x, it cannot equal -8.
3. 49 **5.** 7 **7.** 85 **9.** -45 **11.** $-\frac{3}{2}$
13. no solution **15.** 121 **17.** 8 **19.** 1
21. 6 **23.** no solution **25.** 5
27. The square of the right side is actually $x^2 - 14x + 49$.
29. 12 **31.** 5 **33.** 0, 3 **35.** -1, 3
37. 8 **39.** 4
41. **(a)** 70.5 miles per hour **(b)** 59.8 miles per hour **(c)** 53.9 miles per hour
43. -11, 11 **45.** $-\sqrt{23}, \sqrt{23}$

CHAPTER 8 REVIEW EXERCISES (page 543)

1. -7, 7 **2.** -9, 9 **3.** -14, 14 **4.** -11, 11
5. -15, 15 **6.** -27, 27 **7.** 4 **8.** -6
9. 10 **10.** 3 **11.** not a real number
12. -65 **13.** $\frac{7}{6}$ **14.** $\frac{10}{9}$
15. a must be negative. **16.** 8 **17.** irrational; 4.796
18. rational; 13 **19.** rational; -5 **20.** not a real number **21.** $\sqrt{14}$ **22.** 6 **23.** $5\sqrt{3}$ **24.** 12
25. $-3\sqrt{3}$ **26.** $4\sqrt{3}$ **27.** $4\sqrt{10}$ **28.** -5
29. 12 **30.** 18 **31.** $16\sqrt{6}$ **32.** $25\sqrt{10}$
33. $\frac{3}{2}$ **34.** $-\frac{11}{20}$ **35.** $\frac{\sqrt{3}}{7}$ **36.** $\frac{\sqrt{7}}{13}$ **37.** $\frac{\sqrt{5}}{6}$
38. $\frac{2}{15}$ **39.** $3\sqrt{2}$ **40.** 8 **41.** $2\sqrt{2}$ **42.** p
43. $\sqrt{km}$ **44.** r^9 **45.** x^5y^8 **46.** $x^4\sqrt{x}$ **47.** $\frac{6}{p}$
48. $a^7b^{10}\sqrt{ab}$ **49.** $11x^3y^5$ **50.** Yes, because both approximations are .7071067812. **51.** $2\sqrt{11}$
52. $9\sqrt{2}$ **53.** $21\sqrt{3}$ **54.** $12\sqrt{3}$ **55.** 0
56. $3\sqrt{7}$ **57.** $2\sqrt{3} + 3\sqrt{10}$ **58.** $2\sqrt{2}$
59. $6\sqrt{30}$ **60.** $5\sqrt{x}$ **61.** 0 **62.** $-m\sqrt{5}$
63. $11k^2\sqrt{2n}$ **64.** $\frac{10\sqrt{3}}{3}$ **65.** $\frac{15\sqrt{2}}{2}$ **66.** $\frac{8\sqrt{10}}{5}$
67. $\sqrt{5}$ **68.** $\sqrt{6}$ **69.** $\frac{\sqrt{30}}{15}$ **70.** $\frac{\sqrt{10}}{5}$
71. $\sqrt{10}$ **72.** $\frac{\sqrt{42}}{21}$ **73.** $\frac{r\sqrt{x}}{4x}$ **74.** $\frac{\sqrt[3]{9}}{3}$
75. $\frac{\sqrt[3]{98}}{7}$ **77.** $-\sqrt{15} - 9$ **78.** $3\sqrt{6} + 12$
79. $22 - 16\sqrt{3}$ **80.** $179 + 20\sqrt{7}$ **81.** -2
82. -13 **83.** $2\sqrt{21} - \sqrt{14} + 12\sqrt{2} - 4\sqrt{3}$
84. $-2 + \sqrt{5}$ **85.** $\frac{-2\sqrt{2} - 6}{7}$ **86.** $\frac{-2 + 6\sqrt{2}}{17}$
87. $\frac{-\sqrt{3} + 3}{2}$ **88.** $\frac{-\sqrt{10} + 3\sqrt{5} + \sqrt{2} - 3}{7}$
89. $\frac{2\sqrt{3} + 2 + 3\sqrt{2} + \sqrt{6}}{2}$ **90.** $\frac{3 + 2\sqrt{6}}{3}$
91. $\frac{1 + 3\sqrt{7}}{4}$ **92.** $3 + 4\sqrt{3}$ **93.** no solution
94. no solution **95.** 48
96. 1 **97.** 2 **98.** 6 **99.** -2 **100.** -2
101. 4 **102.** -2 **103.** 3 **104.** -1
105. 9 **106.** $11\sqrt{3}$ **107.** $\frac{11}{t}$ **108.** $\frac{5 - \sqrt{2}}{23}$
109. $\frac{2\sqrt{10}}{5}$ **110.** $5y\sqrt{2}$ **111.** -5
112. $-\sqrt{10} - 5\sqrt{15}$ **113.** $\frac{4r\sqrt{3rs}}{3s}$ **114.** $\frac{2 + \sqrt{13}}{2}$
115. $-7\sqrt{2}$ **116.** $7 - 2\sqrt{10}$ **117.** $166 + 2\sqrt{7}$
118. -11 **119.** $7\sqrt{2}$ **120.** 7 **121.** no solution
122. 8

CHAPTER 8 TEST (page 549)

[8.1] **1.** -14, 14 **2.** **(a)** irrational **(b)** 11.916
[8.2–8.5] **3.** 6 **4.** $-3\sqrt{3}$ **5.** $\frac{8\sqrt{2}}{5}$
6. $2\sqrt[3]{4}$ **7.** $4\sqrt{6}$ **8.** $9\sqrt{7}$ **9.** $-5\sqrt{3x}$
10. $2y\sqrt[3]{4x^2}$ **11.** 31 **12.** $6\sqrt{2} + 2 - 3\sqrt{14} - \sqrt{7}$
13. $11 + 2\sqrt{30}$ **[8.1]** **14.** **(a)** $6\sqrt{2}$ inches

(b) 8.485 inches **[8.4]** **15.** $\frac{5\sqrt{14}}{7}$ **16.** $\frac{\sqrt{6x}}{3x}$
17. $-\sqrt[3]{2}$ **18.** $\frac{-12 - 3\sqrt{3}}{13}$
[8.6] **19.** 3 **20.** $\frac{1}{4}, 1$

CUMULATIVE REVIEW R-8 (page 551)

1. 54 **2.** 6 **3.** 3 **4.** 18 **5.** 15
6. 4.223 **7.** 3 **8.** $y \geq -16$ **9.** $z > 5$
10. length: $17\frac{1}{2}$ meters; width: $10\frac{1}{2}$ meters **11.** $12x^{10}y^2$
12. $\frac{y^{15}}{5832}$ **13.** $3x^3 + 11x^2 - 13$ **14.** $4t^2 - 8t + 5$
15. $(m + 8)(m + 4)$ **16.** $(5t^2 + 6)(5t^2 - 6)$
17. $(6a + 5b)(2a - b)$ **18.** $(9z + 4)^2$
19. 3, 4 **20.** $-2, -1$ **21.** $2, -7$
22. $\frac{x + 1}{x}$ **23.** $(t + 5)(t + 3)$ **24.** -21
25. $\frac{y^2}{(y + 1)(y - 1)}$ **26.** $\frac{-2x - 14}{(x + 3)(x - 1)}$
27. $-4x + 5y = -20$ **28.** $x = 2$

29.

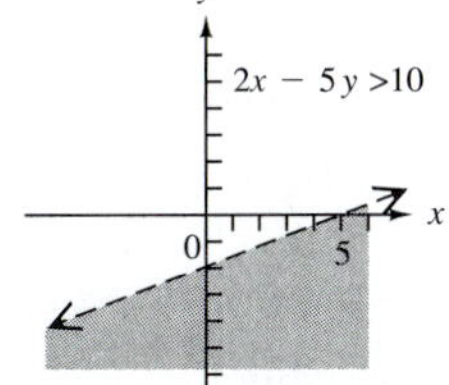

30. $-\frac{5}{6}$ **31.** $(3, -7)$ **32.** infinite number of solutions
33. 15 tens and 5 twenties **34.** $2\sqrt{2}$ **35.** $29\sqrt{3}$
36. $-\sqrt{3} + \sqrt{5}$ **37.** $10xy^2\sqrt{2y}$ **38.** $\frac{1 + \sqrt{3}}{2}$
39. $21 - 5\sqrt{2}$ **40.** 16

CHAPTER 9

SECTION 9.1 (page 557)

1. true **3.** false; $x = 0$ is a solution.
5. true **9.** $-9, 9$ **11.** $-\sqrt{14}, \sqrt{14}$
13. $-4\sqrt{3}, 4\sqrt{3}$ **15.** $-\frac{5}{2}, \frac{5}{2}$
17. $-1.5, 1.5$ **19.** $-\sqrt{3}, \sqrt{3}$ **21.** $-2, 8$
23. no real number solution **25.** $8 + 3\sqrt{3}, 8 - 3\sqrt{3}$
27. $-3, \frac{5}{3}$ **29.** $0, \frac{3}{2}$ **31.** $\frac{5 + \sqrt{30}}{2}, \frac{5 - \sqrt{30}}{2}$
33. $\frac{-1 + 3\sqrt{2}}{3}, \frac{-1 - 3\sqrt{2}}{3}$
35. $-10 + 4\sqrt{3}, -10 - 4\sqrt{3}$
37. $\frac{1 + 4\sqrt{3}}{4}, \frac{1 - 4\sqrt{3}}{4}$ **41.** $-7, 3$
43. 6 inches **45.** $\frac{4 + 4\sqrt{3}}{5}$ **47.** $\frac{3 + \sqrt{6}}{4}$
49. $\left(x - \frac{7}{2}\right)^2$

SECTION 9.2 (page 565)

1. Divide both sides by 4. **3.** 1, 3
5. $-1 + \sqrt{6}, -1 - \sqrt{6}$ **7.** -3
9. 49 **11.** $\frac{25}{4}$ **13.** $\frac{1}{16}$ **15.** (d)
17. $\frac{2 + \sqrt{14}}{2}, \frac{2 - \sqrt{14}}{2}$
19. $-\frac{3}{2}, \frac{1}{2}$ **21.** no real number solution
23. $\frac{-7 + \sqrt{97}}{6}, \frac{-7 - \sqrt{97}}{6}$
25. $-4, 2$ **27.** $1 + \sqrt{6}, 1 - \sqrt{6}$
29. 1 and 5 seconds **31.** $\frac{4 - 3\sqrt{3}}{3}$ **33.** $\frac{2 - \sqrt{5}}{2}$

SECTION 9.3 (page 571)

1. 4; 5; -9 **3.** 3; -4; -2 **5.** 3; 7; 0
7. $\frac{-5 + \sqrt{13}}{6}, \frac{-5 - \sqrt{13}}{6}$
9. $-13, 1$ **11.** 2 **13.** $\frac{-6 + \sqrt{26}}{2}, \frac{-6 - \sqrt{26}}{2}$
15. $-1, \frac{5}{2}$ **17.** $-1, 0$ **19.** $0, \frac{12}{7}$
21. $-2\sqrt{6}, 2\sqrt{6}$ **23.** $-\frac{2}{5}, \frac{2}{5}$
25. $\frac{6 + 2\sqrt{6}}{3}, \frac{6 - 2\sqrt{6}}{3}$
27. no real number solution
29. no real number solution
31. There are no real number solutions.
33. $-\frac{2}{3}, \frac{4}{3}$ **35.** $\frac{-1 + \sqrt{73}}{6}, \frac{-1 - \sqrt{73}}{6}$
37. $1 + \sqrt{2}, 1 - \sqrt{2}$
39. no real number solution
41. $r = \frac{-\pi h \pm \sqrt{\pi^2h^2 + \pi S}}{\pi}$
43. 7.899, 8.899, 11.899

45.

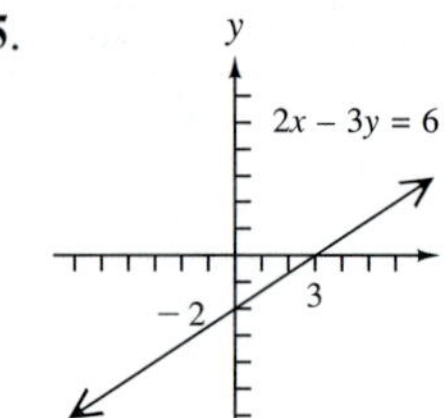

47.

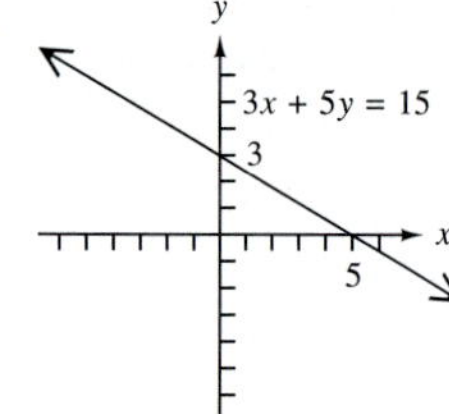

49. 16

SUMMARY EXERCISES ON QUADRATIC EQUATIONS (page 575)

1. $-6, 6$ **3.** $-\frac{10}{9}, \frac{10}{9}$ **5.** $1, 3$ **7.** $4, 5$ **9.** $-\frac{1}{3}, \frac{5}{3}$ **11.** $-17, 5$ **13.** $\frac{7 + 2\sqrt{6}}{3}, \frac{7 - 2\sqrt{6}}{3}$ **15.** no real number solution **17.** $-\frac{1}{2}, 2$ **19.** $-\frac{5}{4}, \frac{3}{2}$ **21.** $1 + \sqrt{2}, 1 - \sqrt{2}$ **23.** $\frac{2}{5}, 4$ **25.** $\frac{-3 + \sqrt{41}}{2}, \frac{-3 - \sqrt{41}}{2}$ **27.** $\frac{1}{4}, 1$ **29.** $\frac{-2 + \sqrt{11}}{3}, \frac{-2 - \sqrt{11}}{3}$ **31.** $\frac{-7 + \sqrt{5}}{4}, \frac{-7 - \sqrt{5}}{4}$ **33.** $\frac{8 + 8\sqrt{2}}{3}, \frac{8 - 8\sqrt{2}}{3}$ **35.** no real number solution **37.** $-\frac{2}{3}, 2$ **39.** $-4, \frac{3}{5}$ **41.** $-\frac{2}{3}, \frac{2}{5}$

SECTION 9.4 (page 581)

3. (0, 0)

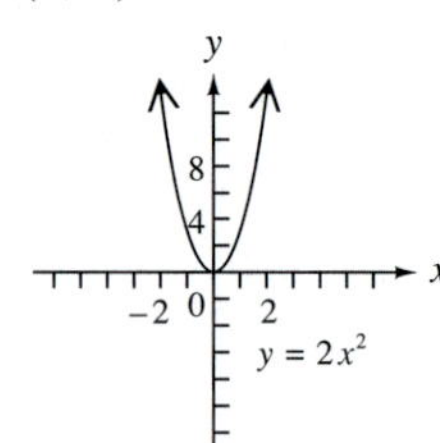

5. (−1, 2)

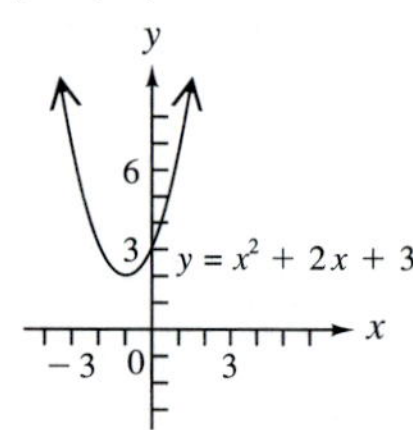

7. (0, −4)

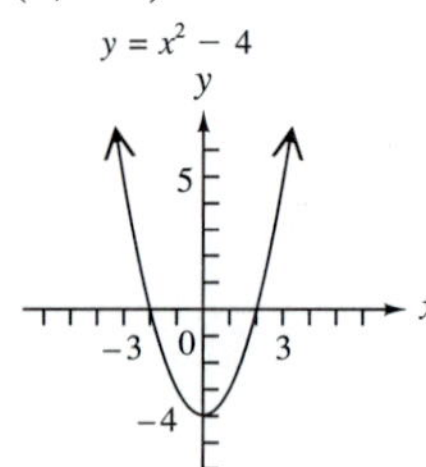

9. (0, 2)

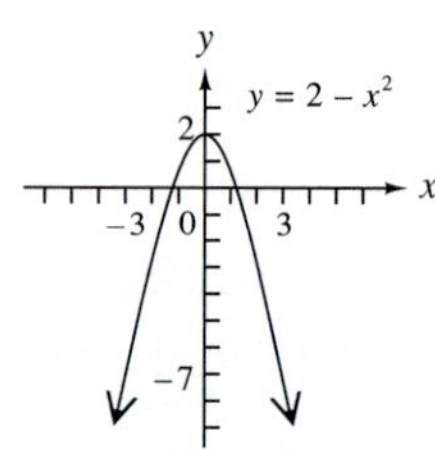

11. (−3, 0)

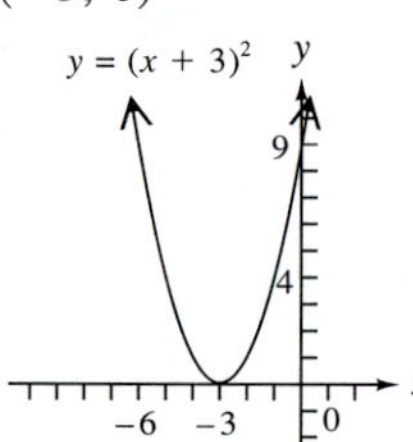

13. (3, 4)

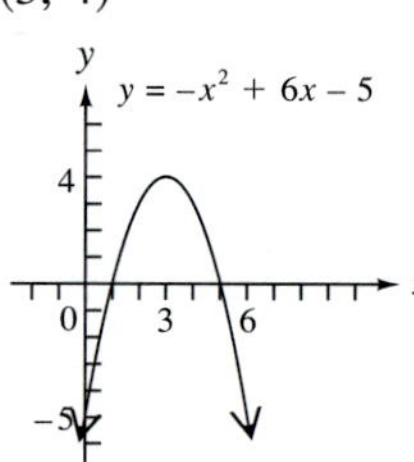

15. (2, 0)

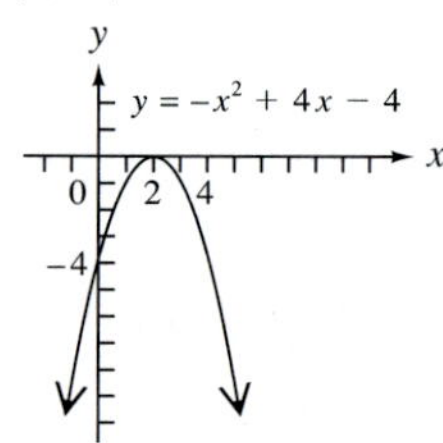

17. If $a > 0$, it opens upward, and if $a < 0$, it opens downward.

CHAPTER 9 REVIEW EXERCISES (page 585)

1. $-12, 12$ **2.** $-\sqrt{37}, \sqrt{37}$ **3.** $-8\sqrt{2}, 8\sqrt{2}$ **4.** $-7, 3$ **5.** $3 + \sqrt{10}, 3 - \sqrt{10}$ **6.** $\frac{-1 + \sqrt{14}}{2}, \frac{-1 - \sqrt{14}}{2}$ **7.** no real number solution **8.** $(ax + b)^2 =$ a number **9.** $-5, -1$ **10.** $-2 + \sqrt{11}, -2 - \sqrt{11}$ **11.** $-1 + \sqrt{6}, -1 - \sqrt{6}$ **12.** $\frac{-4 + \sqrt{22}}{2}, \frac{-4 - \sqrt{22}}{2}$ **13.** $-\frac{2}{5}, 1$ **14.** no real number solution **15.** $2\frac{1}{2}$ seconds **16.** 4 seconds **17.** no real number solution **18.** $\frac{1 + \sqrt{21}}{10}, \frac{1 - \sqrt{21}}{10}$ **19.** $\frac{2 + \sqrt{10}}{2}, \frac{2 - \sqrt{10}}{2}$ **20.** $\frac{-1 + \sqrt{29}}{4}, \frac{-1 - \sqrt{29}}{4}$ **21.** $\frac{-3 + \sqrt{41}}{2}, \frac{-3 - \sqrt{41}}{2}$ **22.** The $-b$ term should be above the fraction bar. **23.** (0, 0) **24.** (1, 0)

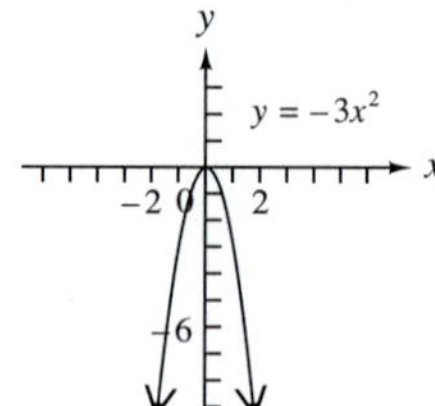

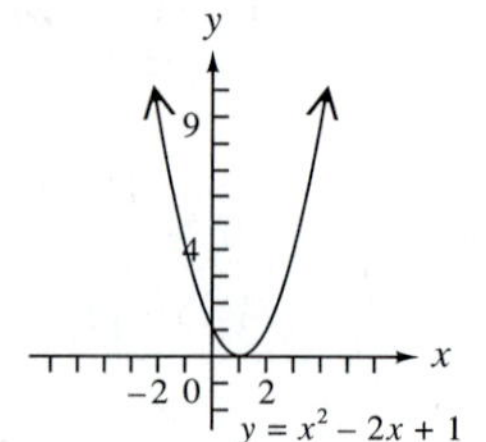

25. (0, 5)

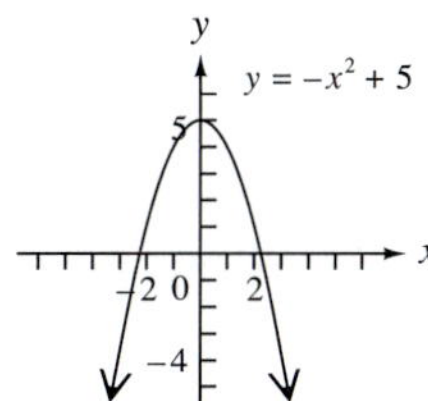

26. (1, 4)

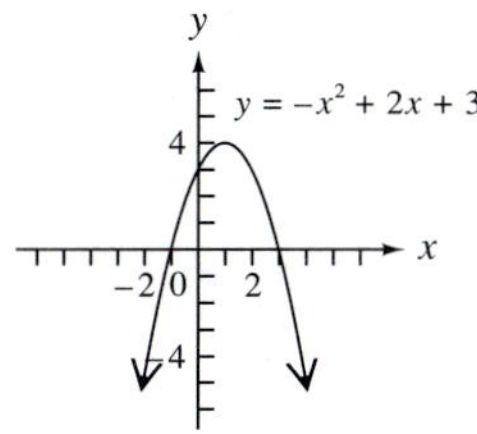

27. $(-2, -2)$

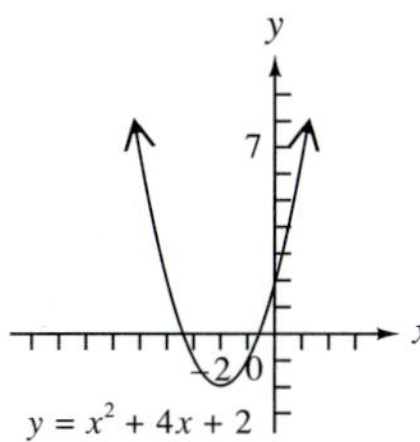

28. $(-4, 0)$

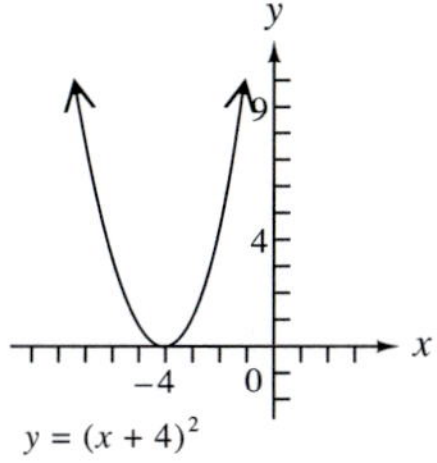

29. $-\frac{11}{2}, 5$ **30.** $-\frac{11}{2}, \frac{9}{2}$

31. $\frac{-1 + \sqrt{21}}{2}, \frac{-1 - \sqrt{21}}{2}$ **32.** $-\frac{3}{2}, \frac{1}{3}$

33. $\frac{-5 + \sqrt{17}}{2}, \frac{-5 - \sqrt{17}}{2}$

34. $-1 + \sqrt{3}, -1 - \sqrt{3}$ **35.** no real number solution **36.** $\frac{9 + \sqrt{41}}{2}, \frac{9 - \sqrt{41}}{2}$

37. $-\frac{5}{3}$ **38.** $-1 + 2\sqrt{2}, -1 - 2\sqrt{2}$

39. $-2 + \sqrt{5}, -2 - \sqrt{5}$ **40.** $-2\sqrt{2}, 2\sqrt{2}$

CHAPTER 9 TEST (page 589)

[9.1] 1. $-\sqrt{39}, \sqrt{39}$ **2.** $-11, 5$

3. $\frac{-3 + 2\sqrt{6}}{4}, \frac{-3 - 2\sqrt{6}}{4}$

[9.2] 4. $2 + \sqrt{10}, 2 - \sqrt{10}$

5. $\frac{-6 + \sqrt{42}}{2}, \frac{-6 - \sqrt{42}}{2}$

[9.3] 6. The quantity must be positive. **7.** $-3, \frac{1}{2}$

8. $\frac{3 + \sqrt{3}}{3}, \frac{3 - \sqrt{3}}{3}$ **9.** no real number solution

10. $\frac{5 + \sqrt{13}}{6}, \frac{5 - \sqrt{13}}{6}$

[9.1–9.3] 11. $1 + \sqrt{2}, 1 - \sqrt{2}$ **12.** $\frac{-1 + 3\sqrt{2}}{2}, \frac{-1 - 3\sqrt{2}}{2}$ **13.** $\frac{11 + \sqrt{89}}{4}, \frac{11 - \sqrt{89}}{4}$

14. 5 **[9.2] 15.** 2 seconds

[9.1] 16. (d) **[9.3] 17.** 6

[9.4] 18. (3, 0)

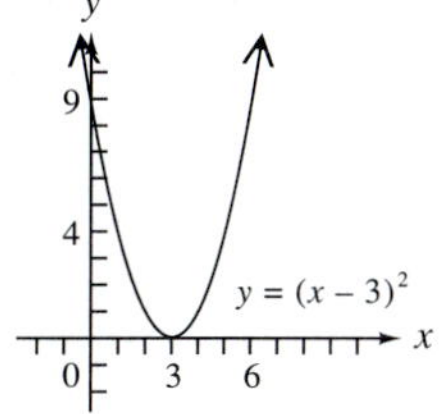

19. $(-1, -3)$

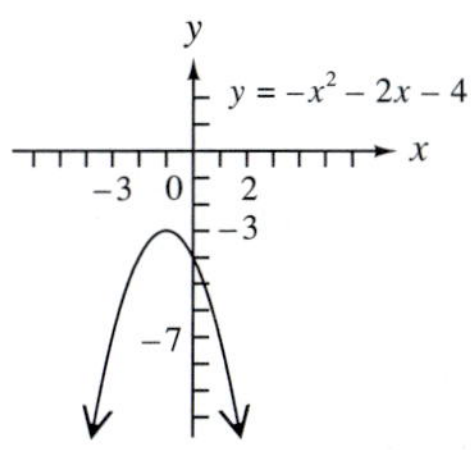

[9.1] 20. The square of a real number cannot be negative.

CUMULATIVE REVIEW R–9 (page 591)

1. 15 **2.** 5 **3.** -2 **4.** $-r + 7$ **5.** $-2k$
6. $19m - 17$ **7.** 3 **8.** 5 **9.** 2 **10.** 133 votes
11. $p = \frac{I}{rt}$ **12.** $L = \frac{P - 2W}{2}$
13. $m > -2$ **14.** $p \leq 4$

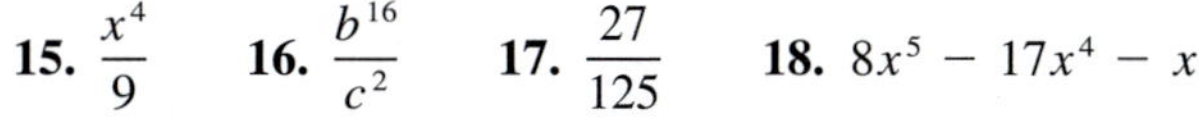

15. $\frac{x^4}{9}$ **16.** $\frac{b^{16}}{c^2}$ **17.** $\frac{27}{125}$ **18.** $8x^5 - 17x^4 - x^2$
19. $2x^4 + x^3 - 19x^2 + 2x + 20$
20. $25t^2 + 90t + 81$ **21.** $3x^2 - 2x + 1$
22. $16x^2(x - 3y)$ **23.** $(4x^2 + 1)(2x + 1)(2x - 1)$
24. $(2a + 1)(a - 3)$ **25.** $(5m - 2)^2$ **26.** $-9, 6$
27. $-1, \frac{4}{3}$ **28.** 50 meters **29.** $\frac{4}{5}$ **30.** $\frac{-k - 1}{k(k - 1)}$
31. $\frac{5a + 2}{(a - 2)^2(a + 2)}$ **32.** $\frac{6x + 1}{3x - 1}$ **33.** $-\frac{15}{7}, 2$

34.

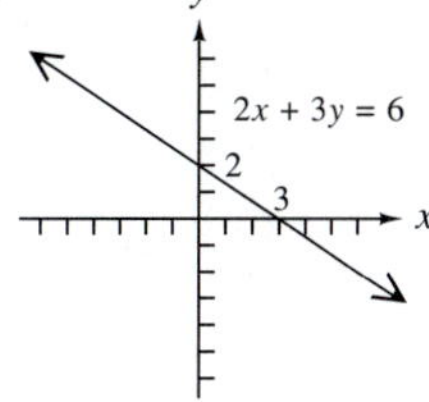

35.

y
3 y = 3
x
0

36.

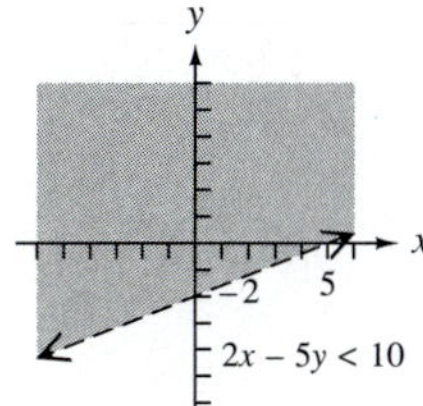

37. $-\frac{1}{3}$ **38.** $2x - y = -3$ **39.** $(-3, 2)$

40. no solution

41.

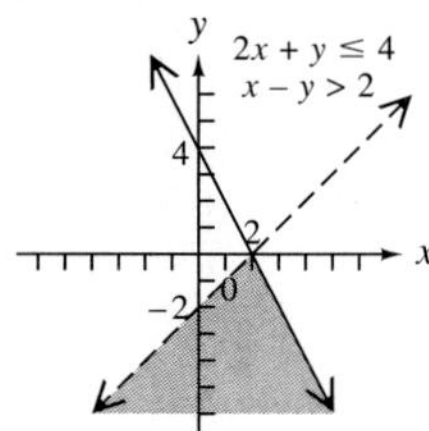

42. $-17, 17$ **43.** 10 **44.** $\frac{6\sqrt{30}}{5}$ **45.** $\frac{\sqrt[3]{28}}{4}$

46. $4\sqrt{5}$ **47.** $-ab\sqrt[3]{2b}$ **48.** 7

49. $\frac{1 + \sqrt{3}}{2}, \frac{1 - \sqrt{3}}{2}$

50. $(0, -4)$

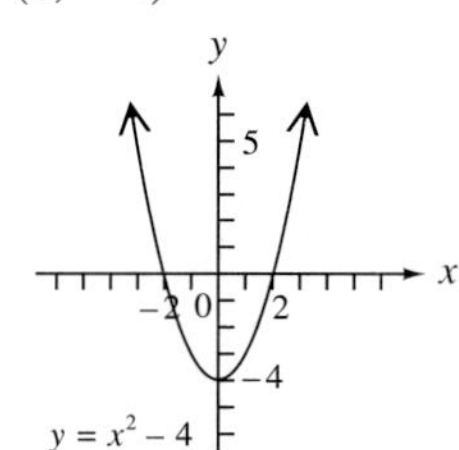

Solutions to Selected Exercises

For the answers to all odd-numbered section exercises, all chapter review exercises, and all chapter tests, see the section beginning on page A-1.

If you would like to see more solutions, you may order the *Student's Solutions Manual for Introductory Algebra* from your college bookstore. It contains solutions to the odd-numbered exercises that do not appear in this section, as well as solutions to all chapter review exercises and chapter tests.

As you are looking at these solutions, remember that many algebraic exercises and problems can be solved in a variety of ways. In this section we provide only one method of solving selected exercises from each section; space does not permit showing other methods that may be equally correct.

If you work the exercise differently but obtain the same answer, then *as long as your steps are mathematically valid,* your work is correct. Mathematical thinking is a creative process, and solving problems in more than one way is an example of this creativity.

CHAPTER R

SECTION R.1 (page 9)

1. In the fraction $\frac{3}{8}$, 3 is the numerator and 8 is the denominator.

5. The answer in a multiplication problem is called the product, and the answer in a division problem is called the quotient.

9. The number 54 is composite since it has factors other than 1 and itself (2, for example).

13. The number 1 is neither prime nor composite, by agreement.

17. $500 = 2 \cdot 250$
$= 2 \cdot 2 \cdot 125$
$= 2 \cdot 2 \cdot 5 \cdot 25$
$= 2 \cdot 2 \cdot 5 \cdot 5 \cdot 5$

21. The number 29 is prime since none of the primes up to and including 5 are factors.

25. $\frac{15}{18} = \frac{3 \cdot 5}{3 \cdot 6} = \frac{3}{3} \cdot \frac{5}{6} = 1 \cdot \frac{5}{6} = \frac{5}{6}$

29. $\frac{144}{120} = \frac{2 \cdot 2 \cdot 2 \cdot 2 \cdot 3 \cdot 3}{2 \cdot 2 \cdot 2 \cdot 3 \cdot 5} = \frac{2}{2} \cdot \frac{2}{2} \cdot \frac{2}{2} \cdot \frac{2}{1} \cdot \frac{3}{3} \cdot \frac{3}{5}$
$= 1 \cdot 1 \cdot 1 \cdot \frac{2}{1} \cdot 1 \cdot \frac{3}{5} = \frac{6}{5}$

33. $\frac{4}{5} \cdot \frac{6}{7} = \frac{4 \cdot 6}{5 \cdot 7} = \frac{24}{35}$

37. $\frac{15}{4} \cdot \frac{8}{25} = \frac{15 \cdot 8}{4 \cdot 25} = \frac{3 \cdot 5 \cdot 4 \cdot 2}{4 \cdot 5 \cdot 5} = \frac{3 \cdot 2}{5} = \frac{6}{5}$

41. $\frac{5}{4} \div \frac{3}{8} = \frac{5}{4} \cdot \frac{8}{3} = \frac{5 \cdot 8}{4 \cdot 3} = \frac{5 \cdot 4 \cdot 2}{4 \cdot 3} = \frac{5 \cdot 2}{3} = \frac{10}{3}$ or $3\frac{1}{3}$

↑ Multiply by the reciprocal of the second fraction.

45. $\frac{3}{4} \div 12 = \frac{3}{4} \cdot \frac{1}{12} = \frac{3 \cdot 1}{4 \cdot 12} = \frac{3 \cdot 1}{4 \cdot 3 \cdot 4} = \frac{1}{4 \cdot 4} = \frac{1}{16}$

↑ The reciprocal of 12

53. $\frac{5}{9} + \frac{1}{3}$

Since $9 = 3 \cdot 3$, the LCD is $3 \cdot 3 = 9$. Use the property of 1 to write $\frac{1}{3}$ with the LCD as denominator.

$$\frac{1}{3} = \frac{1}{3} \cdot \frac{3}{3} = \frac{3}{9}$$

Now add the two like fractions.

$$\frac{5}{9} + \frac{1}{3} = \frac{5}{9} + \frac{3}{9} = \frac{8}{9}$$

57. $\frac{7}{12} - \frac{1}{9}$

Since $12 = 2 \cdot 2 \cdot 3$ and $9 = 3 \cdot 3$, the LCD is $2 \cdot 2 \cdot 3 \cdot 3 = 36$. Use the property of 1 to write each fraction with 36 as denominator.

$$\frac{7}{12} = \frac{7}{12} \cdot \frac{3}{3} = \frac{21}{36} \quad \text{and} \quad \frac{1}{9} \cdot \frac{4}{4} = \frac{4}{36}$$

Now subtract the like fractions.

$$\frac{7}{12} - \frac{1}{9} = \frac{21}{36} - \frac{4}{36} = \frac{17}{36}$$

SOLUTIONS

61. $\frac{5}{3} + \frac{1}{6} - \frac{1}{2}$

Since 2 and 3 are prime and $6 = 2 \cdot 3$, the LCD is $2 \cdot 3 = 6$. Write $\frac{5}{3}$ and $\frac{1}{2}$ as equivalent fractions with 6 as denominator.

$$\frac{5}{3} = \frac{5}{3} \cdot \frac{2}{2} = \frac{10}{6} \quad \text{and} \quad \frac{1}{2} = \frac{1}{2} \cdot \frac{3}{3} = \frac{3}{6}$$

Now add and subtract, then write the answer in lowest terms.

$$\frac{5}{3} + \frac{1}{6} - \frac{1}{2} = \frac{10}{6} + \frac{1}{6} - \frac{3}{6} = \frac{8}{6} = \frac{4 \cdot 2}{3 \cdot 2} = \frac{4}{3} \cdot \frac{2}{2}$$
$$= \frac{4}{3} \cdot 1 = \frac{4}{3}$$

65. The perimeter is the sum of the measures of the 5 sides.

$$196 + 98\frac{3}{4} + 146\frac{1}{2} + 100\frac{7}{8} + 76\frac{5}{8}$$
$$= 196 + \frac{395}{4} + \frac{293}{2} + \frac{807}{8} + \frac{613}{8}$$

Since 2 is prime and $4 = 2 \cdot 2$ and $8 = 2 \cdot 2 \cdot 2$, the LCD is $2 \cdot 2 \cdot 2 = 8$.

$$196 = \frac{196}{1} \cdot \frac{8}{8} = \frac{1568}{8}; \frac{395}{4} = \frac{395}{4} \cdot \frac{2}{2} = \frac{790}{8};$$
$$\frac{293}{2} = \frac{293}{2} \cdot \frac{4}{4} = \frac{1172}{8}$$

Now add the equivalent fractions.

$$\frac{1568}{8} + \frac{790}{8} + \frac{1172}{8} + \frac{807}{8} + \frac{613}{8} = \frac{4950}{8}$$
$$= \frac{2 \cdot 2475}{2 \cdot 4}$$
$$= \frac{2}{2} \cdot \frac{2475}{4}$$
$$= 1 \cdot \frac{2475}{4}$$
$$= 618\frac{3}{4}$$

The perimeter is $618\frac{3}{4}$ feet.

69. Divide the total amount of tomato sauce by the number of servings.

$$2\frac{1}{3} \div 7 = \frac{7}{3} \div 7 = \frac{7}{3} \cdot \frac{1}{7} = \frac{7 \cdot 1}{3 \cdot 7} = \frac{1}{3}$$

$\frac{1}{3}$ cup is needed.

SECTION R.2 (page 19)

1. In the number 367.9412, 3 is in the hundreds place, 6 is in the tens place, 7 is in the units or ones place, 9 is in the tenths place, 4 is in the hundredths place, 1 is in the thousandths place, and 2 is in the ten thousandths place. **(a)** 6 **(b)** 9 **(c)** 1 **(d)** 7 **(e)** 4

5. 84.9×98.3 is approximately $85 \times 100 = 8500$. The answer is (b).

9. Since there are two places to the right of the decimal point, there will be two zeros in the denominator.

$$.64 = \frac{64}{100}$$

13. Since there are three places to the right of the decimal point, there will be three zeros in the denominator.

$$3.805 = \frac{3805}{1000}$$

17. Write the numbers in a column with the decimal points lined up.

$$\begin{array}{r} 28.73 \\ -\ \ 3.12 \\ \hline 25.61 \end{array}$$

21. Attach zeros to make all the numbers the same length.

$$\begin{array}{r} 32.560 \\ 47.356 \\ +\ \ 1.800 \\ \hline 81.716 \end{array}$$

25. Multiply as if the numbers were whole numbers.

$$\begin{array}{rl} 44.3 & \text{1 decimal place} \\ \times\ 5.6 & \text{1 decimal place} \\ \hline 2658 & 1 + 1 = 2 \\ 2215\ \ & \downarrow \\ \hline 248.08 & \text{2 decimal places} \end{array}$$

29. Bring the decimal point straight up and divide as with whole numbers.

$$\begin{array}{r} 7.15 \\ 8\overline{)57.20} \leftarrow \text{Attach zero.} \\ \underline{56} \\ 1\ 2 \\ \underline{8} \\ 40 \\ \underline{40} \\ 0 \end{array}$$

37. Divide 1 by 8.

$$\begin{array}{r} .125 \\ 8\overline{)1.000} \\ \underline{8} \\ 20 \\ \underline{16} \\ 40 \\ \underline{40} \\ 0 \end{array}$$

41. Divide 3 by 16.

$$\begin{array}{r} .1875 \\ 16\overline{)3.0000} \\ \underline{1\,6} \\ 1\,40 \\ \underline{1\,28} \\ 120 \\ \underline{112} \\ 80 \\ \underline{80} \\ 0 \end{array}$$

45. Divide 5 by 9.

$$\begin{array}{r} .5555 \\ 9\overline{)5.0000} \\ \underline{4\,5} \\ 50 \\ \underline{45} \\ 50 \\ \underline{45} \\ 50 \\ \underline{45} \\ 5 \end{array}$$

Since a 5 is always left after the subtraction, this quotient is a repeating decimal.

$$\frac{5}{9} = .\overline{5}$$

Rounded to the nearest thousandth,

$$\frac{5}{9} = .556.$$

53. $117\% = 117 \cdot 1\% = 117 \cdot .01 = 1.17$

57. $.8\% = .8 \cdot 1\% = .8 \cdot .01 = .008$

61. $.004 = .4 \cdot .01 = .4 \cdot 1\% = .4\%$

65. $.3 = 30 \cdot .01 = 30 \cdot 1\% = 30\%$

69. Divide 3 by 2.

$$\begin{array}{r} 1.5 \\ 2\overline{)3.0} \\ \underline{2} \\ 1\,0 \\ \underline{1\,0} \\ 0 \end{array}$$

$$\frac{3}{2} = 1.5 = 150 \cdot .01 = 150 \cdot 1\% = 150\%$$

73. Divide 5 by 16.

$$\begin{array}{r} .3125 \\ 16\overline{)5.0000} \\ \underline{4\,8} \\ 20 \\ \underline{16} \\ 40 \\ \underline{32} \\ 80 \\ \underline{80} \\ 0 \end{array}$$

$$\frac{5}{16} = .3125 = 31.25 \cdot .01 = 31.25 \times 1\% = 31.25\%$$

CHAPTER 1

SECTION 1.1 (page 27)

1. The statement is false. 3^5 means $3 \cdot 3 \cdot 3 \cdot 3 \cdot 3$.

5. $7^2 = \underbrace{7 \cdot 7} = 49$ ← 7 is used as a factor 2 times.

9. $4^3 = \underbrace{4 \cdot 4 \cdot 4} = 64$ ← 4 is used as a factor 3 times.

13. $3^4 = \underbrace{3 \cdot 3 \cdot 3 \cdot 3} = 81$ ← 3 is used as a factor 4 times.

17. $\left(\frac{2}{3}\right)^4 = \underbrace{\frac{2}{3} \cdot \frac{2}{3} \cdot \frac{2}{3} \cdot \frac{2}{3}} = \frac{16}{81}$ ← $\frac{2}{3}$ is used as a factor 4 times.

25. $\frac{1}{4} \cdot \frac{2}{3} + \frac{2}{5} \cdot \frac{11}{3} = \frac{1}{6} + \frac{22}{15}$ Multiply.

$= \frac{5}{30} + \frac{44}{30}$ Get a common denominator.

$= \frac{49}{30}$ Add.

29. $(4.3)(1.2) + (2.1)(8.5) = 5.16 + 17.85$ Multiply.

$= 23.01$ Add.

33. $3^2[(11 + 3) - 4] = 3^2[14 - 4]$ Add inside parentheses.

$= 3^2[10]$ Subtract.

$= 9[10]$ Use the exponent.

$= 90$ Multiply.

37. $\frac{4(6 + 2) + 8(8 - 3)}{6(4 - 2) - 2^2} = \frac{4(8) + 8(5)}{6(2) - 2^2}$ Add and subtract inside parentheses.

$= \frac{4(8) + 8(5)}{6(2) - 4}$ Use the exponent.

$= \frac{32 + 40}{12 - 4}$ Multiply.

$= \frac{72}{8}$ Add and subtract.

$= 9$ Divide.

41. The statement $5 < 6$ is true since 5 "is less than" 6.

45. $17 \le 18 - 1$
$17 \le 17$ Subtract.

The statement is true since $17 = 17$ is true.

49. $6[5 + 3(4 + 2)] \le 70$
$6[5 + 3(6)] \le 70$ Add inside parentheses.
$6[5 + 18] \le 70$ Multiply.
$6[23] \le 70$ Add.
$138 \le 70$ Multiply.

Both $138 < 70$ and $138 = 70$ are false. Because of this, $138 \le 70$ is false.

53. $8 \le 4^2 - 2^2$
$8 \le 16 - 4$ Use the exponents.
$8 \le 12$ Subtract.

Since $8 < 12$, the statement is true.

57. Since "is greater than" indicates $>$ and "minus" indicates $-$, write as $9 > 5 - 4$.

61. Since "is less than or equal to" indicates $\le$, write $2 \le 3$.

65. Three is not equal to six. The statement is true since three and six are not equal.

69. Answers will vary. One example is

$5 + 3 \ge 2 \cdot 2$
$5 + 3 \ge 4$ Multiply on right side.
$8 \ge 4.$ Add on left side.

The statement is true since $8 > 4$.

73. $12 \ge 3$ becomes $3 \le 12$ Exchange numbers. Reverse symbol.

77. $3.12 + 9.34 - 7.33 = 12.46 - 7.33$ Add.
$= 5.13$ Subtract.

SECTION 1.2 (page 33)

1. There is no equals sign, so $3x + 2(x - 4)$ is an expression.

5. Because of the equals sign, $x + y = 3$ is an equation.

9. (a) $x + 9 = 4 + 9$ Let $x = 4$.
$= 13$ Add.

(b) $x + 9 = 6 + 9$ Let $x = 6$.
$= 15$ Add.

13. (a) $4x^2 = 4 \cdot 4^2$ Let $x = 4$.
$= 4 \cdot 16$ Square.
$= 64$ Multiply.

(b) $4x^2 = 4 \cdot 6^2$ Let $x = 6$.
$= 4 \cdot 36$ Square.
$= 144$ Multiply.

17. (a) $\frac{3x - 5}{2x} = \frac{3 \cdot 4 - 5}{2 \cdot 4}$ Let $x = 4$.
$= \frac{12 - 5}{8}$ Multiply.
$= \frac{7}{8}$ Subtract.

(b) $\frac{3x - 5}{2x} = \frac{3 \cdot 6 - 5}{2 \cdot 6}$ Let $x = 6$.
$= \frac{18 - 5}{12}$ Multiply.
$= \frac{13}{12}$ Subtract.

21. (a) $6.459x = 6.459 \cdot 4$ Let $x = 4$.
$= 25.836$ Multiply.

(b) $6.459x = 6.459 \cdot 6$ Let $x = 6$.
$= 38.754$ Multiply.

25. (a) $3(x + 2y) = 3(2 + 2 \cdot 1)$ Replace x with 2 and y with 1.
$= 3(2 + 2)$ Multiply.
$= 3(4)$ Add.
$= 12$ Multiply.

(b) $3(x + 2y) = 3(1 + 2 \cdot 5)$ Replace x with 1 and y with 5.
$= 3(1 + 10)$ Multiply.
$= 3(11)$ Add.
$= 33$ Multiply.

29. (a) $\frac{x}{2} + \frac{y}{3} = \frac{2}{2} + \frac{1}{3}$ Replace x with 2 and y with 1.
$= 1 + \frac{1}{3}$ Use property of one.
$= \frac{3}{3} + \frac{1}{3}$ Get a common denominator.
$= \frac{4}{3}$ Add.

(b) $\frac{x}{2} + \frac{y}{3} = \frac{1}{2} + \frac{5}{3}$ Replace x with 1 and y with 5.
$= \frac{3}{6} + \frac{10}{6}$ Get a common denominator.
$= \frac{13}{6}$ Add.

33. (a) $2y^2 + 5x = 2 \cdot 1^2 + 5 \cdot 2$ Replace x with 2 and y with 1.
$= 2 \cdot 1 + 5 \cdot 2$ Square.
$= 2 + 10$ Multiply.
$= 12$ Add.

(b) $2y^2 + 5x = 2 \cdot 5^2 + 5 \cdot 1$ Replace x with 1 and y with 5.
$= 2 \cdot 25 + 5 \cdot 1$ Square.
$= 50 + 5$ Multiply.
$= 55$ Add.

37. (a) $.841x^2 + .32y^2 = .841 \cdot 2^2 + .32 \cdot 1^2$
Replace x with 2 and y with 1.
$= .841 \cdot 4 + .32 \cdot 1$ Use the exponents.
$= 3.364 + .32$ Multiply.
$= 3.684$ Add.

(b) $.841x^2 + .32y^2 = .841 \cdot 1^2 + .32 \cdot 5^2$
Replace x with 1 and y with 5.
$= .841 \cdot 1 + .32 \cdot 25$ Use the exponents.
$= .841 + 8$ Multiply.
$= 8.841$ Add.

41. "Added to" indicates addition. This phrase translates as $x + 7$.

45. Since a number is subtracted *from* 7, write this as $7 - x$.

49. "Divided by" indicates division, so write this as $\frac{12}{x}$.

57.
$$\begin{aligned} p - 5 &= 12 & \\ 7 - 5 &= 12 & \text{Replace } p \text{ with 7.} \\ 2 &= 12 & \text{False} \end{aligned}$$

The number 7 is not a solution of the equation.

61.
$$\begin{aligned} 2y + 3(y - 2) &= 14 & \\ 2 \cdot 3 + 3(3 - 2) &= 14 & \text{Replace } y \text{ with 3.} \\ 2 \cdot 3 + 3 \cdot 1 &= 14 & \text{Subtract.} \\ 6 + 3 &= 14 & \text{Multiply.} \\ 9 &= 14 & \text{False} \end{aligned}$$

The number 3 is not a solution of the equation.

65.
$$\begin{aligned} 3r^2 - 2 &= 53.47 & \\ 3 \cdot 4.3^2 - 2 &= 53.47 & \text{Replace } r \text{ with 4.3.} \\ 3 \cdot 18.49 - 2 &= 53.47 & \text{Square.} \\ 55.47 - 2 &= 53.47 & \text{Multiply.} \\ 53.47 &= 53.47 & \text{True} \end{aligned}$$

The number 4.3 is a solution of the equation.

69. Let x represent the unknown number and translate as follows.

The sum of a number and 8	is	18.
↓	↓	↓
$x + 8$	$=$	18.

Try each number from the given domain, {0, 2, 4, 6, 8, 10} in turn.

$$\begin{aligned} x + 8 &= 18 & \text{Given equation} \\ 0 + 8 &= 18 & \text{False} \\ 2 + 8 &= 18 & \text{False} \\ 4 + 8 &= 18 & \text{False} \\ 6 + 8 &= 18 & \text{False} \\ 8 + 8 &= 18 & \text{False} \\ 10 + 8 &= 18 & \text{True} \end{aligned}$$

The only solution of $x + 8 = 18$ is 10.

73. Use x to represent the unknown number.

Five	more than	twice a number	is	5.
↓	↓	↓	↓	↓
5	+	$2x$	=	5

Try each number from the given domain, {0, 2, 4, 6, 8, 10}. The solution is 0, since $5 + 2 \cdot 0 = 5$.

SECTION 1.3 (page 43)

1. The statement is true. The set of integers includes the natural numbers, their opposites, and zero. The set of whole numbers is {0, 1, 2, 3, 4, 5, . . .}. Thus, every whole number is an integer.

5. The statement is false. Zero is a whole number that is not positive.

9. The statement is false. All whole numbers are integers. (See Exercise 1.)

13. "Below sea level" indicates a negative number, -8.

17. "Negative balance of trade" indicates a negative number, $-66{,}000{,}000$ (dollars).

21. See the graph in the answer section.

25. -11 is to the left of -4 on the number line, so -11 is the smaller number.

29. -100 is smaller than 0 since any negative number is smaller than 0.

33. The statement $8 < -16$ is false. Any positive number is greater than any negative number.

37. **(a)** The opposite of -2 is found by changing the sign of -2. The opposite of -2 is 2.
(b) The absolute value of -2 is the distance between 0 and -2 on the number line.

$$|-2| = 2$$

The absolute value of -2 is 2.

41. **(a)** The opposite of $-\frac{3}{4}$ is found by changing the sign of $-\frac{3}{4}$. The opposite of $-\frac{3}{4}$ is $\frac{3}{4}$.
(b) The absolute value of $-\frac{3}{4}$ is the distance between 0 and $-\frac{3}{4}$ on the number line.

$$\left|-\frac{3}{4}\right| = \frac{3}{4}$$

The absolute value of $-\frac{3}{4}$ is $\frac{3}{4}$.

45. Since 4 is 4 units from 0 on the number line, $|4| = 4$.

49. Replace $|-14|$ with 14 since -14 is 14 units from 0 on the number line. Thus, $-|-14| = -14$.

57.
$$\begin{aligned} \frac{3}{4} + \frac{1}{6} - \frac{1}{2} &= \frac{9}{12} + \frac{2}{12} - \frac{6}{12} & \text{Get a common denominator.} \\ &= \frac{11}{12} - \frac{6}{12} & \text{Add.} \\ &= \frac{5}{12} & \text{Subtract.} \end{aligned}$$

SECTION 1.4 (page 49)

1. The sum of two *negative* numbers is a *negative* number.

5. To add $6 + (-4)$, find the difference between the absolute values of the numbers. $|6| = 6$ and $|-4| = 4$; $6 - 4 = 2$. Since $|6| > |-4|$, the sum will be positive, and $6 + (-4) = 2$.

9. $-7 + (-3) = -10$ The sum of two negative numbers is negative.

13. $-12.4 + (-3.5) = -15.9$ The sum of two negative numbers is negative.

17. First work inside the brackets.

$$5 + [14 + (-6)] = 5 + 8 = 13$$

21. First work inside the brackets.

$$-3 + [5 + (-2)] = -3 + 3 = 0$$

25. $-\frac{1}{6}+\frac{2}{3}=-\frac{1}{6}+\frac{4}{6}=\frac{3}{6}=\frac{1}{2}$

29. $2\frac{1}{2}+\left(-3\frac{1}{4}\right)=\frac{5}{2}+\left(-\frac{13}{4}\right)$

$=\frac{10}{4}+\left(-\frac{13}{4}\right)=-\frac{3}{4}$

33. $-7.1+[3.3+(-4.9)]=-7.1+(-1.6)=-8.7$

37. $[-5+(-7)]+[-4+(-9)]+[13+(-12)]$
$=-12+(-13)+1=-24$

41. $-11+13=13+(-11)$
$2=2$ True

45. $18+(-6)+(-12)=0$
$12+(-12)=0$
$0=0$ True

49. $\frac{11}{5}+\left(-\frac{6}{11}\right)=-\frac{6}{11}+\frac{11}{5}$

$\frac{121}{55}+\left(-\frac{30}{55}\right)=-\frac{30}{55}+\frac{121}{55}$

$\frac{91}{55}=\frac{91}{55}$ True

53. "Sum" indicates addition.

$-5+12+6=[-5+12]+6=7+6=13$

57. "Sum" and "increased by" indicate addition.

$[-4+(-10)]+12=-14+12=-2$

61. Receiving a check is represented by a positive number and spending by a negative number.

$100+(-53)=47$

The student has \$47 left.

65. The lowest temperature is represented by -5.
The highest temperature is represented by $117+(-5)$, or 112°F.

69. A payment is represented by a positive number and a purchase by a negative number.

$-153+(-14)+60=[-153+(-14)]+60$
$=-167+60=-107$

Jennifer still owes \$107.

In Exercises 73, 77, and 81, mentally replace the variable with values from the given domain $\{-3,-2,-1,0,1,2,3\}$ to find that:

73. -2 is the solution to $x+3=1$, since $-2+3=1$.

77. -1 is the solution to $13+y=12$, since $13+(-1)=12$.

81. 1 is the solution to $r+(-4.6)=-3.6$, since $1+(-4.6)=-3.6$.

85. $(3-1)+(17-14)=2+3=5$

SECTION 1.5 (page 57)

1. We must add the opposite of -8 to -6.

5. $-8-4=-8+(-4)$

9. $6-10=6+(-10)=-4$

13. $-10-6=-10+(-6)=-16$

17. $6-(-13)=6+13=19$

21. $3-(4-6)=3-[4+(-6)]$
$=3-(-2)$
$=3+2$
$=5$

25. $\frac{1}{2}-\left(-\frac{1}{4}\right)=\frac{1}{2}+\frac{1}{4}$
$=\frac{2}{4}+\frac{1}{4}$
$=\frac{3}{4}$

29. $\frac{5}{8}-\left(-\frac{1}{2}-\frac{3}{4}\right)=\frac{5}{8}-\left(-\frac{1}{2}+\left(-\frac{3}{4}\right)\right)$
$=\frac{5}{8}-\left(-\frac{2}{4}+\left(-\frac{3}{4}\right)\right)$
$=\frac{5}{8}-\left(-\frac{5}{4}\right)$
$=\frac{5}{8}+\frac{5}{4}$
$=\frac{5}{8}+\frac{10}{8}$
$=\frac{15}{8}$

33. $-7.4-4.5=-7.4+(-4.5)=-11.9$

37. $[(-3.1)-4.5]-(.8-2.1)=[(-3.1)+(-4.5)]-[.8+(-2.1)]$
$=-7.6-(-1.3)$
$=-7.6+1.3$
$=-6.3$

41. $(-3-8)-(7-4)=[-3+(-8)]-3$
$=-11+(-3)$
$=-14$

45. $-4+[(-6-9)-(-7+4)]=-4+[-6+(-9)-(-3)]$
$=-4+(-15+3)$
$=-4+(-12)$
$=-16$

49. $[-34.99+(6.59-12.25)]-8.33$
$=[-34.99+(6.59+(-12.25))]-8.33$
$=[-34.99+(-5.66)]-8.33$
$=-40.65+(-8.33)$
$=-48.98$

53. "Difference" indicates subtraction.

$4-(-8)=4+8=12$

57. "Sum" indicates addition, and "decreased by" indicates subtraction.

$[9+(-4)]-7=5+(-7)=-2$

61. 10° below -5°F indicates subtraction.

$-5-10=-5+(-10)=-15$

It was -15°F the next day.

65. Owing money is represented by a negative number as is borrowing money.

$$-10 + (-70) = -80$$

Chris' financial status is represented by $-\$80$.

In Exercises 69 and 73, mentally replace the variable with values from the domain $\{-3, -2, -1, 0, 1, 2, 3\}$ to find that:

69. -2 is the solution to $x - 1 = -3$, since $-2 - 1 = -3$.

73. -3 is the solution to $3 - (-x) = 0$, since $3 - [-(-3)] = 0$.

77. In the expression $a + |b|$, a is positive and $|b|$ is positive. Therefore, $a + |b|$ is positive.

81. $(x + y)(2x - y) = (5 + 2)(2 \cdot 5 - 2)$ Replace x with 5 and y with 2.
$= (5 + 2)(10 - 2)$ Multiply.
$= 7 \cdot 8$ Add and subtract.
$= 56$ Multiply.

SECTION 1.6 (page 65)

1. The statement is true.

5. This statement is true, because the product of two negative numbers is a positive number. (See Exercise 1.)

9. $5(-6) = -(5 \cdot 6) = -30$

13. $(-4)(-20) = 80$

17. $\left(-\frac{3}{8}\right)\left(-\frac{20}{9}\right) = \left(\frac{3}{8}\right)\left(\frac{20}{9}\right) = \frac{3 \cdot 20}{8 \cdot 9}$
$= \frac{3 \cdot 4 \cdot 5}{4 \cdot 2 \cdot 3 \cdot 3} = \frac{5}{6}$

21. $(-6)\left(-\frac{1}{4}\right) = (6)\left(\frac{1}{4}\right) = \frac{6 \cdot 1}{1 \cdot 4} = \frac{2 \cdot 3}{2 \cdot 2} = \frac{3}{2}$

25. $-10 - (-4)(2) = -10 - (-8) = -10 + 8 = -2$

29. $-7(3 - 8) = -7[3 + (-8)] = -7(-5) = 35$

33. $(7 - 10)(10 - 4) = [7 + (-10)](6)$
$= (-3)(6) = -18$

37. $3(-5) - (-7) = -15 - (-7) = -15 + 7 = -8$

45. $(2x + y)(3a) = [2(6) + (-4)][3(3)]$ Substitute.
$= [12 + (-4)](9)$ Multiply.
$= (8)(9)$ Add.
$= 72$ Multiply.

49. $(-5 + x)(-3 + y)(3 - a)$
$= (-5 + 6)[-3 + (-4)][3 - 3]$ Substitute.
$= (-5 + 6)[-3 + (-4)][3 + (-3]$
$= (1)(-7)(0)$ Add.
$= 0$ Multiply.

53. $3a^2 - x^2 = 3 \cdot 3^2 - 6^2$ Substitute.
$= 3 \cdot 9 - 36$ Square 3 and 6.
$= 27 - 36$ Multiply.
$= -9$ Subtract.

57. Twice the product of -1 and 6 is subtracted from -4.

$$-4 - 2[(-1)(6)] = -4 - 2(-6)$$
$$= -4 - (-12) = -4 + 12 = 8.$$

61. "Product" indicates multiplication, and "difference" means subtraction.

$$12[9 - (-8)] = 12[9 + 8] = 12(17) = 204$$

In Exercises 65, 69, 73, mentally replace the variable with values from the domain $\{-3, -2, -1, 0, 1, 2, 3\}$ to find that:

65. 3 is the solution to $-2x = -6$, since $-2(3) = -6$.

69. -2 is the solution to $-5x = 10$, since $-5(-2) = 10$.

73. -1 is the solution to $\frac{1}{5}w = -\frac{1}{5}$, since $\frac{1}{5}(-1) = -\frac{1}{5}$.

77. $\frac{3x + 5y}{2} = \frac{3(2) + 5(10)}{2}$ Substitute.
$= \frac{6 + 50}{2}$ Multiply.
$= \frac{56}{2}$ Add.
$= 28$ Divide.

81. $\frac{3x^2 + 2y^2}{10y + 3} = \frac{3 \cdot 2^2 + 2 \cdot 10^2}{10(10) + 3}$ Substitute.
$= \frac{3 \cdot 4 + 2 \cdot 100}{10(10) + 3}$ Square 2 and 10.
$= \frac{12 + 200}{100 + 3}$ Multiply.
$= \frac{212}{103}$ or $2\frac{6}{103}$ Add.

SECTION 1.7 (page 75)

1. The answer is "positive."

5. The answer is "positive."

9. The reciprocal of -5 is $-\frac{1}{5}$ because $-5 \cdot -\frac{1}{5} = 1$.

13. $3 - 3 = 0$
0 has no reciprocal.

17. $.4 = \frac{4}{10} = \frac{2}{5}$
Because $\frac{2}{5} \cdot \frac{5}{2} = 1$, the reciprocal is

$$\frac{5}{2} = 2\frac{1}{2} = 2.5.$$

21. $\frac{-15}{5} = -3$

25. $\frac{-160}{-10} = 16$

29. $\frac{-10.252}{-.4} = 25.63$

33. $(-6.8) \div (-2) = \frac{-6.8}{-2} = 3.4$

37. $\frac{-50}{7 - (-3)} = \frac{-50}{10} = -5$

41. $\frac{-5(-6)}{9 - (-1)} = \frac{30}{10} = 3$

45. $\dfrac{-10(2) + 6(2)}{-3 - (-1)} = \dfrac{-20 + 12}{-2}$ Multiply in numerator. Subtract in denominator.

$= \dfrac{-8}{-2}$ Add in numerator.

$= 4$ Divide.

49. $\dfrac{1^2 + 4^2}{3^2 + 5^2} = \dfrac{1 + 16}{9 + 25}$ Square 1 and 4. Square 3 and 5.

$= \dfrac{17}{34}$ Add.

$= \dfrac{1}{2}$ Express in lowest terms.

53. "Quotient" indicates division. The numerator is -36, and the denominator is -9.

$$\frac{-36}{-9} = 4$$

57. "Divided by" indicates division. The sum of 15 and -3 is the numerator, and the product of 4 and -3 is the denominator.

$$\frac{15 + (-3)}{4(-3)} = \frac{12}{-12} = -1$$

In Exercises 61 and 65, mentally replace the variable with values from the domain $\{-8, -6, -4, -2, 0, 2, 4, 6, 8\}$ to find that:

61. -8 is the solution to $\dfrac{x}{-4} = 2$, since $\dfrac{-8}{-4} = 2$.

65. -4 is the solution to $\dfrac{2x}{2} = -4$, since $\dfrac{2(-4)}{2} = -4$.

69. $\dfrac{x}{3} = -3$

Here, x must be a negative number, since the denominator is positive and the quotient is negative. Since

$$\frac{-9}{3} = -3,$$

the solution is -9.

73. $x + 5 = -5$

Since $-10 + 5 = -5$, the solution is -10.

77. $-\frac{1}{3}(-3) = 1$

SECTION 1.8 (page 87)

1. The statement is true. The sum of 0 and a number leaves the number unchanged.

5. The statement is true. $a + (-a) = 0$

9. This shows the commutative property. Two numbers added in either order give the same result.

13. This shows the commutative property: $ab = ba$.

17. This shows the inverse property: $a + (-a) = 0$.

21. This shows the identity property: $a \cdot 1 = a$.

25. This is an example of the distributive property. The number 6 is "distributed" over x and y.

29. This is an example of the distributive property. The number 5 is "distributed" over $2x$ and $3y$.

37. By the identity property $s + 0 = s$.

41. $(w + 5) + (-3) = w + [5 + (-3)]$ Associative property

$= w + 2$ Add.

45.

$6t + 8 - 6t + 3$

$(6t + 8) + (-6t) + 3$ Order of operations

$(8 + 6t) + (-6t) + 3$ Commutative property

$8 + [6t + (-6t)] + 3$ Associative property

$8 + 0 + 3$ Inverse property

$8 + 3$ Identity property

11 Add.

49.

$\frac{2}{3}x - 11 + 11 - \frac{2}{3}x$

$\left[\frac{2}{3}x + (-11)\right] + 11 + \left(\frac{-2}{3}x\right)$ Order of operations

$\frac{2}{3}x + (-11 + 11) + \left(\frac{-2}{3}x\right)$ Associative property

$\frac{2}{3}x + 0 + \left(\frac{-2}{3}x\right)$ Inverse property

$\frac{2}{3}x + \left(\frac{-2}{3}x\right)$ Identity property

0 Inverse property

53.

$t + (-t) + \frac{1}{2}(2)$

$t + (-t) + 1$ Inverse property

$[t + (-t)] + 1$ Order of operations

$0 + 1$ Inverse property

1 Identity property

57. $4(t + 3) = 4t + 4 \cdot 3$ Distributive property

$= 4t + 12$ Multiply.

61. $-5(y - 4) = -5y + (-5) \cdot (-4)$ Distributive property

$= -5y + 20$ Multiply.

65. $8 \cdot z + 8 \cdot w = 8(z + w)$ Distributive property

69. $8(3r + 4s - 5y) = 8(3r) + 8(4s) - 8(5y)$ Distributive property

$= (8 \cdot 3)r + (8 \cdot 4)s - (8 \cdot 5)y$ Associative property

$= 24r + 32s - 40y$ Multiply.

73. $-5x + x = -5x + 1x$ Identity property

$= (-5 + 1)x$ Distributive property

$= -4x$ Add.

77. $-(-5c - 4d) = -1(-5c - 4d)$ Identity property

$= -1(-5c) + (-1)(-4d)$ Distributive property

$= (-1 \cdot -5)c + (-1 \cdot -4)d$ Associative property

$= 5c + 4d$ Multiply.

81. Answers will vary. For example, "turning on the oven" and "baking a cake."

85. $10 - [-2 - (4 - 6)] = 10 - [-2 - (-2)]$
$= 10 - [-2 + 2]$
$= 10 - 0$
$= 10$

SECTION 1.9 (page 95)

1. The answer is (c).

$6x - 2x = (6 - 2)x$ Distributive property
$= 4x$ Subtract.

5. Since $19 - 8 = 11$, $4r + 19 - 8 = 4r + 11$.

9. $5 + 2(x - 3y) = 5 + 2(x) + 2(-3y)$ Distributive property
$= 5 + 2x - 6y$ Multiply.

13. The numerical coefficient of $14x$ is 14.

17. The numerical coefficient of $5m^2$ is 5.

21. The numerical coefficient of $-x$ is -1.

25. Answers will vary. For example, $-4x$ and $7x$ since $-4x$ has a negative numerical coefficient and $7x$ has a positive numerical coefficient. The sum, $-4x + 7x = 3x$, has a positive numerical coefficient.

29. These terms are unlike. Although both have the variable z, the exponents are not the same.

33. These terms are unlike. x and y do not have the same variable part.

37. $4k + 3 - 2k + 8 + 7k - 16$
$= (4 - 2 + 7)k + 3 + 8 - 16$ Distributive property
$= 9k - 5$ Add and subtract.

41. $-5.3r + 4.9 - 2r + .7 + 3.2r$
$= (-5.3 - 2 + 3.2)r + 4.9 + .7$ Distributive property
$= -4.1r + 5.6$ Add and subtract.

45. $13p + 4(4 - 8p)$
$= 13p + 4(4) + 4(-8p)$ Distributive property
$= 13p + 16 - 32p$ Multiply.
$= -19p + 16$ Combine like terms.

49. $-5(5y - 9) + 3(3y + 6)$
$= -5(5y) - 5(-9) + 3(3y) + 3(6)$ Distributive property
$= -25y + 45 + 9y + 18$ Multiply.
$= -16y + 63$ Combine like terms.

53. $-7.5(2y + 4) - 2.9(3y - 6)$
$= -7.5(2y) - 7.5(4) - 2.9(3y) - 2.9(-6)$ Distributive property
$= -15y - 30 - 8.7y + 17.4$ Multiply.
$= -23.7y - 12.6$ Combine like terms.

57. Since $-7x$ is "subtracted from" the sum $13 + 6x$, write $(13 + 6x) - (-7x) = 13 + 6x + 7x = 13 + 13x$.

61. Wording may vary. One example is "the sum of a number and 2 is subtracted from the product of 9 and the number."

65. The additive inverse or opposite of a number is found by changing the sign of the number. The additive inverse of -15 is 15.

69. $x + 14 + (-14) = x + [14 + (-14)]$ Add -14.
$= x + 0$ Inverse property
$= x$ Identity property

The desired number is -14.

CHAPTER 2

SECTION 2.1 (page 113)

Each of these answers should be checked by substituting in the original equation.

1. $x - 4 = 8$
$x - 4 + 4 = 8 + 4$ Add 4 to each side.
$x = 12$ Combine terms.

5. $t + 2.3 = 8.9$
$t + 2.3 - 2.3 = 8.9 - 2.3$ Subtract 2.3 from each side.
$t = 6.6$ Combine terms.

9. $\frac{9}{7}r - 3 = \frac{2}{7}r$
$\frac{9}{7}r - 3 - \frac{2}{7}r = \frac{2}{7}r - \frac{2}{7}r$ Subtract $\frac{2}{7}r$.
$r - 3 = 0$ Combine terms.
$r - 3 + 3 = 0 + 3$ Add 3.
$r = 3$ Combine terms.

13. $3p + 6 = 10 + 2p$
$3p + 6 - 2p = 10 + 2p - 2p$ Subtract $2p$.
$p + 6 = 10$ Combine terms.
$p + 6 - 6 = 10 - 6$ Subtract 6.
$p = 4$

17. $3x + 9 = 3x + 8$
$3x + 9 - 3x = 3x + 8 - 3x$ Subtract $3x$.
$9 = 8$ False

Therefore, there is no solution.

25. $10x + 5x + 7 - 8 = 12x + 3 + 2x$
$15x - 1 = 14x + 3$ Combine terms.
$15x - 1 - 14x = 14x + 3 - 14x$ Subtract $14x$.
$x - 1 = 3$ Combine terms.
$x - 1 + 1 = 3 + 1$ Add 1.
$x = 4$

29. $5.2q - 4.6 - 7.1q = -2.1 - 1.9q - 2.5$
$-1.9q - 4.6 = -4.6 - 1.9q$ Combine terms.
$-1.9q - 4.6 + 1.9q = -4.6 - 1.9q + 1.9q$ Add $1.9q$.
$-4.6 = -4.6$ Combine terms.
$-4.6 + 4.6 = -4.6 + 4.6$ Add 4.6.
$0 = 0$ True

All real numbers satisfy the equation.

33. $(5y + 6) - (3 + 4y) = 10$
$5y + 6 - 3 - 4y = 10$ Distributive property
$y + 3 = 10$ Combine terms.
$y + 3 - 3 = 10 - 3$ Subtract 3.
$y = 7$

37. $-6(2b + 1) + (13b - 7) = 0$
$-12b - 6 + 13b - 7 = 0$ Distributive property
$b - 13 = 0$ Combine terms.
$b - 13 + 13 = 0 + 13$ Add 13.
$b = 13$

41. $-2(8p + 2) - 3(2 - 7p) = 2(4 + 2p)$
$-16p - 4 - 6 + 21p = 8 + 4p$ Distributive property
$5p - 10 = 8 + 4p$ Combine terms.
$5p - 10 - 4p = 8 + 4p - 4p$ Subtract $4p$.
$p - 10 = 8$ Combine terms.
$p - 10 + 10 = 8 + 10$ Add 10.
$p = 18$

49. $\frac{2}{3}\left(\frac{3}{2}z\right) = \left(\frac{2}{3} \cdot \frac{3}{2}\right)z$ Associative property
$= 1 \cdot z$ Inverse property
$= z$ Identity property

53. Since the coefficient of x is $-\frac{9}{7}$, multiply by the reciprocal, $-\frac{7}{9}$, to get $1x$ or just x.

SECTION 2.2 (page 119)

1. To get just x on the left side, multiply both sides by $\frac{3}{2}$, the reciprocal of $\frac{2}{3}$.
5. To get just x on the left side, multiply both sides by $-\frac{2}{9}$, the reciprocal of $-\frac{9}{2}$.
9. To get just x on the left side, divide both sides by 6, since 6 is the coefficient of x.
13. To get just x on the left side, divide both sides by .12, since .12 is the coefficient of x.

Each of these answers should be checked by substituting in the original equation.

17. $5x = 30$
$\frac{5x}{5} = \frac{30}{5}$ Divide by 5.
$1x = 6$ Divide.
$x = 6$ Multiplicative identity property

21. $3a = -15$
$\frac{3a}{3} = \frac{-15}{3}$ Divide by 3.
$1a = -5$ Divide.
$a = -5$ Multiplicative identity property

25. $-6x = -72$
$\frac{-6x}{-6} = \frac{-72}{-6}$ Divide by -6.
$1x = 12$ Divide.
$x = 12$ Multiplicative identity property

29. $\frac{1}{4}y = -12$
$4 \cdot \frac{1}{4}y = 4(-12)$ Multiply by 4.
$1y = -48$ Multiplicative inverse property
$y = -48$ Multiplicative identity property

33. $-x = -\frac{4}{7}$
$-1 \cdot x = -\frac{4}{7}$ $-x = -1 \cdot x$
$-1(-1 \cdot x) = -1 \cdot \left(-\frac{4}{7}\right)$ Multiply by -1, since $-1 \cdot -1 = 1$.
$x = \frac{4}{7}$ Multiply.

37. $4x + 3x = 21$
$7x = 21$ Combine terms.
$\frac{7x}{7} = \frac{21}{7}$ Divide by 7.
$1x = 3$ Divide.
$x = 3$ Multiplicative identity property

41. $3r - 5r = 10$
$-2r = 10$ Combine terms.
$\frac{-2r}{-2} = \frac{10}{-2}$ Divide by -2.
$1r = -5$ Divide.
$r = -5$ Multiplicative identity property

45. $\frac{2}{3}t = 6$
$\frac{3}{2} \cdot \frac{2}{3}t = \frac{3}{2} \cdot 6$ Multiply by $\frac{3}{2}$.
$1t = \frac{3}{2} \cdot \frac{6}{1}$ Multiplicative inverse property
$t = 9$ Multiplicative identity property

49. $-\frac{7}{9}c = \frac{3}{5}$
$-\frac{9}{7} \cdot \left(-\frac{7}{9}c\right) = -\frac{9}{7} \cdot \frac{3}{5}$ Multiply by $-\frac{9}{7}$.
$1c = -\frac{27}{35}$ Multiplicative inverse property
$c = -\frac{27}{35}$ Multiplicative identity property

57. $-7(4p - 3) + 8 = -7(4p) - 7(-3) + 8$
$= -28p + 21 + 8$
$= -28p + 29$

SECTION 2.3 (page 125)

Each of these answers should be checked by substituting in the original equation.

1. $2h + 4 = 8$
$2h + 4 - 4 = 8 - 4$ Subtract 4.
$2h = 4$ Combine terms.
$\frac{2h}{2} = \frac{4}{2}$ Divide by 2.
$h = 2$ Reduce.

5.
$$10p + 6 = 12p - 4$$
$$10p + 6 - 6 = 12p - 4 - 6 \quad \text{Subtract 6.}$$
$$10p = 12p - 10 \quad \text{Combine terms.}$$
$$10p - 12p = 12p - 10 - 12p \quad \text{Subtract } 12p.$$
$$-2p = -10 \quad \text{Combine terms.}$$
$$\frac{-2p}{-2} = \frac{-10}{-2} \quad \text{Divide by } -2.$$
$$p = 5 \quad \text{Reduce.}$$

9.
$$4(2x - 1) = -6(x + 3)$$
$$8x - 4 = -6x - 18 \quad \text{Distributive property}$$
$$8x - 4 + 4 = -6x - 18 + 4 \quad \text{Add 4.}$$
$$8x = -6x - 14 \quad \text{Combine terms.}$$
$$8x + 6x = -6x - 14 + 6x \quad \text{Add } 6x.$$
$$14x = -14 \quad \text{Combine terms.}$$
$$\frac{14x}{14} = \frac{-14}{14} \quad \text{Divide by 14.}$$
$$x = -1 \quad \text{Reduce.}$$

13.
$$3(2x - 4) = 6(x - 2)$$
$$6x - 12 = 6x - 12 \quad \text{Distributive property}$$
$$6x - 12 - 6x = 6x - 12 - 6x \quad \text{Subtract } 6x.$$
$$-12 = -12 \quad \text{Combine terms.}$$
$$-12 + 12 = -12 + 12 \quad \text{Add 12.}$$
$$0 = 0 \quad \text{True}$$
All real numbers satisfy the equation.

21.
$$-\frac{1}{4}(x - 12) + \frac{1}{2}(x + 2) = x + 4$$
$$4\left(-\frac{1}{4}(x - 12) + \frac{1}{2}(x + 2)\right) = 4(x + 4)$$
Multiply by 4 to clear fractions.
$$4\left(-\frac{1}{4}\right)(x - 12) + 4\left(\frac{1}{2}\right)(x + 2) = 4x + 16$$
Distributive property
$$(-1)(x - 12) + 2(x + 2) = 4x + 16 \quad \text{Multiply.}$$
$$-x + 12 + 2x + 4 = 4x + 16$$
Distributive property
$$x + 16 = 4x + 16$$
Combine terms.
$$x + 16 - 4x = 4x + 16 - 4x$$
Subtract $4x$.
$$-3x + 16 = 16$$
Combine terms.
$$-3x + 16 - 16 = 16 - 16$$
Subtract 16.
$$-3x = 0 \quad \text{Combine terms.}$$
$$\frac{-3x}{-3} = \frac{0}{-3} \quad \text{Divide by } -3.$$
$$x = 0 \quad \text{Reduce.}$$

25.
$$.20(60) + .05(x) = .10(60 + x)$$
$$20(60) + 5(x) = 10(60 + x) \quad \text{Multiply by 100 to clear decimals.}$$
$$1200 + 5x = 600 + 10x \quad \text{Distributive property}$$
$$1200 + 5x - 10x = 600 + 10x - 10x \quad \text{Subtract } 10x.$$
$$1200 - 5x = 600 \quad \text{Combine terms.}$$
$$1200 - 5x - 1200 = 600 - 1200 \quad \text{Subtract 1200.}$$
$$-5x = -600 \quad \text{Combine terms.}$$
$$\frac{-5x}{-5} = \frac{-600}{-5} \quad \text{Divide by } -5.$$
$$x = 120 \quad \text{Reduce.}$$

29.
$$.06(10{,}000) + .08x = .072(10{,}000 + x)$$
$$60(10{,}000) + 80x = 72(10{,}000 + x)$$
Multiply by 1000.
$$600{,}000 + 80x = 720{,}000 + 72x$$
Distributive property
$$600{,}000 + 80x - 72x = 720{,}000 + 72x - 72x$$
Subtract $72x$.
$$600{,}000 + 8x = 720{,}000 \quad \text{Combine terms.}$$
$$600{,}000 + 8x - 600{,}000 = 720{,}000 - 600{,}000$$
Subtract 600,000.
$$8x = 120{,}000 \quad \text{Combine terms.}$$
$$\frac{8x}{8} = \frac{120{,}000}{8} \quad \text{Divide by 8.}$$
$$x = 15{,}000 \quad \text{Reduce.}$$

33.
$$-2(2s - 4) - 8 = -3(4s + 4) - 1$$
$$-4s + 8 - 8 = -12s - 12 - 1 \quad \text{Distributive property}$$
$$-4s = -12s - 13$$
$$-4s + 12s = -12s - 13 + 12s \quad \text{Add } 12s.$$
$$8s = -13$$
$$\frac{8s}{8} = \frac{-13}{8} \quad \text{Divide by 8.}$$
$$s = -\frac{13}{8}$$

37.
$$\frac{1}{2}(x + 2) + \frac{3}{4}(x + 4) = x + 5$$
$$4\left(\frac{1}{2}(x + 2) + \frac{3}{4}(x + 4)\right) = 4(x + 5) \quad \text{Multiply by 4.}$$
$$4\left(\frac{1}{2}\right)(x + 2) + 4\left(\frac{3}{4}\right)(x + 4) = 4x + 20 \quad \text{Distributive property}$$
$$2(x + 2) + 3(x + 4) = 4x + 20$$
$$2x + 4 + 3x + 12 = 4x + 20 \quad \text{Distributive property}$$
$$5x + 16 = 4x + 20$$
$$5x + 16 - 4x = 4x + 20 - 4x$$
Subtract $4x$.
$$x + 16 = 20$$
$$x + 16 - 16 = 20 - 16 \quad \text{Subtract 16.}$$
$$x = 4$$

41.
$$4(x + 8) = 2(2x + 6) + 20$$
$$4x + 32 = 4x + 12 + 20 \quad \text{Distributive property}$$
$$4x + 32 = 4x + 32$$
$$4x + 32 - 32 = 4x + 32 - 32 \quad \text{Subtract 32.}$$
$$4x = 4x$$
$$4x - 4x = 4x - 4x \quad \text{Subtract } 4x.$$
$$0 = 0 \quad \text{True}$$
All real numbers satisfy the equation.

49. "Sum" indicates addition. The sum of a number and twice the number is written as $x + 2x$.

53. "Product" indicates multiplication and "difference" indicates subtraction. The product of 12 and the difference between a number and 9 is written $12(x - 9)$.

SECTION 2.4 (page 133)

1. Choice (c) $6\frac{2}{3}$ is not a reasonable answer in an applied problem that requires finding the number of coins in a jar, since you cannot get $\frac{2}{3}$ of a coin.

5. Let x represent the number.

Double the sum	of 1 added to a number	results in	5 more than the number.
↓	↓	↓	↓
$2 \cdot$	$(x + 1)$	$=$	$x + 5$

Solve the equation.

$$2(x + 1) = x + 5$$
$$2x + 2 = x + 5 \quad \text{Distributive property}$$
$$2x + 2 - x = x + 5 - x \quad \text{Subtract } x.$$
$$x + 2 = 5 \quad \text{Combine terms.}$$
$$x + 2 - 2 = 5 - 2 \quad \text{Subtract 2.}$$
$$x = 3$$

The number is 3.

9. Let x = the number of Republicans
$x + 14$ = the number of Democrats.

The number of Republicans	plus	the number of Democrats	equals	the members of the Senate.
↓	↓	↓	↓	↓
x	$+$	$(x + 14)$	$=$	100

Solve the equation.

$$x + (x + 14) = 100$$
$$2x + 14 = 100 \quad \text{Combine terms.}$$
$$2x + 14 - 14 = 100 - 14 \quad \text{Subtract 14.}$$
$$2x = 86 \quad \text{Combine terms.}$$
$$\frac{2x}{2} = \frac{86}{2} \quad \text{Divide by 2.}$$
$$x = 43$$

There were 43 Republicans and $43 + 14 = 57$ Democrats.

13. Let x = the length of the longer piece of paper
$x - 9$ = the length of the shorter piece of paper.

Longer piece		Shorter piece		Total length
↓		↓		↓
x	$+$	$(x - 9)$	$=$	39

Solve the equation.

$$2x - 9 = 39 \quad \text{Combine terms.}$$
$$2x - 9 + 9 = 39 + 9 \quad \text{Add 9.}$$
$$2x = 48 \quad \text{Combine terms.}$$
$$\frac{2x}{2} = \frac{48}{2} \quad \text{Divide by 2.}$$
$$x = 24$$

The longer piece of paper is 24 inches, and the shorter piece of paper is $24 - 9 = 15$ inches.

17. Let x = the distance of Mercury from the sun (in millions of miles)
$x + 31.2$ = the distance of Venus from the sun
$(x + 31.2) + 25.7$ = the distance of Earth from the sun.

Mercury's distance	Venus' distance	Earth's distance	Total distance
↓	↓	↓	↓
x	$+ (x + 31.2)$	$+ [(x + 31.2) + 25.7]$	$= 196.1$

Solve the equation.

$$3x + 88.1 = 196.1 \quad \text{Combine terms.}$$
$$3x + 88.1 - 88.1 = 196.1 - 88.1 \quad \text{Subtract 88.1.}$$
$$3x = 108 \quad \text{Combine terms.}$$
$$\frac{3x}{3} = \frac{108}{3} \quad \text{Divide by 3.}$$
$$x = 36$$

Mercury is 36 million miles from the sun.

21. Let x = the measure of angle A
x = the measure of angle B (since both A and B have the same measure)
$x + 60$ = measure of angle C.

Angle A measure		Angle B measure		Angle C measure		Total measure
↓		↓		↓		↓
x	$+$	x	$+$	$(x + 60)$	$=$	180

Solve the equation.

$$3x + 60 = 180 \quad \text{Combine terms.}$$
$$3x + 60 - 60 = 180 - 60 \quad \text{Subtract 60.}$$
$$3x = 120 \quad \text{Combine terms.}$$
$$\frac{3x}{3} = \frac{120}{3} \quad \text{Divide by 3.}$$
$$x = 40$$

Angles A and B have measures of 40 degrees, and angle C has a measure of $40 + 60 = 100$ degrees.

25. Let x = the number of centigrams of active ingredient
$95x$ = the number of centigrams of inert ingredient.

Active ingredient		Inert ingredient		Total weight
↓		↓		↓
x	$+$	$95x$	$=$	672

Solve the equation.

$$96x = 672 \quad \text{Combine terms.}$$
$$\frac{96x}{96} = \frac{672}{96} \quad \text{Divide by 96.}$$
$$x = 7$$

There are 7 centigrams of active ingredient and $95(7) = 665$ centigrams of inert ingredient.

29. Let x = the measure of the angle
$180 - x$ = the measure of the angle's supplement
$90 - x$ = the measure of the angle's complement.

Angle's supplement		Three times		Angle's complement		Less 38°
↓		↓		↓		↓
$180 - x$	$=$	3	$\cdot$	$(90 - x)$	$-$	38

Solve the equation.

$$\begin{aligned} 180 - x &= 270 - 3x - 38 && \text{Distributive property} \\ 180 - x &= 232 - 3x && \text{Combine terms.} \\ 180 - x + 3x &= 232 - 3x + 3x && \text{Add } 3x. \\ 180 + 2x &= 232 && \text{Combine terms.} \\ 180 + 2x - 180 &= 232 - 180 && \text{Subtract 180.} \\ 2x &= 52 && \text{Combine terms.} \\ \frac{2x}{2} &= \frac{52}{2} && \text{Divide by 2.} \\ x &= 26 \end{aligned}$$

The angle measures 26 degrees.

33. Let x = the first integer
$x + 1$ = the next consecutive integer

First integer		Next consecutive integer		Sum
↓		↓		↓
x	$+$	$(x + 1)$	$=$	243

Solve the equation.

$$\begin{aligned} 2x + 1 &= 243 && \text{Combine terms.} \\ 2x + 1 - 1 &= 243 - 1 && \text{Subtract 1.} \\ 2x &= 242 && \text{Combine terms.} \\ \frac{2x}{2} &= \frac{242}{2} && \text{Divide by 2.} \\ x &= 121 \end{aligned}$$

The integers are 121 and $121 + 1 = 122$.

37. $LW = (6)(4)$ $\quad L = 6, W = 4$
$= 24$

SECTION 2.5 (page 143)

5. To measure fencing for a yard, use perimeter since you would need to measure the lengths of the sides of the yard.

9. To determine the cost for replacing a linoleum floor with a wood floor, use area since you need to know the measure of the surface covered by the wood.

13. $P = 4s$
$P = 4 \cdot 6$ $\quad$ Let $s = 6$.
$P = 24$

17. $P = a + b + c$
$15 = 3 + 7 + c$ $\quad$ Let $P = 15$, $a = 3$, $b = 7$.
$15 = 10 + c$
$5 = c$ $\quad$ Subtract 10.

21. $I = prt$
$I = (5000)(.025)(7)$ $\quad$ Let $p = 5000$, $r = .025$, $t = 7$.
$I = 875$

25. $C = 2\pi r$
$8.164 = 2(3.14)r$ $\quad$ Let $C = 8.164$, $\pi = 3.14$.
$8.164 = 6.28r$
$1.3 = r$ $\quad$ Divide by 6.28.

29. $V = LWH$
$V = (12)(8)(4)$ $\quad$ Let $L = 12$, $W = 8$, $H = 4$.
$V = 384$

33. $V = \frac{4}{3}\pi r^3$

$V = \frac{4}{3}(3.14)(6^3)$ $\quad$ Let $r = 6$, $\pi = 3.14$.

$V = \frac{4}{3}(3.14)(216)$

$V = 904.32$

37. Use the formula

$$C = 2\pi r,$$

where C is the circumference and r is the radius. Note that the radius is one-half the diameter. Therefore,

$$r = \frac{1}{2}(328) = 164.$$

Now find the circumference by substituting 164 for r in the formula.

$$\begin{aligned} C &= 2(3.14)(164) && r = 164 \\ C &= 1029.92 \end{aligned}$$

The circumference of a circular cross section of the telescope is 1029.92 feet.

41. Substitute the values given in the problem in the formula $V = LWH$ to find the volume of the rectangular set.

$V = LWH$

$V = \left(6\frac{3}{4}\right)(3)\left(1\frac{1}{8}\right)$ $\quad$ Let $L = 6\frac{3}{4}$, $W = 3$, $H = 1\frac{1}{8}$.

$V = \left(\frac{27}{4}\right)(3)\left(\frac{9}{8}\right)$

$V = \frac{729}{32}$ or $22\frac{25}{32}$

The volume of the television set is $22\frac{25}{32}$ cubic inches.

SOLUTIONS

45. In the picture, the two angles are supplementary; thus, their sum equals 180°.

$$(10x + 7) + (7x + 3) = 180$$
$$17x + 10 = 180 \quad \text{Combine terms.}$$
$$17x = 170 \quad \text{Subtract 10.}$$
$$x = 10 \quad \text{Divide by 17.}$$

To find the measures of the angles, replace x with 10 in the two expressions.

$$10x + 7 = 10(10) + 7 = 100 + 7 = 107$$
$$7x + 3 = 7(10) + 3 = 70 + 3 = 73$$

The two angle measures are 107° and 73°.

49. In the figure, the angles are vertical angles which means they have the same measure. Set $11x - 37$ equal to $7x + 27$ and solve.

$$11x - 37 = 7x + 27$$
$$4x - 37 = 27 \quad \text{Subtract } 7x.$$
$$4x = 64 \quad \text{Add 37.}$$
$$x = 16 \quad \text{Divide by 4.}$$

Now replace x with 16 in the first expression.

$$11x - 37 = 11(16) - 37 = 176 - 37 = 139°.$$

The second angle should have the same measure.

$$7x + 27 = 7(16) + 27 = 112 + 27 = 139°.$$

The angles both measure 139.°

53. $d = rt$ Solve for r.

$$\frac{d}{t} = \frac{rt}{t} \quad \text{Divide by } t.$$
$$\frac{d}{t} = r \quad \tfrac{t}{t} = 1;\ 1r = r$$

57.
$$P = a + b + c \quad \text{Solve for } a.$$
$$P - b - c = a + b + c - b - c \quad \text{Subtract } b \text{ and } c.$$
$$P - b - c = a$$

61.
$$A = p + prt \quad \text{Solve for } r.$$
$$A - p = p + prt - p \quad \text{Subtract } p.$$
$$A - p = prt \quad \text{Combine terms.}$$
$$\frac{A - p}{pt} = \frac{prt}{pt} \quad \text{Divide by } pt.$$
$$\frac{A - p}{pt} = r \quad \tfrac{pt}{pt} = 1;\ 1r = r$$

65.
$$y = mx + b \quad \text{Solve for } m.$$
$$y - b = mx + b - b \quad \text{Subtract } b.$$
$$y - b = mx \quad \text{Combine terms.}$$
$$\frac{y - b}{x} = \frac{mx}{x} \quad \text{Divide by } x.$$
$$\frac{y - b}{x} = m \quad \tfrac{x}{x} = 1;\ 1m = m$$

or $m = \dfrac{y}{x} - \dfrac{b}{x}$

69. $.005 = .5 \cdot .01 = .5 \cdot 1\% = .5\%$

SECTION 2.6 (page 153)

1. $38\% \cdot 12 = .38 \cdot 12$
$= 4.56$

5. Let $x =$ the percent in decimal form

What percent	of	30	is	36?
↓	↓	↓	↓	↓
x	$\cdot$	30	$=$	36

$$\frac{x \cdot 30}{30} = \frac{36}{30} \quad \text{Divide by 30.}$$
$$x = 1.2$$
$$x = 120\% \quad \text{Change to percent.}$$

36 is 120% of 30.

9. Let $x =$ the percent in decimal form

.392	is	what percent	of	28?
↓	↓	↓	↓	↓
.392	$=$	x	$\cdot$	28

$$\frac{.392}{28} = \frac{x \cdot 28}{28} \quad \text{Divide by 28.}$$
$$\frac{.392}{28} = x$$
$$.014 = x$$
$$1.4\% = x \quad \text{Change to percent.}$$

.392 is 1.4% of 28.

13. Let $x =$ the number of gold medals.
In the games, 25% of the medals were gold medals.

$$25\% \cdot 20 = x \quad \text{Translate into symbols.}$$
$$(.25)(20) = x \quad \text{Change \% to a decimal.}$$
$$5 = x$$

Since there were 5 gold medals out of 20 total medals, $20 - 5 = 15$ medals were not gold.

17. Let $x =$ the amount saved each month.

$$x = 12\% \cdot 3200 \quad \text{Translate into symbols.}$$
$$x = (.12)(3200) \quad \text{Change \% to a decimal.}$$
$$x = 384$$

Quinhon Dac Ho will save $384 monthly.

21. Since Pittsburgh won 65 games and lost 51 games, they played a total of $65 + 51 = 116$ games. The winning percentage is $\frac{65}{116} = .560$ (to the nearest thousandth).

29. $\dfrac{120}{90} = \dfrac{4 \cdot 30}{3 \cdot 30} = \dfrac{4}{3}$

33. Convert 2 hours to minutes.

$$2 \text{ hours} = 2 \cdot 60$$
$$= 120 \text{ minutes}$$

The ratio of 24 minutes to 2 hours is then

$$\frac{24}{120} = \frac{24}{5 \cdot 24} = \frac{1}{5}.$$

41. $\frac{x}{6} = \frac{18}{4}$

$x \cdot 4 = 6 \cdot 18$ Cross products

$4x = 108$ Multiply.

$\frac{4x}{4} = \frac{108}{4}$ Divide by 4.

$x = 27$

45. $\frac{2r + 8}{4} = \frac{3r - 9}{3}$

$3(2r + 8) = 4(3r - 9)$ Cross products

$6r + 24 = 12r - 36$ Distributive property

$6r + 24 - 12r = 12r - 36 - 12r$ Subtract $12r$.

$-6r + 24 = -36$

$-6r + 24 - 24 = -36 - 24$ Subtract 24.

$-6r = -60$

$\frac{-6r}{-6} = \frac{-60}{-6}$ Divide by -6.

$r = 10$

49. Let x = the cost of 9 gallons of gas.
Set up a proportion; compare gallons to cost in each case.

$\frac{6}{9.72} = \frac{9}{x}$

$6x = 9(9.72)$ Cross products

$6x = 87.48$

$\frac{6x}{6} = \frac{87.48}{6}$ Divide by 6.

$x = 14.58$

9 gallons of gas will cost \$14.58.

53. Let x = the number of fish in Willow Lake. Set up a proportion; compare number tagged to number of fish in each case.

$\frac{7}{350} = \frac{250}{x}$

$7x = 350 \cdot 250$ Cross products

$7x = 87{,}500$

$\frac{7x}{7} = \frac{87{,}500}{7}$ Divide by 7.

$x = 12{,}500$

There are 12,500 fish in Willow Lake.

57. Divide the price by the number of units (ounces) to get the price per unit.

Size	*Price*	*Unit Cost*
$15\frac{1}{2}$ ounces	\$1.19	$\frac{\$1.19}{15\frac{1}{2}} = \$.077$
32 ounces	\$1.69	$\frac{\$1.69}{32} = \$.053$
48 ounces	\$2.69	$\frac{\$2.69}{48} = \$.056$

The best buy is the 32-ounce size at \$1.69.

61. Since -9 is to the left of 6 on a number line, $-9 < 6$.

65. Since $-\frac{2}{3}$ is to the right of $-\frac{7}{8}$ on a number line, $-\frac{2}{3} > -\frac{7}{8}$.

SECTION 2.7 (page 163)

All of the graphs for this section can be found in the answer section.

1. On a number line, place a solid circle on 4 and shade everything to the left.

5. On a number line, place an open circle on 4 and shade everything to the right.

9. On a number line, place an open circle on 0 and a solid circle on 10 and shade everything between them.

13. See answer section.

17. $2k + 3 \geq k + 8$

$2k + 3 - k \geq k + 8 - k$ Subtract k.

$k + 3 \geq 8$ Combine terms.

$k + 3 - 3 \geq 8 - 3$ Subtract 3.

$k \geq 5$ Combine terms.

21. See answer section.

25. $3x < 18$

$\frac{3x}{3} < \frac{18}{3}$ Divide by 3.

$x < 6$

29. $-8t > 24$

$\frac{-8t}{-8} < \frac{24}{-8}$ Divide by -8; reverse symbol.

$t < -3$

33. $-\frac{3}{4}r < -15$

$\left(-\frac{4}{3}\right)\left(-\frac{3}{4}r\right) > \left(-\frac{4}{3}\right)(-15)$ Multiply by $-\frac{4}{3}$; reverse symbol.

$r > 20$

37. $5r + 1 \geq 3r - 9$

$5r + 1 - 3r \geq 3r - 9 - 3r$ Subtract $3r$.

$2r + 1 \geq -9$ Combine terms.

$2r + 1 - 1 \geq -9 - 1$ Subtract 1.

$2r \geq -10$ Combine terms.

$\frac{2r}{2} \geq \frac{-10}{2}$ Divide by 2.

$r \geq -5$

41. $-x + 4 + 7x \leq -2 + 3x + 6$

$6x + 4 \leq 4 + 3x$ Combine terms.

$6x + 4 - 3x \leq 4 + 3x - 3x$ Subtract $3x$.

$3x + 4 \leq 4$ Combine terms.

$3x + 4 - 4 \leq 4 - 4$ Subtract 4.

$3x \leq 0$ Combine terms.

$\frac{3x}{3} \leq \frac{0}{3}$ Divide by 3.

$x \leq 0$

45. $\frac{2}{3}(p + 3) > \frac{5}{6}(p - 4)$

$6 \cdot \frac{2}{3}(p + 3) > 6 \cdot \frac{5}{6}(p - 4)$	Multiply by 6.
$4(p + 3) > 5(p - 4)$	Multiply.
$4p + 12 > 5p - 20$	Distributive property
$4p + 12 - 5p > 5p - 20 - 5p$	Subtract $5p$.
$-p + 12 > -20$	Combine terms.
$-p + 12 - 12 > -20 - 12$	Subtract 12.
$-p > -32$	Combine terms.
$\frac{-p}{-1} < \frac{-32}{-1}$	Divide by -1; reverse symbol.
$p < 32$	

49. $5(2k + 3) - 2(k - 8) > 3(2k + 4) + k - 2$

$10k + 15 - 2k + 16 > 6k + 12 + k - 2$	Distributive property
$8k + 31 > 7k + 10$	Combine terms.
$8k + 31 - 7k > 7k + 10 - 7k$	Subtract $7k$.
$k + 31 > 10$	Combine terms.
$k + 31 - 31 > 10 - 31$	Subtract 31.
$k > -21$	

53. Let x = the score on the third test.

Average	is at least	80.
↓	↓	↓
$\frac{1}{3}(76 + 81 + x)$	$\geq$	80

$3 \cdot \frac{1}{3}(76 + 81 + x) \geq 3 \cdot 80$	Multiply by 3.
$76 + 81 + x \geq 240$	Multiply.
$157 + x \geq 240$	Add.
$157 + x - 157 \geq 240 - 157$	Subtract 157.
$x \geq 83$	

Twylene must receive a score of 83 or more.

57. We are given $R = 60x$ and $C = 50x + 5000$. The product will break even or produce a profit if $R \geq C$.

$60x \geq 50x + 5000$	
$60x - 50x \geq 50x + 5000 - 50x$	Subtract $50x$.
$10x \geq 5000$	Combine terms.
$\frac{10x}{10} \geq \frac{5000}{10}$	Divide by 10.
$x \geq 500$	

The company will break even or make a profit if 500 or more units of bicycle helmets are produced.

61. $2^6 = 2 \cdot 2 \cdot 2 \cdot 2 \cdot 2 \cdot 2 = 64$

65. $-3 + 8 + (-2) = 5 + (-2) = 3$

69.

$3x^2 - 4x + 7 = 3(4)^2 - 4(4) + 7$	Let $x = 4$.
$= 3(16) - 4(4) + 7$	Use the exponent.
$= 48 - 16 + 7$	Multiply.
$= 32 + 7$	Subract.
$= 39$	Add.

CHAPTER 3

SECTION 3.1 (page 187)

1. Since 3 occurs as a factor seven times, the base is 3 and the exponent is 7. The exponential expression is 3^7.

5. Since w occurs as a factor six times, the base is w and the exponent is 6. The exponential expression is w^6.

9. $-7x$ occurs as a factor four times. $-7x$ is the base and the exponent is 4. The exponential expression is $(-7x)^4$.

17. The base is -3 and the exponent is 5.

$$(-3)^5 = (-3)(-3)(-3)(-3)(-3) = -243$$

21. In $-6x^4$, -6 is not part of the base. The base is x and the exponent is 4.

25. $5^2 \cdot 5^6 = 5^{2+6} = 5^8$

29. $(-7)^3(-7)^6 = (-7)^{3+6} = (-7)^9$

33.

$(-8r^4)(7r^3) = -8 \cdot 7 \cdot r^4 \cdot r^3$	Commutative and associative properties
$= -56r^{4+3}$	Product rule
$= -56r^7$	

41.

$-7a^2 + 2a^2 + 10a^2 = (-7 + 2 + 10)a^2$	Distributive property
$= 5a^2$	Add.
$(-7a^2)(2a^2)(10a^2) = (-7 \cdot 2 \cdot 10) \cdot a^2 \cdot a^2 \cdot a^2$	Commutative and associative properties
$= -140a^2a^2a^2$	Multiply.
$= -140a^{2+2+2}$	Product rule
$= -140a^6$	

45. $(t^4)^5 = t^{4 \cdot 5} = t^{20}$ Power rule (a)

49. $(5xy)^5 = 5^5x^5y^5$ Power rule (b)

53.

$(-8^3)^5 = (-1 \cdot 8^3)^5$	
$= (-1)^5(8^3)^5$	Power rule (b)
$= -1 \cdot 8^{3 \cdot 5}$	Power rule (a)
$= -8^{15}$	

57. $\left(\frac{1}{2}\right)^3 = \frac{1^3}{2^3} = \frac{1}{2^3}$ Power rule (c)

61. $\left(\frac{9}{5}\right)^8 = \frac{9^8}{5^8}$ Power rule (c)

65.

$\left(\frac{9}{8}\right)^3 \cdot 9^2 = \frac{9^3}{8^3} \cdot \frac{9^2}{1}$	Power rule (c)
$= \frac{9^3 \cdot 9^2}{8^3}$	Multiply fractions.
$= \frac{9^{3+2}}{8^3}$	Product rule
$= \frac{9^5}{8^3}$	

69.

$(-6p)^4(-6p) = (-6p)^5$	Product rule
$= (-6)^5p^5$	Power rule (b)

73.

$(x^2)^3(x^3)^5 = x^6 \cdot x^{15}$	Power rule (a)
$= x^{21}$	Product rule

77. $(-r^4s)^2(-r^2s^3)^5 = [(-1)r^4s]^2[(-1)r^2s^3]^5$
$= [(-1)^2(r^4)^2s^2][(-1)^5(r^2)^5(s^3)^5]$ Power rule (b)
$= [(-1)^2r^8s^2][(-1)^5r^{10}s^{15}]$ Power rule (a)
$= (-1)^7r^{18}s^{17}$ Product rule
$= -r^{18}s^{17}$ Use exponent.

85. $9 - 12 = 9 + (-12) = -3$

SECTION 3.2 (page 195)

1. By definition, $a^0 = 1$ $(a \neq 0)$, so $9^0 = 1$.
5. $-9^0 = (-9^0) = -1$
9. $\dfrac{0^{10}}{10^0} = \dfrac{0}{1} = 0$
13. $7^0 + 9^0 = 1 + 1 = 2$
17. $\left(\dfrac{1}{2}\right)^{-4} = 2^4 = 16$ $\quad\frac{1}{2}$ and 2 are reciprocals.
21. $(-3)^{-4} = \left(-\dfrac{1}{3}\right)^4 = \dfrac{1}{81}$ $\quad -3$ and $-\frac{1}{3}$ are reciprocals.
25. $(-2)^{-3}$ is negative. A negative base to an odd power will have a negative result.
29. $(\frac{1}{4})^{-2}$ is positive. A nonzero base to an even power will have a positive result.
33. $\dfrac{5^8}{5^5} = 5^{8-5} = 5^3$ Quotient rule
37. $\dfrac{6^{-3}}{6^2} = 6^{-3-2}$ Quotient rule
$= 6^{-5}$
$= \dfrac{1}{6^5}$ Negative to positive exponent rule
41. $\dfrac{1}{6^{-3}} = 6^3$ Negative to positive exponent rule
45. $\dfrac{4^{-3}}{5^{-2}} = \dfrac{5^2}{4^3}$ Negative to positive exponent rule
49. $\dfrac{r^5}{r^{-4}} = r^{5-(-4)} = r^9$ Quotient rule
53. $\dfrac{(7^4)^3}{7^9} = \dfrac{7^{12}}{7^9}$ Power rule (a)
$= 7^{12-9}$ Quotient rule
$= 7^3$
57. $\dfrac{(3x)^{-2}}{(4x)^{-3}} = \dfrac{(4x)^3}{(3x)^2}$ Negative to positive exponent rule
$= \dfrac{4^3x^3}{3^2x^2}$ Power rule (b)
$= \dfrac{4^3x^{3-2}}{3^2}$ Quotient rule
$= \dfrac{4^3x}{3^2}$
61. $(6x)^4(6x)^{-3} = (6x)^{4+(-3)}$ Product rule
$= (6x)^1$ or $6x$
65. $10{,}000(.5687) = 5687$

SECTION 3.3 (page 199)

1. 4.56×10^3 is in scientific notation since $1 \leq |4.56| < 10$ and the exponent is an integer.
5. $.8 \times 10^2$ is not in scientific notation since $|.8|$ is not greater than or equal to 1 and less than 10.

$.8 \times 10^2 = (8. \times 10^{-1}) \times 10^2$ Move 1 place to the right
$= 8 \times (10^{-1} \times 10^2)$ Associative property
$= 8 \times 10^1$ Product rule

13. Move the decimal point 4 places so it is to the right of the first nonzero digit.

8.2350 4 places

Since the number got smaller, multiply by a positive power of 10.

$$82{,}350 = 8.2350 \times 10^4 \quad \text{or} \quad 8.235 \times 10^4$$

(Note that the final zero need not be written.)
17. Move the decimal point 3 places so it is to the right of the first nonzero digit.

002.03 3 places

Since the number got larger, the exponent must be negative.

$$-.00203 = -2.03 \times 10^{-3}$$

21. Since the exponent is positive, make 5.677 larger by moving the decimal point 12 places to the right. We need to add 9 zeros.

$$5.677 \times 10^{12} = 5{,}677{,}000{,}000{,}000$$

25. Since the exponent is negative, move the decimal point 4 places to the left.

$$7.8 \times 10^{-4} = .00078$$

29. $(2 \times 10^8) \times (3 \times 10^3) = (2 \times 3)(10^8 \times 10^3)$ Commutative and associative properties
$= 6 \times 10^{11}$ Product rule
$= 600{,}000{,}000{,}000$ Write without exponents.
33. $(3 \times 10^{-4}) \times (2 \times 10^8) = (3 \times 2)(10^{-4} \times 10^8)$
$= 6 \times 10^4 = 60{,}000$
37. $\dfrac{8 \times 10^3}{2 \times 10^2} = \dfrac{8}{2} \times \dfrac{10^3}{10^2} = 4 \times 10^1$
41. $6.35 \times 10^{11} = 635{,}000{,}000{,}000$ Move 11 places to the right.
45. $2.3 \times 10^{-4} = .00023$ Move 4 places to the left.
$6 \times 10^{-3} = .006$ Move 3 places to the left.
49. $6(2x - 3) + 5 = 6(2x) - 6(3) + 5$ Distributive property
$= 12x - 18 + 5$ Multiply.
$= 12x - 13$ Add.

SECTION 3.4 (page 207)

1. $6x^4$ has one term with a coefficient of 6.

5. $-19r^2 - r$ has two terms. The coefficients are -19 and -1.

9. $-3m^5 + 5m^5 = (-3 + 5)m^5 = 2m^5$

13. The polynomial $.2m^5 - .5m^2$ cannot be simplified. The two terms are unlike, since the exponents on the variables are different.

17. $-4p^7 + 8p^7 + 5p^9 = (-4 + 8)p^7 + 5p^9$
$= 4p^7 + 5p^9$ or $5p^9 + 4p^7$

21. Polynomials can be binomials or they can be monomials, trinomials, or have more than 3 terms, so the statement is sometimes true.

25. A trinomial has three terms. A binomial has two terms. The statement is never true.

29. The polynomial is already simplified. In descending powers of the variable, it is written $6m^5 + 5m^4 - 7m^3 - 3m^2$. The degree is 5. It is none of these (4 terms).

33. $.8x^4 - .3x^4 - .5x^4 + 7 = (.8 - .3 - .5)x^4 + 7$
$= 0x^4 + 7 = 7$

The degree is 0, it is a monomial (one term).

37.
$$\begin{array}{rr} & 12x^4 - x^2 \\ - & 8x^4 + 3x^2 \\ \hline \end{array} \rightarrow \begin{array}{rr} & 12x^4 - x^2 \\ + & -8x^4 - 3x^2 \\ \hline & 4x^4 - 4x^2 \end{array}$$

41.
$$\begin{array}{rr} - & 12m^3 - 8m^2 + 6m + 7 \\ & -3m^3 + 5m^2 - 2m - 4 \\ \hline \end{array} \rightarrow$$

$$\begin{array}{rr} & 12m^3 - 8m^2 + 6m + 7 \\ + & 3m^2 - 5m^2 + 2m + 4 \\ \hline & 15m^3 - 13m^2 + 8m + 11 \end{array}$$

45. $(2r^2 + 3r - 12) + (6r^2 + 2r)$
$= 2r^2 + 3r - 12 + 6r^2 + 2r$
$= 2r^2 + 6r^2 + 3r + 2r - 12$
$= 8r^2 + 5r - 12$

49. $(16x^3 - x^2 + 3x) + (-12x^3 + 3x^2 + 2x)$
$= 16x^3 - x^2 + 3x - 12x^3 + 3x^2 + 2x$
$= 16x^3 - 12x^3 - x^2 + 3x^2 + 3x + 2x$
$= 4x^3 + 2x^2 + 5x$

53. $[(8m^2 + 4m - 7) - (2m^3 - 5m + 2)]$
$- (m^2 + m + 1)$
$= [(8m^2 + 4m - 7) + (-2m^3 + 5m - 2)]$
$+ (-m^2 - m - 1)$
$= 8m^2 + 4m - 7 - 2m^3 + 5m - 2 - m^2 - m - 1$
$= - 2m^3 + 8m^2 - m^2 + 4m + 5m - m - 7$
$- 2 - 1$
$= -2m^3 + 7m^2 + 8m - 10$

57. To find the perimeter, add the lengths of the sides.

$3t^2 + 2t + 7 + 5t^2 + 2 + 6t + 4$
$= 3t^2 + 5t^2 + 2t + 6t + 7 + 2 + 4$
$= 8t^2 + 8t + 13$

65. $(2c^4d + 3c^2d^2 - 4d^2) - (c^4d + 8c^2d^2 - 5d^2)$
$= 2c^4d + 3c^2d^2 - 4d^2 - c^4d - 8c^2d^2 + 5d^2$
$= c^4d - 5c^2d^2 + d^2$

The terms have degree 5, 4, and 2, so the degree of the polynomial is 5, the highest of these.

69. $-8n^3(3n^4) = (-8 \cdot 3)(n^3 \cdot n^4)$
$= -24n^7$

SECTION 3.5 (page 215)

1. Associative property (grouping changed)
Commutative property (order changed)
Associative property (grouping changed)

5. $-2m(3m + 2) = (-2m)(3m) + (-2m)(2)$ Distributive property
$= -6m^2 - 4m$ Multiply monomials.

9. $2y^5(3 + 2y + 5y^4) = 2y^5(3) + 2y^5(2y) + 2y^5(5y^4)$ Distributive property
$= 6y^5 + 4y^6 + 10y^9$ Multiply monomials.

13. $(4r + 1)(2r - 3) = (4r)(2r) + (1)(2r) + (4r)(-3) + (1)(-3)$
$= 8r^2 + 2r + (-12r) + (-3)$
$= 8r^2 - 10r - 3$

17.
$$\begin{array}{rl} 3q + 1 & \\ 3q + 1 & \\ \hline 3q + 1 & \leftarrow 1(3q + 1) \\ 9q^2 + 3q & \leftarrow 3q(3q + 1) \\ \hline 9q^2 + 6q + 1 & \end{array}$$

21.
$$\begin{array}{rl} -3t + 4 & \\ t + 6 & \\ \hline -18t + 24 & \leftarrow 6(-3t + 4) \\ -3t^2 + 4t & \leftarrow t(-3t + 4) \\ \hline -3t^2 - 14t + 24 & \end{array}$$

29.
$$\begin{array}{r} 5m^3 - 4m^2 + m - 5 \\ 4m + 3 \\ \hline 15m^3 - 12m^2 + 3m - 15 \\ 20m^4 - 16m^3 + 4m^2 - 20m \\ \hline 20m^4 - m^3 - 8m^2 - 17m - 15 \end{array}$$

33. $(x + 12)^2 = (x + 12)(x + 12)$
$= (x + 12)(x) + (x + 12)(12)$
$= x(x) + 12(x) + x(12) + 12(12)$
$= x^2 + 12x + 12x + 144$
$= x^2 + 24x + 144$

37.
$$\begin{array}{r} 5x - 2y \\ 5x - 2y \\ \hline -10xy + 4y^2 \\ 25x^2 - 10xy \\ \hline 25x^2 - 20xy + 4y^2 \end{array}$$

41. Since $(r + 1)^4 = (r + 1)(r + 1)(r + 1)(r + 1)$, first find $(r + 1)(r + 1)$.

$$(r + 1)(r + 1) = r^2 + r + r + 1 = r^2 + 2r + 1$$

To multiply this result by $(r + 1)(r + 1)$, multiply $(r^2 + 2r + 1)(r^2 + 2r + 1)$.

$$\begin{array}{r} r^2 + 2r + 1 \\ r^2 + 2r + 1 \\ \hline r^2 + 2r + 1 \\ 2r^3 + 4r^2 + 2r \\ r^4 + 2r^3 + r^2 \\ \hline r^4 + 4r^3 + 6r^2 + 4r + 1 \end{array}$$

45. The two numbers are 2 and 4, since $2 \cdot 4 = 8$ and $2 + 4 = 6$.

SECTION 3.6 (page 221)

1. **(a)** $2x(x) = 2x^2$
(b) $2x(-5) = -10x$
(c) $3(x) = 3x$
(d) $3(-5) = -15$
(e) $-10x + 3x = -7x$
(f) $2x^2 - 7x - 15$

5. $(w - 4)(w - 6)$

$$\begin{aligned} &\quad\ \ \text{F} \qquad\quad \text{O} \qquad\qquad \text{I} \qquad\qquad \text{L} \\ &= (w)(w) + (w)(-6) + (-4)(w) + (-4)(-6) \\ &= w^2 - 6w - 4w + 24 \\ &= w^2 - 10w + 24 \end{aligned}$$

9. $(2x - 1)(5x + 3)$

$$\begin{aligned} &\quad\ \ \text{F} \qquad\quad \text{O} \qquad\qquad \text{I} \qquad\qquad \text{L} \\ &= (2x)(5x) + (2x)(3) + (-1)(5x) + (-1)(3) \\ &= 10x^2 + 6x - 5x - 3 \\ &= 10x^2 + x - 3 \end{aligned}$$

13. $(3m + 7)(3m + 5) = 3m(3m) + 3m(5) + 7(3m) + 7(5)$
$= 9m^2 + 15m + 21m + 35$
$= 9m^2 + 36m + 35$

17. $(-5 + 6z)(3 - z) = -5(3) - 5(-z) + 6z(3) + 6z(-z)$
$= -15 + 5z + 18z - 6z^2$
$= -15 + 23z - 6z^2$

21. $(3w + 2z)(9w - z) = 3w(9w) + 3w(-z) + 2z(9w) + 2z(-z)$
$= 27w^2 - 3wz + 18wz - 2z^2$
$= 27w^2 + 15wz - 2z^2$

25. $(x - .3)(x + .8) = x(x) + x(.8) - .3(x) - .3(.8)$
$= x^2 + .8x - .3x - .24$
$= x^2 + .5x - .24$

29. **(a)** $(2x)^2 = 2^2x^2 = 4x^2$
(b) $2(2x)(3) = 2 \cdot 2 \cdot 3x = 12x$
(c) $3^2 = 9$
(d) $4x^2 + 12x + 9$

33. Use the rule for the square of a binomial.

$$\begin{aligned} (a - c)^2 &= a^2 - 2 \cdot a \cdot c + c^2 \\ &= a^2 - 2ac + c^2 \end{aligned}$$

37. $(8t + 7s)^2 = (8t)^2 + 2(8t)(7s) + (7s)^2$
$= 64t^2 + 112ts + 49s^2$

41. **(a)** $(7x)(7x) = 49x^2$
(b) $(7x)(-3y) + (3y)(7x) = -21xy + 21xy = 0$
(c) $(3y)(-3y) = -9y^2$
(d) $49x^2 - 9y^2$
The sum found in (b) is omitted because it is 0. Adding 0, the identity element for addition, would not change the answer.

45. $(2w + 5)(2w - 5) = (2w)^2 - 5^2 = 4w^2 - 25$

49. $(2x^2 - 5)(2x^2 + 5) = (2x^2)^2 - 5^2 = 4x^4 - 25$

53. $\dfrac{6x^5}{2x} = \dfrac{6}{2}x^{5-1} = 3x^4$

SECTION 3.7 (page 225)

1. $\dfrac{12x^5}{-2x} = \dfrac{12}{-2}x^{5-1} = -6x^4$

9. $$\begin{aligned} \frac{10m^5 - 16m^4 + 8m^3}{2m} &= \frac{10m^5}{2m} - \frac{16m^4}{2m} + \frac{8m^3}{2m} \\ &= 5m^4 - 8m^3 + 4m^2 \end{aligned}$$

13. $$\begin{aligned} \frac{m^5 - 4m^2 + 8}{2m} &= \frac{m^5}{2m} - \frac{4m^2}{2m} + \frac{8}{2m} \\ &= \frac{m^4}{2} - 2m + \frac{4}{m} \end{aligned}$$

17. $$\begin{aligned} \frac{3x^2 + 15x^3 - 27x^4}{3x^2} &= \frac{3x^2}{3x^2} + \frac{15x^3}{3x^2} - \frac{27x^4}{3x^2} \\ &= 1 + 5x - 9x^2 \end{aligned}$$

21. $$\begin{aligned} \frac{4x^4 + 3x^3 + 2x}{3x^2} &= \frac{4x^4}{3x^2} + \frac{3x^3}{3x^2} + \frac{2x}{3x^2} \\ &= \frac{4x^2}{3} + x + \frac{2}{3x} \end{aligned}$$

25. $$\begin{aligned} \frac{2m^5 - 6m^4 + 8m^2}{-2m^3} &= \frac{2m^5}{-2m^3} - \frac{6m^4}{-2m^3} + \frac{8m^2}{-2m^3} \\ &= -m^2 + 3m - \frac{4}{m} \end{aligned}$$

29. $$\begin{aligned} &\frac{120x^{11} - 60x^{10} + 140x^9 - 100x^8}{10x^{12}} \\ &= \frac{120x^{11}}{10x^{12}} - \frac{60x^{10}}{10x^{12}} + \frac{140x^9}{10x^{12}} - \frac{100x^8}{10x^{12}} \\ &= \frac{12}{x} - \frac{6}{x^2} + \frac{14}{x^3} - \frac{10}{x^4} \end{aligned}$$

33. $\dfrac{(\quad)}{5x^3} = 3x^2 - 7x + 7$

Multiply the quotient and the divisor to find the missing polynomial.

$$\begin{aligned} (5x^3)(3x^2 - 7x + 7) &= 5x^3(3x^2) + 5x^3(-7x) + 5x^3(7) \\ &= 15x^5 - 35x^4 + 35x^3 \end{aligned}$$

37.
$$\begin{array}{r} 8k^2 + 9k + 3 \\ -2k + 1 \\ \hline 8k^2 + 9k + 3 \\ -16k^3 - 18k^2 - 6k \quad\;\; \\ \hline -16k^3 - 10k^2 + 3k + 3 \end{array}$$

SECTION 3.8 (page 231)

1.
$$\begin{array}{r} x + 2 \\ x - 3\overline{)x^2 - \ \ x - 6} \\ \underline{x^2 - 3x} \\ 2x - 6 \\ \underline{2x - 6} \\ 0 \end{array}$$

The remainder is 0. The quotient is $x + 2$.

5.
$$\begin{array}{r} p - 4 \\ p + 6\overline{)p^2 + 2p + 20} \\ \underline{p^2 + 6p} \\ -4p + 20 \\ \underline{-4p - 24} \\ 44 \end{array}$$

The remainder is 44. The quotient is $p - 4 + \frac{44}{p + 6}$.

9.
$$\begin{array}{r} 6m - 1 \\ 2m - 3\overline{)12m^2 - 20m + 3} \\ \underline{12m^2 - 18m} \\ -2m + 3 \\ \underline{-2m + 3} \\ 0 \end{array}$$

The remainder is 0. The quotient is $6m - 1$.

13.
$$\begin{array}{r} 4x^2 - 7x + 3 \\ 2x + 1\overline{)8x^3 - 10x^2 - x + 3} \\ \underline{8x^3 + 4x^2} \\ -14x^2 - x \\ \underline{-14x^2 - 7x} \\ 6x + 3 \\ \underline{6x + 3} \\ 0 \end{array}$$

The remainder is 0. The quotient is $4x^2 - 7x + 3$.

17.
$$\begin{array}{r} 3y^2 - 2y + 2 \\ y + 1\overline{)3y^3 + y^2 + 0y + 2} \\ \underline{3y^3 + 3y^2} \\ -2y^2 + 0y \\ \underline{-2y^2 - 2y} \\ 2y + 2 \\ \underline{2y + 2} \\ 0 \end{array}$$

The remainder is 0. The quotient is $3y^2 - 2y + 2$.

21.
$$\begin{array}{r} x^2 \qquad + 1 \\ x^2 + 0x - 2\overline{)x^4 + 0x^3 - x^2 + 0x - 2} \\ \underline{x^4 + 0x^3 - 2x^2} \\ x^2 + 0x - 2 \\ \underline{x^2 + 0x - 2} \\ 0 \end{array}$$

The remainder is 0. The quotient is $x^2 + 1$.

25.
$$\begin{array}{r} x^3 + 3x^2 - x + 5 \\ 2x^2 + 3x + 1\overline{)2x^5 + 9x^4 + 8x^3 + 10x^2 + 14x + 5} \\ \underline{2x^5 + 3x^4 + x^3} \\ 6x^4 + 7x^3 + 10x^2 \\ \underline{6x^4 + 9x^3 + 3x^2} \\ -2x^3 + 7x^2 + 14x \\ \underline{-2x^3 - 3x^2 - x} \\ 10x^2 + 15x + 5 \\ \underline{10x^2 + 15x + 5} \\ 0 \end{array}$$

The remainder is 0. The quotient is $x^3 + 3x^2 - x + 5$.

29. Positive integer factors of 18 are 1, 2, 3, 6, 9, and 18. Each divides 18 with no remainder.

CHAPTER 4

SECTION 4.1 (page 249)

1. $12 = 2^2 \cdot 3 \qquad 16 = 2^4$
Take each factor the least number of times it appears in both factored forms. The greatest common factor is $2^2 = 4$.

5. $18 = 2 \cdot 3^2 \qquad 24 = 2^3 \cdot 3 \qquad 36 = 2^2 \cdot 3^2$
$48 = 2^4 \cdot 3$
Take each factor the least number of times it appears in all the factored forms. The greatest common factor is $2 \cdot 3 = 6$.

13. $30x^3 = 2 \cdot 3 \cdot 5 \cdot x^3 \qquad 40x^6 = 2^3 \cdot 5 \cdot x^6$
$50x^7 = 2 \cdot 5^2 \cdot x^7$
Take each factor the least number of times it appears in all the factored forms. The greatest common factor is $2 \cdot 5 \cdot x^3 = 10x^3$.

17. $-x^4y^3 = -1 \cdot x^4 \cdot y^3 \qquad -xy^2 = -1 \cdot x \cdot y^2$
Take each factor the least number of times it appears in both factored forms. The greatest common factor is $x \cdot y^2 = xy^2$. (The factors of -1 are -1 and 1. Since $1 > -1$, the greatest common factor is $1xy^2$ or xy^2.)

21. $-xy$ is a common factor of $-x^4y^3$ and $-xy^2$. When $-xy$ is multiplied by x^3y^2, the result is $-x^4y^3$.

25. Since $3x^2 = 3x \cdot x$, place x in the blank.

29. Since $-8z^9 = -4z^5 \cdot 2z^4$, place $2z^4$ in the blank.

33. Since $-14x^4y^3 = 2xy(-7x^3y^2)$, place $-7x^3y^2$ in the blank.

37. The greatest common factor for $10a^2$ and $20a$ is $10a$.

$$10a^2 - 20a = 10a \cdot a - 10a \cdot 2 = 10a(a - 2)$$

41. $11w^3 - 100$ has no common factor except 1.

45. The greatest common factor for $13y^8 + 26y^4 - 39y^2$ is $13y^2$.

$$\begin{aligned} 13y^8 + 26y^4 - 39y^2 &= 13y^2 \cdot y^6 + 13y^2 \cdot 2y^2 \\ &\quad -13y^2 \cdot 3 \\ &= 13y^2(y^6 + 2y^2 - 3) \end{aligned}$$

49. The greatest common factor is a^3.

$$\begin{aligned} a^5 + 2a^3b^2 - 3a^5b^2 + 4a^4b^3 &= a^3 \cdot a^2 + a^3 \cdot 2b^2 \\ &\quad - a^3 \cdot 3a^2b^2 + a^3 \cdot 4ab^3 \\ &= a^3(a^2 + 2b^2 \\ &\quad - 3a^2b^2 + 4ab^3) \end{aligned}$$

53.
$$\begin{aligned} a^2 - 2a + 5a - 10 &= (a^2 - 2a) + (5a - 10) && \text{Group the terms.} \\ &= a(a - 2) + 5(a - 2) && \text{Factor each group.} \\ &= (a - 2)(a + 5) && \text{Factor out } a - 2. \end{aligned}$$

57.
$$\begin{aligned} &18r^2 + 12ry - 3xr - 2xy \\ &= (18r^2 + 12ry) - (3xr + 2xy) && \text{Group the terms.} \\ &= 6r(3r + 2y) - x(3r + 2y) && \text{Factor each group.} \\ &= (3r + 2y)(6r - x) && \text{Factor out } 3r + 2y. \end{aligned}$$

61. $1 - a + ab - b = (1 - a) + (ab - b)$ Group the terms.

$= 1(1 - a) - b(-a + 1)$ Factor each group.

$= 1(1 - a) - b(1 - a)$

Must be same.

$= (1 - a)(1 - b)$ Factor out $1 - a$.

69. F O I L

$(x + 2)(x + 7) = x \cdot x + x \cdot 7 + 2 \cdot x + 2 \cdot 7$

$= x^2 + 7x + 2x + 14$

$= x^2 + 9x + 14$

SECTION 4.2 (page 255)

1.

	Factors of 12	Sum of factors	
	12, 1	$12 + 1 = 13$	
	−12, −1	$-12 + (-1) = -13$	
	6, 2	$6 + 2 = 8$	
	−6, −2	$-6 + (-2) = -8$	
→	4, 3	$4 + 3 = 7$	Sum is 7.
	−4, −3	$-4 + (-3) = -7$	

5.

	Factors of 27	Sum of factors	
→	27, 1	$27 + 1 = 28$	Sum is 28.
	−27, −1	$-27 + (-1) = -28$	
	9, 3	$9 + 3 = 12$	
	−9, −3	$-9 + (-3) = -12$	

9. $x^2 - 12x + 32$

	Factors of 32	Sum of factors	
	1, 32	$1 + 32 = 33$	
	−1, −32	$-1 + (-32) = -33$	
	2, 16	$2 + 16 = 18$	
	−2, −16	$-2 + (-16) = -18$	
	4, 8	$4 + 8 = 12$	
→	−4, −8	$-4 + (-8) = -12$	Sum is −12.

Choice (c) is the correct factored form.

13. $x^2 + 15x + 44 = (x + 4)(\quad)$

To complete the factoring, use 11 since $4 \cdot 11 = 44$ and $4 + 11 = 15$, the coefficient of the middle term.

$$x^2 + 15x + 44 = (x + 4)(x + 11)$$

17. $y^2 - 2y - 15 = (y + 3)(\quad)$

To complete the factoring, use −5 since $3(-5) = -15$ and $3 + (-5) = -2$, the coefficient of the middle term.

$$y^2 - 2y - 15 = (y + 3)(y - 5)$$

21. $y^2 - 7y - 18 = (y + 2)(\quad)$

To complete the factoring, use −9 since $2(-9) = -18$ and $2 + (-9) = -7$, the coefficient of the middle term.

$$y^2 - 7y - 18 = (y + 2)(y - 9)$$

25. $b^2 + 8b + 15$

Only positive integers need be considered since all signs in the trinomial are positive.

	Factors of 15	Sum of factors
	1, 15	$1 + 15 = 16$
→	3, 5	$3 + 5 = 8$

From the list, 3 and 5 are the required integers, so

$$b^2 + 8b + 15 = (b + 3)(b + 5).$$

29. $y^2 - 8y + 15$

	Factors of 15	Sum of factors
	1, 15	$1 + 15 = 16$
	−1, −15	$-1 + (-15) = -16$
	3, 5	$3 + 5 = 8$
→	−3, −5	$-3 + (-5) = -8$

The required integers are −3 and −5, so

$$y^2 - 8y + 15 = (y - 5)(y - 3).$$

33. $t^2 - 8t + 16$

	Factors of 16	Sum of factors
	1, 16	$1 + 16 = 17$
	−1, −16	$-1 + (-16) = -17$
	2, 8	$2 + 8 = 10$
	−2, −8	$-2 + (-8) = -10$
	4, 4	$4 + 4 = 8$
→	−4, −4	$-4 + (-4) = -8$

$$t^2 - 8t + 16 = (t - 4)(t - 4) = (t - 4)^2$$

37. $t^2 - tz - 6z^2$

	Factors of $-6z^2$	Sum of factors
	$-2z, 3z$	$-2z + 3z = z$
→	$2z, -3z$	$2z + (-3z) = -z$
	$6z, -z$	$6z + (-z) = 5z$
	$-6z, z$	$-6z + z = -5z$

$$t^2 - tz - 6z^2 = (t + 2z)(t - 3z)$$

41. $v^2 - 11vw + 30w^2$

Factors of 30w²	*Sum of factors*
$30w, w$	$30w + w = 31w$
$-30w, -w$	$-30w + (-w) = -31w$
$15w, 2w$	$15w + 2w = 17w$
$-15w, -2w$	$-15w + (-2w) = -17w$
$10w, 3w$	$10w + 3w = 13w$
$-10w, -3w$	$-10w + (-3w) = -13w$
$5w, 6w$	$5w + 6w = 11w$
→ $-5w, -6w$	$-5w + (-6w) = -11w$

$v^2 - 11vw + 30w^2 = (v - 5w)(v - 6w)$

45. $2t^3 + 8t^2 + 6t$

$= 2t \cdot t^2 + 2t \cdot 4t + 2t \cdot 3$ $2t$ is the greatest common factor.

$= 2t(t^2 + 4t + 3)$ Distributive property

$= 2t(t + 1)(t + 3)$ $1 \cdot 3 = 3$; $1 + 3 = 4$

49. $m^3n - 10m^2n^2 + 24mn^3$

$= mn \cdot m^2 - mn \cdot 10mn + mn \cdot 24n^2$

mn is the greatest common factor.

$= mn(m^2 - 10mn + 24n^2)$ Distributive property

$= mn(m - 6n)(m - 4n)$

$-6(-4) = 24$; $-6 + (-4) = -10$

53. The product of $(a + 9)(a + 4)$ is $a^2 + 13a + 36$. Thus, the polynomial $a^2 + 13a + 36$ can be factored to give $(a + 9)(a + 4)$.

57. $(5z + 2)(3z - 2) = 5z(3z) + 5z(-2) + 2(3z) + 2(-2)$

$= 15z^2 - 10z + 6z - 4$

$= 15z^2 - 4z - 4$

SECTION 4.3 (page 261)

1. Find the product of each choice.

$$(2x - 1)(x + 1) = 2x^2 + x - 1$$

$$(2x + 1)(x - 1) = 2x^2 - x - 1$$

Choice (b) is the factored form.

5. Find the product of each choice.

$$(4k + m)(k + 3m) = 4k^2 + 13km + 3m^2$$

$$(4k + 3m)(k + m) = 4k^2 + 7km + 3m^2$$

Choice (a) is the factored form.

9. $2x^2 + 6x - 8 = 2 \cdot x^2 + 2 \cdot 3x - 2 \cdot 4$

$= 2(x^2 + 3x - 4)$ Factor out 2.

$= 2(x + 4)(x - 1)$ $4(-1) = -4$; $4 + (-1) = 3$

17. $3a^2 + 10a + 7 = 3a^2 + 3a + 7a + 7$ $10a = 3a + 7a$

$= (3a^2 + 3a) + (7a + 7)$ Group the terms.

$= 3a(a + 1) + 7(a + 1)$ Factor each group.

$= (a + 1)(3a + 7)$ Factor out $a + 1$.

21. $15m^2 + m - 2 = 15m^2 + 6m - 5m - 2$

$= m = 6m - 5m$

$= (15m^2 + 6m) + (-5m - 2)$

Group the terms.

$= 3m(5m + 2) - 1(5m + 2)$

Factor each group.

$= (5m + 2)(3m - 1)$

Factor out $5m + 2$.

25. $20x^2 + 11x - 3 = 20x^2 + 15x - 4x - 3$

$11x = 15x - 4x$

$= (20x^2 + 15x) + (-4x - 3)$

Group the terms.

$= 5x(4x + 3) - 1(4x + 3)$

Factor each group.

$= (4x + 3)(5x - 1)$

Factor out $4x + 3$.

29. $20y^2 + 39y - 11 = 20y^2 + 44y - 5y - 11$

$39y = 44y - 5y$

$= (20y^2 + 44y) + (-5y - 11)$

Group the terms.

$= 4y(5y + 11) - 1(5y + 11)$

Factor each group.

$= (5y + 11)(4y - 1)$

Factor out $5y + 11$.

33. $24x^2 - 42x + 9 = 3 \cdot 8x^2 - 3 \cdot 14x + 3 \cdot 3$

$= 3(8x^2 - 14x + 3)$ Factor out 3.

$= 3(8x^2 - 12x - 2x + 3)$

$-14x = -12x - 2x$

$= 3[(8x^2 - 12x) + (-2x + 3)]$

Group the terms.

$= 3[4x(2x - 3) - 1(2x - 3)]$

Factor each group.

$= 3(2x - 3)(4x - 1)$

Factor out $2x - 3$.

37. $2m^3 + 2m^2 - 40m = 2m \cdot m^2 + 2m \cdot m - 2m \cdot 20$

$= 2m(m^2 + m - 20)$

Factor out $2m$.

$= 2m(m + 5)(m - 4)$

$5(-4) = -20$; $5 + (-4) = 1$

41. $18x^5 + 15x^4 - 75x^3$

$= 3x^3 \cdot 6x^2 + 3x^3 \cdot 5x - 3x^3 \cdot 25$

$= 3x^3(6x^2 + 5x - 25)$

Factor out $3x^3$.

$= 3x^3(6x^2 + 15x - 10x - 25)$

$5x = 15x - 10x$

$= 3x^3[(6x^2 + 15x) + (-10x - 25)]$

Group the terms.

$= 3x^3[3x(2x + 5) - 5(2x + 5)]$

Factor each group.

$= 3x^3(2x + 5)(3x - 5)$

Factor out $2x + 5$.

45. $12p^2 + 7pq - 12q^2 = 12p^2 + 16pq - 9pq - 12q^2$

$7pq = 16pq - 9pq$

$= (12p^2 + 16pq) + (-9pq - 12q^2)$

Group the terms.

$= 4p(3p + 4q) - 3q(3p + 4q)$

Factor each group.

$= (3p + 4q)(4p - 3q)$

Factor out $3p + 4q$.

49. $6a^2 - 7ab - 5b^2 = 6a^2 - 10ab + 3ab - 5b^2$ $\quad -7ab = -10ab + 3ab$
$= (6a^2 - 10ab) + (3ab - 5b^2)$ Group the terms.
$= 2a(3a - 5b) + b(3a - 5b)$ Factor each group.
$= (3a - 5b)(2a + b)$ Factor out $3a - 5b$.

53. $5 - 6x + x^2 = 5 - 5x - x + x^2$ $\quad -6x = -5x - x$
$= (5 - 5x) + (-x + x^2)$ Group the terms.
$= 5(1 - x) - x(1 - x)$ Factor each group.
$= (1 - x)(5 - x)$ Factor out $1 - x$.

57. $-x^2 - 4x + 21 = -1(x^2 + 4x - 21)$ Factor out -1.
$= -1(x + 7)(x - 3)$
$7(-3) = -21;\ 7 + (-3) = 4$

61. $-2a^2 - 5ab - 2b^2 = -1(2a^2 + 5ab + 2b^2)$ Factor out -1.
$= -1(2a^2 + 4ab + ab + 2b^2)$ $\quad 5ab = 4ab + ab$
$= -1[(2a^2 + 4ab) + (ab + 2b^2)]$ Group the terms.
$= -1[2a(a + 2b) + b(a + 2b)]$ Factor each group.
$= -1(a + 2b)(2a + b)$ Factor out $(a + 2b)$.

65. $(7p + 3)(7p - 3) = (7p)^2 - 3^2$
$= 49p^2 - 9$

69. $(3t + 4)^2 = (3t)^2 + 2(3t)(4) + 4^2$
$= 9t^2 + 24t + 16$

SECTION 4.4 (page 269)

5. $y^2 - 25 = y^2 - 5^2$
$= (y + 5)(y - 5)$

9. $36m^2 - \frac{16}{25} = (6m)^2 - \left(\frac{4}{5}\right)^2$
$= \left(6m + \frac{4}{5}\right)\left(6m - \frac{4}{5}\right)$

13. $196p^2 - 225 = (14p)^2 - 15^2$
$= (14p + 15)(14p - 15)$

17. $100x^2 + 49$ is prime because we can't factor the sum of two squares with no common factor.

21. $x^4 - 1 = (x^2)^2 - 1^2$
$= (x^2 + 1)(x^2 - 1)$
$= (x^2 + 1)(x^2 - 1^2)$
$= (x^2 + 1)(x + 1)(x - 1)$
Factor $(x^2 - 1)$.

29. $x^2 - 8x + 16 = x^2 - 2 \cdot 4 \cdot x + 4^2$
$= (x - 4)^2$

33. $x^2 - 1.0x + .25 = x^2 - 2(.5)x + .5^2$
$= (x - .5)^2$

37. $16x^2 - 40x + 25 = (4x)^2 - 2(4x)(5) + 5^2$
$= (4x - 5)^2$

41. $64x^2 + 48xy + 9y^2 = (8x)^2 + 2(8x)(3y) + (3y)^2$
$= (8x + 3y)^2$

49. $m - 4 = 0$
$m - 4 + 4 = 0 + 4$
$m = 4$

53. $9x - 6 = 0$
$9x - 6 + 6 = 0 + 6$
$9x = 6$
$\frac{9x}{9} = \frac{6}{9}$
$x = \frac{2}{3}$

SUMMARY EXERCISES ON FACTORING (page 271)

1. $32m^9 + 16m^5 + 24m^3 = 8m^3 \cdot 4m^6 + 8m^3 \cdot 2m^2 + 8m^3 \cdot 3$
$= 8m^3(4m^6 + 2m^2 + 3)$ Factor out $8m^3$.

5. $6z^2 + 31z + 5 = 6z^2 + 30z + z + 5$ $\quad 31z = 30z + z$
$= (6z^2 + 30z) + (z + 5)$ Group the terms.
$= 6z(z + 5) + 1(z + 5)$ Factor each group.
$= (z + 5)(6z + 1)$ Factor out $z + 5$.

9. $16x + 20 = 4 \cdot 4x + 4 \cdot 5$
$= 4(4x + 5)$ Factor out 4.

13. $m^2 + 2m - 15$
Look for a pair of numbers whose product is -15 and whose sum is 2. The numbers are -3 and 5, so

$$m^2 + 2m - 15 = (m - 3)(m + 5).$$

17. $z^2 - 12z + 36 = z^2 - 2(6)z + 6^2$ Perfect square trinomial
$= (z - 6)^2$

21. $6y^2 - 6y - 12 = 6 \cdot y^2 - 6 \cdot y - 6 \cdot 2$
$= 6(y^2 - y - 2)$ Factor out 6.
$= 6(y - 2)(y + 1)$
$(-2) \cdot 1 = -2;\ -2 + 1 = -1$

25. $k^2 + 9$ is prime because we can't factor the sum of two squares with no common factor.

29. $4k^2 - 12k + 9 = (2k)^2 - 2(2k)(3) + 3^2$ Perfect square trinomial
$= (2k - 3)^2$

33. $3k^3 - 12k^2 - 15k = 3k \cdot k^2 - 3k \cdot 4k - 3k \cdot 5$
$= 3k(k^2 - 4k - 5)$ Factor out $3k$.
$= 3k(k + 1)(k - 5)$
$1 \cdot (-5) = -5;\ 1 + (-5) = -4$

37. $36y^6 - 42y^5 - 120y^4 = 6y^4 \cdot 6y^2 - 6y^4 \cdot 7y - 6y^4 \cdot 20$
$= 6y^4(6y^2 - 7y - 20)$ Factor out $6y^4$.
$= 6y^4(6y^2 - 15y + 8y - 20)$ $-7y = -15y + 8y$
$= 6y^4[(6y^2 - 15y) + (8y - 20)]$ Group the terms.
$= 6y^4[3y(2y - 5) + 4(2y - 5)]$ Factor each group.
$= 6y^4(2y - 5)(3y + 4)$ Factor out $2y - 5$.

41. $54m^2 - 24z^2 = 6 \cdot 9m^2 - 6 \cdot 4z^2$
$= 6(9m^2 - 4z^2)$ Factor out 6.
$= 6[(3m)^2 - (2z)^2]$
$= 6(3m + 2z)(3m - 2z)$ Difference of squares

45. $m^2 - 81 = m^2 - 9^2$
$= (m + 9)(m - 9)$ Difference of squares

49. $m^2 - 4m + 4 = m^2 - 2(2)(m) + 2^2$ Perfect square trinomial
$= (m - 2)^2$

53. $12p^2 + pq - 6q^2 = 12p^2 + 9pq - 8pq - 6q^2$ $pq = 9pq - 8pq$
$= (12p^2 + 9pq) + (-8pq - 6q^2)$ Group the terms.
$= 3p(4p + 3q) - 2q(4p + 3q)$ Factor each group.
$= (4p + 3q)(3p - 2q)$ Factor out $4p + 3q$.

57. $100a^2 - 81y^2 = (10a)^2 - (9y)^2$
$= (10a + 9y)(10a - 9y)$ Difference of squares

SECTION 4.5 (page 279)

Each of these answers should be checked by substituting in the original equation.

1. $(x - 4)(x + 5) = 0$
$x - 4 = 0$ or $x + 5 = 0$ Zero-factor property
$x = 4$ or $x = -5$ Solve each equation.

5. $t(t + 4) = 0$
$t = 0$ or $t + 4 = 0$ Zero-factor property
$t = 0$ or $t = -4$ Solve each equation.

13. $y^2 + 3y + 2 = 0$
$(y + 2)(y + 1) = 0$ Factor.
$y + 2 = 0$ or $y + 1 = 0$ Zero-factor property
$y = -2$ or $y = -1$ Solve each equation.

17. $x^2 = 24 - 5x$
$x^2 + 5x - 24 = 0$ Add $5x$ and subtract 24.
$(x + 8)(x - 3) = 0$ Factor.
$x + 8 = 0$ or $x - 3 = 0$ Zero-factor property
$x = -8$ or $x = 3$ Solve each equation.

21. $z^2 = -2 - 3z$
$z^2 + 3z + 2 = 0$ Add $3z$ and 2.
$(z + 2)(z + 1) = 0$ Factor.
$z + 2 = 0$ or $z + 1 = 0$ Zero-factor property
$z = -2$ or $z = -1$

25. $3x^2 + 5x - 2 = 0$
$(3x - 1)(x + 2) = 0$ Factor.
$3x - 1 = 0$ or $x + 2 = 0$ Zero-factor property
$3x = 1$
$x = \frac{1}{3}$ or $x = -2$

29. $9s^2 + 12s = -4$
$9s^2 + 12s + 4 = 0$ Add 4.
$(3s + 2)^2 = 0$ Factor.
$3s + 2 = 0$ Zero-factor property
$3s = -2$
$s = -\frac{2}{3}$

33. $16k^2 - 49 = 0$
$(4k + 7)(4k - 7) = 0$ Factor.
$4k + 7 = 0$ or $4k - 7 = 0$ Zero-factor property
$4k = -7$ $4k = 7$
$k = -\frac{7}{4}$ or $k = \frac{7}{4}$

37. Because $(-11)^2 = 121$, -11 is another solution. The correct way to solve the equation is as follows:

$n^2 = 121$
$n^2 - 121 = 0$ Subtract 121.
$(n + 11)(n - 11) = 0$ Factor.
$n + 11 = 0$ or $n - 11 = 0$ Zero-factor property
$n = -11$ or $n = 11$

41. $6r^2 = 3r$
$6r^2 - 3r = 0$ Subtract $3r$.
$3r(2r - 1) = 0$ Factor out $3r$.
$3r = 0$ or $2r - 1 = 0$ Zero-factor property
$2r = 1$
$r = 0$ or $r = \frac{1}{2}$

45. $z(2z + 7) = 4$
$2z^2 + 7z = 4$ Multiply.
$2z^2 + 7z - 4 = 0$ Subtract 4.
$(2z - 1)(z + 4) = 0$ Factor.
$2z - 1 = 0$ or $z + 4 = 0$ Zero-factor property
$2z = 1$
$z = \frac{1}{2}$ or $z = -4$

49. $3x(x + 1) = (2x + 3)(x + 1)$
$3x^2 + 3x = 2x^2 + 5x + 3$ Multiply.
$x^2 - 2x - 3 = 0$ Subtract $2x^2$, $5x$, and 3.
$(x - 3)(x + 1) = 0$ Factor.
$x - 3 = 0$ or $x + 1 = 0$ Zero-factor property
$x = 3$ or $x = -1$

53. $(2x + 7)(x^2 + 2x - 3) = 0$

$(2x + 7)(x + 3)(x - 1) = 0$ Factor $x^2 + 2x - 3$.

$2x + 7 = 0$ or $x + 3 = 0$ or $x - 1 = 0$ Zero-factor property

$2x = -7$

$x = -\frac{7}{2}$ or $x = -3$ or $x = 1$

57. Let x = number of counties in California

$x + 9$ = number of counties in Florida.

California counties	plus	Florida counties	equals	total counties.
↓	↓	↓	↓	↓
x	$+$	$(x + 9)$	$=$	125

Solve the equation.

$2x + 9 = 125$ Combine terms.

$2x = 116$ Subtract 9.

$x = 58$ Divide by 2.

California has 58 counties, and Florida has $58 + 9 = 67$ counties.

SECTION 4.6 (page 285)

1. $A = LW$

Area of a rectangle

$80 = (x + 8)(x - 8)$ Let $A = 80$, $L = x + 8$, $W = x - 8$.

$80 = x^2 - 64$ Multiply.

$0 = x^2 - 144$ Subtract 80.

$0 = (x + 12)(x - 12)$ Factor.

$x + 12 = 0$ or $x - 12 = 0$ Zero-factor property

$x = -12$ or $x = 12$

The solution cannot be $x = -12$ since, when substituted, $-12 + 8$ and $-12 - 8$ give negative answers and length cannot be negative. Thus, $x = 12$ and

$$L = x + 8 = 12 + 8 = 20$$

$$W = x - 8 = 12 - 8 = 4.$$

The length is 20 units, and the width is 4 units.

5. $A = \pi r^2$

Area of a circle

$36\pi = \pi(x + 2)^2$ Let $A = 36\pi$, $r = x + 2$.

$36 = (x + 2)^2$ Divide by π.

$36 = x^2 + 4x + 4$ Multiply.

$0 = x^2 + 4x - 32$ Subtract 36.

$0 = (x + 8)(x - 4)$ Factor.

$x + 8 = 0$ or $x - 4 = 0$ Zero-factor property

$x = -8$ or $x = 4$

The solution cannot be $x = -8$ or a negative radius $-8 + 2$ would result. Thus, $x = 4$ and

$$r = x + 2 = 4 + 2 = 6.$$

The radius is 6 units.

9. Let w = the width of the original screen

$w + 3$ = the length of the original screen

$(w + 3) + 1$ or $w + 4$ = the length of the new screen.

Length × width (area) of new screen	equals	length × width (area) of original screen	increased by	8.
↓	↓	↓	↓	↓
$w(w + 4)$	$=$	$w(w + 3)$	$+$	8

Simplify this equation and solve it.

$w^2 + 4w = w^2 + 3w + 8$ Distributive property

$w = 8$ Subtract w^2 and $3w$.

The width of the original screen is 8 inches, and the length is $w + 3 = 8 + 3 = 11$ inches.

13. Let h = the height of the triangle

$2h + 2$ = the base of the triangle.

$A = \frac{1}{2}bh$ Area of triangle

Area	equals	half the base	times	the height.
↓	↓	↓	↓	↓
30	$=$	$\frac{1}{2}(2h + 2)$	$\cdot$	h

Let $A = 30$, $b = 2h + 2$.

Simplify this equation and solve it.

$60 = (2h + 2)h$ Multiply by 2.

$60 = 2h^2 + 2h$ Distributive property

$0 = 2h^2 + 2h - 60$ Subtract 60.

$0 = 2(h^2 + h - 30)$ Factor out 2.

$0 = 2(h + 6)(h - 5)$ Factor.

$h + 6 = 0$ or $h - 5 = 0$ Zero-factor property

$h = -6$ or $h = 5$

The solution $h = -6$ must be discarded since a triangle cannot have a negative height. Thus,

$$h = 5 \text{ and } 2h + 2 = 2(5) + 2 = 12.$$

The height is 5 inches, and the base is 12 inches.

17. Let x = the first integer

$x + 1$ = the next consecutive integer.

The product of two consecutive integers	is	their sum	plus	11.
↓	↓	↓	↓	↓
$x(x + 1)$	$=$	$(x + (x + 1))$	$+$	11

Simplify this equation and solve it.

$$x^2 + x = x + x + 1 + 11 \quad \text{Multiply.}$$
$$x^2 + x = 2x + 12 \quad \text{Add.}$$
$$x^2 - x - 12 = 0 \quad \text{Subtract } 2x \text{ and } 12.$$
$$(x - 4)(x + 3) = 0 \quad \text{Factor.}$$
$$x - 4 = 0 \quad \text{or} \quad x + 3 = 0 \quad \text{Zero-factor property}$$
$$x = 4 \quad \text{or} \quad x = -3$$

The two integers are 4 and $x + 1 = 4 + 1 = 5$ or -3 and $x + 1 = -3 + 1 = -2$.

21. Let x = the first even integer
$x + 2$ = the second consecutive even integer
$(x + 2) + 2$ or $x + 4$ = the third consecutive even integer.

The sum of the squares of the smaller two integers	equals	the square of the largest integer.
↓	↓	↓
$x^2 + (x + 2)^2$	$=$	$(x + 4)^2$

Simplify this equation and solve it.

$$x^2 + x^2 + 4x + 4 = x^2 + 8x + 16 \quad \text{Square both quantities.}$$
$$2x^2 + 4x + 4 = x^2 + 8x + 16 \quad \text{Combine terms.}$$
$$x^2 - 4x - 12 = 0 \quad \text{Subtract } x^2, 8x, \text{ and } 16.$$
$$(x - 6)(x + 2) = 0 \quad \text{Factor.}$$
$$x - 6 = 0 \quad \text{or} \quad x + 2 = 0 \quad \text{Zero-factor property}$$
$$x = 6 \quad \text{or} \quad x = -2$$

The three numbers are 6 and $x + 2 = 6 + 2 = 8$ and $x + 4 = 6 + 4 = 10$ or -2 and $x + 2 = -2 + 2 = 0$ and $x + 4 = -2 + 4 = 2$.

25. Let x = the length of the ladder.
$x - 4$ = the distance from the bottom of the ladder to the building
$x - 2$ = the distance on the side of the building where the top of the ladder is.

Substitute into the Pythagorean formula.

a^2	$+$	b^2	$=$	c^2
↓		↓		↓
$(x - 2)^2$	$+$	$(x - 4)^2$	$=$	x^2

$$x^2 - 4x + 4 + x^2 - 8x + 16 = x^2 \quad \text{Square both quantities.}$$
$$2x^2 - 12x + 20 = x^2 \quad \text{Combine terms.}$$
$$x^2 - 12x + 20 = 0 \quad \text{Subtract } x^2.$$
$$(x - 10)(x - 2) = 0 \quad \text{Factor.}$$
$$x - 10 = 0 \quad \text{or} \quad x - 2 = 0 \quad \text{Zero-factor property}$$
$$x = 10 \quad \text{or} \quad x = 2$$

The solution cannot be 2 or a negative distance results. Thus, $x = 10$ and the top of the ladder reaches $x - 2 = 10 - 2 = 8$ feet up the side of the building.

29. (a)
$$d = 16t^2$$
$$d = 16(3)^2 \quad \text{Let } t = 3.$$
$$d = 16(9) \quad \text{Use the exponent.}$$
$$d = 144 \quad \text{Multiply.}$$

The object would fall 144 feet.

(b)
$$d = 16t^2$$
$$1600 = 16t^2 \quad \text{Let } d = 1600.$$
$$0 = 16t^2 - 1600 \quad \text{Subtract 1600.}$$
$$0 = 16(t^2 - 100) \quad \text{Factor out 16.}$$
$$0 = 16(t + 10)(t - 10) \quad \text{Factor.}$$
$$t + 10 = 0 \quad \text{or} \quad t - 10 = 0 \quad \text{Zero-factor property}$$
$$t = -10 \quad \text{or} \quad t = 10$$

Discard $t = -10$ since time cannot be negative. It would take an object 10 seconds to fall 1600 feet.

33. Each side of the shaded region would be $y - x - x = y - 2x$. The area of a square is $A = s^2$. The area would be $(y - 2x)^2$ or $y^2 - 4xy + 4x^2$.

37. $-\dfrac{42}{14} = -\dfrac{2 \cdot 3 \cdot 7}{2 \cdot 7} = -\dfrac{2}{2} \cdot \dfrac{3}{1} \cdot \dfrac{7}{7} = -1 \cdot 3 \cdot 1 = -3$

CHAPTER 5

SECTION 5.1 (page 309)

1. The denominator $5y$ will be zero when $y = 0$, so $\frac{2}{5y}$ is undefined for $y = 0$.

5. To find the numbers that make the denominator 0, we must solve

$$m^2 + m - 6 = 0.$$
$$(m + 3)(m - 2) = 0 \quad \text{Factor.}$$
$$m + 3 = 0 \quad \text{or} \quad m - 2 = 0 \quad \text{Zero-factor property}$$
$$m = -3 \qquad m = 2$$

The expression $\dfrac{m + 2}{m^2 + m - 6}$ is undefined for $m = -3$ or $m = 2$.

9. (a) Let $x = 2$.

$$\frac{5x - 2}{4x} = \frac{5 \cdot 2 - 2}{4 \cdot 2} = \frac{10 - 2}{8} = \frac{8}{8} = 1$$

(b) Let $x = -3$.

$$\frac{5x - 2}{4x} = \frac{5(-3) - 2}{4(-3)} = \frac{-15 - 2}{-12} = \frac{-17}{-12} = \frac{17}{12}$$

13. (a)
$$\frac{(-3x)^2}{4x + 12} = \frac{(-3 \cdot 2)^2}{4 \cdot 2 + 12} \quad \text{Let } x = 2.$$
$$= \frac{(-6)^2}{8 + 12} \quad \text{Simplify.}$$
$$= \frac{36}{20}$$
$$= \frac{9}{5} \quad \text{Lowest terms.}$$

(b) $\dfrac{(-3x)^2}{4x + 12} = \dfrac{(-3(-3))^2}{4(-3) + 12}$ Let $x = -3$.

$= \dfrac{(9)^2}{-12 + 12}$ Simplify.

$= \dfrac{(9)^2}{0}$

The expression is undefined when $x = -3$ because of the 0 in the denominator.

21. $\dfrac{4(y - 2)}{10(y - 2)} = \dfrac{2 \cdot 2(y - 2)}{5 \cdot 2(y - 2)}$ Factor.

$= \dfrac{2}{5}$ Fundamental property

25. $\dfrac{7m + 14}{5m + 10} = \dfrac{7(m + 2)}{5(m + 2)}$ Factor.

$= \dfrac{7}{5}$ Fundamental property

29. $\dfrac{12m^2 - 3}{8m - 4} = \dfrac{3(4m^2 - 1)}{4(2m - 1)} = \dfrac{3(2m + 1)(2m - 1)}{4(2m - 1)}$ Factor.

$= \dfrac{3(2m + 1)}{4}$ Fundamental property

33. $\dfrac{9r^2 - 4s^2}{9r + 6s} = \dfrac{(3r + 2s)(3r - 2s)}{3(3r + 2s)}$ Factor.

$= \dfrac{3r - 2s}{3}$ Fundamental property

37. $\dfrac{2x^2 - 3x - 5}{2x^2 - 7x + 5} = \dfrac{(2x - 5)(x + 1)}{(2x - 5)(x - 1)}$ Factor.

$= \dfrac{x + 1}{x - 1}$ Fundamental property

41. $\dfrac{m^2 - 1}{1 - m} = \dfrac{(m + 1)(m - 1)}{-1(m - 1)}$ Factor.

$= \dfrac{m + 1}{-1}$ Fundamental property

$= -(m + 1)$

45. The appropriate choice is (d) since

$$\frac{-(3x + 4)}{7} = \frac{-3x - 4}{7} \neq \frac{4 - 3x}{7}.$$

49. $\dfrac{6}{15} \cdot \dfrac{25}{3} = \dfrac{6 \cdot 25}{15 \cdot 3} = \dfrac{2 \cdot 3 \cdot 5 \cdot 5}{3 \cdot 5 \cdot 3} = \dfrac{10}{3}$

SECTION 5.2 (page 315)

1. $\dfrac{15a^2}{14} \cdot \dfrac{7}{5a} = \dfrac{3 \cdot 5 \cdot a \cdot a \cdot 7}{2 \cdot 7 \cdot 5 \cdot a}$ Multiply and factor.

$= \dfrac{3 \cdot a(5 \cdot 7 \cdot a)}{2(5 \cdot 7 \cdot a)}$

$= \dfrac{3a}{2}$ Lowest terms

5. $\dfrac{2(c + d)}{3} \cdot \dfrac{18}{6(c + d)^2} = \dfrac{3 \cdot 3 \cdot 2 \cdot 2(c + d)}{3 \cdot 3 \cdot 2(c + d)(c + d)}$ Multiply and factor.

$= \dfrac{2}{c + d}$ Lowest terms

9. Invert the rational expression to find the reciprocal. The reciprocal of $\dfrac{r^2 + rp}{7}$ is $\dfrac{7}{r^2 + rp}$.

13. $\dfrac{9z^4}{3z^5} \div \dfrac{3z^2}{5z^3} = \dfrac{9z^4}{3z^5} \cdot \dfrac{5z^3}{3z^2}$ Multiply by the reciprocal of the second expression.

$= \dfrac{9 \cdot 5z^7}{3 \cdot 3z^7}$ Multiply.

$= 5$ Lowest terms

17. $\dfrac{3}{2y - 6} \div \dfrac{6}{y - 3} = \dfrac{3}{2(y - 3)} \div \dfrac{6}{y - 3}$ Factor.

$= \dfrac{3}{2(y - 3)} \cdot \dfrac{y - 3}{6}$ Multiply by the reciprocal of the second expression.

$= \dfrac{3(y - 3)}{3 \cdot 4(y - 3)}$

$= \dfrac{1}{4}$ Lowest terms

21. $\dfrac{5x - 15}{3x + 9} \cdot \dfrac{4x + 12}{6x - 18}$

$= \dfrac{5(x - 3)}{3(x + 3)} \cdot \dfrac{4(x + 3)}{6(x - 3)}$ Factor.

$= \dfrac{5 \cdot 4 \cdot (x - 3)(x + 3)}{3 \cdot 6 \cdot (x - 3)(x + 3)}$ Multiply.

$= \dfrac{10}{9}$ Lowest terms

25. $\dfrac{27 - 3z}{4} \cdot \dfrac{12}{2z - 18}$

$= \dfrac{3(9 - z)}{4} \cdot \dfrac{12}{2(z - 9)}$ Factor.

$= \dfrac{3 \cdot -1(z - 9)}{4} \cdot \dfrac{12}{2(z - 9)}$ Factor out -1.

$= \dfrac{-3 \cdot 12(z - 9)}{4 \cdot 2(z - 9)}$ Multiply.

$= -\dfrac{9}{2}$ Lowest terms

29. $\dfrac{p^2 + 4p - 5}{p^2 + 7p + 10} \div \dfrac{p - 1}{p + 4}$

$= \dfrac{p^2 + 4p - 5}{p^2 + 7p + 10} \cdot \dfrac{p + 4}{p - 1}$ Multiply by the reciprocal.

$= \dfrac{(p + 5)(p - 1) \cdot (p + 4)}{(p + 5)(p + 2) \cdot (p - 1)}$ Factor and multiply.

$= \dfrac{p + 4}{p + 2}$ Lowest terms

33. $\dfrac{2k^2 + 3k - 2}{6k^2 - 7k + 2} \cdot \dfrac{4k^2 - 5k + 1}{k^2 + k - 2}$

$= \dfrac{(2k-1)(k+2)}{(3k-2)(2k-1)} \cdot \dfrac{(4k-1)(k-1)}{(k+2)(k-1)}$ Factor.

$= \dfrac{(2k-1)(k+2)(4k-1)(k-1)}{(3k-2)(2k-1)(k+2)(k-1)}$ Multiply.

$= \dfrac{4k-1}{3k-2}$ Lowest terms

37. $\left(\dfrac{x^2 + 10x + 25}{x^2 + 10x} \cdot \dfrac{10x}{x^2 + 15x + 50}\right) \div \dfrac{x+5}{x+10}$

$= \left(\dfrac{(x+5)(x+5)}{x(x+10)} \cdot \dfrac{10x}{(x+5)(x+10)}\right) \div \dfrac{x+5}{x+10}$ Factor.

$= \left(\dfrac{(x+5)(x+5) \cdot 10x}{x(x+10)(x+5)(x+10)}\right) \cdot \dfrac{x+10}{x+5}$ Multiply by the reciprocal.

$= \dfrac{10x(x+5)^2(x+10)}{x(x+5)^2(x+10)^2}$ Multiply.

$= \dfrac{10}{x+10}$ Lowest terms

41. $108 = 12 \cdot 9$
$= 4 \cdot 3 \cdot 3 \cdot 3$
$= 2 \cdot 2 \cdot 3 \cdot 3 \cdot 3$
$= 2^2 \cdot 3^3$

45. $84q^3 = 2^2 \cdot 3 \cdot 7 \cdot q^3$
$90q^6 = 2 \cdot 3^2 \cdot 5 \cdot q^6$
Take each factor the least number of times it appears in both factored forms. The greatest common factor is $2 \cdot 3 \cdot q^3 = 6q^3$.

SECTION 5.3 (page 321)

1. Factor each denominator.

$$15 = 3 \cdot 5; \quad 21 = 3 \cdot 7; \quad 24 = 2^3 \cdot 3$$

Take each factor the greatest number of times it appears. The least common denominator is $2^3 \cdot 3 \cdot 5 \cdot 7 = 840$.

5. The denominators are x^4 and x^7. Use the highest exponent on x. The least common denominator is x^7.

9. Factor each denominator.

$$21r^3 = 3 \cdot 7 \cdot r^3$$
$$12r^5 = 2^2 \cdot 3 \cdot r^5$$

Take each factor the greatest number of times it appears, and use the highest exponent on r. The least common denominator is $2^2 \cdot 3 \cdot 7 \cdot r^5 = 84r^5$.

13. Factor each denominator.

$$28m^2 = 2^2 \cdot 7 \cdot m^2$$
$$12m - 20 = 2^2(3m - 5)$$

The least common denominator is $2^2 \cdot 7m^2(3m - 5) = 28m^2(3m - 5)$.

17. The denominators, $c - d$ and $d - c$, are opposites of each other since

$$c - d = -d + c = -1(d - c).$$

Either $c - d$ or $d - c$ can be used as the least common denominator.

21. Factor each denominator.

$$p^2 + 8p + 15 = (p + 5)(p + 3)$$
$$p^2 - 3p - 18 = (p - 6)(p + 3)$$

The least common denominator is $(p + 3)(p + 5)(p - 6)$.

25. $\dfrac{-5}{k} = \dfrac{\quad}{9k}$
To get a denominator of $9k$, multiply numerator and denominator by 9.

$$\frac{9(-5)}{9 \cdot k} = \frac{-45}{9k}$$

29. $\dfrac{5t^2}{6r} = \dfrac{\quad}{42r^4}$
To get a denominator of $42r^4$, multiply numerator and denominator by $7r^3$.

$$\frac{5t^2 \cdot 7r^3}{6r \cdot 7r^3} = \frac{35t^2r^3}{42r^4}$$

33. $\dfrac{-4t}{3t - 6} = \dfrac{\quad}{6t - 12}$

$\dfrac{-4t}{3(t-2)} = \dfrac{\quad}{6(t-2)}$ Factor both denominators.

$\dfrac{2(-4t)}{2 \cdot 3(t-2)} = \dfrac{-8t}{6(t-2)}$ Multiply numerator and denominator by 2.

or $\dfrac{-8t}{6t - 12}$

37. $\dfrac{2(b-1)}{b^2 + b} = \dfrac{\quad}{b^3 + 3b^2 + 2b}$

$\dfrac{2(b-1)}{b(b+1)} = \dfrac{\quad}{b(b^2 + 3b + 2)}$ Factor denominators.

$\dfrac{2(b-1)}{b(b+1)} = \dfrac{\quad}{b(b+1)(b+2)}$ Factor the trinomial.

$\dfrac{2(b-1) \cdot (b+2)}{b(b+1) \cdot (b+2)} = \dfrac{2(b-1)(b+2)}{b^3 + 3b^2 + 2b}$ Multiply numerator and denominator by $b + 2$.

41. $\dfrac{1}{2} + \dfrac{7}{8} = \dfrac{4}{8} + \dfrac{7}{8}$ Get a common denominator.

$= \dfrac{11}{8}$ Add numerators.

45. $\dfrac{4}{3} - \dfrac{1}{4} = \dfrac{16}{12} - \dfrac{3}{12}$ Get a common denominator.

$= \dfrac{13}{12}$ Subtract numerators.

SECTION 5.4 (page 329)

5. $\frac{a+b}{2} - \frac{a-b}{2} = \frac{(a+b)-(a-b)}{2}$ Subtract numerators.

$= \frac{a+b-a+b}{2}$ Distributive property

$= \frac{2b}{2}$ Combine terms.

$= b$ Lowest terms

9. $\frac{y^2-3y}{y+3} + \frac{-18}{y+3} = \frac{y^2-3y-18}{y+3}$ Add numerators.

$= \frac{(y-6)(y+3)}{y+3}$ Factor.

$= y - 6$ Fundamental property

13. Rewrite the fractions with the least common denominator, 15.

$$\frac{z}{5} + \frac{1}{3} = \frac{z \cdot 3}{5 \cdot 3} + \frac{1 \cdot 5}{3 \cdot 5}$$

$$= \frac{3z}{15} + \frac{5}{15}$$

$= \frac{3z+5}{15}$ Add numerators.

17. Rewrite the fractions with the least common denominator, $4x$.

$$-\frac{3}{4} - \frac{1}{2x} = -\frac{3 \cdot x}{4 \cdot x} - \frac{2 \cdot 1}{2 \cdot 2x}$$

$= \frac{-3x-2}{4x}$ Subtract numerators.

21. Rewrite the fractions with the least common denominator, $3b$.

$$\frac{b+3}{b} + \frac{b+7}{3b} = \frac{3 \cdot (b+3)}{3 \cdot b} + \frac{b+7}{3b}$$

$= \frac{3b+9}{3b} + \frac{b+7}{3b}$ Distributive property

$= \frac{4b+16}{3b}$ Add numerators.

$= \frac{4(b+4)}{3b}$ Factor.

25. Rewrite each fraction with the least common denominator, $p(p-2)$.

$$\frac{1}{p-2} - \frac{3}{p} = \frac{1 \cdot p}{(p-2) \cdot p} - \frac{3 \cdot (p-2)}{p \cdot (p-2)}$$

$= \frac{p-3(p-2)}{p(p-2)}$ Subtract numerators.

$= \frac{p-3p+6}{p(p-2)}$ Distributive property

$= \frac{-2p+6}{p(p-2)}$ Combine terms.

$= \frac{-2(p-3)}{p(p-2)}$ Factor.

29. The least common denominator is either $1 - y$ or $y - 1$.

$$\frac{-1}{1-y} - \frac{3}{y-1} = \frac{-1 \cdot -1}{-1 \cdot (1-y)} - \frac{3}{y-1} = \frac{1-3}{y-1} = \frac{-2}{y-1}$$

or

$$\frac{-1}{1-y} - \frac{3}{y-1} = \frac{-1}{1-y} - \frac{-1 \cdot 3}{-1 \cdot (y-1)} = \frac{-1}{1-y} - \frac{(-3)}{1-y} = \frac{-1+3}{1-y} = \frac{2}{1-y}$$

33. Factor the second denominator.

$$\frac{2m}{m-n} - \frac{5m+n}{2m-2n} = \frac{2m}{m-n} - \frac{5m+n}{2(m-n)}$$

The least common denominator is $2(m - n)$.

$$= \frac{2 \cdot 2m}{2(m-n)} - \frac{5m+n}{2(m-n)}$$

$= \frac{4m-(5m+n)}{2(m-n)}$ Subtract numerators.

$= \frac{4m-5m-n}{2(m-n)}$ Distribute the negative sign.

$= \frac{-m-n}{2(m-n)}$ Combine terms.

$= \frac{-(m+n)}{2(m-n)}$ Factor.

37. Factor the denominators.

$$\frac{1}{a^2-1} - \frac{a-1}{a^2+3a-4}$$

$$= \frac{1}{(a+1)(a-1)} - \frac{a-1}{(a+4)(a-1)}$$

SOLUTIONS

The least common denominator is $(a + 1)(a - 1)(a + 4)$.

$$= \frac{1 \cdot (a + 4)}{(a + 1)(a - 1) \cdot (a + 4)} - \frac{(a - 1) \cdot (a + 1)}{(a + 4)(a - 1) \cdot (a + 1)}$$

$$= \frac{a + 4 - (a^2 - 1)}{(a + 1)(a - 1)(a + 4)}$$ Multiply each numerator and subtract.

$$= \frac{a + 4 - a^2 + 1}{(a + 1)(a - 1)(a + 4)}$$ Distributive property

$$= \frac{-a^2 + a + 5}{(a + 1)(a - 1)(a + 4)}$$ Combine terms.

41. Factor the denominators.

$$\frac{4y - 1}{2y^2 + 5y - 3} - \frac{y + 3}{6y^2 + y - 2} = \frac{4y - 1}{(2y - 1)(y + 3)} - \frac{y + 3}{(3y + 2)(2y - 1)}$$

The least common denominator is $(2y - 1)(y + 3)(3y + 2)$.

$$= \frac{(4y - 1) \cdot (3y + 2)}{(2y - 1)(y + 3) \cdot (3y + 2)} - \frac{(y + 3) \cdot (y + 3)}{(3y + 2)(2y - 1) \cdot (y + 3)}$$

$$= \frac{12y^2 + 5y - 2 - (y^2 + 6y + 9)}{(2y - 1)(y + 3)(3y + 2)}$$ Multiply each numerator and subtract.

$$= \frac{12y^2 + 5y - 2 - y^2 - 6y - 9}{(2y - 1)(y + 3)(3y + 2)}$$ Distributive property

$$= \frac{11y^2 - y - 11}{(2y - 1)(y + 3)(3y + 2)}$$ Combine terms.

45. $$\frac{\frac{3}{2} - \frac{5}{4}}{\frac{7}{4} + \frac{1}{3}} = \frac{\frac{3}{2} \cdot \frac{2}{2} - \frac{5}{4}}{\frac{7}{4} \cdot \frac{3}{3} + \frac{1}{3} \cdot \frac{4}{4}}$$ Work in numerator and denominator separately.

$$= \frac{\frac{6}{4} - \frac{5}{4}}{\frac{21}{12} + \frac{4}{12}}$$

$$= \frac{\frac{1}{4}}{\frac{25}{12}}$$

Rewrite as a division problem.

$$\frac{1}{4} \div \frac{25}{12} = \frac{1}{4} \cdot \frac{12}{25} = \frac{3}{25}$$

SECTION 5.5 (page 335)

1. The fraction bar represents division.

Only one method is shown for each exercise in this section.

5. $$\frac{\frac{p}{q^2}}{\frac{p^2}{q}} = \frac{p}{q^2} \div \frac{p^2}{q} = \frac{p}{q^2} \cdot \frac{q}{p^2} = \frac{p \cdot q}{q^2 \cdot p^2} = \frac{1}{pq}$$

9. $$\frac{\frac{4a^4b^3}{3a}}{\frac{2ab^4}{b^2}} = \frac{4a^4b^3}{3a} \div \frac{2ab^4}{b^2} = \frac{4a^4b^3}{3a} \cdot \frac{b^2}{2ab^4} = \frac{4a^4b^3 \cdot b^2}{3a \cdot 2ab^4} = \frac{4a^4b^5}{3 \cdot 2a^2b^4} = \frac{2a^2b}{3}$$

13. $$\frac{\frac{2}{x} - 3}{\frac{2 - 3x}{2}} = \frac{2x\left(\frac{2}{x} - 3\right)}{2x\left(\frac{2 - 3x}{2}\right)}$$ Multiply numerator and denominator by the LCD $= 2x$.

$$= \frac{2x\left(\frac{2}{x}\right) - 2x(3)}{2x\left(\frac{2 - 3x}{2}\right)}$$ Distributive property

$$= \frac{4 - 6x}{2x - 3x^2}$$

$$= \frac{2(2 - 3x)}{x(2 - 3x)}$$ Factor.

$$= \frac{2}{x}$$ Lowest terms

17. $$\frac{a - \frac{5}{a}}{a + \frac{1}{a}} = \frac{a\left(a - \frac{5}{a}\right)}{a\left(a + \frac{1}{a}\right)}$$ Multiply numerator and denominator by the LCD $= a$.

$$= \frac{a^2 - 5}{a^2 + 1}$$

21. $$\frac{\dfrac{t}{t+2}}{\dfrac{4}{t^2-4}} = \frac{t}{t+2} \div \frac{4}{t^2-4}$$

$$= \frac{t}{t+2} \cdot \frac{t^2-4}{4}$$

$$= \frac{t \cdot (t+2)(t-2)}{(t+2) \cdot 4} \qquad \text{Multiply and factor.}$$

$$= \frac{t(t-2)}{4} \qquad \text{Lowest terms}$$

25. The least common denominator is $(m-1)(m+2)(m-3)$.

$$\frac{\dfrac{1}{m-1} + \dfrac{2}{m+2}}{\dfrac{2}{m+2} - \dfrac{1}{m-3}}$$

$$= \frac{\left(\dfrac{1}{m-1} + \dfrac{2}{m+2}\right) \cdot (m-1)(m+2)(m-3)}{\left(\dfrac{2}{m+2} - \dfrac{1}{m-3}\right) \cdot (m-1)(m+2)(m-3)}$$

Multiply numerator and denominator by LCD.

$$= \frac{1(m+2)(m-3) + 2(m-1)(m-3)}{2(m-1)(m-3) - 1(m-1)(m+2)}$$

Distributive property

$$= \frac{m^2 - m - 6 + 2m^2 - 8m + 6}{2m^2 - 8m + 6 - m^2 - m + 2}$$

$$= \frac{3m^2 - 9m}{m^2 - 9m + 8} \qquad \text{Combine terms.}$$

$$= \frac{3m(m-3)}{(m-1)(m-8)} \qquad \text{Factor.}$$

29. $4x - 5 = 8x + 3$

$-5 = 4x + 3$ Subtract $4x$.

$-8 = 4x$ Subtract 3.

$\dfrac{-8}{4} = \dfrac{4x}{4}$ Divide by 4.

$-2 = x$

33. $6 - (4 - 3y) = 9$

$6 - 4 + 3y = 9$ Distributive property

$3y + 2 = 9$ Combine terms.

$3y = 7$ Subtract 2.

$\dfrac{3y}{3} = \dfrac{7}{3}$ Divide by 3.

$y = \dfrac{7}{3}$

SECTION 5.6 (page 341)

1. The appropriate choice is (d) because if 3 is substituted for y

$$\frac{y-1}{2} - \frac{y-3}{4} = \frac{3-1}{2} - \frac{3-3}{4} = \frac{2}{2} - \frac{0}{4} = 1,$$

and this satisfies the equation. (The other three choices do not satisfy the equation.)

Each of the answers should be checked by substituting in the original equation.

5. $\dfrac{5}{y} + 4 = \dfrac{2}{y}$

$y\left(\dfrac{5}{y} + 4\right) = y\left(\dfrac{2}{y}\right)$ Multiply both sides by the LCD, y.

$y\left(\dfrac{5}{y}\right) + y \cdot 4 = y\left(\dfrac{2}{y}\right)$ Distributive property

$5 + 4y = 2$

$4y = -3$ Subtract 5.

$y = -\dfrac{3}{4}$ Divide by 4.

9. $\dfrac{3x}{5} - 6 = x$

$5\left(\dfrac{3x}{5} - 6\right) = 5 \cdot x$ Multiply both sides by the LCD, 5.

$5\left(\dfrac{3x}{5}\right) - 5 \cdot 6 = 5x$ Distributive property

$3x - 30 = 5x$

$-30 = 2x$ Subtract $3x$.

$-15 = x$ Divide by 2.

13. $\dfrac{z-1}{4} = \dfrac{z+3}{3}$

$12\left(\dfrac{z-1}{4}\right) = 12\left(\dfrac{z+3}{3}\right)$ Multiply both sides by the LCD, 12.

$3(z-1) = 4(z+3)$

$3z - 3 = 4z + 12$ Distributive property

$-3 = z + 12$ Subtract $3z$.

$-15 = z$ Subtract 12.

17. $\dfrac{2x+3}{x} = \dfrac{3}{2}$

$2x\left(\dfrac{2x+3}{x}\right) = 2x\left(\dfrac{3}{2}\right)$ Multiply both sides by the LCD, $2x$.

$2(2x+3) = 3x$

$4x + 6 = 3x$ Distributive property

$x + 6 = 0$ Subtract $3x$.

$x = -6$ Subtract 6.

21. $$\frac{q+2}{3} + \frac{q-5}{5} = \frac{7}{3}$$

$$15\left(\frac{q+2}{3} + \frac{q-5}{5}\right) = 15\left(\frac{7}{3}\right) \quad \text{Multiply both sides by the LCD, 15.}$$

$$15\left(\frac{q+2}{3}\right) + 15\left(\frac{q-5}{5}\right) = 15\left(\frac{7}{3}\right) \quad \text{Distributive property}$$

$$\begin{aligned} 5(q+2) + 3(q-5) &= 5 \cdot 7 \\ 5q + 10 + 3q - 15 &= 35 && \text{Distributive property} \\ 8q - 5 &= 35 && \text{Combine terms.} \\ 8q &= 40 && \text{Add 5.} \\ q &= 5 && \text{Divide by 8.} \end{aligned}$$

25. $$\frac{3m}{5} - \frac{3m-2}{4} = \frac{1}{5}$$

$$20\left(\frac{3m}{5} - \frac{3m-2}{4}\right) = 20\left(\frac{1}{5}\right) \quad \text{Multiply both sides by the LCD, 20.}$$

$$\begin{aligned} 4(3m) - 5(3m-2) &= 4 \cdot 1 && \text{Distributive property} \\ 12m - 15m + 10 &= 4 && \text{Distributive property; multiply.} \\ -3m + 10 &= 4 && \text{Combine terms.} \\ -3m &= -6 && \text{Subtract 10.} \\ m &= 2 && \text{Divide by } -3. \end{aligned}$$

29. $$\frac{3}{x-1} + \frac{2}{4x-4} = \frac{7}{4}$$

$$\frac{3}{x-1} + \frac{2}{4(x-1)} = \frac{7}{4} \quad \text{Factor denominator.}$$

$$4(x-1)\left(\frac{3}{x-1} + \frac{2}{4(x-1)}\right) = 4(x-1)\left(\frac{7}{4}\right)$$

Multiply both sides by the LCD, $4(x - 1)$.

$$\begin{aligned} 12 + 2 &= (x-1)(7) && \text{Distributive property} \\ 14 &= 7x - 7 && \text{Add; distributive property} \\ 21 &= 7x && \text{Add 7.} \\ 3 &= x && \text{Divide by 7.} \end{aligned}$$

33. $$\frac{2}{m} = \frac{m}{5m+12}$$

$$m(5m+12)\left(\frac{2}{m}\right) = m(5m+12)\left(\frac{m}{5m+2}\right)$$

Multiply by the LCD, $m(5m + 12)$.

$$\begin{aligned} 10m + 24 &= m^2 \\ -m^2 + 10m + 24 &= 0 && \text{Subtract } m^2. \\ m^2 - 10m - 24 &= 0 && \text{Multiply by } -1. \\ (m-12)(m+2) &= 0 && \text{Factor.} \end{aligned}$$

$$m - 12 = 0 \quad \text{or} \quad m + 2 = 0 \quad \text{Zero-factor property}$$

$$m = 12 \qquad\qquad m = -2$$

The solutions are -2 and 12, since both numbers satisfy the original equation.

37. $$\frac{3y}{y^2+5y+6} = \frac{5y}{y^2+2y-3} - \frac{2}{y^2+y-2}$$

Factor the denominators.

$$\frac{3y}{(y+3)(y+2)} = \frac{5y}{(y+3)(y-1)} - \frac{2}{(y+2)(y-1)}$$

Multiply each side of the equation by the least common denominator, $(y + 3)(y + 2)(y - 1)$.

$$(y+3)(y+2)(y-1)\left(\frac{3y}{(y+3)(y+2)}\right) = (y+3)(y+2)(y-1) \cdot \left(\frac{5y}{(y+3)(y-1)} - \frac{2}{(y+2)(y-1)}\right)$$

$$\begin{aligned} 3y(y-1) &= 5y(y+2) - 2(y+3) \\ 3y^2 - 3y &= 5y^2 + 10y - 2y - 6 \\ 3y^2 - 3y &= 5y^2 + 8y - 6 \\ -2y^2 - 11y + 6 &= 0 \\ 2y^2 + 11y - 6 &= 0 && \text{Multiply by } -1. \\ (2y-1)(y+6) &= 0 && \text{Factor.} \end{aligned}$$

$$2y - 1 = 0 \quad \text{or} \quad y + 6 = 0 \quad \text{Zero-factor property}$$

$$2y = 1 \qquad\qquad y = -6$$

$$y = \frac{1}{2}$$

The solutions are -6 and $\frac{1}{2}$, since both numbers satisfy the original equation.

41. Multiply each side of the equation by the least common denominator, $4(z + 1)(z - 1)$.

$$\frac{2}{z-1} - \frac{5}{4} = \frac{-1}{z+1}$$

$$4(z+1)(z-1) \cdot \frac{2}{z-1} - 4(z+1)(z-1) \cdot \frac{5}{4} = 4(z+1)(z-1) \cdot \frac{-1}{z+1}$$

$$\begin{aligned} 8(z+1) - 5(z^2-1) &= -4(z-1) \\ 8z + 8 - 5z^2 + 5 &= -4z + 4 \\ -5z^2 + 12z + 9 &= 0 && \text{Add } 4z\text{; subtract 4.} \\ 5z^2 - 12z - 9 &= 0 && \text{Multiply by } -1. \\ (5z+3)(z-3) &= 0 && \text{Factor.} \end{aligned}$$

$$5z + 3 = 0 \quad \text{or} \quad z - 3 = 0 \quad \text{Zero-factor property}$$

$$5z = -3 \qquad\qquad z = 3$$

$$z = -\frac{3}{5}$$

Both numbers check, so the solutions are $-\frac{3}{5}$ and 3.

45. $m = \dfrac{kF}{a}$ Solve for a.

$a \cdot m = a\left(\dfrac{kF}{a}\right)$ Multiply both sides by the LCD, a.

$am = kF$

$a = \dfrac{kF}{m}$ Divide by m.

49. $h = \dfrac{2A}{B + b}$ Solve for A.

$(B + b)\,h = (B + b) \cdot \dfrac{2A}{B + b}$ Multiply by the LCD, $B + b$.

$h(B + b) = 2A$

$\dfrac{h(B + b)}{2} = A$ Divide by 2.

53. Use the formula for distance, rate, and time: $d = rt$. Solve for r.

$d = rt$

$288 = rt$ Let $d = 288$.

$\dfrac{288}{t} = r$ Divide by t.

or $r = \dfrac{288}{t}$

Her rate is $\frac{288}{t}$ miles per hour.

SUMMARY EXERCISES ON RATIONAL EXPRESSIONS (page 345)

1. $\dfrac{4}{p} + \dfrac{6}{p} = \dfrac{4 + 6}{p} = \dfrac{10}{p}$

5. $\dfrac{2y^2 + y - 6}{2y^2 - 9y + 9} \cdot \dfrac{y^2 - 2y - 3}{y^2 - 1}$

$= \dfrac{(2y - 3)(y + 2)(y - 3)(y + 1)}{(2y - 3)(y - 3)(y + 1)(y - 1)}$ Multiply; factor.

$= \dfrac{y + 2}{y - 1}$ Lowest terms

9. $\dfrac{4}{p + 2} + \dfrac{1}{3p + 6} = \dfrac{4}{p + 2} + \dfrac{1}{3(p + 2)}$ Factor.

$= \dfrac{3 \cdot 4}{3(p + 2)} + \dfrac{1}{3(p + 2)}$ Get a common denominator.

$= \dfrac{12 + 1}{3(p + 2)}$

$= \dfrac{13}{3(p + 2)}$ Add fractions.

13. $\dfrac{5}{4z} - \dfrac{2}{3z} = \dfrac{3 \cdot 5}{3 \cdot 4z} - \dfrac{4 \cdot 2}{4 \cdot 3z}$ Get a common denominator.

$= \dfrac{15}{12z} - \dfrac{8}{12z}$

$= \dfrac{15 - 8}{12z}$

$= \dfrac{7}{12z}$ Subtract.

SECTION 5.7 (page 355)

1. Let x represent the number. Then

One-third	of	a number	is	2 more than	one-sixth	of	the same number.
↓	↓	↓	↓	↓	↓	↓	↓
$\frac{1}{3}$	$\cdot$	x	$=$	$2 +$	$\frac{1}{6}$	$\cdot$	x

$\dfrac{1}{3}x = 2 + \dfrac{1}{6}x$

$6\left(\dfrac{1}{3}x\right) = 6(2) + 6\left(\dfrac{1}{6}x\right)$ Multiply each term by 6.

$2x = 12 + x$

$x = 12$ Subtract x.

The number is 12.

5. Let x represent the total revenue in millions of dollars. Then $\frac{1}{2}x$ represents that part of total revenue provided by tomatoes and $\frac{1}{4}x$ represents that part of total revenue provided by oils, which gives the following equation.

Tomato revenue	plus	oil revenue	equals	$74.8 million.
↓	↓	↓	↓	↓
$\frac{1}{2}x$	$+$	$\frac{1}{4}x$	$=$	74.8

$4\left(\dfrac{1}{2}x\right) + 4\left(\dfrac{1}{4}x\right) = 4(74.8)$ Multiply by 4.

$2x + x = 299.2$

$3x = 299.2$

$x = 99.7$ (rounded) Divide by 3.

Revenue is approximately $99.7 million.

SOLUTIONS

9. Use the formula for distance, rate, and time: $d = rt$. Let $d = 50$ and $t = 2.08$. Solve for r.

$$d = rt$$
$$50 = 2.08r$$
$$\frac{50}{2.08} = r \quad \text{Divide by 2.08.}$$
$$r = 24.04 \text{ (rounded)}$$

His average speed was 24.04 kilometers per hour.

13. Let t represent McDermott's time. Then $t - .8$ represents Hussein's time, since he completed the race in .8 of an hour less time. We fill in the chart as follows, realizing that the rate column is filled in by using the formula $r = \frac{d}{t}$.

	d	r	t
Hussein	26	$\frac{26}{t - .8}$	$t - .8$
McDermott	26	$\frac{26}{t}$	t

McDermott's speed was .73 times Hussein's speed.

$$\frac{26}{t} = .73 \cdot \left(\frac{26}{t - .8}\right)$$

$$\frac{26}{t} = \frac{18.98}{t - .8} \quad \text{Multiply.}$$

$$t(t - .8)\left(\frac{26}{t}\right) = t(t - .8)\left(\frac{18.98}{t - .8}\right) \quad \text{Multiply by } t(t - .8).$$

$$26(t - .8) = 18.98t$$
$$26t - 20.8 = 18.98t \quad \text{Distributive property}$$
$$-20.8 = -7.02t \quad \text{Subtract } 26t.$$
$$2.9630 \approx t \quad \text{Divide by } -7.02.$$

(rounded)

To find each runner's speed, substitute 2.9630 for t.

$$\text{Hussein's speed} = \frac{26}{t - .8} = \frac{26}{2.9630 - .8} \approx 12.02 \text{ (rounded)}$$

$$\text{McDermott's speed} = \frac{26}{t} = \frac{26}{2.9630} \approx 8.77 \text{ (rounded)}$$

Hussein's speed was approximately 12.02 miles per hour, and McDermott's speed was approximately 8.77 miles per hour.

17. Let x represent the speed of the boat in still water. We fill in the chart as follows, realizing that the time column is filled in by using the formula $t = \frac{d}{r}$.

	d	r	t
against current	20	$x - 4$	$\frac{20}{x - 4}$
with current	60	$x + 4$	$\frac{60}{x + 4}$

Times are equal.

Since the times are equal we get the following equation.

$$\frac{20}{x - 4} = \frac{60}{x + 4}$$

$$(x + 4)(x - 4)\frac{20}{x - 4} = (x + 4)(x - 4)\frac{60}{x + 4}$$

Multiply by $(x + 4)(x - 4)$.

$$20(x + 4) = 60(x - 4)$$
$$20x + 80 = 60x - 240 \quad \text{Distributive property}$$
$$320 = 40x \quad \text{Add 240; subtract } 20x.$$
$$8 = x \quad \text{Divide by 40.}$$

The speed of the boat in still water is 8 miles per hour.

21. Let x represent the number of hours it will take for Geraldo and Luisa to do a day's laundry working together. Since Geraldo can do the laundry in 8 hours, his rate alone is $\frac{1}{8}$ job per hour. Also, since Luisa can do the job alone in 9 hours her rate is $\frac{1}{9}$ job per hour. We fill in the chart as follows:

	Rate	*Time working together*	*Fractional part of the job done when working together*
Geraldo	$\frac{1}{8}$	x	$\frac{1}{8}x$
Luisa	$\frac{1}{9}$	x	$\frac{1}{9}x$

Sum is 1 whole job.

Since together Geraldo and Luisa complete 1 whole job, we must add their individual fractional parts and set the sum equal to 1.

$$\frac{1}{8}x + \frac{1}{9}x = 1$$

$$72\left(\frac{1}{8}x\right) + 72\left(\frac{1}{9}x\right) = 72(1) \quad \text{Multiply by 72.}$$

$$9x + 8x = 72$$
$$17x = 72 \quad \text{Combine terms.}$$
$$x = \frac{72}{17} \quad \text{Divide by 17.}$$

or 4.24 hours (rounded)

It will take Geraldo and Luisa about 4.24 hours to do a day's laundry if they work together.

25. Let x represent the number of hours it will take the experienced employee to enter the data. Then $2x$ represents the number of hours it will take the new employee (the experienced employee takes less time). The experienced employee's rate is $\frac{1}{x}$ job per hour and the new employee's rate is $\frac{1}{2x}$ job per hour. We fill in the chart as follows.

	Rate	*Time working together*	*Fractional part of the job done when working together*
Experienced employee	$\frac{1}{x}$	2	$\frac{2}{x}$
New employee	$\frac{1}{2x}$	2	$\frac{1}{x}$

Sum is 1 whole job.

Since together the two employees complete the whole job, we add their individual fractional parts and set the sum equal to 1.

$$\frac{2}{x} + \frac{1}{x} = 1$$

$$x\left(\frac{2}{x}\right) + x\left(\frac{1}{x}\right) = x(1) \quad \text{Multiply by } x.$$

$$2 + 1 = x$$

$$3 = x$$

Working alone it will take the experienced employee 3 hours to enter the data.

29. If the interest varies directly as the rate of interest, and if I represents the interest and r the rate of interest, then there is a constant k such that

$$I = kr.$$

Find k by letting $I = 48$ and $r = 5\% = .05$.

$$48 = k(.05)$$

$$960 = k$$

Since $I = kr$ and $k = 960$,

$$I = 960r.$$

If $r = 4.2\% = .042$, then

$$I = 960(.042) = 40.32.$$

The interest is \$40.32.

33. If the pressure varies directly as the depth, and if P represents the pressure and d the depth, then there is a constant k such that

$$P = kd.$$

Find k by letting $P = 50$ and $d = 10$.

$$50 = k \cdot 10$$

$$5 = k$$

Since $P = kd$ and $k = 5$,

$$P = 5d.$$

If $d = 20$, then

$$P = 5 \cdot 20 = 100.$$

The pressure is 100 pounds per square inch.

37. $y = 5x + 3$

(a) Let $x = -2$.

$$y = 5(-2) + 3 = -10 + 3 = -7$$

(b) Let $x = 4$.

$$y = 5(4) + 3 = 20 + 3 = 23$$

41. $2x - 5y = 10$

(a) Let $x = -2$.

$$2(-2) - 5y = 10$$

$$-4 - 5y = 10$$

$$-5y = 14$$

$$y = -\frac{14}{5}$$

(b) Let $x = 4$.

$$2(4) - 5y = 10$$

$$8 - 5y = 10$$

$$-5y = 2$$

$$y = -\frac{2}{5}$$

CHAPTER 6

SECTION 6.1 (page 381)

1. The symbol (x, y) does represent an ordered pair, because parentheses are used. The symbols $[x, y]$ and $\{x, y\}$ do not since square brackets and braces are used, not parentheses.

5. All solutions of the equation $y = 3$ have a y-value of 3, so the missing number is 3.

9. Substitute 4 for p and 2 for q in the equation.

$$2p - q = 6$$

$$2(4) - 2 = 6$$

$$8 - 2 = 6$$

$$6 = 6 \quad \text{True}$$

The result is true, so (4, 2) is a solution of $2p - q = 6$.

13. Substitute 2 for x and 6 for y in the equation.

$$y = 3x$$

$$6 = 3(2)$$

$$6 = 6 \quad \text{True}$$

The result is true, so (2, 6) is a solution of $y = 3x$.

17. Substitute -6 for x in the equation; y does not appear.

$$x + 4 = 0$$

$$-6 + 4 = 0$$

$$-2 = 0 \quad \text{False}$$

The result is false, so $(-6, 2)$ is not a solution of $x + 4 = 0$.

21. Given $y = 2x + 7$.

$$y = 2(0) + 7 \quad \text{Let } x = 0.$$
$$y = 0 + 7$$
$$y = 7$$

The ordered pair is (0, 7).

25. Given $y = -4x - 4$.

$$y = -4(0) - 4 \quad \text{Let } x = 0.$$
$$y = 0 - 4$$
$$y = -4$$

The ordered pair is (0, −4).

29. $y = -4x - 4$
$y = -4(10) - 4 \quad$ Let $x = 10$.
$y = -40 - 4$
$y = -44$
The ordered pair is (10, −44).

33. Given $y = 2x + 6$.

$$y = 2(3) + 6 \quad \text{Let } x = 3.$$
$$y = 6 + 6$$
$$y = 12$$

The ordered pair is (3, 12).

$$y = 2x + 6$$
$$y = 2(0) + 6 \quad \text{Let } x = 0.$$
$$y = 0 + 6$$
$$y = 6$$

The ordered pair is (0, 6).

$$y = 2x + 6$$
$$y = 2(-1) + 6 \quad \text{Let } x = -1.$$
$$y = -2 + 6$$
$$y = 4$$

The ordered pair is (−1, 4).

37. Given $2x + y = 5$.

$$2(0) + y = 5 \quad \text{Let } x = 0.$$
$$0 + y = 5$$
$$y = 5$$

The ordered pair is (0, 5).

$$2x + y = 5$$
$$2x + 0 = 5 \quad \text{Let } y = 0.$$
$$2x = 5$$
$$x = \frac{5}{2} \quad \text{Divide by 2.}$$

The ordered pair is $(\frac{5}{2}, 0)$.

$$2x + y = 5$$
$$2(12) + y = 5 \quad \text{Let } x = 12.$$
$$24 + y = 5$$
$$y = -19 \quad \text{Subtract 24.}$$

The ordered pair is (12, −19).

41. Given $3u - 5v = -15$.
If $u = 0$,
then $3u - 5v = -15$
becomes $3(0) - 5v = -15$.
$0 - 5v = -15$
$-5v = -15$
$v = 3$
If $v = 0$,
then $3u - 5v = -15$
becomes $3u - 5(0) = -15$.
$3u - 0 = -15$
$3u = -15$
$u = -5$
If $v = -6$,
then $3u - 5v = -15$
becomes $3u - 5(-6) = -15$.
$3u + 30 = -15$
$3u = -45$
$u = -15$

The completed table of ordered pairs is as follows:

u	0	−5	−15
v	3	0	−6

45. Given $x = -9$. No matter what value y has, the x-value will always be −9. The completed table of ordered pairs is as follows:

x	−9	−9	−9
y	6	2	−3

49. Given $x - 8 = 0$.
$x = 8$
No matter what value y has, the x-value will always be 8. The completed table of ordered pairs is as follows:

x	8	8	8
y	8	3	0

53. C is 5 units to the left of the origin and 4 units above the origin. The ordered pair is (−5, 4).

57. We must have

$$a + 1 = 4 \quad \text{and} \quad b - 2 = 7$$
$$a = 3 \qquad\qquad b = 9.$$

The value of a is 3, and the value of b is 9.

For Exercises 61–77, see the graphs in the answer section.

69. If the point with coordinates (x, y) is in quadrant III, then x is negative and y is negative.

73. Given $x - 2y = 6$.

If $x = 0$, then $x - 2y = 6$ becomes $0 - 2y = 6$.
$-2y = 6$
$y = -3$

If $y = 0$, then $x - 2y = 6$ becomes $x - 2(0) = 6$.
$x - 0 = 6$
$x = 6$

If $x = 2$, then $x - 2y = 6$ becomes $2 - 2y = 6$.
$-2y = 4$
$y = -2$

If $y = -1$, then $x - 2y = 6$ becomes $x - 2(-1) = 6$.
$x + 2 = 6$
$x = 4$

The completed table of ordered pairs follows.

x	y
0	−3
6	0
2	−2
4	−1

77. Given $y + 4 = 0$.
$y = -4$
No matter what value x has, the y value will always be -4. The ordered pairs are $(0, -4)$, $(5, -4)$, $(-2, -4)$, and $(-3, -4)$. The completed table of ordered pairs follows.

x	y
0	−4
5	−4
−2	−4
−3	−4

81. $3x + 6 = 9$
$3x = 3$ Subtract 6.
$x = 1$ Divide by 3.

85. $2x - 8 = 0$
$2x = 8$ Add 8.
$x = 4$ Divide by 2.

SECTION 6.2 (page 393)

See the answer section in your text for all graphs.

9. Given $y = \frac{2}{3}x + 1$.

If $x = 0$,
$$y = \frac{2}{3}(0) + 1$$
$y = 1$.

If $x = 3$,
$$y = \frac{2}{3}(3) + 1$$
$y = 2 + 1$
$y = 3$.

If $x = -3$,
$$y = \frac{2}{3}(-3) + 1$$
$y = -2 + 1$
$y = -1$.

The ordered pairs are $(0, 1)$, $(3, 3)$, and $(-3, -1)$.

13. Let $x = 0$ to find the y-intercept, and let $y = 0$ to find the x-intercept.

If $x = 0$,
$2x - 3y = 24$
$2(0) - 3y = 24$
$-3y = 24$
$y = -8$.

If $y = 0$,
$2x - 3y = 24$
$2x - 3(0) = 24$
$2x = 24$
$x = 12$.

The y-intercept is $(0, -8)$; the x-intercept is $(12, 0)$.

21. Find two ordered pairs. We will find the intercepts.

If $x = 0$,
$x - y = 4$
$0 - y = 4$,
$y = -4$.

If $y = 0$,
$x - y = 4$
$x - 0 = 4$
$x = 4$.

Two points on the line are $(0, -4)$ and $(4, 0)$. Plot these points and draw the line through them.

25. Find two ordered pairs. We will find the intercepts.

If $x = 0$,
$3y = 4x + 12$
$3y = 4(0) + 12$
$3y = 12$
$y = 4$.

If $y = 0$,
$3y = 4x + 12$
$3(0) = 4x + 12$
$0 = 4x + 12$
$-12 = 4x$
$x = -3$.

Two points are $(0, 4)$ and $(-3, 0)$. Plot these points and draw the line through them.

29. Find two ordered pairs. There is only one intercept for this equation, so we choose $x = 0$ and $x = 2$.

If $x = 0$,
$y - 2x = 0$
$y - 2(0) = 0$.
$y = 0$

If $x = 2$,
$y - 2x = 0$
$y - 2(2) = 0$
$y - 4 = 0$
$y = 4$.

Two points are $(0, 0)$ and $(2, 4)$. Plot these points and draw the line through them.

33. If $y + 1 = 0$, then $y = -1$. This is the equation of a horizontal line with y-intercept $(0, -1)$.

37. **(a)** The initial value is the y-intercept, \$30,000.
(b) The depreciation is the change in value, y, from year 0 to year 3. Use the corresponding y-values: \$30,000 − \$15,000 = \$15,000.
(c) The yearly depreciation is the change in y for 1 year's change in x. This is the definition of slope, so find the slope. Use the points (0, 30,000) and (3, 15,000).

$$m = \frac{y_2 - y_1}{x_2 - x_1} = \frac{30{,}000 - 15{,}000}{0 - 3} = \frac{15{,}000}{-3} = -5000.$$

The yearly depreciation is \$5000. The negative slope indicates that the value is decreasing.

41. $\frac{-2 + 2}{-9 - 4} = \frac{0}{-13} = 0$

SOLUTIONS

SECTION 6.3 (page 403)

See the answer section in your text for all graphs.

9. The given points are (3, 2) and (−1, −4). Let $(x_1, y_1) = (3, 2)$ and $(x_2, y_2) = (-1, -4)$. Then

$$m = \frac{y_2 - y_1}{x_2 - x_1} = \frac{-4 - 2}{-1 - 3} = \frac{-6}{-4} = \frac{3}{2}.$$

13. The given points are (−2, 3) and (−2, −3). Let $(x_1, y_1) = (-2, 3)$ and $(x_2, y_2) = (-2, -3)$. Then

$$m = \frac{y_2 - y_1}{x_2 - x_1} = \frac{-3 - 3}{-2 - (-2)} = \frac{-6}{0}.$$

The slope is undefined because of the 0 denominator.

17. Let $(x_1, y_1) = (4, -1)$ and $(x_2, y_2) = (-2, -8)$. Then

$$m = \frac{y_2 - y_1}{x_2 - x_1} = \frac{-8 - (-1)}{-2 - 4} = \frac{-7}{-6} = \frac{7}{6}.$$

21. Let $(x_1, y_1) = (6, -5)$ and $(x_2, y_2) = (-12, -5)$. Then

$$m = \frac{y_2 - y_1}{x_2 - x_1} = \frac{-5 - (-5)}{-12 - 6} = \frac{0}{-18} = 0.$$

25. Let $(x_1, y_1) = (3.1, 2.6)$ and $(x_2, y_2) = (1.6, 2.1)$. Then

$$m = \frac{y_2 - y_1}{x_2 - x_1} = \frac{2.1 - 2.6}{1.6 - 3.1} = \frac{-.5}{-1.5} = \frac{1}{3}.$$

29. Solve the equation for y.

$$2y = -x + 4$$

$$y = -\frac{1}{2}x + 2 \quad \text{Divide by 2.}$$

The coefficient of x gives the slope, so the slope is $-\frac{1}{2}$.

33. Solve the equation for y.

$$-6x + 4y = 4$$

$$4y = 6x + 4 \quad \text{Add } 6x.$$

$$y = \frac{3}{2}x + 1 \quad \text{Divide by 4.}$$

The slope is $\frac{3}{2}$, the coefficient of x.

37. **(a)** The slope is negative because the graph falls from left to right.
(b) The y-value of the y-intercept is negative because the graph intersects the y-axis below the origin.

41. **(a)** The slope is zero because the line is horizontal.
(b) The y-value of the y-intercept is positive because the graph intersects the y-axis above the origin.

45. Solve each equation for y.

$$8x - 9y = 6 \qquad 8x + 6y = -5$$

$$-9y = -8x + 6 \qquad 6y = -8x - 5$$

$$y = \frac{8}{9}x - \frac{2}{3} \qquad y = -\frac{4}{3}x - \frac{5}{6}$$

The slopes are given by the coefficients of x. Since the two slopes are $\frac{8}{9}$ and $-\frac{4}{3}$, the lines are neither parallel nor perpendicular.

49. Since the slope is the ratio of the vertical rise to the horizontal run, the slope is $\frac{6}{20} = \frac{3}{10}$.

SECTION 6.4 (page 411)

1. Count to see that $\frac{\text{rise}}{\text{run}} = \frac{3}{1} = 3$, so the slope is 3. The y-intercept is (0, −3). Use the slope-intercept form $y = mx + b$ to get $y = 3x - 3$.

5. Given $m = 4$ and $b = -3$, use $y = mx + b$ to get $y = 4x - 3$.

13. Locate (1, −5) on your grid. Since $\frac{\text{rise}}{\text{run}} = -\frac{2}{5} = \frac{-2}{5}$, count 2 units down (−2) and 5 units to the right (5) to get another point on the line. Draw a line through the two points.

17. Locate (3, 2) on your grid. Because $m = 0$, the line is a horizontal line through the point (3, 2).

25. Let $(x_1, y_1) = (3, -10)$; $m = -2$. Use the form

$$y - y_1 = m(x - x_1).$$

$$y - (-10) = -2(x - 3)$$

$$y + 10 = -2x + 6 \quad \text{Distributive property}$$

$$2x + y + 10 = 6 \quad \text{Add } 2x.$$

$$2x + y = -4 \quad \text{Subtract 10.}$$

29. Let $(x_1, y_1) = (6, -3)$; $m = -\frac{4}{5}$. Use the form

$$y - y_1 = m(x - x_1).$$

$$y - (-3) = -\frac{4}{5}(x - 6)$$

$$5(y + 3) = -4(x - 6) \quad \text{Multiply by 5 to clear fractions.}$$

$$5y + 15 = -4x + 24 \quad \text{Distributive property}$$

$$4x + 5y + 15 = 24 \quad \text{Add } 4x.$$

$$4x + 5y = 9 \quad \text{Subtract 15.}$$

33. Given (8, 5) and (9, 6), first find the slope of the line.

$$m = \frac{6 - 5}{9 - 8} = \frac{1}{1} = 1$$

Now use the point (8, 5) and $m = 1$ in the point-slope form.

$$y - y_1 = m(x - x_1)$$

$$y - 5 = 1(x - 8)$$

$$y - 5 = x - 8$$

$$-x + y - 5 = -8 \quad \text{Subtract } x.$$

$$-x + y = -3 \quad \text{Add 5.}$$

$$x - y = 3 \quad \text{Multiply by } -1.$$

37. First find the slope using the given points (0, −2) and (−3, 0).

$$m = \frac{0 - (-2)}{-3 - 0} = \frac{2}{-3} = -\frac{2}{3}$$

Now use the point $(-3, 0)$ and $m = -\frac{2}{3}$ in the point-slope form.

$$y - y_1 = m(x - x_1)$$
$$y - 0 = -\frac{2}{3}(x - (-3))$$
$$3y = -2(x + 3) \quad \text{Multiply by 3.}$$
$$3y = -2x - 6 \quad \text{Distributive property}$$
$$2x + 3y = -6 \quad \text{Add } 2x.$$

41. If $m = .25$ and $b = 400$, substitute these values into $y = mx + b$ to get

$y = .25x + 400$ as the cost equation.

(a) Let $x = 100$.

$$y = .25(100) + 400$$
$$y = 25 + 400$$
$$y = 425$$

The cost is \$425.

(b) Let $y = 775$.

$$775 = .25x + 400$$
$$375 = .25x \quad \text{Subtract 400.}$$
$$1500 = x \quad \text{Divide by .25.}$$

1500 snow cones will have a total cost of \$775.

45. If $C = 0$ and $F = 32$, a corresponding ordered pair will be (0, 32). Similarly, if $C = 100$ and $F = 212$, the ordered pair will be (100, 212).

49. $3x + 8 > -1$

$$3x > -9 \quad \text{Subtract 8.}$$
$$x > -3 \quad \text{Divide by 3.}$$

See the graph in the answer section of your text.

SECTION 6.5 (page 421)

1. Substitute the x- and y-values into the inequality.

(a)
$$3x - 4y < 12$$
$$3(2) - 4(6) < 12 \quad \text{Let } x = 2 \text{ and } y = 6.$$
$$6 - 24 < 12$$
$$-18 < 12 \quad \text{True}$$

Yes, (2, 6) is a solution.

(b)
$$3x - 4y < 12$$
$$3(4) - 4(0) < 12 \quad \text{Let } x = 4 \text{ and } y = 0.$$
$$12 - 0 < 12$$
$$12 < 12 \quad \text{False}$$

No, (4, 0) is not a solution.

See the answer section in your text for all graphs.

5. Graph $x + y \geq 4$. Choose the point (0, 0), and test it in the inequality.

$$x + y \geq 4$$
$$0 + 0 \geq 4 \quad \text{Let } x = 0 \text{ and } y = 0.$$
$$0 \geq 4 \quad \text{False}$$

Therefore, shade the region not containing (0, 0).

9. Graph $-3x + 4y > 12$. Choose the point (0, 0), and test it in the inequality.

$$-3x + 4y > 12$$
$$-3(0) + 4(0) > 12 \quad \text{Let } x = 0 \text{ and } y = 0.$$
$$0 + 0 > 12$$
$$0 > 12 \quad \text{False}$$

Therefore, shade the region not containing (0, 0).

13. Graph $x \leq 3y$. Choose the point (0, 2), and test it in the inequality.

$$x \leq 3y$$
$$0 \leq 3(2) \quad \text{Let } x = 0 \text{ and } y = 2.$$
$$0 \leq 6 \quad \text{True}$$

Therefore, shade the region containing (0, 2).

17. For $x + y \leq 5$, draw the line $x + y = 5$. This line goes through the points (0, 5) and (5, 0). The line should be solid because of the $\leq$ sign. Choose the point (0, 0) as a test point. Since $0 + 0 \leq 5$ is true, shade the region containing (0, 0).

21. For $2x + 3y > -6$, draw the line $2x + 3y = -6$. This line goes through the points $(0, -2)$ and $(-3, 0)$. The line should be dashed because of the $>$ sign. Choose the point (0, 0) as a test point. Since $2(0) + 3(0) > -6$ is true, shade the region containing (0, 0).

25. For $x \leq -2$, draw the vertical line $x = -2$. This line goes through the points $(-2, 0)$ and $(-2, 4)$. The line should be solid because of the $\leq$ sign. Choose the point (0, 0) as a test point. Since $0 \leq -2$ is false, shade the region not containing (0, 0).

29. For $y \geq 4x$, draw the line $y = 4x$. This line goes through the points (0, 0) and (1, 4). The line should be solid because of the $\geq$ sign. Choose the point (2, 2) as a test point. Since $2 \geq 4(2)$ is false, shade the region not containing (2, 2).

33. There are many ordered pairs that satisfy all three inequalities. Examples of some are (0, 6000), (6000, 0), (2000, 2000), (1500, 1500), (2000, 3000), (1000, 3000), and (3000, 1000). To graph the region that satisfies all the inequalities, draw the solid line $x + y = 6000$. Shade the region below this line, above the x-axis ($y = 0$), and to the right of the y-axis ($x = 0$).

37. Add in columns.

$$\begin{array}{r} -3x + 2y \\ \underline{3x - 2y} \\ 0 + 0 \end{array}$$

The sum is 0.

CHAPTER 7

SECTION 7.1 (page 443)

1. The ordered pair (1, −2) does make the first equation true, but if we replace x with 1 and y with −2 in the second equation, we get $2(1) + (-2) = 4$, or $0 = 4$, which is false. Because the ordered pair does not satisfy both equations in the system, it is not a solution of the system.

5. Decide whether or not (2, −3) is a solution of the system by substituting 2 for x and −3 for y in each equation.

$$\begin{aligned} x + y &= -1 \\ 2 + (-3) &= -1 \\ -1 &= -1 \quad \text{True} \end{aligned} \qquad \begin{aligned} 2x + 5y &= 19 \\ 2(2) + 5(-3) &= 19 \\ 4 + (-15) &= 19 \\ -11 &= 19 \quad \text{False} \end{aligned}$$

The ordered pair (2, −3) is not a solution, because it does not satisfy the second equation.

9. Decide whether (7, −2) is a solution of the system by substituting 7 for x and −2 for y in each equation.

$$\begin{aligned} 4x &= 26 - y \\ 4(7) &= 26 - (-2) \\ 28 &= 26 + 2 \\ 28 &= 28 \quad \text{True} \end{aligned} \qquad \begin{aligned} 3x &= 29 + 4y \\ 3(7) &= 29 + 4(-2) \\ 21 &= 29 - 8 \\ 21 &= 21 \quad \text{True} \end{aligned}$$

Since (7, −2) satisfies both equations, it is a solution of the system.

13. From the graph, the ordered pair that is a solution of the system is in the second quadrant. The ordered pair (2, 2) is in the first quadrant, (−2, −2) is in the third quadrant and (2, −2) is in the fourth quadrant. Choice (b), (−2, 2), is the only ordered pair given that is in the second quadrant.

17. Graph both lines on the same axes. For $x + y = 4$, use the intercepts, (0, 4) and (4, 0). For $y - x = 4$, use the intercepts, (0, 4) and (−4, 0). These lines intersect at (0, 4). See the answer section in your text for the graph.

21. Graph both lines on the same axes. For $3x - 2y = -3$, use the points (−1, 0) and (1, 3). For $-3x - y = -6$, use the intercepts (0, 6) and (2, 0). These lines intersect at (1, 3).

25. Graph both lines on the same axes. For $3x - 4y = 24$, use the intercepts (8, 0) and (0, −6). For $y = -\frac{3}{2}x + 3$, use the intercepts (0, 3) and (2, 0). These lines intersect at (4, −3).

29. Answers will vary. One example would be

$$\begin{aligned} x + y &= 1 \\ x - y &= -5, \end{aligned}$$

because both equations are satisfied when $x = -2$ and $y = 3$. Graph both lines on the same axes. For $x + y = 1$, use the intercepts, (0, 1) and (1, 0). For $x - y = -5$, use the intercepts, (0, 5) and (−5, 0). These lines intersect at (−2, 3).

33. Graph each line. The two lines are parallel, so there is no solution.

37. Graph each line. The two lines are parallel, so there is no solution.

41. Add in columns.

$$\begin{array}{r} -3x + 7y \\ \underline{3x - 7y} \\ 0 + 0 \end{array}$$

The sum is 0.

SECTION 7.2 (page 453)

Answers should be checked by substituting in the original equations.

1. If we use the addition method on each system, choice (c) is the only system in which a variable will be eliminated. If we add the equations in (c) we get $3x = 17$. (The variable y was eliminated.)

5. $x + y = 2$
$2x - y = -5$
Add the equations, getting $3x = -3$ or $x = -1$. Replace x with −1 in either equation. If we use $x + y = 2$, we get $-1 + y = 2$ or $y = 3$. The solution is (−1, 3).

9. $3x + 2y = 0$
$-3x - y = 3$
Add the equations, getting $y = 3$. Replace y with 3 in either equation. If we use $3x + 2y = 0$, we get $3x + 2(3) = 0$ or $3x + 6 = 0$ or $x = -2$. The solution is (−2, 3).

13. $2x - y = 12$
$3x + 2y = -3$
Multiply each side of the first equation by 2 and add.

$$\begin{aligned} 4x - 2y &= 24 && \text{Multiply by 2.} \\ \underline{3x + 2y} &\underline{= -3} \\ 7x \quad &= 21 && \text{Add.} \\ x &= 3 \end{aligned}$$

Let $x = 3$ in the second equation to get

$$\begin{aligned} 3x + 2y &= -3 \\ 3(3) + 2y &= -3 && \text{Let } x = 3. \\ 9 + 2y &= -3 && \text{Multiply.} \\ 2y &= -12 && \text{Subtract 9.} \\ y &= -6. && \text{Divide by 2.} \end{aligned}$$

The solution is (3, −6).

17. $x + 4y = 16$
$3x + 5y = 20$
Multiply each side of the first equation by -3 and add.

$$\begin{array}{rll} -3x - 12y &= -48 & \text{Multiply by } -3. \\ 3x + 5y &= 20 & \\ \hline -7y &= -28 & \text{Add.} \\ y &= 4 & \text{Divide by } -7. \end{array}$$

Let $y = 4$ in the first equation to get

$$\begin{aligned} x + 4y &= 16 \\ x + 4(4) &= 16 && \text{Let } y = 4. \\ x + 16 &= 16 && \text{Multiply.} \\ x &= 0. && \text{Subtract 16.} \end{aligned}$$

The solution is (0, 4).

21. $3x - 2y = 22$
$-5x + 4y = -36$
One way to proceed is to multiply the first equation by 2, then add.

$$\begin{array}{rll} 6x - 4y &= 44 & \text{Multiply by 2.} \\ -5x + 4y &= -36 & \\ \hline x &= 8 & \text{Add.} \end{array}$$

Replace x with 8 in either equation to find that $y = 1$. The solution is (8, 1).

29. $2x + 3y = 0$
$7y - 29 = 5x$
Rewrite the second equation in standard form.

$$\begin{aligned} 7y - 29 &= 5x \\ -29 &= 5x - 7y && \text{Subtract } 7y. \\ 5x - 7y &= -29 \end{aligned}$$

One way to proceed is to multiply the first equation by 7 to get $14x + 21y = 0$, and the second equation by 3 to get $15x - 21y = -87$. Add these two equations to get $29x = -87$ or $x = -3$. Replace x with -3 in either of the original equations to find $y = 2$, giving the solution $(-3, 2)$.

33. $3x = 3 + 2y$
$-\frac{4}{3}x + y = \frac{1}{3}$

Begin by writing the first equation as $3x - 2y = 3$. One way to proceed is to multiply the second equation by 3 to get $-4x + 3y = 1$. Multiply the first equation by 4 to get $12x - 8y = 12$, and multiply the second equation by 3 to get $-12x + 9y = 3$. Add these equations to get $y = 15$. Replace y with 15 in one of the original equations to find $x = 11$, giving the solution (11, 15).

37. $-x + 3y = 4$
$-2x + 6y = 8$
Multiply the first equation by -2 to get $2x - 6y = -8$. Adding this equation to the second gives $0 = 0$. This true result means that the system has an infinite number of solutions. The equations of the system represent the same line.

41. $6x + 3y = 0$
$-18x - 9y = 0$
Begin by multiplying the first equation by 3 to get $18x + 9y = 0$. Adding this equation to the second equation gives $0 = 0$. This true result means that the system has an infinite number of solutions. The equations of the system represent the same line.

45.
$$\begin{aligned} -2(y - 2) + 5y &= -5 \\ -2y + 4 + 5y &= -5 && \text{Distributive property} \\ 3y + 4 &= -5 && \text{Combine terms.} \\ 3y &= -9 && \text{Subtract 4.} \\ y &= -3 && \text{Divide by 3.} \end{aligned}$$

49. $4x - 2\left(\frac{1 - 3x}{2}\right) = 6$

$$\begin{aligned} 4x - (1 - 3x) &= 6 && \text{Multiply.} \\ 4x - 1 + 3x &= 6 && \text{Distributive property} \\ 7x - 1 &= 6 && \text{Combine terms.} \\ 7x &= 7 && \text{Add 1.} \\ x &= 1 && \text{Divide by 7.} \end{aligned}$$

SECTION 7.3 (page 461)

Answers should be checked by substituting in the original equations.

5. $3x + 2y = 27$
$x = y + 4$
The second equation gives x in terms of y. Substitute $y + 4$ for x in the first equation.

$$\begin{aligned} 3x + 2y &= 27 \\ 3(y + 4) + 2y &= 27 && \text{Let } x = y + 4. \\ 3y + 12 + 2y &= 27 && \text{Distributive property} \\ 5y + 12 &= 27 && \text{Combine terms.} \\ 5y &= 15 && \text{Subtract 12.} \\ y &= 3 && \text{Divide by 5.} \end{aligned}$$

To find x, substitute $y = 3$ in the equation $x = y + 4$ to get $x = 3 + 4 = 7$. The solution is (7, 3).

9. $3x + 4 = -y$
$2x + y = 0$
Solve either equation for y. If we solve the first equation for y, we get $y = -3x - 4$. Now substitute this value for x in the second equation.

$$\begin{aligned} 2x + y &= 0 \\ 2x + (-3x - 4) &= 0 && \text{Let } y = -3x - 4. \\ -x - 4 &= 0 && \text{Combine terms.} \\ -x &= 4 && \text{Add 4.} \\ x &= -4 && \text{Multiply by } -1. \end{aligned}$$

To find y, substitute $x = -4$ in $y = -3x - 4$ to get $y = -3(-4) - 4 = 12 - 4 = 8$. The solution is $(-4, 8)$.

13. $3x - y = 5$
$y = 3x - 5$
The second equation says that $y = 3x - 5$. Replace y with $3x - 5$ in the first equation to get

$$3x - (3x - 5) = 5$$
$$3x - 3x + 5 = 5$$
$$5 = 5. \quad \text{True}$$

This true result means that the system has an infinite number of solutions. The equations of the system represent the same line.

17. $2x + 8y = 3$
$x = 8 - 4y$
The second equation says that $x = 8 - 4y$. Replace x with $8 - 4y$ in the first equation to get

$$2(8 - 4y) + 8y = 3$$
$$16 - 8y + 8y = 3$$
$$16 = 3. \quad \text{False}$$

This false statement means that the system has no solution. The equations of the system represent parallel lines.

21. $4x - 3y = -8$
$x + 3y = 13$
(a) Add the two equations to get $5x = 5$ or $x = 1$. Replace x with 1 in either of the original equations to find $y = 4$, giving the solution (1, 4).
(b) One way to proceed is to solve the second equation for x to get $x = -3y + 13$. Replace x with $-3y + 13$ in the first equation to get

$$4(-3y + 13) - 3y = -8$$
$$-12y + 52 - 3y = -8$$
$$-15y + 52 = -8$$
$$-15y = -60$$
$$y = 4.$$

Now replace y with 4 in $x = -3y + 13$ to get $x = -3(4) + 13 = -12 + 13 = 1$. The solution is (1, 4).

In Exercises 25–37, only one method is shown.

25. $2x - 8y + 3y + 2 = 5y + 16$
$8x - 2y = 4x + 28$
Combine terms in each equation to get $2x - 10y = 14$ for the first equation and $4x - 2y = 28$ for the second equation. Multiply the first equation by -2 to get $-4x + 20y = -28$. Add this result to the second equation to get $18y = 0$ or $y = 0$. Replace y with 0 in either equation to get $x = 7$. The solution is (7, 0).

29. $5x + y = 12 - x - 7y$
$3x + 2y = 10 - 6x - 10y$
Rearrange and combine terms in each equation to get $6x + 8y = 12$ for the first equation and $9x + 12y = 10$ for the second equation. Multiply the first of these equations by -3 and the second by 2 to get $-18x - 24y = -36$ and $18x + 24y = 20$. Add these two equations to get $0 = -16$. This false statement means that the system has no solution.

33. $x + \frac{1}{3}y = y - 2$

$\frac{1}{4}x + y = x + y$

Rearrange terms and multiply the first equation by 3.

$$3\left(x + \frac{1}{3}y\right) = 3(y - 2)$$
$$3x + y = 3y - 6 \quad \text{Distributive property}$$
$$3x - 2y = -6 \quad \text{Subtract } 3y.$$

Rearrange terms and multiply the second equation by 4.

$$4\left(\frac{1}{4}x + y\right) = 4(x + y)$$
$$x + 4y = 4x + 4y \quad \text{Distributive property}$$
$$-3x + 0 = 0 \quad \text{Subtract } 4x \text{ and } 4y.$$
$$-3x = 0$$
$$x = 0$$

Replace x with 0 in $3x - 2y = -6$ to get $-2y = -6$ or $y = 3$. The solution is (0, 3).

37. $\frac{x}{3} - \frac{3y}{4} = -\frac{1}{2}$

$\frac{x}{6} + \frac{y}{8} = \frac{3}{4}$

Multiply the first equation by 12.

$$12\left(\frac{x}{3} - \frac{3y}{4}\right) = 12\left(-\frac{1}{2}\right)$$
$$4x - 9y = -6$$

Multiply the second equation by 24.

$$24\left(\frac{x}{6} + \frac{y}{8}\right) = 24\left(\frac{3}{4}\right)$$
$$4x + 3y = 18$$

Multiply $4x - 9y = -6$ by -1 and add to $4x + 3y = 18$.

$$\begin{aligned} -4x + 9y &= 6 \\ 4x + 3y &= 18 \\ \hline 12y &= 24 \\ y &= 2 \end{aligned}$$

Replace y with 2 in either equation to get $x = 3$. The solution is (3, 2).

41. From Exercise 40, the United States earned 108 medals. Let x represent the number of gold medals won. Then x also represents the number of bronze medals earned since the team had the same number of gold and bronze medals. Then $x - 3$ represents the number of silver medals won. Add the number of each type of medal to get the total number of medals, 108.

$$x + x + (x - 3) = 108$$
$$3x - 3 = 108 \quad \text{Combine terms.}$$
$$3x = 111 \quad \text{Add 3.}$$
$$x = 37 \quad \text{Divide by 3.}$$

If $x = 37$, then $x - 3 = 37 - 3 = 34$. The United States earned 37 gold medals, 34 silver medals, and 37 bronze medals.

SECTION 7.4 (page 469)

5. Let x represent the number of people who lived in Harding County and let y represent those who lived in Los Alamos County. Since a total of 1096 people lived in these counties, one equation is

$$x + y = 1096.$$

Harding County had 878 more people than Los Alamos, so another equation is

$$x = y + 878.$$

We must solve the system

$$x + y = 1096$$
$$x = y + 878.$$

Write the second equation as $x - y = 878$. Add this to the first equation to get $2x = 1974$ or $x = 987$. Replace x with 987 in either equation to get $y = 109$. There were 987 people in Harding County and 109 people in Los Alamos County.

9. Fill in the chart as follows, realizing that the entries in the "total value" column were found by multiplying the denomination of the bill by the number of bills.

Denomination of bill	*Number of bills*	*Total value (in dollars)*
\$1	x	x
\$10	y	$10y$
Totals	74	\$326

The total number of bills is 74, so

$$x + y = 74.$$

Since the total value is \$326, the right-hand column leads to

$$x + 10y = 326.$$

These two equations give the system:

$$x + y = 74$$
$$x + 10y = 326.$$

Multiply the first equation by -1 to get $-x - y = -74$. Add this to the second equation to get $9y = 252$ or $y = 28$. Replace y with 28 in either equation to get $x = 46$. The clerk has 46 one-dollar bills and 28 ten-dollar bills.

13. Let x represent the amount invested at 5% and y represent the amount invested at 4%. Make a chart.

Amount invested	*Interest rate (as a decimal)*	*Interest income (yearly)*
x	.05	$.05x$
y	.04	$.04y$
		350

Since Maria has invested twice as much at 5% as at 4% one equation is

$$x = 2y.$$

Her total interest income is \$350, so a second equation is

$$.05x + .04y = 350.$$

Solve the system by substitution. Replace x with $2y$ in the second equation to get

$$.05(2y) + .04y = 350$$
$$.14y = 350$$
$$y = 2500.$$

Replace y with 2500 in $x = 2y$ to get $x = 2(2500) = 5000$. Maria has \$5000 invested at 5% and \$2500 invested at 4%.

17. Let x represent the number of liters of 40% solution and y represent the number of liters of 70% solution. Make a chart.

Liters of solution	*Percent (as a decimal)*	*Liters of pure dye*
x	.40	$.40x$
y	.70	$.70y$
120	.50	$.50(120) = 60$

Write two equations:

$$x + y = 120$$

and

$$.40x + .70y = 60.$$

Solve this system. Multiply the second equation by 100 (clear the decimals) to get $40x + 70y = 6000$. Now,

SOLUTIONS

solve the first equation for x to get $x = 120 - y$, and substitute $120 - y$ for x in the second equation.

$$\begin{aligned} 40(120 - y) + 70y &= 6000 \\ 4800 - 40y + 70y &= 6000 \\ 4800 + 30y &= 6000 \\ 30y &= 1200 \\ y &= 40 \end{aligned}$$

To find x, substitute 40 for y in $x = 120 - y$ to get $x = 120 - 40 = 80$. Thus, 80 liters of 40% solution and 40 liters of 70% solution are needed.

21. Let x represent the number of barrels of \$40 pickles and y represent the number of barrels of \$60 pickles. Make a chart.

Barrels of pickles	*Price per barrel (in dollars)*	*Total price (in dollars)*
x	40	$40x$
y	60	$60y$
50	48	$48(50) = 2400$

Write two equations: $x + y = 50$ and $40x + 60y = 2400$. Solve this system. Solve the first equation for x to get $x = 50 - y$ and substitute $50 - y$ for x in the second equation.

$$\begin{aligned} 40(50 - y) + 60y &= 2400 \\ 2000 - 40y + 60y &= 2400 \\ 2000 + 20y &= 2400 \\ 20y &= 400 \\ y &= 20 \end{aligned}$$

To find x, substitute 20 for y in $x = 50 - y$ to get $x = 50 - 20 = 30$. One should mix 30 barrels of \$40 pickles and 20 barrels of \$60 pickles.

25. Let x represent the plane's speed in still air and y the wind speed. The rate of the plane with the wind is $x + y$, so one equation is $x + y = 500$. The rate of the plane against the wind is $x - y$, so another equation is $x - y = 440$. Solve the system by adding to eliminate y.

$$\begin{aligned} x + y &= 500 \\ x - y &= 440 \\ \hline 2x \quad &= 940 \\ x \quad &= 470 \end{aligned}$$

Substitute 470 for x in the first equation to get $y = 30$. The speed of the plane in still air is 470 miles per hour. The wind speed is 30 miles per hour.

29. Let x represent Roberto's speed and y represent Juana's speed. Draw two diagrams.

Roberto → Juana →

30 miles

Roberto's distance is 30 miles more than Juana's when he overtakes her in 60 hours.

Roberto	Juana
time: 60 hours	time: 60 hours
rate: x	rate: y
distance: $60x$	distance: $60y$

Thus, one equation is $60x = 60y + 30$, or $60x - 60y = 30$.

Roberto →← Juana

30 miles

Roberto and Juana meet after 5 hours. The sum of their distances is 30 miles.

Roberto	Juana
time: 5 hours	time: 5 hours
rate: x	rate: y
distance: $5x$	distance: $5y$

Therefore, another equation is $5x + 5y = 30$. Solve the system:

$$\begin{aligned} 60x - 60y &= 30 \\ 5x + 5y &= 30. \end{aligned}$$

Multiply the second equation by -12 to get $-60x - 60y = -360$. Add this to the first equation to get $-120y = -330$ or $y = 2\frac{3}{4}$. Replace y with $2\frac{3}{4}$ in either equation to get $x = 3\frac{1}{4}$. Roberto's rate is $3\frac{1}{4}$ miles per hour, and Juana's rate is $2\frac{3}{4}$ miles per hour.

33. *Step 1* Graph $y = -3x + 2$.
This graph is a straight line with intercepts $(0, 2)$ and $(\frac{2}{3}, 0)$. Draw this as a solid line because of the $\geq$ symbol.

Step 2 Use $(0, 0)$ as a test point.

$y \geq -3x + 2$	Original inequality
$0 \geq -3(0) + 2$	Let $x = 0$ and $y = 0$.
$0 \geq 0 + 2$	
$0 \geq 2$	False

Step 3 Since the statement is false, shade the region that does not contain $(0, 0)$. See the graph in the answer section.

SECTION 7.5 (page 477)

See the answer section in your text for all graphs.

5. $4x + 5y \geq 20$
$x - 2y \leq 5$

Step 1 Graph the line $4x + 5y = 20$. If $x = 0$, then $y = 4$, giving the ordered pair (0, 4). If $y = 0$, then $x = 5$, giving the ordered pair (5, 0). Plot these two points and draw the solid line through them.

Step 2 Use (0, 0) as a test point.

$4x + 5y \geq 20$	Original inequality
$0 + 0 \geq 20$	Substitute $x = 0$, $y = 0$.
$0 \geq 20$	False

Step 3 Since the inequality is false for (0, 0), shade the region on the side of the line that does not contain (0, 0).

Repeat the same steps for the second inequality.

Step 1 Graph the line $x - 2y = 5$. Use the points (5, 0) and (1, −2). Plot these two points and draw the solid line through them.

Step 2 Use (0, 0) as a test point.

$x - 2y \leq 5$	Original inequality
$0 - 0 \leq 5$	Substitute $x = 0$, $y = 0$.
$0 \leq 5$	True

Step 3 Since the inequality is true for (0, 0), shade the region on the side of the line containing (0, 0). The solution is the area where the two shaded regions overlap, or the darkest shaded region.

9. $y \leq 2x - 5$
$x < 3y + 2$
Graph the line $y = 2x - 5$. Make it a solid line. Use (0, 0) as a test point.

$y \leq 2x - 5$	Original inequality
$0 \leq 2 \cdot 0 - 5$	Substitute $x = 0$, $y = 0$.
$0 \leq 0 - 5$	
$0 \leq -5$	False

Shade the region that does not contain (0, 0). Graph the line $x = 3y + 2$. Make it a dashed line. Use (0, 0) as a test point.

$x < 3y + 2$	Original inequality
$0 < 3 \cdot 0 + 2$	Substitute $x = 0$, $y = 0$.
$0 < 0 + 2$	
$0 < 2$	True

Shade the region containing (0, 0). The solution is the overlapped portion of the shaded areas.

13. $x \leq 2y + 3$
$x + y < 0$
Graph the line $x = 2y + 3$. Make the line solid. Use (0, 0) as a test point.

$x \leq 2y + 3$	Original inequality
$0 \leq 2 \cdot 0 + 3$	Substitute $x = 0$, $y = 0$.
$0 \leq 0 + 3$	
$0 \leq 3$	True

Shade the region containing (0, 0). Graph the line $x + y = 0$. Make the line dashed. Use a test point not on the line, for example, (1, 0).

$x + y < 0$	Original inequality
$1 + 0 < 0$	Substitute $x = 1$, $y = 0$.
$1 < 0$	False

Shade the region not containing (1, 0). The solution is the overlapped portion of the shaded regions.

17. $x - 3y \leq 6$
$x \geq -4$
Graph the line $x - 3y = 6$. Make it a solid line. Use (0, 0) for a test point. The inequality $x - 3y \leq 6$ is true at (0, 0), so shade the region containing (0, 0). Graph the line $x = -4$. Make it a solid line. Use (0, 0) for a test point. The inequality $x \geq -4$ is true at (0, 0), so shade the region containing (0, 0). The solution is the overlapped portion of the shaded regions.

21. $16^2 = 16 \cdot 16 = 256$

CHAPTER 8

SECTION 8.1 (page 503)

1. A positive number has two square roots. One is positive, and the other is negative.

5. A negative number has one real cube root; this root is negative.

9. The square roots of 144 are 12 and −12, since $(12)(12) = 144$ and $(-12)(-12) = 144$.

13. The square roots of 900 are 30 and −30, since $(30)(30) = 900$ and $(-30)(-30) = 900$.

17. $(-\sqrt{19})^2 = (-\sqrt{19})(-\sqrt{19}) = 19$

21. a must be positive because the square root of a negative number is not a real number and $\sqrt{0} = 0$.

25. $\sqrt{49} = \sqrt{7 \cdot 7} = 7$

29. $$-\sqrt{\frac{144}{121}} = -\sqrt{\frac{2 \cdot 2 \cdot 2 \cdot 2 \cdot 3 \cdot 3}{11 \cdot 11}} = -\left(\frac{2 \cdot 2 \cdot 3}{11}\right) = -\frac{12}{11}$$

33. The number $\sqrt{25}$ is rational; $\sqrt{25} = \sqrt{5 \cdot 5} = 5$.

37. The number $-\sqrt{64}$ is rational; $-\sqrt{64} = -\sqrt{2 \cdot 2 \cdot 2 \cdot 2 \cdot 2 \cdot 2} = -(2 \cdot 2 \cdot 2) = -8$.

41. The number $\sqrt{-29}$ is not a real number since there is no real number whose square is −29.

45. $c^2 = a^2 + b^2$
$c^2 = 8^2 + 15^2$ Let $a = 8$ and $b = 15$.
$c^2 = 64 + 225$ Square.
$c^2 = 289$
$c = \sqrt{289} = \sqrt{17 \cdot 17} = 17$

49. $c^2 = a^2 + b^2$
$c^2 = 11^2 + 4^2$ Let $a = 11$ and $b = 4$.
$c^2 = 121 + 16$ Square.
$c^2 = 137$
$c = \sqrt{137} \approx 11.705$

53. Use the Pythagorean formula. Let b = the vertical distance of the kite above Margaret's hand.

$$a^2 + b^2 = c^2$$
$$60^2 + b^2 = 100^2 \quad \text{Let } a = 60 \text{ and } c = 100.$$
$$3600 + b^2 = 10{,}000 \quad \text{Square.}$$
$$b^2 = 6400 \quad \text{Subtract 3600.}$$
$$b = \sqrt{6400} = 80$$

The kite is 80 feet above her hand.

57. Use the Pythagorean formula.

$$a^2 + b^2 = c^2$$
$$5^2 + 8^2 = x^2 \quad \text{Let } a = 5 \text{ and } b = 8.$$
$$25 + 64 = x^2 \quad \text{Square.}$$
$$89 = x^2$$
$$\sqrt{89} = x$$
$$9.434 \approx x$$

61. $\sqrt[3]{1000} = \sqrt[3]{2 \cdot 2 \cdot 2 \cdot 5 \cdot 5 \cdot 5} = 2 \cdot 5 = 10$

65. $\sqrt[3]{-27} = \sqrt[3]{(-3)(-3)(-3)} = -3$

69. $\sqrt[4]{-1}$ is not a real number since a fourth power of a nonzero real number must be positive.

73. $72 = 2 \cdot 36$
$= 2 \cdot 2 \cdot 18$
$= 2 \cdot 2 \cdot 2 \cdot 9$
$= 2 \cdot 2 \cdot 2 \cdot 3 \cdot 3$
$= 2^3 \cdot 3^2$

SECTION 8.2 (page 511)

1. The statement is true by the product rule for radicals.

5. The statement is false. In general, $\sqrt{x^2} = |x|$, so $\sqrt{(-6)^2} = |-6| = 6$.

9. $\sqrt{6} \cdot \sqrt{15} = \sqrt{6 \cdot 15} = \sqrt{2 \cdot 3 \cdot 3 \cdot 5} = 3\sqrt{2 \cdot 5}$
$= 3\sqrt{10}$

13. $\sqrt{13} \cdot \sqrt{r} = \sqrt{13 \cdot r} = \sqrt{13r}$

17. $\sqrt{45} = \sqrt{9 \cdot 5} = \sqrt{9} \cdot \sqrt{5} = 3\sqrt{5}$

21. $\sqrt{75} = \sqrt{25 \cdot 3} = \sqrt{25} \cdot \sqrt{3} = 5\sqrt{3}$

25. $-\sqrt{700} = -\sqrt{100 \cdot 7} = -\sqrt{100} \cdot \sqrt{7} = -10\sqrt{7}$

29. $\sqrt{3} \cdot \sqrt{18} = \sqrt{3 \cdot 18} = \sqrt{3 \cdot 3 \cdot 6} = 3\sqrt{6}$

33. $\sqrt{12} \cdot \sqrt{30} = \sqrt{12 \cdot 30} = \sqrt{360} = \sqrt{36 \cdot 10}$
$= \sqrt{36} \cdot \sqrt{10} = 6\sqrt{10}$

37. $\sqrt{\dfrac{16}{225}} = \dfrac{\sqrt{16}}{\sqrt{225}} = \dfrac{4}{15}$

41. $\sqrt{\dfrac{5}{7}} \cdot \sqrt{35} = \dfrac{\sqrt{5}}{\sqrt{7}} \cdot \dfrac{\sqrt{35}}{1} = \dfrac{\sqrt{5 \cdot 35}}{\sqrt{7}} = \dfrac{\sqrt{5 \cdot 5 \cdot 7}}{\sqrt{7}}$
$= \dfrac{5\sqrt{7}}{\sqrt{7}} = 5$

45. $\dfrac{30\sqrt{10}}{5\sqrt{2}} = \dfrac{30}{5}\sqrt{\dfrac{10}{2}} = 6\sqrt{5}$

49. $\sqrt{y^4} = \sqrt{y^2 \cdot y^2} = y^2$

53. $\sqrt{400x^6} = \sqrt{20 \cdot 20 \cdot x^3 \cdot x^3} = 20x^3$

57. $\sqrt{x^6y^{12}} = \sqrt{x^3 \cdot x^3 \cdot y^6 \cdot y^6} = x^3y^6$

61. $\sqrt[3]{54} = \sqrt[3]{27 \cdot 2} = \sqrt[3]{27} \cdot \sqrt[3]{2} = 3\sqrt[3]{2}$

65. $\sqrt[4]{80} = \sqrt[4]{16 \cdot 5} = \sqrt[4]{16} \cdot \sqrt[4]{5} = 2\sqrt[4]{5}$

69. $\sqrt[3]{-\dfrac{216}{125}} = \dfrac{\sqrt[3]{-216}}{\sqrt[3]{125}} = \dfrac{-6}{5} = -\dfrac{6}{5}$

73. $4x + 7 - 9x + 12 = 4x - 9x + 7 + 12$
$= -5x + 19$

SECTION 8.3 (page 517)

1. $2\sqrt{3} + 4\sqrt{3} = (2 + 4)\sqrt{3} = 6\sqrt{3}$
The distributive property is used in the first step, where $a = \sqrt{3}$, $b = 2$, and $c = 4$.

5. $14\sqrt{7} - 19\sqrt{7} = (14 - 19)\sqrt{7} = -5\sqrt{7}$

9. $6\sqrt{7} - \sqrt{7} = 6\sqrt{7} - 1\sqrt{7} = (6 - 1)\sqrt{7} = 5\sqrt{7}$

13. $5\sqrt{72} - 3\sqrt{50} = 5\sqrt{36 \cdot 2} - 3\sqrt{25 \cdot 2}$
$= 5 \cdot \sqrt{36} \cdot \sqrt{2} - 3 \cdot \sqrt{25} \cdot \sqrt{2}$
$= 5 \cdot 6 \cdot \sqrt{2} - 3 \cdot 5 \cdot \sqrt{2}$
$= 30\sqrt{2} - 15\sqrt{2}$
$= 15\sqrt{2}$

17. $5\sqrt{7} - 3\sqrt{28} + 6\sqrt{63}$
$= 5\sqrt{7} - 3\sqrt{4 \cdot 7} + 6\sqrt{9 \cdot 7}$
$= 5\sqrt{7} - 3 \cdot \sqrt{4} \cdot \sqrt{7} + 6 \cdot \sqrt{9} \cdot \sqrt{7}$
$= 5\sqrt{7} - 3 \cdot 2 \cdot \sqrt{7} + 6 \cdot 3 \cdot \sqrt{7}$
$= 5\sqrt{7} - 6\sqrt{7} + 18\sqrt{7}$
$= 17\sqrt{7}$

21. $4\sqrt{50} + 3\sqrt{12} - 5\sqrt{45}$
$= 4\sqrt{25 \cdot 2} + 3\sqrt{4 \cdot 3} - 5\sqrt{9 \cdot 5}$
$= 4 \cdot \sqrt{25} \cdot \sqrt{2} + 3 \cdot \sqrt{4} \cdot \sqrt{3} - 5 \cdot \sqrt{9} \cdot \sqrt{5}$
$= 4 \cdot 5 \cdot \sqrt{2} + 3 \cdot 2 \cdot \sqrt{3} - 5 \cdot 3 \cdot \sqrt{5}$
$= 20\sqrt{2} + 6\sqrt{3} - 15\sqrt{5}$

25. $\sqrt{3} \cdot \sqrt{7} + 4\sqrt{21} = \sqrt{3 \cdot 7} + 4\sqrt{21}$
$= \sqrt{21} + 4\sqrt{21}$
$= 5\sqrt{21}$

29. $\sqrt{9x} + \sqrt{49x} - \sqrt{25x} = \sqrt{9} \cdot \sqrt{x} + \sqrt{49} \cdot \sqrt{x}$
$- \sqrt{25} \cdot \sqrt{x}$
$= 3\sqrt{x} + 7\sqrt{x} - 5\sqrt{x}$
$= 5\sqrt{x}$

33. $3\sqrt{8x^2} - 4x\sqrt{2} - x\sqrt{8}$
$= 3\sqrt{4x^2 \cdot 2} - 4x\sqrt{2} - x\sqrt{4 \cdot 2}$
$= 3 \cdot \sqrt{4x^2} \cdot \sqrt{2} - 4x\sqrt{2} - x \cdot \sqrt{4} \cdot \sqrt{2}$
$= 3 \cdot 2x \cdot \sqrt{2} - 4x\sqrt{2} - x \cdot 2 \cdot \sqrt{2}$
$= 6x\sqrt{2} - 4x\sqrt{2} - 2x\sqrt{2}$
$= 0$

37. $2\sqrt{125x^2z} + 8x\sqrt{80z} = 2\sqrt{25x^2 \cdot 5z} + 8x\sqrt{16 \cdot 5z}$
$= 2\sqrt{25x^2}\sqrt{5z} + 8x\sqrt{16}\sqrt{5z}$
$= 2 \cdot 5x\sqrt{5z} + 8x \cdot 4 \cdot \sqrt{5z}$
$= 10x\sqrt{5z} + 32x\sqrt{5z}$
$= 42x\sqrt{5z}$

41. $6\sqrt[3]{8p^2} - 2\sqrt[3]{27p^2} = 6 \cdot \sqrt[3]{8} \cdot \sqrt[3]{p^2}$
$- 2 \cdot \sqrt[3]{27} \cdot \sqrt[3]{p^2}$
$= 6 \cdot 2 \cdot \sqrt[3]{p^2} - 2 \cdot 3 \cdot \sqrt[3]{p^2}$
$= 12\sqrt[3]{p^2} - 6\sqrt[3]{p^2}$
$= 6\sqrt[3]{p^2}$

49. $(-\sqrt{37})^2 = (-\sqrt{37})(-\sqrt{37}) = 37$

SECTION 8.4 (page 523)

5. $\frac{8}{\sqrt{2}} = \frac{8 \cdot \sqrt{2}}{\sqrt{2} \cdot \sqrt{2}} = \frac{8\sqrt{2}}{2} = 4\sqrt{2}$

9. $\frac{7\sqrt{3}}{\sqrt{5}} = \frac{7\sqrt{3} \cdot \sqrt{5}}{\sqrt{5} \cdot \sqrt{5}} = \frac{7\sqrt{15}}{5}$

13. $\frac{16}{\sqrt{27}} = \frac{16}{\sqrt{9 \cdot 3}} = \frac{16}{\sqrt{9} \cdot \sqrt{3}} = \frac{16}{3\sqrt{3}} = \frac{16 \cdot \sqrt{3}}{3\sqrt{3} \cdot \sqrt{3}}$
$= \frac{16\sqrt{3}}{9}$

17. $\frac{63}{\sqrt{45}} = \frac{63 \cdot \sqrt{5}}{\sqrt{45} \cdot \sqrt{5}} = \frac{63\sqrt{5}}{\sqrt{9 \cdot 5 \cdot 5}} = \frac{63\sqrt{5}}{15} = \frac{21\sqrt{5}}{5}$

21. $\frac{-\sqrt{80}}{\sqrt{6}} = \frac{-\sqrt{80} \cdot \sqrt{6}}{\sqrt{6} \cdot \sqrt{6}} = \frac{-\sqrt{8 \cdot 10 \cdot 2 \cdot 3}}{6}$
$= \frac{-\sqrt{16 \cdot 30}}{6} = \frac{-\sqrt{16} \cdot \sqrt{30}}{6}$
$= \frac{-4\sqrt{30}}{6} = \frac{-2\sqrt{30}}{3}$

25. $\sqrt{\frac{13}{5}} = \frac{\sqrt{13}}{\sqrt{5}} = \frac{\sqrt{13} \cdot \sqrt{5}}{\sqrt{5} \cdot \sqrt{5}} = \frac{\sqrt{65}}{5}$

29. $\sqrt{\frac{21}{7}} \cdot \sqrt{\frac{21}{8}} = \frac{\sqrt{21}}{\sqrt{7}} \cdot \frac{\sqrt{21}}{\sqrt{8}} = \frac{21}{\sqrt{7 \cdot 2 \cdot 4}} = \frac{21}{2\sqrt{14}}$
$= \frac{21 \cdot \sqrt{14}}{2 \cdot \sqrt{14} \cdot \sqrt{14}}$
$= \frac{21\sqrt{14}}{2 \cdot 14} = \frac{3\sqrt{14}}{4}$

33. $\sqrt{\frac{2}{9}} \cdot \sqrt{\frac{9}{2}} = \frac{\sqrt{2}}{\sqrt{9}} \cdot \frac{\sqrt{9}}{\sqrt{2}} = \frac{\sqrt{18}}{\sqrt{18}} = 1$

37. $\sqrt{\frac{4x^3}{y}} = \frac{\sqrt{4x^3}}{\sqrt{y}} = \frac{\sqrt{4x^2} \cdot \sqrt{x}}{\sqrt{y}} = \frac{2x\sqrt{x}}{\sqrt{y}}$
$= \frac{2x\sqrt{x} \cdot \sqrt{y}}{\sqrt{y} \cdot \sqrt{y}} = \frac{2x\sqrt{xy}}{y}$

41. $\sqrt{\frac{9a^2r^5}{7t}} = \frac{\sqrt{9a^2r^5}}{\sqrt{7t}} = \frac{\sqrt{9a^2r^4 \cdot r}}{\sqrt{7t}} = \frac{\sqrt{9a^2r^4} \cdot \sqrt{r}}{\sqrt{7t}}$
$= \frac{3ar^2\sqrt{r}}{\sqrt{7t}} = \frac{3ar^2\sqrt{r} \cdot \sqrt{7t}}{\sqrt{7t} \cdot \sqrt{7t}} = \frac{3ar^2\sqrt{7rt}}{7t}$

45. $\sqrt[3]{\frac{3}{2}} = \frac{\sqrt[3]{3}}{\sqrt[3]{2}} = \frac{\sqrt[3]{3} \cdot \sqrt[3]{2^2}}{\sqrt[3]{2} \cdot \sqrt[3]{2^2}} = \frac{\sqrt[3]{3 \cdot 2^2}}{\sqrt[3]{2 \cdot 2^2}} = \frac{\sqrt[3]{12}}{\sqrt[3]{2^3}}$
$= \frac{\sqrt[3]{12}}{2}$

49. $\sqrt[3]{\frac{3}{4y^2}} = \frac{\sqrt[3]{3}}{\sqrt[3]{4y^2}} = \frac{\sqrt[3]{3} \cdot \sqrt[3]{2y}}{\sqrt[3]{4y^2} \cdot \sqrt[3]{2y}}$
$= \frac{\sqrt[3]{6y}}{\sqrt[3]{8y^3}} = \frac{\sqrt[3]{6y}}{2y}$

53. $(4x + 7)(8x - 3)$
$= (4x)(8x) + (4x)(-3) + 7(8x) + 7(-3)$
$= 32x^2 - 12x + 56x - 21$
$= 32x^2 + 44x - 21$

57. $(p + q)(a - m)$
$= p(a) + p(-m) + q(a) + q(-m)$
$= pa - pm + qa - qm$

SECTION 8.5 (page 529)

1. $\sqrt{49} + \sqrt{36} = 7 + 6 = 13$

5. $\sqrt{2}(\sqrt{32} - \sqrt{8}) = \sqrt{2} \cdot \sqrt{32} - \sqrt{2} \cdot \sqrt{8}$
$= \sqrt{64} - \sqrt{16} = 8 - 4 = 4$

9. $3\sqrt{5} + 2\sqrt{45} = 3\sqrt{5} + 2\sqrt{9 \cdot 5}$
$= 3\sqrt{5} + 2 \cdot \sqrt{9} \cdot \sqrt{5}$
$= 3\sqrt{5} + 2 \cdot 3 \cdot \sqrt{5}$
$= 3\sqrt{5} + 6\sqrt{5}$
$= 9\sqrt{5}$

13. $\sqrt{5}(\sqrt{3} - \sqrt{7}) = \sqrt{5} \cdot \sqrt{3} - \sqrt{5} \cdot \sqrt{7}$
$= \sqrt{5 \cdot 3} - \sqrt{5 \cdot 7}$
$= \sqrt{15} - \sqrt{35}$

17. $3\sqrt{14} \cdot \sqrt{2} - \sqrt{28} = 3\sqrt{14 \cdot 2} - \sqrt{28}$
$= 3\sqrt{28} - \sqrt{28}$
$= 2\sqrt{28}$
$= 2\sqrt{4 \cdot 7}$
$= 2 \cdot \sqrt{4} \cdot \sqrt{7}$
$= 2 \cdot 2 \cdot \sqrt{7}$
$= 4\sqrt{7}$

21. $(5\sqrt{7} - 2\sqrt{3})(3\sqrt{7} + 4\sqrt{3})$
$= 5\sqrt{7}(3\sqrt{7}) + 5\sqrt{7}(4\sqrt{3}) - 2\sqrt{3}(3\sqrt{7})$
$- 2\sqrt{3}(4\sqrt{3})$
$= 15 \cdot 7 + 20\sqrt{21} - 6\sqrt{21} - 8 \cdot 3$
$= 105 + 20\sqrt{21} - 6\sqrt{21} - 24$
$= 81 + 14\sqrt{21}$

25. $(5 - \sqrt{2})(5 + \sqrt{2}) = (5)^2 - (\sqrt{2})^2$
$= 25 - 2$
$= 23$

29. $(\sqrt{2} + \sqrt{3})(\sqrt{6} - \sqrt{2})$
$= \sqrt{2}(\sqrt{6}) - \sqrt{2}(\sqrt{2}) + \sqrt{3}(\sqrt{6}) + \sqrt{3}(-\sqrt{2})$
$= \sqrt{12} - 2 + \sqrt{18} - \sqrt{6}$
$= \sqrt{4 \cdot 3} - 2 + \sqrt{9 \cdot 2} - \sqrt{6}$
$= \sqrt{4} \cdot \sqrt{3} - 2 + \sqrt{9} \cdot \sqrt{2} - \sqrt{6}$
$= 2\sqrt{3} - 2 + 3\sqrt{2} - \sqrt{6}$

33. $(\sqrt{5} + \sqrt{30})(\sqrt{6} + \sqrt{3})$
$= \sqrt{5}(\sqrt{6}) + \sqrt{5}(\sqrt{3}) + \sqrt{30}(\sqrt{6}) + \sqrt{30}(\sqrt{3})$
$= \sqrt{30} + \sqrt{15} + \sqrt{180} + \sqrt{90}$
$= \sqrt{30} + \sqrt{15} + \sqrt{36 \cdot 5} + \sqrt{9 \cdot 10}$
$= \sqrt{30} + \sqrt{15} + \sqrt{36} \cdot \sqrt{5} + \sqrt{9} \cdot \sqrt{10}$
$= \sqrt{30} + \sqrt{15} + 6\sqrt{5} + 3\sqrt{10}$

37. $\frac{1}{3 + \sqrt{2}} = \frac{1(3 - \sqrt{2})}{(3 + \sqrt{2})(3 - \sqrt{2})}$
$= \frac{3 - \sqrt{2}}{3^2 - (\sqrt{2})^2} = \frac{3 - \sqrt{2}}{9 - 2}$
$= \frac{3 - \sqrt{2}}{7}$

41. $\dfrac{\sqrt{2}}{2-\sqrt{2}} = \dfrac{\sqrt{2}(2+\sqrt{2})}{(2-\sqrt{2})(2+\sqrt{2})} = \dfrac{2\sqrt{2}+2}{2^2-(\sqrt{2})^2}$
$= \dfrac{2\sqrt{2}+2}{4-2} = \dfrac{2\sqrt{2}+2}{2} = \dfrac{2(\sqrt{2}+1)}{2}$
$= \sqrt{2}+1$ or $1+\sqrt{2}$

45. $\dfrac{\sqrt{12}}{\sqrt{3}+1} = \dfrac{\sqrt{12}(\sqrt{3}-1)}{(\sqrt{3}+1)(\sqrt{3}-1)} = \dfrac{\sqrt{36}-\sqrt{12}}{(\sqrt{3})^2-1^2}$
$= \dfrac{6-\sqrt{12}}{3-1} = \dfrac{6-\sqrt{12}}{2} = \dfrac{6-\sqrt{4\cdot 3}}{2}$
$= \dfrac{6-\sqrt{4}\cdot\sqrt{3}}{2}$
$= \dfrac{6-2\sqrt{3}}{2} = \dfrac{2(3-\sqrt{3})}{2} = 3-\sqrt{3}$

49. $\dfrac{6\sqrt{11}-12}{6} = \dfrac{6(\sqrt{11}-2)}{6} = \sqrt{11}-2$

53. $\dfrac{12-\sqrt{40}}{4} = \dfrac{12-\sqrt{4\cdot 10}}{4} = \dfrac{12-\sqrt{4}\cdot\sqrt{10}}{4}$
$= \dfrac{12-2\sqrt{10}}{4}$
$= \dfrac{2(6-\sqrt{10})}{2\cdot 2} = \dfrac{6-\sqrt{10}}{2}$

57. $(k-1) = (k-1)^2$
$k-1 = k^2-2k+1$
$0 = k^2-3k+2$
$0 = (k-1)(k-2)$
$k-1=0$ or $k-2=0$
$k=1$ $\quad k=2$

SECTION 8.6 (page 537)

Each of these answers should be checked by substituting in the original equation.

5. $\sqrt{y+2} = 3$
$(\sqrt{y+2})^2 = 3^2$
$y+2 = 9$
$y = 7$
This answer checks in the original equation, so the solution is 7.

9. $\sqrt{4-t} = 7$
$(\sqrt{4-t})^2 = 7^2$
$4-t = 49$
$-t = 45$
$t = -45$
This answer checks in the original equation, so the solution is -45.

13. $\sqrt{3x-8} = -2$
$(\sqrt{3x-8})^2 = (-2)^2$
$3x-8 = 4$
$3x = 12$
$x = 4$
Since this answer does not check in the original equation, there is no solution.

17. $\sqrt{10x-8} = 3\sqrt{x}$
$(\sqrt{10x-8})^2 = (3\sqrt{x})^2$
$(\sqrt{10x-8})^2 = 3^2(\sqrt{x})^2$
$10x-8 = 9x$
$x-8 = 0$
$x = 8$
Since this answer checks in the original equation, the solution is 8.

21. $\sqrt{3x-5} = \sqrt{2x+1}$
$(\sqrt{3x-5})^2 = (\sqrt{2x+1})^2$
$3x-5 = 2x+1$
$x-5 = 1$
$x = 6$
This answer checks in the original equation, so the solution is 6.

25. $7x = \sqrt{49x^2+2x-10}$
$(7x)^2 = (\sqrt{49x^2+2x-10})^2$
$7^2x^2 = (\sqrt{49x^2+2x-10})^2$
$49x^2 = 49x^2+2x-10$
$0 = 2x-10$
$10 = 2x$
$5 = x$
This answer checks in the original equation, so the solution is 5.

29. $\sqrt{2x+1} = x-7$
$(\sqrt{2x+1})^2 = (x-7)^2$
$2x+1 = x^2-14x+49$
$0 = x^2-16x+48$
$0 = (x-4)(x-12)$
$x-4=0$ or $x-12=0$
$x=4$ or $x=12$
Check both of these answers in the original equation. The only solution is 12.

33. $\sqrt{5x+1}-1 = x$
$\sqrt{5x+1} = x+1$
$(\sqrt{5x+1})^2 = (x+1)^2$
$5x+1 = x^2+2x+1$
$0 = x^2-3x$
$0 = x(x-3)$
$x=0$ or $x-3=0$
$x=0$ or $x=3$
Since both of these answers check in the original equation, the solutions are 0 and 3.

37. $x-4-\sqrt{2x} = 0$
$x-4 = \sqrt{2x}$
$(x-4)^2 = (\sqrt{2x})^2$
$x^2-8x+16 = 2x$
$x^2-10x+16 = 0$
$(x-8)(x-2) = 0$
$x-8=0$ or $x-2=0$
$x=8$ or $x=2$
Check both of these answers in the original equation. The only solution is 8.

41. $s = 30\sqrt{\dfrac{a}{p}}$

(a) $s = 30\sqrt{\dfrac{862}{156}}$ Let $a = 862$ and $p = 156$.

$s \approx 70.5$ miles per hour Use a calculator.

(b) $s = 30\sqrt{\frac{382}{96}}$ Let $a = 382$ and $p = 96$.

$s \approx 59.8$ miles per hour Use a calculator.

(c) $s = 30\sqrt{\frac{84}{26}}$ Let $a = 84$ and $p = 26$.

$s \approx 53.9$ miles per hour Use a calculator.

45. The square roots of 23 are $\sqrt{23}$ and $-\sqrt{23}$, since $\sqrt{23} \cdot \sqrt{23} = 23$ and $(-\sqrt{23})(-\sqrt{23}) = 23$.

CHAPTER 9

SECTION 9.1 (page 557)

1. The statement is true. If $k > 0$, the solutions to $x^2 = k$ are $x = \sqrt{k}$ and $x = -\sqrt{k}$.

5. The statement is true. If k is prime, then it is positive and the solutions to $x^2 = k$ are $x = \sqrt{k}$ and $x = -\sqrt{k}$. If k is prime then $\sqrt{k}$ is an irrational number.

9. If $x^2 = 81$, then by the square root property,

$$x = \sqrt{81} = 9 \quad \text{or} \quad x = -\sqrt{81} = -9.$$

The solutions are 9 and -9, which may be written ± 9.

13. If $t^2 = 48$, then $t = \sqrt{48}$ or $t = -\sqrt{48}$. Write $\sqrt{48}$ in simplest form.

$$\sqrt{48} = \sqrt{16} \cdot \sqrt{3} = 4\sqrt{3}.$$

The solutions are $4\sqrt{3}$ and $-4\sqrt{3}$, or $\pm 4\sqrt{3}$.

17. If $z^2 = 2.25$, then by the square root property,

$$z = \sqrt{2.25} = 1.5 \quad \text{or} \quad z = -\sqrt{2.25} = -1.5.$$

The solutions are 1.5 and $-$ 1.5, or ± 1.5.

21. $(x - 3)^2 = 25$

$x - 3 = 5$ or $x - 3 = -5$ Square root property

$x = 8$ or $x = -2$ Add 3.

The solutions are -2 and 8.

25. $(x - 8)^2 = 27$

$x - 8 = \sqrt{27}$ or $x - 8 = -\sqrt{27}$ Square root property

Now simplify the radical: $\sqrt{27} = \sqrt{9} \cdot \sqrt{3} = 3\sqrt{3}$.

$x - 8 = 3\sqrt{3}$ or $x - 8 = -3\sqrt{3}$

$x = 8 + 3\sqrt{3}$ or $x = 8 - 3\sqrt{3}$ Add 8.

The solutions are $8 + 3\sqrt{3}$ and $8 - 3\sqrt{3}$.

29. $(4x - 3)^2 = 9$

$4x - 3 = 3$ or $4x - 3 = -3$ Square root property

$4x = 6$ or $4x = 0$ Add 3.

$x = \frac{6}{4} = \frac{3}{2}$ or $x = 0$ Divide by 4.

The solutions are 0 and $\frac{3}{2}$.

33. $(3k + 1)^2 = 18$

$3k + 1 = \sqrt{18}$ or

$3k + 1 = -\sqrt{18}$ Square root property

Now simplify the radical: $\sqrt{18} = \sqrt{9} \cdot \sqrt{2} = 3\sqrt{2}$.

$3k + 1 = 3\sqrt{2}$ or $3k + 1 = -3\sqrt{2}$

$3k = -1 + 3\sqrt{2}$ or $3k = -1 - 3\sqrt{2}$ Subtract 1.

$k = \frac{-1 + 3\sqrt{2}}{3}$ or $k = \frac{-1 - 3\sqrt{2}}{3}$ Divide by 3.

The solutions are $\frac{-1 + 3\sqrt{2}}{3}$ and $\frac{-1 - 3\sqrt{2}}{3}$.

37. $(4k - 1)^2 - 48 = 0$

$(4k - 1)^2 = 48$ Add 48.

$4k - 1 = \sqrt{48}$ or $4k - 1 = -\sqrt{48}$ Square root property

Now simplify the radical: $\sqrt{48} = \sqrt{16} \cdot \sqrt{3} = 4\sqrt{3}$.

$4k - 1 = 4\sqrt{3}$ or $4k - 1 = -4\sqrt{3}$

$4k = 1 + 4\sqrt{3}$ or $4k = 1 - 4\sqrt{3}$ Add 1.

$k = \frac{1 + 4\sqrt{3}}{4}$ or $k = \frac{1 - 4\sqrt{3}}{4}$ Divide by 4.

The solutions are $\frac{1 + 4\sqrt{3}}{4}$ and $\frac{1 - 4\sqrt{3}}{4}$.

41. $x^2 + 4x + 4 = 25$

$(x + 2)^2 = 25$ Factor.

$x + 2 = 5$ or $x + 2 = -5$ Square root property

$x = 3$ or $x = -7$ Subtract 2.

The solutions are -7 and 3.

45. $\frac{4}{5} + \sqrt{\frac{48}{25}} = \frac{4}{5} + \frac{\sqrt{48}}{\sqrt{25}} = \frac{4}{5} + \frac{\sqrt{16} \cdot \sqrt{3}}{5}$

$= \frac{4}{5} + \frac{4\sqrt{3}}{5} = \frac{4 + 4\sqrt{3}}{5}$

49. Since $(x)^2 = x^2$ and $\frac{49}{4} = (\frac{7}{2})^2$, try $(x - \frac{7}{2})^2$. Since the middle term is $2(x)(-\frac{7}{2}) = -7x$, $x^2 - 7x + \frac{49}{4} = (x - \frac{7}{2})^2$.

SECTION 9.2 (page 565)

1. In order to solve the equation $4x^2 + 8x = 3$, one should first divide both sides of the equation by 4. The coefficient of the squared term must be 1.

5. Solve $x^2 + 2x - 5 = 0$.
Rewrite the equation with only variable terms on one side.

$x^2 + 2x = 5$ Add 5.

Take half the coefficient of x and square it.

$$\frac{1}{2}(2) = 1 \quad \text{and} \quad 1^2 = 1.$$

Add 1 to each side of the equation.

$$x^2 + 2x + 1 = 5 + 1 \quad \text{Add 1.}$$
$$(x + 1)^2 = 6 \quad \text{Factor; combine terms.}$$
$$x + 1 = \sqrt{6} \quad \text{or} \quad x + 1 = -\sqrt{6} \quad \text{Square root property}$$
$$x = -1 + \sqrt{6} \quad x = -1 - \sqrt{6} \quad \text{Subtract 1.}$$

The solutions are $-1 + \sqrt{6}$ and $-1 - \sqrt{6}$.

9. Take half the coefficient of y and square it.

$$\frac{1}{2}(14) = 7 \quad \text{and} \quad 7^2 = 49$$

The answer is 49.

13. Take half the coefficient of r, and square it.

$$\frac{1}{2}\left(\frac{1}{2}\right) = \frac{1}{4} \quad \text{and} \quad \left(\frac{1}{4}\right)^2 = \frac{1}{16}$$

The answer is $\frac{1}{16}$.

17. Solve $2x^2 - 4x - 5 = 0$.
Divide each term by 2, the coefficient of x^2.

$$x^2 - 2x - \frac{5}{2} = 0$$

Rewrite the equation with only variable terms on one side.

$$x^2 - 2x = \frac{5}{2} \quad \text{Add } \tfrac{5}{2}.$$

Square half the coefficient of x.

$$\frac{1}{2}(-2) = -1 \quad \text{and} \quad (-1)^2 = 1.$$

Add 1 to both sides of the equation.

$$x^2 - 2x + 1 = \frac{5}{2} + 1 \quad \text{Add 1.}$$
$$(x - 1)^2 = \frac{7}{2} \quad \text{Factor; combine terms.}$$
$$x - 1 = \sqrt{\frac{7}{2}} \quad \text{or} \quad x - 1 = -\sqrt{\frac{7}{2}}$$

Square root property

Simplify the radical:

$$\sqrt{\frac{7}{2}} = \frac{\sqrt{7}}{\sqrt{2}} = \frac{\sqrt{7}}{\sqrt{2}} \cdot \frac{\sqrt{2}}{\sqrt{2}} = \frac{\sqrt{14}}{2}.$$

$$x - 1 = \frac{\sqrt{14}}{2} \quad \text{or} \quad x - 1 = -\frac{\sqrt{14}}{2}$$
$$x = 1 + \frac{\sqrt{14}}{2} \quad \text{or} \quad x = 1 - \frac{\sqrt{14}}{2} \quad \text{Add 1.}$$
$$x = \frac{2}{2} + \frac{\sqrt{14}}{2} \quad \text{or} \quad x = \frac{2}{2} - \frac{\sqrt{14}}{2}$$
$$x = \frac{2 + \sqrt{14}}{2} \quad \text{or} \quad x = \frac{2 - \sqrt{14}}{2}$$

The solutions are $\frac{2 + \sqrt{14}}{2}$ and $\frac{2 - \sqrt{14}}{2}$.

21. Solve $2p^2 - 2p + 3 = 0$.

$$p^2 - p + \frac{3}{2} = 0 \quad \text{Divide by 2.}$$
$$p^2 - p = -\frac{3}{2} \quad \text{Subtract } \tfrac{3}{2}.$$

Square half the coefficient of p.

$$\frac{1}{2}(-1) = -\frac{1}{2} \quad \text{and} \quad \left(-\frac{1}{2}\right)^2 = \frac{1}{4}$$

Add $\frac{1}{4}$ to each side of the equation.

$$p^2 - p + \frac{1}{4} = -\frac{3}{2} + \frac{1}{4} \quad \text{Add } \tfrac{1}{4}.$$
$$\left(p - \frac{1}{2}\right)^2 = -\frac{5}{4} \quad \text{Factor; combine terms.}$$

The square root of $-\frac{5}{4}$ is not a real number, so the equation has no real number solution.

25. Solve $(x + 3)(x - 1) = 5$.

$$x^2 + 2x - 3 = 5 \quad \text{Multiply.}$$
$$x^2 + 2x = 8 \quad \text{Add 3.}$$

Square half the coefficient of x.

$$\frac{1}{2}(2) = 1 \quad \text{and} \quad 1^2 = 1$$

Add 1 to both sides of the equation.

$$x^2 + 2x + 1 = 8 + 1 \quad \text{Add 1.}$$
$$(x + 1)^2 = 9 \quad \text{Factor; combine terms.}$$
$$x + 1 = 3 \quad \text{or} \quad x + 1 = -3 \quad \text{Square root property}$$
$$x = 2 \quad \text{or} \quad x = -4 \quad \text{Subtract 1.}$$

The solutions are -4 and 2.

29. $s = -16t^2 + 96t$
$80 = -16t^2 + 96t$ Let $s = 80$.
We must find t when $s = 80$.

$$-5 = t^2 - 6t \quad \text{Divide by } -16.$$
$$t^2 - 6t = -5 \quad \text{Reverse the sides.}$$
$$t^2 - 6t + 9 = -5 + 9 \quad \text{Add } [\tfrac{1}{2}(-6)]^2 = 9.$$
$$(t - 3)^2 = 4 \quad \text{Factor; combine terms.}$$
$$t - 3 = 2 \quad \text{or} \quad t - 3 = -2 \quad \text{Square root property}$$
$$t = 5 \quad \text{or} \quad t = 1 \quad \text{Add 3.}$$

The object will reach a height of 80 feet at 1 and 5 seconds.

33. $\dfrac{6-\sqrt{45}}{6}=\dfrac{6-\sqrt{9}\cdot\sqrt{5}}{6}=\dfrac{6-3\sqrt{5}}{6}$

$=\dfrac{3(2-\sqrt{5})}{3\cdot 2}=\dfrac{2-\sqrt{5}}{2}$

SECTION 9.3 (page 571)

1. In $4x^2+5x-9=0$, the coefficient of the x^2 term is 4, so $a=4$. The coefficient of the x term is 5, so $b=5$. The constant is -9, so $c=-9$. (Note that one side of the equation was equal to 0 before we started.)

5. Rewrite the equation as $3x^2+7x=0$. The value of a is 3, while $b=7$. No constant term is given, so $c=0$.

9. Rewrite the equation as

$$k^2+12k-13=0.$$

This matches the standard form, so $a=1$, $b=12$, and $c=-13$. Substitute these values in the quadratic formula.

$$k=\frac{-b\pm\sqrt{b^2-4ac}}{2a}$$

$$k=\frac{-12\pm\sqrt{12^2-4(1)(-13)}}{2(1)}\qquad \text{Let } a=1,\ b=12,\ c=-13.$$

$$k=\frac{-12\pm\sqrt{144+52}}{2}$$

$$k=\frac{-12\pm\sqrt{196}}{2}$$

$$k=\frac{-12\pm 14}{2}\qquad \sqrt{196}=14$$

$$k=\frac{-12+14}{2}=\frac{2}{2}=1\quad\text{or}\quad k=\frac{-12-14}{2}=\frac{-26}{2}=-13$$

The solutions are -13 and 1.

13. Rewrite the equation in standard form as

$$2x^2+12x+5=0.$$

Substitute $a=2$, $b=12$, and $c=5$ in the quadratic formula.

$$x=\frac{-b\pm\sqrt{b^2-4ac}}{2a}$$

$$x=\frac{-12\pm\sqrt{12^2-4(2)(5)}}{2(2)}\qquad \text{Let } a=2,\ b=12,\ c=5.$$

$$x=\frac{-12\pm\sqrt{144-40}}{4}$$

$$x=\frac{-12\pm\sqrt{104}}{4}=\frac{-12\pm\sqrt{4}\cdot\sqrt{26}}{4}$$

$$x=\frac{-12\pm 2\sqrt{26}}{4}=\frac{2(-6\pm\sqrt{26})}{2\cdot 2}=\frac{-6\pm\sqrt{26}}{2}$$

The solutions are $\dfrac{-6+\sqrt{26}}{2}$ and $\dfrac{-6-\sqrt{26}}{2}$.

17. The equation

$$6x^2+6x=0$$

matches the standard form, so $a=6$, $b=6$, and $c=0$. Substitute these values in the quadratic formula.

$$x=\frac{-b\pm\sqrt{b^2-4ac}}{2a}$$

$$x=\frac{-6\pm\sqrt{6^2-4(6)(0)}}{2(6)}\qquad \text{Let } a=6,\ b=6,\ c=0.$$

$$x=\frac{-6\pm\sqrt{36-0}}{12}$$

$$x=\frac{-6\pm 6}{12}$$

$$x=\frac{-6+6}{12}=\frac{0}{12}=0\quad\text{or}\quad x=\frac{-6-6}{12}=\frac{-12}{12}=-1.$$

The solutions are -1 and 0.

21. The equation

$$x^2-24=0$$

matches the standard form, so $a=1$, $b=0$, and $c=-24$. Substitute these values in the quadratic formula.

$$x=\frac{-b\pm\sqrt{b^2-4ac}}{2a}$$

$$x=\frac{-0\pm\sqrt{0^2-4(1)(-24)}}{2(1)}\qquad \text{Let } a=1,\ b=0,\ c=-24.$$

$$x=\frac{\pm\sqrt{96}}{2}=\frac{\pm\sqrt{16}\cdot\sqrt{6}}{2}$$

$$x=\frac{\pm 4\sqrt{6}}{2}=\pm 2\sqrt{6}$$

The solutions are $-2\sqrt{6}$ and $2\sqrt{6}$.

25. Rewrite the equation to match the standard form:

$$3x^2-12x+4=0.$$

Substitute the values $a=3$, $b=-12$, and $c=4$ in the quadratic formula.

$$x=\frac{-b\pm\sqrt{b^2-4ac}}{2a}$$

$$x=\frac{-(-12)\pm\sqrt{(-12)^2-4(3)(4)}}{2(3)}\qquad \text{Let } a=3,\ b=-12,\ c=4.$$

$$x=\frac{12\pm\sqrt{144-48}}{6}$$

$$x=\frac{12\pm\sqrt{96}}{6}=\frac{12\pm\sqrt{16}\cdot\sqrt{6}}{6}$$

$$x=\frac{12\pm 4\sqrt{6}}{6}=\frac{2(6\pm 2\sqrt{6})}{2\cdot 3}=\frac{6\pm 2\sqrt{6}}{3}$$

The solutions are $\dfrac{6+2\sqrt{6}}{3}$ and $\dfrac{6-2\sqrt{6}}{3}$.

29. The equation

$$2x^2 + x + 5 = 0$$

matches the standard form, so substitute the values $a = 2$, $b = 1$, and $c = 5$ in the quadratic formula.

$$x = \frac{-b \pm \sqrt{b^2 - 4ac}}{2a}$$

$$x = \frac{-1 \pm \sqrt{(1)^2 - 4(2)(5)}}{2(2)} \qquad \text{Let } a = 2,\ b = 1,\ c = 5.$$

$$x = \frac{-1 \pm \sqrt{1 - 40}}{4}$$

$$x = \frac{-1 \pm \sqrt{-39}}{4}$$

Since $\sqrt{-39}$ is not a real number, there is no real number solution.

33. To clear fractions in

$$\frac{3}{2}k^2 - k - \frac{4}{3} = 0,$$

multiply by 6 to get

$$9k^2 - 6k - 8 = 0.$$

Use $a = 9$, $b = -6$, and $c = -8$ in the quadratic formula.

$$k = \frac{-b \pm \sqrt{b^2 - 4ac}}{2a}$$

$$k = \frac{-(-6) \pm \sqrt{(-6)^2 - 4(9)(-8)}}{2(9)} \qquad \text{Let } a = 9,\ b = -6,\ c = -8.$$

$$k = \frac{6 \pm \sqrt{36 + 288}}{18}$$

$$k = \frac{6 \pm \sqrt{324}}{18}$$

$$k = \frac{6 \pm 18}{18} \qquad \sqrt{324} = 18$$

$$k = \frac{6 + 18}{18} = \frac{24}{18} = \frac{4}{3} \quad \text{or} \quad k = \frac{6 - 18}{18} = \frac{-12}{18} = -\frac{2}{3}$$

The solutions are $-\frac{2}{3}$ and $\frac{4}{3}$.

37. Multiply by 10 to clear decimals in

$$.5x^2 = x + .5.$$

This gives

$$5x^2 = 10x + 5.$$

$$5x^2 - 10x - 5 = 0 \qquad \text{Standard form}$$

Substitute $a = 5$, $b = -10$, and $c = -5$ in the quadratic formula.

$$x = \frac{-b \pm \sqrt{b^2 - 4ac}}{2a}$$

$$x = \frac{-(-10) \pm \sqrt{(-10)^2 - 4(5)(-5)}}{2(5)} \qquad \text{Let } a = 5,\ b = -10,\ c = -5.$$

$$x = \frac{10 \pm \sqrt{100 + 100}}{10}$$

$$x = \frac{10 \pm \sqrt{200}}{10} = \frac{10 \pm \sqrt{100} \cdot \sqrt{2}}{10}$$

$$x = \frac{10 \pm 10\sqrt{2}}{10} = \frac{10(1 \pm \sqrt{2})}{10} = 1 \pm \sqrt{2}$$

The solutions are $1 + \sqrt{2}$ and $1 - \sqrt{2}$.

41. Write the equation $S = 2\pi rh + \pi r^2$ in standard form (r is the variable).

$$\pi r^2 + 2\pi hr - S = 0$$

Use $a = \pi$, $b = 2\pi h$, and $c = -S$ in the quadratic formula.

$$r = \frac{-b \pm \sqrt{b^2 - 4ac}}{2a}$$

$$r = \frac{-2\pi h \pm \sqrt{(2\pi h)^2 - 4(\pi)(-S)}}{2(\pi)} \qquad \text{Let } a = \pi,\ b = 2\pi h,\ c = -S.$$

$$r = \frac{-2\pi h \pm \sqrt{4\pi^2 h^2 + 4\pi S}}{2\pi}$$

$$r = \frac{-2\pi h \pm \sqrt{4(\pi^2 h^2 + \pi S)}}{2\pi}$$

$$r = \frac{-2\pi h \pm 2\sqrt{\pi^2 h^2 + \pi S}}{2\pi}$$

$$r = \frac{-\pi h \pm \sqrt{\pi^2 h^2 + \pi S}}{\pi} \qquad \text{Divide numerator and denominator by 2.}$$

The solutions are $\dfrac{-\pi h + \sqrt{\pi^2 h^2 + \pi S}}{\pi}$ and $\dfrac{-\pi h - \sqrt{\pi^2 h^2 + \pi S}}{\pi}$.

45. To graph $2x - 3y = 6$, find the intercepts. If $x = 0$ then $y = -2$, which gives the point $(0, -2)$. If $y = 0$ then $x = 3$, which gives the point $(3, 0)$. Plot these points and draw the line through them. See the graph in the answer section.

49.

$$\begin{aligned} 2x^2 - x + 1 &= 2(3)^2 - 3 + 1 \qquad \text{Let } x = 3. \\ &= 2 \cdot 9 - 3 + 1 \\ &= 18 - 3 + 1 \\ &= 16 \end{aligned}$$

SUMMARY EXERCISES ON QUADRATIC EQUATIONS (PAGE 575)

Only one method is shown for solving each of these exercises.

1. $s^2 = 36$

$s^2 - 36 = 0$

$(s + 6)(s - 6) = 0$ Factor.

$s + 6 = 0$ or $s - 6 = 0$ Zero-factor property

$s = -6$ or $s = 6$

The solutions are -6 and 6.

5. $z^2 - 4z + 3 = 0$

$(z - 3)(z - 1) = 0$ Factor.

$z - 3 = 0$ or $z - 1 = 0$ Zero-factor property

$z = 3$ or $z = 1$

The solutions are 1 and 3.

9. $(3k - 2)^2 = 9$

$3k - 2 = \sqrt{9}$ or $3k - 2 = -\sqrt{9}$ Square root property

$3k - 2 = 3$ or $3k - 2 = -3$

$3k = 5$ or $3k = -1$ Add 2.

$k = \frac{5}{3}$ or $k = -\frac{1}{3}$ Divide by 3.

The solutions are $-\frac{1}{3}$ and $\frac{5}{3}$.

13. $(3r - 7)^2 = 24$

$3r - 7 = \sqrt{24}$ or $3r - 7 = -\sqrt{24}$ Square root property

Now simplify the radical: $\sqrt{24} = \sqrt{4} \cdot \sqrt{6} = 2\sqrt{6}$

$3r - 7 = 2\sqrt{6}$ or $3r - 7 = -2\sqrt{6}$

$3r = 7 + 2\sqrt{6}$ or $3r = 7 - 2\sqrt{6}$ Add 7.

$r = \frac{7 + 2\sqrt{6}}{3}$ or $r = \frac{7 - 2\sqrt{6}}{3}$ Divide by 3.

The solutions are $\frac{7 + 2\sqrt{6}}{3}$ and $\frac{7 - 2\sqrt{6}}{3}$.

17. $-2x^2 = -3x - 2$

$2x^2 - 3x - 2 = 0$ Standard form

$(2x + 1)(x - 2) = 0$ Factor.

$2x + 1 = 0$ or $x - 2 = 0$ Zero-factor property

$2x = -1$ or $x = 2$

$x = -\frac{1}{2}$

The solutions are $-\frac{1}{2}$ and 2.

21. $0 = -x^2 + 2x + 1$

$x^2 - 2x - 1 = 0$ Standard form

This will not factor, so use the quadratic formula with $a = 1$, $b = -2$, and $c = -1$.

$$x = \frac{-b \pm \sqrt{b^2 - 4ac}}{2a}$$

$$x = \frac{-(-2) \pm \sqrt{(-2)^2 - 4(1)(-1)}}{2(1)} \quad \text{Let } a = 1, b = -2, c = -1.$$

$$x = \frac{2 \pm \sqrt{4 + 4}}{2}$$

$$x = \frac{2 \pm \sqrt{8}}{2} = \frac{2 \pm 2\sqrt{2}}{2} = \frac{2(1 \pm \sqrt{2})}{2}$$

$$x = 1 \pm \sqrt{2}$$

The solutions are $1 + \sqrt{2}$ and $1 - \sqrt{2}$.

25. $(x + 2)(x + 1) = 10$

$x^2 + 3x + 2 = 10$ Use FOIL on the left.

$x^2 + 3x - 8 = 0$ Standard form

Use $a = 1$, $b = 3$, and $c = -8$ in the quadratic formula.

$$x = \frac{-b \pm \sqrt{b^2 - 4ac}}{2a}$$

$$x = \frac{-3 \pm \sqrt{(3)^2 - 4(1)(-8)}}{2(1)} \quad \text{Let } a = 1, b = 3, c = -8.$$

$$x = \frac{-3 \pm \sqrt{9 + 32}}{2}$$

$$x = \frac{-3 \pm \sqrt{41}}{2}$$

The solutions are $\frac{-3 + \sqrt{41}}{2}$ and $\frac{-3 - \sqrt{41}}{2}$.

29. $3m(3m + 4) = 7$

$9m^2 + 12m = 7$ Distributive property

$9m^2 + 12m - 7 = 0$ Standard form

Use $a = 9$, $b = 12$, and $c = -7$ in the quadratic formula.

$$m = \frac{-b \pm \sqrt{b^2 - 4ac}}{2a}$$

$$m = \frac{-12 \pm \sqrt{12^2 - 4(9)(-7)}}{2(9)} \quad \text{Let } a = 9, b = 12, c = -7.$$

$$m = \frac{-12 \pm \sqrt{144 + 252}}{18}$$

$$m = \frac{-12 \pm \sqrt{396}}{18} = \frac{-12 \pm \sqrt{36} \cdot \sqrt{11}}{18}$$

$$m = \frac{-12 \pm 6\sqrt{11}}{18} = \frac{6(-2 \pm \sqrt{11})}{6 \cdot 3} = \frac{-2 \pm \sqrt{11}}{3}$$

The solutions are $\frac{-2 + \sqrt{11}}{3}$ and $\frac{-2 - \sqrt{11}}{3}$.

33. $9k^2 = 16(3k + 4)$

$9k^2 = 48k + 64$ Distributive property

$9k^2 - 48k - 64 = 0$ Standard form

Use the quadratic formula with $a = 9$, $b = -48$, and $c = -64$.

$$k = \frac{-b \pm \sqrt{b^2 - 4ac}}{2a}$$

$$k = \frac{-(-48) \pm \sqrt{(-48)^2 - 4(9)(-64)}}{2(9)} \quad \text{Let } a = 9, b = -48, c = -64.$$

$$k = \frac{48 \pm \sqrt{2304 + 2304}}{18}$$

$$k = \frac{48 \pm \sqrt{4608}}{18} = \frac{48 \pm \sqrt{2304} \cdot \sqrt{2}}{18}$$

$$k = \frac{48 \pm 48\sqrt{2}}{18} = \frac{6(8 \pm 8\sqrt{2})}{6 \cdot 3} = \frac{8 \pm 8\sqrt{2}}{3}$$

The solutions are $\frac{8 + 8\sqrt{2}}{3}$ and $\frac{8 - 8\sqrt{2}}{3}$.

37. $-3x^2 + 4x = -4$

$3x^2 - 4x - 4 = 0$ Standard form

$(3x + 2)(x - 2) = 0$ Factor.

$3x + 2 = 0$ or $x - 2 = 0$ Zero-factor property

$3x = -2$ or $x = 2$

$x = -\frac{2}{3}$

The solutions are $-\frac{2}{3}$ and 2.

41. $k^2 - \frac{4}{15} = -\frac{4}{15}k$

$15k^2 - 4 = -4k$ Multiply by 15 to clear fractions.

$15k^2 + 4k - 4 = 0$ Standard form

$(5k - 2)(3k + 2) = 0$ Factor.

$5k - 2 = 0$ or $3k + 2 = 0$ Zero-factor property

$5k = 2$ or $3k = -2$

$k = \frac{2}{5}$ or $k = -\frac{2}{3}$

The solutions are $-\frac{2}{3}$ and $\frac{2}{5}$.

SECTION 9.4 (page 581)

See the answer section in your text for all graphs.

5. The graph of $y = x^2 + 2x + 3$ is a parabola. The x-value of the vertex is $-\frac{b}{2a}$. Here $a = 1$ and $b = 2$, so

$$x = -\frac{b}{2a} = -\frac{2}{2(1)} = -\frac{2}{2} = -1$$

and

$$y = (-1)^2 + 2(-1) + 3 = 1 - 2 + 3 = 2.$$

The vertex is $(-1, 2)$. Because the coefficient of x^2 is 1 (positive), the graph opens upward and has the same shape as that of $y = x^2$. Some additional points on the graph are $(0, 3)$, $(1, 6)$, $(2, 11)$, and $(-2, 3)$.

9. The graph of $y = 2 - x^2$ is a parabola. In this equation, $a = -1$ and $b = 0$, so the x-value of the vertex is

$$x = -\frac{b}{2a} = -\frac{0}{2(-1)} = 0$$

and

$$y = 2 - (0)^2 = 2.$$

The vertex is $(0, 2)$. Since the coefficient of x^2 is -1, the graph opens downward and has the same shape as that of $y = x^2$. Some additional points on the graph are $(-2, -2)$, $(-1, 1)$, $(1, 1)$, and $(2, -2)$.

13. The graph of $y = -x^2 + 6x - 5$ is a parabola. In this equation, $a = -1$ and $b = 6$, so the x-value of the vertex is

$$x = -\frac{b}{2a} = -\frac{6}{2(-1)} = \frac{6}{2} = 3$$

and

$$y = -(3)^2 + 6(3) - 5 = -9 + 18 - 5 = 4.$$

The vertex is $(3, 4)$. Since the coefficient of x^2 is -1, the graph opens downward and has the same shape as that of $y = x^2$. Some additional points on the graph are $(0, -5)$, $(1, 0)$, $(5, 0)$, and $(2, 3)$.

17. If $a > 0$, then the coefficient of x^2 is positive and the parabola opens upward. If $a < 0$, then the coefficient of x^2 is negative and the parabola opens downward.

Index

V

W

X

Y

Z